SPECIAL INDIAN EDITION

Engineering Electromagnetics

Eighth Edition

About the Authors

William H Hayt Jr (deceased) received his BS and MS at Purdue University and his PhD from the University of Illinois. After spending four years in industry, Professor Hayt joined the faculty of Purdue University, where he served as Professor and Head of the School of Electrical Engineering, and as Professor Emeritus after retiring in 1986. Professor Hayt's professional society memberships included Eta Kappa Nu, Tau Beta Pi, Sigma Xi, Sigma Delta Chi, Fellow of IEEE, ASEE, and NAEB. While at Purdue, he received numerous teaching awards, including the university's Best Teacher Award. He is also listed in *Purdue's Book of Great Teachers*, a permanent wall display in the Purdue Memorial Union, dedicated on 23 April, 1999. The book bears the names of the inaugural group of 225 faculty members, past and present, who have devoted their lives to excellence in teaching and scholarship. They were chosen by their students and their peers as Purdue's finest educators.

A native of Los Angeles, California, **John A Buck** received his MS and PhD in Electrical Engineering from the University of California at Berkeley in 1977 and 1982, and his BS in Engineering from UCLA in 1975. In 1982, he joined the faculty of the School of Electrical and Computer Engineering at Georgia Tech, where he has remained for the past 32 years. His research areas and publications have centered within the fields of ultrafast switching, nonlinear optics, and optical fiber communications. He is the author of the graduate text *Fundamentals of Optical Fibers*, which is now in its second edition. When not glued to his computer or confined to the lab, Dr. Buck enjoys music, hiking, and photography.

M Jaleel Akhtar received his PhD (Dr.-Ing.) in Electrical Engineering from the Otto-von-Guericke University of Magdeburg, Magdeburg, Germany, in 2003. He was a Post-Doctoral Research Scientist and a Project Leader with the Institute for Pulsed Power and Microwave Technology, Karlsruhe Institute of Technology, Karlsruhe, Germany, from 2003 to 2009, where he was involved in the microwave material modelling, processing and testing. He joined the Department of Electrical Engineering, Indian Institute of Technology (IIT) Kanpur, Kanpur, India in 2009, where he is currently an Associate Professor. He also worked as a Scientist with the Central Electronics Engineering Research Institute, Pilani, India, from 1994 to 1997, where he was involved in the design and development of high-power microwave tubes.

He has authored/co-authored more than 80 papers in various peer-reviewed international journals and conference proceedings as well as books and book chapters. He holds one patent on coplanar waveguide-based RF sensors. His current research interests include microwave imaging and nondestructive testing, RF sensors, design and testing of artificial dielectrics and functional materials for electromagnetic stealth applications, UWB antennas for imaging, and design of RF filters and components.

Dr Akhtar is a Fellow of the Institution of Electronics and Telecommunication Engineers (IETE), New Delhi, India; Senior Member of the Institute of Electrical and Electronics Engineers (IEEE), USA; and a Life Member of the Indian Physics Association and the Indo-French Technical Association. He was the recipient of the Computer Simulation Technology (CST) University Publication Award in 2009 from the CST AG, Darmstadt, Germany.

SPECIAL INDIAN EDITION

Engineering Electromagnetics

Eighth Edition

William H Hayt, Jr
Late Emeritus Professor
Purdue University, Indiana, USA

John A Buck
Georgia Institute of Technology, Georgia, USA

Adapted by

M Jaleel Akhtar
Department of Electrical Engineering
Indian Institute of Technology (IIT) Kanpur, India

McGraw Hill Education (India) Private Limited
CHENNAI

McGraw Hill Education Offices
Chennai New York St Louis San Francisco Auckland Bogotá Caracas
Kuala Lumpur Lisbon London Madrid Mexico City Milan Montreal
San Juan Santiago Singapore Sydney Tokyo Toronto

McGraw Hill Education (India) Private Limited

Engineering Electromagnetics, 8/e (SIE)

Indian Adaptation done by arrangement with McGraw-Hill Global Education Holdings, LLC, New York.

Sales Territories: India, Pakistan, Nepal, Bangladesh, Sri Lanka and Bhutan.

Eleventh reprint 2017
RQACRDLMRYLRD

Special Indian Edition 2014

Print Edition
ISBN (13 digits): 978-93-392-0327-6
ISBN (10 digits): 93-392-0327-5

E-book Edition
ISBN (13 digits): 978-93-392-0328-3
ISBN (10 digits): 93-392-0328-3

12 13 14 15 16 RAJ 21 20 19 18 17

Managing Director: *Kaushik Bellani*
Head—Higher Education (Publishing and Marketing): *Vibha Mahajan*
Publishing Manager—SEM & Tech Ed.: *Shalini Jha*
Manager—Production Systems: *Satinder S Baveja*
General Manager—Production: *Rajender P Ghansela*
Manager—Production: *Reji Kumar*

Published by McGraw Hill Education (India) Private Limited,
444/1, Sri Ekambara Naicker Industrial Estate, Alapakkam, Porur, Chennai 600116, Tamil Nadu, India

Typeset at Script Makers, 19, A-1B Block, DDA Market, Paschim Vihar, New Delhi 110063, and Printed and Bound in India at Rajkamal Electric Press, Plot No. 2, Phase – IV, Sonepat, Haryana.

Visit us at: www.mheducation.co.in

Contents

Preface to the Adapted Edition

It is my privilege to carry out the adaptation of this classic book, *Engineering Electromagnetics* by Hayt and Buck. This book has been followed for the first course in electromagnetic theory in undergraduate classes throughout the world for many decades. During the last decade, education on electromagnetic engineering has undergone a lot of changes due to the advent of many advanced computational and simulation tools. The focus of electromagnetics education, especially for engineers, has shifted towards the transmission-line-based approach rather than the more conventional field-theory approach. The transmission-line concepts are not limited to the RF and microwave frequency range any more, but they have also now become equally important to the field of high-speed digital circuits due to the requirement of carrying digital signals at very high data rates.

Hence, it is now vital for any student of electrical engineering to learn the basic concepts of electromagnetics and transmission lines irrespective of whether he/she wants to work in the microelectronics field or the computer industry. As a teacher of electromagnetics, my experience has been that students can learn the concepts easily if the practical usage of these theoretical concepts is explained to them. Similarly, a number of solved examples on each topic provide students an opportunity to learn the subject in a better way, thereby giving them more command on the subject. The earlier edition of this book did not contain many solved examples in chapters related to the electrodynamics and the transmission lines. Another observation was that the earlier version of this book did not cover in detail waveguides and resonators. These topics are, however, part of the undergraduate electrical engineering curriculum of most academic institutions of India. The adaptation of *Engineering Electromagnetics* has been done mainly to suit the contents of this book to the syllabi and the teaching style followed in most teaching institutions of the Indian subcontinent.

Salient Features

- **Updated contents**: The content follows an application-oriented approach. A new chapter on Waveguides and Resonators has been added. This chapter provides a detailed analysis of the parallel-plate waveguide, the rectangular waveguide, and the rectangular cavity resonator. The chapter explains various concepts from the application point of view, and includes some real-world examples. Traditionally, the book is well known for its comprehensive coverage of Electrostatics and Magnetostatics, together with excellent treatment of Vector Analytics in Electromagnetics.
- **Rich pedagogy**: A total of more than 40 new thought-provoking solved examples are added to various chapters of the book. This is in addition to the drill problems and solved examples of the earlier edition. The total pedagogy count now stands at
  - Examples: 111
  - Drill Problems: 121
  - Review Problems: 474

– Diagrams:261
– Images: 7

Adaptation Changes

The main focus of the adaptation has been on the electrodynamics part and the topics related to transmission lines. Hence, no changes have been made in this book in Chapter 1 to Chapter 9, which mainly deal with Electrostatics, Dielectric and Magnetic Materials, and the Steady Magnetic Field. The changes in the adapted edition are mainly done on topics dealing with the time-varying electromagnetic fields starting from Chapter 11.

I have tried to put some value addition to the book without changing its basic flavor. Mainly in **Chapter 11** to **Chapter 13**, a number of solved examples have been added at appropriate places without affecting the continuity of the text. A detailed explanation is provided in all the solved examples, which also helps the reader learn some new concepts not covered in the main text of that particular chapter. These solved examples can also be effectively used by instructors in tutorial classes to emphasize various points related to a particular topic. The rectangular waveguides and resonators were not widely covered in the earlier editions of this book. In view of this, an altogether new chapter **(Chapter 15)** on Waveguides and Resonators is added in the adapted edition. The new chapter provides a detailed analysis of the parallel-plate waveguide, the rectangular waveguide and the rectangular cavity resonator. Details of the changes have been mentioned below.

In **Chapter 11**, two examples have been added to find out the distributed parameters and some basic wave characteristics of the transmission line. The third example in this chapter puts emphasis on the fact that a piece of transmission line can quite effectively be used as a circuit element to show either inductive or capacitive behavior in the UHF frequency range. The fourth example shows how mismatching at the load end affects the power flow along the line. The next four new examples in this chapter deal with the Smith chart, explaining various concepts and showing relevant applications of the chart. One example is specially added here to show the procedure of matching a transmission line with a load using a stub of different characteristic impedance. The usefulness of a quarter-wave transformer to determine the physical length of a line is shown in one of the examples. The last two examples in this chapter are related to transient phenomenon where the concept of TDR (time domain reflectometer), as a practical tool to locate faults along the line, is explained.

Chapter 12 is augmented by adding a number of solved examples explaining the procedure to analyze the electromagnetic fields of a plane wave in both instantaneous and phasor forms. Firstly, two examples related to the propagation of waves in free space are added where it is shown that a plane wave can be uniquely defined either in terms of the electric field or the magnetic field which automatically leads to determination of all the wave characteristics. A number of examples are then included in the chapter to describe the wave-propagation phenomenon in dielectrics, where various parameters such as the displacement current, the conduction current, the power density, the attenuation constant, etc., are determined. The last three new examples of this chapter mainly deal with finding the polarization of a plane wave. It is shown, with the help of these examples, how a wave can be characterized as linearly, circularly or elliptically polarized starting from the first principle.

In **Chapter 13**, a number of examples are added to show the propagation behavior of a plane wave when it encounters a discontinuity along the propagation path leading to partial reflection and partial transmission. In the first few examples, the two-layered medium is considered where the wave is assumed to be incident normally from the first medium. A number of parameters such as the reflection coefficient, the transmission coefficient, the reflected power, the transmitted power, etc., are determined. One new example is added for the three-layered medium, where the concept of anti-reflection coating and its dependence on various parameters such as the dielectric constant, the frequency, etc., is explained. Lastly, few examples are added in this chapter related to the oblique incidence in order to show the usefulness of Snell's law and the Poynting theorem in a general situation.

Chapter 14 of this book is revisited, and a number of changes have been made. While the basic theory of the waveguide operation including the plane-wave analysis of the parallel-plate waveguide is retained, the wave-equation approach to analyze the parallel-plate line is removed. Similarly, the rectangular waveguide section, which was described very briefly, is taken out from this chapter. The general topics such as the optical fiber and the basic antenna principle are retained for advanced reading.

Chapter 15 is an altogether new chapter to the adapted edition of this book. This chapter is focused on waveguides and resonators, where the wave-equation approach is used to analyze various guiding structures and cavities. A very systematic approach starting from Maxwell's equations is given here to obtain the electric and magnetic field components corresponding to any arbitrary TE or TM mode. The field expressions for TE and TM modes are derived inside a number of waveguide geometries and resonant structures by applying appropriate boundary conditions in each case. The chapter starts with the parallel-plate waveguide, where the width of both the plates is considered quite large so that the transverse dependence of all the field components is limited to only one direction. The electric and magnetic field expressions for any arbitrary TE and TM mode supported by the parallel-plate geometry are derived starting from the wave equation after applying the boundary conditions on both the plates. The concept of the cut-off frequency and the modified propagation constant inside the guided media is explained. It is shown that the TEM wave, which is the lowest order TM mode, is the dominant mode of the parallel-plate waveguide. After the parallel-plate line, the rectangular waveguides, which are single-conductor hollow metallic guides, most commonly being used in the microwave frequency range, are discussed in detail. It is explained how the method of separation of variables combined with the boundary condition on all the four metallic walls can be used to determine the longitudinal component of the magnetic and the electric field in case of TE and TM modes, respectively. The transverse components of the electric and magnetic fields for both TE and TM cases are derived in terms of the corresponding longitudinal component using Maxwell's equations. The concept of dominant mode in the rectangular waveguide is explained, and various parameters such as the cut-off frequency, the wave impedance, the phase and group velocities, the attenuation constant, etc., are defined. The usage of rectangular waveguides including some modern-day application is provided.

The last section of this chapter deals with resonators, where rectangular cavity resonators are primarily discussed. It is shown how a piece of rectangular waveguide can be easily converted into a rectangular cavity. The fields inside the rectangular cavity are derived starting from the field expressions of the corresponding rectangular waveguide, and applying additional boundary conditions along the longitudinal direction. The expressions for the resonant frequency and the quality factor of the rectangular cavity are derived, and the practical application of the rectangular cavity for dielectric measurement is emphasized. A number of solved examples are provided throughout this new chapter to help readers understand various concepts. The first three examples of the chapter deal with the parallel-plate waveguide where the cut-off frequencies, the phase and group velocities, and the wave impedances of various modes are computed. The next few examples are devoted to rectangular waveguides which help readers understand the design procedure of the waveguide geometry. It is shown how the cross-sectional dimensions of the waveguide can be chosen in order to obtain the highest possible bandwidth for the dominant-mode propagation. The last two examples of the chapter are focused on the rectangular cavity, which help readers determine various parameters of the cavity such as the resonant frequency, the quality factor, the stored energy, the dissipated power, etc.

How to Use the Book

The adapted edition of the book can now be used in more than one semester of any undergraduate engineering course. Chapters 1 to 9, mostly dealing with the statics part, can be used in the introductory course on electromagnetic theory, while Chapters 10 to 15 can be used in the second-level undergraduate course on electromagnetics and microwaves. It is, however, difficult to provide the exact demarcation between the two level courses as it would also depend upon the instructor and the course design of a particular institution.

Chapter 15 is written in such a way that the instructors can directly switch to it after Chapter 13. It would, however, be better if Sections 14.1 and 14.2 are briefly discussed before switching to Chapter 15. Section 14.3 provides the analysis of a parallel-plate waveguide using the plane-wave model which is independent of the approach used in Chapter 15 for the analysis of parallel-plate lines. Similarly, Sections 14.4, 14.5, and 14.6, dealing with dielectric waveguides, the optical fiber and the antenna principle can be skipped for later reading without losing any continuity, thus allowing the reader to directly progress to Chapter 15.

Web Supplements

The following material can be accessed at *http://www.mhhe.com/hayt/ee8*

For Instructors: Solution Manual, PowerPoint Lecture Slides

For Students: Links to Virtual Labs *http://www.iitk.ac.in/mimt_lab/vlab/*

The text is also accompanied by a CD containing animations and interactive quizzes.

Acknowledgements

It is my humble duty at this stage to acknowledge all those persons who have, directly or indirectly, contributed to the successful completion of this project. First of all, I would like to thank my colleagues of the RF and Microwaves group at IIT Kanpur, Professors A Biswas, A R Harish and K V Srivastava. Their cooperation provided me a chance to teach the electromagnetic theory undergraduate course in the department, which certainly helped me a lot in bringing out the new edition of this book. I also appreciate efforts of my graduate students, especially Abhishek, Nilesh, Shaji, and Zubair, who provided assistance in making illustrations of the book. I am indebted to the authorities of my host institute, Indian Institute of Technology (IIT) Kanpur, for providing me the appropriate atmosphere in many ways to carry out the adaptation of this book.

This project would not have been completed without the active support and cooperation from the McGraw Hill Education (India) project team throughout this adaptation process. I would especially like to thank Koyel Ghosh for initiating the project and motivating me at various stages, and Satinder Singh Baveja, P L Pandita, and Sohini Mukherjee for typesetting and copyediting the whole text at a short notice.

I would like to take this opportunity to thank Dr J A Buck for editing the previous editions of this book, which certainly made my job easier. The overall adaptation process certainly took more time than expected, and due to this reason, I was also not able to complete all the tasks which I had earlier planned. However, I hope, many more changes can be incorporated in any future edition if enthusiasm is shown by readers of the book.

I express sincere thanks to all those reviewers who took out time to review the content. Their names are given below:

Dharmendra Singh	*Indian Institute of Technology (IIT) Roorkee Roorkee, Uttarakhand*
S P Singh	*Indian Institute of Technology (Banaras Hindu University)—IIT (BHU) Varanasi, Uttar Pradesh*
Prem Chaddha	*Delhi Technological University, New Delhi*
Rajesh Khanna	*Thapar University, Patiala, Punjab*
Ghanshyam Singh	*Malaviya National Institute of Technology Jaipur, Rajasthan*
Satish K Jain	*Shri Govindram Seksaria Institute of Technology and Science (SGSITS) Indore, Madhya Pradesh*
Sushrut Das	*Indian School of Mines (ISM) Dhanbad, Jharkhand*
Jibendu Sekhar Roy	*KIIT University, Bhubaneswar, Odisha*
Sangram Mohapatra	*C V Raman College of Engineering Bhubaneswar, Odisha*
Tanushree Bose	*Sikkim Manipal University, Majitar, Sikkim*
I Kathir	*Mepco Schlenck Engineering College, Sivakasi, Tamil Nadu*
E S Gopi	*National Institute of Technology (NIT) Tiruchirappalli Tiruchirappalli, Tamil Nadu*
N V S N Sarma	*National Institute of Technology (NIT) Warangal Warangal, Andhra Pradesh*
A V Narasimhadhan	*National Institute of Technology (NIT) Surathkal Surathkal, Karnataka*
Deepak Ghodgaonkar	*Dhirubhai Ambani Institute of Information and Communication Technology, Gandhinagar, Gujarat*

Finally, I would like to acknowledge the support of my parents and other family members whose blessings have always been with me at every stage of my life. I would specially like to dedicate this book to my father who is not with me anymore but his memories are still alive.

Feedback Request

Although I have tried my best to make the adapted edition of this book more interesting and useful, I am of the opinion that there is always some room for improvement. Hence, as an author of this book, I invite suggestions and feedback from instructors as well as students in order to improve any future edition of this book. I would be also happy to clarify any point or reply to any specific queries related to this book. I can be reached by e-mail at *mjakhtar@iitk.ac.in*

You can also visit the website of this book given at the back cover for some advanced teaching resources.

M Jaleel Akhtar

Indian Institute of Technology Kanpur

April 2014

Publisher's Note

McGraw Hill Education (India) invites suggestions and comments from you, all of which can be sent to *info.india@mheducation.com* (kindly mention the title and author name in the subject line).

Piracy-related issues may also be reported.

Preface to the Seventh Edition

Preparing a new edition of a textbook is an odd mixture of toil and gratification to their limits. In the midst of long hours and endless minutiae, the acts of incorporating new ideas that augment those already there or that displace those that may have become tiresome provide feelings of relief and accomplishment. Through the effort, there is hope and a growing belief that the new book will be better and more useful.

In the case of electromagnetics, the core subject matter never changes, and so it could be argued that previous treatments that have proven successful are probably best left alone. This was my philosophy when preparing the sixth edition. In this new seventh edition, I have taken a few more liberties. The older topics, present since the first edition, were reassessed, and a few were either dropped or moved to new locations. These changes were made sparingly, as my intent was to improve the flow of material while attempting to avoid anything that would damage the classic appeal and success of Hayt's original work as it has existed for nearly fifty years.

In recent years, many electromagnetics core courses have shifted their emphasis in the direction of transmission-line theory, in a manner consistent with the rise of computer engineering as a primary component within electrical engineering curricula. This has resulted in the most significant change in our new edition, which is a substantially rewritten (and now independent) chapter on transmission lines. This (the former Chapter 13) is now Chapter 11, and it precedes the chapters on electromagnetic waves. In Chapter 11, transmission lines are treated entirely within the context of circuit theory; wave phenomena are introduced and used exclusively in the form of voltages and currents. Line losses are now covered, along with a more detailed development of the wave equation. Inductance and capacitance concepts are treated as known parameters, and so there is no reliance on any other chapter. This allows transmission lines to be covered as the initial topic in a course, if desired. Field concepts and parameter computation in lines are still present, but now they appear in the early part of Chapter 14, where they play the additional roles of helping to introduce waveguiding concepts while adding perspective to the waveguiding problem. The specific cases of planar, coaxial, and two-wire lines within different frequency regimes are treated as in the earlier editions, and a new section on microstrip line has been added. This material can be covered after Chapter 12 and does not require Chapter 13.

The electromagnetic waves chapters, now 12 and 13 (formerly 11 and 12), retain their independence from transmission-line theory in that one can progress from Chapter 10 directly to Chapter 12. In this way, wave phenomena are introduced from first principles, but within the context of the uniform plane wave. Chapter 12 refers to Chapter 11 in places where the latter may give additional perspective, along with a little more detail. Nevertheless, all the necessary material for learning plane waves without first studying transmission-line waves is present in Chapter 12, should the student or instructor wish to proceed in that order.

The discussion of plane wave reflection and dispersion in Chapter 13 moves directly into Chapter 14, in which waveguiding fundamentals are covered in the light of plane wave reflection models, as well as through direct solution of the wave equation. This chapter retains its original content from the sixth edition, but it now includes an expanded section on optical fibers in addition to the one on transmission-line structures previously mentioned. The last part of Chapter 14 covers basic radiation concepts, a carryover from earlier editions.

The restructuring of the earlier chapters includes the division of the former Chapter 5 (Conductors, Dielectrics, and Capacitance) into two chapters (now 5 and 6) that deal separately with conductors and capacitors. The previous Chapter 6 (which covered field plotting and numerical techniques) has been eliminated, but some of its material has been retained in other chapters. Curvilinear square mapping and discussions of current analogies are now part of the new capacitance chapter (6), and the old section on iterative solution is now part of the Laplace and Poisson equation development in Chapter 7.

A major new supplement to this edition is a CD containing computer demonstrations and interactive programs developed by Natalia Nikolova of McMaster University, and Vikram Jandhyala and Indranil Chowdhury of the University of Washington. Their excellent contributions are geared to the text, and CD icons appear in the margins whenever an exercise that pertains to the narrative exists. In addition, quizzes are provided on the CD to aid in further study. Numerous animations (including a few of my own) are present that help in visualizing many of the phenomena described in the text.

Approximately 40 percent of the problems in the sixth edition have been replaced. In addition to many new problems, I have included several excellent ``classic'' problems of Bill Hayt's that appeared in the early editions. I decided to revive what I felt were the best and most relevant of these. The drill problems have been reworked and errors have been corrected.

Apart from these changes, the theme of the text is the same as it has been since the first edition of 1958. An inductive approach is used that is consistent with the historical development. In it, the experimental laws are presented as individual concepts that are later unified in Maxwell's equations. After the first chapter on vector analysis, additional mathematical tools are introduced in the text on an as-needed basis. Throughout every edition, as well as this one, the primary goal has been to enable students to learn independently. Numerous examples, drill problems (usually having multiple parts), the end-of-chapter problems, and the material on the media disk, are provided to facilitate this. Answers to the drill problems are given below each problem. Answers to odd-numbered end-of-chapter problems are found in Appendix E. A solutions manual will also be available to instructors. This material, along with the media suite contents and other teaching resources, is also available at the book's website, *http://www.mhhe.com/haytbuck*. **COSMOS** (Complete Online Solutions Manual Operating System), available to instructors on CD-ROM, contains the entire book problem set, enhanced to include any referenced images or text, as well as the entire solution set for the book. This application will assist instructors in organizing, distributing, and tracking problem sets as they are assigned. Also acknowledged are ANSOFT and Faustus Scientific Corp.

The book contains more than enough material for a one-semester course. As is evident, statics concepts are emphasized and occur first in the presentation. In a course that places more emphasis on dynamics, the transmission lines chapter can be covered initially as mentioned, or at any point in the course. The statics material can be covered more rapidly by omitting Chapter 1 (assigned to be read as a review) and skipping Sections 2.6, 5.5, 5.6, 6.5, 6.6, 7.4 through 7.6, 8.6, 8.7, 9.3 through 9.6, 9.8, and 10.5. A more streamlined presentation of plane waves can be accomplished by omitting Sections 12.5, 13.5, and 13.6. Chapter 14 is intended as an advanced topics chapter, in which the development of waveguide and antenna concepts occurs through the application of the methods learned in earlier chapters, therefore helping to solidify that knowledge. It may also serve as a bridge between the basic course and more advanced courses that follow it.

ACKNOWLEDGMENTS

I am deeply indebted to many students and colleagues who have provided feedback and encouragement prior to and during the preparation of this new edition. In the initial review process, many thoughtful and valuable insights were provided by

Raviraj Sadanand Adve, *University of Toronto*
Jonathan S Bagby, *Florida Atlantic University*
Arun V Bakshi, *College of Engineering, Pimpri, India*
Shanker Balasubramaniam, *Michigan State University*
N Scott Barker, *University of Virginia*
Vikram Jandhyala, *University of Washington*
Brian A Lail, *University of Central Florida*
Sharad R Laxpati, *University of Illinois–Chicago*
Reinhold Ludwig, *Worcester Polytechnic Institute*
Masoud Mostafavi, *San Jose State University*
Natalia K Nikolova, *McMaster University*
J Scott Tyo, University of New Mexico
Kathleen L Virga, *University of Arizona*
Clive Woods, *Iowa State University*

Their remarks and suggestions affected many aspects of the final product. Errors and inconsistencies in the text and in some of the problems were pointed out in detail in several communications by William Thompson, Jr, Pennsylvania State University. At Georgia Tech, Shannon Madison provided assistance in proofing the drill problems, and Diana Fouts is responsible for the cover illustration and design.

In the four years since the last edition went to press, I received many e-mails from a host of people with questions and suggestions, often about portions of the text that, on further reflection, could have been clearer. It was these acts of calling my attention to the details that have perhaps been of most value in improving the final product. I regret not having been able to answer every message, but they were all considered and acted upon as appropriate. As then, I invite future correspondence, and can be reached at *john.buck@ece.gatech.edu*.

Finally, I acknowledge the project team at McGraw-Hill, whose enthusiasm, encouragement, and support were indispensable. Of these most especially, Michelle Flomenhoft and Carlise Stembridge held everything together and made it all happen. I value my association with them. As with the previous revision, time was too short to complete everything that I had wanted. I am certain that my enthusiasm will be high for a chance at an eighth edition, once rested, and with patience regained by my wife and daughters. The girls, too young to understand why Dad remains hunched over a computer all weekend, have hopefully not grown too old to want him back. I dedicate this book to them.

John A Buck
Marietta, GA
September 2004

1

Vector Analysis

Vector analysis is a mathematical subject which is much better taught by mathematicians than by engineers. Most junior and senior engineering students, however, have not had the time (or perhaps the inclination) to take a course in vector analysis, although it is likely that many elementary vector concepts and operations were introduced in the calculus sequence. These fundamental concepts and operations are covered in this chapter, and the time devoted to them now should depend on past exposure.

The viewpoint here is also that of the engineer or physicist and not that of the mathematician in that proofs are indicated rather than rigorously expounded, and the physical interpretation is stressed. It is easier for engineers to take a more rigorous and complete course in the mathematics department after they have been presented with a few physical pictures and applications.

Vector analysis is a mathematical shorthand. It has some new symbols, some new rules, and a pitfall here and there like most new fields, and it demands concentration, attention, and practice. The drill problems, first met at the end of Section 1.4, should be considered an integral part of the text and should all be worked. They should not prove to be difficult if the material in the accompanying section of the text has been thoroughly understood. It takes a little longer to "read" the chapter this way, but the investment in time will produce a surprising interest.

1.1 SCALARS AND VECTORS

The term *scalar* refers to a quantity whose value may be represented by a single (positive or negative) real number. The x, y, and z we used in basic algebra are scalars, and the quantities they represent are scalars. If we speak of a body falling a distance L in a time t, or the temperature T at any point in a bowl of soup whose coordinates are x, y, and z, then L, t, T, x, y, and z are all scalars. Other scalar quantities are mass, density, pressure (but not force), volume, and volume resistivity. Voltage is also a scalar quantity, although the complex representation of a sinusoidal voltage, an artificial procedure, produces a *complex scalar*, or *phasor*, which requires two real numbers for its representation, such as amplitude and phase angle, or real part and imaginary part.

A *vector* quantity has both a magnitude[1] and a direction in space. We shall be concerned with two- and three-dimensional spaces only, but vectors may be defined in n-dimensional space in more advanced applications. Force, velocity, acceleration, and a straight line from the positive to the negative terminal of a storage battery are examples of vectors. Each quantity is characterized by both a magnitude and a direction.

[1] We adopt the convention that magnitude infers absolute value; the magnitude of any quantity is therefore always positive.

We shall be mostly concerned with scalar and vector *fields*. A field (scalar or vector) may be defined mathematically as some function of that vector which connects an arbitrary origin to a general point in space. We usually find it possible to associate some physical effect with a field, such as the force on a compass needle in the earth's magnetic field, or the movement of smoke particles in the field defined by the vector velocity of air in some region of space. Note that the field concept invariably is related to a region. Some quantity is defined at every point in a region. Both *scalar fields* and *vector fields* exist. The temperature throughout the bowl of soup and the density at any point in the earth are examples of scalar fields. The gravitational and magnetic fields of the earth, the voltage gradient in a cable, and the temperature gradient in a soldering-iron tip are examples of vector fields. The value of a field varies in general with both position and time.

In this book, as in most others using vector notation, vectors will be indicated by boldface type, for example, **A**. Scalars are printed in italic type, for example, *A*. When writing longhand or using a typewriter, it is customary to draw a line or an arrow over a vector quantity to show its vector character. (CAUTION: This is the first pitfall. Sloppy notation, such as the omission of the line or arrow symbol for a vector, is the major cause of errors in vector analysis.)

1.2 VECTOR ALGEBRA

With the definition of vectors and vector fields now accomplished, we may proceed to define the rules of vector arithmetic, vector algebra, and (later) vector calculus. Some of the rules will be similar to those of scalar algebra, some will differ slightly, and some will be entirely new and strange. This is to be expected, for a vector represents more information than does a scalar, and the multiplication of two vectors, for example, will be more involved than the multiplication of two scalars.

The rules are those of a branch of mathematics which is firmly established. Everyone "plays by the same rules," and we, of course, are merely going to look at and interpret these rules. However, it is enlightening to consider ourselves pioneers in the field. We are making our own rules, and we can make any rules we wish. The only requirement is that the rules be self-consistent. Of course, it would be nice if the rules agreed with those of scalar algebra where possible, and it would be even nicer if the rules enabled us to solve a few practical problems.

To begin, the addition of vectors follows the parallelogram law, and this is easily, if inaccurately, accomplished graphically. Figure 1.1 shows the sum of two vectors, **A** and **B**. It is easily seen that $\mathbf{A} + \mathbf{B} = \mathbf{B} + \mathbf{A}$, or that vector addition obeys the commutative law. Vector addition also obeys the associative law,

$$\mathbf{A} + (\mathbf{B} + \mathbf{C}) = (\mathbf{A} + \mathbf{B}) + \mathbf{C}$$

Fig. 1.1 Two vectors may be added graphically either by drawing both vectors from a common origin and completing the parallelogram or by beginning the second vector from the head of the first and completing the triangle; either method is easily extended to three or more vectors.

Note that when a vector is drawn as an arrow of finite length, its location is defined to be at the tail end of the arrow.

Coplanar vectors are vectors lying in a common plane, such as those shown in Figure 1.1. Both lie in the plane of the paper and may be added by expressing each vector in terms of "horizontal" and "vertical" components and then adding the corresponding components.

Vectors in three dimensions may likewise be added by expressing the vectors in terms of three components and adding the corresponding components. Examples of this process of addition will be given after vector components are discussed in Section 1.4.

The rule for the subtraction of vectors follows easily from that for addition, for we may always express $\mathbf{A} - \mathbf{B}$ as $\mathbf{A} + (-\mathbf{B})$; the sign, or direction, of the second vector is reversed, and this vector is then added to the first by the rule for vector addition.

Vectors may be multiplied by scalars. The magnitude of the vector changes, but its direction does not when the scalar is positive, although it reverses direction when multiplied by a negative scalar. Multiplication of a vector by a scalar also obeys the associative and distributive laws of algebra, leading to

$$(r + s)(\mathbf{A} + \mathbf{B}) = r(\mathbf{A} + \mathbf{B}) + s(\mathbf{A} + \mathbf{B}) = r\mathbf{A} + r\mathbf{B} + s\mathbf{A} + s\mathbf{B}$$

Division of a vector by a scalar is merely multiplication by the reciprocal of that scalar.

The multiplication of a vector by a vector is discussed in Sections 1.6 and 1.7.

Two vectors are said to be equal if their difference is zero, or $\mathbf{A} = \mathbf{B}$ if $\mathbf{A} - \mathbf{B} = 0$.

In our use of vector fields we shall always add and subtract vectors which are defined at the same point. For example, the *total* magnetic field about a small horseshoe magnet will be shown to be the sum of the fields produced by the earth and the permanent magnet; the total field at any point is the sum of the individual fields at that point.

If we are not considering a vector *field*, however, we may add or subtract vectors which are not defined at the same point. For example, the sum of the gravitational force acting on a 150 lb_f (pound-force) man at the North Pole and that acting on a 175 lb_f man at the South Pole may be obtained by shifting each force vector to the South Pole before addition. The resultant is a force of 25 lb_f directed toward the center of the earth at the South Pole; if we wanted to be difficult, we could just as well describe the force as 25 lb_f directed *away* from the center of the earth (or "upward") at the North Pole.[2]

1.3 THE RECTANGULAR COORDINATE SYSTEM

In order to describe a vector accurately, some specific lengths, directions, angles, projections, or components must be given. There are three simple methods of doing this, and about eight or ten other methods which are useful in very special cases. We are going to use only the three simple methods, and the simplest of these is the *rectangular*, or *rectangular cartesian* coordinate system.

In the rectangular coordinate system we set up three coordinate axes mutually at right angles to each other and call them the x, y, and z axes. It is customary to choose a *right-handed* coordinate system, in which a rotation (through the smaller angle) of the x axis into the y axis would cause a right-handed screw to progress in the direction of the z axis. If the right hand is used, then the

---

[2] A few students have argued that the force might be described at the equator as being in a "northerly" direction. They are right, but enough is enough.

thumb, forefinger, and middle finger may be identified, respectively, as the x, y, and z axes. Figure 1.2a shows a right-handed rectangular coordinate system.

A point is located by giving its x, y, and z coordinates. These are, respectively, the distances from the origin to the intersection of a perpendicular dropped from the point to the x, y, and z axes. An alternative method of interpreting coordinate values, and a method corresponding to that which *must* be used in all other coordinate systems, is to consider the point as being at the common intersection of three surfaces, the planes $x =$ constant, $y =$ constant, and $z =$ constant, the constants being the coordinate values of the point.

Figure 1.2b shows the points P and Q whose coordinates are $(1, 2, 3)$ and $(2, -2, 1)$, respectively. Point P is therefore located at the common point of intersection of the planes $x = 1$, $y = 2$, and $z = 3$, while point Q is located at the intersection of the planes $x = 2$, $y = -2$, $z = 1$.

As we encounter other coordinate systems in Sections 1.8 and 1.9, we should expect points to be located at the common intersection of three surfaces, not necessarily planes, but still mutually perpendicular at the point of intersection.

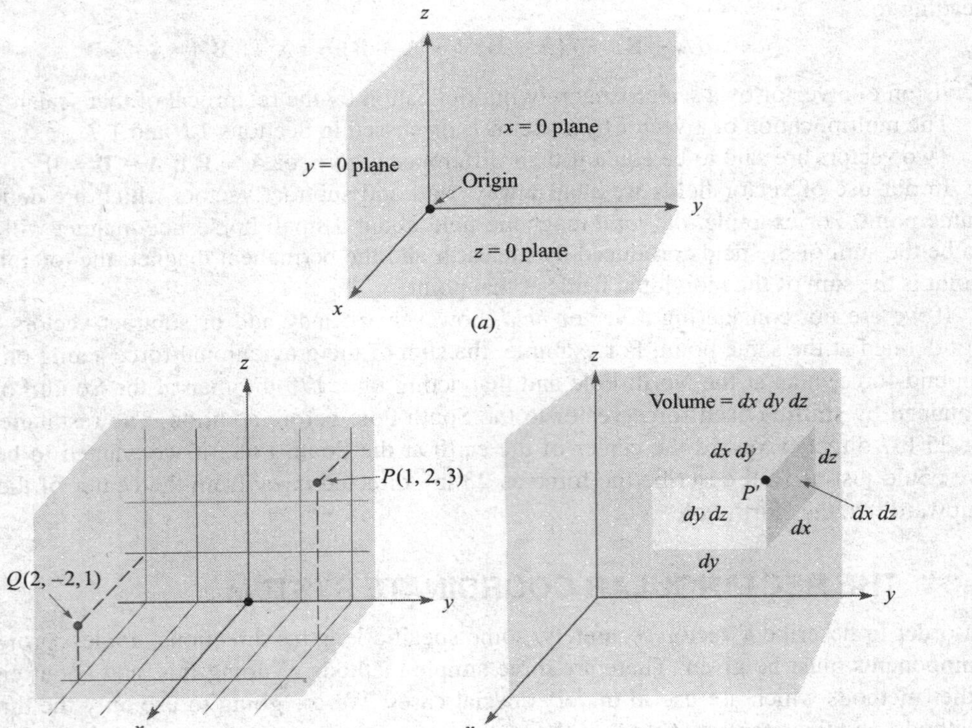

Fig. 1.2 (a) A right-handed rectangular coordinate system. If the curved fingers of the right hand indicate the direction through which the x axis is turned into coincidence with the y axis, the thumb shows the direction of the z axis. (b) The location of points $P(1, 2, 3)$ and $Q(2, -2, 1)$. (c) The differential volume element in rectangular coordinates; dx, dy, and dz are, in general, independent differentials.

If we visualize three planes intersecting at the general point P, whose coordinates are $x, y,$ and z, we may increase each coordinate value by a differential amount and obtain three slightly displaced planes intersecting at point P', whose coordinates are $x + dx$, $y + dy$, and $z + dz$. The six planes define a rectangular parallelepiped whose volume is $dv = dxdydz$; the surfaces have differential areas dS of $dxdy$, $dydz$, and $dzdx$. Finally, the distance dL from P to P' is the diagonal of the parallelepiped and has a length of $\sqrt{(dx)^2 + (dy)^2 + (dz)^2}$. The volume element is shown in Figure 1.2c; point P' is indicated, but point P is located at the only invisible corner.

All this is familiar from trigonometry or solid geometry and as yet involves only scalar quantities. We shall begin to describe vectors in terms of a coordinate system in the next section.

1.4 VECTOR COMPONENTS AND UNIT VECTORS

To describe a vector in the rectangular coordinate system, let us first consider a vector $\mathbf{r}$ extending outward from the origin. A logical way to identify this vector is by giving the three *component vectors*, lying along the three coordinate axes, whose vector sum must be the given vector. If the component vectors of the vector $\mathbf{r}$ are $\mathbf{x}$, $\mathbf{y}$, and $\mathbf{z}$, then $\mathbf{r} = \mathbf{x} + \mathbf{y} + \mathbf{z}$. The component vectors are shown in Figure 1.3a. Instead of one vector, we now have three, but this is a step forward because the three vectors are of a very simple nature; each is always directed along one of the coordinate axes.

In other words, the component vectors have magnitudes which depend on the given vector (such as $\mathbf{r}$), but they each have a known and constant direction. This suggests the use of *unit vectors* having unit magnitude, by definition, and directed along the coordinate axes in the direction of the increasing coordinate values. We shall reserve the symbol $\mathbf{a}$ for a unit vector and identify the direction of the unit vector by an appropriate subscript. Thus $\mathbf{a}_x$, $\mathbf{a}_y$, and $\mathbf{a}_z$ are the unit vectors in the rectangular coordinate system.[3] They are directed along the $x, y,$ and z axes, respectively, as shown in Figure 1.3b.

If the component vector $\mathbf{y}$ happens to be two units in magnitude and directed toward increasing values of y, we should then write $\mathbf{y} = 2\mathbf{a}_y$. A vector $\mathbf{r}_P$ pointing from the origin to point $P(1, 2, 3)$ is written $\mathbf{r}_P = \mathbf{a}_x + 2\mathbf{a}_y + 3\mathbf{a}_z$. The vector from P to Q may be obtained by applying the rule of vector addition. This rule shows that the vector from the origin to P plus the vector from P to Q is equal to the vector from the origin to Q. The desired vector from $P(1, 2, 3)$ to $Q(2, -2, 1)$ is therefore

$$\begin{aligned}\mathbf{R}_{PQ} &= \mathbf{r}_Q - \mathbf{r}_P = (2-1)\mathbf{a}_x + (-2-2)\mathbf{a}_y + (1-3)\mathbf{a}_z \\ &= \mathbf{a}_x - 4\mathbf{a}_y - 2\mathbf{a}_z\end{aligned}$$

The vectors $\mathbf{r}_P$, $\mathbf{r}_Q$, and $\mathbf{R}_{PQ}$ are shown in Figure 1.3c.

This last vector does not extend outward from the origin, as did the vector $\mathbf{r}$ we initially considered. However, we have already learned that vectors having the same magnitude and pointing in the same direction are equal, so we see that to help our visualization processes we are at liberty to slide any vector over to the origin before determining its component vectors. Parallelism must, of course, be maintained during the sliding process.

If we are discussing a force vector $\mathbf{F}$, or indeed any vector other than a displacement-type vector such as $\mathbf{r}$, the problem arises of providing suitable letters for the three component vectors.

[3] The symbols $\mathbf{i}, \mathbf{j}$, and $\mathbf{k}$ are also commonly used for the unit vectors in rectangular coordinates.

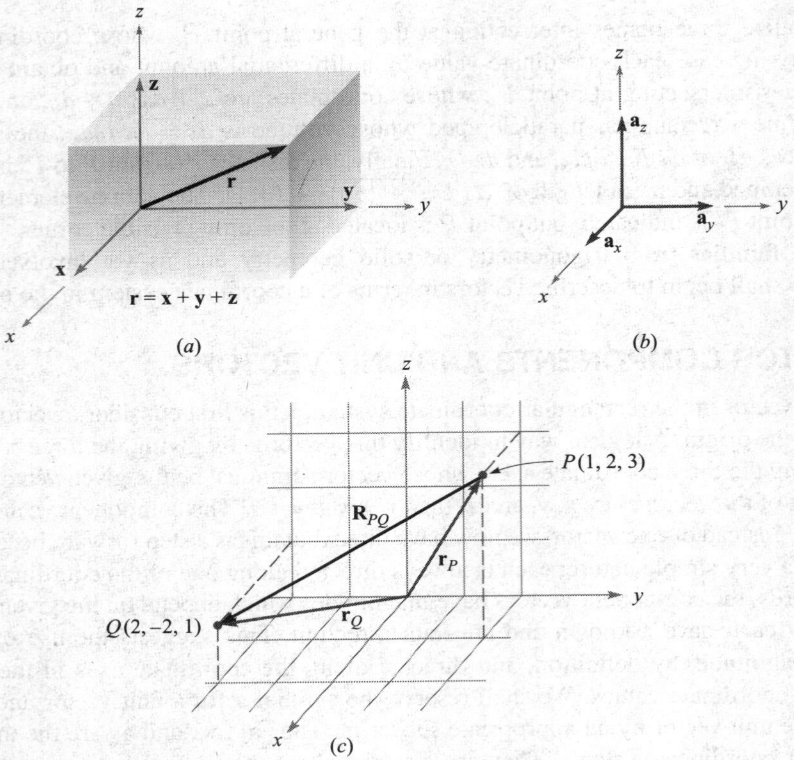

Fig. 1.3 (a) The component vectors **x**, **y**, and **z** of vector **r**. (b) The unit vectors of the rectangular coordinate system have unit magnitude and are directed toward increasing values of their respective variables. (c) The vector $\mathbf{R}_{PQ}$ is equal to the vector difference $\mathbf{r}_Q - \mathbf{r}_P$.

It would not do to call them **x**, **y**, and **z**, for these are displacements, or directed distances, and are measured in meters (abbreviated m) or some other unit of length. The problem is most often avoided by using *component scalars*, simply called *components*, F_x, F_y, and F_z. The components are the signed magnitudes of the component vectors. We may then write $\mathbf{F} = F_x\mathbf{a}_x + F_y\mathbf{a}_y + F_z\mathbf{a}_z$. The component vectors are $F_x\mathbf{a}_x$, $F_y\mathbf{a}_y$, and $F_z\mathbf{a}_z$.

Any vector **B** then may be described by $\mathbf{B} = B_x\mathbf{a}_x + B_y\mathbf{a}_y + B_z\mathbf{a}_z$. The magnitude of **B** written $|\mathbf{B}|$ or simply B, is given by

$$|\mathbf{B}| = \sqrt{B_x^2 + B_y^2 + B_z^2} \tag{1}$$

Each of the three coordinate systems we discuss will have its three fundamental and mutually perpendicular unit vectors which are used to resolve any vector into its component vectors. However, unit vectors are not limited to this application. It is often helpful to be able to write a unit vector having a specified direction. This is easily done, for a unit vector in a given direction is merely a vector in that direction divided by its magnitude. A unit vector in the **r** direction is $\mathbf{r}/\sqrt{x^2+y^2+z^2}$, and a unit vector in the direction of the vector **B** is

$$\mathbf{a}_B = \frac{\mathbf{B}}{\sqrt{B_x^2+B_y^2+B_z^2}} = \frac{\mathbf{B}}{|\mathbf{B}|} \tag{2}$$

EXAMPLE 1.1

Specify the unit vector extending from the origin toward the point $G(2, -2, -1)$.

Solution. We first construct the vector extending from the origin to point G,

$$\mathbf{G} = 2\mathbf{a}_x - 2\mathbf{a}_y - \mathbf{a}_z$$

We continue by finding the magnitude of **G**,

$$|\mathbf{G}| = \sqrt{(2)^2+(-2)^2+(-1)^2} = 3$$

and finally expressing the desired unit vector as the quotient,

$$\mathbf{a}_G = \frac{\mathbf{G}}{|\mathbf{G}|} = \tfrac{2}{3}\mathbf{a}_x - \tfrac{2}{3}\mathbf{a}_y - \tfrac{1}{3}\mathbf{a}_z = 0.667\mathbf{a}_x - 0.667\mathbf{a}_y - 0.333\mathbf{a}_z$$

A special identifying symbol is desirable for a unit vector so that its character is immediately apparent. Symbols which have been used are $\mathbf{u}_B, \mathbf{a}_B, \mathbf{1_B}$, or even **b**. We shall consistently use the lowercase **a** with an appropriate subscript.

[NOTE: Throughout the text, drill problems appear following sections in which a new principle is introduced in order to allow students to test their understanding of the basic fact itself. The problems are useful in gaining familiarity with new terms and ideas and should all be worked. More general problems appear at the ends of the chapters. The answers to the drill problems are given in the same order as the parts of the problem.]

D1.1 Given points $M(-1, 2, 1)$, $N(3, -3, 0)$, and $P(-2, -3, -4)$, find: (*a*) $\mathbf{R}_{MN}$; (*b*) $\mathbf{R}_{MN} + \mathbf{R}_{MP}$; (*c*) $|\mathbf{r}_M|$; (*d*) $\mathbf{a}_{MP}$; (*e*) $|2\mathbf{r}_P - 3\mathbf{r}_N|$.

Ans. $4\mathbf{a}_x - 5\mathbf{a}_y - \mathbf{a}_z$; $3\mathbf{a}_x - 10\mathbf{a}_y - 6\mathbf{a}_z$; 2.45; $-0.14\mathbf{a}_x - 0.7\mathbf{a}_y - 0.7\mathbf{a}_z$; 15.56

1.5 THE VECTOR FIELD

We have already defined a vector field as a vector function of a position vector. In general, the magnitude and direction of the function will change as we move throughout the region, and the value of the vector function must be determined using the coordinate values of the point in question. Since we have considered only the rectangular coordinate system, we should expect the vector to be a function of the variables x, y, and z.

If we again represent the position vector as **r**, then a vector field **G** can be expressed in functional notation as **G**(**r**); a scalar field T is written as T(**r**).

If we inspect the velocity of the water in the ocean in some region near the surface where tides and currents are important, we might decide to represent it by a velocity vector which is in any direction, even up or down. If the z axis is taken as upward, the x axis in a northerly direction, the y axis to the west, and the origin at the surface, we have a right-handed coordinate system and may write the velocity vector as $\mathbf{v} = v_x\mathbf{a}_x + v_y\mathbf{a}_y + v_z\mathbf{a}_z$, or $\mathbf{v}(\mathbf{r}) = v_x(\mathbf{r})\mathbf{a}_x + v_y(\mathbf{r})\mathbf{a}_y + v_z(\mathbf{r})\mathbf{a}_z$; each of the components v_x, v_y, and v_z may be a function of the three variables x, y, and z. If the problem is simplified by assuming that we are in some portion of the Gulf Stream where the water is moving only to the north, then v_y, and v_z are zero. Further simplifying assumptions might be made if the velocity falls off with depth and changes very slowly as we move north, south, east, or west. A suitable expression could be $\mathbf{v} = 2e^{z/100}\mathbf{a}_x$. We have a velocity of 2 m/s (meters per second) at the surface and a velocity of 0.368×2, or 0.736 m/s, at a depth of 100 m ($z = -100$), and the velocity continues to decrease with depth; in this example the vector velocity has a constant direction.

While the preceding example is fairly simple and only a rough approximation to a physical situation, a more exact expression would be correspondingly more complex and difficult to interpret. We shall come across many fields in our study of electricity and magnetism which are simpler than the velocity example, an example in which only the component and one variable were involved (the x component and the variable z). We shall also study more complicated fields, and methods of interpreting these expressions physically will be discussed then.

D1.2 A vector field $\mathbf{S}$ is expressed in rectangular coordinates as $\mathbf{S} = \{125/[(x-1)^2 + (y-2)^2 + (z+1)^2]\}\{(x-1)\mathbf{a}_x + (y-2)\mathbf{a}_y + (z+1)\mathbf{a}_z\}$. (*a*) Evaluate $\mathbf{S}$ at $P(2, 4, 3)$. (*b*) Determine a unit vector that gives the direction of $\mathbf{S}$ at P. (*c*) Specify the surface $f(x, y, z)$ on which $|\mathbf{S}| = 1$.

Ans. $5.95\mathbf{a}_x + 11.90\mathbf{a}_y + 23.8\mathbf{a}_z$; $0.218\mathbf{a}_x + 0.436\mathbf{a}_y + 0.873\mathbf{a}_z$; $\sqrt{(x-1)^2 + (y-2)^2 + (z+1)^2} = 125$

1.6 THE DOT PRODUCT

We now consider the first of two types of vector multiplication. The second type will be discussed in the following section.

Given two vectors **A** and **B**, the *dot product*, or *scalar product*, is defined as the product of the magnitude of **A**, the magnitude of **B**, and the cosine of the smaller angle between them,

$$\mathbf{A} \cdot \mathbf{B} = |\mathbf{A}|\,|\mathbf{B}| \cos\theta_{AB} \tag{3}$$

The dot appears between the two vectors and should be made heavy for emphasis. The dot, or scalar, product is a scalar, as one of the names implies, and it obeys the commutative law,

$$\mathbf{A} \cdot \mathbf{B} = \mathbf{B} \cdot \mathbf{A} \tag{4}$$

for the sign of the angle does not affect the cosine term. The expression $\mathbf{A} \cdot \mathbf{B}$ is read "**A** dot **B**."

Perhaps the most common application of the dot product is in mechanics, where a constant force **F** applied over a straight displacement **L** does an amount of work $FL\cos\theta$, which is more easily written $\mathbf{F} \cdot \mathbf{L}$. We might anticipate one of the results of Chapter 4 by pointing out that if the force varies along the path, integration is necessary to find the total work, and the result becomes

$$\text{Work} = \int \mathbf{F} \cdot d\mathbf{L}$$

Another example might be taken from magnetic fields, a subject about which we shall have a lot more to say later. The total flux Φ crossing a surface of area S is given by BS if the magnetic flux density B is perpendicular to the surface and uniform over it. We define a *vector surface* **S** as having the usual area for its magnitude and having a direction *normal* to the surface (avoiding for the moment the problem of which of the two possible normals to take). The flux crossing the surface is then $\mathbf{B} \cdot \mathbf{S}$. This expression is valid for any direction of the uniform magnetic flux density. However, if the flux density is not constant over the surface, the total flux is $\Phi = \int \mathbf{B} \cdot d\mathbf{S}$. Integrals of this general form appear in Chapter 3 when we study electric flux density.

Finding the angle between two vectors in three-dimensional space is often a job we would prefer to avoid, and for that reason the definition of the dot product is usually not used in its basic form. A more helpful result is obtained by considering two vectors whose rectangular components are given, such as $\mathbf{A} = A_x\mathbf{a}_x + A_y\mathbf{a}_y + A_z\mathbf{a}_z$ and $\mathbf{B} = B_x\mathbf{a}_x + B_y\mathbf{a}_y + B_z\mathbf{a}_z$. The dot product also obeys the distributive law, and, therefore, $\mathbf{A} \cdot \mathbf{B}$ yields the sum of nine scalar terms, each involving the dot product of two unit vectors. Since the angle between two different unit vectors of the rectangular coordinate system is 90°, we then have

$$\mathbf{a}_x \cdot \mathbf{a}_y = \mathbf{a}_y \cdot \mathbf{a}_x = \mathbf{a}_x \cdot \mathbf{a}_z = \mathbf{a}_z \cdot \mathbf{a}_x = \mathbf{a}_y \cdot \mathbf{a}_z = \mathbf{a}_z \cdot \mathbf{a}_y = 0$$

The remaining three terms involve the dot product of a unit vector with itself, which is unity, giving finally

$$\mathbf{A} \cdot \mathbf{B} = A_x B_x + A_y B_y + A_z B_z \tag{5}$$

which is an expression involving no angles.

A vector dotted with itself yields the magnitude squared, or

$$\mathbf{A} \cdot \mathbf{A} = A^2 = |\mathbf{A}|^2 \tag{6}$$

and any unit vector dotted with itself is unity,

$$\mathbf{a}_A \cdot \mathbf{a}_A = 1$$

One of the most important applications of the dot product is that of finding the component of a vector in a given direction. Referring to Figure 1.4*a*, we can obtain the component (scalar) of **B** in the direction specified by the unit vector **a** as

$$\mathbf{B} \cdot \mathbf{a} = |\mathbf{B}|\,|\mathbf{a}| \cos\theta_{Ba} = |\mathbf{B}| \cos\theta_{Ba}$$

The sign of the component is positive if $0 \leq \theta_{Ba} \leq 90°$ and negative whenever $90° \leq \theta_{Ba} \leq 180°$.

In order to obtain the component *vector* of **B** in the direction of **a**, we simply multiply the component (scalar) by **a**, as illustrated by Figure 1.4*b*. For example, the component of **B** in the direction of $\mathbf{a}_x$ is $\mathbf{B} \cdot \mathbf{a}_x = B_x$, and the component vector is $B_x\mathbf{a}_x$, or $(\mathbf{B} \cdot \mathbf{a}_x)\mathbf{a}_x$. Hence, the problem of finding the component of a vector in any desired direction becomes the problem of finding a unit vector in that direction, and that we can do.

The geometrical term *projection* is also used with the dot product. Thus, $\mathbf{B} \cdot \mathbf{a}$ is the projection of **B** in the **a** direction.

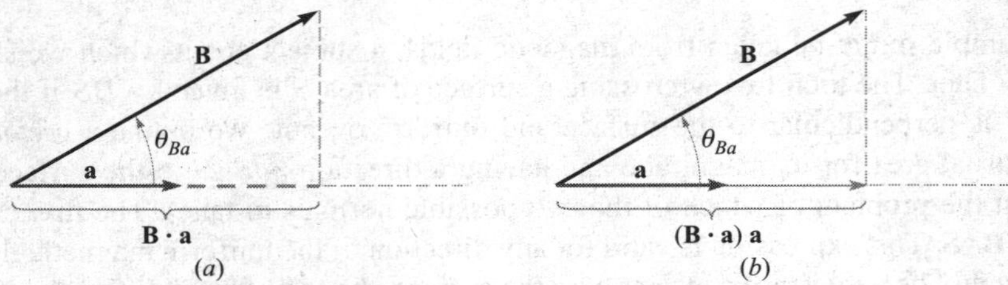

Fig. 1.4 (*a*) The scalar component of **B** in the direction of the unit vector **a** is **B** · **a**. (*b*) The vector component of **B** in the direction of the unit vector **a** is (**B** · **a**)**a**.

EXAMPLE 1.2

In order to illustrate these definitions and operations, let us consider the vector field $\mathbf{G} = y\mathbf{a}_x - 2.5x\mathbf{a}_y + 3\mathbf{a}_z$ and the point $Q(4, 5, 2)$. We wish to find: **G** at Q; the scalar component of **G** at Q in the direction of $\mathbf{a}_N = \frac{1}{3}(2\mathbf{a}_x + \mathbf{a}_y - 2\mathbf{a}_z)$; the vector component of **G** at Q in the direction of $\mathbf{a}_N$; and finally, the angle θ_{Ga} between $\mathbf{G}(\mathbf{r}_Q)$ and $\mathbf{a}_N$.

Solution. Substituting the coordinates of point Q into the expression for **G**, we have

$$\mathbf{G}(\mathbf{r}_Q) = 5\mathbf{a}_x - 10\mathbf{a}_y + 3\mathbf{a}_z$$

Next we find the scalar component. Using the dot product, we have

$$\mathbf{G}\cdot\mathbf{a}_N = (5\mathbf{a}_x - 10\mathbf{a}_y + 3\mathbf{a}_z)\cdot\tfrac{1}{3}(2\mathbf{a}_x + \mathbf{a}_y - 2\mathbf{a}_z) = \tfrac{1}{3}(10 - 10 - 6) = -2$$

The vector component is obtained by multiplying the scalar component by the unit vector in the direction of $\mathbf{a}_N$,

$$(\mathbf{G}\cdot\mathbf{a}_N)\mathbf{a}_N = -(2)\tfrac{1}{3}(2\mathbf{a}_x + \mathbf{a}_y - 2\mathbf{a}_z) = -1.333\mathbf{a}_x - 0.667\mathbf{a}_y + 1.333\mathbf{a}_z$$

The angle between $\mathbf{G}(\mathbf{r}_Q)$ and $\mathbf{a}_N$ is found from

$$\mathbf{G}\cdot\mathbf{a}_N = |\mathbf{G}|\cos\theta_{Ga}$$

$$-2 = \sqrt{25 + 100 + 9}\cos\theta_{Ga}$$

and

$$\theta_{Ga} = \cos^{-1}\frac{-2}{\sqrt{134}} = 99.9^\circ$$

D1.3 The three vertices of a triangle are located at $A(6, -1, 2)$, $B(-2, 3, -4)$, and $C(-3, 1, 5)$. Find: (*a*) $\mathbf{R}_{AB}$; (*b*) $\mathbf{R}_{AC}$; (*c*) the angle θ_{BAC} at vertex A; (*d*) the (vector) projection of $\mathbf{R}_{AB}$ on $\mathbf{R}_{AC}$.

Ans. $-8\mathbf{a}_x + 4\mathbf{a}_y - 6\mathbf{a}_z$; $-9\mathbf{a}_x + 2\mathbf{a}_y + 3\mathbf{a}_z$; 53.6°; $-5.94\mathbf{a}_x + 1.319\mathbf{a}_y + 1.979\mathbf{a}_z$

1.7 THE CROSS PRODUCT

Given two vectors **A** and **B**, we shall now define the *cross product*, or *vector product*, of **A** and **B**, written with a cross between the two vectors as $\mathbf{A} \times \mathbf{B}$ and read "**A** cross **B**." The cross product $\mathbf{A} \times \mathbf{B}$ is a vector; the magnitude of $\mathbf{A} \times \mathbf{B}$ is equal to the product of the magnitudes of **A**, **B**, and the sine of the smaller angle between **A** and **B**; the direction of $\mathbf{A} \times \mathbf{B}$ is perpendicular to the plane containing **A** and **B** and is along that one of the two possible perpendiculars which is in the direction of advance of a right-handed screw as **A** is turned into **B**. This direction is illustrated in Figure 1.5. Remember that either vector may be moved about at will, maintaining its direction constant, until the two vectors have a "common origin." This determines the plane containing both. However, in most of our applications we shall be concerned with vectors defined at the same point.

As an equation we can write

$$\mathbf{A} \times \mathbf{B} = \mathbf{a}_N |\mathbf{A}|\,|\mathbf{B}| \sin\theta_{AB} \tag{7}$$

where an additional statement, such as that given above, is still required to explain the direction of the unit vector $\mathbf{a}_N$. The subscript stands for "normal."

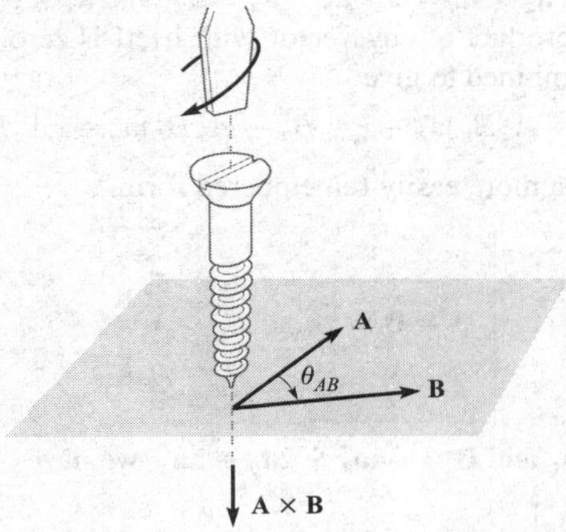

Fig. 1.5 The direction of $\mathbf{A} \times \mathbf{B}$ is in the direction of advance of a right-handed screw as **A** is turned into **B**.

Reversing the order of the vectors **A** and **B** results in a unit vector in the opposite direction, and we see that the cross product is not commutative, for $\mathbf{B} \times \mathbf{A} = -(\mathbf{A} \times \mathbf{B})$. If the definition of the cross product is applied to the unit vectors $\mathbf{a}_x$ and $\mathbf{a}_y$, we find $\mathbf{a}_x \times \mathbf{a}_y = \mathbf{a}_z$, for each vector has unit magnitude, the two vectors are perpendicular, and the rotation of $\mathbf{a}_x$ into $\mathbf{a}_y$ indicates the positive z direction by the definition of a right-handed coordinate system. In a similar way $\mathbf{a}_y \times \mathbf{a}_z = \mathbf{a}_x$, and $\mathbf{a}_z \times \mathbf{a}_x = \mathbf{a}_y$. Note the alphabetic symmetry. As long as the three vectors $\mathbf{a}_x$, $\mathbf{a}_y$, and $\mathbf{a}_z$ are written in order (and assuming that $\mathbf{a}_x$ follows $\mathbf{a}_z$, like three elephants in a circle holding tails, so that we could also write $\mathbf{a}_y$, $\mathbf{a}_z$, $\mathbf{a}_x$ or $\mathbf{a}_z$, $\mathbf{a}_x$, $\mathbf{a}_y$), then the cross and equal sign may be placed in either of the two vacant spaces. As a matter of fact, it is now simpler to define a right-handed rectangular coordinate system by saying that $\mathbf{a}_x \times \mathbf{a}_y = \mathbf{a}_z$.

A simple example of the use of the cross product may be taken from geometry or trigonometry. To find the area of a parallelogram, the product of the lengths of two adjacent sides is multiplied by the sine of the angle between them. Using vector notation for the two sides, we then may express the (scalar) area as the *magnitude* of $\mathbf{A} \times \mathbf{B}$, or $|\mathbf{A} \times \mathbf{B}|$.

The cross product may be used to replace the right-hand rule familiar to all electrical engineers. Consider the force on a straight conductor of length $\mathbf{L}$, where the direction assigned to $\mathbf{L}$ corresponds to the direction of the steady current I, and a uniform magnetic field of flux density $\mathbf{B}$ is present. Using vector notation, we may write the result neatly as $\mathbf{F} = I\mathbf{L} \times \mathbf{B}$. This relationship will be obtained later in Chapter 9.

The evaluation of a cross product by means of its definition turns out to be more work than the evaluation of the dot product from its definition, for not only must we find the angle between the vectors, but we must also find an expression for the unit vector $\mathbf{a}_N$. This work may be avoided by using rectangular components for the two vectors $\mathbf{A}$ and $\mathbf{B}$ and expanding the cross product as a sum of nine simpler cross products, each involving two unit vectors,

$$\begin{aligned}\mathbf{A} \times \mathbf{B} = &A_x B_x \mathbf{a}_x \times \mathbf{a}_x + A_x B_y \mathbf{a}_x \times \mathbf{a}_y + A_x B_z \mathbf{a}_x \times \mathbf{a}_z \\ &+ A_y B_x \mathbf{a}_y \times \mathbf{a}_x + A_y B_y \mathbf{a}_y \times \mathbf{a}_y + A_y B_z \mathbf{a}_y \times \mathbf{a}_z \\ &+ A_z B_x \mathbf{a}_z \times \mathbf{a}_x + A_z B_y \mathbf{a}_z \times \mathbf{a}_y + A_z B_z \mathbf{a}_z \times \mathbf{a}_z\end{aligned}$$

We have already found that $\mathbf{a}_x \times \mathbf{a}_y = \mathbf{a}_z$, $\mathbf{a}_y \times \mathbf{a}_z = \mathbf{a}_x$, and $\mathbf{a}_z \times \mathbf{a}_x = \mathbf{a}_y$. The three remaining terms are zero, for the cross product of any vector with itself is zero, since the included angle is zero. These results may be combined to give

$$\mathbf{A} \times \mathbf{B} = (A_y B_z - A_z B_y)\mathbf{a}_x + (A_z B_x - A_x B_z)\mathbf{a}_y + (A_x B_y - A_y B_x)\mathbf{a}_z \tag{8}$$

or written as a determinant in a more easily remembered form,

$$\mathbf{A} \times \mathbf{B} = \begin{vmatrix} \mathbf{a}_x & \mathbf{a}_y & \mathbf{a}_z \\ A_x & A_y & A_z \\ B_x & B_y & B_z \end{vmatrix} \tag{9}$$

Thus, if $\mathbf{A} = 2\mathbf{a}_x - 3\mathbf{a}_y + \mathbf{a}_z$ and $\mathbf{B} = -4\mathbf{a}_x - 2\mathbf{a}_y + 5\mathbf{a}_z$, we have

$$\begin{aligned}\mathbf{A} \times \mathbf{B} &= \begin{vmatrix} \mathbf{a}_x & \mathbf{a}_y & \mathbf{a}_z \\ 2 & -3 & 1 \\ -4 & -2 & 5 \end{vmatrix} \\ &= [(-3)(5) - (1(-2)]\mathbf{a}_x - [(2)(5) - (1)(-4)]\mathbf{a}_y + [(2)(-2) - (-3)(-4)]\mathbf{a}_z \\ &= -13\mathbf{a}_x - 14\mathbf{a}_y - 16\mathbf{a}_z\end{aligned}$$

D1.4 The three vertices of a triangle are located at $A(6, -1, 2)$, $B(-2, 3, -4)$, and $C(-3, 1, 5)$. Find: (*a*) $\mathbf{R}_{AB} \times \mathbf{R}_{AC}$; (*b*) the area of the triangle; (*c*) a unit vector perpendicular to the plane in which the triangle is located.

Ans. $24\mathbf{a}_x + 78\mathbf{a}_y + 20\mathbf{a}_z$; 42.0; $0.286\mathbf{a}_x + 0.928\mathbf{a}_y + 0.238\mathbf{a}_z$

1.8 OTHER COORDINATE SYSTEMS: CIRCULAR CYLINDRICAL COORDINATES

The rectangular coordinate system is generally the one in which students prefer to work every problem. This often means a lot more work for the student, because many problems possess a type of symmetry which pleads for a more logical treatment. It is easier to do now, once and for all, the work required to become familiar with cylindrical and spherical coordinates, instead of applying an equal or greater effort to every problem involving cylindrical or spherical symmetry later. With this future saving of labor in mind, we shall take a careful and unhurried look at cylindrical and spherical coordinates.

The circular cylindrical coordinate system is the three-dimensional version of the polar coordinates of analytic geometry. In the two-dimensional polar coordinates, a point was located in a plane by giving its distance ρ from the origin, and the angle ϕ between the line from the point to the origin and an arbitrary radial line, taken as $\phi = 0$.[4] A three-dimensional coordinate system, circular cylindrical coordinates, is obtained by also specifying the distance z of the point from an arbitrary $z = 0$ reference plane which is perpendicular to the line $\rho = 0$. For simplicity, we usually refer to circular cylindrical coordinates simply as cylindrical coordinates. This will not cause any confusion in reading this book, but it is only fair to point out that there are such systems as elliptic cylindrical coordinates, hyperbolic cylindrical coordinates, parabolic cylindrical coordinates, and others.

We no longer set up three axes as with rectangular coordinates, but we must instead consider any point as the intersection of three mutually perpendicular surfaces. These surfaces are a circular cylinder ($\rho =$ constant), a plane ($\phi =$ constant), and another plane ($z =$ constant). This corresponds to the location of a point in a rectangular coordinate system by the intersection of three planes ($x =$ constant, $y =$ constant, and $z =$ constant). The three surfaces of circular cylindrical coordinates are shown in Figure 1.6a. Note that three such surfaces may be passed through any point, unless it lies on the z axis, in which case one plane suffices.

Three unit vectors must also be defined, but we may no longer direct them along the "coordinate axes," for such axes exist only in rectangular coordinates. Instead, we take a broader view of the unit vectors in rectangular coordinates and realize that they are directed toward increasing coordinate values and are perpendicular to the surface on which that coordinate value is constant (i.e., the unit vector $\mathbf{a}_x$ is normal to the plane $x =$ constant and points toward larger values of x). In a corresponding way we may now define three unit vectors in cylindrical coordinates, $\mathbf{a}_\rho$, $\mathbf{a}_\phi$, and $\mathbf{a}_z$.

The unit vector $\mathbf{a}_\rho$ at a point $P(\rho_1, \phi_1, z_1)$ is directed radially outward, normal to the cylindrical surface $\rho = \rho_1$. It lies in the planes $\phi = \phi_1$ and $z = z_1$. The unit vector $\mathbf{a}_\phi$ is normal to the plane $\phi = \phi_1$, points in the direction of increasing ϕ, lies in the plane $z = z_1$, and is tangent to the cylindrical surface $\rho = \rho_1$. The unit vector $\mathbf{a}_z$ is the same as the unit vector $\mathbf{a}_z$ of the rectangular coordinate system. Figure 1.6b shows the three vectors in cylindrical coordinates.

In rectangular coordinates, the unit vectors are not functions of the coordinates. Two of the unit vectors in cylindrical coordinates, $\mathbf{a}_\rho$ and $\mathbf{a}_\phi$, however, *do* vary with the coordinate ϕ, since their

[4] The two variables of polar coordinates are commonly called r and θ. With three coordinates, however, it is more common to use ρ for the radius variable of cylindrical coordinates and r for the (different) radius variable of spherical coordinates. Also, the angle variable of cylindrical coordinates is customarily called ϕ because everyone uses θ for a different angle in spherical coordinates. The angle ϕ is common to both cylindrical and spherical coordinates. See?

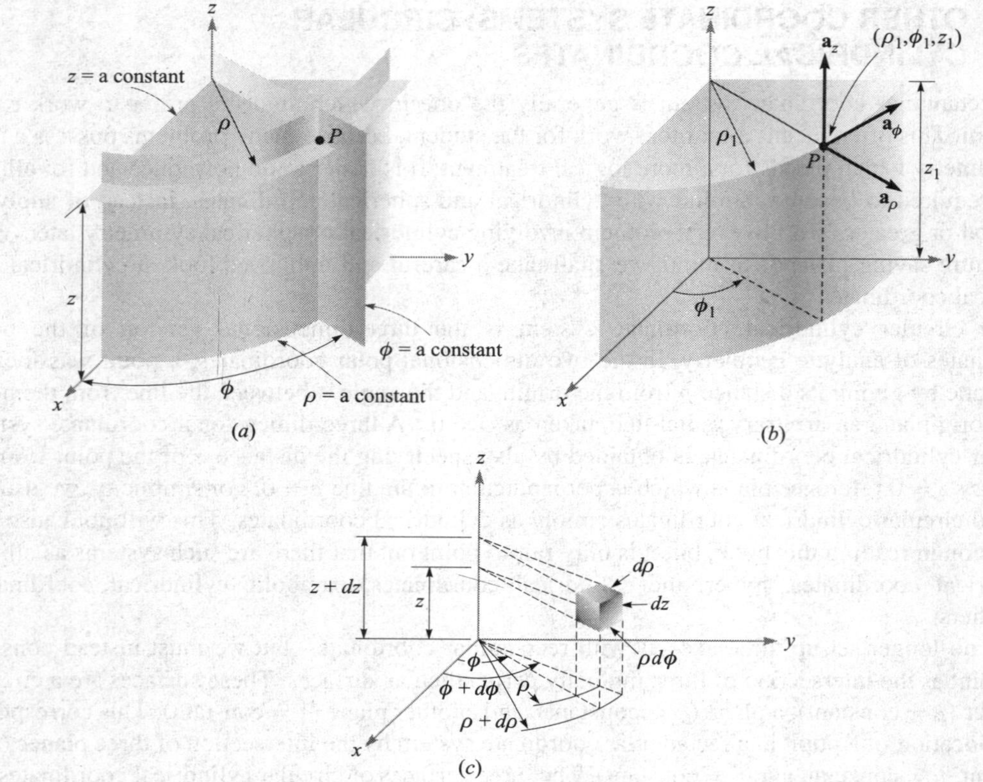

Fig. 1.6 (*a*) The three mutually perpendicular surfaces of the circular cylindrical coordinate system. (*b*) The three unit vectors of the circular cylindrical coordinate system. (*c*) The differential volume unit in the circular cylindrical coordinate system; $d\rho$, $\rho d\phi$, and dz are all elements of length.

directions change. In integration or differentiation with respect to ϕ, then, $\mathbf{a}_\rho$ and $\mathbf{a}_\phi$ must not be treated as constants.

The unit vectors are again mutually perpendicular, for each is normal to one of the three mutually perpendicular surfaces, and we may define a right-handed cylindrical coordinate system as one in which $\mathbf{a}_\rho \times \mathbf{a}_\phi = \mathbf{a}_z$, or (for those who have flexible fingers) as one in which the thumb, forefinger, and middle finger point in the direction of increasing ρ, ϕ, and z, respectively.

A differential volume element in cylindrical coordinates may be obtained by increasing ρ, ϕ, and z by the differential increments $d\rho$, $d\phi$, and dz. The two cylinders of radius ρ and $\rho + d\rho$, the two radial planes at angles ϕ and $\phi + d\phi$, and the two "horizontal" planes at "elevations" z and $z + dz$ now enclose a small volume, as shown in Figure 1.6*c*, having the shape of a truncated wedge. As the volume element becomes very small, its shape approaches that of a rectangular parallelepiped having sides of length $d\rho$, $\rho d\phi$ and dz. Note that $d\rho$ and dz are dimensionally lengths, but $d\phi$ is not; $\rho d\phi$ is the length. The surfaces have areas of $\rho\, d\rho\, d\phi$, $d\rho\, dz$, and $\rho\, d\phi\, dz$, and the volume becomes $\rho\, d\rho\, d\phi\, dz$.

The variables of the rectangular and cylindrical coordinate systems are easily related to each other. With reference to Figure 1.7, we see that

$$x = \rho \cos\phi$$
$$y = \rho \sin\phi \tag{10}$$
$$z = z$$

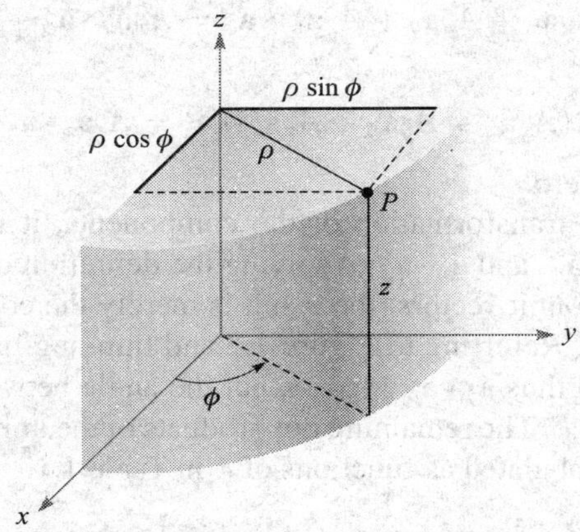

Fig. 1.7 The relationship between the rectangular variables x, y, z and the cylindrical coordinate variables ρ, ϕ, z. There is no change in the variable z between the two systems.

From the other viewpoint, we may express the cylindrical variables in terms of x, y, and z:

$$\rho = \sqrt{x^2 + y^2} \quad (\rho \geq 0)$$
$$\phi = \tan^{-1}\frac{y}{x} \tag{11}$$
$$z = z$$

We shall consider the variable ρ to be positive or zero, thus using only the positive sign for the radical in (11). The proper value of the angle ϕ is determined by inspecting the signs of x and y. Thus, if $x = -3$ and $y = 4$, we find that the point lies in the second quadrant so that $\rho = 5$ and $\phi = 126.9°$. For $x = 3$ and $y = -4$, we have $\phi = -53.1°$ or $306.9°$, whichever is more convenient.

Using (10) or (11), scalar functions given in one coordinate system are easily transformed into the other system.

A vector function in one coordinate system, however, requires two steps in order to transform it to another coordinate system, because a different set of component vectors is generally required. That is, we may be given a rectangular vector

$$\mathbf{A} = A_x\mathbf{a}_x + A_y\mathbf{a}_y + A_z\mathbf{a}_z$$

where each component is given as a function of x, y, and z, and we need a vector in cylindrical coordinates

$$\mathbf{A} = A_\rho\mathbf{a}_\rho + A_\phi\mathbf{a}_\phi + A_z\mathbf{a}_z$$

where each component is given as a function of ρ, ϕ, and z.

To find any desired component of a vector, we recall from the discussion of the dot product that a component in a desired direction may be obtained by taking the dot product of the vector and a unit vector in the desired direction. Hence,

$$A_\rho = \mathbf{A} \cdot \mathbf{a}_\rho \qquad \text{and} \qquad A_\phi = \mathbf{A} \cdot \mathbf{a}_\phi$$

Expanding these dot products, we have

$$A_\rho = (A_x\mathbf{a}_x + A_y\mathbf{a}_y + A_z\mathbf{a}_z)\cdot\mathbf{a}_\rho = A_x\mathbf{a}_x\cdot\mathbf{a}_\rho + A_y\mathbf{a}_y\cdot\mathbf{a}_\rho \tag{12}$$

$$A_\phi = (A_x\mathbf{a}_x + A_y\mathbf{a}_y + A_z\mathbf{a}_z)\cdot\mathbf{a}_\phi = A_x\mathbf{a}_x\cdot\mathbf{a}_\phi + A_y\mathbf{a}_y\cdot\mathbf{a}_\phi \tag{13}$$

and

$$A_z = (A_x\mathbf{a}_x + A_y\mathbf{a}_y + A_z\mathbf{a}_z)\cdot\mathbf{a}_z = A_z\mathbf{a}_z\cdot\mathbf{a}_z = A_z \tag{14}$$

since $\mathbf{a}_z \cdot \mathbf{a}_\rho$ and $\mathbf{a}_z \cdot \mathbf{a}_\phi$ are zero.

In order to complete the transformation of the components, it is necessary to know the dot products $\mathbf{a}_x \cdot \mathbf{a}_\rho$, $\mathbf{a}_y \cdot \mathbf{a}_\rho$, $\mathbf{a}_x \cdot \mathbf{a}_\phi$, and $\mathbf{a}_y \cdot \mathbf{a}_\phi$. Applying the definition of the dot product, we see that since we are concerned with unit vectors, the result is merely the cosine of the angle between the two unit vectors in question. Referring to Figure 1.7 and thinking mightily, we identify the angle between $\mathbf{a}_x$ and $\mathbf{a}_\rho$ as ϕ, and thus $\mathbf{a}_x \cdot \mathbf{a}_\rho = \cos\phi$, but the angle between $\mathbf{a}_y$ and $\mathbf{a}_\rho$ is $90° - \phi$, and $\mathbf{a}_y \cdot \mathbf{a}_\rho = \cos(90° - \phi) = \sin\phi$. The remaining dot products of the unit vectors are found in a similar manner, and the results are tabulated as functions of ϕ in Table 1.1.

Table 1.1 Dot products of unit vectors in cylindrical and rectangular coordinate systems

	$\mathbf{a}_\rho$	$\mathbf{a}_\phi$	$\mathbf{a}_z$
$\mathbf{a}_x\cdot$	$\cos\phi$	$-\sin\phi$	0
$\mathbf{a}_y\cdot$	$\sin\phi$	$\cos\phi$	0
$\mathbf{a}_z\cdot$	0	0	1

Transforming vectors from rectangular to cylindrical coordinates or vice versa is therefore accomplished by using (10) or (11) to change variables, and by using the dot products of the unit vectors given in Table 1.1 to change components. The two steps may be taken in either order.

EXAMPLE 1.3

Transform the vector $\mathbf{B} = y\mathbf{a}_x - x\mathbf{a}_y + z\mathbf{a}_z$ into cylindrical coordinates.

Solution. The new components are

$$B_\rho = \mathbf{B}\cdot\mathbf{a}_\rho = y(\mathbf{a}_x\cdot\mathbf{a}_\rho) - x(\mathbf{a}_y\cdot\mathbf{a}_\rho)$$
$$= y\cos\phi - x\sin\phi = \rho\sin\phi\cos\phi - \rho\cos\phi\sin\phi = 0$$
$$B_\phi = \mathbf{B}\cdot\mathbf{a}_\phi = y(\mathbf{a}_x\cdot\mathbf{a}_\phi) - x(\mathbf{a}_y\cdot\mathbf{a}_\phi)$$
$$= -y\sin\phi - x\cos\phi = -\rho\sin^2\phi - \rho\cos^2\phi = -\rho$$

Thus,

$$\mathbf{B} = -\rho\mathbf{a}_\phi + z\mathbf{a}_z$$

D1.5 (*a*) Give the rectangular coordinates of the point $C(\rho = 4.4, \phi = -115°, z = 2)$. (*b*) Give the cylindrical coordinates of the point $D(x = -3.1, y = 2.6, z = -3)$. (*c*) Specify the distance from C to D.

Ans. $C(x = -1.860, y = -3.99, z = 2)$; $D(\rho = 4.05, \phi = 140.0°, z = -3)$; 8.36

D1.6 Transform to cylindrical coordinates: (*a*) $\mathbf{F} = 10\mathbf{a}_x - 8\mathbf{a}_y + 6\mathbf{a}_z$ at point $P(10, -8, 6)$; (*b*) $\mathbf{G} = (2x + y)\mathbf{a}_x - (y - 4x)\mathbf{a}_y$ at point $Q(\rho, \phi, z)$. (*c*) Give the rectangular components of the vector $\mathbf{H} = 20\mathbf{a}_\rho - 10\mathbf{a}_\phi + 3\mathbf{a}_z$ at $P(x = 5, y = 2, z = -1)$.

Ans. $12.81\mathbf{a}_\rho + 6\mathbf{a}_z$; $(2\rho\cos^2\phi - \rho\sin^2\phi + 5\rho\sin\phi\cos\phi)\mathbf{a}_\rho + (4\rho\cos^2\phi - \rho\sin^2\phi - 3\rho\sin\phi\cos\phi)\mathbf{a}_\phi$; $H_x = 22.3, H_y = -1.857, H_z = 3$

1.9 THE SPHERICAL COORDINATE SYSTEM

We have no two-dimensional coordinate system to help us understand the three-dimensional spherical coordinate system, as we have for the circular cylindrical coordinate system. In certain respects we can draw on our knowledge of the latitude-and-longitude system of locating a place on the surface of the earth, but usually we consider only points on the surface and not those below or above ground.

Let us start by building a spherical coordinate system on the three rectangular axes (Figure 1.8*a*). We first define the distance from the origin to any point as r. The surface $r =$ constant is a sphere.

The second coordinate is an angle θ between the z axis and the line drawn from the origin to the point in question. The surface $\theta =$ constant is a cone, and the two surfaces, cone and sphere, are everywhere perpendicular along their intersection, which is a circle of radius $r\sin\theta$. The coordinate θ corresponds to latitude, except that latitude is measured from the equator and θ is measured from the "North Pole."

The third coordinate ϕ is also an angle and is exactly the same as the angle ϕ of cylindrical coordinates. It is the angle between the x axis and the projection in the $z = 0$ plane of the line drawn from the origin to the point. It corresponds to the angle of longitude, but the angle ϕ increases to the "east." The surface $\phi =$ constant is a plane passing through the $\theta = 0$ line (or the z axis).

We should again consider any point as the intersection of three mutually perpendicular surfaces—a sphere, a cone, and a plane—each oriented in the manner just described. The three surfaces are shown in Figure 1.8*b*.

Three unit vectors may again be defined at any point. Each unit vector is perpendicular to one of the three mutually perpendicular surfaces and oriented in that direction in which the coordinate increases. The unit vector $\mathbf{a}_r$ is directed radially outward, normal to the sphere $r =$ constant, and lies in the cone $\theta =$ constant and the plane $\phi =$ constant. The unit vector $\mathbf{a}_\theta$ is normal to the conical surface, lies in the plane, and is tangent to the sphere. It is directed along a line of "longitude" and points "south." The third unit vector $\mathbf{a}_\phi$ is the same as in cylindrical coordinates, being normal to the plane and tangent to both the cone and sphere. It is directed to the "east."

The three unit vectors are shown in Figure 1.8*c*. They are, of course, mutually perpendicular, and a right-handed coordinate system is defined by causing $\mathbf{a}_r \times \mathbf{a}_\theta = \mathbf{a}_\phi$. Our system is right-handed,

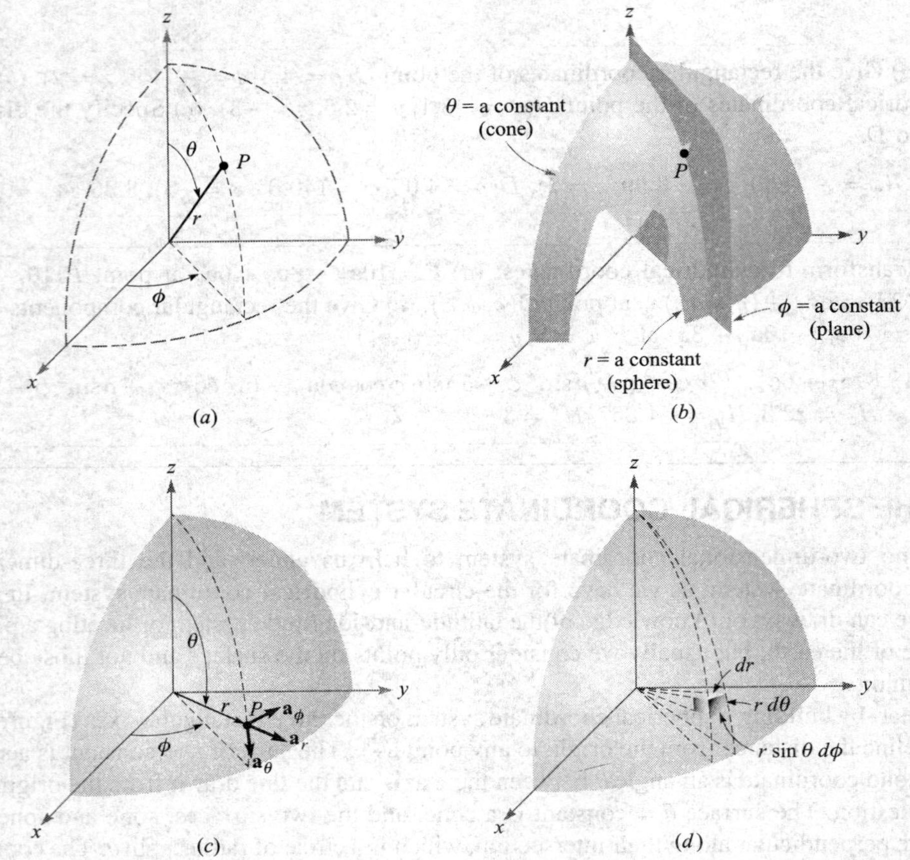

Fig. 1.8 (*a*) The three spherical coordinates. (*b*) The three mutually perpendicular surfaces of the spherical coordinate system. (*c*) The three unit vectors of spherical coordinates: $\mathbf{a}_r \times \mathbf{a}_\theta = \mathbf{a}_\phi$. (*d*) The differential volume element in the spherical coordinate system.

as an inspection of Figure 1.8*c* will show, on application of the definition of the cross product. The right-hand rule serves to identify the thumb, forefinger, and middle finger with the direction of increasing r, θ, and ϕ, respectively. (Note that the identification in cylindrical coordinates was with ρ, ϕ, and z, and in rectangular coordinates with x, y, and z). A differential volume element may be constructed in spherical coordinates by increasing r, θ, and ϕ by dr, $d\theta$, and $d\phi$, as shown in Figure 1.8*d*. The distance between the two spherical surfaces of radius r and $r + dr$ is dr; the distance between the two cones having generating angles of θ and $\theta + d\theta$ is $r d\theta$; and the distance between the two radial planes at angles ϕ and $\phi + d\phi$ is found to be $r \sin\theta d\phi$, after a few moments of trigonometric thought. The surfaces have areas of $r\,dr\,d\theta$, $r \sin\theta\,dr\,d\phi$, and $r^2 \sin\theta\,d\theta\,d\phi$, and the volume is $r^2 \sin\theta\,dr\,d\theta\,d\phi$.

The transformation of scalars from the rectangular to the spherical coordinate system is easily made by using Figure 1.8*a* to relate the two sets of variables:

$$x = r \sin\theta \cos\phi$$
$$y = r \sin\theta \sin\phi \qquad (15)$$
$$z = r \cos\theta$$

The transformation in the reverse direction is achieved with the help of

$$r = \sqrt{x^2 + y^2 + z^2} \qquad (r \geq 0)$$
$$\theta = \cos^{-1} \frac{z}{\sqrt{x^2 + y^2 + z^2}} \qquad (0^\circ \leq \theta \leq 180^\circ) \qquad (16)$$
$$\phi = \tan^{-1} \frac{y}{x}$$

The radius variable r is nonnegative, and θ is restricted to the range from 0° to 180°, inclusive. The angles are placed in the proper quadrants by inspecting the signs of x, y, and z.

The transformation of vectors requires the determination of the products of the unit vectors in rectangular and spherical coordinates. We work out these products from Figure 1.8c and a pinch of trigonometry. Since the dot product of any spherical unit vector with any rectangular unit vector is the component of the spherical vector in the direction of the rectangular vector, the dot products with $\mathbf{a}_z$ are found to be

$$\mathbf{a}_z \cdot \mathbf{a}_r = \cos\theta$$
$$\mathbf{a}_z \cdot \mathbf{a}_\theta = -\sin\theta$$
$$\mathbf{a}_z \cdot \mathbf{a}_\phi = 0$$

The dot products involving $\mathbf{a}_x$ and $\mathbf{a}_y$ require first the projection of the spherical unit vector on the xy plane and then the projection onto the desired axis. For example, $\mathbf{a}_r \cdot \mathbf{a}_x$ is obtained by projecting $\mathbf{a}_r$ onto the xy plane, giving $\sin\theta$, and then projecting $\sin\theta$ on the x axis, which yields $\sin\theta\cos\phi$. The other dot products are found in a like manner, and all are shown in Table 1.2.

Table 1.2 Dot products of unit vectors in spherical and rectangular coordinate systems

	$\mathbf{a}_r$	$\mathbf{a}_\theta$	$\mathbf{a}_\phi$
$\mathbf{a}_x\cdot$	$\sin\theta\cos\phi$	$\cos\theta\cos\phi$	$-\sin\phi$
$\mathbf{a}_y\cdot$	$\sin\theta\sin\phi$	$\cos\theta\sin\phi$	$\cos\phi$
$\mathbf{a}_z\cdot$	$\cos\theta$	$-\sin\theta$	0

EXAMPLE 1.4

We illustrate this transformation procedure by transforming the vector field $\mathbf{G} = (xz/y)\mathbf{a}_x$ into spherical components and variables.

Solution. We find the three spherical components by dotting $\mathbf{G}$ with the appropriate unit vectors, and we change variables during the procedure:

$$G_r = \mathbf{G} \cdot \mathbf{a}_r = \frac{xz}{y} \mathbf{a}_x \cdot \mathbf{a}_r = \frac{xz}{y} \sin\theta \cos\phi$$

$$= r\sin\theta\cos\theta\frac{\cos^2\phi}{\sin\phi}$$

$$G_\theta = \mathbf{G}\cdot\mathbf{a}_\theta = \frac{xz}{y}\mathbf{a}_x\cdot\mathbf{a}_\theta = \frac{xz}{y}\cos\theta\cos\phi$$

$$= r\cos^2\theta\frac{\cos^2\phi}{\sin\phi}$$

$$G\phi = \mathbf{G}\cdot\mathbf{a}_\phi = \frac{xz}{y}\mathbf{a}_x\cdot\mathbf{a}_\phi = \frac{xz}{y}(-\sin\phi)$$

$$= -r\cos\theta\cos\phi$$

Collecting these results, we have

$$\mathbf{G} = r\cos\theta\cos\phi\,(\sin\theta\cot\phi\,\mathbf{a}_r + \cos\theta\cot\phi\,\mathbf{a}_\theta - \mathbf{a}_\phi)$$

Appendix A describes the general curvilinear coordinate system of which the rectangular, circular cylindrical, and spherical coordinate systems are special cases. The first section of this appendix could well be scanned now.

D1.7 Given the two points, $C(-3, 2, 1)$ and $D(r = 5, \theta = 20°, \phi = -70°)$, find: (*a*) the spherical coordinates of C; (*b*) the rectangular coordinates of D; (*c*) the distance from C to D.

Ans. $C(r = 3.74, \theta = 74.5°, \phi = 146.3°)$; $D(x = 0.585, y = -1.607, z = 4.70)$; 6.29

D1.8 Transform the following vectors to spherical coordinates at the points given: (*a*) $10\mathbf{a}_x$ at $P(x = -3, y = 2, z = 4)$; (*b*) $10\mathbf{a}_y$ at $Q(\rho = 5, \phi = 30°, z = 4)$; (*c*) $10\mathbf{a}_z$ at $M(r = 4, \theta = 110°, \phi = 120°)$.

Ans. $-5.57\mathbf{a}_r - 6.18\mathbf{a}_\theta - 5.55\mathbf{a}_\phi$; $3.90\mathbf{a}_r + 3.12\mathbf{a}_\theta + 8.66\mathbf{a}_\phi$; $-3.42\mathbf{a}_r - 9.40\mathbf{a}_\theta$

References

1. Grossman, S. I. *Calculus*. 3d ed. Orlando, Fla.: Academic Press and Harcourt Brace Jovanovich, 1984. Vector algebra and cylindrical and spherical coordinates appear in Chapter 17, and vector calculus is introduced in Chapter 20.
2. Spiegel, M. R. *Vector Analysis*. Schaum Outline Series. New York: McGraw-Hill, 1959. A large number of examples and problems with answers are provided in this concise, inexpensive member of an outline series.
3. Swokowski, E. W. *Calculus with Analytic Geometry*. 3d ed. Boston: Prindle, Weber, & Schmidt, 1984. Vector algebra and the cylindrical and spherical coordinate systems are discussed in Chapter 14, and vector calculus appears in Chapter 18.
4. Thomas, G. B., Jr., and R. L. Finney: *Calculus and Analytic Geometry*. 6th ed. Reading, Mass.: Addison-Wesley Publishing Company, 1984. Vector algebra and the three coordinate systems we use are discussed in Chapter 13. Other vector operations are discussed in Chapters 15 and 17.

CHAPTER 1 Problems

Quizzes

1.1 Given the vectors $\mathbf{M} = -10\mathbf{a}_x + 4\mathbf{a}_y - 8\mathbf{a}_z$ and $\mathbf{N} = 8\mathbf{a}_x + 7\mathbf{a}_y - 2\mathbf{a}_z$, find: (*a*) a unit vector in the direction of $-\mathbf{M} + 2\mathbf{N}$; (*b*) the magnitude of $5\mathbf{a}_x + \mathbf{N} - 3\mathbf{M}$; (*c*) $|\mathbf{M}||2\mathbf{N}|(\mathbf{M} + \mathbf{N})$.

1.2 The three vertices of a triangle are located at $A(-1, 2, 5)$, $B(-4, -2, -3)$, and $C(1, 3, -2)$. (*a*) Find the length of the perimeter of the triangle. (*b*) Find a unit vector that is directed from the midpoint of the side AB to the midpoint of the side BC. (*c*) Show that this unit vector multiplied by a scalar is equal to the vector from A to C and that the unit vector is therefore parallel to AC.

1.3 The vector from the origin to point A is given as $(6, -2, -4)$, and the unit vector directed from the origin toward point B is $(2, -2, 1)/3$. If points A and B are ten units apart, find the coordinates of point B.

1.4 A circle, centered at the origin with a radius of 2 units, lies in the xy plane. Determine the unit vector in rectangular components that lies in the xy plane, is tangent to the circle at $(\sqrt{3}, 1, 0)$, and is in the general direction of increasing values of y.

1.5 A vector field is specified as $\mathbf{G} = 24xy\mathbf{a}_x + 12(x^2 + 2)\mathbf{a}_y + 18z^2\mathbf{a}_z$. Given two points, $P(1, 2, -1)$ and $Q(-2, 1, 3)$, find: (*a*) $\mathbf{G}$ at P; (*b*) a unit vector in the direction of $\mathbf{G}$ at Q; (*c*) a unit vector directed from Q toward P; (*d*) the equation of the surface on which $|\mathbf{G}| = 60$.

1.6 If $\mathbf{a}$ is a unit vector in a given direction, B is a scalar constant, and $\mathbf{r} = x\mathbf{a}_x + y\mathbf{a}_y + z\mathbf{a}_z$, describe the surface $\mathbf{r} \cdot \mathbf{a} = B$. What is the relation between the unit vector $\mathbf{a}$ and the scalar B to this surface? (HINT: Consider first a simple example with $\mathbf{a} = \mathbf{a}_x$ and $B = 1$, and then consider any $\mathbf{a}$ and B.)

1.7 Given the vector field $\mathbf{E} = 4zy^2 \cos 2x\mathbf{a}_x + 2zy \sin 2x\mathbf{a}_y + y^2 \sin 2x\mathbf{a}_z$ for the region $|x|$, $|y|$, and $|z|$ less than 2, find: (*a*) the surfaces on which $E_y = 0$; (*b*) the region in which $E_y = E_z$; (*c*) the region in which $\mathbf{E} = 0$.

1.8 Demonstrate the ambiguity that results when the cross product is used to find the angle between two vectors by finding the angle between $\mathbf{A} = 3\mathbf{a}_x - 2\mathbf{a}_y + 4\mathbf{a}_z$ and $\mathbf{B} = 2\mathbf{a}_x + \mathbf{a}_y - 2\mathbf{a}_z$. Does this ambiguity exist when the dot product is used?

1.9 A field is given as $\mathbf{G} = [25/(x^2 + y^2)](x\mathbf{a}_x + y\mathbf{a}_y)$. Find: (*a*) a unit vector in the direction of $\mathbf{G}$ at $P(3, 4, -2)$; (*b*) the angle between $\mathbf{G}$ and $\mathbf{a}_x$ at P; (*c*) the value of the double integral on the plane $y = 7$.

1.10 By expressing diagonals as vectors and using the definition of the dot product, find the smaller angle between any two diagonals of a cube, where each diagonal connects diametrically opposite corners, and passes through the center of the cube.

1.11 Given the points $M(0.1, -0.2, -0.1)$, $N(-0.2, 0.1, 0.3)$, and $P(0.4, 0, 0.1)$, find: (*a*) the vector $\mathbf{R}_{MN}$; (*b*) the dot product $\mathbf{R}_{MN} \cdot \mathbf{R}_{MP}$; (*c*) the scalar projection of $\mathbf{R}_{MN}$ on $\mathbf{R}_{MP}$; (*d*) the angle between $\mathbf{R}_{MN}$ and $\mathbf{R}_{MP}$.

1.12 Show that the vector fields $\mathbf{A} = \rho \cos \phi \mathbf{a}_\rho + \rho \sin \phi \mathbf{a}_\phi + \rho \mathbf{a}_z$ and $\mathbf{B} = \rho \cos \phi \mathbf{a}_\rho + \rho \sin \phi \mathbf{a}_\phi - \rho \mathbf{a}_z$ are everywhere perpendicular to each other.

1.13 (*a*) Find the vector component of $\mathbf{F} = 10\mathbf{a}_x - 6\mathbf{a}_y + 5\mathbf{a}_z$ that is parallel to $\mathbf{G} = 0.1\mathbf{a}_x + 0.2\mathbf{a}_y + 0.3\mathbf{a}_z$. (*b*) Find the vector component of **F** that is perpendicular to **G**. (*c*) Find the vector component of **G** that is perpendicular to **F**.

1.14 Show that the vector fields $\mathbf{A} = \mathbf{a}_r(\sin 2\theta)/r^2 + 2\mathbf{a}_\theta(\sin\theta)/r^2$ and $\mathbf{B} = r\cos\theta\mathbf{a}_r + r\mathbf{a}_\theta$ are everywhere parallel to each other.

1.15 Three vectors extending from the origin are given as $\mathbf{r}_1 = (7, 3, -2)$, $\mathbf{r}_2 = (-2, 7, -3)$, and $\mathbf{r}_3 = (0, 2, 3)$. Find: (*a*) a unit vector perpendicular to both $\mathbf{r}_1$ and $\mathbf{r}_2$; (*b*) a unit vector perpendicular to the vectors $\mathbf{r}_1 - \mathbf{r}_2$ and $\mathbf{r}_2 - \mathbf{r}_3$; (*c*) the area of the triangle defined by $\mathbf{r}_1$ and $\mathbf{r}_2$; (*d*) the area of the triangle defined by the heads of $\mathbf{r}_1$, $\mathbf{r}_2$, and $\mathbf{r}_3$.

1.16 The vector field $\mathbf{E} = (B/\rho)\mathbf{a}_\rho$ where B is a constant, is to be translated such that it originates at the line, $x = 2, y = 0$. Write the translated form of **E** in rectangular components.

1.17 Point $A(-4, 2, 5)$ and the two vectors, $\mathbf{R}_{AM} = (20, 18 - 10)$ and $\mathbf{R}_{AN} = (-10, 8, 15)$, define a triangle. (*a*) Find a unit vector perpendicular to the triangle. (*b*) Find a unit vector in the plane of the triangle and perpendicular to $\mathbf{R}_{AN}$. (*c*) Find a unit vector in the plane of the triangle that bisects the interior angle at A.

1.18 Transform the vector field $\mathbf{H} = (A/\rho)\mathbf{a}_\phi$, where A is a constant, from cylindrical coordinates to spherical coordinates.

1.19 (*a*) Express the field $\mathbf{D} = (x^2 + y^2)^{-1}(x\mathbf{a}_x + y\mathbf{a}_y)$ in cylindrical components and cylindrical variables; (*b*) Evaluate **D** at the point where $\rho = 2$, $\phi = 0.2\pi$, and $z = 5$, expressing the result in cylindrical and rectangular components.

1.20 A cylinder of radius a, centered on the z axis, rotates about the z axis at angular velocity Ω rad/s. The rotation direction is counter-clockwise when looking in the positive z direction. (*a*) Using cylindrical components, write an expression for the velocity field, **v**, which gives the tangential velocity at any point within the cylinden; (*b*) convert your result from part *a* to spherical components; (*c*) convert to rectangular components.

1.21 Express in cylindrical components: (*a*) the vector from $C(3, 2, -7)$ to $D(-1, -4, 2)$; (*b*) a unit vector at D directed toward C; (*c*) a unit vector at D directed toward the origin.

1.22 A sphere of radius a, centered at the origin, rotates about the z axis at angular velocity Ω rad/s. The rotation direction is clockwise when one is looking in the positive z direction. (*a*) Using spherical components, write an expression for the velocity field, **v**, that gives the tangential velocity at any point within the sphere; (*b*) convert to rectangular components.

1.23 The surfaces $\rho = 3$, $\rho = 5$, $\phi = 100°$, $\phi = 130°$, $z = 3$, and $z = 4.5$ define a closed surface. (*a*) Find the enclosed volume; (*b*) Find the total area of the enclosing surface; (*c*) Find the total length of the twelve edges of the surfaces; (*d*) Find the length of the longest straight line that lies entirely within the volume.

1.24 Express the field $\mathbf{E} = A\mathbf{a}_r/r^2$ in (*a*) rectangular components; (*b*) cylindrical components.

1.25 Given point $P(r = 0.8, \theta = 30°, \phi = 45°)$, and $\mathbf{E} = 1/r^2\ (\cos\phi\mathbf{a}_r + \sin\phi/\sin\theta\,\mathbf{a}_\phi)$; (*a*) Find **E** at P; (*b*) Find $|\mathbf{E}|$ at P; (*c*) Find a unit vector in the direction of **E** at P.

1.26 Express the uniform vector field $\mathbf{F} = 5\mathbf{a}_x$ in (*a*) cylindrical components; (*b*) spherical components.

1.27 The surfaces $r = 2$ and 4, $\theta = 30°$ and $50°$, and $\phi = 20°$ and $60°$ identify a closed surface. (*a*) Find the enclosed volume; (*b*) Find the total area of the enclosing surface; (*c*) Find the total length of the twelve edges of the surface; (*d*) Find the length of the longest straight line that lies entirely within the surface.

1.28 Express the vector field, $\mathbf{G} = 8\sin\phi\,\mathbf{a}_\theta$ in (*a*) rectangular components; (*b*) cylindrical components.

1.29 Express the unit vector $\mathbf{a}_x$ in spherical components at the point: (*a*) $r = 2, \theta = 1$ rad, $\phi = 0.8$ rad; (*b*) $x = 3, y = 2, z = -1$; (*c*) $\rho = 2.5, \phi = 0.7$ rad, $z = 1.5$.

1.30 At point $B(5, 120^\circ, 75^\circ)$ a vector field has the value $\mathbf{A} = -12\mathbf{a}_r - 5\mathbf{a}_\theta + 15\mathbf{a}_\phi$. Find the vector component of **A** that is: (*a*) normal to the surface $r = 5$; (*b*) tangent to the surface $r = 5$; (*c*) tangent to the cone $\theta = 120^\circ$. (*d*) Find a unit vector that is perpendicular to **A** and tangent to the cone $\theta = 120^\circ$.

2

Coulomb's Law and Electric Field Intensity

Now that we have formulated a new language in the first chapter, we shall establish a few basic principles of electricity and attempt to describe them in terms of it. If we had used vector calculus for several years and already had a few correct ideas about electricity and magnetism, we might jump in now with both feet and present a handful of equations, including Maxwell's equations and a few other auxiliary equations, and proceed to describe them physically by virtue of our knowledge of vector analysis. This is perhaps the ideal way, starting with the most general results and then showing that Ohm's, Gauss's, Coulomb's, Faraday's, Ampère's, Biot-Savart's, Kirchhoff's, and a few less familiar laws are all special cases of these equations. It is philosophically satisfying to have the most general result and to feel that we can obtain the results for any special case at will. However, such a jump would lead to many frantic cries of "Help" and not a few drowned students.

Instead we shall present at decent intervals the experimental laws mentioned above, expressing each in vector notation, and we will use these laws to solve a number of simple problems. In this way our familiarity with both vector analysis and electric and magnetic fields will gradually increase, and by the time we have finally reached our handful of general equations, little additional explanation will be required. The entire field of electromagnetic theory is then open to us, and we may use Maxwell's equations to describe wave propagation, radiation from antennas, skin effect, waveguides and transmission lines, and optical fibers, and even to obtain a new insight into the ordinary power transformer.

In this chapter we shall restrict our attention to *static* electric fields in *vacuum* or *free space*. For all practical purposes, our results will be applicable to air and other gases. Other materials will be introduced in Chapter 6, and time-varying fields will be introduced in Chapter 10.

We shall begin by describing a quantitative experiment performed in the seventeenth century.

2.1 THE EXPERIMENTAL LAW OF COULOMB

Records from at least 600 B.C. show evidence of the knowledge of static electricity. The Greeks were responsible for the term *electricity*, derived from their word for amber, and they spent many leisure hours rubbing a small piece of amber on their sleeves and observing how it would then attract pieces of fluff and stuff. However, their main interest lay in philosophy and logic, not in experimental science, and it was many centuries before the attracting effect was considered to be anything other than magic or a "life force."

Dr. Gilbert, physician to Her Majesty the Queen of England, was the first to do any true experimental work with this effect, and in 1600 he stated that glass, sulfur, amber, and other materials which he named would "not only draw to themselves straws and chaff, but all metals, wood, leaves, stone, earths, even water and oil."

Shortly thereafter an officer in the French Army Engineers, Colonel Charles Coulomb, performed an elaborate series of experiments using a delicate torsion balance, invented by himself, to determine quantitatively the force exerted between two objects, each having a static charge of electricity. His published result is now known to many high school students and bears a great similarity to Newton's gravitational law (discovered about a hundred years earlier). Coulomb stated that the force between two very small objects separated in a vacuum or free space by a distance which is large compared to their size is proportional to the charge on each and inversely proportional to the square of the distance between them, or

$$F = k\frac{Q_1 Q_2}{R^2}$$

where Q_1 and Q_2 are the positive or negative quantities of charge, R is the separation, and k is a proportionality constant. If the International System of Units[1] (SI) is used, Q is measured in coulombs (C), R is in meters (m), and the force should be newtons (N). This will be achieved if the constant of proportionality k is written as

$$k = \frac{1}{4\pi\epsilon_0}$$

The factor 4π will appear in the denominator of Coulomb's law but will not appear in the more useful equations (including Maxwell's equations) which we shall obtain with the help of Coulomb's law. The new constant ϵ_0 is called the *permittivity of free space* and has the magnitude, measured in farads per meter (F/m),

$$\epsilon_0 = 8.854 \times 10^{-12} \doteq \frac{1}{36\pi} 10^{-9} \text{ F/m} \tag{1}$$

The quantity ϵ_0 is not dimensionless, for Coulomb's law shows that it has the label $C^2/N \cdot m^2$. We shall later define the farad and show that it has the dimensions $C^2/N \cdot m$; we have anticipated this definition by using the unit F/m in equation (1).

Coulomb's law is now

$$F = \frac{Q_1 Q_2}{4\pi\epsilon_0 R^2} \tag{2}$$

Not all SI units are as familiar as the English units we use daily, but they are now standard in electrical engineering and physics. The newton is a unit of force that is equal to 0.2248 lb_f, and is the force required to give a 1-kilogram (kg) mass an acceleration of 1 meter per second per second (m/s^2). The coulomb is an extremely large unit of charge, for the smallest known quantity of charge is that of the electron (negative) or proton (positive), given in mks units as 1.602×10^{-19} C; hence a negative charge of one coulomb represents about 6×10^{18} electrons.[2] Coulomb's law

[1] The International System of Units (an mks system) is described in Appendix B. Abbreviations for the units are given in Table B.1. Conversions to other systems of units are given in Table B.2, while the prefixes designating powers of ten in SI appear in Table B.3.

[2] The charge and mass of an electron and other physical constants are tabulated in Table C.4 of Appendix C.

shows that the force between two charges of one coulomb each, separated by one meter, is 9×10^9 N, or about one million tons. The electron has a rest mass of 9.109×10^{-31} kg and has a radius of the order of magnitude of 3.8×10^{-15} m. This does not mean that the electron is spherical in shape, but merely serves to describe the size of the region in which a slowly moving electron has the greatest probability of being found. All other known charged particles, including the proton, have larger masses, and larger radii, and occupy a probabilistic volume larger than does the electron.

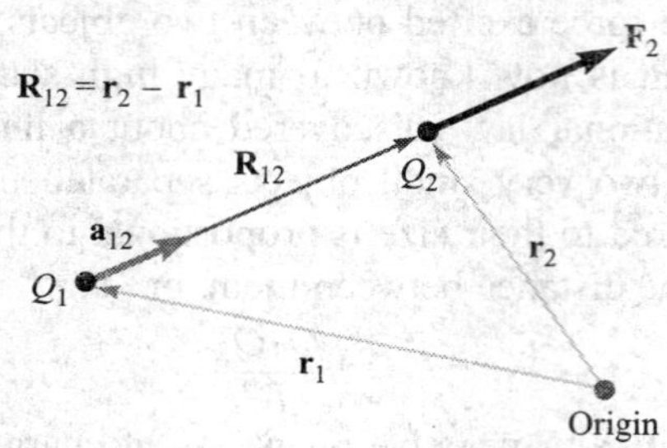

Fig. 2.1 If Q_1 and Q_2 have like signs, the vector force $\mathbf{F}_2$ on Q_2 is in the same direction as the vector $\mathbf{R}_{12}$.

In order to write the vector form of (2), we need the additional fact (furnished also by Colonel Coulomb) that the force acts along the line joining the two charges and is repulsive if the charges are alike in sign and attractive if they are of opposite sign. Let the vector $\mathbf{r}_1$ locate Q_1 while $\mathbf{r}_2$ locates Q_2. Then the vector $\mathbf{R}_{12} = \mathbf{r}_2 - \mathbf{r}_1$ represents the directed line segment from Q_1 to Q_2, as shown in Figure 2.1. The vector $\mathbf{F}_2$ is the force on Q_2 and is shown for the case where Q_1 and Q_2 have the same sign. The vector form of Coulomb's law is

$$\mathbf{F}_2 = \frac{Q_1 Q_2}{4\pi\epsilon_0 R_{12}^2}\mathbf{a}_{12} \tag{3}$$

where $\mathbf{a}_{12}$ = a unit vector in the direction of R_{12}, or

$$\mathbf{a}_{12} = \frac{\mathbf{R}_{12}}{|\mathbf{R}_{12}|} = \frac{\mathbf{R}_{12}}{R_{12}} = \frac{\mathbf{r}_2 - \mathbf{r}_1}{|\mathbf{r}_2 - \mathbf{r}_1|} \tag{4}$$

EXAMPLE 2.1

Let us illustrate the use of the vector form of Coulomb's law by locating a charge of $Q_1 = 3 \times 10^{-4}$ C at $M(1, 2, 3)$ and a charge of $Q_2 = -10^{-4}$ C at $N(2, 0, 5)$ in a vacuum. We desire the force exerted on Q_2 by Q_1.

Solution. We shall make use of (3) and (4) to obtain the vector force. The vector $\mathbf{R}_{12}$ is

$$\mathbf{R}_{12} = \mathbf{r}_2 - \mathbf{r}_1 = (2-1)\mathbf{a}_x + (0-2)\mathbf{a}_y + (5-3)\mathbf{a}_z = \mathbf{a}_x - 2\mathbf{a}_y + 2\mathbf{a}_z$$

leading to $|\mathbf{R}_{12}| = 3$, and the unit vector, $\mathbf{a}_{12} = \frac{1}{3}(\mathbf{a}_x - 2\mathbf{a}_y + 2\mathbf{a}_z)$. Thus,

$$\mathbf{F}_2 = \frac{3 \times 10^{-4}(-10^{-4})}{4\pi(1/36\pi)10^{-9} \times 3^2}\left(\frac{\mathbf{a}_x - 2\mathbf{a}_y + 2\mathbf{a}_z}{3}\right)$$

$$= -30\left(\frac{\mathbf{a}_x - 2\mathbf{a}_y + 2\mathbf{a}_z}{3}\right) \text{ N}$$

The magnitude of the force is 30 N (or about 7 lb_f), and the direction is specified by the unit vector, which has been left in parentheses to display the magnitude of the force. The force on Q_2 may also be considered as three component forces,

$$\mathbf{F}_2 = -10\mathbf{a}_x + 20\mathbf{a}_y - 20\mathbf{a}_z$$

The force expressed by Coulomb's law is a mutual force, for each of the two charges experiences a force of the same magnitude, although of opposite direction. We might equally well have written

$$\mathbf{F}_1 = -\mathbf{F}_2 = \frac{Q_1 Q_2}{4\pi\epsilon_0 R_{12}^2}\mathbf{a}_{21} = -\frac{Q_1 Q_2}{4\pi\epsilon_0 R_{12}^2}\mathbf{a}_{12} \tag{5}$$

Coulomb's law is linear, for if we multiply Q_1 by a factor n, the force on Q_2 is also multiplied by the same factor n. It is also true that the force on a charge in the presence of several other charges is the sum of the forces on that charge due to each of the other charges acting alone.

D2.1 A charge $Q_A = -20\,\mu\text{C}$ is located at $A(-6, 4, 7)$, and a charge $Q_B = 50\,\mu\text{C}$ is at $B(5, 8, -2)$ in free space. If distances are given in meters, find: *(a)* $\mathbf{R}_{AB}$; *(b)* R_{AB}. Determine the vector force exerted on Q_A by Q_B if $\epsilon_0 =$: *(c)* $10^{-9}/(36\pi)$ F/m; *(d)* 8.854×10^{-12} F/m.

Ans. $11\mathbf{a}_x + 4\mathbf{a}_y - 9\mathbf{a}_z$ m; 14.76 m; $30.76\mathbf{a}_x + 11.184\mathbf{a}_y - 25.16\mathbf{a}_z$ mN; $30.72\mathbf{a}_x + 11.169\mathbf{a}_y - 25.13\mathbf{a}_z$ mN

2.2 ELECTRIC FIELD INTENSITY

If we now consider one charge fixed in position, say Q_1, and move a second charge slowly around, we note that there exists everywhere a force on this second charge; in other words, this second charge is displaying the existence of a force *field*. Call this second charge a test charge Q_t. The force on it is given by Coulomb's law,

$$\mathbf{F}_t = \frac{Q_1 Q_t}{4\pi\epsilon_0 R_{1t}^2}\mathbf{a}_{1t}$$

Writing this force as a force per unit charge gives

$$\frac{\mathbf{F}_t}{Q_t} = \frac{Q_1}{4\pi\epsilon_0 R_{1t}^2}\mathbf{a}_{1t} \tag{6}$$

Interactives

The quantity on the right side of (6) is a function only of Q_1 and the directed line segment from Q_1 to the position of the test charge. This describes a vector field and is called the *electric field intensity*.

We define the electric field intensity as the vector force on a unit positive test charge. We would not *measure* it experimentally by finding the force on a 1-C test charge, however, for this would probably cause such a force on Q_1 as to change the position of that charge.

Electric field intensity must be measured by the unit newtons per coulomb—the force per unit charge. Again anticipating a new dimensional quantity, the *volt* (V), to be presented in Chapter 4 and having the label of joules per coulomb (J/C) or newton-meters per coulomb (N · m/C), we shall at once measure electric field intensity in the practical units of volts per meter (V/m).

Using a capital letter **E** for electric field intensity, we have finally

$$\mathbf{E} = \frac{\mathbf{F}_t}{Q_t} \tag{7}$$

$$\mathbf{E} = \frac{Q_1}{4\pi\epsilon_0 R_{1t}^2}\mathbf{a}_{1t} \tag{8}$$

Equation (7) is the defining expression for electric field intensity, and (8) is the expression for the electric field intensity due to a single point charge Q_1 in a vacuum. In the succeeding sections we shall obtain and interpret expressions for the electric field intensity due to more complicated arrangements of charge, but now let us see what information we can obtain from (8), the field of a single point charge.

First, let us dispense with most of the subscripts in (8), reserving the right to use them again any time there is a possibility of misunderstanding:

$$\mathbf{E} = \frac{Q}{4\pi\epsilon_0 R^2}\mathbf{a}_R \tag{9}$$

We should remember that R is the magnitude of the vector **R**, the directed line segment from the point at which the point charge Q is located to the point at which **E** is desired, and $\mathbf{a}_R$ is a unit vector in the **R** direction.[3]

Let us arbitrarily locate Q_1 at the center of a spherical coordinate system. The unit vector $\mathbf{a}_R$ then becomes the radial unit vector $\mathbf{a}_r$, and R is r. Hence

$$\mathbf{E} = \frac{Q_1}{4\pi\epsilon_0 r^2}\mathbf{a}_r \tag{10}$$

or

$$E_r = \frac{Q_1}{4\pi\epsilon_0 r^2}$$

The field has a single radial component, and its inverse-square-law relationship is quite obvious.

Writing these expressions in rectangular coordinates for a charge Q at the origin, we have $\mathbf{R} = \mathbf{r} = x\mathbf{a}_x + y\mathbf{a}_y + z\mathbf{a}_z$ and $\mathbf{a}_R = \mathbf{a}_r = (x\mathbf{a}_x + y\mathbf{a}_y + z\mathbf{a}_z)/\sqrt{x^2+y^2+z^2}$; therefore,

$$\mathbf{E} = \frac{Q}{4\pi\epsilon_0(x^2+y^2+z^2)}\left(\frac{x}{\sqrt{x^2+y^2+z^2}}\mathbf{a}_x + \frac{y}{\sqrt{x^2+y^2+z^2}}\mathbf{a}_y + \frac{z}{\sqrt{x^2+y^2+z^2}}\mathbf{a}_z\right) \tag{11}$$

This expression no longer shows immediately the simple nature of the field, and its complexity is the price we pay for solving a problem having spherical symmetry in a coordinate system with which we may (temporarily) have more familiarity.

---

[3] We firmly intend to avoid confusing r and $\mathbf{a}_r$ with R and $\mathbf{a}_R$. The first two refer specifically to the spherical coordinate system, whereas R and $\mathbf{a}_R$ do not refer to any coordinate system—the choice is still available to us.

Without vector analysis, the information contained in (11) would have to be expressed in three equations, one for each component, and in order to obtain the equation we would have to break up the magnitude of the electric field intensity into the three components by finding the projection on each coordinate axis. Using vector notation, this is done automatically when we write the unit vector.

If we consider a charge which is *not* at the origin of our coordinate system, the field no longer possesses spherical symmetry (nor cylindrical symmetry, unless the charge lies on the z axis), and we might as well use rectangular coordinates. For a charge Q located at the source point $\mathbf{r}' = x'\mathbf{a}_x + y'\mathbf{a}_y + z'\mathbf{a}_z$, as illustrated in Figure 2.2, we find the field at a general field point $\mathbf{r} = x\mathbf{a}_x + y\mathbf{a}_y + z\mathbf{a}_z$ by expressing $\mathbf{R}$ as $\mathbf{r} - \mathbf{r}'$, and then

$$\mathbf{E}(\mathbf{r}) = \frac{Q}{4\pi\epsilon_0|\mathbf{r}-\mathbf{r}'|^2}\frac{\mathbf{r}-\mathbf{r}'}{|\mathbf{r}-\mathbf{r}'|} = \frac{Q(\mathbf{r}-\mathbf{r}')}{4\pi\epsilon_0|\mathbf{r}-\mathbf{r}'|^3}$$

$$= \frac{Q[(x-x')\mathbf{a}_x + (y-y')\mathbf{a}_y + (z-z')\mathbf{a}_z]}{4\pi\epsilon_0[(x-x')^2+(y-y')^2+(z-z')^2]^{3/2}} \tag{12}$$

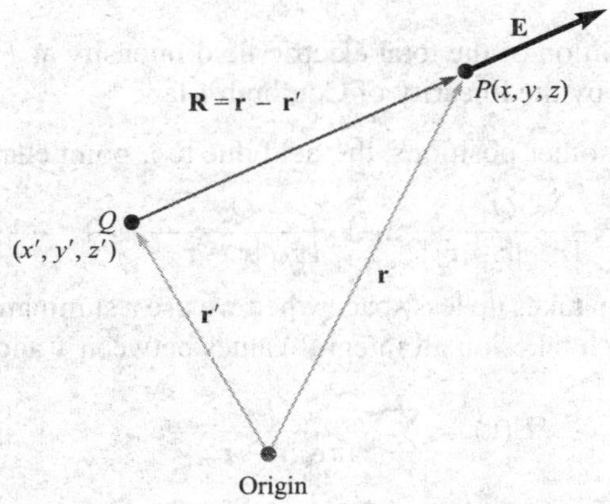

Fig. 2.2 The vector $\mathbf{r}'$ locates the point charge Q, the vector $\mathbf{r}$ identifies the general point in space $P(x, y, z)$, and the vector $\mathbf{R}$ from Q to $P(x, y, z)$ is then $\mathbf{R} = \mathbf{r} - \mathbf{r}'$.

Earlier, we defined a vector field as a vector function of a position vector, and this is emphasized by letting $\mathbf{E}$ be symbolized in functional notation by $\mathbf{E}(\mathbf{r})$.

Equation (11) is merely a special case of (12), where $x' = y' = z' = 0$.

Since the coulomb forces are linear, the electric field intensity due to two point charges, Q_1 at $\mathbf{r}_1$ and Q_2 at $\mathbf{r}_2$, is the sum of the forces on Q_t caused by Q_1 and Q_2 acting alone, or

$$\mathbf{E}(\mathbf{r}) = \frac{Q_1}{4\pi\epsilon_0|\mathbf{r}-\mathbf{r}_1|^2}\mathbf{a}_1 + \frac{Q_2}{4\pi\epsilon_0|\mathbf{r}-\mathbf{r}_2|^2}\mathbf{a}_2$$

where $\mathbf{a}_1$ and $\mathbf{a}_2$ are unit vectors in the direction of $(\mathbf{r} - \mathbf{r}_1)$ and $(\mathbf{r} - \mathbf{r}_2)$, respectively. The vectors $\mathbf{r}, \mathbf{r}_1, \mathbf{r}_2, \mathbf{r} - \mathbf{r}_1, \mathbf{r} - \mathbf{r}_2, \mathbf{a}_1$, and $\mathbf{a}_2$ are shown in Figure 2.3.

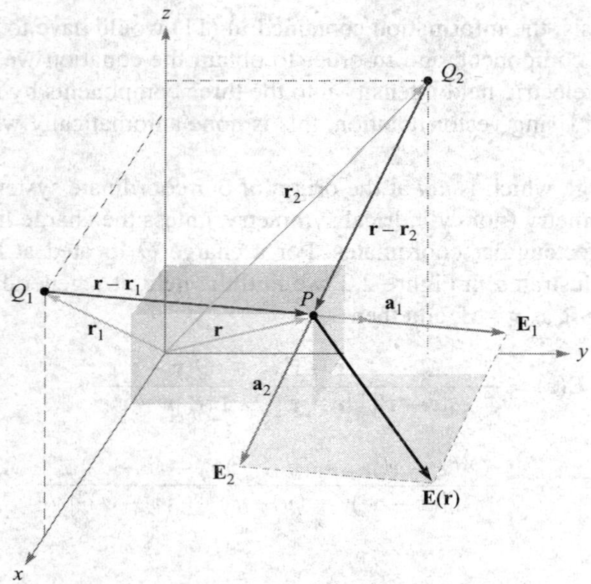

Fig. 2.3 The vector addition of the total electric field intensity at P due to Q_1 and Q_2 is made possible by the linearity of Coulomb's law.

If we add more charges at other positions, the field due to n point charges is

$$\mathbf{E}(\mathbf{r}) = \frac{Q_1}{4\pi\epsilon_0|\mathbf{r} - \mathbf{r}_1|^2}\mathbf{a}_1 + \frac{Q_2}{4\pi\epsilon_0|\mathbf{r} - \mathbf{r}_2|^2}\mathbf{a}_2 + \cdots + \frac{Q_n}{4\pi\epsilon_0|\mathbf{r} - \mathbf{r}_n|^2}\mathbf{a}_n \tag{13}$$

Interactives

This expression takes up less space when we use a summation sign $\sum$ and a summing integer m which takes on all integral values between 1 and n,

$$\mathbf{E}(\mathbf{r}) = \sum_{m=1}^{n} \frac{Q_m}{4\pi\epsilon_0|\mathbf{r} - \mathbf{r}_m|^2}\mathbf{a}_m \tag{14}$$

When expanded, (14) is identical with (13), and students unfamiliar with summation signs should check that result.

EXAMPLE 2.2

In order to illustrate the application of (13) or (14), let us find $\mathbf{E}$ at $P(1, 1, 1)$ caused by four identical 3-nC (nanocoulomb) charges located at $P_1(1, 1, 0)$, $P_2(-1, 1, 0)$, $P_3(-1, -1, 0)$, and $P_4(1, -1, 0)$, as shown in Figure 2.4.

Solution. We find that $\mathbf{r} = \mathbf{a}_x + \mathbf{a}_y + \mathbf{a}_z$, $\mathbf{r}_1 = \mathbf{a}_x + \mathbf{a}_y$, and thus $\mathbf{r} - \mathbf{r}_1 = \mathbf{a}_z$. The magnitudes are: $|\mathbf{r} - \mathbf{r}_1| = 1$, $|\mathbf{r} - \mathbf{r}_2| = \sqrt{5}$, $|\mathbf{r} - \mathbf{r}_3| = 3$, and $|\mathbf{r} - \mathbf{r}_4| = \sqrt{5}$. Since $Q/4\pi\epsilon_0 = 3 \times 10^{-9}/(4\pi \times 8.854 \times 10^{-12}) = 26.96\ \text{V} \cdot \text{m}$, we may now use (13) or (14) to obtain

$$\mathbf{E} = 26.96 \left[\frac{\mathbf{a}_z}{1}\frac{1}{1^2} + \frac{2\mathbf{a}_x + \mathbf{a}_z}{\sqrt{5}}\frac{1}{(\sqrt{5})^2} + \frac{2\mathbf{a}_x + 2\mathbf{a}_y + \mathbf{a}_z}{3}\frac{1}{3^2} + \frac{2\mathbf{a}_y + \mathbf{a}_z}{\sqrt{5}}\frac{1}{(\sqrt{5})^2} \right]$$

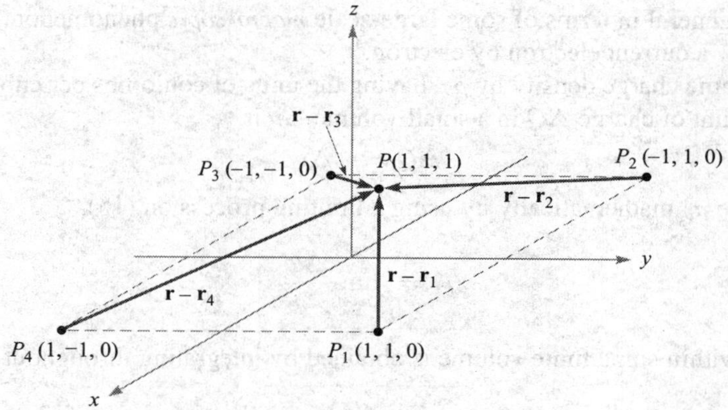

Fig. 2.4 A symmetrical distribution of four identical 3-nC point charges produces a field at P, $\mathbf{E} = 6.82\mathbf{a}_x + 6.82\mathbf{a}_y + 32.8\mathbf{a}_z$ V/m.

or

$$\mathbf{E} = 6.82\mathbf{a}_x + 6.82\mathbf{a}_y + 32.8\mathbf{a}_z \text{ V/m}$$

D2.2 A charge of $-0.3\,\mu$C is located at $A(25, -30, 15)$ (in cm), and a second charge of $0.5\,\mu$C is at $B(-10, 8, 12)$ cm. Find **E** at: (*a*) the origin; (*b*) $P(15, 20, 50)$ cm.

Ans. $92.3\mathbf{a}_x - 77.6\mathbf{a}_y - 94.2\mathbf{a}_z$ kV/m; $11.9\mathbf{a}_x - 0.519\mathbf{a}_y + 12.4\mathbf{a}_z$ kV/m

D2.3 Evaluate the sums: (*a*) $\sum_{m=0}^{5} \frac{1+(-1)^m}{m^2+1}$; (*b*) $\sum_{m=1}^{4} \frac{(0.1)^m + 1}{(4+m^2)^{1.5}}$

Ans. 2.52; 0.176

2.3 FIELD DUE TO A CONTINUOUS VOLUME CHARGE DISTRIBUTION

Illustrations

If we now visualize a region of space filled with a tremendous number of charges separated by minute distances, such as the space between the control grid and the cathode in the electron-gun assembly of a cathode-ray tube operating with space charge, we see that we can replace this distribution of very small particles with a smooth continuous distribution described by a *volume charge density*, just as we describe water as having a density of 1 g/cm^3 (gram per cubic centimeter) even though it consists of atomic- and molecular-sized particles. We can do this only if we are uninterested in the small irregularities (or ripples) in the field as we move from electron to electron or if we care little that the mass of the water actually increases in small but finite steps as each new molecule is added.

This is really no limitation at all, because the end results for electrical engineers are almost always in terms of a current in a receiving antenna, a voltage in an electronic circuit, or a charge on a capacitor, or in general in terms of some large-scale *macroscopic* phenomenon. It is very seldom that we must know a current electron by electron.[4]

We denote volume charge density by ρ_ν, having the units of coulombs per cubic meter (C/m^3). The small amount of charge ΔQ in a small volume $\Delta\nu$ is

$$\Delta Q = \rho_\nu \Delta\nu \tag{15}$$

and we may define ρ_ν mathematically by using a limiting process on (15),

$$\rho_\nu = \lim_{\Delta\nu \to 0} \frac{\Delta Q}{\Delta\nu} \tag{16}$$

The total charge within some finite volume is obtained by integrating throughout that volume,

$$Q = \int_{\text{vol}} \rho_\nu \, d\nu \tag{17}$$

Only one integral sign is customarily indicated, but the differential $d\nu$ signifies integration throughout a volume, and hence a triple integration. Fortunately, we may be content for the most part with no more than the indicated integration, for multiple integrals are very difficult to evaluate in all but the most symmetrical problems.

EXAMPLE 2.3

As an example of the evaluation of a volume integral, we shall find the total charge contained in a 2-cm length of the electron beam shown in Figure 2.5.

Solution. From the illustration, we see that the charge density is

$$\rho_\nu = -5 \times 10^{-6} e^{-10^5 \rho z} \text{ C/m}^2$$

The volume differential in cylindrical coordinates is given in Section 1.8; therefore,

$$Q = \int_{0.02}^{0.04} \int_0^{2\pi} \int_0^{0.01} -5 \times 10^{-6} e^{-10^5 \rho z} \rho \, d\rho \, d\phi \, dz$$

We integrate first with respect to ϕ since it is so easy,

$$Q = \int_{0.02}^{0.04} \int_0^{0.01} -10^{-5} \pi e^{-10^5 \rho z} \rho \, d\rho \, dz$$

and then with respect to z, because this will simplify the last integration with respect to ρ,

$$Q = \int_0^{0.01} \left(\frac{-10^{-5}\pi}{-10^5 \rho} e^{-10^5 \rho z} \rho \, d\rho \right)_{z=0.02}^{z=0.04}$$

$$= \int_0^{0.01} -10^{-5}\pi (e^{-2000\rho} - e^{-4000\rho}) d\rho$$

[4] A study of the noise generated by electrons in semiconductors and resistors, however, requires just such an examination of the charge through statistical analysis.

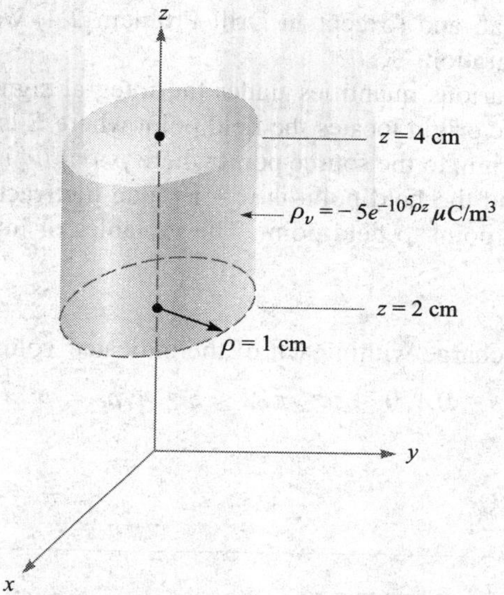

Fig. 2.5 The total charge contained within the right circular cylinder may be obtained by evaluating $Q = \int_{\text{vol}} \rho_\nu d\nu$.

Finally,

$$Q = -10^{-10}\pi \left(\frac{e^{-2000\rho}}{-2000} - \frac{e^{-4000\rho}}{-4000} \right)_0^{0.01}$$

$$Q = -10^{-10}\pi \left(\frac{1}{2000} - \frac{1}{4000} \right) = \frac{-\pi}{40} = 0.0785 \text{ pC}$$

where pC indicates picocoulombs.

Incidentally, we may use this result to make a rough estimate of the electron-beam current. If we assume these electrons are moving at a constant velocity of 10 percent of the velocity of light, this 2-cm-long packet will have moved 2 cm in $\frac{2}{3}$ns, and the current is about equal to

$$\frac{\Delta Q}{\Delta t} = \frac{-(\pi/40)10^{-12}}{(2/3)10^{-9}}$$

or approximately 118 μA.

The incremental contribution to the electric field intensity at **r** produced by an incremental charge ΔQ at $\mathbf{r}'$ is

$$\Delta \mathbf{E}(\mathbf{r}) = \frac{\Delta Q}{4\pi\epsilon_0 |\mathbf{r} - \mathbf{r}'|^2} \frac{\mathbf{r} - \mathbf{r}'}{|\mathbf{r} - \mathbf{r}'|} = \frac{\rho_\nu \Delta \nu}{4\pi\epsilon_0 |\mathbf{r} - \mathbf{r}'|^2} \frac{\mathbf{r} - \mathbf{r}'}{|\mathbf{r} - \mathbf{r}'|}$$

If we sum the contributions of all the volume charge in a given region and let the volume element $\Delta\nu$ approach zero as the number of these elements becomes infinite, the summation becomes an integral,

$$\mathbf{E}(\mathbf{r}) = \int_{\text{vol}} \frac{\rho_\nu(\mathbf{r}')\, d\nu'}{4\pi\epsilon_0 |\mathbf{r} - \mathbf{r}'|^2} \frac{\mathbf{r} - \mathbf{r}'}{|\mathbf{r} - \mathbf{r}'|} \tag{18}$$

This is again a triple integral, and (except in Drill Problem 2.4) we shall do our best to avoid actually performing the integration.

The significance of the various quantities under the integral sign of (18) might stand a little review. The vector $\mathbf{r}$ from the origin locates the field point where $\mathbf{E}$ is being determined, while the vector $\mathbf{r}'$ extends from the origin to the source point where $\rho_v(\mathbf{r}')dv'$ is located. The scalar distance between the source point and the field point is $|\mathbf{r} - \mathbf{r}'|$, and the fraction $(\mathbf{r} - \mathbf{r}')/|\mathbf{r} - \mathbf{r}'|$ is a unit vector directed from source point to field point. The variables of integration are x', y', and z' in rectangular coordinates.

D2.4 Calculate the total charge within each of the indicated volumes: (*a*) $0.1 \le |x|, |y|, |z| \le 0.2$: $\rho_\nu = \dfrac{1}{x^3y^3z^3}$; (*b*) $0 \le \rho \le 0.1, 0 \le \phi \le \pi, 2 \le z \le 4$; $\rho_\nu = \rho^2 z^2 \sin 0.6\phi$; (*c*) universe: $\rho_\nu = e^{-2r}/r^2$.

Ans. 0; 1.018 mC; 6.28 C

2.4 FIELD OF A LINE CHARGE

Up to this point we have considered two types of charge distribution, the point charge and charge distributed throughout a volume with a density ρ_ν C/m^3. If we now consider a filamentlike distribution of volume charge density, such as a very fine, sharp beam in a cathode-ray tube or a charged conductor of very small radius, we find it convenient to treat the charge as a line charge of density ρ_L C/m. In the case of the electron beam the charges are in motion and it is true that we do not have an electrostatic problem. However, if the electron motion is steady and uniform (a dc beam) and if we ignore for the moment the magnetic field which is produced, the electron beam may be considered to be composed of stationary electrons, for snapshots taken at any time will show the same charge distribution.

Let us assume a straight-line charge extending along the z axis in a cylindrical coordinate system from $-\infty$ to ∞, as shown in Figure 2.6. We desire the electric field intensity $\mathbf{E}$ at any and every point resulting from a *uniform* line charge density ρ_L.

Symmetry should always be considered first in order to determine two specific factors: (1) with which coordinates the field does *not* vary, and (2) which components of the field are *not* present. The answers to these questions then tell us which components are present and with which coordinates they *do* vary.

Referring to Figure 2.6, we realize that as we move around the line charge, varying ϕ while keeping ρ and z constant, the line charge appears the same from every angle. In other words, azimuthal symmetry is present, and no field component may vary with ϕ.

Again, if we maintain ρ and ϕ constant while moving up and down the line charge by changing z, the line charge still recedes into infinite distance in both directions and the problem is unchanged. This is axial symmetry and leads to fields which are not functions of z.

If we maintain ϕ and z constant and vary ρ, the problem changes, and Coulomb's law leads us to expect the field to become weaker as ρ increases. Hence, by a process of elimination we are led to the fact that the field varies only with ρ.

Now, which components are present? Each incremental length of line charge acts as a point charge and produces an incremental contribution to the electric field intensity which is directed

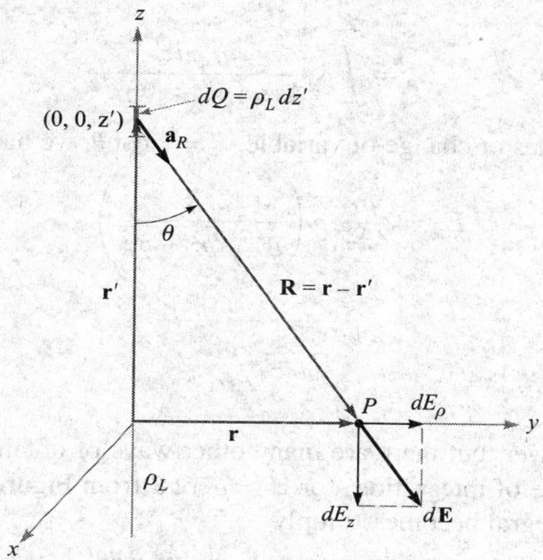

Fig. 2.6 The contribution $d\mathbf{E} = dE_\rho \mathbf{a}_\rho + dE_z \mathbf{a}_z$ to the electric field intensity produced by an element of charge $dQ = \rho_L dz'$ located a distance z' from the origin. The linear charge density is uniform and extends along the entire z axis.

away from the bit of charge (assuming a positive line charge). No element of charge produces a ϕ component of electric intensity; E_ϕ is zero. However, each element does produce an E_ρ and an E_z component, but the contribution to E_z by elements of charge which are equal distances above and below the point at which we are determining the field will cancel.

We therefore have found that we have only an E_ρ component and it varies only with ρ. Now to find this component.

We choose a point $P(0, y, 0)$ on the y axis at which to determine the field. This is a perfectly general point in view of the lack of variation of the field with ϕ and z. Applying (12) to find the incremental field at P due to the incremental charge $dQ = \rho_L dz'$, we have

$$d\mathbf{E} = \frac{\rho_L dz'(\mathbf{r} - \mathbf{r}')}{4\pi\epsilon_0 |\mathbf{r} - \mathbf{r}'|^3}$$

where

$$\mathbf{r} = y\mathbf{a}_y = \rho\mathbf{a}_\rho$$

$$\mathbf{r}' = z'\mathbf{a}_z$$

and

$$\mathbf{r} - \mathbf{r}' = \rho\mathbf{a}_\rho - z'\mathbf{a}_z$$

Therefore,

$$d\mathbf{E} = \frac{\rho_L dz'(\rho\mathbf{a}_\rho - z'\mathbf{a}_z)}{4\pi\epsilon_0(\rho^2 + z'^2)^{3/2}}$$

Since only the $\mathbf{E}_\rho$ component is present, we may simplify:

$$dE_\rho = \frac{\rho_L \rho dz'}{4\pi\epsilon_0(\rho^2 + z'^2)^{3/2}}$$

and

$$E_\rho = \int_{-\infty}^{\infty} \frac{\rho_L \rho dz'}{4\pi\epsilon_0(\rho^2 + z'^2)^{3/2}}$$

Integrating by integral tables or change of variable, $z' = \rho \cot\theta$, we have

$$E_\rho = \frac{\rho_L}{4\pi\epsilon_0}\rho \left(\frac{1}{\rho^2} \frac{z'}{\sqrt{\rho^2 + z'^2}} \right)_{-\infty}^{\infty}$$

and

$$E_\rho = \frac{\rho_L}{2\pi\epsilon_0 \rho} \tag{19}$$

This is the desired answer, but there are many other ways of obtaining it. We might have used the angle θ as our variable of integration, for $z' = \rho\cot\theta$ from Figure 2.6 and $dz' = -\rho\csc^2\theta\, d\theta$. Since $R = \rho\csc\theta$, our integral becomes, simply,

$$dE_\rho = \frac{\rho_L dz'}{4\pi\epsilon_0 R^2}\sin\theta = -\frac{\rho_L \sin\theta\, d\theta}{4\pi\epsilon_0\rho}$$

$$E_\rho = -\frac{\rho_L}{4\pi\epsilon_0\rho}\int_{\pi}^{0}\sin\theta\, d\theta = \frac{\rho_L}{4\pi\epsilon_0\rho}\cos\theta\Big]_{\pi}^{0}$$

$$= \frac{\rho_L}{2\pi\epsilon_0\rho}$$

Here the integration was simpler, but some experience with problems of this type is necessary before we can unerringly choose the simplest variable of integration at the beginning of the problem.

We might also have considered (18) as our starting point,

$$\mathbf{E} = \int_{\text{vol}} \frac{\rho_\nu\, dv'(\mathbf{r} - \mathbf{r}')}{4\pi\epsilon_0 |\mathbf{r} - \mathbf{r}'|^3}$$

letting $\rho_\nu\, dv' = \rho_L\, dz'$ and integrating along the line which is now our "volume" containing all the charge. Suppose we do this and forget everything we have learned from the symmetry of the problem. Choose point P now at a general location (ρ, ϕ, z) (Figure 2.7) and write

$$\mathbf{r} = \rho\mathbf{a}_\rho + z\mathbf{a}_z$$

$$\mathbf{r}' = z'\mathbf{a}_z$$

$$\mathbf{R} = \mathbf{r} - \mathbf{r}' = \rho\mathbf{a}_\rho + (z - z')\mathbf{a}_z$$

$$R = \sqrt{\rho^2 + (z - z')^2}$$

$$\mathbf{a}_R = \frac{\rho\mathbf{a}_\rho + (z - z')\mathbf{a}_z}{\sqrt{\rho^2 + (z - z')^2}}$$

$$\mathbf{E} = \int_{-\infty}^{\infty} \frac{\rho_L dz'[\rho\mathbf{a}_\rho + (z - z')\mathbf{a}_z]}{4\pi\epsilon_0[\rho^2 + (z - z')^2]^{3/2}}$$

$$= \frac{\rho_L}{4\pi\epsilon_0}\left\{ \int_{-\infty}^{\infty} \frac{\rho dz'\mathbf{a}_\rho}{[\rho^2 + (z - z')^2]^{3/2}} + \int_{-\infty}^{\infty} \frac{(z - z')\, dz'\mathbf{a}_z}{[\rho^2 + (z - z')^2]^{3/2}} \right\}$$

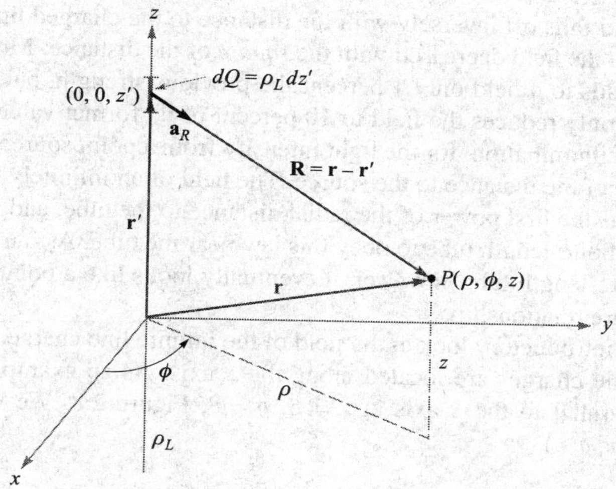

Fig. 2.7 The geometry of the problem for the field about an infinite line charge leads to more difficult integrations when symmetry is ignored.

Before integrating a vector expression, we must know whether or not a vector under the integral sign (here the unit vectors $\mathbf{a}_\rho$ and $\mathbf{a}_z$) varies with the variable of integration (here dz'). If it does not, then it is a constant and may be removed from within the integral, leaving a scalar which may be integrated by normal methods. Our unit vectors, of course, cannot change in magnitude, but a change in direction is just as troublesome. Fortunately, the direction of $\mathbf{a}_\rho$ does not change with z' (nor with ρ, but it does change with ϕ), and $\mathbf{a}_z$ is constant always.

Hence we remove the unit vectors from the integrals and again integrate with tables or by changing variables,

$$\mathbf{E} = \frac{\rho L}{4\pi\epsilon_0}\left\{\mathbf{a}_\rho \int_{-\infty}^{\infty} \frac{\rho\, dz'}{[\rho^2 + (z - z')^2]^{3/2}} + \mathbf{a}_z \int_{-\infty}^{\infty} \frac{(z - z')\, dz'}{[\rho^2 + (z - z')^2]^{3/2}}\right\}$$

$$= \frac{\rho_L}{4\pi\epsilon_0}\left\{\left[\mathbf{a}_\rho \rho \frac{1}{\rho^2} \frac{-(z - z')}{\sqrt{\rho^2 + (z - z')^2}}\right]_{-\infty}^{\infty} + \left[\mathbf{a}_z \frac{1}{\sqrt{\rho^2 + (z - z')^2}}\right]_{-\infty}^{\infty}\right\}$$

$$= \frac{\rho_L}{4\pi\epsilon_0}\left[\mathbf{a}_\rho \frac{2}{\rho} + \mathbf{a}_z(0)\right] = \frac{\rho_L}{2\pi\epsilon_0 \rho}\mathbf{a}_\rho$$

Again we obtain the same answer, as we should, for there is nothing wrong with the method except that the integration was harder and there were two integrations to perform. This is the price we pay for neglecting the consideration of symmetry and plunging doggedly ahead with mathematics. Look before you integrate.

Other methods for solving this basic problem will be discussed later after we introduce Gauss's law and the concept of potential.

Now let us consider the answer itself,

Illustrations

$$\mathbf{E} = \frac{\rho_L}{2\pi\epsilon_0 \rho}\mathbf{a}_\rho \tag{20}$$

We note that the field falls off inversely with the distance to the charged line, as compared with the point charge, where the field decreased with the *square* of the distance. Moving ten times as far from a point charge leads to a field only 1 percent the previous strength, but moving ten times as far from a line charge only reduces the field to 10 percent of its former value. An analogy can be drawn with a source of illumination, for the light intensity from a point source of light also falls off inversely as the square of the distance to the source. The field of an infinitely long fluorescent tube thus decays inversely as the first power of the radial distance to the tube, and we should expect the light intensity about a finite-length tube to obey this law near the tube. As our point recedes farther and farther from a finite-length tube, however, it eventually looks like a point source, and the field obeys the inverse-square relationship.

Before leaving this introductory look at the field of the infinite line charge, we should recognize the fact that not all line charges are located along the z axis. As an example, let us consider an infinite line charge parallel to the z axis at $x = 6$, $y = 8$, Figure 2.8. We wish to find **E** at the general field point $P(x, y, z)$.

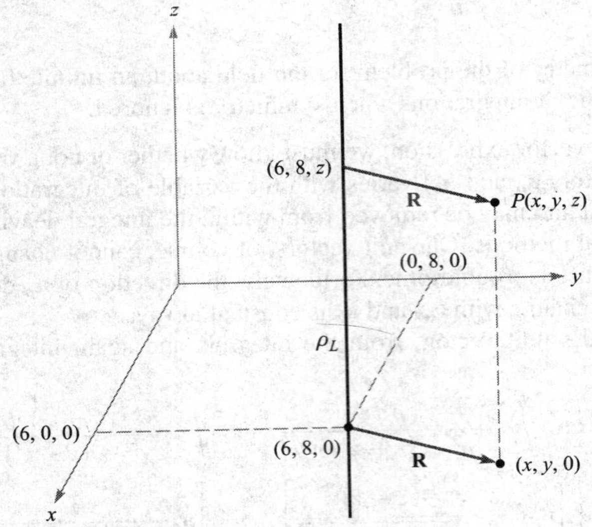

Fig. 2.8 A point $P(x, y, z)$ is identified near an infinite uniform line charge located at $x = 6, y = 8$.

We replace ρ in (20) by the radial distance between the line charge and point, $P, R = \sqrt{(x-6)^2 + (y-8)^2}$, and let $\mathbf{a}_\rho$ be $\mathbf{a}_R$. Thus,

$$\mathbf{E} = \frac{\rho_L}{2\pi\epsilon_0\sqrt{(x-6)^2 + (y-8)^2}}\mathbf{a}_R$$

where

$$\mathbf{a}_R = \frac{\mathbf{R}}{|\mathbf{R}|} = \frac{(x-6)\mathbf{a}_x + (y-8)\mathbf{a}_y}{\sqrt{(x-6)^2 + (y-8)^2}}$$

Therefore,

$$\mathbf{E} = \frac{\rho_L}{2\pi\epsilon_0}\frac{(x-6)\mathbf{a}_x + (y-8)\mathbf{a}_y}{(x-6)^2 + (y-8)^2}$$

We again note that the field is not a function of z.

In Section 2.6 we shall describe how fields may be sketched, and we will use the field of the line charge as one example.

D2.5 Infinite uniform line charges of 5 nC/m lie along the (positive and negative) x and y axes in free space. Find **E** at: (*a*) $P_A(0, 0, 4)$; (*b*) $P_B(0, 3, 4)$.

Ans. $45\mathbf{a}_z$ V/m; $10.8\mathbf{a}_y + 36.9\mathbf{a}_z$ V/m

2.5 FIELD OF A SHEET OF CHARGE

Another basic charge configuration is the infinite sheet of charge having a uniform density of ρ_S C/m^2. Such a charge distribution may often be used to approximate that found on the conductors of a strip transmission line or a parallel-plate capacitor. As we shall see in Chapter 5, static charge resides on conductor surfaces and not in their interiors; for this reason, ρ_S is commonly known as *surface charge density*. The charge-distribution family now is complete—point, line, surface, and volume, or Q, ρ_L, ρ_S, and ρ_ν.

Let us place a sheet of charge in the yz plane and again consider symmetry (Figure 2.9). We see first that the field cannot vary with y or with z, and then we see that the y and z components arising from differential elements of charge symmetrically located with respect to the point at which we evaluate the field will cancel. Hence only E_x is present, and this component is a function of x alone. We are again faced with a choice of many methods by which to evaluate this component, and this time we shall use but one method and leave the others as exercises for a quiet Sunday afternoon.

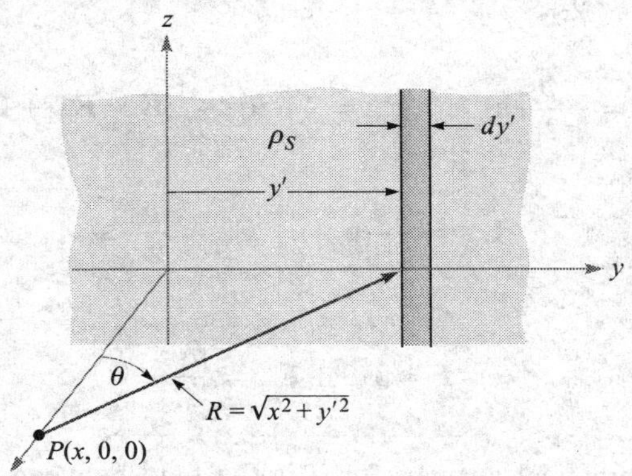

Fig. 2.9 An infinite sheet of charge in the yz plane, a general point P on the x axis, and the differential-width line charge used as the element in determining the field at P by $d\mathbf{E} = \rho_S\, dy'\mathbf{a}_R/(2\pi\varepsilon_0 R)$.

Let us use the field of the infinite line charge (19) by dividing the infinite sheet into differential-width strips. One such strip is shown in Figure 2.9. The line charge density, or charge per unit length, is $\rho_L = \rho_S\, dy'$, and the distance from this line charge to our general point P on the x axis is $R = \sqrt{x^2 + y'^2}$. The contribution to E_x at P from this differential-width strip is then

$$dE_x = \frac{\rho_S\, dy'}{2\pi\epsilon_0\sqrt{x^2 + y'^2}}\cos\theta = \frac{\rho_S}{2\pi\epsilon_0}\frac{x\,dy'}{x^2 + y'^2}$$

Adding the effects of all the strips,

$$E_x = \frac{\rho_S}{2\pi\epsilon_0}\int_{-\infty}^{\infty}\frac{x\,dy'}{x^2+y'^2} = \frac{\rho_S}{2\pi\epsilon_0}\tan^{-1}\frac{y'}{x}\bigg]_{-\infty}^{\infty} = \frac{\rho_S}{2\epsilon_0}$$

If the point P were chosen on the negative x axis, then

$$E_x = -\frac{\rho_S}{2\epsilon_0}$$

for the field is always directed away from the positive charge. This difficulty in sign is usually overcome by specifying a unit vector $\mathbf{a}_N$, which is normal to the sheet and directed outward, or away from it. Then

$$\mathbf{E} = \frac{\rho_S}{2\epsilon_0}\mathbf{a}_N \tag{21}$$

This is a startling answer, for the field is constant in magnitude and direction. It is just as strong a million miles away from the sheet as it is right off the surface. Returning to our light analogy, we see that a uniform source of light on the ceiling of a very large room leads to just as much illumination on a square foot on the floor as it does on a square foot a few inches below the ceiling. If you desire greater illumination on this subject, it will do you no good to hold the book closer to such a light source.

If a second infinite sheet of charge, having a *negative* charge density $-\rho_S$, is located in the plane $x = a$, we may find the total field by adding the contribution of each sheet. In the region $x > a$,

$$\mathbf{E}_+ = \frac{\rho_S}{2\epsilon_0}\mathbf{a}_x \qquad \mathbf{E}_- = -\frac{\rho_S}{2\epsilon_0}\mathbf{a}_x \qquad \mathbf{E} = \mathbf{E}_+ + \mathbf{E}_- = 0$$

and for $x < 0$,

$$\mathbf{E}_+ = -\frac{\rho_S}{2\epsilon_0}\mathbf{a}_x \qquad \mathbf{E}_- = \frac{\rho_S}{2\epsilon_0}\mathbf{a}_x \qquad \mathbf{E} = \mathbf{E}_+ + \mathbf{E}_- = 0$$

and when $0 < x < a$,

$$\mathbf{E}_+ = \frac{\rho_S}{2\epsilon_0}\mathbf{a}_x \qquad \mathbf{E}_- = \frac{\rho_S}{2\epsilon_0}\mathbf{a}_x$$

and

$$\mathbf{E} = \mathbf{E}_+ + \mathbf{E}_- = \frac{\rho_S}{\epsilon_0}\mathbf{a}_x \tag{22}$$

This is an important practical answer, for it is the field between the parallel plates of an air capacitor, provided the linear dimensions of the plates are very much greater than their separation and provided also that we are considering a point well removed from the edges. The field outside the capacitor, while not zero, as we found for the preceding ideal case, is usually negligible.

D2.6 Three infinite uniform sheets of charge are located in free space as follows: 3 nC/m^2 at $z = -4$, 6 nC/m^2 at $z = 1$, and -8 nC/m^2 at $z = 4$. Find $\mathbf{E}$ at the point: (*a*) $P_A(2, 5, -5)$; (*b*) $P_B(4, 2, -3)$; (*c*) $P_C(-1, -5, 2)$; (*d*) $P_D(-2, 4, 5)$.

Ans. $-56.5\mathbf{a}_z$; $283\mathbf{a}_z$; $961\mathbf{a}_z$; $56.5\mathbf{a}_z$ all V/m

2.6 STREAMLINES AND SKETCHES OF FIELDS

Illustrations

We now have vector equations for the electric field intensity resulting from several different charge configurations, and we have had little difficulty in interpreting the magnitude and direction of the field from the equations. Unfortunately, this simplicity cannot last much longer, for we have solved most of the simple cases and our new charge distributions must lead to more complicated expressions for the fields and more difficulty in visualizing the fields through the equations. However, it is true that one picture would be worth about a thousand words, if we just knew what picture to draw.

Consider the field about the line charge,

$$\mathbf{E} = \frac{\rho_L}{2\pi\epsilon_0\rho}\mathbf{a}_\rho$$

Figure 2.10*a* shows a cross-sectional view of the line charge and presents what might be our first effort at picturing the field—short line segments drawn here and there having lengths proportional to the magnitude of **E** and pointing in the direction of **E**. The figure fails to show the symmetry with respect to ϕ, so we try again in Figure 2.10*b* with a symmetrical location of the line segments. The real trouble now appears—the longest lines must be drawn in the most crowded region, and this also plagues us if we use line segments of equal length but of a thickness which is proportional to **E** (Figure 2.10*c*). Other schemes include drawing shorter lines to represent stronger fields (inherently misleading) and using intensity of color or different colors to represent stronger fields.

For the present, let us be content to show only the *direction* of **E** by drawing continuous lines from the charge which are everywhere tangent to **E**. Figure 2.10*d* shows this compromise. A symmetrical distribution of lines (one every 45°) indicates azimuthal symmetry, and arrowheads should be used to show direction.

These lines are usually called *streamlines*, although other terms such as flux lines and direction lines are also used. A small positive test charge placed at any point in this field and free to move would accelerate in the direction of the streamline passing through that point. If the field represented the velocity of a liquid or a gas (which, incidentally, would have to have a source at $\rho = 0$), small suspended particles in the liquid or gas would trace out the streamlines.

We shall find out later that a bonus accompanies this streamline sketch, for the magnitude of the field can be shown to be inversely proportional to the spacing of the streamlines for some important special cases. The closer they are together, the stronger is the field. At that time we shall also find an easier, more accurate method of making that type of streamline sketch.

If we attempted to sketch the field of the point charge, the variation of the field into and away from the page would cause essentially insurmountable difficulties; for this reason sketching is usually limited to two-dimensional fields.

In the case of the two-dimensional field let us arbitrarily set $E_z = 0$. The streamlines are thus confined to planes for which z is constant, and the sketch is the same for any such plane. Several streamlines are shown in Figure 2.11, and the E_x and E_y components are indicated at a general point. Since it is apparent from the geometry that

$$\frac{E_y}{E_x} = \frac{dy}{dx} \tag{23}$$

a knowledge of the functional form of E_x and E_y (and the ability to solve the resultant differential equation) will enable us to obtain the equations of the streamlines.

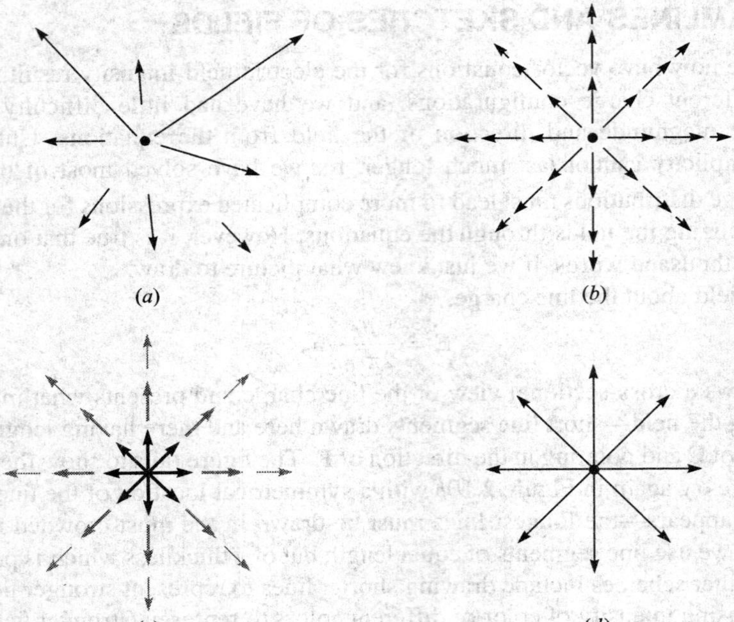

Fig. 2.10 (*a*) One very poor sketch, (*b*) and (*c*) two fair sketches, and (*d*) the usual form of streamline sketch. In the last form, the arrows show the direction of the field at every point along the line, and the spacing of the lines is inversely proportional to the strength of the field.

As an illustration of this method, consider the field of the uniform line charge with $\rho_L = 2\pi\epsilon_0$,

$$\mathbf{E} = \frac{1}{\rho}\mathbf{a}_\rho$$

In rectangular coordinates,

$$\mathbf{E} = \frac{x}{x^2+y^2}\mathbf{a}_x + \frac{y}{x^2+y^2}\mathbf{a}_y$$

Thus we form the differential equation

$$\frac{dy}{dx} = \frac{E_y}{E_x} = \frac{y}{x} \quad \text{or} \quad \frac{dy}{y} = \frac{dx}{x}$$

Therefore,

$$\ln y = \ln x + C_1 \quad \text{or} \quad \ln y = \ln x + \ln C$$

from which the equations of the streamlines are obtained,

$$y = Cx$$

If we want to find the equation of one particular streamline, say that one passing through $P(-2, 7, 10)$, we merely substitute the coordinates of that point into our equation and evaluate C. Here, $7 = C(-2)$, and $C = -3.5$, so that $y = -3.5x$.

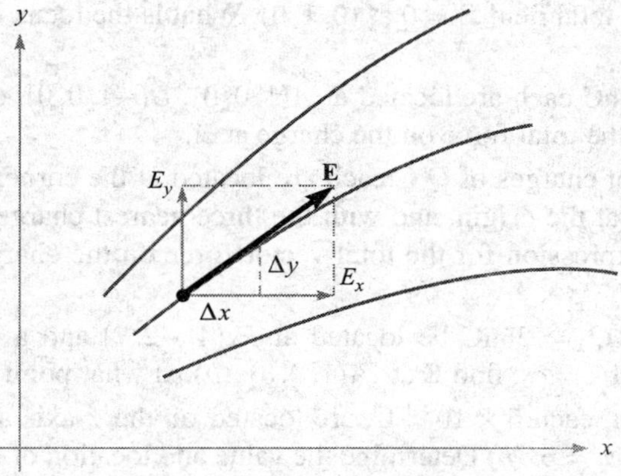

Fig. 2.11 The equation of a streamline is obtained by solving the differential equation $E_y/E_x = dy/dx$.

Each streamline is associated with a specific value of C, and the radial lines shown in Figure 2.10d are obtained when $C = 0, 1, -1$, and $1/C = 0$.

The equations of streamlines may also be obtained directly in cylindrical or spherical coordinates. A spherical coordinate example will be examined in Section 4.7.

D2.7 Find the equation of that streamline that passes through the point $P(1, 4, -2)$ in the field $\mathbf{E} =$: (a) $\dfrac{-8x}{y}\mathbf{a}_x + \dfrac{4x^2}{y^2}\mathbf{a}_y$; (b) $2e^{5x}[y(5x+1)\mathbf{a}_x + x\mathbf{a}_y]$.

Ans. $x^2 + 2y^2 = 33$; $y^2 = 15.7 + 0.4x - 0.08\ln(5x+1)$

References

1. Boast, W. B. *Vector Fields*. New York: Harper and Row, 1964. This book contains numerous examples and sketches of fields.
2. Della Torre, E., and Longo, C. L. *The Electromagnetic Field*. Boston: Allyn and Bacon, 1969. The authors introduce all of electromagnetic theory with a careful and rigorous development based on a single experimental law—that of Coulomb. It begins in Chapter 1.
3. Schelkunoff, S. A. *Electromagnetic Fields*. New York: Blaisdell Publishing Company, 1963. Many of the physical aspects of fields are discussed early in this text without advanced mathematics.

CHAPTER 2 Problems

Quizzes

2.1 Four 10nC positive charges are located in the $z = 0$ plane at the corners of a square 8 cm on a side. A fifth 10nC positive charge is located at a point 8 cm distant from the other charges. Calculate the magnitude of the total force on this fifth charge for $\epsilon = \epsilon_0$.

2.2 Two point charges of Q_1 coulombs each are located at $(0, 0, 1)$ and $(0, 0, -1)$. Determine the locus of the possible positions of a third charge Q_2 where Q_2 may be any positive or negative value, such that the total field $\mathbf{E} = 0$ at $(0, 1, 0)$. What is the locus if the two original charges are Q_1 and $-Q_1$?

2.3 Point charges of 50nC each are located at $A(1, 0, 0)$, $B(-1, 0, 0)$, $C(0, 1, 0)$, and $D(0, -1, 0)$ in free space. Find the total force on the charge at A.

2.4 Eight identical point charges of Q C each are located at the corners of a cube of side length a, with one charge at the origin, and with the three nearest charges at $(a, 0, 0)$, $(0, a, 0)$, and $(0, 0, a)$. Find an expression for the total vector force on the charge at $P(a, a, a)$, assuming free space.

2.5 Let a point charge $Q_1 = 25$nC be located at $P_1(4, -2, 7)$ and a charge $Q_2 = 60$ nC be at $P_2(-3, 4, -2)$. (*a*) If $\epsilon = \epsilon_0$, find $\mathbf{E}$ at $P_3(1, 2, 3)$. (*b*) At what point on the y axis is $E_x = 0$?

2.6 Three point charges, each 5×10^{-9} C, are located on the x axis at $x = -1, 0$, and 1 in free space. (*a*) Find $\mathbf{E}$ at $x = 5$. (*b*) Determine the value and location of the equivalent single point charge that would produce the same field at very large distances. (*c*) Determine $\mathbf{E}$ at $x = 5$, using the approximation of (*b*).

2.7 A $2\,\mu$C point charge is located at $A(4, 3, 5)$ in free space. Find E_ρ, E_ϕ, and E_z at $P(8, 12, 2)$.

2.8 A crude device for measuring charge consists of two small insulating spheres of radius a, one of which is fixed in position. The other is movable along the x axis and is subject to a restraining force kx, where k is a spring constant. The uncharged spheres are centered at $x = 0$ and $x = d$, the latter fixed. If the spheres are given equal and opposite charges of Q coulombs, obtain the expression by which Q may be found as a function of x. Determine the maximum charge that can be measured in terms of ϵ_0, k, and d, and state the separation of the spheres then. What happens if a larger charge is applied?

2.9 A 100 nC point charge is located at $A(-1, 1, 3)$ in free space. (*a*) Find the locus of all points $P(x, y, z)$ at which $E_x = 500$ V/m. (*b*) Find y_1 if $P(-2, y_1, 3)$ lies on that locus.

2.10 A positive test charge is used to explore the field of a single positive point charge Q at $P(a, b, c)$. If the test charge is placed at the origin, the force on it is in the direction $0.5\mathbf{a}_x - 0.5\sqrt{3}\mathbf{a}_y$, and when the test charge is moved to $(1, 0, 0)$, the force is in the direction of $0.6\mathbf{a}_x - 0.8\mathbf{a}_y$. Find a, b, and c.

2.11 A charge Q_0 located at the origin in free space produces a field for which $E_z = 1$ kV/m at point $P(-2, 1, -1)$. (*a*) Find Q_0. Find $\mathbf{E}$ at $M(1, 6, 5)$ in (*b*) rectangular coordinates; (*c*) cylindrical coordinates; (*d*) spherical coordinates.

2.12 Electrons are in random motion in a fixed region in space. During any 1μs interval, the probability of finding an electron in a subregion of volume 10^{-15} m^2 is 0.27. What volume charge density, appropriate for such time durations, should be assigned to that subregion?

2.13 A uniform volume charge density of $0.2\,\mu$C/m^3 is present throughout the spherical shell extending from $r = 3$ cm to $r = 5$ cm. If $\rho_\nu = 0$ elsewhere, find: (*a*) the total charge present throughout the shell, and (*b*) r_1 if half the total charge is located in the region 3 cm $< r < r_1$.

2.14 The charge density varies with radius in a cylindrical coordinate system as $\rho_\nu = \rho_0/(\rho^2 + a^2)^2$ C/m^3. Within what distance from the z axis does half the total charge lie?

2.15 A spherical volume having a $2\,\mu$m radius contains a uniform volume charge density of 10^{15} C/m^3. (*a*) What total charge is enclosed in the spherical volume? (*b*) Now assume that a large region contains one of these little spheres at every corner of a cubical grid 3 mm on a side, and that there is no charge between the spheres. What is the average volume charge density throughout this large region?

2.16 Within a region of free space, charge density is given as $\rho_\nu = \rho_0 r/a$ C/m^3, where ρ_0 and a are constants. Find the total charge lying within: (*a*) the sphere, $r \leq a$; (*b*) the cone, $r \leq a$, $0 \leq \theta \leq 0.1\pi$; (*c*) the region, $r \leq a$, $0 \leq \theta \leq 0.1\pi$, $0 \leq \phi \leq 0.2\pi$.

2.17 A uniform line charge of 16 nC/m is located along the line defined by $y = -2$, $z = 5$. If $\epsilon = \epsilon_0$: (*a*) find **E** at $P(1, 2, 3)$. (*b*) find **E** at that point in the $z = 0$ plane where the direction of **E** is given by $(1/3)\mathbf{a}_y - (2/3)\mathbf{a}_z$.

2.18 An infinite uniform line charge $\rho_L = 2$ nC/m lies along the x axis in free space, while point charges of 8 nC each are located at $(0, 0, 1)$ and $(0, 0, -1)$. (*a*) Find **E** at $(2, 3, -4)$. (*b*) To what value should ρ_L be changed to cause **E** to be zero at $(0, 0, 3)$?

2.19 A uniform line charge of 2 μC/m is located on the z axis. Find **E** in rectangular coordinates at $P(1, 2, 3)$ if the charge exists from: (*a*) $-\infty < z < \infty$; (*b*) $-4 \leq z \leq 4$.

2.20 The portion of the z axis for which $|z| < 2$ carries a nonuniform line charge density of $10|z|$ nC/m, and $\rho_L = 0$ elsewhere. Determine **E** in free space at: (*a*) $(0, 0, 4)$; (*b*) $(0, 4, 0)$.

2.21 Two identical uniform line charges, with $\rho_l = 75$ nC/m, are located in free space at $x = 0$, $y = \pm 0.4$ m. What force per unit length does each line charge exert on the other?

2.22 Two identical uniform sheet charges with $\rho_s = 100$ nC/m^2 are located in free space at $z = \pm 2.0$ cm. What force per unit area does each sheet exert on the other?

2.23 Given the surface charge density, $\rho_s = 2\,\mu$C/m^2, existing in the region $\rho < 0.2$ m, $z = 0$, and is zero elsewhere, find **E** at: (*a*) $P_A(\rho = 0, z = 0.5)$; (*b*) $P_B(\rho = 0, z = -0.5)$.

2.24 For the charged disk of Problem 2.23, show that: (*a*) the field along the z axis reduces to that of an infinite sheet charge at small values of z; (*b*) the z axis field reduces to that of a point charge at large values of z.

2.25 Find **E** at the origin if the following charge distributions are present in free space: point charge, 12 nC, at $P(2, 0, 6)$; uniform line charge density, 3 nC/m, at $x = -2$, $y = 3$; uniform surface charge density, 0.2 nC/m^2 at $x = 2$.

2.26 An electric dipole (discussed in detail in Section 4.7) consists of two point charges of equal and opposite magnitude $\pm Q$ spaced by distance d. With the charges along the z axis at positions $z = \pm d/2$ (with the positive charge at the positive z location), the electric field in spherical coordinates is given by $\mathbf{E}(r, \theta) = [Qd/(4\pi\epsilon_0 r^3)][2\cos\theta\mathbf{a}_r + \sin\theta\mathbf{a}_\theta]$, where $r >> d$. Using rectangular coordinates, determine expressions for the vector force on a point charge of magnitude q: (*a*) at $(0, 0, z)$; (*b*) at $(0, y, 0)$.

2.27 Given the electric field $\mathbf{E} = (4x - 2y)\mathbf{a}_x - (2x + 4y)\mathbf{a}_y$, find: (*a*) the equation of the streamline that passes through the point $P(2, 3, -4)$; (*b*) a unit vector specifying the direction of **E** at $Q(3, -2, 5)$.

2.28 A field is given as $\mathbf{E} = 2xz^2\mathbf{a}_x + 2z(x^2 + 1)\mathbf{a}_z$. Find the equation of the streamline passing through the point $(1, 3, -1)$.

2.29 If $\mathbf{E} = 20e^{-5y}(\cos 5x\mathbf{a}_x - \sin 5x\mathbf{a}_y)$, find: (*a*) $|\mathbf{E}|$ at $P(\pi/6, 0.1, 2)$; (*b*) a unit vector in the direction of **E** at P; (*c*) the equation of the direction line passing through P.

2.30 For fields that do not vary with z in cylindrical coordinates, the equations of the streamlines are obtained by solving the differential equation $E_p/E_\phi = dp/(\rho d\phi)$. Find the equation of the line passing through the point $(2, 30°, 0)$ for the field $\mathbf{E} = \rho\cos 2\phi\mathbf{a}_\rho - \rho\sin 2\phi\mathbf{a}_\phi$.

3

Electric Flux Density, Gauss's Law, and Divergence

After drawing a few of the fields described in the previous chapter and becoming familiar with the concept of the streamlines which show the direction of the force on a test charge at every point, it is difficult to avoid giving these lines a physical significance and thinking of them as *flux* lines. No physical particle is projected radially outward from the point charge, and there are no steel tentacles reaching out to attract or repel an unwary test charge, but as soon as the streamlines are drawn on paper there seems to be a picture showing "something" is present.

It is very helpful to invent an *electric flux* which streams away symmetrically from a point charge and is coincident with the streamlines and to visualize this flux wherever an electric field is present.

This chapter introduces and uses the concept of electric flux and electric flux density to solve again several of the problems presented in Chapter 2. The work here turns out to be much easier, and this is due to the extremely symmetrical problems which we are solving.

3.1 ELECTRIC FLUX DENSITY

About 1837 the Director of the Royal Society in London, Michael Faraday, became very interested in static electric fields and the effect of various insulating materials on these fields. This problem had been bothering him during the past ten years when he was experimenting in his now famous work on induced electromotive force, which we shall discuss in Chapter 10. With that subject completed, he had a pair of concentric metallic spheres constructed, the outer one consisting of two hemispheres that could be firmly clamped together. He also prepared shells of insulating material (or dielectric material, or simply *dielectric*) which would occupy the entire volume between the concentric spheres. We shall not make immediate use of his findings about dielectric materials, for we are restricting our attention to fields in free space until Chapter 6. At that time we shall see that the materials he used will be classified as ideal dielectrics.

His experiment, then, consisted essentially of the following steps:

1. With the equipment dismantled, the inner sphere was given a known positive charge.
2. The hemispheres were then clamped together around the charged sphere with about 2 cm of dielectric material between them.
3. The outer sphere was discharged by connecting it momentarily to ground.

4. The outer space was separated carefully, using tools made of insulating material in order not to disturb the induced charge on it, and the negative induced charge on each hemisphere was measured.

Faraday found that the total charge on the outer sphere was equal in *magnitude* to the original charge placed on the inner sphere and that this was true regardless of the dielectric material separating the two spheres. He concluded that there was some sort of "displacement" from the inner sphere to the outer which was independent of the medium, and we now refer to this flux as *displacement*, *displacement* flux, or simply *electric flux*.

Faraday's experiments also showed, of course, that a larger positive charge on the inner sphere induced a correspondingly larger negative charge on the outer sphere, leading to a direct proportionality between the electric flux and the charge on the inner sphere. The constant of proportionality is dependent on the system of units involved, and we are fortunate in our use of SI units, because the constant is unity. If electric flux is denoted by Ψ (psi) and the total charge on the inner sphere by Q, then for Faraday's experiment

$$\Psi = Q$$

and the electric flux Ψ is measured in coulombs.

We can obtain more quantitative information by considering an inner sphere of radius a and an outer sphere of radius b, with charges of Q and $-Q$, respectively (Figure 3.1). The paths of electric flux Ψ extending from the inner sphere to the outer sphere are indicated by the symmetrically distributed streamlines drawn radially from one sphere to the other.

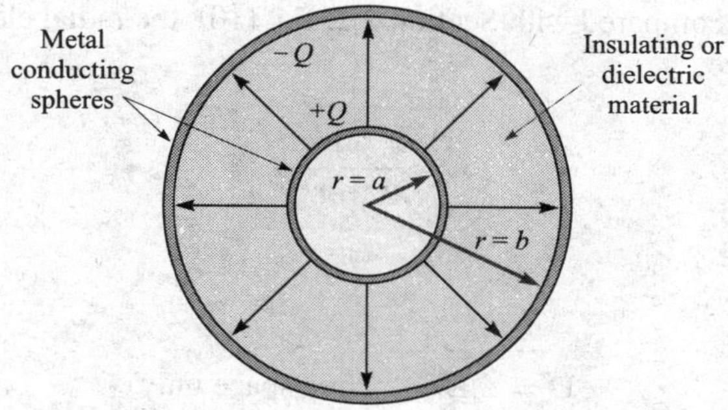

Fig. 3.1 The electric flux in the region between a pair of charged concentric spheres. The direction and magnitude of **D** are not functions of the dielectric between the spheres.

At the surface of the inner sphere, Ψ coulombs of electric flux are produced by the charge $Q(= \Psi)$ coulombs distributed uniformly over a surface having an area of $4\pi a^2$ m^2. The density of the flux at this surface is $\Psi/4\pi a^2$ or $Q/4\pi a^2$ C/m^2, and this is an important new quantity.

Electric flux density, measured in coulombs per square meter (sometimes described as "lines per square meter," for each line is due to one coulomb), is given the letter **D**, which was originally chosen because of the alternate names of *displacement flux density* or *displacement density*. Electric flux density is more descriptive, however, and we shall use the term consistently.

The electric flux density **D** is a vector field and is a member of the "flux density" class of vector fields, as opposed to the "force fields" class, which includes the electric field intensity **E**. The direction of **D** at a point is the direction of the flux lines at that point, and the magnitude is given by the number of flux lines crossing a surface normal to the lines divided by the surface area.

Referring again to Figure 3.1, the electric flux density is in the radial direction and has a value of

$$\mathbf{D}\Big|_{r=a} = \frac{Q}{4\pi a^2}\mathbf{a}_r \qquad \text{(inner sphere)}$$

$$\mathbf{D}\Big|_{r=b} = \frac{Q}{4\pi b^2}\mathbf{a}_r \qquad \text{(outer sphere)}$$

and at a radial distance r, where $a \le r \le b$,

$$\mathbf{D} = \frac{Q}{4\pi r^2}\mathbf{a}_r$$

If we now let the inner sphere become smaller and smaller, while still retaining a charge of Q, it becomes a point charge in the limit, but the electric flux density at a point r meters from the point charge is still given by

$$\mathbf{D} = \frac{Q}{4\pi r^2}\mathbf{a}_r \tag{1}$$

for Q lines of flux are symmetrically directed outward from the point and pass through an imaginary spherical surface of area $4\pi r^2$.

This result should be compared with Section 2.2, Eq. (10), the radial electric field intensity of a point charge in free space,

$$\mathbf{E} = \frac{Q}{4\pi\epsilon_0 r^2}\mathbf{a}_r$$

In free space, therefore,

$$\mathbf{D} = \epsilon_0\mathbf{E} \qquad \text{(free space only)} \tag{2}$$

Although (2) is applicable only to a vacuum, it is not restricted solely to the field of a point charge. For a general volume charge distribution in free space

$$\mathbf{E} = \int_{\text{vol}} \frac{\rho_\nu\, dv}{4\pi\epsilon_0 R^2}\mathbf{a}_R \qquad \text{(free space only)} \tag{3}$$

where this relationship was developed from the field of a single point charge. In a similar manner, (1) leads to

$$\mathbf{D} = \int_{\text{vol}} \frac{\rho_\nu\, dv}{4\pi R^2}\mathbf{a}_R \tag{4}$$

and (2) is therefore true for any free-space charge configuration; we shall consider (2) as defining **D** in free space.

As a preparation for the study of dielectrics later, it might be well to point out now that, for a point charge embedded in an infinite ideal dielectric medium, Faraday's results show that (1) is still applicable, and thus so is (4). Equation (3) is not applicable, however, and so the relationship between **D** and **E** will be slightly more complicated than (2).

Since **D** is directly proportional to **E** in free space, it does not seem that it should really be necessary to introduce a new symbol. We do so for several reasons. First, **D** is associated with the flux concept, which is an important new idea. Second, the **D** fields we obtain will be a little simpler than the corresponding **E** fields, since ϵ_0 does not appear. And, finally, it helps to become a little familiar with **D** before it is applied to dielectric materials in Chapter 6.

Let us consider a simple numerical example to illustrate these new quantities and units.

EXAMPLE 3.1

We wish to find **D** in the region about a uniform line charge of 8 nC/m lying along the z axis in free space.

Solution. The **E** field is

$$\mathbf{E} = \frac{\rho_L}{2\pi\epsilon_0\rho}\mathbf{a}_\rho = \frac{8\times10^{-9}}{2\pi(8.854\times10^{-12})\rho}\mathbf{a}_\rho = \frac{143.8}{\rho}\mathbf{a}_\rho \text{ V/m}$$

At $\rho = 3\text{m}$, $\mathbf{E} = 47.9\mathbf{a}_\rho$ V/m.

Associated with the **E** field, we find

$$\mathbf{D} = \frac{\rho_L}{2\pi\rho}\mathbf{a}_\rho = \frac{8\times10^{-9}}{2\pi\rho}\mathbf{a}_\rho = \frac{1.273\times10^{-9}}{\rho}\mathbf{a}_\rho \text{ C/m}^2$$

The value at $\rho = 3$ m is $\mathbf{D} = 0.424\mathbf{a}_\rho$ nC/m.

The total flux leaving a 5-m length of the line charge is equal to the total charge on that length, or $\Psi = 40$ nC.

D3.1 Given a 60-μC point charge located at the origin, find the total electric flux passing through: (a) that portion of the sphere $r = 26$ cm bounded by $0 < \theta < \frac{\pi}{2}$ and $0 < \phi < \frac{\pi}{2}$; (b) the closed surface defined by $\rho = 26$ cm and $z = \pm 26$ cm; (c) the plane $z = 26$ cm.

Ans. 7.5 μC; 60 μC; 30 μC

D3.2 Calculate **D** in rectangular coordinates at point $P(2, -3, 6)$ produced by: (a) a point charge $Q_A = 55$ mC at $Q(-2, 3, -6)$; (b) a uniform line charge $\rho_{LB} = 20$ mC/m on the x axis; (c) a uniform surface charge density $\rho_{SC} = 120\ \mu\text{C/m}^2$ on the plane $z = -5$ m.

Ans. $6.38\mathbf{a}_x - 9.57\mathbf{a}_y + 19.14\mathbf{a}_z\ \mu\text{C/m}^2$; $-212\mathbf{a}_y + 424\mathbf{a}_z\ \mu\text{C/m}^2$; $60\mathbf{a}_z\ \mu\text{C/m}^2$

3.2 GAUSS'S LAW

The results of Faraday's experiments with the concentric spheres could be summed up as an experimental law by stating that the electric flux passing through any imaginary spherical surface lying between the two conducting spheres is equal to the charge enclosed within that imaginary surface. This enclosed charge is distributed on the surface of the inner sphere, or it might be concentrated as a point charge at the center of the imaginary sphere. However, since one coulomb of electric flux is produced by one coulomb of charge, the inner conductor might just as well have been a cube or a brass door key and the total induced charge on the outer sphere would still be the same. Certainly the flux density would change from its previous symmetrical distribution to some unknown configuration, but $+Q$ coulombs on any inner conductor would produce an induced charge of $-Q$ coulombs on the surrounding sphere. Going one step further, we could now replace the two outer hemispheres by an empty (but completely closed) soup can. Q coulombs on the brass door key would produce $\Psi = Q$ lines of electric flux and would induce $-Q$ coulombs on the tin can[1]

These generalizations of Faraday's experiment lead to the following statement, which is known as *Gauss's law*:

> The electric flux passing through any closed surface is equal to the total charge enclosed by that surface.

The contribution of Gauss, one of the greatest mathematicians the world has ever produced, was actually not in stating the law as we have, but in providing a mathematical form for this statement, which we shall now obtain.

Let us imagine a distribution of charge, shown as a cloud of point charges in Figure 3.2, surrounded by a closed surface of any shape. The closed surface may be the surface of some real material, but more generally it is any closed surface we wish to visualize. If the total charge is Q, then Q coulombs of electric flux will pass through the enclosing surface. At every point on the surface the electric-flux-density vector $\mathbf{D}$ will have some value $\mathbf{D}_S$, where the subscript S merely reminds us that $\mathbf{D}$ must be evaluated at the surface, and $\mathbf{D}_S$ will in general vary in magnitude and direction from one point on the surface to another.

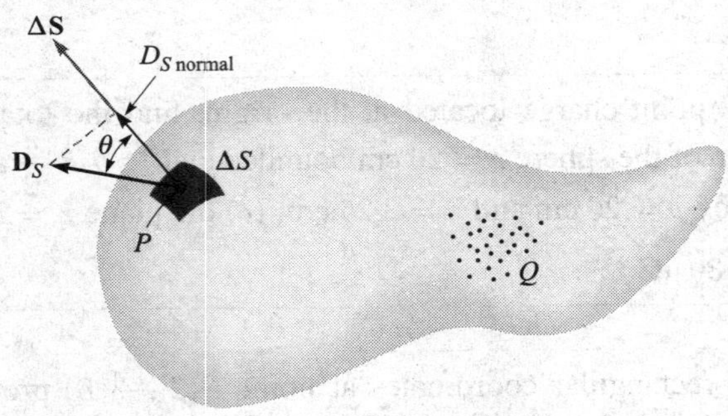

Fig. 3.2 The electric flux density $\mathbf{D}_S$ at P due to charge Q. The total flux passing through ΔS is $\mathbf{D}_S \cdot \Delta\mathbf{S}$.

We must now consider the nature of an incremental element of the surface. An incremental element of area ΔS is very nearly a portion of a plane surface, and the complete description of

[1] If it were a perfect insulator, the soup could even be left in the can without any difference in the results.

this surface element requires not only a statement of its magnitude ΔS but also of its orientation in space. In other words, the incremental surface element is a vector quantity. The only unique direction which may be associated with $\Delta\mathbf{S}$ is the direction of the normal to that plane which is tangent to the surface at the point in question. There are, of course, two such normals, and the ambiguity is removed by specifying the outward normal whenever the surface is closed and "outward" has a specific meaning.

At any point P consider an incremental element of surface ΔS and let $\mathbf{D}_S$ make an angle θ with $\Delta\mathbf{S}$, as shown in Figure 3.2. The flux crossing ΔS is then the product of the normal component of $\mathbf{D}_S$ and $\Delta\mathbf{S}$,

$$\Delta\Psi = \text{flux crossing } \Delta S = D_{S,norm}\Delta S = D_S \cos\theta \Delta S = \mathbf{D}_S \cdot \Delta\mathbf{S}$$

where we are able to apply the definition of the dot product developed in Chapter 1.

The *total* flux passing through the closed surface is obtained by adding the differential contributions crossing each surface element $\Delta\mathbf{S}$,

$$\Psi = \int d\Psi = \oint_{\substack{\text{closed}\\ \text{surface}}} \mathbf{D}_S \cdot d\mathbf{S}$$

Illustrations

The resultant integral is a *closed surface integral*, and since the surface element $d\mathbf{S}$ always involves the differentials of two coordinates, such as $dx\,dy$, $\rho\, d\phi\ d\rho$, or $r^2 \sin\theta\, d\theta\, d\phi$, the integral is a double integral. Usually only one integral sign is used for brevity, and we shall always place an S below the integral sign to indicate a surface integral, although this is not actually necessary since the differential $d\mathbf{S}$ is automatically the signal for a surface integral. One last convention is to place a small circle on the integral sign itself to indicate that the integration is to be performed over a *closed* surface. Such a surface is often called a *gaussian surface*. We then have the mathematical formulation of Gauss's law,

$$\Psi = \oint_S \mathbf{D}_S \cdot d\mathbf{S} = \text{charge enclosed} = Q \tag{5}$$

The charge enclosed might be several point charges, in which case

$$Q = \Sigma Qn$$

or a line charge,

$$Q = \int \rho_L\, dL$$

or a surface charge,

$$Q = \int_S \rho_S dS \qquad \text{(not necessarily a closed surface)}$$

or a volume charge distribution,

$$Q = \int_{\text{vol}} \rho_\nu\, dv$$

The last form is usually used, and we should agree now that it represents any or all of the other forms. With this understanding, Gauss's law may be written in terms of the charge distribution as

$$\oint_S \mathbf{D}_S \cdot d\mathbf{S} = \int_{\text{vol}} \rho_\nu \, dv \qquad (6)$$

a mathematical statement meaning simply that the total electric flux through any closed surface is equal to the charge enclosed.

To illustrate the application of Gauss's law, let us check the results of Faraday's experiment by placing a point charge Q at the origin of a spherical coordinate system (Figure 3.3) and by choosing our closed surface as a sphere of radius a. The electric field intensity of the point charge has been found to be

$$\mathbf{E} = \frac{Q}{4\pi\epsilon_0 r^2}\mathbf{a}_r$$

and since

$$\mathbf{D} = \epsilon_0 \mathbf{E}$$

we have, as before,

$$\mathbf{D} = \frac{Q}{4\pi r^2}\mathbf{a}_r$$

At the surface of the sphere,

$$\mathbf{D}_S = \frac{Q}{4\pi a^2}\mathbf{a}_r$$

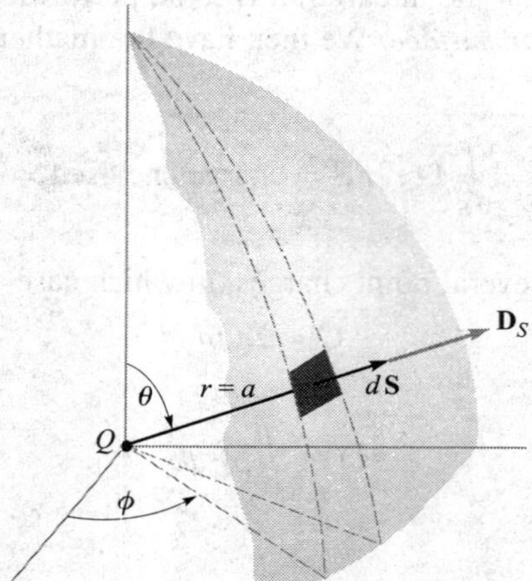

Fig. 3.3 Application of Gauss's law to the field of a point charge Q on a spherical closed surface of radius a. The electric flux density D is everywhere normal to the spherical surface and has a constant magnitude at every point on it.

The differential element of area on a spherical surface is, in spherical coordinates from Chapter 1,

$$dS = r^2 \sin\theta \, d\theta \, d\phi = a^2 \sin\theta \, d\theta \, d\phi$$

or

$$d\mathbf{S} = a^2 \sin\theta \, d\theta \, d\phi \, \mathbf{a}_r$$

The integrand is

$$\mathbf{D}_S \cdot d\mathbf{S} = \frac{Q}{4\pi a^2} a^2 \sin\theta \, d\theta \, d\phi \mathbf{a}_r \cdot \mathbf{a}_r = \frac{Q}{4\pi} \sin\theta \, d\theta \, d\phi$$

leading to the closed surface integral

$$\int_{\phi=0}^{\phi=2\pi} \int_{\theta=\phi}^{\theta=\pi} \frac{Q}{4\pi} \sin\theta \, d\theta \, d\phi$$

where the limits on the integrals have been chosen so that the integration is carried over the entire surface of the sphere once.[2] Integrating gives

$$\int_0^{2\pi} \frac{Q}{4\pi} \left(-\cos\theta \right)_0^\pi d\phi = \int_0^{2\pi} \frac{Q}{2\pi} d\phi = Q$$

Animations

and we obtain a result showing that Q coulombs of electric flux are crossing the surface, as we should since the enclosed charge is Q coulombs.

The following section contains examples of the application of Gauss's law to problems of a simple symmetrical geometry with the object of finding the electric field intensity.

D3.3 Given the electric flux density, $\mathbf{D} = 0.3r^2\mathbf{a}_r$ nC/m^2 in free space: (*a*) find $\mathbf{E}$ at point $P(r = 2, \theta = 25^\circ, \phi = 90^\circ)$; (*b*) find the total charge within the sphere $r = 3$; (*c*) find the total electric flux leaving the sphere $r = 4$.

Ans. 135.5$\mathbf{a}_r$ V/m; 305 nC; 965 nC

D3.4 Calculate the total electric flux leaving the cubical surface formed by the six planes $x, y, z = \pm 5$ if the charge distribution is: (*a*) two point charges, 0.1 μC at $(1, -2, 3)$ and $\frac{1}{7}$ μC at $(-1, 2, -2)$; (*b*) a uniform line charge of π μC/m at $x = -2$, $y = 3$; (*c*) a uniform surface charge of 0.1 μC/m^2 on the plane $y = 3x$.

Ans. 0.243 μC; 31.4 μC; 10.54 μC

3.3 APPLICATION OF GAUSS'S LAW: SOME SYMMETRICAL CHARGE DISTRIBUTIONS

Let us now consider how we may use Gauss's law,

$$Q = \oint_S \mathbf{D}_S \cdot d\mathbf{S}$$

to determine $\mathbf{D}_S$ if the charge distribution is known. This is an example of an integral equation in which the unknown quantity to be determined appears inside the integral.

The solution is easy if we are able to choose a closed surface which satisfies two conditions:

[2] Note that if θ and ϕ both cover the range from 0 to 2π, the spherical surface is covered twice.

1. $\mathbf{D}_S$ is everywhere either normal or tangential to the closed surface, so that $\mathbf{D}_S \cdot d\mathbf{S}$ becomes either $D_S dS$ or zero, respectively.
2. On that portion of the closed surface for which $\mathbf{D}_S \cdot d\mathbf{S}$ is not zero, D_S = constant.

This allows us to replace the dot product with the product of the scalars D_S and dS and then to bring D_S outside the integral sign. The remaining integral is then $\int_S dS$ over that portion of the closed surface which $\mathbf{D}_S$ crosses normally, and this is simply the area of this section of that surface.

Only a knowledge of the symmetry of the problem enables us to choose such a closed surface, and this knowledge is obtained easily by remembering that the electric field intensity due to a positive point charge is directed radially outward from the point charge.

Let us again consider a point charge Q at the origin of a spherical coordinate system and decide on a suitable closed surface which will meet the two requirements previously listed. The surface in question is obviously a spherical surface, centered at the origin and of any radius r. $\mathbf{D}_S$ is everywhere normal to the surface; D_S has the same value at all points on the surface.

Then we have, in order,

$$\begin{aligned} Q &= \oint_S \mathbf{D}_S \cdot d\mathbf{S} = \oint_{\text{sph}} D_S dS \\ &= D_S \oint_{\text{sph}} dS = D_S \int_{\phi=0}^{\phi=2\pi} \int_{\theta=0}^{\theta=\pi} r^2 \sin\theta \, d\theta \, d\phi \\ &= 4\pi r^2 D_S \end{aligned}$$

and hence

$$D_S = \frac{Q}{4\pi r^2}$$

Since r may have any value and since $\mathbf{D}_S$ is directed radially outward,

$$\mathbf{D} = \frac{Q}{4\pi r^2}\mathbf{a}_r \qquad \mathbf{E} = \frac{Q}{4\pi\epsilon_0 r^2}\mathbf{a}_r$$

which agrees with the results of Chapter 2. The example is a trivial one, and the objection could be raised that we had to know that the field was symmetrical and directed radially outward before we could obtain an answer. This is true, and that leaves the inverse-square-law relationship as the only check obtained from Gauss's law. The example does, however, serve to illustrate a method which we may apply to other problems, including several to which Coulomb's law is almost incapable of supplying an answer.

Are there any other surfaces which would have satisfied our two conditions? The student should determine that such simple surfaces as a cube or a cylinder do not meet the requirements.

As a second example, let us reconsider the uniform line charge distribution ρ_L lying along the z axis and extending from $-\infty$ to $+\infty$. We must first obtain a knowledge of the symmetry of the field, and we may consider this knowledge complete when the answers to these two questions are known:

1. With which coodinates does the field vary (or of what variables is D a function)?
2. Which components of $\mathbf{D}$ are present?

These same questions were asked when we used Coulomb's law to solve this problem in Section 2.5. We found then that the knowledge obtained from answering them enabled us to make a much simpler integration. The problem could have been (and was) worked without any consideration of symmetry, but it was more difficult.

In using Gauss's law, however, it is not a question of using symmetry to simplify the solution, for the application of Gauss's law depends on symmetry, and *if we cannot show that symmetry exists then we cannot use Gauss's law* to obtain a solution. The preceding two questions now become "musts."

From our previous discussion of the uniform line charge, it is evident that only the radial component of D is present, or

$$\mathbf{D} = D_\rho \mathbf{a}_\rho$$

and this component is a function of ρ only.

$$D_\rho = f(\rho)$$

The choice of a closed surface is now simple, for a cylindrical surface is the only surface to which D_ρ is everywhere normal, and it may be closed by plane surfaces normal to the z axis. A closed right circular cylinder of radius ρ extending from $z = 0$ to $z = L$ is shown in Figure 3.4.

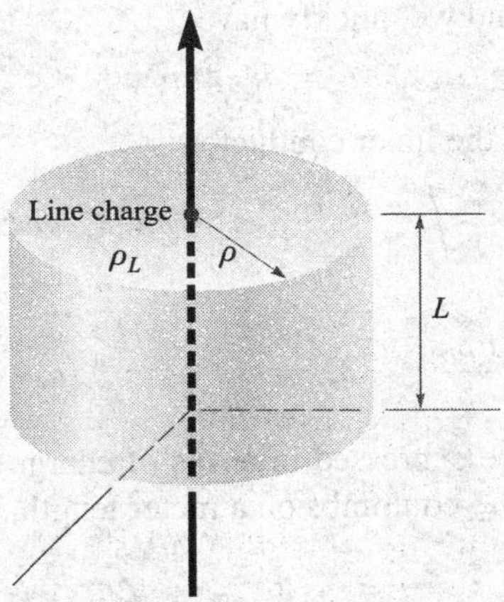

Fig. 3.4 The gaussian surface for an infinite uniform line charge is a right circular cylinder of length L and radius ρ. $\mathbf{D}$ is constant in magnitude and everywhere perpendicular to the cylindrical surface; $\mathbf{D}$ is parallel to the end faces.

We apply Gauss's law,

$$Q = \oint_{\text{cyl}} \mathbf{D}_S \cdot d\mathbf{S} = D_S \int_{\text{sides}} dS + 0 \int_{\text{top}} dS + 0 \int_{\text{bottom}} dS$$

$$= D_S \int_{z=0}^{L} \int_{\phi=0}^{2\pi} \rho \, d\phi \, dz = D_S 2\pi\rho L$$

and obtain

$$D_S = D_\rho = \frac{Q}{2\pi\rho L}$$

In terms of the charge density ρ_L, the total charge enclosed is

$$Q = \rho_L L$$

giving

$$D_\rho = \frac{\rho_L}{2\pi\rho}$$

or

$$E_\rho = \frac{\rho_L}{2\pi\epsilon_0\rho}$$

Comparison with Section 2.4, Eq. (20), shows that the correct result has been obtained and with much less work. Once the appropriate surface has been chosen, the integration usually amounts only to writing down the area of the surface at which **D** is normal.

The problem of a coaxial cable is almost identical with that of the line charge and is an example which is extremely difficult to solve from the standpoint of Coulomb's law. Suppose that we have two coaxial cylindrical conductors, the inner of radius a and the outer of radius b, each infinite in extent (Figure 3.5). We shall assume a charge distribution of ρ_S on the outer surface of the inner conductor.

Symmetry considerations show us that only the D_ρ component is present and that it can be a function only of ρ. A right circular cylinder of length L and radius ρ, where $a < \rho < b$, is necessarily chosen as the gaussian surface, and we quickly have

$$Q = D_S 2\pi\rho L$$

The total charge on a length L of the inner conductor is

$$Q = \int_{z=0}^{L} \int_{\phi=0}^{2\pi} \rho_S a \, d\phi \, dz = 2\pi a L \rho_S$$

from which we have

$$D_S = \frac{a\rho_S}{\rho} \qquad \mathbf{D} = \frac{a\rho_S}{\rho}\mathbf{a}_\rho \qquad (a < \rho < b)$$

Illustrations

This result might be expressed in terms of charge per unit length because the inner conductor has $2\pi a\rho_S$ coulombs on a meter length, and hence, letting $\rho_L = 2\pi a\rho_S$,

$$\mathbf{D} = \frac{\rho_L}{2\pi\rho}\mathbf{a}_\rho$$

and the solution has a form identical with that of the infinite line charge.

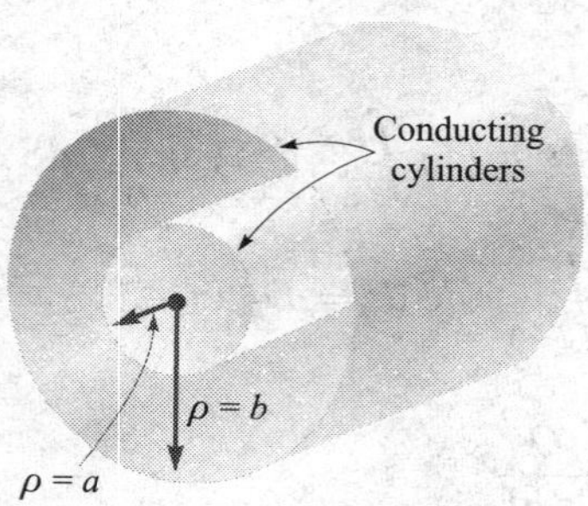

Fig. 3.5 The two coaxial cylindrical conductors forming a coaxial cable provide an electric flux density within the cylinders, given by $D_\rho = a\rho_S/\rho$.

Since every line of electric flux starting from the charge on the inner cylinder must terminate on a negative charge on the inner surface of the outer cylinder, the total charge on that surface must be

$$Q_{\text{outer cyl}} = -2\pi a L \rho_{S,\text{inner cyl}}$$

and the surface charge on the outer cylinder is found as

$$2\pi b L \rho_{S,\text{outer cyl}} = -2\pi a L \rho_{S,\text{inner cyl}}$$

or

$$\rho_{S,\text{outer cyl}} = -\frac{a}{b}\rho_{S,\text{inner cyl}}$$

What would happen if we should use a cylinder of radius ρ, $\rho > b$, for the gaussian surface? The total charge enclosed would then be zero, for there are equal and opposite charges on each conducting cylinder. Hence

$$0 = D_S 2\pi\rho L \qquad (\rho > b)$$

$$D_S = 0 \qquad (\rho > b)$$

An identical result would be obtained for $\rho < a$. Thus the coaxial cable or capacitor has no external field (we have proved that the outer conductor is a "shield"), and there is no field within the center conductor.

Our result is also useful for a *finite* length of coaxial cable, open at both ends, provided the length L is many times greater than the radius b so that the unsymmetrical conditions at the two ends do not appreciably affect the solution. Such a device is also termed a *coaxial capacitor*. Both the coaxial cable and the coaxial capacitor will appear frequently in the work that follows.

Perhaps a numerical example can illuminate some of these results.

EXAMPLE 3.2

Let us select a 50-cm length of coaxial cable having an inner radius of 1 mm and an outer radius of 4 mm. The space between conductors is assumed to be filled with air. The total charge on the inner conductor is 30 nC. We wish to know the charge density on each conductor, and the **E** and **D** fields.

Solution. We begin by finding the surface charge density on the inner cylinder,

$$\rho_{S,\text{inner cyl}} = \frac{Q_{\text{inner cyl}}}{2\pi a L} = \frac{30 \times 10^{-9}}{2\pi(10^{-3})(0.5)} = 9.55\ \mu\text{C/m}^2$$

The negative charge density on the inner surface of the outer cylinder is

$$\rho_{S,\text{outer cyl}} = \frac{Q_{\text{outer cyl}}}{2\pi b L} = \frac{-30 \times 10^{-9}}{2\pi(4 \times 10^{-3})(0.5)} = -2.39\ \mu\text{C/m}^2$$

The internal fields may therefore be calculated easily:

$$D_\rho = \frac{a\rho_S}{\rho} = \frac{10^{-3}(9.55 \times 10^{-6})}{\rho} = \frac{9.55}{\rho}\ \text{nC/m}^2$$

and

$$E_\rho = \frac{D_\rho}{\epsilon_0} = \frac{9.55 \times 10^{-9}}{8.854 \times 10^{-12}\rho} = \frac{1079}{\rho}\ \text{V/m}$$

Both of these expressions apply to the region where $1 < \rho < 4$ mm. For $\rho < 1$ mm or $\rho > 4$ mm, **E** and **D** are zero.

D3.5 A point charge of 0.25 μC is located at $r = 0$, and uniform surface charge densities are located as follows: 2 mC/m² at $r = 1$ cm, and −0.6 mC/m² at $r = 1.8$ cm. Calculate **D** at: (a) $r = 0.5$ cm; (b) $r = 1.5$ cm; (c) $r = 2.5$ cm. (d) What uniform surface charge density should be established at $r = 3$ cm to cause $\mathbf{D} = 0$ at $r = 3.5$ cm?

Ans. $796\mathbf{a}_r$ μC/m²; $977\mathbf{a}_r$ μC/m²; $40.8\mathbf{a}_r$ μC/m²; −28.3 μC/m²

3.4 APPLICATION OF GAUSS'S LAW: DIFFERENTIAL VOLUME ELEMENT

We are now going to apply the methods of Gauss's law to a slightly different type of problem—one which does not possess any symmetry at all. At first glance it might seem that our case is hopeless, for without symmetry a simple gaussian surface cannot be chosen such that the normal component of **D** is constant or zero everywhere on the surface. Without such a surface, the integral cannot be evaluated. There is only one way to circumvent these difficulties, and that is to choose such a very small closed surface that **D** is *almost* constant over the surface, and the small change in **D** may be adequately represented by using the first two terms of the Taylor's-series expansion for **D**. The result will become more nearly correct as the volume enclosed by the gaussian surface decreases, and we intend eventually to allow this volume to approach zero.

This example also differs from the preceding ones in that we shall not obtain the value of **D** as our answer, but will instead receive some extremely valuable information about the way **D** varies in the region of our small surface. This leads directly to one of Maxwell's four equations, which are basic to all electromagnetic theory.

Let us consider any point P, shown in Figure 3.6, located by a rectangular coordinate system. The value of **D** at the point P may be expressed in rectangular components, $\mathbf{D}_0 = D_{x0}\mathbf{a}_x + D_{y0}\mathbf{a}_y + D_{z0}\mathbf{a}_z$. We choose as our closed surface the small rectangular box, centered at P, having sides of lengths Δx, Δy, and Δz, and apply Gauss's law,

$$\oint_S \mathbf{D} \cdot d\mathbf{S} = Q$$

In order to evaluate the integral over the closed surface, the integral must be broken up into six integrals, one over each face,

$$\oint_S \mathbf{D} \cdot d\mathbf{S} = \int_{\text{front}} + \int_{\text{back}} + \int_{\text{left}} + \int_{\text{right}} + \int_{\text{top}} + \int_{\text{bottom}}$$

Consider the first of these in detail. Since the surface element is very small, **D** is essentially constant (over *this* portion of the entire closed surface) and

$$\int_{\text{front}} \doteq \mathbf{D}_{\text{front}} \cdot \Delta\mathbf{S}_{\text{front}}$$

$$\doteq \mathbf{D}_{\text{front}} \cdot \Delta y\, \Delta z\, \mathbf{a}_x$$

$$\doteq D_{\text{x,front}} \Delta y\, \Delta z$$

where we have only to approximate the value of D_x at this front face. The front face is at a distance of $\Delta x/2$ from P, and hence

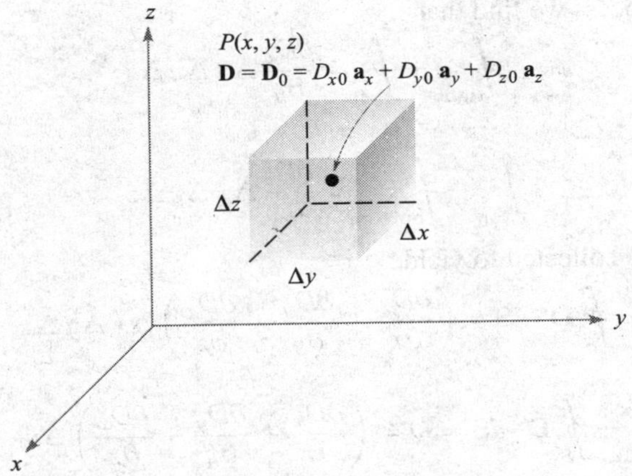

Fig. 3.6 A differential-sized gaussian surface about the point P is used to investigate the space rate of change of D in the neighborhood of P.

$$D_{x,\text{front}} \doteq D_{x0} + \frac{\Delta x}{2} \times \text{rate of change of } D_x \text{with } x$$

$$\doteq D_{x0} + \frac{\Delta x}{2}\frac{\partial D_x}{\partial x}$$

where D_{x0} is the value of D_x at P, and where a partial derivative must be used to express the rate of change of D_x with x, since D_x in general also varies with y and z. This expression could have been obtained more formally by using the constant term and the term involving the first derivative in the Taylor's-series expansion for D_x in the neighborhood of P.

We have now

$$\int_{\text{front}} \doteq \left(D_{x0} + \frac{\Delta x}{2}\frac{\partial D_x}{\partial x}\right)\Delta y\,\Delta z$$

Consider now the integral over the back surface,

$$\int_{\text{back}} \doteq \mathbf{D}_{\text{back}} \cdot \Delta\mathbf{S}_{\text{back}}$$

$$\doteq \mathbf{D}_{\text{back}} \cdot (-\Delta y\,\Delta z\,\mathbf{a}_x)$$

$$\doteq -D_{x,\text{back}}\Delta y\,\Delta z$$

and

$$D_{x,\text{back}} \doteq D_{x0} - \frac{\Delta x}{2}\frac{\partial D_x}{\partial x}$$

giving

$$\int_{\text{back}} \doteq \left(-D_{x0} + \frac{\Delta x}{2}\frac{\partial D_x}{\partial x}\right)\Delta y\,\Delta z$$

If we combine these two integrals, we have

$$\int_{\text{front}} + \int_{\text{back}} \doteq \frac{\partial D_x}{\partial x}\Delta x\,\Delta y\,\Delta z$$

By exactly the same process we find that

$$\int_{\text{right}} + \int_{\text{left}} \doteq \frac{\partial D_y}{\partial y} \Delta x \, \Delta y \, \Delta z$$

and

$$\int_{\text{top}} + \int_{\text{bottom}} \doteq \frac{\partial D_z}{\partial z} \Delta x \, \Delta y \, \Delta z$$

and these results may be collected to yield

$$\oint_S \mathbf{D} \cdot d\mathbf{S} \doteq \left(\frac{\partial D_x}{\partial x} + \frac{\partial D_y}{\partial y} + \frac{\partial D_z}{\partial z} \right) \Delta x \, \Delta y \, \Delta z$$

or

$$\oint_S \mathbf{D} \cdot d\mathbf{S} = Q \doteq \left(\frac{\partial D_x}{\partial x} + \frac{\partial D_y}{\partial y} + \frac{\partial D_z}{\partial z} \right) \Delta \nu \tag{7}$$

The expression is an approximation which becomes better as $\Delta \nu$ becomes smaller, and in the following section we shall let the volume $\Delta \nu$ approach zero. For the moment, we have applied Gauss's law to the closed surface surrounding the volume element $\Delta \nu$ and have as a result the approximation (7) stating that

$$\text{Charge enclosed in volume } \Delta \nu \doteq \left(\frac{\partial D_x}{\partial x} + \frac{\partial D_y}{\partial y} + \frac{\partial D_z}{\partial z} \right) \times \text{volume } \Delta \nu \tag{8}$$

EXAMPLE 3.3

Find an approximate value for the total charge enclosed in an incremental volume of 10^{-9} m^3 located at the origin, if $\mathbf{D} = e^{-x} \sin y \, \mathbf{a}_x - e^{-x} \cos y \, \mathbf{a}_y + 2z\mathbf{a}_z$ C/m^2.

Solution. We first evaluate the three partial derivatives in (8):

$$\frac{\partial D_x}{\partial x} = -e^{-x} \sin y$$

$$\frac{\partial D_y}{\partial y} = e^{-x} \sin y$$

$$\frac{\partial D_z}{\partial z} = 2$$

At the origin, the first two expressions are zero, and the last is 2. Thus, we find that the charge enclosed in a small volume element there must be approximately $2\Delta \nu$. If $\Delta \nu$ is 10^{-9} m^3, then we have enclosed about 2 nC.

D3.6 In free space, let $\mathbf{D} = 8xyz^4\mathbf{a}_x + 4x^2z^4\mathbf{a}_y + 16x^2yz^3\mathbf{a}_z$ pC/m^2. (*a*) Find the total electric flux passing through the rectangular surface $z = 2$, $0 < x < 2$, $1 < y < 3$, in the $\mathbf{a}_z$ direction. (*b*) Find $\mathbf{E}$ at $P(2, -1, 3)$. (*c*) Find an approximate value for the total charge contained in an incremental sphere located at $P(2, -1, 3)$ and having a volume of 10^{-12} m^3.

Ans. 1365 pC; $-146.4\mathbf{a}_x + 146.4\mathbf{a}_y - 195.2\mathbf{a}_z$ V/m; -2.38×10^{-21} C

3.5 DIVERGENCE

We shall now obtain an exact relationship from (7), by allowing the volume element $\Delta\nu$ to shrink to zero. We write this equation as

$$\left(\frac{\partial D_x}{\partial x} + \frac{\partial D_y}{\partial y} + \frac{\partial D_z}{\partial z}\right) \doteq \frac{\oint_S \mathbf{D} \cdot d\mathbf{S}}{\Delta\nu} = \frac{Q}{\Delta\nu}$$

or, as a limit

$$\left(\frac{\partial D_x}{\partial x} + \frac{\partial D_y}{\partial y} + \frac{\partial D_z}{\partial z}\right) = \lim_{\Delta\nu \to 0} \frac{\oint_S \mathbf{D} \cdot d\mathbf{S}}{\Delta\nu} = \lim_{\Delta\nu \to 0} \frac{Q}{\Delta\nu}$$

where the approximation has been replaced by an equality. It is evident that the last term is the volume charge density ρ_ν, and hence that

$$\left(\frac{\partial D_x}{\partial x} + \frac{\partial D_y}{\partial y} + \frac{\partial D_z}{\partial z}\right) = \lim_{\Delta\nu \to 0} \frac{\oint_S \mathbf{D} \cdot d\mathbf{S}}{\Delta\nu} = \rho_\nu \tag{9}$$

This equation contains too much information to discuss all at once, and we shall write it as two separate equations,

$$\left(\frac{\partial D_x}{\partial x} + \frac{\partial D_y}{\partial y} + \frac{\partial D_z}{\partial z}\right) = \lim_{\Delta\nu \to 0} \frac{\oint_S \mathbf{D} \cdot d\mathbf{S}}{\Delta\nu} \tag{10}$$

and

$$\left(\frac{\partial D_x}{\partial x} + \frac{\partial D_y}{\partial y} + \frac{\partial D_z}{\partial z}\right) = \rho_\nu \tag{11}$$

where we shall save (11) for consideration in the next section.

Equation (10) does not involve charge density, and the methods of the previous section could have been used on any vector $\mathbf{A}$ to find $\oint_S \mathbf{A} \cdot d\mathbf{S}$ for a small closed surface, leading to

$$\left(\frac{\partial A_x}{\partial x} + \frac{\partial A_y}{\partial y} + \frac{\partial A_z}{\partial z}\right) = \lim_{\Delta\nu \to 0} \frac{\oint_S \mathbf{A} \cdot d\mathbf{S}}{\Delta\nu} \tag{12}$$

where $\mathbf{A}$ could represent velocity, temperature gradient, force, or any other vector field.

Interactives

This operation appeared so many times in physical investigations in the last century that it received a descriptive name, *divergence*. The divergence of $\mathbf{A}$ is defined as

$$\text{Divergence of } \mathbf{A} = \text{div } \mathbf{A} = \lim_{\Delta\nu \to 0} \frac{\oint_S \mathbf{A} \cdot d\mathbf{S}}{\Delta\nu} \tag{13}$$

and is usually abbreviated div $\mathbf{A}$. The physical interpretation of the divergence of a vector is obtained by describing carefully the operations implied by the right-hand side of (13), where we shall consider $\mathbf{A}$ to be a member of the flux-density family of vectors in order to aid the physical interpretation.

> The divergence of the vector flux density $\mathbf{A}$ is the outflow of flux from a small closed surface per unit volume as the volume shrinks to zero.

The physical interpretation of divergence afforded by this statement is often useful in obtaining qualitative information about the divergence of a vector field without resorting to a mathematical investigation. For instance, let us consider the divergence of the velocity of water in a bathtub after the drain has been opened. The net outflow of water through *any* closed surface lying entirely within the water must be zero, for water is essentially incompressible, and the water entering and leaving different regions of the closed surface must be equal. Hence the divergence of this velocity is zero.

If, however, we consider the velocity of the air in a tire which has just been punctured by a nail, we realize that the air is expanding as the pressure drops, and that consequently there is a net outflow from any closed surface lying within the tire. The divergence of this velocity is therefore greater than zero.

A positive divergence for any vector quantity indicates a *source* of that vector quantity at that point. Similarly, a negative divergence indicates a *sink*. Since the divergence of the water velocity above is zero, no source or sink exists.[3] The expanding air, however, produces a positive divergence of the velocity, and each interior point may be considered a source.

Writing (10) with our new term, we have

$$\text{div } \mathbf{D} = \left(\frac{\partial D_x}{\partial x} + \frac{\partial D_y}{\partial y} + \frac{\partial D_z}{\partial z} \right) \tag{14}$$

This expression is again of a form which does not involve the charge density. It is the result of applying the definition of divergence (13) to a differential volume element in *rectangular coordinates*.

If a differential volume unit $\rho \, d\rho \, d\phi \, dz$ in cylindrical coordinates, or $r^2 \sin\theta \, dr d\theta \, d\phi$ in spherical coordinates, had been chosen, expressions for divergence involving the components of the vector in the particular coordinate system and involving partial derivatives with respect to the variables of that system would have been obtained. These expressions are obtained in Appendix A and are given here for convenience:

$$\text{div } \mathbf{D} = \frac{\partial D_x}{\partial x} + \frac{\partial D_y}{\partial y} + \frac{\partial D_z}{\partial z} \qquad \text{(rectangular)} \tag{15}$$

$$\text{div } \mathbf{D} = \frac{1}{\rho}\frac{\partial}{\partial \rho}(\rho D_\rho) + \frac{1}{\rho}\frac{\partial D_\phi}{\partial \phi} + \frac{\partial D_z}{\partial z} \qquad \text{(cylindrical)} \tag{16}$$

$$\text{div } \mathbf{D} = \frac{1}{r^2}\frac{\partial}{\partial r}(r^2 D_r) + \frac{1}{r\sin\theta}\frac{\partial}{\partial \theta}(\sin\theta \, D_\theta) + \frac{1}{r\sin\theta}\frac{\partial D_\phi}{\partial \phi} \qquad \text{(spherical)} \tag{17}$$

These relationships are also shown towards the end of the book for easy reference.

It should be noted that the divergence is an operation which is performed on a vector, but that the result is a scalar. We should recall that, in a somewhat similar way, the dot or scalar product was a multiplication of two vectors which yielded a scalar.

For some reason it is a common mistake on meeting divergence for the first time to impart a vector quality to the operation by scattering unit vectors around in the partial derivatives.

[3] Having chosen a differential element of volume within the water, the gradual decrease in water level with time will eventually cause the volume element to lie above the surface of the water. At the instant the surface of the water intersects the volume element, the divergence is positive and the small volume is a source. This complication is avoided above by specifying an integral point.

Divergence merely tells us *how much* flux is leaving a small volume on a per-unit-volume basis; no direction is associated with it.

We can illustrate the concept of divergence by continuing with the example at the end of Section 3.4.

EXAMPLE 3.4

Find div $\mathbf{D}$ at the origin if $\mathbf{D} = e^{-x} \sin y\, \mathbf{a}_x - e^{-x} \cos y\, \mathbf{a}_y + 2z\mathbf{a}_z$.

Solution. We use (14) or (15) to obtain

$$\text{div } \mathbf{D} = \frac{\partial D_x}{\partial x} + \frac{\partial D_y}{\partial y} + \frac{\partial D_z}{\partial z} = -e^{-x} \sin y + e^{-x} \sin y + 2 = 2$$

The value is the constant 2, regardless of location.

If the units of $\mathbf{D}$ are C/m^2, then the units of div $\mathbf{D}$ are C/m^3. This is a volume charge density, a concept discussed in the next section.

D3.7 In each of the following parts, find a numerical value for div $\mathbf{D}$ at the point specified: (*a*) $\mathbf{D} = (2xyz - y^2)\mathbf{a}_x + (x^2z - 2xy)\mathbf{a}_y + x^2y\mathbf{a}_z$ C/m^2 at $P_A(2, 3, -1)$; (*b*) $\mathbf{D} = 2\rho z^2 \sin^2\phi\, \mathbf{a}_\rho + \rho z^2 \sin 2\phi\, \mathbf{a}_\phi + 2\rho^2 z \sin^2\phi\, \mathbf{a}_z$ C/m^2 at $P_B(\rho = 2, \phi = 110°, z = -1)$; (*c*) $\mathbf{D} = 2r \sin\theta \cos\phi\, \mathbf{a}_r + r\cos\theta\cos\phi\, \mathbf{a}_\theta - r \sin\phi\, \mathbf{a}_\phi$ C/m^2 at $P_C(r = 1.5, \theta = 30°, \phi = 50°)$.

Ans. −10.00; 9.06; 1.29

3.6 MAXWELL'S FIRST EQUATION (ELECTROSTATICS)

We now wish to consolidate the gains of the last two sections and to provide an interpretation of the divergence operation as it relates to electric flux density. The expressions developed there may be written as

$$\text{div } \mathbf{D} = \lim_{\Delta\nu \to 0} \frac{\oint_S \mathbf{D} \cdot d\mathbf{S}}{\Delta\nu} \tag{18}$$

$$\text{div } \mathbf{D} = \frac{\partial D_x}{\partial x} + \frac{\partial D_y}{\partial y} + \frac{\partial D_z}{\partial z} \tag{19}$$

and

$$\text{div } \mathbf{D} = \rho_\nu \tag{20}$$

The first equation is the definition of divergence, the second is the result of applying the definition to a differential volume element in rectangular coordinates, giving us an equation by which the divergence of a vector expressed in rectangular coordinates may be evaluated, and the third is merely (11) written using the new term div $\mathbf{D}$. Equation (20) is almost an obvious result if we have achieved any familiarity at all with the concept of divergence as defined by (18), for given Gauss's law,

$$\oint_S \mathbf{D} \cdot d\mathbf{S} = Q$$

per unit volume

$$\frac{\oint_S \mathbf{D} \cdot d\mathbf{S}}{\Delta\nu} = \frac{Q}{\Delta\nu}$$

As the volume shrinks to zero,

$$\lim_{\Delta\nu \to 0} \frac{\oint_S \mathbf{D} \cdot d\mathbf{S}}{\Delta\nu} = \lim_{\Delta\nu \to 0} \frac{Q}{\Delta\nu}$$

we should see div **D** on the left and volume charge density on the right,

$$\text{div}\, \mathbf{D} = \rho_\nu \tag{20}$$

This is the first of Maxwell's four equations as they apply to electrostatics and steady magnetic fields, and it states that the electric flux per unit volume leaving a vanishingly small volume unit is exactly equal to the volume charge density there. This equation is aptly called the *point form of Gauss's law*. Gauss's law relates the flux leaving any closed surface to the charge enclosed, and Maxwell's first equation makes an identical statement on a per-unit-volume basis for a vanishingly small volume, or at a point. Because that the divergence may be expressed as the sum of three partial derivatives, Maxwell's first equation is also described as the differential-equation form of Gauss's law, and conversely, Gauss's law is recognized as the integral form of Maxwell's first equation.

As a specific illustration, let us consider the divergence of **D** in the region about a point charge Q located at the origin. We have the field

$$\mathbf{D} = \frac{Q}{4\pi r^2}\mathbf{a}_r$$

and make use of (17), the expression for divergence in spherical coordinates given in Section 3.5:

$$\text{div}\, \mathbf{D} = \frac{1}{r^2}\frac{\partial}{\partial r}(r^2 D_r) + \frac{1}{r\sin\theta}\frac{\partial}{\partial\theta}(D_\theta \sin\theta) + \frac{1}{r\sin\theta}\frac{\partial D_\phi}{\partial\phi}$$

Since D_θ and D_ϕ are zero, we have

$$\text{div}\, \mathbf{D} = \frac{1}{r^2}\frac{d}{dr}\left(r^2 \frac{Q}{4\pi r^2}\right) = 0 \qquad (\text{if } r \neq 0)$$

Thus, $\rho_\nu = 0$ everywhere except at the origin, where it is infinite.

The divergence operation is not limited to electric flux density; it can be applied to any vector field. We shall apply it to several other electromagnetic fields in the coming chapters.

D3.8 Determine an expression for the volume charge density associated with each **D** field following: (*a*) $\mathbf{D} = \frac{4xy}{z}\mathbf{a}_x + \frac{2x^2}{z}\mathbf{a}_y - \frac{2x^2y}{z^2}\mathbf{a}_z$; (*b*) $\mathbf{D} = z\sin\phi\,\mathbf{a}_\rho + z\cos\phi\,\mathbf{a}_\phi + \rho\sin\phi\,\mathbf{a}_z$; (*c*) $\mathbf{D} = \sin\theta\sin\phi\,\mathbf{a}_r + \cos\theta\sin\phi\,\mathbf{a}_\theta + \cos\phi\,\mathbf{a}_\phi$.

Ans. $\frac{4y}{z^3}(x^2 + z^2)$; 0; 0.

3.7 THE VECTOR OPERATOR ∇ AND THE DIVERGENCE THEOREM

If we remind ourselves again that divergence is an operation on a vector yielding a scalar result, just as the dot product of two vectors gives a scalar result, it seems possible that we can find something which may be dotted formally with **D** to yield the scalar

$$\frac{\partial D_x}{\partial x} + \frac{\partial D_y}{\partial y} + \frac{\partial D_z}{\partial z}$$

Obviously, this cannot be accomplished by using a dot *product*; the process must be a dot *operation*.

With this in mind, we define the del *operator* ∇ as a *vector operator*,

$$\nabla = \frac{\partial}{\partial x}\mathbf{a}_x + \frac{\partial}{\partial y}\mathbf{a}_y + \frac{\partial}{\partial z}\mathbf{a}_z \tag{21}$$

Similar *scalar operators* appear in several methods of solving differential equations where we often let D replace d/dx, D^2 replace d^2/dx^2, and so forth.[4] We agree on defining ∇ (pronounced "del") that it shall be treated in every way as an ordinary vector with the one important exception that partial derivatives result instead of products of scalars.

Consider $\nabla \cdot \mathbf{D}$, signifying

$$\nabla \cdot \mathbf{D} = \left(\frac{\partial}{\partial x}\mathbf{a}_x + \frac{\partial}{\partial y}\mathbf{a}_y + \frac{\partial}{\partial z}\mathbf{a}_z\right) \cdot (D_x\mathbf{a}_x + D_y\mathbf{a}_y + D_z\mathbf{a}_z)$$

We first consider the dot products of the unit vectors, discarding the six zero terms and having left

$$\nabla \cdot \mathbf{D} = \frac{\partial}{\partial x}(D_x) + \frac{\partial}{\partial y}(D_y) + \frac{\partial}{\partial z}(D_z)$$

where the parentheses are now removed by operating or differentiating:

$$\nabla \cdot \mathbf{D} = \frac{\partial D_x}{\partial x} + \frac{\partial D_y}{\partial y} + \frac{\partial D_z}{\partial z}$$

This is recognized as the divergence of $\mathbf{D}$, so that we have

$$\text{div}\,\mathbf{D} = \nabla \cdot \mathbf{D} = \frac{\partial D_x}{\partial x} + \frac{\partial D_y}{\partial y} + \frac{\partial D_z}{\partial z}$$

The use of $\nabla \cdot \mathbf{D}$ is much more prevalent than that of div $\mathbf{D}$, although both usages have their advantages. Writing $\nabla \cdot \mathbf{D}$ allows us to obtain simply and quickly the correct partial derivatives, but only in rectangular coordinates, as we shall see. On the other hand, div $\mathbf{D}$ is an excellent reminder of the physical interpretation of divergence. We shall use the operator notation $\nabla \cdot \mathbf{D}$ from now on to indicate the divergence operation.

The vector operator ∇ is used not only with divergence, but also with several other very important operations that appear later. One of these is ∇u, where u is any scalar field, and leads to

$$\nabla u = \left(\frac{\partial}{\partial x}\mathbf{a}_x + \frac{\partial}{\partial y}\mathbf{a}_y + \frac{\partial}{\partial z}\mathbf{a}_z\right) u = \frac{\partial u}{\partial x}\mathbf{a}_x + \frac{\partial u}{\partial y}\mathbf{a}_y + \frac{\partial u}{\partial z}\mathbf{a}_z$$

The ∇ operator does not have a specific form in other coordinate systems. If we are considering $\mathbf{D}$ in cylindrical coordinates, then $\nabla \cdot \mathbf{D}$ still indicates the divergence of $\mathbf{D}$, or

$$\nabla \cdot \mathbf{D} = \frac{1}{\rho}\frac{\partial}{\partial \rho}(\rho D_\rho) + \frac{1}{\rho}\frac{\partial D_\phi}{\partial \phi} + \frac{\partial D_z}{\partial z}$$

where this expression has been taken from Section 3.5. We have no form for ∇ itself to help us obtain this sum of partial derivatives. This means that ∇u, as yet unnamed but easily written in rectangular coordinates, cannot be expressed by us at this time in cylindrical coordinates. Such an expression will be obtained when ∇u is defined in Chapter 4.

---

[4] This scalar operator D, which will not appear again, is not to be confused with the electric flux density.

We shall close our discussion of divergence by presenting a theorem which will be needed several times in later chapters, the *divergence theorem*. This theorem applies to any vector field for which the appropriate partial derivatives exist, although it is easiest for us to develop it for the electric flux density. We have actually obtained it already and now have little more to do than point it out and name it, for starting from Gauss's law,

$$\oint_S \mathbf{D} \cdot d\mathbf{S} = Q$$

and letting

$$Q = \int_{\text{vol}} \rho_\nu dv$$

and then replacing ρ_ν by its equal,

$$\nabla \cdot \mathbf{D} = \rho_\nu$$

we have

$$\oint_S \mathbf{D} \cdot d\mathbf{S} = Q = \int_{\text{vol}} \rho_\nu dv = \int_{\text{vol}} \nabla \cdot \mathbf{D}\, dv$$

The first and last expressions constitute the divergence theorem,

$$\oint_S \mathbf{D} \cdot d\mathbf{S} = \int_{\text{vol}} \nabla \cdot \mathbf{D}\, dv \tag{22}$$

which may be stated as follows:

> The integral of the normal component of any vector field over a closed surface is equal to the integral of the divergence of this vector field throughout the volume enclosed by the closed surface.

Again, we emphasize that the divergence theorem is true for any vector field, although we have obtained it specifically for the electric flux density **D**, and we shall have occasion later to apply it to several different fields. Its benefits derive from the fact that it relates a triple integration *throughout some volume* to a double integration *over the surface* of that volume. For example, it is much easier to look for leaks in a bottle full of some agitated liquid by an inspection of the surface than by calculating the velocity at every internal point.

The divergence theorem becomes obvious physically if we consider a volume ν, shown in cross section in Figure 3.7, which is surrounded by a closed surface S. Division of the volume into a number of small compartments of differential size and consideration of one cell show that the flux diverging from such a cell *enters*, or *converges* on, the adjacent cells unless the cell contains a portion of the outer surface. In summary, the divergence of the flux density throughout a volume leads, then, to the same result as determining the net flux crossing the enclosing surface.

Let us consider an example to illustrate the divergence theorem.

EXAMPLE 3.5

Evaluate both sides of the divergence theorem for the field $\mathbf{D} = 2xy\mathbf{a}_x + x^2\mathbf{a}_y$ C/m^2 and the rectangular parellelepiped formed by the planes $x = 0$ and 1, $y = 0$ and 2, and $z = 0$ and 3.

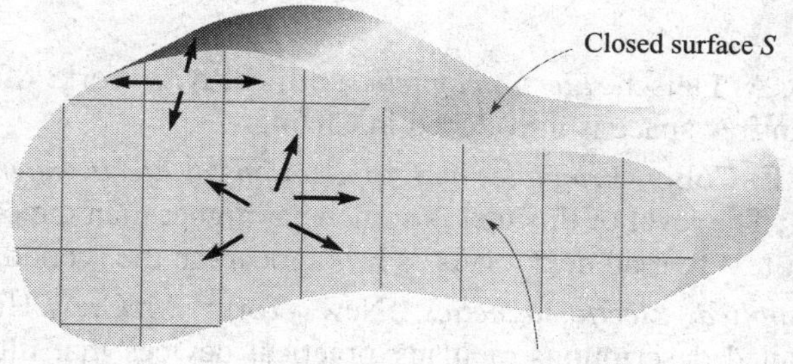

Fig. 3.7 The divergence theorem states that the total flux crossing the closed surface is equal to the integral of the divergence of the flux density throughout the enclosed volume. The volume is shown here in cross section.

Solution. Evaluating the surface integral first, we note that $\mathbf{D}$ is parallel to the surfaces at $z = 0$ and $z = 3$, so $\mathbf{D} \cdot d\mathbf{S} = 0$ there. For the remaining four surfaces we have

$$\oint_S \mathbf{D} \cdot d\mathbf{S} = \int_0^3 \int_0^2 (\mathbf{D})_{x=0} \cdot (-dy\, dz\, \mathbf{a}_x) + \int_0^3 \int_0^2 (\mathbf{D})_{x=1} \cdot (dy\, dz\, \mathbf{a}_x)$$

$$+ \int_0^3 \int_0^1 (\mathbf{D})_{y=0} \cdot (-dx\, dz\, \mathbf{a}_y) + \int_0^3 \int_0^1 (\mathbf{D})_{y=2} \cdot (dx\, dz\, \mathbf{a}_y)$$

$$= -\int_0^3 \int_0^2 (D_x)_{x=0} dy\, dz + \int_0^3 \int_0^2 (D_x)_{x=1} dy\, dz$$

$$- \int_0^3 \int_0^1 (D_y)_{y=0} dx\, dz + \int_0^3 \int_0^1 (D_y)_{y=2} dx\, dz$$

However, $(D_x)_{x=0} = 0$, and $(D_y)_{y=0} = (D_y)_{y=2}$, which leaves only

$$\oint_S \mathbf{D} \cdot d\mathbf{S} = \int_0^3 \int_0^2 (D_x)_{x=1} dy\, dz = \int_0^3 \int_0^2 2y\, dy\, dz = \int_0^3 4\, dz = 12$$

Since

$$\nabla \cdot \mathbf{D} = \frac{\partial}{\partial x}(2xy) + \frac{\partial}{\partial y}(x^2) = 2y$$

the volume integral becomes

$$\int_{\text{vol}} \nabla \cdot \mathbf{D}\, dv = \int_0^3 \int_0^2 \int_0^1 2y\, dx\, dy\, dz = \int_0^3 \int_0^2 2y\, dy\, dz = \int_0^3 4\, dz = 12$$

and the check is accomplished. Remembering Gauss's law, we see that we have also determined that a total charge of 12 C lies within this parallelepiped.

D3.9 Given the field $\mathbf{D} = 6\rho \sin \frac{1}{2}\phi\, \mathbf{a}_\rho + 1.5\rho \cos \frac{1}{2}\phi\, \mathbf{a}_\phi$ C/m^2, evaluate both sides of the divergence theorem for the region bounded by $\rho = 2$, $\phi = 0$, $\phi = \pi$, $z = 0$, and $z = 5$.

Ans. 225; 225

References

1. Kraus, J. D., and D. A. Fleisch. *Electromagnetics.* 5th ed. New York: McGraw-Hill, 1999. The static electric field in free space is introduced in Chapter 2.
2. Plonsey, R., and R. E. Collin. *Principles and Applications of Electromagnetic Fields.* New York: McGraw-Hill, 1961. The level of this text is somewhat higher than the one we are reading now, but it is an excellent text to read next. Gauss's law appears in the second chapter.
3. Plonus, M. A. *Applied Electromagnetics.* New York: McGraw-Hill, 1978. This book contains rather detailed descriptions of many practical devices that illustrate electromagnetic applications. For example, see the discussion of xerography on pp. 95–98 as an electrostatics application.
4. Skilling, H. H. *Fundamentals of Electric Waves.* 2d ed. New York: John Wiley & Sons, 1948. The operations of vector calculus are well illustrated. Divergence is discussed on pp. 22 and 38. Chapter 1 is interesting reading.
5. Thomas, G. B., Jr., and R. L. Finney. (see Suggested References for Chapter 1). The divergence theorem is developed and illustrated from several different points of view on pp. 976–980.

CHAPTER 3 Problems

Quizzes

3.1 An empty metal paint can is placed on a marble table, the lid is removed, and both parts are discharged (honorably) by touching them to ground. An insulating nylon thread is glued to the center of the lid, and a penny, a nickel, and a dime are glued to the thread so that they are not touching each other. The penny is given a charge of +5 nC, and the nickel and dime are discharged. The assembly is lowered into the can so that the coins hang clear of all walls, and the lid is secured. The outside of the can is again touched momentarily to ground. The device is carefully disassembled with insulating gloves and tools. (*a*) What charges are found on each of the five metallic pieces? (*b*) If the penny had been given a charge of +5 nC, the dime a charge of −2 nC, and the nickel a charge of −1 nC, what would the final charge arrangement have been?

3.2 A point charge of 20 nC is located at $(4, -1, 3)$, and a uniform line charge of −25 nC/m lies along the intersection of the planes $x = -4$ and $z = 6$. (*a*) Calculate **D** at $(3, -1, 0)$. (*b*) How much electric flux leaves the surface of a sphere of radius 5, centered at the origin? (*c*) Repeat part *b* if the radius of the sphere is 10.

3.3 The cylindrical surface $\rho = 8$ cm contains the surface charge density, $\rho_S = 5e^{-20|z|}$ nC/m^2. (*a*) What is the total amount of charge present? (*b*) How much electric flux leaves the surface $\rho = 8$ cm, 1 cm $< z <$ 5 cm, $30^\circ < \phi < 90^\circ$?

3.4 In cylindrical coordinates, let $\mathbf{D} = (\rho\mathbf{a}_\rho + z\mathbf{a}_z)/[4\pi(\rho^2 + z^2)^{1.5}]$. Determine the total flux leaving: (*a*) the infinitely-long cylindrical surface $\rho = 7$; (*b*) the finite cylinder, $\rho = 7, |z| \leq 10$.

3.5 Let $\mathbf{D} = 4xy\mathbf{a}_x + 2(x^2 + z^2)\mathbf{a}_y + 4yz\mathbf{a}_z$ C/m^2 and evaluate surface integrals to find the total charge enclosed in the rectangular parallelepiped $0 < x < 2, 0 < y < 3, 0 < z < 5$ m.

3.6 In free space, a volume charge of constant density $\rho_\nu = \rho_0$ exists within the region $-\infty < x < \infty, -\infty < y < \infty$, and $-d/2 < z < d/2$. Find **D** and **E** everywhere.

3.7 Volume charge density is located in free space as $\rho_\nu = 2e^{-1000r}$ nC/m^3 for $0 < r < 1$ mm, and $\rho_\nu = 0$ elsewhere. (a) Find the total charge enclosed by the spherical surface $r = 1$ mm. (b) By using Gauss's law, calculate the value of D_r on the surface $r = 1$ mm.

3.8 Use Gauss's law in integral form to show that an inverse distance field in spherical coordinates, $\mathbf{D} = A\mathbf{a}_r/r$, where A is a constant, requires every spherical shell of 1 m thickness to contain $4\pi A$ coulombs of charge. Does this indicate a continuous charge distribution? If so, find the charge density variation with r.

3.9 A uniform volume charge density of 80 μC/m^3 is present throughout the region 8 mm $< r <$ 10 mm. Let $\rho_\nu = 0$ for $0 < r < 8$ mm. (a) Find the total charge inside the spherical surface $r = 10$ mm. (b) Find D_r at $r = 10$ mm. (c) If there is no charge for $r > 10$ mm, find D_r at $r = 20$ mm.

3.10 Volume charge density varies in spherical coordinates as $\rho_\nu = (\rho_0 \sin \pi r)/r^2$, where ρ_0 is a constant. Find the surfaces on which $\mathbf{D} = 0$.

3.11 In cylindrical coordinates, let $\rho_\nu = 0$ for $\rho < 1$ mm, $\rho_\nu = 2\sin(2000\,\pi\rho)$ nC/m^3 for 1 mm $< \rho <$ 1.5 mm, and $\rho_\nu = 0$ for $\rho > 1.5$ mm. Find $\mathbf{D}$ everywhere.

3.12 The sun radiates a total power of about 2×10^{26} watts (W). If we imagine the sun's surface to be marked off in latitude and longitude and assume uniform radiation, (a) what power is radiated by the region lying between latitude 50° N and 60° N and longitude 12° W and 27° W? (b) What is the power density on a spherical surface 93,000,000 miles from the sun in W/m^2?

3.13 Spherical surfaces at $r = 2$, 4, and 6 m carry uniform surface charge densities of 20 nC/m^2, -4 nC/m^2, and ρ_{S0}, respectively. (a) Find $\mathbf{D}$ at $r = 1$, 3, and 5 m. (b) Determine ρ_{S0} such that $\mathbf{D} = 0$ at $r = 7$ m.

3.14 A light source inside a translucent sphere of 20 cm diameter causes a light flux density at the spherical surface $1{,}000\cos^2(\theta/2)\mathbf{a}_r$ lumens/m^2. (a) In what direction is the flux density a maximum? (b) Determine the angle $\theta = \theta_0$ at which the flux density is one-half its maximum value. (c) Determine the angle $\theta = \theta_1$ such that one-half the total light flux is emitted within the cone $\theta < \theta_1$.

3.15 Volume charge density is located as follows: $\rho_\nu = 0$ for $\rho < 1$ mm and for $\rho > 2$ mm, $\rho_\nu = 4\rho$ μC/m^3 for $1 < \rho < 2$ mm. (a) Calculate the total charge in the region $0 < \rho < \rho_1$, $0 < z < L$, where $1 < \rho_1 < 2$ mm. (b) Use Gauss's law to determine D_ρ at $\rho = \rho_1$. (c) Evaluate D_ρ at $\rho = 0.8$ mm, 1.6 mm, and 2.4 mm.

3.16 In spherical coordinates, a volume charge density $\rho_\nu = 10e^{-2r}$ C/m^3 is present. (a) Determine $\mathbf{D}$. (b) Check your result of part a by evaluating $\nabla \cdot \mathbf{D}$.

3.17 A cube is defined by $1 < x, y, z < 1.2$. If $\mathbf{D} = 2x^2y\mathbf{a}_x + 3x^2y^2\mathbf{a}_y$ C/m^2. (a) Apply Gauss's law to find the total flux leaving the closed surface of the cube. (b) Evaluate $\nabla \cdot \mathbf{D}$ at the center of the cube. (c) Estimate the total charge enclosed within the cube by using Eq. (8).

3.18 State whether the divergence of the following vector fields is positive, negative, or zero: (a) the thermal energy flow in J/(m^2 − s) at any point in a freezing ice cube; (b) the current density in A/m^2 in a bus bar carrying direct current; (c) the mass flow rate in kg/(m^2 − s) below the surface of water in a basin, in which the water is circulating clockwise as viewed from above.

3.19 A spherical surface of radius 3 mm is centered at $P(4, 1, 5)$ in free space. Let $\mathbf{D} = x\mathbf{a}_x$ C/m^2. Use the results of Section 3.4 to estimate the net electric flux leaving the spherical surface.

3.20 Suppose that an electric flux density in cylindrical coordinates is of the form $\mathbf{D} = D_\rho \mathbf{a}_\rho$. Describe the dependence of the charge density ρ_v on coordinates ρ, ϕ, and z if (*a*) $D_\rho = f(\phi, z)$; (*b*) $D_\rho = (1/\rho)f(\phi, z)$; (*c*) $D_\rho = f(\rho)$.

3.21 Calculate $\nabla \cdot \mathbf{D}$ at the point specified if (*a*) $\mathbf{D} = (1/z^2)[10xyz\ \mathbf{a}_x + 5x^2z\ \mathbf{a}_y + (2z^3 - 5x^2y)\ \mathbf{a}_z]$ at $P(-2, 3, 5)$; (*b*) $\mathbf{D} = 5z^2\ \mathbf{a}_\rho + 10\rho z\ \mathbf{a}_z$ at $P(3, -45°, 5)$; (*c*) $\mathbf{D} = 2r \sin\theta \sin\phi\ \mathbf{a}_r + r\cos\theta \sin\phi\ \mathbf{a}_\theta + r\cos\phi\ \mathbf{a}_\phi$ at $P(3, 45°, -45°)$.

3.22 (*a*) A flux density field is given as $\mathbf{F}_1 = 5\mathbf{a}_z$. Evaluate the outward flux of $\mathbf{F}_1$ through the hemispherical surface, $r = a, 0 < \theta < \pi/2, 0 < \phi < 2\pi$. (*b*) What simple observation would have saved a lot of work in part *a*? (*c*) Now suppose the field is given by $\mathbf{F}_2 = 5z\mathbf{a}_z$. Using the appropriate surface integrals, evaluate the net outward flux of $\mathbf{F}_2$ through the closed surface consisting of the hemisphere of part *a* and its circular base in the *xy* plane. (*d*) Repeat part *c* by using the divergence theorem and an appropriate volume integral.

3.23 (*a*) A point charge Q lies at the origin. Show that div $\mathbf{D}$ is zero everywhere except at the origin. (*b*) Replace the point charge with a uniform volume charge density ρ_{v0} for $0 < r < a$. Relate ρ_{v0} to Q and a so that the total charge is the same. Find div $\mathbf{D}$ everywhere.

3.24 (*a*) A uniform line charge density ρ_L lies along the z axis. Show that $\nabla \cdot \mathbf{D} = 0$ everywhere except on the line charge. (*b*) Replace the line charge with a uniform volume charge density ρ_0 for $0 < \rho < a$. Relate ρ_0 to ρ_L so that the charge per unit length is the same. Then find $\nabla \cdot \mathbf{D}$ everywhere.

3.25 Within the spherical shell, $3 < r < 4$ m, the electric flux density is given as $\mathbf{D} = 5(r-3)^3\ \mathbf{a}_r$ C/m^2. (*a*) What is the volume charge density at $r = 4$? (*b*) What is the electric flux density at $r = 4$? (*c*) How much electric flux leaves the sphere $r = 4$? (*d*) How much charge is contained within the sphere $r = 4$?

3.26 If we have a perfect gas of mass density ρ_m kg/m^3, and we assign a velocity $\mathbf{U}$ m/s to each differential element, then the mass flow rate is $\rho_m\mathbf{U}$ kg/(m^2 − s). Physical reasoning then leads to the *continuity equation*, $\nabla \cdot (\rho_m \mathbf{U}) = -\partial\rho_m/\partial t$. (*a*) Explain in words the physical interpretation of this equation. (*b*) Show that $\oint_s \rho_m \mathbf{U} \cdot d\mathbf{S} = -dM/dt$, where M is the total mass of the gas within the constant closed surface S, and explain the physical significance of the equation.

3.27 Let $\mathbf{D} = 5.00r^2\mathbf{a}_r$ mC/m^2 for $r \le 0.08$ m and $\mathbf{D} = 0.205\ \mathbf{a}_r/r^2$ μC/m^2 for $r \ge 0.08$ m. (*a*) Find ρ_v for $r = 0.06$ m. (*b*) Find ρ_v for $r = 0.1$ m. (*c*) What surface charge density could be located at $r = 0.08$ m to cause $\mathbf{D} = 0$ for $r > 0.08$ m?

3.28 Repeat Problem 3.8, but use $\nabla \cdot \mathbf{D} = \rho_v$ and take an appropriate volume integral.

3.29 In the region of free space that includes the volume, $2 < x, y, z < 3$, $\mathbf{D} = \frac{2}{z^2}(yz\ \mathbf{a}_x + xz\ \mathbf{a}_y - 2xy\ \mathbf{a}_z)$ C/m^2. (*a*) Evaluate the volume integral side of the divergence theorem for the volume defined here. (*b*) Evaluate the surface integral side for the corresponding closed surface.

3.30 Let $\mathbf{D} = 20\rho^2\ \mathbf{a}_\rho$ nC/m^2. (*a*) What is the volume charge density at the point $P(0.5, 60°, 2)$? (*b*) Use two different methods to find the amount of charge lying within the closed surface bounded by $\rho = 3, 0 \le z \le 2$.

3.31 Given the flux density $\mathbf{D} = \frac{16}{r}\cos(2\theta)\ \mathbf{a}_\theta$ C/m^2, use two different methods to find the total charge within the region $1 < r < 2$ m, $1 < \theta < 2$ rad, $1 < \phi < 2$ rad.

4

Energy and Potential

In Chapters 2 and 3 we became acquainted with Coulomb's law and its use in finding the electric field about several simple distributions of charge, and also with Gauss's law and its application in determining the field about some symmetrical charge arrangements. The use of Gauss's law was invariably easier for these highly symmetrical distributions because the problem of integration always disappeared when the proper closed surface was chosen.

However, if we had attempted to find a slightly more complicated field, such as that of two unlike point charges separated by a small distance, we would have found it impossible to choose a suitable gaussian surface and obtain an answer. Coulomb's law, however, is more powerful and enables us to solve problems for which Gauss's law is not applicable. The application of Coulomb's law is laborious, detailed, and often quite complex, the reason for this being precisely the fact that the electric field intensity, a vector field, must be found directly from the charge distribution. Three different integrations are needed in general, one for each component, and the resolution of the vector into components usually adds to the complexity of the integrals.

Certainly it would be desirable if we could find some as yet undefined scalar function with a single integration and then determine the electric field from this scalar by some simple straightforward procedure, such as differentiation.

This scalar function does exist and is known as the *potential* or *potential field*. We shall find that it has a very real physical interpretation and is more familiar to most of us than is the electric field which it will be used to find.

We should expect, then, to be equipped soon with a third method of finding electric fields—a single scalar integration, although not always as simple as we might wish, followed by a pleasant differentiation.

The remaining difficult portion of the task, the integration, we intend to remove in Chapter 7.

4.1 ENERGY EXPENDED IN MOVING A POINT CHARGE IN AN ELECTRIC FIELD

The electric field intensity was defined as the force on a unit test charge at that point at which we wish to find the value of this vector field. If we attempt to move the test charge against the electric field, we have to exert a force equal and opposite to that exerted by the field, and this requires us to expend energy or do work. If we wish to move the charge in the direction of the field, our energy expenditure turns out to be negative; we do not do the work, the field does.

Suppose we wish to move a charge Q a distance $d\mathbf{L}$ in an electric field $\mathbf{E}$. The force on Q arising from the electric field is

$$\mathbf{F}_E = Q\mathbf{E} \tag{1}$$

where the subscript reminds us that this force arises from the field. The component of this force in the direction $d\mathbf{L}$ which we must overcome is

$$F_{EL} = \mathbf{F} \cdot \mathbf{a}_L = Q\mathbf{E} \cdot \mathbf{a}_L$$

where $\mathbf{a}_L$ = a unit vector in the direction of $d\mathbf{L}$.

The force which we must apply is equal and opposite to the force associated with the field,

$$F_{\text{appl}} = -Q\mathbf{E} \cdot \mathbf{a}_L$$

and our expenditure of energy is the product of the force and distance. That is,

Differential work done by external source moving $Q = -Q\mathbf{E} \cdot \mathbf{a}_L dL = -Q\mathbf{E} \cdot d\mathbf{L}$

or

$$dW = -Q\mathbf{E} \cdot d\mathbf{L} \tag{2}$$

where we have replaced $\mathbf{a}_L dL$ by the simpler expression $d\mathbf{L}$.

This differential amount of work required may be zero under several conditions determined easily from (2). There are the trivial conditions for which $\mathbf{E}$, Q, or $d\mathbf{L}$ is zero, and a much more important case in which $\mathbf{E}$ and $d\mathbf{L}$ are perpendicular. Here the charge is moved always in a direction at right angles to the electric field. We can draw on a good analogy between the electric field and the gravitational field, where, again, energy must be expended to move against the field. Sliding a mass around with constant velocity on a frictionless surface is an effortless process if the mass is moved along a constant elevation contour; positive or negative work must be done in moving it to a higher or lower elevation, respectively.

Interactives

Returning to the charge in the electric field, the work required to move the charge a finite distance must be determined by integrating, "work done by elecr. field in moving a charge".

$$W = -Q \int_{\text{init}}^{\text{final}} \mathbf{E} \cdot d\mathbf{L} \tag{3}$$

where the path must be specified before the integral can be evaluated. The charge is assumed to be at rest at both its initial and final positions.

This definite integral is basic to field theory, and we shall devote the following section to its interpretation and evaluation.

D4.1 Given the electric field $\mathbf{E} = \frac{1}{z^2}(8xyz\mathbf{a}_x + 4x^2z\mathbf{a}_y - 4x^2y\mathbf{a}_z)$ V/m, find the differential amount of work done in moving a 6-nC charge a distance of 2 μm, starting at $P(2, -2, 3)$ and proceeding in the direction $\mathbf{a}_L$ =:(a) $-\frac{6}{7}\mathbf{a}_x + \frac{3}{7}\mathbf{a}_y + \frac{2}{7}\mathbf{a}_z$; (b) $\frac{6}{7}\mathbf{a}_x - \frac{3}{7}\mathbf{a}_y - \frac{2}{7}\mathbf{a}_z$; (c) $\frac{3}{7}\mathbf{a}_x + \frac{6}{7}\mathbf{a}_y$

Ans. −149.3 fJ; 149.3 fJ; 0

4.2 THE LINE INTEGRAL

The integral expression for the work done in moving a point charge Q from one position to another, Eq. (3), is an example of a line integral, which in vector-analysis notation always takes the form of the integral along some prescribed path of the dot product of a vector field and a differential vector path length $d\mathbf{L}$.

Without using vector analysis we should have to write

$$W = -Q \int_{\text{init}}^{\text{final}} E_L dL$$

where E_L = component of **E** along $d\mathbf{L}$.

A line integral is like many other integrals which appear in advanced analysis, including the surface integral appearing in Gauss's law, in that it is essentially descriptive. We like to look at it much more than we like to work it out. It tells us to choose a path, break it up into a large number of very small segments, multiply the component of the field along each segment by the length of the segment, and then add the results for all the segments. This is a summation, of course, and the integral is obtained exactly only when the number of segments becomes infinite.

This procedure is indicated in Figure 4.1, where a path has been chosen from an initial position B to a final position[1] A and a *uniform electric field* selected for simplicity. The path is divided into six segments, $\Delta\mathbf{L}_1$, $\Delta\mathbf{L}_2$, . . ., $\Delta\mathbf{L}_6$, and the components of **E** along each segment are denoted by E_{L1}, E_{L2}, . . ., E_{L6}. The work involved in moving a charge Q from B to A is then approximately

$$W = -Q(E_{L1}\Delta L_1 + E_{L2}\Delta L_2 + \cdots + E_{L6}\Delta L_6)$$

or, using vector notation,

$$W = -Q(\mathbf{E}_1 \cdot \Delta\mathbf{L}_1 + \mathbf{E}_2 \cdot \Delta\mathbf{L}_2 + \cdots + \mathbf{E}_6 \cdot \Delta\mathbf{L}_6)$$

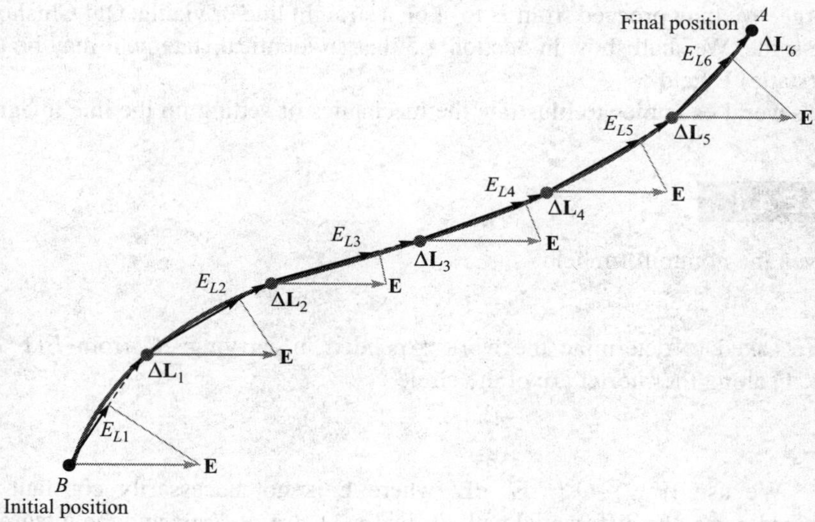

Fig. 4.1 A graphical interpretation of a line integral in a uniform field. The line integral of **E** between points B and A is independent of the path selected, even in a nonuniform field; this result is not, in general, true for time-varying fields.

and since we have assumed a uniform field,

$$\mathbf{E}_1 = \mathbf{E}_2 = \cdots = \mathbf{E}_6$$

$$W = -Q\mathbf{E} \cdot (\Delta\mathbf{L}_1 + \Delta\mathbf{L}_2 + \cdots + \Delta\mathbf{L}_6)$$

[1] The final position is given the designation A to correspond with the convention for potential difference, as discussed in the following section.

What is this sum of vector segments in the preceding parentheses? Vectors add by the parallelogram law, and the sum is just the vector directed from the initial point B to the final point A, $\mathbf{L}_{BA}$. Therefore,

$$W = -Q\mathbf{E} \cdot \mathbf{L}_{BA} \qquad \text{(uniform } \mathbf{E}\text{)} \tag{4}$$

Remembering the summation interpretation of the line integral, this result for the uniform field can be obtained rapidly now from the integral expression

$$W = -Q \int_B^A \mathbf{E} \cdot d\mathbf{L} \tag{5}$$

as applied to a uniform field

$$W = -Q\mathbf{E} \cdot \int_B^A d\mathbf{L}$$

where the last integral becomes $\mathbf{L}_{BA}$ and

$$W = -Q\mathbf{E} \cdot \mathbf{L}_{BA} \qquad \text{(uniform } \mathbf{E}\text{)}$$

For this special case of a uniform electric field intensity, we should note that the work involved in moving the charge depends only on Q, $\mathbf{E}$, and $\mathbf{L}_{BA}$, a vector drawn from the initial to the final point of the path chosen. It does not depend on the particular path we have selected along which to carry the charge. We may proceed from B to A on a straight line or via the Old Chisholm Trail; the answer is the same. We shall show in Section 4.5 that an identical statement may be made for any nonuniform (static) $\mathbf{E}$ field.

Let us use several examples to illustrate the mechanics of setting up the line integral appearing in (5).

EXAMPLE 4.1

We are given the nonuniform field

$$\mathbf{E} = y\mathbf{a}_x + x\mathbf{a}_y + 2\mathbf{a}_z$$

and we are asked to determine the work expended in carrying 2C from $B(1\ ,\ 0\ ,\ 1)$ to $A(0.8, 0.6, 1)$ along the shorter arc of the circle

$$x^2 + y^2 = 1 \quad z = 1$$

Solution. We use $W = -Q \int_B^A \mathbf{E} \cdot d\mathbf{L}$, where $\mathbf{E}$ is not necessarily constant. Working in rectangular coordinates, the differential path $d\mathbf{L}$ is $dx\mathbf{a}_x + dy\mathbf{a}_y + dz\mathbf{a}_z$, and the integral becomes

$$\begin{aligned} W &= -Q \int_B^A \mathbf{E} \cdot d\mathbf{L} \\ &= -2 \int_B^A (y\mathbf{a}_x + x\mathbf{a}_y + 2\mathbf{a}_z) \cdot (dx\mathbf{a}_x + dy\,\mathbf{a}_y + dz\,\mathbf{a}_z) \\ &= -2 \int_1^{0.8} y\,dx - 2 \int_0^{0.6} x\,dy - 4 \int_1^1 dz \end{aligned}$$

where the limits on the integrals have been chosen to agree with the initial and final values of the appropriate variable of integration. Using the equation of the circular path (and selecting the sign of the radical which is correct for the quadrant involved), we have

$$W = -2\int_1^{0.8} \sqrt{1-x^2}\,dx - 2\int_0^{0.6} \sqrt{1-y^2}\,dy - 0$$

$$= -\left[x\sqrt{1-x^2} + \sin^{-1} x\right]_1^{0.8} - \left[y\sqrt{1-y^2} + \sin^{-1} y\right]_0^{0.6}$$

$$= -(0.48 + 0.927 - 0 - 1.571) - (0.48 + 0.644 - 0 - 0) = -0.96 \text{ J}$$

EXAMPLE 4.2

Again find the work required to carry 2C from B to A in the same field, but this time use the straight-line path from B to A.

Solution. We start by determining the equations of the straight line. Any two of the following three equations for planes passing through the line are sufficient to define the line:

$$y - y_B = \frac{y_A - y_B}{x_A - x_B}(x - x_B)$$

$$z - z_B = \frac{z_A - z_B}{y_A - y_B}(y - y_B)$$

$$x - x_B = \frac{x_A - x_B}{z_A - z_B}(z - z_B)$$

From the first equation we have

$$y = -3(x - 1)$$

and from the second we obtain

$$z = 1$$

Thus,

$$W = -2\int_1^{0.8} y\,dx - 2\int_0^{0.6} x\,dy - 4\int_1^1 dz$$

$$= 6\int_1^{0.8} (x-1)\,dx - 2\int_0^{0.6} \left(1 - \frac{y}{3}\right) dy = -0.96 \text{ J}$$

This is the same answer we found using the circular path between the same two points, and it again demonstrates the statement (unproved) that the work done is independent of the path taken in any electrostatic field.

It should be noted that the equations of the straight line show that $dy = -3\,dx$ and $dx = -3\,dy$. These substitutions may be made in the first two integrals, along with a change in limits, and the answer may be obtained by evaluating the new integrals. This method is often simpler if the integrand is a function of only one variable.

Note that the expressions for $d\mathbf{L}$ in our three coordinate systems use the differential lengths obtained in Chapter 1 (rectangular in Section 1.3, cylindrical in Section 1.8, and spherical in Section 1.9):

$$d\mathbf{L} = dx\,\mathbf{a}_x + dy\,\mathbf{a}_y + dz\,\mathbf{a}_z \qquad \text{(rectangular)} \tag{6}$$

$$d\mathbf{L} = d\rho\,\mathbf{a}_\rho + \rho d\phi \mathbf{a}_\phi + dz\,\mathbf{a}_z \qquad \text{(cylindrical)} \tag{7}$$

$$d\mathbf{L} = dr\,\mathbf{a}_r + r\,d\theta\,\mathbf{a}_\theta + r\sin\theta\,d\phi\,\mathbf{a}_\phi \qquad \text{(spherical)} \tag{8}$$

The interrelationships among the several variables in each expression are determined from the specific equations for the path.

As a final example illustrating the evaluation of the line integral, let us investigate several paths which we might take near an infinite line charge. The field has been obtained several times and is entirely in the radial direction,

$$\mathbf{E} = E_\rho \mathbf{a}_\rho = \frac{\rho_L}{2\pi\epsilon_0 \rho}\mathbf{a}_\rho$$

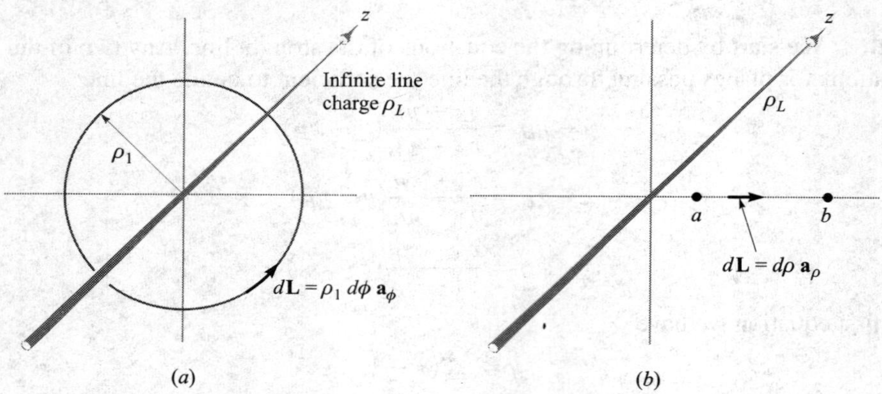

Fig. 4.2 (*a*) A circular path and (*b*) a radial path along which a charge of Q is carried in the field of an infinite line charge. No work is expected in the former case.

Let us first find the work done in carrying the positive charge Q about a circular path of radius ρ_b centered at the line charge, as illustrated in Figure 4.2*a*. Without lifting a pencil, we see that the work must be nil, for the path is always perpendicular to the electric field intensity, or the force on the charge is always exerted at right angles to the direction in which we are moving it. For practice, however, let us set up the integral and obtain the answer.

The differential element $d\mathbf{L}$ is chosen in cylindrical coordinates, and the circular path selected demands that $d\rho$ and dz be zero, so $d\mathbf{L} = \rho_1 d\phi \mathbf{a}_\phi$. The work is then

$$W = -Q\int_{\text{init}}^{\text{final}} \frac{\rho_L}{2\pi\epsilon_0\rho_1}\mathbf{a}_\rho \cdot \rho_1 d\phi\,\mathbf{a}\phi$$

$$= -Q\int_0^{2\pi} \frac{\rho_L}{2\pi\epsilon_0} d\phi\,\mathbf{a}_\rho \cdot \mathbf{a}_\phi = 0$$

Let us now carry the charge from $\rho = a$ to $\rho = b$ along a radial path (Figure 4.2 *b*). Here $d\mathbf{L} = d\rho\mathbf{a}_\rho$ and

$$W = -Q\int_{\text{init}}^{\text{final}} \frac{\rho_L}{2\pi\epsilon_0\rho}\mathbf{a}_\rho \cdot d\rho\mathbf{a}_\rho = -Q\int_a^b \frac{\rho_L}{2\pi\epsilon_0}\frac{d\rho}{\rho}$$

or

$$W = -\frac{Q\rho_L}{2\pi\epsilon_0}\ln\frac{b}{a}$$

Since b is larger than a, $\ln(b/a)$ is positive, and we see that the work done is negative, indicating that the external source that is moving the charge receives energy.

One of the pitfalls in evaluating line integrals is a tendency to use too many minus signs when a charge is moved in the direction of a *decreasing* coordinate value. This is taken care of completely by the limits on the integral, and no misguided attempt should be made to change the sign of $d\mathbf{L}$. Suppose we carry Q from b to a (Figure 4.2 *b*). We still have $d\mathbf{L} = d\rho\mathbf{a}_\rho$ and show the different direction by recognizing $\rho = b$ as the initial point and $\rho = a$ as the final point,

$$W = -Q\int_b^a \frac{\rho_L}{2\pi\epsilon_0}\frac{d\rho}{\rho} = \frac{Q\rho_L}{2\pi\epsilon_0}\ln\frac{b}{a}$$

This is the negative of the previous answer and is obviously correct.

D4.2 Calculate the work done in moving a 4-C charge from $B(1,0,0)$ to $A\ (0, 2, 0)$ along the path $y = 2 - 2x$, $z = 0$ in the field $\mathbf{E} =$: (*a*) $5\mathbf{a}_x$ V/m; (*b*) $5x\mathbf{a}_x$ V/m; (*c*) $5x\mathbf{a}_x + 5\,y\,\mathbf{a}_y$ V/m.

Ans. 20 J; 10 J; – 30 J

D4.3 We shall see later that a time-varying **E** field need not be conservative. (If it is not conservative, the work expressed by Eq. (3) may be a function of the path used.) Let $\mathbf{E} = y\mathbf{a}_x$ V/m at a certain instant of time, and calculate the work required to move a 3-C charge from (1, 3, 5) to (2, 0, 3) along the straight-line segments joining: (*a*) (1, 3, 5) to (2, 3, 5) to (2, 0, 5) to (2, 0, 3); (*b*) (1, 3, 5) to (1, 3, 3) to (1, 0, 3) to (2, 0, 3).

Ans. – 9 J; 0

4.3 DEFINITION OF POTENTIAL DIFFERENCE AND POTENTIAL

Illustrations

We are now ready to define a new concept from the expression for the work done by an external source in moving a charge Q from one point to another in an electric field E, "Potential difference and work".

$$W = -Q\int_{\text{init}}^{\text{final}} \mathbf{E}\cdot d\mathbf{L}$$

In much the same way as we defined the electric field intensity as the force on a *unit* test charge, we now define *potential difference* V as the work done (by an external source) in moving a *unit* positive charge from one point to another in an electric field,

$$\text{Potential difference} = V = -\int_{\text{init}}^{\text{final}} \mathbf{E}\cdot d\mathbf{L} \qquad (9)$$

We shall have to agree on the direction of movement, as implied by our language, and we do this by stating that V_{AB} signifies the potential difference between points A and B and is the work done in moving the unit charge from B (last named) to A (first named). Thus, in determining V_{AB}, B is the initial point and A is the final point. The reason for this somewhat peculiar definition will become clearer shortly, when it is seen that the initial point B is often taken at infinity, whereas the final point A represents the fixed position of the charge; point A is thus inherently more significant.

Potential difference is measured in joules per coulomb, for which the *volt* is defined as a more common unit, abbreviated as V. Hence the potential difference between points A and B is

$$V_{AB} = -\int_B^A \mathbf{E} \cdot d\mathbf{L} \text{ V} \tag{10}$$

and V_{AB} is positive if work is done in carrying the positive charge from B to A.

From the line-charge example of Section 4.2 we found that the work done in taking a charge Q from $\rho = b$ to $\rho = a$ was

$$W = \frac{Q\rho_L}{2\pi\epsilon_0} \ln \frac{b}{a}$$

Thus, the potential difference between points at $\rho = a$ and $\rho = b$ is

$$V_{ab} = \frac{W}{Q} = \frac{\rho_L}{2\pi\epsilon_0} \ln \frac{b}{a} \tag{11}$$

We can try out this definition by finding the potential difference between points A and B at radial distances r_A and r_B from a point charge Q. Choosing an origin at Q,

$$\mathbf{E} = E_r \mathbf{a}_r = \frac{Q}{4\pi\epsilon_0 r^2} \mathbf{a}_r$$

and

$$d\mathbf{L} = dr\, \mathbf{a}_r$$

we have

$$V_{AB} = -\int_B^A \mathbf{E} \cdot d\mathbf{L} = -\int_{r_B}^{r_A} \frac{Q}{4\pi\epsilon_0 r^2}\, dr = \frac{Q}{4\pi\epsilon_0} \left(\frac{1}{r_A} - \frac{1}{r_B} \right) \tag{12}$$

If $r_B > r_A$, the potential difference V_{AB} is positive, indicating that energy is expended by the external source in bringing the positive charge from r_B to r_A. This agrees with the physical picture showing the two like charges repelling each other.

It is often convenient to speak of the *potential*, or *absolute potential*, of a point, rather than the potential difference between two points, but this means only that we agree to measure every potential difference with respect to a specified reference point which we consider to have zero potential. Common agreement must be reached on the zero reference before a statement of the potential has any significance. A person having one hand on the deflection plates of a cathode-ray tube which are "at a potential of 50 V" and the other hand on the cathode terminal would probably be too shaken up to understand that the cathode is not the zero reference, but that all potentials in that circuit are customarily measured with respect to the metallic shield about the tube. The cathode may be several thousands of volts negative with respect to the shield.

Perhaps the most universal zero reference point in experimental or physical potential measurements is "ground," by which we mean the potential of the surface region of the earth itself. Theoretically, we usually represent this surface by an infinite plane at zero potential, although some large-scale problems, such as those involving propagation across the Atlantic Ocean, require a spherical surface at zero potential.

Another widely used reference "point" is infinity. This usually appears in theoretical problems approximating a physical situation in which the earth is relatively far removed from the region in which we are interested, such as the static field near the wing tip of an airplane that has acquired a charge in flying through a thunderhead, or the field inside an atom. Working with the *gravitational* potential field on earth, the zero reference is normally taken at sea level; for an interplanetary mission, however, the zero reference is more conveniently selected at infinity.

A cylindrical surface of some definite radius may occasionally be used as a zero reference when cylindrical symmetry is present and infinity proves inconvenient. In a coaxial cable the outer conductor is selected as the zero reference for potential. And, of course, there are numerous special problems, such as those for which a two-sheeted hyperboloid or an oblate spheroid must be selected as the zero-potential reference, but these need not concern us immediately.

If the potential at point A is V_A and that at B is V_B, then

$$V_{AB} = V_A - V_B \tag{13}$$

where we necessarily agree that V_A and V_B shall have the same zero reference point.

D4.4 An electric field is expressed in rectangular coordinates by $\mathbf{E} = 6x^2\mathbf{a}_x + 6y\mathbf{a}_y + 4\mathbf{a}_z$ V/m. Find: (*a*) V_{MN} if points M and N are specified by $M(2, 6, -1)$ and $N(-3, -3, 2)$; (*b*) V_M if $V = 0$ at $Q(4, -2, -35)$; (*c*) V_N if $V = 2$ at $P(1, 2, -4)$.

Ans. −139.0 V; −120.0 V; 19.0 V

4.4 THE POTENTIAL FIELD OF A POINT CHARGE

In Section 4.3 we found an expression (12) for the potential difference between two points located at $r = r_A$ and $r = r_B$ in the field of a point charge Q placed at the origin,

$$V_{AB} = \frac{Q}{4\pi\epsilon_0}\left(\frac{1}{r_A} - \frac{1}{r_B}\right) = V_A - V_B \tag{14}$$

It was assumed that the two points lay on the same radial line or had the same θ and ϕ coordinate values, allowing us to set up a simple path on this radial line along which to carry our positive charge. We now should ask whether different θ and ϕ coordinate values for the initial and final position will affect our answer and whether we could choose more complicated paths between the two points without changing the results. Let us answer both questions at once by choosing two general points A and B (Figure 4.3) at radial distances of r_A and r_B, and any values for the other coordinates.

The differential path length $d\mathbf{L}$ has r, θ, and ϕ components, and the electric field has only a radial component. Taking the dot product then leaves us only

$$V_{AB} = -\int_{r_B}^{r_A} E_r\,dr = -\int_{r_B}^{r_A} \frac{Q}{4\pi\epsilon_0 r^2}\,dr = \frac{Q}{4\pi\epsilon_0}\left(\frac{1}{r_A} - \frac{1}{r_B}\right)$$

We obtain the same answer and see, therefore, that the potential difference between two points in the field of a point charge depends only on the distance of each point from the charge and does not depend on the particular path used to carry our unit charge from one point to the other.

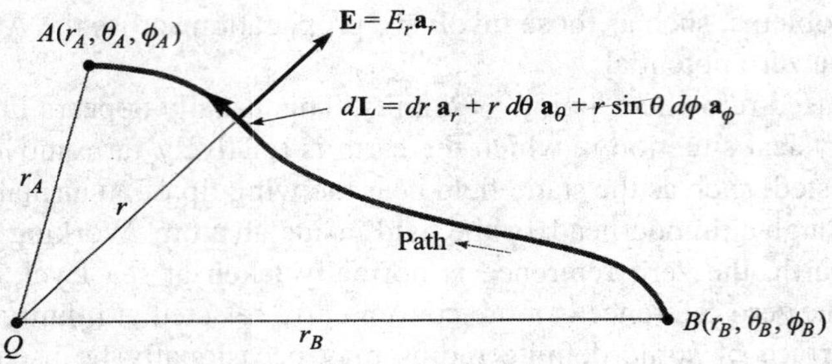

Fig. 4.3 A general path between general points B and A in the field of a point charge Q at the origin. The potential difference V_{AB} is independent of the path selected.

How might we conveniently define a zero reference for potential? The simplest possibility is to let $V = 0$ at infinity. If we let the point at $r = r_B$ recede to infinity the potential at r_A becomes

$$V_A = \frac{Q}{4\pi\epsilon_0 r_A}$$

or, since there is no reason to identify this point with the A subscript,

$$V = \frac{Q}{4\pi\epsilon_0 r} \tag{15}$$

This expression defines the potential at any point distant r from a point charge Q at the origin, the potential at infinite radius being taken as the zero reference. Returning to a physical interpretation, we may say that $Q/4\pi\epsilon_0 r$ joules of work must be done in carrying a 1-C charge from infinity to any point r meters from the charge Q.

A convenient method to express the potential without selecting a specific zero reference entails identifying r_A as r once again and letting $Q/4\pi\epsilon_0 r_B$ be a constant. Then

$$V = \frac{Q}{4\pi\epsilon_0 r} + C_1 \tag{16}$$

and C_1 may be selected so that $V = 0$ at any desired value of r. We could also select the zero reference indirectly by electing to let V be V_0 at $r = r_0$.

It should be noted that the *potential difference* between two points is not a function of C_1.

Equation(15) or (16) represents the potential field of a point charge. The potential is a scalar field and does not involve any unit vectors.

Let us now define an *equipotential surface* as a surface composed of all those points having the same value of potential. No work is involved in moving a unit charge around on an equipotential surface, for, by definition, there is no potential difference between any two points on this surface.

The equipotential surfaces in the potential field of a point charge are spheres centered at the point charge.

Illustrations

An inspection of the form of the potential field of a point charge shows that it is an inverse-distance field, whereas the electric field intensity was found to be an inverse-square-law relationship. A similar result occurs for the gravitational force field of a point mass (inverse-square law) and the gravitational potential field (inverse distance). The gravitational force exerted by the earth on an object one million miles from it is four times that exerted on the same object two million miles away. The kinetic energy given to a freely falling object starting from the end of the universe with zero velocity, however, is only twice as much at one million miles as it is at two million miles.

D4.5 A 15-nC point charge is at the origin in free space. Calculate V_1 if point P_1 is located at $P_1(-2, 3, -1)$ and: (*a*) $V = 0$ at $(6, 5, 4)$; (*b*) $V = 0$ at infinity; (*c*) $V = 5$ V at $(2, 0, 4)$.

Ans. 20.67 V; 36.0 V; 10.89 V

4.5 THE POTENTIAL FIELD OF A SYSTEM OF CHARGES: CONSERVATIVE PROPERTY

The potential at a point has been defined as the work done in bringing a unit positive charge from the zero reference to the point, and we have suspected that this work, and hence the potential, is independent of the path taken. If it were not, potential would not be a very useful concept.

Let us now prove our assertion. We shall do so by beginning with the potential field of the single point charge for which we showed, in Section 4.4, the independence with regard to the path, noting that the field is linear with respect to charge so that superposition is applicable. It will then follow that the potential of a system of charges has a value at any point which is independent of the path taken in carrying the test charge to that point.

Thus the potential field of a single point charge, which we shall identify as Q_1 and locate at $\mathbf{r}_1$, involves only the distance $|\mathbf{r} - \mathbf{r}_1|$ from Q_1 to the point at $\mathbf{r}$ where we are establishing the value of the potential. For a zero reference at infinity, we have

$$V(\mathbf{r}) = \frac{Q_1}{4\pi\epsilon_0|\mathbf{r} - \mathbf{r}_1|}$$

The potential arising from two charges, Q_1 at $\mathbf{r}_1$ and Q_2 at $\mathbf{r}_2$, is a function only of $|\mathbf{r} - \mathbf{r}_1|$ and $|\mathbf{r} - \mathbf{r}_2|$, the distances from Q_1 and Q_2 to the field point, respectively.

$$V(\mathbf{r}) = \frac{Q_1}{4\pi\epsilon_0|\mathbf{r} - \mathbf{r}_1|} + \frac{Q_2}{4\pi\epsilon_0|\mathbf{r} - \mathbf{r}_2|}$$

Continuing to add charges, we find that the potential arising from n point charges is

$$V(\mathbf{r}) = \frac{Q_1}{4\pi\epsilon_0|\mathbf{r} - \mathbf{r}_1|} + \frac{Q_2}{4\pi\epsilon_0|\mathbf{r} - \mathbf{r}_2|} + \cdots + \frac{Q_n}{4\pi\epsilon_0|\mathbf{r} - \mathbf{r}_n|}$$

or

$$V(\mathbf{r}) = \sum_{m=1}^{n} \frac{Q_m}{4\pi\epsilon_0|\mathbf{r} - \mathbf{r}_m|} \tag{17}$$

If each point charge is now represented as a small element of a continuous volume charge distribution $\rho_\nu \Delta\nu$, then

$$V(\mathbf{r}) = \frac{\rho_\nu(\mathbf{r}_1)\Delta\nu_1}{4\pi\epsilon_0|\mathbf{r} - \mathbf{r}_1|} + \frac{\rho_\nu(\mathbf{r}_2)\Delta\nu_2}{4\pi\epsilon_0|\mathbf{r} - \mathbf{r}_2|} + \cdots + \frac{\rho_\nu(\mathbf{r}_n)\Delta\nu_n}{4\pi\epsilon_0|\mathbf{r} - \mathbf{r}_n|}$$

As we allow the number of elements to become infinite, we obtain the integral expression

$$V(\mathbf{r}) = \int_{\text{vol}} \frac{\rho_v(\mathbf{r}')dv'}{4\pi\epsilon_0|\mathbf{r} - \mathbf{r}'|} \tag{18}$$

We have come quite a distance from the potential field of the single point charge, and it might be helpful to examine (18) and refresh ourselves as to the meaning of each term. The potential $V(\mathbf{r})$ is determined with respect to a zero reference potential at infinity and is an exact measure of the work done in bringing a unit charge from infinity to the field point at $\mathbf{r}$ where we are finding the potential. The volume charge density $\rho_y(\mathbf{r}')$ and differential volume element dv' combine to represent a differential amount of charge $\rho_\nu(\mathbf{r}')\,dv'$ located at $\mathbf{r}'$. The distance $|\mathbf{r}-\mathbf{r}'|$ is that distance from the source point to the field point. The integral is a multiple (volume) integral.

If the charge distribution takes the form of a line charge or a surface charge, the integration is along the line or over the surface:

$$V(\mathbf{r}) = \int \frac{\rho_L(\mathbf{r}')dL'}{4\pi\epsilon_0|\mathbf{r}-\mathbf{r}'|} \tag{19}$$

$$V(\mathbf{r}) = \int_S \frac{\rho_S(\mathbf{r}')dS'}{4\pi\epsilon_0|\mathbf{r}-\mathbf{r}'|} \tag{20}$$

The most general expression for potential is obtained by combining (17), (18), (19), and (20).

These integral expressions for potential in terms of the charge distribution should be compared with similar expressions for the electric field intensity, such as (18) in Section 2.3:

$$\mathbf{E}(\mathbf{r}) = \int_{\text{vol}} \frac{\rho_\nu(r')dv'}{4\pi\epsilon_0|\mathbf{r}-\mathbf{r}'|^2}\frac{\mathbf{r}-\mathbf{r}'}{|\mathbf{r}-\mathbf{r}'|}$$

The potential again is inverse distance, and the electric field intensity, inverse-square law. The latter, of course, is also a vector field.

To illustrate the use of one of these potential integrals, let us find V on the z axis for a uniform line charge ρ_L in the form of a ring, $\rho = a$, in the $z = 0$ plane, as shown in Figure 4.4. Working with (19), we have $dL' = ad\phi'$, $\mathbf{r} = z\mathbf{a}_z$, $\mathbf{r}' = a\mathbf{a}_\rho$, $|\mathbf{r}-\mathbf{r}'| = \sqrt{a^2+z^2}$, and

$$V = \int_0^{2\pi} \frac{\rho_L a d\phi'}{4\pi\epsilon_0\sqrt{a^2+z^2}} = \frac{\rho_L a}{2\epsilon_0\sqrt{a^2+z^2}}$$

For a zero reference at infinity, then:

1. The potential arising from a single point charge is the work done in carrying a unit positive charge from infinity to the point at which we desire the potential, and the work is independent of the path chosen between those two points.
2. The potential field in the presence of a number of point charges is the sum of the individual potential fields arising from each charge.
3. The potential arising from a number of point charges or any continuous charge distribution may therefore be found by carrying a unit charge from infinity to the point in question along any path we choose.

   In other words, the expression for potential (zero reference at infinity),

$$V_A = -\int_\infty^A \mathbf{E}\cdot d\mathbf{L}$$

or potential difference,

$$V_{AB} = V_A - V_B = -\int_B^A \mathbf{E}\cdot d\mathbf{L}$$

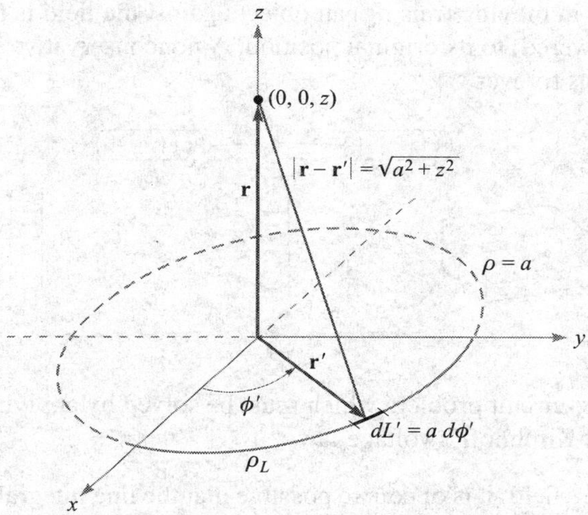

Fig. 4.4 The potential field of a ring of uniform line charge density is easily obtained from $V = \int \rho_L(\mathbf{r}')dL'/(4\pi\epsilon_0|\mathbf{r} - \mathbf{r}'|)$.

is not dependent on the path chosen for the line integral, regardless of the source of the **E** field.

This result is often stated concisely by recognizing that no work is done in carrying the unit charge around any *closed path*, or

$$\oint \mathbf{E} \cdot d\mathbf{L} = 0 \tag{21}$$

A small circle is placed on the integral sign to indicate the closed nature of the path. This symbol also appeared in the formulation of Gauss' law, where a closed *surface* integral was used.

Equation (21) is true for *static* fields, but we shall see in Chapter 10 that Faraday demonstrated it was incomplete when time-varying magnetic fields were present. One of Maxwell's greatest contributions to electromagnetic theory was in showing that a time-varying electric field produces a magnetic field, and therefore we should expect to find later that (21) is not correct when either **E** or the magnetic field varies with time.

Restricting our attention to the static case where **E** does not change with time, consider the dc circuit shown in Figure 4.5. Two points, A and B, are marked, and (21) states that no work is involved in carrying a unit charge from A through R_2 and R_3 to B and back to A through R_1, or that the sum of the potential differences around any closed path is zero.

Equation (21) is therefore just a more general form of Kirchhoff's circuital law for voltages, more general in that we can apply it to any region where an electric field exists and we are not restricted to a conventional circuit composed of wires, resistances, and batteries. Equation (21) must be amended before we can apply it to time-varying fields. We shall take care of this in Chapter 10, and in Chapter 13 we will then be able to establish the general form of Kirchhoff's voltage law for circuits in which currents and voltages vary with time.

Any field that satisfies an equation of the form of (21), (i.e., where the closed line integral of the field is zero) is said to be a *conservative field*. The name arises from the fact that no work is done (or that energy is *conserved*) around a closed path. The gravitational field is also conservative, for any energy expended in moving (raising) an object against the field is recovered exactly, when the object is returned (lowered) to its original position. A nonconservative gravitational field could solve our energy problems forever.

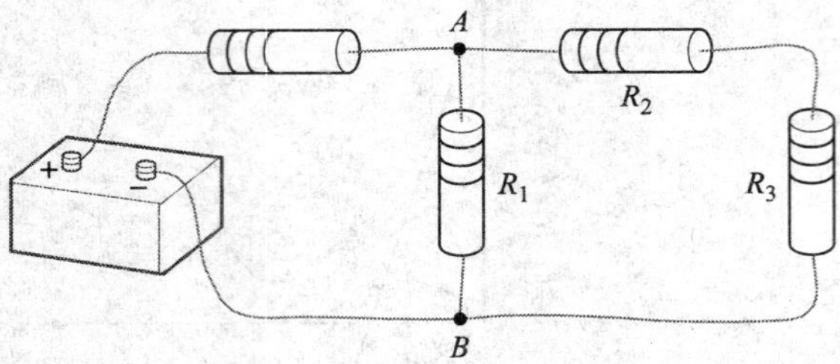

Fig. 4.5 A simple dc-circuit problem which must be solved by applying $\oint \mathbf{E} \cdot d\mathbf{L} = 0$ in the form of Kirchhoff's voltage law.

Given a *nonconservative* field, it is of course possible that the line integral may be zero for certain closed paths. For example, consider the force field, $\mathbf{F} = \sin \pi\rho_\phi \mathbf{a}_\phi$. Around a circular path of radius $\rho = \rho_1$, we have $d\mathbf{L} = \rho d\phi \mathbf{a}_\phi$, and

$$\oint \mathbf{F} \cdot d\mathbf{L} = \int_0^{2\pi} \sin \pi\rho_1 \mathbf{a}_\phi \cdot \rho_1 d\phi\, \mathbf{a}_\phi = \int_0^{2\pi} \rho_1 \sin \pi\rho_1 d\phi$$
$$= 2\pi\rho_1 \sin \pi\rho_1$$

The integral is zero if $\rho_1 = 1, 2, 3, \ldots$, etc., but it is not zero for other values of ρ_1, or for most other closed paths, and the given field is not conservative. A conservative field must yield a zero value for the line integral around every possible closed path.

D4.6 If we take the zero reference for potential at infinity, find the potential at (0, 0, 2) caused by this charge configuration in free space: (*a*) 12 nC/m on the line $\rho = 2.5$ m, $z = 0$; (*b*) point charge of 18 nC at (1, 2, − 1); (*c*) 12 nC/m on the line $y = 2.5$, $z = 0$.

Ans. 529 V; 43.2 V; 67.4 V

4.6 POTENTIAL GRADIENT

Interactives

We now have two methods of determining potential, one directly from the electric field intensity by means of a line integral, and another from the basic charge distribution itself by a volume integral. Neither method is very helpful in determining the fields in most practical problems, however, for as we shall see later, neither the electric field intensity nor the charge distribution is very often known. Preliminary information is much more apt to consist of a description of two equipotential surfaces, such as the statement that we have two parallel conductors of circular cross section at potentials of 100 and −100 V. Perhaps we wish to find the capacitance between the conductors, or the charge and current distribution on the conductors from which losses may be calculated.

These quantities may be easily obtained from the potential field, and our immediate goal will be a simple method of finding the electric field intensity from the potential.

We already have the general line-integral relationship between these quantities,

$$V = -\int \mathbf{E} \cdot d\mathbf{L} \tag{22}$$

but this is much easier to use in the reverse direction: given **E**, find V.

However, (22) may be applied to a very short element of length $\Delta\mathbf{L}$ along which **E** is essentially constant, leading to an incremental potential difference ΔV,

$$\Delta V \doteq -\mathbf{E} \cdot d\mathbf{L} \tag{23}$$

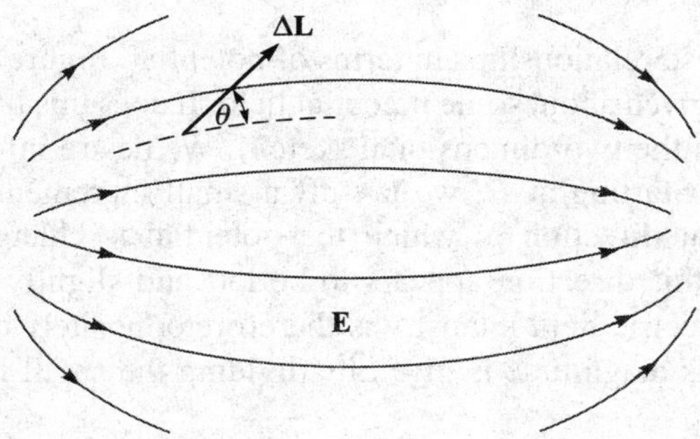

Fig. 4.6 A vector incremental element of length $\Delta\mathbf{L}$ is shown making an angle of θ with an E field, indicated by its streamlines. The sources of the field are not shown.

Let us see first if we can determine any new information about the relation of V to **E** from this equation. Consider a general region of space, as shown in Figure 4.6, in which **E** and V both change as we move from point to point. Equation (23) tells us to choose an incremental vector element of length $\Delta\mathbf{L} = \Delta L \mathbf{a}_L$ and multiply its magnitude by the component of **E** in the direction of $\mathbf{a}_L$ (one interpretation of the dot product) to obtain the small potential difference between the final and initial points of $\Delta\mathbf{L}$.

If we designate the angle between $\Delta\mathbf{L}$ and **E** as θ, then

$$\Delta V \doteq -E\Delta L \cos\theta$$

We now wish to pass to the limit and consider the derivative *dV*/*dL*. To do this, we need to show that V may be interpreted as a *function* $V(x, y, z)$. So far, V is merely the result of the line integral (22). If we assume a specified starting point or zero reference and then let our end point be (x, y, z), we know that the result of the integration is a unique function of the end point (x, y, z) because **E** is a conservative field. Therefore V is a single-valued function $V(x, y, z)$. We may then pass to the limit and obtain

$$\frac{dV}{dL} = -E\cos\theta$$

In which direction should $\Delta\mathbf{L}$ be placed to obtain a maximum value of ΔV? Remember that **E** is a definite value at the point at which we are working and is independent of the direction of $\Delta\mathbf{L}$. The magnitude ΔL is also constant, and our variable is $\mathbf{a}_L$, the unit vector showing the direction of $\Delta\mathbf{L}$. It is obvious that the maximum positive increment of potential, $\Delta V_{\max}$, will occur when $\cos\theta$

is -1, or $\Delta\mathbf{L}$ points in the direction *opposite* to $\mathbf{E}$. For this condition,

$$\left.\frac{dV}{dL}\right|_{\max} = E$$

This little exercise shows us two characteristics of the relationship between $\mathbf{E}$ and V at any point:

1. The magnitude of the electric field intensity is given by the maximum value of the rate of change of potential with distance.
2. This maximum value is obtained when the direction of the distance increment is opposite to $\mathbf{E}$ or, in other words, the direction of $\mathbf{E}$ is *opposite* to the direction in which the potential is *increasing* the most rapidly.

Let us now illustrate these relationships in terms of potential. Figure 4.7 is intended to show the information we have been given about some potential field. It does this by showing the equipotential surfaces (shown as lines in the two-dimensional sketch). We desire information about the electric field intensity at point P. Starting at P, we lay off a small incremental distance $\Delta\mathbf{L}$ in various directions, hunting for that direction in which the potential is changing (increasing) the most rapidly. From the sketch, this direction appears to be left and slightly upward. From our second characteristic above, the electric field intensity is therefore oppositely directed, or to the right and slightly downward at P. Its magnitude is given by dividing the small increase in potential by the small element of length.

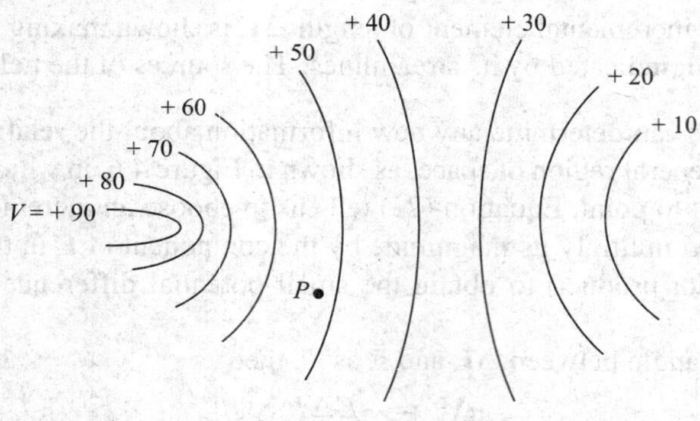

Fig. 4.7 A potential field is shown by its equipotential surfaces. At any point the E field is normal to the equipotential surface passing through that point and is directed toward the more negative surfaces.

It seems likely that the direction in which the potential is increasing the most rapidly is perpendicular to the equipotentials (in the direction of *increasing* potential), and this is correct, for if $\Delta\mathbf{L}$ is directed along an equipotential, $\Delta V = 0$ by our definition of an equipotential surface. But then

$$\Delta V = -\mathbf{E} \cdot \Delta\mathbf{L} = 0$$

and since neither $\mathbf{E}$ nor $\Delta\mathbf{L}$ is zero, $\mathbf{E}$ must be perpendicular to this $\Delta\mathbf{L}$ or perpendicular to the equipotentials.

Since the potential field information is more likely to be determined first, let us describe the direction of $\Delta\mathbf{L}$ which leads to a maximum increase in potential mathematically in terms of the potential field rather than the electric field intensity. We do this by letting $\mathbf{a}_N$ be a unit vector normal to the equipotential surface and directed toward the higher potentials. The electric field intensity is then expressed in terms of the potential,

$$\mathbf{E} = -\left.\frac{dV}{dL}\right|_{\max} \mathbf{a}_N \tag{24}$$

which shows that the magnitude of $\mathbf{E}$ is given by the maximum space rate of change of V and the direction of $\mathbf{E}$ is *normal* to the equipotential surface (in the direction of *decreasing* potential).

Since $dV/dL\,|_{\max}$ occurs when $\Delta\mathbf{L}$ is in the direction of $\mathbf{a}_N$, we may remind ourselves of this fact by letting

$$\left.\frac{dV}{dL}\right|_{\max} = \frac{dV}{dN}$$

and

$$\mathbf{E} = -\frac{dV}{dN}\mathbf{a}_N \tag{25}$$

Equation (24) or (25) serves to provide a physical interpretation of the process of finding the electric field intensity from the potential. Both are descriptive of a general procedure, and we do not intend to use them directly to obtain quantitative information. This procedure leading from V to $\mathbf{E}$ is not unique to this pair of quantities, however, but has appeared as the relationship between a scalar and a vector field in hydraulics, thermodynamics, and magnetics, and indeed in almost every field to which vector analysis has been applied.

The operation on V by which $-\mathbf{E}$ is obtained is known as the *gradient*, and the gradient of a scalar field T is defined as

$$\text{Gradient of } T = \text{grad } T = \frac{dT}{dN}\mathbf{a}_N \tag{26}$$

where $\mathbf{a}_N$ is a unit vector normal to the equipotential surfaces, and that normal is chosen which points in the direction of increasing values of T.

Using this new term, we now may write the relationship between V and $\mathbf{E}$ as

$$\mathbf{E} = -\text{grad } V \tag{27}$$

Since we have shown that V is a unique function of x, y, and z, we may take its total differential

$$dV = \frac{\partial V}{\partial x}dx + \frac{\partial V}{\partial y}dy + \frac{\partial V}{\partial z}dz$$

But we also have

$$dV = -\mathbf{E}\cdot d\mathbf{L} = -E_x dx - E_y dy - E_z dz$$

Since both expressions are true for any *dx, dy*, and *dz*, then

$$E_x = -\frac{\partial V}{\partial x}$$

$$E_y = -\frac{\partial V}{\partial y}$$

$$E_z = -\frac{\partial V}{\partial z}$$

These results may be combined vectorially to yield

$$\mathbf{E} = -\left(\frac{\partial V}{\partial x}\mathbf{a}_x + \frac{\partial V}{\partial y}\mathbf{a}_y + \frac{\partial V}{\partial z}\mathbf{a}_z\right) \tag{28}$$

and comparison of (27) and (28) provides us with an expression which may be used to evaluate the gradient in rectangular coordinates,

$$\text{grad}\, V = \frac{\partial V}{\partial x}\mathbf{a}_x + \frac{\partial V}{\partial y}\mathbf{a}_y + \frac{\partial V}{\partial z}\mathbf{a}_z \tag{29}$$

The gradient of a scalar is a vector, and old quizzes show that the unit vectors which are often incorrectly added to the divergence expression appear to be those which were incorrectly removed from the gradient. Once the physical interpretation of the gradient, expressed by (26), is grasped as showing the maximum space rate of change of a scalar quantity and *the direction in which this maximum occurs*, the vector nature of the gradient should be self-evident.

The vector operator

$$\nabla = \frac{\partial}{\partial x}\mathbf{a}_x + \frac{\partial}{\partial y}\mathbf{a}_y + \frac{\partial}{\partial z}\mathbf{a}_z$$

may be used formally as an operator on a scalar, T, ∇T, producing

$$\nabla T = \frac{\partial T}{\partial x}\mathbf{a}_x + \frac{\partial T}{\partial y}\mathbf{a}_y + \frac{\partial T}{\partial z}\mathbf{a}_z$$

from which we see that

$$\nabla T = \text{grad}\, T$$

This allows us to use a very compact expression to relate $\mathbf{E}$ and V,

$$\mathbf{E} = -\nabla V \tag{30}$$

Interactives

The gradient may be expressed in terms of partial derivatives in other coordinate systems through the application of its definition (26). These expressions are derived in Appendix A and repeated here for convenience when dealing with problems having cylindrical or spherical symmetry. They also appear inside the back cover.

$$\nabla V = \frac{\partial V}{\partial x}\mathbf{a}_x + \frac{\partial V}{\partial y}\mathbf{a}_y + \frac{\partial V}{\partial z}\mathbf{a}_z \qquad \text{(rectangular)} \tag{31}$$

$$\nabla V = \frac{\partial V}{\partial \rho}\mathbf{a}_\rho + \frac{1}{\rho}\frac{\partial V}{\partial \phi}\mathbf{a}_\phi + \frac{\partial V}{\partial z}\mathbf{a}_z \qquad \text{(cylindrical)} \tag{32}$$

$$\nabla V = \frac{\partial V}{\partial r}\mathbf{a}_r + \frac{1}{r}\frac{\partial V}{\partial \theta}\mathbf{a}_\theta + \frac{1}{r\sin\theta}\frac{\partial V}{\partial \phi}\mathbf{a}_\phi \qquad \text{(spherical)} \tag{33}$$

Note that the denominator of each term has the form of one of the components of $d\mathbf{L}$ in that coordinate system, except that partial differentials replace ordinary differentials; for example, r sin θ $d\phi$ becomes r sin θ $\partial\phi$.

Let us now hasten to illustrate the gradient concept with an example.

EXAMPLE 4.3

Given the potential field, $V = 2x^2y - 5z$, and a point $P(-4, 3, 6)$, we wish to find several numerical values at point P: the potential V, the electric field intensity $\mathbf{E}$, the direction of $\mathbf{E}$, the electric flux density $\mathbf{D}$, and the volume charge density ρ_ν.

Solution. The potential at $P(-4, 5, 6)$ is

$$V_P = 2(-4)^2(3) - 5(6) = 66 \text{ V}$$

Next, we may use the gradient operation to obtain the electric field intensity,

$$\mathbf{E} = -\nabla V = -4xy\mathbf{a}_x - 2x^2\mathbf{a}_y + 5\mathbf{a}_z \text{ V/m}$$

The value of $\mathbf{E}$ at point P is

$$\mathbf{E}_P = 48\mathbf{a}_x - 32\mathbf{a}_y + 5\mathbf{a}_z \text{ V/m}$$

and

$$|\mathbf{E}_P| = \sqrt{48^2 + (-32)^2 + 5^2} = 57.9 \text{ V/m}$$

The direction of $\mathbf{E}$ at P is given by the unit vector

$$\mathbf{a}_{E,P} = (48\mathbf{a}_x - 32\mathbf{a}_y + 5\mathbf{a}_z)/57.9$$
$$= 0.829\mathbf{a}_x - 0.553\mathbf{a}_y + 0.086\mathbf{a}_z$$

If we assume these fields exist in free space, then

$$D = \epsilon_0\mathbf{E} = -35.4xy\mathbf{a}_x - 17.71x^2\mathbf{a}_y + 44.3\mathbf{a}_z \text{ pC/m}^3$$

Finally, we may use the divergence relationship to find the volume charge density that is the source of the given potential field,

$$\rho_v = \nabla \cdot \mathbf{D} = -35.4y \text{ pC/m}^3$$

At P, $\rho_\nu = -106.2$ pC/m^3.

D4.7 A portion of a two-dimensional ($E_z = 0$) potential field is shown in Figure 4.8. The grid lines are 1 mm apart in the actual field. Determine approximate values for $\mathbf{E}$ in rectangular coordinates at: (*a*) *a*; (*b*) *b*; (*c*) *c*.

Ans. $-1075\mathbf{a}_y$ V/m; $-600\mathbf{a}_x - 700\mathbf{a}_y$ V/m; $-500\mathbf{a}_x - 650\mathbf{a}_y$ V/m

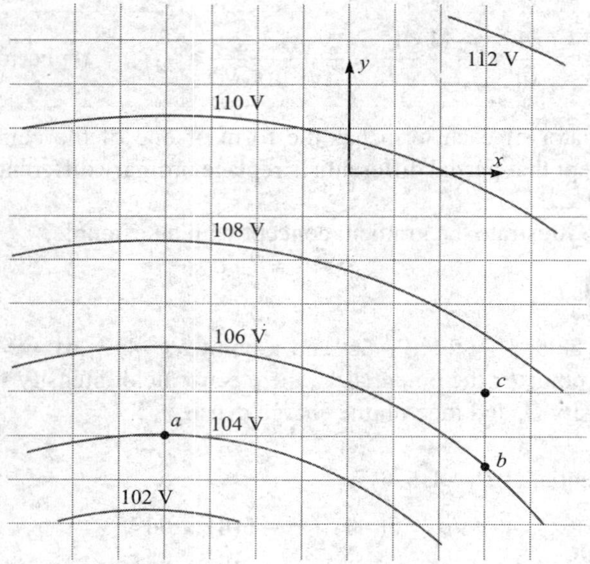

Fig. 4.8 See Problem D4.7.

D4.8 Given the potential field in cylindrical coordinates, $V = \frac{100}{z^2+1}\rho\cos\phi$ V, and point P at $\rho = 3$ m, $\phi = 60°$, $z = 2$ m, find values at P for: (*a*) V; (*b*) $\mathbf{E}$; (*c*) E; (*d*) dV/dN; (*e*) $\mathbf{a}_N$; (*f*) ρ_ν in free space.

Ans. 30.0 V; $-10.00\mathbf{a}_\rho + 17.3\mathbf{a}_\phi + 24.0\mathbf{a}_z$ V/m; 31.2 V/m; 31.2 V/m; $0.32\mathbf{a}_\rho - 0.55\mathbf{a}_\phi - 0.77\mathbf{a}_z$; -234 pC/m^3.

4.7 THE DIPOLE

The dipole fields which we shall develop in this section are quite important because they form the basis for the behavior of dielectric materials in electric fields, as discussed in part of Chapter 6, as well as justifying the use of images, as described in Section 5.5 of Chapter 6. Moreover, this development will serve to illustrate the importance of the potential concept presented in this chapter.

An *electric dipole*, or simply a *dipole*, is the name given to two point charges of equal magnitude and opposite sign, separated by a distance which is small compared to the distance to the point P at which we want to know the electric and potential fields. The dipole is shown in Figure 4.9 a. The distant point P is described by the spherical coordinates r, θ, and $\phi = 90°$, in view of the azimuthal symmetry. The positive and negative point charges have separation d and rectangular coordinates $(0, 0, \frac{1}{2}d)$and $(0, 0, -\frac{1}{2}d)$, respectively.

So much for the geometry. What would we do next? Should we find the total electric field intensity by adding the known fields of each point charge? Would it be easier to find the total potential field first? In either case, having found one, we shall find the other from it before calling the problem solved.

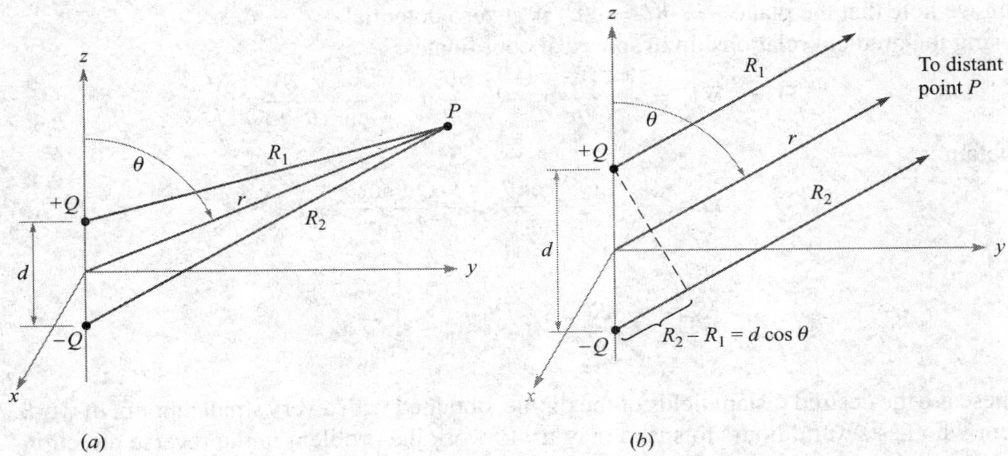

Fig. 4.9 (a) The geometry of the problem of an electric dipole. The dipole moment $p = Qd$ is in the a_z direction. (b) For a distant point P, R_1 is essentially parallel to R_2, and we find that $R_2 - R_1 = d\cos\theta$.

If we choose to find **E** first, we shall have two components to keep track of in spherical coordinates (symmetry shows E_ϕ is zero), and then the only way to find V from **E** is by use of the line integral. This last step includes establishing a suitable zero reference for potential, since the line integral gives us only the potential difference between the two points at the ends of the integral path.

On the other hand, the determination of V first is a much simpler problem. This is because we find the potential as a function of position by simply adding the scalar potentials from the two charges. The position-dependent vector magnitude and direction of **E** are subsequently evaluated with relative ease by taking the negative gradient of V.

Choosing this simpler method, we let the distances from Q and $-Q$ to P be R_1 and R_2, respectively, and write the total potential as

$$V = \frac{Q}{4\pi\epsilon_0}\left(\frac{1}{R_1} - \frac{1}{R_2}\right) = \frac{Q}{4\pi\epsilon_0}\frac{R_2 - R_1}{R_1 R_2}$$

Note that the plane $z = 0$, midway between the two point charges, is the locus of points for which $R_1 = R_2$, and is therefore at zero potential, as are all points at infinity.

For a distant point, $R_1 = R_2$, and the $R_1 R_2$ product in the denominator may be replaced by r^2. The approximation may not be made in the numerator, however, without obtaining the trivial answer that the potential field approaches zero as we go very far away from the dipole. Coming back a little closer to the dipole, we see from Figure 4.9b that $R_2 - R_1$ may be approximated very easily if R_1 and R_2 are assumed to be parallel,

$$R_2 - R_1 \doteq d\cos\theta$$

The final result is then

$$V = \frac{Qd\cos\theta}{4\pi\epsilon_0 r^2} \tag{34}$$

Again we note that the plane $z = 0(\theta = 90°)$ is at zero potential.

Using the gradient relationship in spherical coordinates,

$$\mathbf{E} = -\nabla V = -\left(\frac{\partial V}{\partial r}\mathbf{a}_r + \frac{1}{r}\frac{\partial V}{\partial \theta}\mathbf{a}_\theta + \frac{1}{r\sin\theta}\frac{\partial V}{\partial \phi}\mathbf{a}_\phi\right)$$

we obtain

$$\mathbf{E} = -\left(-\frac{Qd\cos\theta}{2\pi\epsilon_0 r^3}\mathbf{a}_r - \frac{Qd\sin\theta}{4\pi\epsilon_0 r^3}\mathbf{a}_\theta\right) \tag{35}$$

or

$$\mathbf{E} = \frac{Qd}{4\pi\epsilon_0 r^3}(2\cos\theta\,\mathbf{a}_r + \sin\theta\,\mathbf{a}_\theta) \tag{36}$$

These are the desired distant fields of the dipole, obtained with a very small amount of work. Any student who has several hours to spend may try to work the problem in the reverse direction—the authors consider the process too long and detailed to include, even for effect.

To obtain a plot of the potential field, we may choose a dipole such that $Qd/(4\pi\epsilon_0) = 1$, and then $\cos\theta = Vr^2$. The colored lines in Figure 4.10 indicate equipotentials for which $V = 0, +0.2,$

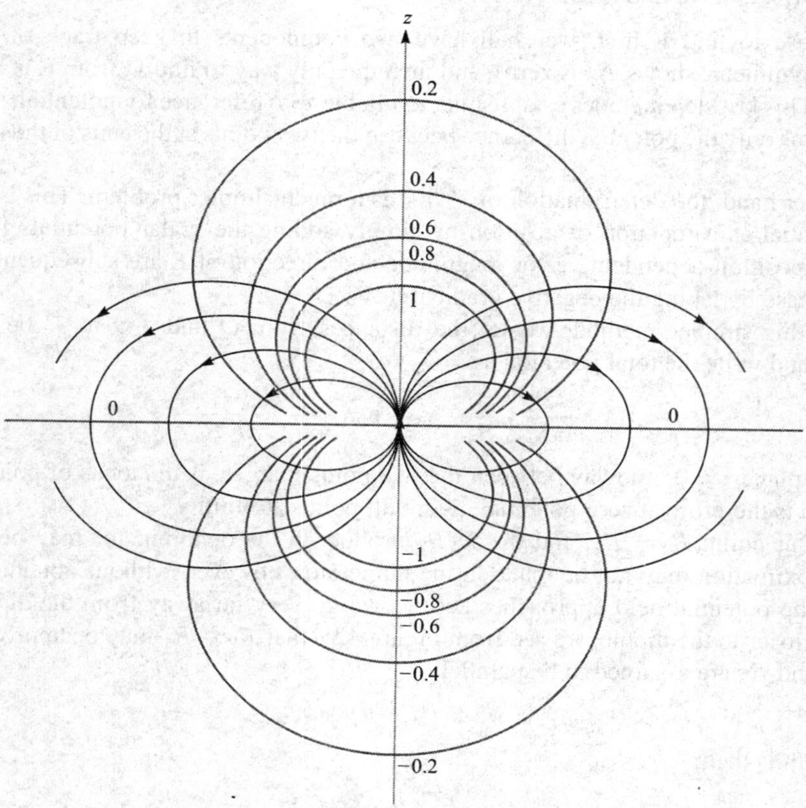

Fig. 4.10 The electrostatic field of a point dipole with its moment in the $\mathbf{a}_z$ direction. Six equipotential surfaces are labeled with relative values of V.

+ 0.4, + 0.6, + 0.8, and + 1, as indicated. The dipole axis is vertical, with the positive charge on the top. The streamlines for the electric field are obtained by applying the methods of Section 2.6 in spherical coordinates,

$$\frac{E_\theta}{E_r} = \frac{r\,d\theta}{dr} = \frac{\sin\theta}{2\cos\theta}$$

or

$$\frac{dr}{r} = 2\cot\theta\,d\theta$$

from which we obtain

$$r = C_1 \sin^2\theta$$

The black streamlines shown in Figure 4.10 are for $C_1 = 1$, 1 .5 , 2 , and 2.5.

The potential field of the dipole, Eq. (34), may be simplified by making use of the dipole moment. Let us first identify the vector length directed from $-Q$ to $+Q$ as $\mathbf{d}$ and then define the *dipole moment* as $Q\mathbf{d}$ and assign it the symbol $\mathbf{p}$. Thus

$$\mathbf{p} = Q\mathbf{d} \tag{37}$$

The units of $\mathbf{p}$ are C · m .

Since $\mathbf{d} \cdot \mathbf{a}_r = d\cos\theta$, we then have

$$V = \frac{\mathbf{p} \cdot \mathbf{a}_r}{4\pi\epsilon_0 r^2} \tag{38}$$

This result may be generalized as

$$V = \frac{1}{4\pi\epsilon_0|\mathbf{r} - \mathbf{r}'|^2}\mathbf{p} \cdot \frac{\mathbf{r} - \mathbf{r}'}{|\mathbf{r} - \mathbf{r}'|} \tag{39}$$

where $\mathbf{r}$ locates the field point P, and $\mathbf{r}'$ determines the dipole center. Equation (39) is independent of any coordinate system.

The dipole moment $\mathbf{p}$ will appear again when we discuss dielectric materials. Since it is equal to the product of the charge and the separation, neither the dipole moment nor the potential will change as Q increases and $\mathbf{d}$ decreases, provided the product remains constant. The limiting case of a *point dipole* is achieved when we let $\mathbf{d}$ approach zero and Q approach infinity such that the product $\mathbf{p}$ is finite.

Turning our attention to the resultant fields, it is interesting to note that the potential field is now proportional to the inverse *square* of the distance, and the electric field intensity is proportional to the inverse *cube* of the distance from the dipole. Each field falls off faster than the corresponding field for the point charge, but this is no more than we should expect because the opposite charges appear to be closer together at greater distances and to act more like a single point charge of 0 C.

Symmetrical arrangements of larger numbers of point charges produce fields proportional to the inverse of higher and higher powers of r. These charge distributions are called *multipoles*, and they are used ininfinite series to approximate more unwieldy charge configurations.

D4.9 An electric dipole located at the origin in free space has a moment $\mathbf{p} = 3\mathbf{a}_x - 2\mathbf{a}_y + \mathbf{a}_z$ nC·m. (*a*) Find V at $P_A(2, 3, 4)$. (*b*) Find V at $r = 2.5$, $\theta = 30°$, $\phi = 40°$.

Ans. 0.23 V; 1.97 V

D4.10 A dipole of moment $\mathbf{p} = 6\mathbf{a}_z$ nC· m is located at the origin in free space. (*a*) Find V at $P(r = 4, \theta = 20°, \phi = 0°)$. (*b*) Find $\mathbf{E}$ at P.

Ans. 3.17 V; $1.58\mathbf{a}_r + 0.29\mathbf{a}_\theta$ V/m

4.8 ENERGY DENSITY IN THE ELECTROSTATIC FIELD

We have introduced the potential concept by considering the work done, or energy expended, in moving a point charge around in an electric field, and now we must tie up the loose ends of that discussion by tracing the energy flow one step further.

Bringing a positive charge from infinity into the field of another positive charge requires work, the work being done by the external source moving the charge. Let us imagine that the external source carries the charge up to a point near the fixed charge and then holds it there. Energy must be conserved, and the energy expended in bringing this charge into position now represents potential energy, for if the external source released its hold on the charge, it would accelerate away from the fixed charge, acquiring kinetic energy of its own and the capability of doing work.

In order to find the potential energy present in a system of charges, we must find the work done by an external source in positioning the charges.

We may start by visualizing an empty universe. Bringing a charge Q_1 from infinity to any position requires no work, for there is no field present.[2] The positioning of Q_2 at a point in the field of Q_1 requires an amount of work given by the product of the charge Q_2 and the potential at that point due to Q_1. We represent this potential as $V_{2,1}$, where the first subscript indicates the location and the second subscript the source. That is, $V_{2,1}$ is the potential at the location of Q_2 due to Q_1. Then

$$\text{Work to position } Q_2 = Q_2 V_{2,1}$$

Similarly, we may express the work required to position each additional charge in the field of all those already present:

$$\text{Work to position } Q_3 = Q_3 V_{3,1} + Q_3 V_{3,2}$$

$$\text{Work to position } Q_4 = Q_4 V_{4,1} + Q_4 V_{4,2} + Q_4 V_{4,3}$$

and so forth. The total work is obtained by adding each contribution:

$$\text{Total positioning work} = \text{potential energy of field}$$

$$= W_E = Q_2 V_{2,1} + Q_3 V_{3,1} + Q_3 V_{3,2} + Q_4 V_{4,1} + Q_4 V_{4,2} + Q_4 V_{4,3} + \cdots \tag{40}$$

Noting the form of a representative term in the preceding equation,

$$Q_3 V_{3,1} = Q_3 \frac{Q_1}{4\pi\epsilon_0 R_{13}} = Q_1 \frac{Q_3}{4\pi\epsilon_0 R_{31}}$$

where R_{13} and R_{31} each represent the scalar distance between Q_1 and Q_3, we see that it might equally well have been written as $Q_1 V_{1,3}$. If each term of the total energy expression is replaced by

[2] However, somebody in the workshop at infinity had to do an infinite amount of work to create the point charge in the first place! How much energy is required to bring two half-charges into coincidence to make a unit charge?

its equal, we have

$$W_E = Q_1V_{1,2} + Q_1V_{1,3} + Q_2V_{2,3} + Q_1V_{1,4} + Q_2V_{2,4} + Q_3V_{3,4} + \cdots \tag{41}$$

Adding the two energy expressions (40) and (41) gives us a chance to simplify the result a little:

$$\begin{aligned} 2W_E = Q_1(&V_{1,2} + V_{1,3} + V_{1,4} + \cdots) \\ &+Q_2(V_{2,1} + V_{2,3} + V_{2,4} + \cdots) \\ &+Q_3(V_{3,1} + V_{3,2} + V_{3,4} + \cdots) \\ &+\cdots \end{aligned}$$

Each sum øf potentials in parentheses is the combined potential due to all the charges except for the charge at the point where this combined potential is being found. In other words,

$$V_{1,2} + V_{1,3} + V_{1,4} + \cdots = V_1$$

the potential at the location of Q_1 due to the presence of $Q_2, Q_3, \ldots$. We therefore have

$$W_E = \frac{1}{2}(Q_1V_1 + Q_2V_2 + Q_3V_3 + \cdots) = \frac{1}{2}\sum_{m=1}^{m=N} Q_m V_m \tag{42}$$

In order to obtain an expression for the energy stored in a region of continuous charge distribution, each charge is replaced by $\rho_\nu\, dv$, and the summation becomes an integral,

$$W_E = \frac{1}{2}\int_{\text{vol}} \rho_\nu V\, dv \tag{43}$$

Equations (42) and (43) allow us to find the total potential energy present in a system of point charges or distributed volume charge density. Similar expressions may be easily written in terms of line or surface charge density. Usually we prefer to use (43) and let it represent all the various types of charge which may have to be considered. This may always be done by considering point charges, line charge density, or surface charge density to be continuous distributions of volume charge density over very small regions. We shall illustrate such a procedure with an example shortly.

Before we undertake any interpretation of this result, we should consider a few lines of more difficult vector analysis and obtain an expression equivalent to (43) but written in terms of **E** and **D**.

We begin by making the expression a little bit longer. Using Maxwell's first equation, replace ρ_ν by its equal $\nabla \cdot \mathbf{D}$ and make use of a vector identity which is true for any scalar function V and any vector function **D**,

$$\nabla \cdot (V\mathbf{D}) \equiv V(\nabla \cdot \mathbf{D}) + \mathbf{D} \cdot (\nabla V) \tag{44}$$

This may be proved readily by expansion in rectangular coordinates. We then have, successively,

$$W_E = \frac{1}{2}\int_{\text{vol}} \rho_\nu V\, dv = \frac{1}{2}\int_{\text{vol}} (\nabla \cdot \mathbf{D})V\, dv$$

$$= \frac{1}{2}\int_{\text{vol}} [\nabla \cdot (V\mathbf{D}) - \mathbf{D} \cdot (\nabla V)\, dv$$

Using the divergence theorem from Chapter 3, the first volume integral of the last equation is changed into a closed surface integral, where the closed surface surrounds the volume considered.

This volume, first appearing in(43), must contain *every* charge, and there can then be no charges outside of the volume. We may therefore consider the volume as *infinite* in extent if we wish. We have

$$W_E = \frac{1}{2}\oint_S (V\mathbf{D}) \cdot d\mathbf{S} - \frac{1}{2}\int_{\text{vol}} \mathbf{D} \cdot (\nabla V)\, dv$$

The surface integral is equal to zero, for over this closed surface surrounding the universe we see that V is approaching zero at least as rapidly as $1/r$ (the charges look like point charges from there), and $\mathbf{D}$ is approaching zero at least as rapidly as $1/r^2$. The integrand therefore approaches zero at least as rapidly as $1/r^3$, while the differential element of surface, looking more and more like a portion of a sphere, is increasing only as r^2. Consequently, in the limit as $r \to \infty$, the integrand and the integral both approach zero. Substituting $\mathbf{E} = -\nabla V$ in the remaining volume integral, we have our answer,

$$W_E = \frac{1}{2}\int_{\text{vol}} \mathbf{D} \cdot \mathbf{E}\, dv = \frac{1}{2}\int_{\text{vol}} \epsilon_0 E^2\, dv \tag{45}$$

Let us now use this last expression to calculate the energy stored in the electrostatic field of a section of a coaxial cable or capacitor of length L. We found in Section 3.3 that

$$D_\rho = \frac{a\rho_S}{\rho}$$

Hence,

$$\mathbf{E} = \frac{a\rho_S}{\epsilon_0 \rho}\mathbf{a}_\rho$$

where ρ_S is the surface charge density on the inner conductor, whose radius is a. Thus,

$$W_E = \frac{1}{2}\int_0^L \int_0^{2\pi} \int_a^b \epsilon_0 \frac{a^2 \rho_S^2}{\epsilon_0^2 \rho^2}\rho\, d\rho\, d\phi\, dz = \frac{\pi L a^2 \rho_S^2}{\epsilon_0}\ln\frac{b}{a}$$

This same result may be obtained from (43). We choose the outer conductor as our zero-potential reference, and the potential of the inner cylinder is then

$$V_a = -\int_b^a E_\rho d\rho = -\int_b^a \frac{a\rho_S}{\epsilon_0 \rho} d\rho = \frac{a\rho_S}{\epsilon_0}\ln\frac{b}{a}$$

The surface charge density ρ_S at $\rho = a$ can be interpreted as a volume charge density $\rho_\nu = \rho_S/t$, extending from $\rho = a - \frac{1}{2}t$ to $\rho = a + \frac{1}{2}t$, where $t << a$. The integrand in (43) is therefore zero everywhere between the cylinders (where the volume charge density is zero), as well as at the outer cylinder (where the potential is zero). The integration is therefore performed only within the thin cylindrical shell at $\rho = a$,

$$W_E = \frac{1}{2}\int_{\text{vol}} \rho_\nu V\, dV = \frac{1}{2}\int_0^L \int_0^{2\pi} \int_{a-t/2}^{a+t/2} \frac{\rho_S}{t} a \frac{\rho_S}{\epsilon_0}\ln\frac{b}{a}\rho\, d\rho\, d\phi\, dz$$

from which

$$W_E = \frac{a^2 \rho_S^2 \ln(b/a)}{\epsilon_0}\pi L$$

once again.

This expression takes on a more familiar form if we recognize the total charge on the inner conductor as $Q = 2\pi a L \rho_S$. Combining this with the potential difference between the cylinders, V_a, we see that

$$W_E = \frac{1}{2} Q V_a$$

which should be familiar as the energy stored in a capacitor.

The question of where the energy is stored in an electric field has not yet been answered. Potential energy can never be pinned down precisely in terms of physical location. Someone lifts a pencil, and the pencil acquires potential energy. Is the energy stored in the molecules of the pencil, in the gravitational field between the pencil and the earth, or in some obscure place? Is the energy in a capacitor stored in the charges themselves, in the field, or where? No one can offer any proof for his or her own private opinion, and the matter of deciding may be left to the philosophers.

Electromagnetic field theory makes it easy to believe that the energy of an electric field or a charge distribution is stored in the field itself, for if we take (45), an exact and rigorously correct expression,

$$W_E = \frac{1}{2} \int_{\text{vol}} \mathbf{D} \cdot \mathbf{E} \, dv$$

and write it on a differential basis,

$$dW_E = \frac{1}{2} \mathbf{D} \cdot \mathbf{E} \, dv$$

or

$$\frac{dW_E}{dv} = \frac{1}{2} \mathbf{D} \cdot \mathbf{E} \tag{46}$$

we obtain a quantity $\frac{1}{2}\mathbf{D} \cdot \mathbf{E}$, which has the dimensions of an energy density, or joules per cubic meter. We know that if we integrate this energy density over the entire field-containing volume, the result is truly the total energy present, but we have no more justification for saying that the energy stored in each differential volume element dv is $\mathbf{D} \cdot \mathbf{E} \, dv$ than we have for looking at (43) and saying that the stored energy is $\frac{1}{2}\rho_\nu V dv$. The interpretation afforded by (46), however, is a convenient one, and we shall use it until proved wrong.

D4.11 Find the energy stored in free space for the region 2 mm $< r <$ 3 mm, $0 < \theta < 90°$, $0 < \phi < 90°$, given the potential field $V =$: (a) $\dfrac{200}{r}$ V; (b) $\dfrac{300 \cos\theta}{r^2}$ V.

Ans. 46.4 μJ; 36.7 J

References

1. Attwood, S. S. *Electric and Magnetic Fields*. 3d ed. New York: John Wiley & Sons, 1949. There are a large number of well-drawn field maps of various charge distributions, including the dipole field. Vector analysis is not used.
2. Skilling, H. H. (See Suggested References for Chapter 3.) Gradient is described on pp. 19–21.
3. Thomas, G. B., Jr., and R. L. Finney. (See Suggested References for Chapter 1.) The directional derivative and the gradient are presented on pp. 823–30.

CHAPTER 4 Problems

Quizzes

4.1 The value of **E** at $P(\rho = 2,\ \phi = 40°,\ z = 3)$ is given as $\mathbf{E} = 100\mathbf{a}_\rho - 200\mathbf{a}_\phi + 300\mathbf{a}_z$ V/m. Determine the incremental work required to move a 20 μC charge a distance of 6 μm: (*a*) in the direction of $\mathbf{a}_\rho$; (*b*) in the direction of $\mathbf{a}_\phi$; (*c*) in the direction of $\mathbf{a}_z$; (*d*) in the direction of **E**; (*e*) in the direction of $\mathbf{G} = 2\mathbf{a}_x \cdot 3\mathbf{a}_y + 4\mathbf{a}_z$.

4.2 An electric field is given as $\mathbf{E} = -10e^y(\sin 2z\mathbf{a}_x + x\sin 2z\mathbf{a}_y + 2x\cos 2z\mathbf{a}_z)$ V/m. (*a*) Find **E** at $P(5, 0, \pi/12)$. (*b*) How much work is done in moving a charge of 2 nC an incremental distance of 1 mm from P in the direction of $\mathbf{a}_x$? (*c*) of $\mathbf{a}_y$? (*d*) of $\mathbf{a}_z$? (*e*) of $(\mathbf{a}_x + \mathbf{a}_y + \mathbf{a}_z)$?

4.3 If $\mathbf{E} = 120\mathbf{a}_\rho$V/m, find the incremental amount of work done in moving a 50-μC charge a distance of 2 mm from: (*a*) $P(1, 2, 3)$ toward $Q(2, 1, 4)$; (*b*) $Q(2, 1, 4)$ toward $P(1, 2, 3)$.

4.4 It is found that the energy expended in carrying a charge of 4 μC from the origin to $(x, 0, 0)$ along the x axis is directly proportional to the square of the path length. If $E_x = 7$ V/m at (1, 0, 0), determine E_x on the x axis as a function of x.

4.5 Compute the value of $\int_A^P \mathbf{G} \cdot d\mathbf{L}$ for $\mathbf{G} = 2y\mathbf{a}_x$ with $A(1 - 1, 2)$ and $P(2, 1, 2)$ using the path: (*a*) straight-line segments $A(1, -1, 2)$ to $B(1, 1, 2)$ to $P(2, 1, 2)$; (*b*) straight-line segments $A(1, -1, 2)$ to $C(2, -1, 2)$ to $P(2, 1, 2)$.

4.6 Determine the work done in carrying a 2-μC charge from (2, 1, −1) to (8, 2, −1) in the field $\mathbf{E} = y\mathbf{a}_x + x\mathbf{a}_y$ along (*a*) the parabola $x = 2y^2$, (*b*) the hyperbola $x = 8/(7.3y)$; (*c*) the straight line $x = 6y - 4$.

4.7 Let $\mathbf{G} = 3xy^2\mathbf{a}_x + 2z\mathbf{a}_y$ Given an initial point $P(2, 1, 1)$ and a final point $Q(4, 3, 1)$ find $\int \mathbf{G} \cdot d\mathbf{L}$ using the path: (*a*) straight line: $y = x - 1$, $z = 1$; (*b*) parabola: $6y = x^2 + 2$, $z = 1$.

4.8 Given $\mathbf{E} = -x\mathbf{a}_x + y\mathbf{a}_y$, find the work involved in moving a unit positive charge on a circular arc, the circle centered at the origin, from $x = a$ to $x = y = a/\sqrt{2}$.

4.9 A uniform surface charge density of 20 nC/m^2 is present on the spherical surface $r = 06$ cm in free space. (*a*) Find the absolute potential at $P(r = 1$ cm, $\theta = 25°, \phi = 50°)$. (*b*) Find$V_{AB}$, given points $A(r = 2$ cm, $\theta = 30°$, $\phi = 60°)$ and $B(r = 3$ cm, $\theta = 45°$, $\phi = 90°)$.

4.10 Express the potential field of an infinite line charge (*a*) with zero reference at $\rho = \rho_0$; (*b*) with $V = V_0$ at $\rho = \rho_0$. (*c*) Can the zero reference be placed at infinity? Why?

4.11 Let a uniform surface charge density of 5 nC/m^2 be present at the $z = 0$ plane, a uniform line charge density of 8 nC/m be located at $x = 0$, $z = 4$, and a point charge of 2μC be present at $P(2, 0, 0)$. If $V = 0$ at $M(0, 0, 5)$, find V at $N(1, 2, 3)$.

4.12 In spherical coordinates, $\mathbf{E} = 2r/(r^2 + a^2)^2\mathbf{a}_r$ V/m. Find the potential at any point, using the reference (*a*)$V = 0$ at infinity; (*b*) $V = 0$ at $r = 0$; (*c*) $V = 100$ V at $r = a$.

4.13 Three identical point charges of 4 pC each are located at the corners of an equilateral triangle 0.5 mm on a side in free space. How much work must be done to move one charge to a point equidistant from the other two and on the line joining them?

4.14 Given the electric field $\mathbf{E} = (y + 1)\mathbf{a}_x + (x - 1)\mathbf{a}_y + 2\mathbf{a}_z$ find the potential difference between the points (*a*) (2, −2, −1) and (0, 0, 0); (*b*) (3, 2, -1) and (−2, −3, 4).

4.15 Two uniform line charges, 8 nC/m each, are located at $x = 1$, $z = 2$, and at $x = -1$, $y = 2$ in free space. If the potential at the origin is 100 V, find V at $P(4, 1, 3)$.

4.16 The potential at any point in space is given in cylindrical coordinates by $V = (k/\rho^2)\cos(b\phi)$ V/m, where k and b are constants. (a) Where is the zero reference for potential? (b) Find the vector electric field intensity at any point (ρ, ϕ, z).

4.17 Uniform surface charge densities of 6 and 2 nC/m^2 are present at $\rho = 2$ and 6 cm, respectively, in free space. Assume $V = 0$ at $\rho = 4$ cm, and calculate V at: (a) $\rho = 5$ cm; (b) $\rho = 7$ cm.

4.18 Find the potential at the origin produced by a line charge $\rho_L = kx/(x^2 + a^2)$ extending along the x axis from $x = a$ to $+\infty$, where $a > 0$. Assume a zero reference at infinity.

4.19 The annular surface 1 cm $< \rho <$ 3 cm, $z = 0$, carries the nonuniform surface charge density $\rho_s = 5\rho$ nC/m^2. Find V at $P(0, 0, 2$ cm) if $V = 0$ at infinity.

4.20 A point charge Q is located at the origin. Express the potential in both rectangular and cylindrical coordinates, and use the gradient operation in that coordinate system to find the electric field intensity. The result may be checked by conversion to spherical coordinates.

4.21 Let $V = 2xy^2z^3 + 3\ln(x^2 + 2y^2 + 3z^2)$ V in free space. Evaluate each of the following quantities at $P(3, 2, -1)$: (a) V; (b) $|V|$; (c) $\mathbf{E}$; (d) $|\mathbf{E}|$; (e) $\mathbf{a}_N$; (f) $\mathbf{D}$.

4.22 A certain potential field is given in spherical coordinates by $V = V_0(r/a)\sin\theta$. Find the total charge contained within the region $r < a$.

4.23 It is known that the potential is given as $V = 80\rho^{0.6}$ V. Assuming free space conditions, find: (a) $\mathbf{E}$; (b) the volume charge density at $\rho = 0.5$ m; (c) the total charge lying within the closed surface $\rho = 0.6$, $0 < z < 1$.

4.24 The surface defined by the equation $x^3 + y^2 + z = 1000$, where x, y, and z are positive, is an equipotential surface on which the potential is 200 V. If $|\mathbf{E}| = 50$ V/m at the point $P(7, 25, 32)$ on the surface, find $\mathbf{E}$ there.

4.25 Within the cylinder $\rho = 2$, $0 < z < 1$, the potential is given by $V = 100 + 50\rho + 150\rho\sin\phi$ V. (a) Find V, $\mathbf{E}$, $\mathbf{D}$, and ρ_ν at $P(1, 60°, 05)$ in free space. (b) How much charge lies within the cylinder?

4.26 Let us assume that we have a very thin, square, imperfectly conducting plate 2 m on a side, located in the plane $z = 0$ with one corner at the origin such that it lies entirely within the first quadrant. The potential at any point in the plate is given as $V = -e^{-x}\sin y$. (a) An electron enters the plate at $x = 0$, $y = \pi/3$ with zero initial velocity; in what direction is its initial movement? (b) Because of collisions with the particles in the plate, the electron achieves a relatively low velocity and little acceleration (the work that the field does on it is converted largely into heat). The electron therefore moves approximately along a streamline. Where does it leave the plate and in what direction is it moving at the time?

4.27 Two point charges, 1 nC at (0, 0, 0.1) and -1 nC at (0, 0, -0.1), are in free space. (a) Calculate V at $P(0.3, 0, 0.4)$, (b) Calculate $|\mathbf{E}|$ at P. (c) Now treat the two charges as a dipole at the origin and find V at P.

4.28 Use the electric field intensity of the dipole [Section 4.7, Eq. (36)] to find the difference in potential between points at θ_a and θ_b, each point having the same r and ϕ coordinates. Under what conditions does the answer agree with Eq. (34), for the potential at θ_a?

4.29 A dipole having a moment $\mathbf{p} = 3\mathbf{a}_x - 5\mathbf{a}_y + 10\mathbf{a}_z$ nC· m is located at $Q(1, 2 - 4)$ in free space. Find V at $P(2, 3, 4)$.

4.30 A dipole for which $\mathbf{p} = 10\epsilon_0\mathbf{a}_z$ C· m is located at the origin. What is the equation of the surface on which $E_z = 0$ but $\mathbf{E} \neq 0$?

4.31 A potential field in free space is expressed as $V = 20/(xyz)$ V. (*a*) Find the total energy stored within the cube $1 < x, y, z < 2$. (*b*) What value would be obtained by assuming a uniform energy density equal to the value at the center of the cube?

4.32 (*a*) Using Eq. (36), find the energy stored in the dipole field in the region $r > a$. (*b*) Why can we not let a approach zero as a limit?

4.33 A copper sphere of radius 4 cm carries a uniformly distributed total charge of 5 μC in free space. (*a*) Use Gauss's law to find **D** external to the sphere. (*b*) Calculate the total energy stored in the electrostatic field. (*c*) Use $W_E = Q^2/(2C)$ to calculate the capacitance of the isolated sphere.

4.34 A sphere of radius a contains volume charge of uniform density ρ_0 C/m^3. Find the total stored energy by applying (*a*) Eq. (43); (*b*) Eq. (45).

4.35 Four 0.8 nC point charges are located in free space at the corners of a square 4 cm on a side. (*a*) Find the total potential energy stored. (*b*) A fifth 0.8 nC charge is installed at the center of the square. Again find the total stored energy.

5

Current and Conductors

In this chapter and in Chapter 6, we will apply the laws and methods that we have learned to some of the materials with which an engineer must work. After defining current and current density and developing the fundamental continuity equation, we will consider a conducting material and present Ohm's law in both its microscopic and macroscopic forms. With these results we may calculate resistance values for a few of the simpler geometric forms that resistors may assume. Conditions which must be met at conductor boundaries are next obtained, and this knowledge enables us to introduce the use of images. We conclude the chapter by considering the properties of a general semiconductor.

In Chapter 6, we will investigate the polarization of dielectric materials and define relative permittivity, or the dielectric constant—an important engineering parameter. Having both conductors and dielectrics, we may then put them together to form capacitors. Most of the knowledge from the previous chapters will be required to determine the capacitance of the structures that will be considered.

The fundamental electromagnetic principles on which resistors and capacitors depend form the subject of this chapter and Chapter 6; the inductor will be introduced in Chapter 9.

5.1 CURRENT AND CURRENT DENSITY

Electric charges in motion constitute a *current*. The unit of current is the ampere (A), defined as a rate of movement of charge passing a given reference point (or crossing a given reference plane) of one coulomb per second. Current is symbolized by I, and therefore

$$I = \frac{dQ}{dt} \tag{1}$$

Current is thus defined as the motion of positive charges, even though conduction in metals takes place through the motion of electrons, as we shall see shortly.

In field theory we are usually interested in events occurring at a point rather than within some large region, and we shall find the concept of *current density*, measured in amperes per square meter (A/m^2), more useful. Current density is a vector[1] represented by **J**.

The increment of current ΔI crossing an incremental surface ΔS normal to the current density is

$$\Delta I = J_N \Delta S$$

[1] Current is not a vector, for it is easy to visualize a problem in which a total current I in a conductor of nonuniform cross section (such as a sphere) may have a different direction at each point of a given cross section. Current in an exceedingly fine wire, or a *filamentary current,* is occasionally defined as a vector, but we usually prefer to be consistent and give the direction to the filament, or path, and not to the current.

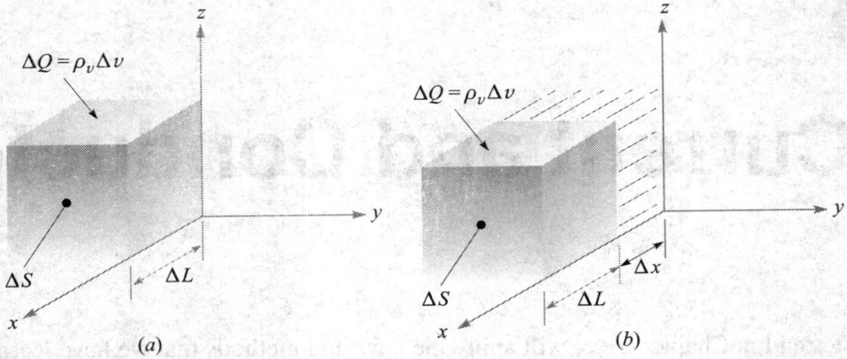

Fig. 5.1 An increment of charge, $\Delta Q = \rho_\nu \Delta S \Delta L$, which moves a distance Δx in a time Δt, produces a component of current density in the limit of $J_x = \rho_\nu \nu_x$.

and in the case where the current density is not perpendicular to the surface,

$$\Delta I = \mathbf{J} \cdot \Delta \mathbf{S}$$

Total current is obtained by integrating,

$$I = \int_S \mathbf{J} \cdot d\mathbf{S} \tag{2}$$

Current density may be related to the velocity of volume charge density at a point. Consider the element of charge $\Delta Q = \rho_\nu \Delta \nu = \rho_\nu \Delta S \Delta L$, as shown in Figure 5.1*a*. To simplify the explanation, let us assume that the charge element is oriented with its edges parallel to the coordinate axes and that it possesses only an x component of velocity. In the time interval Δt, the element of charge has moved a distance Δx, as indicated in Figure 5.1*b*.We have therefore moved a charge $\Delta Q = \rho_\nu \Delta S \Delta x$ through a reference plane perpendicular to the direction of motion in a time increment Δt, and the resultant current is

$$\Delta I = \frac{\Delta Q}{\Delta t} = \rho_\nu \Delta S \frac{\Delta x}{\Delta t}$$

As we take the limit with respect to time, we have

$$\Delta I = \rho_\nu \Delta S \nu_x$$

where ν_x represents the x component of the velocity $\mathbf{v}$.[2] In terms of current density, we find

$$J_x = \rho_\nu \nu_x$$

and in general

$$J = \rho_\nu \mathbf{v} \tag{3}$$

This last result shows very clearly that charge in motion constitutes a current. We call this type of current a *convection current*, and $\mathbf{J}$ or $\rho_\nu \mathbf{v}$ is the *convection current density*. Note that the convection current density is related linearly to charge density as well as to velocity. The mass rate of flow of

[2] The lowercase ν is used both for volume and velocity. Note, however, that velocity always appears as a vector $\mathbf{v}$, a component ν_x, or a magnitude $|\mathbf{v}|$, while volume appears only in differential form as $d\nu$ or $\Delta \nu$.

cars (cars per square foot per second) in the Holland Tunnel could be increased either by raising the density of cars per cubic foot, or by going to higher speeds, if the drivers were capable of doing so.

D5.1 Given the vector current density $\mathbf{J} = 10\,\rho^2 z\mathbf{a}_\rho - 4\rho\cos^2\phi\mathbf{a}_\phi$ mA/m^2: (*a*) find the current density at $P(\rho = 3,\ \phi = 30°,\ z = 2)$; (*b*) determine the total current flowing outward through the circular band $\rho = 3,\ 0 < \phi < 2\pi,\ 2 < z < 2.8$.

Ans. $180\mathbf{a}_\rho - 9\mathbf{a}_\phi$ mA/m^2; 3.26 A

5.2 CONTINUITY OF CURRENT

Although we are supposed to be studying static fields at this time, the introduction of the concept of current is logically followed by a discussion of the conservation of charge and the continuity equation. The principle of conservation of charge states simply that charges can be neither created nor destroyed, although equal amounts of positive and negative charge may be *simultaneously* created, obtained by separation, destroyed, or lost by recombination.

The continuity equation follows from this principle when we consider any region bounded by a closed surface. The current through the closed surface is

$$I = \oint_S \mathbf{J} \cdot d\mathbf{S}$$

and this *outward flow* of positive charge must be balanced by a decrease of positive charge (or perhaps an increase of negative charge) within the closed surface. If the charge inside the closed surface is denoted by Q_i, then the rate of decrease is $-dQ_i/dt$ and the principle of conservation of charge requires

$$I = \oint_S \mathbf{J} \cdot d\mathbf{S} = -\frac{dQ_i}{dt} \qquad (4)$$

It might be well to answer here an often-asked question. "Isn't there a sign error? I thought $I = dQ/dt$." The presence or absence of a negative sign depends on what current and charge we consider. In circuit theory we usually associate the current flow *into* one terminal of a capacitor with the time rate of increase of charge on that plate. The current of (4), however, is an *outward-flowing* current.

Equation (4) is the integral form of the continuity equation, and the differential, or point, form is obtained by using the divergence theorem to change the surface integral into a volume integral:

$$\int_s \mathbf{J} \cdot d\mathbf{S} = \int_{\text{vol}} (\nabla \cdot \mathbf{J})\,d\nu$$

We next represent the enclosed charge Q_i by the volume integral of the charge density,

$$\int_{\text{vol}} (\nabla \cdot \mathbf{J})\,d\nu = -\frac{d}{dt}\int_{\text{vol}} \rho_\nu\,d\nu$$

If we agree to keep the surface constant, the derivative becomes a partial derivative and may appear within the integral,

$$\int_{\text{vol}} (\nabla \cdot \mathbf{J})\,d\nu = \int_{\text{vol}} -\frac{\partial \rho_\nu}{\partial t}\,d\nu$$

Since the expression is true for any volume, however small, it is true for an incremental volume,

$$(\nabla \cdot \mathbf{J})\Delta\nu = -\frac{\partial \rho_\nu}{\partial t}\Delta\nu$$

from which we have our point form of the continuity equation,

$$(\nabla\cdot\mathbf{J}) = -\frac{\partial \rho_\nu}{\partial t} \tag{5}$$

Remembering the physical interpretation of divergence, this equation indicates that the current, or charge per second, diverging from a small volume per unit volume is equal to the time rate of decrease of charge per unit volume at every point.

As a numerical example illustrating some of the concepts from the last two sections, let us consider a current density that is directed radially outward and decreases exponentially with time,

$$\mathbf{J} = \frac{1}{r}e^{-t}\mathbf{a}_r \ \text{A/m}^2$$

Selecting an instant of time $t = 1$ s, we may calculate the total outward current at $r = 5$ m:

$$I = J_r S = \left(\tfrac{1}{5}e^{-1}\right)(4\pi 5^2) = 23.1 \text{ A}$$

At the same instant, but for a slightly larger radius, $r = 6$ m, we have

$$I = J_r S = \left(\tfrac{1}{6}e^{-1}\right)(4\pi 6^2) = 27.7 \text{ A}$$

Thus, the total current is larger at $r = 6$ than it is at $r = 5$.

To see why this happens, we need to look at the volume charge density and the velocity. We use the continuity equation first:

$$-\frac{\partial \rho_\nu}{\partial t} = \nabla\cdot\mathbf{J} = \nabla\cdot\left(\frac{1}{r}e^{-t}\mathbf{a}_r\right) = \frac{1}{r^2}\frac{\partial}{\partial r}\left(r^2\frac{1}{r}e^{-t}\right) = \frac{1}{r^2}e^{-t}$$

We next seek the volume charge density by integrating with respect to t. Since ρ_ν is given by a partial derivative with respect to time, the "constant" of integration may be a function of r:

$$\rho_\nu = -\int \frac{1}{r^2}e^{-t}dt + K(r) = \frac{1}{r^2}e^{-t} + K(r)$$

If we assume that $\rho_\nu \to 0$ as $t \to \infty$, then $K(r) = 0$, and

$$\rho_\nu = \frac{1}{r^2}e^{-t} \ \text{C/m}^3$$

We may now use $\mathbf{J} = \rho_\nu \, \mathbf{v}$ to find the velocity,

$$\nu_r = \frac{J_r}{\rho_\nu} = \frac{\frac{1}{r}e^{-t}}{\frac{1}{r^2}e^{-t}} = r \text{ m/s}$$

The velocity is greater at $r = 6$ than it is at $r = 5$, and we see that some (unspecified) force is accelerating the charge density in an outward direction.

In summary, we have a current density that is inversely proportional to r, a charge density that is inversely proportional to r^2, and a velocity and total current that are proportional to r. All quantities vary as e^{-t}.

D5.2 Current density is given in cylindrical coordinates as $\mathbf{J} = -10^6 z^{1.5} \mathbf{a}_z$ A/m^2 in the region $0 \leq \rho \leq 20\mu$ m; for $\rho \geq 20\mu$ m, $\mathbf{J} = 0$. (*a*) Find the total current crossing the surface $z = 0.1$ m in the $\mathbf{a}_z$ direction. (*b*) If the charge velocity is 2×10^6 m/s at $z = 0.1$ m, find ρ_ν there. (*c*) If the volume charge density at $z = 0.15$ m is -2000 C/m^3, find the charge velocity there.

Ans. -39.7μA; -15.8 mC/m^3; 29.0 m/s

5.3 METALLIC CONDUCTORS

Physicists today describe the behavior of the electrons surrounding the positive atomic nucleus in terms of the total energy of the electron with respect to a zero reference level for an electron at an infinite distance from the nucleus. The total energy is the sum of the kinetic and potential energies, and since energy must be given to an electron to pull it away from the nucleus, the energy of every electron in the atom is a negative quantity. Even though the picture has some limitations, it is convenient to associate these energy values with orbits surrounding the nucleus, the more negative energies corresponding to orbits of smaller radius. According to the quantum theory, only certain discrete energy levels, or energy states, are permissible in a given atom, and an electron must therefore absorb or emit discrete amounts of energy, or quanta, in passing from one level to another. A normal atom at absolute zero temperature has an electron occupying every one of the lower energy shells, starting outward from the nucleus and continuing until the supply of electrons is exhausted.

In a crystalline solid, such as a metal or a diamond, atoms are packed closely together, many more electrons are present, and many more permissible energy levels are available because of the interaction forces between adjacent atoms. We find that the energies which may be possessed by electrons are grouped into broad ranges, or "bands," each band consisting of very numerous, closely spaced, discrete levels. At a temperature of absolute zero, the normal solid also has every level occupied, starting with the lowest and proceeding in order until all the electrons are located. The electrons with the highest (least negative) energy levels, the valence electrons, are located in the *valence band*. If there are permissible higher-energy levels in the valence band, or if the valence band merges smoothly into a *conduction band*, then additional kinetic energy may be given to the valence electrons by an external field, resulting in an electron flow. The solid is called a *metallic conductor*. The filled valence band and the unfilled conduction band for a conductor at absolute zero temperature are suggested by the sketch in Figure 5.2*a*.

If, however, the electron with the greatest energy occupies the top level in the valence band and a gap exists between the valence band and the conduction band, then the electron cannot accept additional energy in small amounts, and the material is an insulator. This band structure is indicated in Figure 5.2*b*. Note that if a relatively large amount of energy can be transferred to the electron, it may be sufficiently excited to jump the gap into the next band where conduction can occur easily. Here the insulator breaks down.

An intermediate condition occurs when only a small "forbidden region" separates the two bands, as illustrated by Figure 5.2*c*. Small amounts of energy in the form of heat, light, or an electric field may raise the energy of the electrons at the top of the filled band and provide the basis for conduction. These materials are insulators which display many of the properties of conductors and are called *semiconductors*.

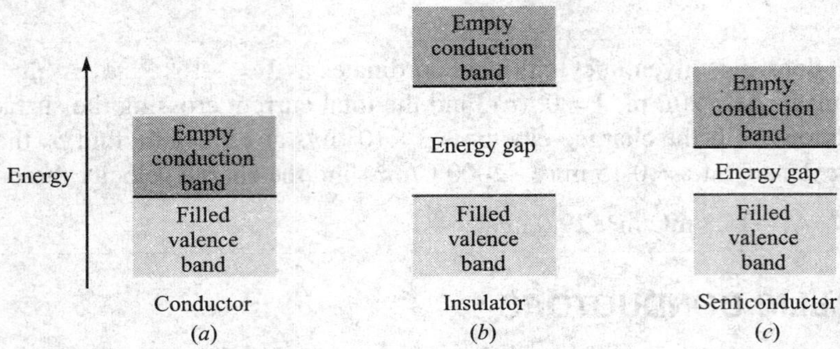

Fig. 5.2 The energy-band structure in three different types of materials at 0 K. (*a*) The conductor exhibits no energy gap between the valence and conduction bands. (*b*) The insulator shows a large energy gap. (*c*) The semiconductor has only a small energy gap.

Let us first consider the conductor. Here the valence electrons, or *conduction,* or *free,* electrons, move under the influence of an electric field. With a field **E**, an electron having a charge $Q = -e$ will experience a force

$$\mathbf{F} = -e\mathbf{E}$$

In free space the electron would accelerate and continuously increase its velocity (and energy); in the crystalline material the progress of the electron is impeded by continual collisions with the thermally excited crystalline lattice structure, and a constant average velocity is soon attained. This velocity $\mathbf{v}_d$ is termed the *drift velocity*, and it is linearly related to the electric field intensity by the *mobility* of the electron in the given material. We designate mobility by the symbol μ (mu), so that

$$\mathbf{v}_d = -\mu_e \mathbf{E} \tag{6}$$

where μ_e is the mobility of an electron and is positive by definition. Note that the electron velocity is in a direction opposite to the direction of **E**. Equation (6) also shows that mobility is measured in the units of square meters per volt-second; typical values[3] are 0.0012 for aluminum, 0.0032 for copper, and 0.0056 for silver.

For these good conductors a drift velocity of a few inches per second is sufficient to produce a noticeable temperature rise and can cause the wire to melt if the heat cannot be quickly removed by thermal conduction or radiation.

Substituting (6) into Eq. (3) of Section 5.1, we obtain

$$\boxed{\mathbf{J} = -\rho_e \mu_e \mathbf{E}} \tag{7}$$

where ρ_e is the free-electron charge density, a negative value. The total charge density ρ_ν is zero because equal positive and negative charge is present in the neutral material. The negative value of ρ_e and the minus sign lead to a current density **J** that is in the same direction as the electric field intensity **E**.

[3] Wert and Thomson, p. 238, listed in the References at the end of this chapter.

The relationship between **J** and **E** for a metallic conductor, however, is also specified by the conductivity σ (sigma),

Illustrations

$$\mathbf{J} = \sigma \mathbf{E} \tag{8}$$

where σ is measured is siemens[4] per meter (S/m). One siemens (1 S) is the basic unit of conductance in the SI system and is defined as one ampere per volt. Formerly, the unit of conductance was called the mho and was symbolized by an *inverted* Ω.

Just as the siemens honors the Siemens brothers, the reciprocal unit of resistance which we call the ohm (1 Ω is one volt per ampere) honors Georg Simon Ohm, a German physicist who first described the current-voltage relationship implied by (8). We call this equation the *point form of Ohm's law*; we shall look at the more common form of Ohm's law shortly.

First, however, it is informative to note the conductivity of several metallic conductors; typical values (in siemens per meter) are 3.82×10^7 for aluminum, 5.80×10^7 for copper, and 6.17×10^7 for silver. Data for other conductors may be found in Appendix C. On seeing data such as these, it is only natural to assume that we are being presented with *constant* values; this is essentially true. Metallic conductors obey Ohm's law quite faithfully, and it is a *linear* relationship; the conductivity is constant over wide ranges of current density and electric field intensity. Ohm's law and the metallic conductors are also described as *isotropic*, or having the same properties in every direction. A material which is not isotropic is called *anisotropic*, and we shall mention such a material in Chapter 6.

The conductivity is a function of temperature, however. The resistivity, which is the reciprocal of the conductivity, varies almost linearly with temperature in the region of room temperature, and for aluminum, copper, and silver it increases about 0.4 percent for a 1 K rise in temperature.[5] For several metals the resistivity drops abruptly to zero at a temperature of a few kelvin; this property is termed *superconductivity*. Copper and silver are not superconductors, although aluminum is (for temperatures below 1.14 K).

If we now combine (7) and (8), the conductivity may be expressed in terms of the charge density and the electron mobility,

$$\sigma = -\rho_e \mu_e \tag{9}$$

From the definition of mobility (6), it is now satisfying to note that a higher temperature infers a greater crystalline lattice vibration, more impeded electron progress for a given electric field strength, lower drift velocity, lower mobility, lower conductivity from (9), and higher resistivity as stated.

The application of Ohm's law in point form to a macroscopic (visible to the naked eye) region leads to a more familiar form. Initially, let us assume that **J** and **E** are *uniform*, as they are in the cylindrical region shown in Figure 5.3. Since they are uniform,

$$I = \int_S \mathbf{J} \cdot d\mathbf{S} = JS \tag{10}$$

[4] This is the family name of two German-born brothers, Karl Wilhelm and Werner von Siemens, who were famous engineer-inventors in the nineteenth century. Karl became a British subject and was knighted, becoming Sir William Siemens.

[5] Copious temperature data for conducting materials are available in the *Standard Handbook for Electrical Engineers,* listed among the References at the end of this chapter.

and

$$V_{ab} = -\int_b^a \mathbf{E} \cdot d\mathbf{L} = -\mathbf{E} \cdot \int_b^a d\mathbf{L} = -\mathbf{E} \cdot \mathbf{L}_{ba} = \mathbf{E} \cdot \mathbf{L}_{ab} \tag{11}$$

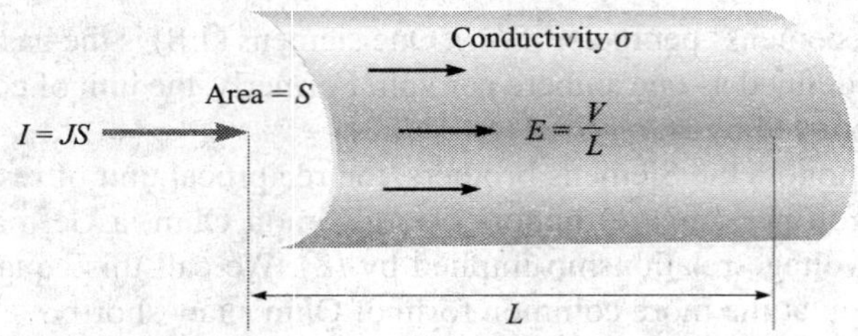

Fig. 5.3 Uniform current density **J** and electric field intensity **E** in a cylindrical region of length L and cross-sectional area S. Here $V = IR$, where $R = L/\sigma S$.

or

$$V = EL$$

Thus

$$J = \frac{I}{S} = \sigma E = \sigma \frac{V}{L}$$

or

$$V = \frac{L}{\sigma S} I$$

The ratio of the potential difference between the two ends of the cylinder to the current entering the more positive end, however, is recognized from elementary circuit theory as the *resistance* of the cylinder, and therefore

$$V = IR \tag{12}$$

where

$$R = \frac{L}{\sigma S} \tag{13}$$

Equation (12) is, of course, known as *Ohm's law*, and (13) enables us to compute the resistance R, measured in ohms (abbreviated as Ω), of conducting objects which possess uniform fields. If the fields are not uniform, the resistance may still be defined as the ratio of V to I, where V is the potential difference between two specified equipotential surfaces in the material and I is the total current crossing the more positive surface into the material. From the general integral relationships in (10) and (11), and from Ohm's law (8), we may write this general expression for resistance when the fields are nonuniform,

$$R = \frac{V_{ab}}{I} = \frac{-\int_b^a \mathbf{E} \cdot d\mathbf{L}}{\int_S \sigma \mathbf{E} \cdot d\mathbf{S}} \tag{14}$$

The line integral is taken between two equipotential surfaces in the conductor, and the surface integral is evaluated over the more positive of these two equipotentials. We cannot solve these nonuniform problems at this time, but we should be able to solve several of them after perusing Chapters 6 and 7.

EXAMPLE 5.1

As an example of the determination of the resistance of a cylinder, let us find the resistance of a 1-mile length of #16 copper wire, which has a diameter of 0.0508 in.

Solution. The diameter of the wire is $0.0508 \times 0.0254 = 1.291 \times 10^{-3}$ m, the area of the cross section is $\pi(1.291 \times 10^{-3}/2)^2 = 1.308 \times 10^{-6}\ \text{m}^2$, and the length is 1609 m. Using a conductivity of 5.80×10^7 S/m, the resistance of the wire is therefore

$$R = \frac{1609}{(5.80 \times 10^7)(1.308 \times 10^{-6})} = 21.2\ \Omega$$

This wire can safely carry about 10 A dc, corresponding to a current density of $10/(1.308 \times 10^{-6}) = 7.65 \times 10^6\ \text{A/m}^2$, or 7.65 A/mm^2. With this current the potential difference between the two ends of the wire is 212 V, the electric field intensity is 0.312 V/m, the drift velocity is 0.000422 m/s, or a little more than one furlong a week, and the free-electron charge density is $-1.81 \times 10^{10}\ \text{C/m}^3$, or about one electron in a cube two angstroms on a side.

D5.3 Find the magnitude of the current density in a sample of silver for which $\sigma = 6.17 \times 10^7$ S/m and $\mu_e = 0.0056\ \text{m}^2/\text{V}\cdot\text{s}$ if: (*a*) the drift velocity is 1.5 μm/s; (*b*) the electric field intensity is 1 mV/m; (*c*) the sample is a cube 2.5 mm on a side having a voltage of 0.4 mV between opposite faces; (*d*) the sample is a cube 2.5 mm on a side carrying a total current of 0.5 A.

Ans. 16.5 kA/m^2; 61.7 kA/m^2; 9.9 MA/m^2; 80.0 kA/m^2

D5.4 A copper conductor has a diameter of 0.6 in. and it is 1200 ft long. Assume that it carries a total dc current of 50 A. (*a*) Find the total resistance of the conductor. (*b*) What current density exists in it? (*c*) What is the dc voltage between the conductor ends? (*d*) How much power is dissipated in the wire?

Ans. 0.035 Ω; 2.74×10^5 A/m^2; 1.73 V; 86.4 W

5.4 CONDUCTOR PROPERTIES AND BOUNDARY CONDITIONS

Once again we must temporarily depart from our assumed static conditions and let time vary for a few microseconds to see what happens when the charge distribution is suddenly unbalanced within a conducting material. Let us suppose, for the sake of the argument, that there suddenly appear a number of electrons in the interior of a conductor. The electric fields set up by these electrons are not counteracted by any positive charges, and the electrons therefore, begin to accelerate away from each other. This continues until the electrons reach the surface of the conductor or until a number of electrons equal to the number injected have reached the surface.

Here the outward progress of the electrons is stopped, for the material surrounding the conductor is an insulator not possessing a convenient conduction band. No charge may remain within the conductor. If it did, the resulting electric field would force the charges to the surface.

Hence the final result within a conductor is zero charge density, and a surface charge density resides on the exterior surface. This is one of the two characteristics of a good conductor.

The other characteristic, stated for static conditions in which no current may flow, follows directly from Ohm's law: the electric field intensity within the conductor is zero. Physically, we see that if an electric field were present, the conduction electrons would move and produce a current, thus leading to a nonstatic condition.

Summarizing for electrostatics, no charge and no electric field may exist at any point *within* a conducting material. Charge may, however, appear on the surface as a surface charge density, and our next investigation concerns the fields *external* to the conductor.

We wish to relate these external fields to the charge on the surface of the conductor. The problem is a simple one, and we may first talk our way to the solution with little mathematics.

If the external electric field intensity is decomposed into two components, one tangential and one normal to the conductor surface, the tangential component is seen to be zero. If it were not zero, a tangential force would be applied to the elements of the surface charge, resulting in their motion and nonstatic conditions. Since static conditions are assumed, the tangential electric field intensity and electric flux density are zero.

Gauss's law answers our questions concerning the normal component. The electric flux leaving a small increment of surface must be equal to the charge residing on that incremental surface. The flux cannot penetrate into the conductor, for the total field there is zero. It must then leave the surface normally. Quantitatively, we may say that the electric flux density in coulombs per square meter leaving the surface normally is equal to the surface charge density in coulombs per square meter, or $D_N = \rho_S$.

If we use some of our previously derived results in making a more careful analysis (and incidentally introducing a general method which we must use later), we should set up a boundary between a conductor and free space (Figure 5.4) showing tangential and normal components of **D** and **E** on the free-space side of the boundary. Both fields are zero in the conductor. The tangential field may be determined by applying Section 4.5, Eq. (21),

$$\oint \mathbf{E} \cdot d\mathbf{L} = 0$$

around the small closed path *abcda*. The integral must be broken up into four parts

$$\int_a^b + \int_b^c + \int_c^d + \int_d^a = 0$$

Remembering that **E** = 0 within the conductor, we let the length from a to b or c to d be Δw and from b to c or d to a be Δh, and obtain

$$E_t \Delta w - E_{N,\text{ at } b} \tfrac{1}{2} \Delta h + E_{N,\text{ at } a} \tfrac{1}{2} \Delta h = 0$$

As we allow Δh to approach zero, keeping Δw small but finite, it makes no difference whether or not the normal fields are equal at a and b, for Δh causes these products to become negligibly small. Hence

$$E_t \Delta w = 0$$

and therefore,

$$E_t = 0$$

The condition on the normal field is found most readily by considering D_N rather than E_N and choosing a small cylinder as the gaussian surface. Let the height be Δh and the area of the top and

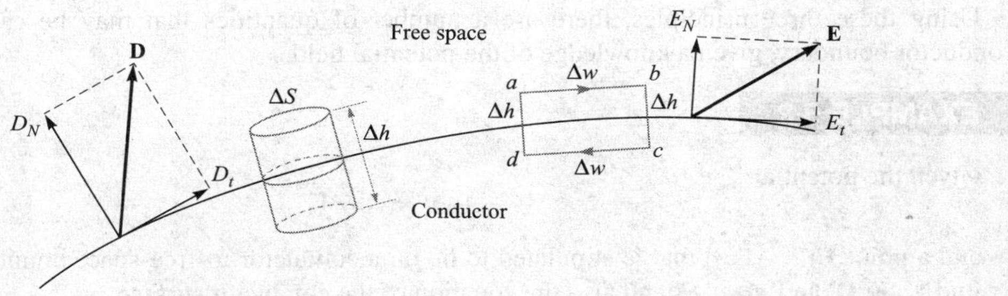

Fig. 5.4 An appropriate closed path and gaussian surface are used to determine boundary conditions at a boundary between a conductor and free space; $E_t = 0$ and $D_N = \rho S$.

bottom faces be ΔS. Again we shall let Δh approach zero. Using Gauss's law,

$$\oint_S \mathbf{D} \cdot d\mathbf{S} = Q$$

we integrate over the three distinct surfaces

$$\int_{\text{top}} + \int_{\text{bottom}} + \int_{\text{sides}} = Q$$

and find that the last two are zero (for different reasons). Then

$$D_N \Delta S = Q = \rho_S \Delta S$$

or

$$D_N = \rho_S$$

These are the desired *boundary conditions* for the conductor-to-free-space boundary in electrostatics,

$$D_t = E_t = 0 \tag{15}$$

$$D_N = \epsilon_0 E_N = \rho_S \tag{16}$$

The electric flux leaves the conductor in a direction normal to the surface, and the value of the electric flux density is numerically equal to the surface charge density.

An immediate and important consequence of a zero tangential electric field intensity is the fact that a conductor surface is an equipotential surface. The evaluation of the potential difference between any two points on the surface by the line integral leads to a zero result, because the path may be chosen on the surface itself where $\mathbf{E} \cdot d\mathbf{L} = 0$.

To summarize the principles which apply to conductors in electrostatic fields, we may state that

1. The static electric field intensity inside a conductor is zero.
2. The static electric field intensity at the surface of a conductor is everywhere directed normal to that surface.
3. The conductor surface is an equipotential surface.

Using these three principles, there are a number of quantities that may be calculated at a conductor boundary, given a knowledge of the potential field.

EXAMPLE 5.2

Given the potential,

$$V = 100(x^2 - y^2)$$

and a point $P(2, -1, 3)$ that is stipulated to lie on a conductor-to-free-space boundary, let us find V, **E**, **D**, and ρ_S at P, and also the equation of the conductor surface.

Solution. The potential at point P is

$$V_P = 100[2^2 - (-1)^2] = 300 \text{ V}$$

Since the conductor is an equipotential surface, the potential at the entire surface must be 300 V. Moreover, if the conductor is a solid object, then the potential everywhere in and on the conductor is 300 V, for $\mathbf{E} = 0$ within the conductor.

The equation representing the locus of all points having a potential of 300 V is

$$300 = 100(x^2 - y^2)$$

or

$$x^2 - y^2 = 3$$

This is therefore the equation of the conductor surface; it happens to be a hyperbolic cylinder, as shown in Figure 5.5. Let us assume arbitrarily that the solid conductor lies above and to the right of the equipotential surface at point P, while free space is down and to the left.

Next, we find **E** by the gradient operation,

$$\mathbf{E} = -100\nabla(x^2 - y^2) = -200x\mathbf{a}_x + 200y\mathbf{a}_y$$

At point P,

$$\mathbf{E}_p = -400\mathbf{a}_x - 200\mathbf{a}_y \text{ V/m}$$

Since $\mathbf{D} = \epsilon_0\mathbf{E}$, we have

$$\mathbf{D}_P = 8.854 \times 10^{-12}\mathbf{E}_P = -3.54\mathbf{a}_x - 1.771\mathbf{a}_y \text{ nC/m}^2$$

The field is directed downward and to the left at P; it is normal to the equipotential surface. Therefore,

$$D_N = |\mathbf{D}_P| = 3.96 \text{ nC/m}^2$$

Thus, the surface charge density at P is

$$\rho_{S,P} = D_N = 3.96 \text{ nC/m}^2$$

Note that if we had taken the region to the left of the equipotential surface as the conductor, the **E** field would *terminate* on the surface charge and we would let $\rho_S = -3.96$ nC/m^2.

EXAMPLE 5.3

Finally, let us determine the equation of the streamline passing through P.

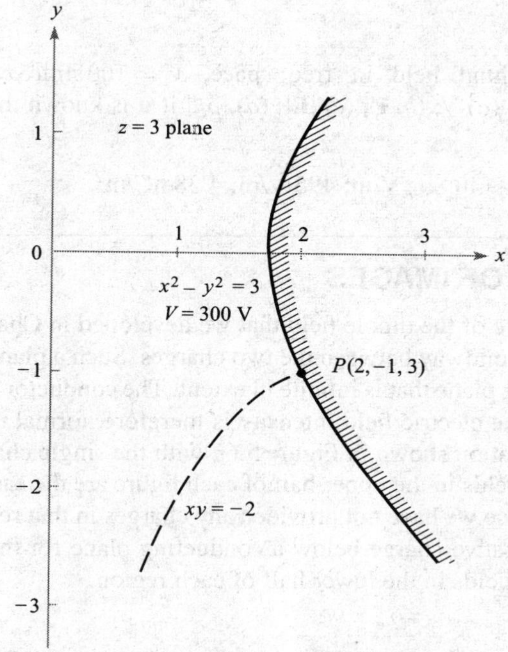

Fig. 5.5 Given point $P(2, -1, 3)$ and the potential field, $V = 100(x^2 - y^2)$, we find the equipotential surface through P is $x^2 - y^2 = 3$, and the streamline through P is $xy = -2$.

Solution. We see that

$$\frac{E_y}{E_x} = \frac{200y}{-200x} = -\frac{y}{x} = \frac{dy}{dx}$$

Thus,

$$\frac{dy}{y} + \frac{dx}{x} = 0$$

and

$$\ln y + \ln x = C_1$$

Therefore,

$$xy = C_2$$

The line (or surface) through P is obtained when $C_2 = (2)(-1) = -2$. Thus, the streamline is the trace of another hyperbolic cylinder,

$$xy = -2$$

This is also shown on Figure 5.5.

D5.5 Given the potential field in free space, $V = 100 \sinh 5x \sin 5y$ V, and a point $P(0.1, 0.2, 0.3)$, find at P: (*a*) V; (*b*) **E**; (*c*) $|\mathbf{E}|$; (*d*) $|\rho_S|$ if it is known that P lies on a conductor surface.

Ans. 43.8 V; $-474\mathbf{a}_x - 140.8\mathbf{a}_y$ V/m; 495 V/m; 4.38 nC/m^2

5.5 THE METHOD OF IMAGES

One important characteristic of the dipole field that we developed in Chapter 4 is the infinite plane at zero potential that exists midway between the two charges. Such a plane may be represented by a vanishingly thin conducting plane that is infinite in extent. The conductor is an equipotential surface at a potential $V = 0$, and the electric field intensity is therefore normal to the surface. Thus, if we replace the dipole configuration shown in Figure 5.6a with the single charge and conducting plane shown in Figure 5.6b, the fields in the upper half of each figure are the same. Below the conducting plane, all fields are zero since we have not provided any charges in that region. Of course, we might also substitute a single negative charge below a conducting plane for the dipole arrangement and obtain equivalence for the fields in the lower half of each region.

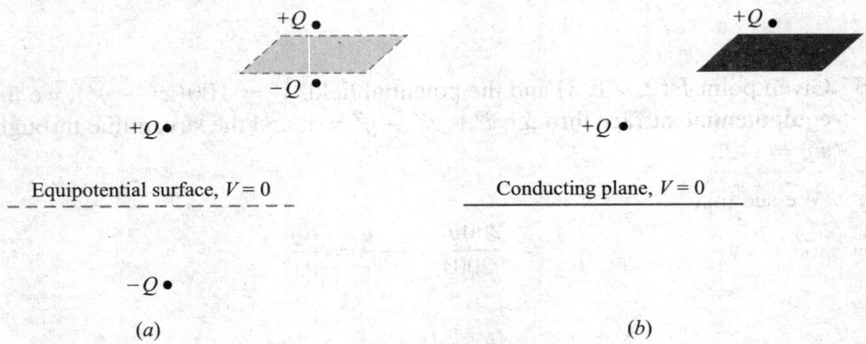

Fig. 5.6 (a) Two equal but opposite charges may be replaced by (b) a single charge and a conducting plane without affecting the fields above the $V = 0$ surface.

If we approach this equivalence from the opposite point of view, we begin with a single charge above a perfectly conducting plane and then see that we may maintain the same fields above the plane by removing the plane and locating a negative charge at a symmetrical location below the plane. This charge is called the *image* of the original charge, and it is the negative of that value.

If we can do this once, linearity allows us to do it again and again, and thus *any* charge configuration above an infinite ground plane may be replaced by an arrangement composed of the given charge configuration, its image, and no conducting plane. This is suggested by the two illustrations of Figure 5.7. In many cases, the potential field of the new system is much easier to find since it does not contain the conducting plane with its unknown surface charge distribution.

As an example of the use of images, let us find the surface charge density at $P(2, 5, 0)$ on the conducting plane $z = 0$ if there is a line charge of 30 nC/m located at $x = 0$, $z = 3$, as shown in Figure 5.8a. We remove the plane and install an image line charge of -30 nC/m at $x = 0$, $z = -3$,

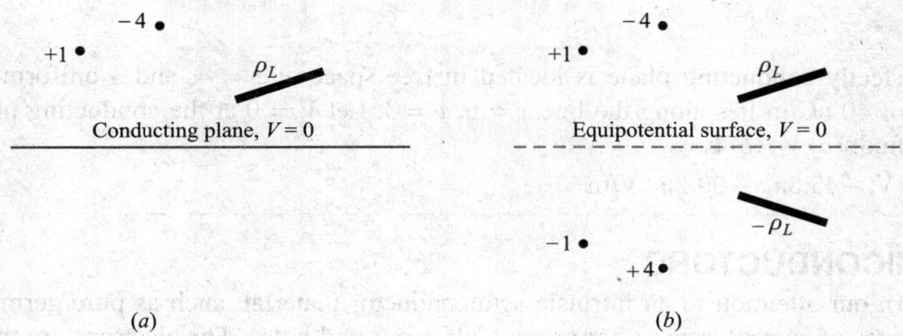

Fig. 5.7 (a) A given charge configuration above an infinite conducting plane may be replaced by (b) the given charge configuration plus the image configuration, without the conducting plane.

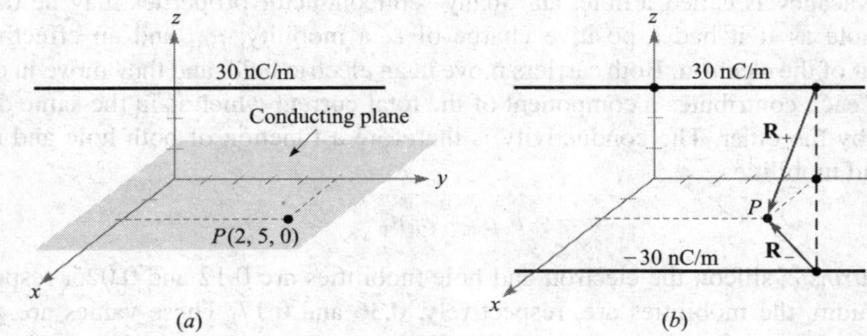

Fig. 5.8 (a) A line charge above a conducting plane. (b) The conductor is removed, and the image of the line charge is added.

as illustrated in Figure 5.8b. The field at P may now be obtained by superposition of the known fields of the line charges. The radial vector from the positive line charge to P is $\mathbf{R}_+ = 2\mathbf{a}_x - 3\mathbf{a}_z$, while $\mathbf{R}_- = 2\mathbf{a}_x + 3\mathbf{a}_z$. Thus, the individual fields are

$$\mathbf{E}_+ = \frac{\rho_L}{2\pi\epsilon_0 R_+}\mathbf{a}_{R+} = \frac{30 \times 10^{-9}}{2\pi\epsilon_0\sqrt{13}}\,\frac{2\mathbf{a}_x - 3\mathbf{a}_z}{\sqrt{13}}$$

and

$$\mathbf{E}_- = \frac{30 \times 10^{-9}}{2\pi\epsilon_0\sqrt{13}}\,\frac{2\mathbf{a}_x + 3\mathbf{a}_z}{\sqrt{13}}$$

Adding these results, we have

$$\mathbf{E} = \frac{-180 \times 10^{-9}\mathbf{a}_z}{2\pi\epsilon_0(13)} = -249\mathbf{a}_z \text{ V/m}$$

This then is the field at (or just above) P in both the configurations of Figure 5.8, and it is certainly satisfying to note that the field is normal to the conducting plane, as it must be. Thus, $\mathbf{D} = \epsilon_0\mathbf{E} = -2.20\mathbf{a}_z$ nC/m^2, and since this is directed *toward* the conducting plane, ρ_S is negative and has a value of -2.20 nC/m^2 at P.

D5.6 A perfectly conducting plane is located in free space at $x = 4$, and a uniform infinite line charge of 40 nC/m lies along the line $x = 6$, $y = 3$. Let $V = 0$ at the conducting plane. At $P(7, -1, 5)$ find: (*a*) V; (*b*) **E**.

Ans. 317 V; $-45.3\mathbf{a}_x - 99.2\mathbf{a}_y$ V/m

5.6 SEMICONDUCTORS

If we now turn our attention to an intrinsic semiconductor material, such as pure germanium or silicon, two types of current carriers are present, electrons and holes. The electrons are those from the top of the filled valence band which have received sufficient energy (usually thermal) to cross the relatively small forbidden band into the conduction band. The forbidden-band energy gap in typical semiconductors is of the order of one electronvolt. The vacancies left by these electrons represent unfilled energy states in the valence band which may also move from atom to atom in the crystal. The vacancy is called a hole, and many semiconductor properties may be described by treating the hole as if it had a positive charge of e, a mobility, μ_h, and an effective mass comparable to that of the electron. Both carriers move in an electric field, and they move in opposite directions; hence each contributes a component of the total current which is in the same direction as that provided by the other. The conductivity is therefore a function of both hole and electron concentrations and mobilities,

$$\sigma = -\rho_e \mu_e + \rho_h \mu_h \tag{17}$$

For pure, or *intrinsic*, silicon the electron and hole mobilities are 0.12 and 0.025, respectively, while for germanium, the mobilities are, respectively, 0.36 and 0.17. These values are given in square meters per volt-second and range from 10 to 100 times as large as those for aluminum, copper, silver, and other metallic conductors.[6] These mobilities are given for a temperature of 300 K.

The electron and hole concentrations depend strongly on temperature. At 300 K the electron and hole volume charge densities are both 0.0024 C/m^3 in magnitude in intrinsic silicon and 3.0 C/m^3 in intrinsic germanium. These values lead to conductivities of 0.000 35 S/m in silicon and 1.6 S/m in germanium. As temperature increases, the mobilities decrease, but the charge densities increase very rapidly. As a result, the conductivity of silicon increases by a factor of 10 as the temperature increases from 300 to about 330 K and decreases by a factor of 10 as the temperature drops from 300 to about 275 K. Note that the conductivity of the intrinsic semiconductor increases with temperature, while that of a metallic conductor decreases with temperature; this is one of the characteristic differences between the metallic conductors and the intrinsic semiconductors.

Intrinsic semiconductors also satisfy the point form of Ohm's law; that is, the conductivity is reasonably constant with current density and with the direction of the current density.

The number of charge carriers and the conductivity may both be increased dramatically by adding very small amounts of impurities. *Donor* materials provide additional electrons and form *n-type* semiconductors, while *acceptors* furnish extra holes and form *p-type* materials. The process is known as *doping*, and a donor concentration in silicon as low as one part in 10^7 causes an increase in conductivity by a factor of 10^5.

[6] Mobility values for semiconductors are given in Refs. 2, 3, and 5 listed at the end of this chapter.

The range of value of the conductivity is extreme as we go from the best insulating materials to semiconductors and the finest conductors. In siemens per meter, σ ranges from 10^{-17} for fused quartz, $10^{-7}$for poor plastic insulators, and roughly unity for semiconductors to almost 10^8 for metallic conductors at room temperature. These values cover the remarkably large range of some 25 orders of magnitude.

D5.7 Using the values given in this section for the electron and hole mobilities in silicon at 300 K, and assuming hole and electron charge densities are 0.0029 C/m^{3}and -0.0029 C/m^3, respectively, find: (*a*) the component of the conductivity due to holes; (*b*) the component of the conductivity due to electrons; (*c*) the conductivity.

Ans. 7.25 μS/m; 348 μS/m; 421 μS/m

References

1. Adler, K. H., A. C. Smith, and R. L. Longini. *Introduction to Semiconductor Physics.* New York: John Wiley & Sons, 1964. Semiconductor theory is treated at an undergraduate level.
2. Dekker, A. J. *Electrical Engineering Materials.* Englewood Cliffs, N.J.: Prentice-Hall, 1959. This admirable little book covers dielectrics, conductors, semiconductors, and magnetic materials.
3. Fink, D. G., and H. W. Beaty. *Standard Handbook for Electrical Engineers.* 12th ed. New York: McGraw-Hill, 1987.
4. Maxwell, J. C. *A Treatise on Electricity and Magnetism.* 3d ed. New York: Oxford University Press, 1904, or an inexpensive paperback edition, Dover Publications, New York, 1954.
5. Wert, C. A., and R. M. Thomson. *Physics of Solids.* 2d ed. New York: McGraw-Hill, 1970. This is an advanced undergraduate-level text that covers metals, semiconductors, and dielectrics.

CHAPTER 5 Problems

Quizzes

5.1 Given the current density $\mathbf{J} = -10^4[\sin(2x)e^{-2y}\mathbf{a}_x + \cos(2x)e^{-2y}\mathbf{a}_y]$ kA/m^2: (*a*) Find the total current crossing the plane $y = 1$ in the $\mathbf{a}_y$ direction in the region $0 < x < 1, 0 < z < 2$. (*b*) Find the total current leaving the region $0 < x, y < 1, 2 < z < 3$ by integrating $\mathbf{J} \cdot d\mathbf{S}$ over the surface of the cube. (*c*) Repeat part (*b*), but use the divergence theorem.

5.2 A certain current density is given in cylindrical coordinates as $\mathbf{J} = 100e^{-2z}(\rho\mathbf{a}_\rho + \mathbf{a}_z)$ A/m^2. Find the total current passing through each of these surfaces: (*a*) $z = 0$, $0 \le \rho \le 1$, in the $\mathbf{a}_z$ direction; (*b*) $z = 1$, $0 \le \rho \le 1$, in the $\mathbf{a}_z$ direction; (*c*) closed cylinder defined by $0 \le z \le 1$, $0 \le \rho \le 1$, in an outward direction.

5.3 Let $\mathbf{J} = 400\sin\theta/(r^2 + 4)\mathbf{a}_r$ A/m^2.(*a*) Find the total current flowing through that portion of the spherical surface $r = 0.8$, bounded by $0.1\pi < \theta < 0.3\pi$, $0 < \phi < 2\pi$. (*b*) Find the average value of $\mathbf{J}$ over the defined area.

5.4 Assume that a uniform electron beam of circular cross section with radius of 0.2 mm is generated by a cathode at $x = 0$ and collected by an anode at $x = 20$ cm. The velocity of the electrons varies with x as $v_x = 10^8 x^{0.5}$ m/s, with x in meters. If the current density at the anode is 10^4 A/m^2, find the volume charge density and the current density as functions of x.

5.5 Let $\mathbf{J} = 25/\rho\mathbf{a}_\rho - 20/(\rho^2 + 0.01)\mathbf{a}_z$ A/m^2. (*a*) Find the total current crossing the plane $z = 0.2$ in the $\mathbf{a}_z$ direction for $\rho < 0.4$. (*b*) Calculate $\partial\rho_v/\partial t$. (*c*) Find the outward current crossing the closed surface defined by $\rho = 0.01$, $\rho = 0.4$, $z = 0$, and $z = 0.2$. (*d*) Show that the divergence theorem is satisified for **J** and the surface specified in part (*c*).

5.6 The current density in a certain region is approximated by $\mathbf{J} = (0.1/r)\exp(-10^6 t)\ \mathbf{a}_r$ A/m^2 in spherical coordinates. (*a*) At $t = 1\ \mu$ s, how much current is crossing the surface $r = 5$? (*b*) Repeat for $r = 6$. (*c*) Use the continuity equation to find $\rho_\nu(r, t)$ assuming that $\rho_\nu \to 0$ as $t \to \infty$. (*d*) Find an expression for the velocity of the charge density.

5.7 Assuming that there is no transformation of mass to energy or vice versa, it is possible to write a continuity equation for mass. (*a*) If we use the continuity equation for charge as our model, what quantities correspond to **J** and ρ_ν? (*b*) Given a cube 1 cm on a side, experimental data show that the rates at which mass is leaving each of the six faces are 10.25, −9.85, 1.75, −2.00, −4.05, and 4.45 mg/s. If we assume that the cube is an incremental volume element, determine an approximate value for the time rate of change of density at its center.

5.8 The conductivity of carbon is about 3×10^4 S/m. (*a*) What size and shape sample of carbon has a conductance of 3×10^4 S? (*b*) What is the conductance if every dimension of the sample found in part (*a*) is halved?

5.9 (*a*) Using data tabulated in Appendix C, calculate the required diameter for a 2-m-long nichrome wire that will dissipate an average power of 450 W when 120 V rms at 60 Hz is applied to it. (*b*) Calculate the rms current density in the wire.

5.10 A solid wire of conductivity a_1 and radius a has a jacket of material having conductivity σ_2, and its inner radius is a and outer radius is b. Show that the ratio of the current densities in the two materials is independent of a and b.

5.11 Two perfectly conducting cylindrical surfaces of length l are located at $\rho = 3$ and $\rho = 5$ cm. The total current passing radially outward through the medium between the cylinders is 3 A dc. (*a*) Find the voltage and resistance between the cylinders, and **E** in the region between the cylinders, if a conducting material having $\sigma = 0.05$ S/m is present for $3 < \rho < 5$ cm. (*b*) Show that integrating the power dissipated per unit volume over the volume gives the total dissipated power.

5.12 Two identical conducting plates, each having area A, are located at $z = 0$ and $z = d$. The region between plates is filled with a material having z-dependent conductivity, $\sigma(z) = \sigma_0 e^{-z/d}$, where σ_0 is a constant. Voltage V_0 is applied to the plate at $z = d$; the plate at $z = 0$ is at zero potential. Find, in terms of the given parameters: (*a*) the resistance of the material; (*b*) the total current flowing between plates; (*c*) the electric field intensity **E** within the material.

5.13 A hollow cylindrical tube with a rectangular cross section has external dimensions of 0.5 in. by 1 in. and a wall thickness of 0.05 in. Assume that the material is brass, for which $\sigma = 1.5 \times 10^7$ S/m. A current of 200 A dc is flowing down the tube. (*a*) What voltage drop is present acrossa1m length of the tube? (*b*) Find the voltage drop if the interior of the tube is filled with a conducting material for which $\sigma = 1.5 \times 10^5$ S/m.

5.14 A rectangular conducting plate lies in the *xy* plane, occupying the region $0 < x < a$, $0 < y < b$. An identical conducting plate is positioned directly above and parallel to the first, at $z = d$. The region between plates is filled with material having conductivity $\sigma(x) = \sigma_0 e^{-x/a}$, where σ_0 is a constant. Voltage V_0 is applied to the plate at $z = d$; the plate at $z = 0$ is at zero potential. Find, in terms of the given parameters: (*a*) the electric field intensity **E** within the material; (*b*) the total current flowing between plates; (*c*) the resistance of the material.

5.15 Let $V = 10(\rho + 1)z^2 \cos\phi$ V in free space. (*a*) Let the equipotential surface $V = 20$ V define a conductor surface. Find the equation of the conductor surface. (*b*) Find ρ and **E** at that point on the conductor surface where $\phi = 0.2\pi$ and $z = 1.5$. (*c*) Find $|\rho_S|$ at that point.

5.16 In cylindrical coordinates, $V = 1000\rho^2$. (*a*) If the region $0.1 < \rho < 0.3$ m is free space while the surfaces $\rho = 0.1$ and $\rho = 0.3$ m are conductors, specify the surface charge density on each conductor. (*b*) What is the total charge in a 1m length of the free space region, $0.1 < \rho < 0.3$ (not including the conductors)? (*c*) What is the total charge in a 1 m length, including both surface charges?

5.17 Given the potential field $V = 100xz/(x^2 + 4)$ V in free space: (*a*) Find **D** at the surface $z = 0$. (*b*) Show that the $z = 0$ surface is an equipotential surface. (*c*) Assume that the $z = 0$ surface is a conductor and find the total charge on that portion of the conductor defined by $0 < x < 2$, $-3 < y < 0$.

5.18 A potential field is given as $V = 100 \ln\{[(x+1)^2 + y^2]/[(x-1)^2 + y^2]\}$ V. It is known that point $P(2, 1, 1)$ is on a conductor surface and that the conductor lies in free space. At P, find a unit vector normal to the surface and also the value of the surface charge density on the conductor.

5.19 Let $V = 20x^2yz - 10z^2$ V in free space. (*a*) Determine the equations of the equipotential surfaces on which $V = 0$ and 60 V. (*b*) Assume these are conducting surfaces and find the surface charge density at that point on the $V = 60$ V surface where $x = 2$ and $z = 1$. It is known that $0 \le V \le 60$ V is the field-containing region. (*c*) Give the unit vector at this point that is normal to the conducting surface and directed toward the $V = 0$ surface.

5.20 Two point charges of $-100\pi\mu$ C are located at (2, −1, 0) and (2, 1, 0). The surface $x = 0$ is a conducting plane. (*a*) Determine the surface charge density at the origin. (*b*) Determine ρ_S at $P(0, h, 0)$.

5.21 Let the surface $y = 0$ be a perfect conductor in free space. Two uniform infinite line charges of 30 nC/m each are located at $x = 0$, $y = 1$, and $x = 0$, $y = 2$. (*a*) Let $V = 0$ at the plane $y = 0$, and find V at $P(1, 2, 0)$. (*b*) Find **E** at P.

5.22 The line segment $x = 0$, $-1 \le y \le 1$, $z = 1$, carries a linear charge density $\rho_L = \pi|y|\mu$ C/m. Let $z = 0$ be a conducting plane and determine the surface charge density at: (*a*) (0, 0, 0); (*b*) (0, 1, 0).

5.23 A dipole with $\mathbf{p} = 0.1\mathbf{a}_z \mu$ C · m is located at $A(1, 0, 0)$ in free space, and the $x = 0$ plane is perfectly conducting. (*a*) Find V at $P(2, 0, 1)$. (*b*) Find the equation of the 200 V equipotential surface in rectangular coordinates.

5.24 At a certain temperature, the electron and hole mobilities in intrinsic germanium are given as 0.43 and 0.21 m^2/V · s, respectively. If the electron and hole concentrations are both 2.3×10^{19} m^{-3}, find the conductivity at this temperature.

5.25 Electron and hole concentrations increase with temperature. For pure silicon, suitable expressions are $\rho_h = -\rho_e = 6200T^{1.5}e^{-7000/T}$ C/m^3. The functional dependence of the mobilities on temperature is given by $\mu_h = 2.3 \times 10^5 T^{-2.7}$ m^2/V · s and $\mu_e = 2.1 \times 10^5 T^{-2.5}$ m^2/V · s, where the temperature, T, is in degrees Kelvin. Find σ at: (*a*) 0° C; (*b*) 40°C; (*c*) 80° C.

5.26 A semiconductor sample has a rectangular cross section 1.5 by 2.0 mm, and a length of 11.0 mm. The material has electron and hole densities of 1.8×10^{18} and 3.0×10^{15} m^{-3}, respectively. If $\mu_e = 0.082$ m^2/V · s and $\mu_h = 0.0021$ m^2/V · s, find the resistance offered between the end faces of the sample.

6 Dielectrics and Capacitance

Having investigated the properties of conducting media in Chapter 5, we now turn our attention to insulating materials, or dielectrics. Such materials differ from conductors in that ideally, there is no free charge that can be transported within them to produce conduction current. Instead, all charge is confined to molecular or lattice sites by coulomb forces. An applied electric field has the effect of displacing the charges slightly, leading to the formation of ensembles of electric dipoles. The extent to which this occurs is measured by the relative permittivity, or dielectric constant. The polarization of the medium may modify the electric field, whose magnitude and direction may differ from the values that it would have in a different dielectric or in free space. Boundary conditions for the fields at interfaces between dielectrics are developed in order to evaluate these differences.

The charge displacement principle constitutes an energy storage mechanism that becomes useful in the construction of capacitors. Furthermore, the response of the dielectric to time-varying fields, particularly electromagnetic waves, is extremely important to the understanding of many physical phenomena and to the development of useful devices, as we will consider in later chapters. It should also be noted that most materials will possess both dielectric and conductive properties; that is, a material considered a dielectric may be slightly conductive, and a material that is mostly conductive may be slightly polarizable. These departures from the ideal cases lead to some interesting behavior, particularly as to effects on wave propagation, as will be seen later.

Following the discussion of dielectric materials, capacitors are considered—particularly with regard to what capacitance can be obtained from a given combination of conductors and dielectrics. The goal is to present the methods for calculating capacitance for a number of basic cases, including transmission line geometries, and to be able to make judgments on how capacitance will be altered by changes in materials or their configuration. Both analytic and graphical methods for determining capacitance are discussed. Additional analytic procedures and numerical methods that can be applied to capacitance computation will be developed in Chapter 7.

6.1 THE NATURE OF DIELECTRIC MATERIALS

Although we have mentioned insulators and dielectric materials, we do not as yet have any quantitative relationships in which they are involved. We shall soon see, however, that a dielectric in an electric field can be viewed as a free-space arrangement of microscopic electric dipoles which are composed of positive and negative charges whose centers do not quite coincide.

These are not free charges, and they cannot contribute to the conduction process. Rather, they are bound in place by atomic and molecular forces and can only shift positions slightly in response

Interactives

to external fields. They are called *bound* charges, in contrast to the free charges that determine conductivity. The bound charges can be treated as any other sources of the electrostatic field. If we did not wish to, therefore, we would not need to introduce the dielectric constant as a new parameter or to deal with permittivities different from the permittivity of free space; however, the alternative would be to consider *every charge within a piece of dielectric material*. This is too great a price to pay for using all our previous equations in an unmodified form, and we shall therefore spend some time theorizing about dielectrics in a qualitative way; introducing polarization $\mathbf{P}$, permittivity ϵ, and relative permittivity ϵ_r; and developing some quantitative relationships involving these new quantities.

The characteristic which all dielectric materials have in common, whether they are solid, liquid, or gas, and whether or not they are crystalline in nature, is their ability to store electric energy. This storage takes place by means of a shift in the relative positions of the internal, bound positive and negative charges against the normal molecular and atomic forces.

This displacement against a restraining force is analogous to lifting a weight or stretching a spring and represents potential energy. The source of the energy is the external field, the motion of the shifting charges resulting perhaps in a transient current through a battery which is producing the field.

The actual mechanism of the charge displacement differs in the various dielectric materials. Some molecules, termed *polar* molecules, have a permanent displacement existing between the centers of "gravity" of the positive and negative charges, and each pair of charges acts as a dipole. Normally the dipoles are oriented in a random way throughout the interior of the material, and the action of the external field is to align these molecules, to some extent, in the same direction. A sufficiently strong field may even produce an additional displacement between the positive and negative charges.

A *nonpolar* molecule does not have this dipole arrangement until after a field is applied. The negative and positive charges shift in opposite directions against their mutual attraction and produce a dipole which is aligned with the electric field.

Either type of dipole may be described by its dipole moment $\mathbf{p}$, as developed in Section 4.7, Eq. (37),

$$\mathbf{p} = Q\mathbf{d} \tag{1}$$

where Q is the positive one of the two bound charges composing the dipole, and $\mathbf{d}$ is the vector from the negative to the positive charge. We note again that the units of $\mathbf{p}$ are coulomb-meters.

If there are n dipoles per unit volume and we deal with a volume $\Delta\nu$, then there are $n\,\Delta\nu$ dipoles, and the total dipole moment is obtained by the vector sum,

$$\mathbf{p}_{\text{total}} = \sum_{i=1}^{n\,\Delta\nu} \mathbf{p}_i$$

If the dipoles are aligned in the same general direction, $\mathbf{p}_{\text{total}}$ may have a significant value. However, a random orientation may cause $\mathbf{p}_{\text{total}}$ to be essentially zero.

We now define the polarization $\mathbf{P}$ as the *dipole moment per unit volume*,

$$\mathbf{P} = \lim_{\Delta\nu \to 0} \frac{1}{\Delta\nu} \sum_{i=1}^{n\,\Delta\nu} \mathbf{p}_i \tag{2}$$

with units of coulombs per square meter. We shall treat $\mathbf{P}$ as a typical continuous field, even though it is obvious that it is essentially undefined at points within an atom or molecule. Instead, we should think of its value at any point as an average value taken over a sample volume $\Delta\nu$—large enough to contain many molecules ($n\,\Delta\nu$ in number), but yet sufficiently small to be considered incremental in concept.

Our immediate goal is to show that the bound-volume charge density acts like the free-volume charge density in producing an external field; we shall obtain a result similar to Gauss's law.

To be specific, let us assume that we have a dielectric containing nonpolar molecules. No molecule has a dipole moment, and $\mathbf{P} = 0$ throughout the material. Somewhere in the interior of the dielectric we select an incremental surface element $\Delta\mathbf{S}$, as shown in Figure 6.1*a*, and apply an electric field $\mathbf{E}$. The electric field produces a moment $\mathbf{p} = Q\mathbf{d}$ in each molecule, such that $\mathbf{p}$ and $\mathbf{d}$ make an angle θ with $\Delta\mathbf{S}$, as indicated in Figure 6.1*b*.

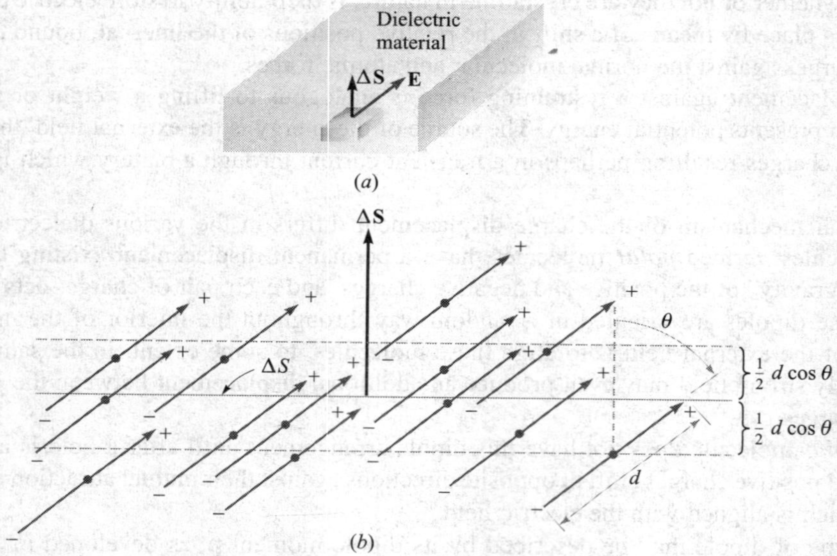

Fig. 6.1 (a) An incremental surface element ΔS is shown in the interior of a dielectric in which an electric field $\mathbf{E}$ is present. (b) The nonpolar molecules form dipole moments $\mathbf{p}$ and a polarization $\mathbf{P}$. There is a net transfer of bound charge across ΔS.

Now let us inspect the movement of bound charges across $\Delta\mathbf{S}$. Each of the charges associated with the creation of a dipole must have moved a distance $\frac{1}{2}d\cos\theta$ in the direction perpendicular to $\Delta\mathbf{S}$. Thus, any positive charges initially lying below the surface $\Delta\mathbf{S}$ and within the distance $\frac{1}{2}d\cos\theta$ of the surface must have crossed $\Delta\mathbf{S}$ going upward. Also, any negative charges initially lying above the surface and within that distance ($\frac{1}{2}d\cos\theta$) from $\Delta\mathbf{S}$ must have crossed $\Delta\mathbf{S}$ going downward. Therefore, since there are n molecules/m^3, the net total charge which crosses the elemental surface in an upward direction is equal to $nQd\cos\theta\Delta\mathbf{S}$, or

$$\Delta Q_b = nQ\mathbf{d}\cdot\Delta\mathbf{S}$$

where the subscript on Q_b reminds us that we are dealing with a bound charge and not a free charge. In terms of the polarization, we have

$$\Delta Q_b = \mathbf{P}\cdot\Delta\mathbf{S}$$

If we interpret $\Delta\mathbf{S}$ as an element of a *closed* surface inside the dielectric material, then the direction of $\Delta\mathbf{S}$ is outward, and the net increase in the bound charge *within* the closed surface is obtained through the integral

$$Q_b = -\oint_S \mathbf{P} \cdot d\mathbf{S} \tag{3}$$

This last relationship has some resemblance to Gauss's law, and we may now generalize our definition of electric flux density so that it applies to media other than free space. We first write Gauss's law in terms of $\epsilon_0\mathbf{E}$ and Q_T, the *total* enclosed charge, bound plus free:

$$Q_T = \oint_S \epsilon_0\mathbf{E} \cdot d\mathbf{S} \tag{4}$$

where

$$Q_T = Q_b + Q$$

and Q is the total *free* charge enclosed by the surface S. Note that the free charge appears without subscript since it is the most important type of charge and will appear in Maxwell's equations.

Combining these last three equations, we obtain an expression for the free charge enclosed,

$$Q = Q_T - Q_b = \oint_S (\epsilon_0\mathbf{E} + \mathbf{P}) \cdot d\mathbf{S} \tag{5}$$

We may now define $\mathbf{D}$ in more general terms than we did in Chapter 3,

$$\mathbf{D} = \epsilon_0\mathbf{E} + \mathbf{P} \tag{6}$$

There is thus an added term to D which appears when polarizable material is present. Thus,

$$Q = \oint_S \mathbf{D} \cdot d\mathbf{S} \tag{7}$$

where Q is the free charge enclosed.

Utilizing the several volume charge densities, we have

$$Q_b = \int_\nu \rho_b \, dv$$

$$Q = \int_\nu \rho_\nu \, dv$$

$$Q_T = \int_\nu \rho_T \, dv$$

With the help of the divergence theorem, we may therefore transform (3), (4), and (7) into the equivalent divergence relationships,

$$\nabla \cdot \mathbf{P} = -\rho_b$$

$$\nabla \cdot \epsilon_0\mathbf{E} = \rho_T$$

$$\nabla \cdot \mathbf{D} = \rho_\nu \tag{8}$$

We shall emphasize only (7) and (8), the two expressions involving the free charge, in the work that follows.

In order to make any real use of these new concepts, it is necessary to know the relationship between the electric field intensity $\mathbf{E}$ and the polarization $\mathbf{P}$ which results. This relationship will, of course, be a function of the type of material, and we shall essentially limit our discussion to those isotropic materials for which **E** and **P** are linearly related. In an isotropic material the vectors **E** and **P** are always parallel, regardless of the orientation of the field. Although most engineering dielectrics are linear for moderate-to-large field strengths and are also isotropic, single crystals may be anisotropic. The periodic nature of crystalline materials causes dipole moments to be formed most easily along the crystal axes, and not necessarily in the direction of the applied field.

In *ferroelectric* materials the relationship between **P** and **E** not only is nonlinear, but also shows hysteresis effects; that is, the polarization produced by a given electric field intensity depends on the past history of the sample. Important examples of this type of dielectric are barium titanate, often used in ceramic capacitors, and Rochelle salt.

The linear relationship between **P** and **E** is

$$\mathbf{P} = \chi_e \epsilon_0 \mathbf{E} \tag{9}$$

where χ_e (chi) is a dimensionless quantity called the *electric susceptibility* of the material.

Using this relationship in (6), we have

$$\mathbf{D} = \epsilon_0 \mathbf{E} + \chi_e \epsilon_0 \mathbf{E} = (\chi_e + 1)\epsilon_0 \mathbf{E}$$

The expression within the parentheses is now defined as

$$\epsilon_r = \chi_e + 1 \tag{10}$$

This is another dimensionless quantity, and it is known as the *relative permittivity*, or *dielectric constant* of the material. Thus,

$$\mathbf{D} = \epsilon_0 \epsilon_r \mathbf{E} = \epsilon \mathbf{E} \tag{11}$$

where

$$\epsilon = \epsilon_0 \epsilon_r \tag{12}$$

and ϵ is the *permittivity*. The dielectric constants are given for some representative materials in Appendix C.

Anisotropic dielectric materials cannot be described in terms of a simple susceptibility or permittivity parameter. Instead, we find that each component of **D** may be a function of every component of **E**, and $\mathbf{D} = \epsilon \mathbf{E}$ becomes a matrix equation where **D** and **E** are each 3×1 column matrices and ϵ is a 3×3 square matrix. Expanding the matrix equation gives

$$D_x = \epsilon_{xx} E_x + \epsilon_{xy} E_y + \epsilon_{xz} E_z$$

$$D_y = \epsilon_{yx} E_x + \epsilon_{yy} E_y + \epsilon_{yz} E_z$$

$$D_z = \epsilon_{zx} E_x + \epsilon_{zy} E_y + \epsilon_{zz} E_z$$

Note that the elements of the matrix depend on the selection of the coordinate axes in the anisotropic material. Certain choices of axis directions lead to simpler matrices.[1]

D and **E** (and **P**) are no longer parallel, and although $\mathbf{D} = \epsilon_0 \mathbf{E} + \mathbf{P}$ remains a valid equation for anisotropic materials, we may continue to use $\mathbf{D} = \epsilon \mathbf{E}$ only by interpreting it as a matrix equation.

---

[1] A more complete discussion of this matrix may be found in the Ramo, Whinnery, and Van Duzer reference listed at the end of this chapter.

We shall concentrate our attention on linear isotropic materials and reserve the general case for a more advanced text.

In summary, then, we now have a relationship between **D** and **E** which depends on the dielectric material present,

$$\mathbf{D} = \epsilon \mathbf{E} \tag{11}$$

where

$$\epsilon = \epsilon_0 \epsilon_r \tag{12}$$

This electric flux density is still related to the free charge by either the point or integral form of Gauss's law:

$$\nabla \cdot \mathbf{D} = \rho_\nu \tag{8}$$

$$\oint_S \mathbf{D} \cdot d\mathbf{S} = Q \tag{7}$$

The use of the relative permittivity, as indicated by (12), makes consideration of the polarization, dipole moments, and bound charge unnecessary. However, when anisotropic or nonlinear materials must be considered, the relative permittivity, in the simple scalar form that we have discussed, is no longer applicable.

Let us now illustrate these new concepts with a numerical example.

EXAMPLE 6.1

We locate a slab of Teflon in the region $0 \le x \le a$, and assume free space where $x < 0$ and $x > a$. Outside the Teflon there is a uniform field $\mathbf{E}_{\text{out}} = E_0 \mathbf{a}_x$ V/m. We seek values for **D**, **E**, and **P** everywhere.

Solution. The dielectric constant of Teflon is 2.1, and thus the electric susceptibility is 1.1.

Outside the slab, we have immediately $\mathbf{D}_{\text{out}} = \epsilon_0 E_0 \mathbf{a}_x$. Also since there is no dielectric material there, $\mathbf{P}_{\text{out}} = 0$. Now, any of the last four or five equations will enable us to relate the several fields inside the material to each other. Thus

$$\mathbf{D}_{\text{in}} = 2.1 \epsilon_0 \mathbf{E}_{\text{in}} \qquad (0 \le x \le a)$$
$$\mathbf{P}_{\text{in}} = 1.1 \epsilon_0 \mathbf{E}_{\text{in}} \qquad (0 \le x \le a)$$

As soon as we establish a value for any of these three fields within the dielectric, the other two can be found immediately. The difficulty lies in crossing over the boundary from the known fields external to the dielectric to the unknown ones within it. To do this we need a boundary condition, and this is the subject of the next exciting section. We shall complete this example then.

In the remainder of this text we shall describe polarizable materials in terms of **D** and ϵ rather than **P** and χ_e. We shall limit our discussion to isotropic materials.

D6.1 A slab of dielectric material has a relative dielectric constant of 3.8 and contains a uniform electric flux density of 8 nC/m^2. If the material is lossless, find: (*a*) E; (*b*) P; (*c*) the average number of dipoles per cubic meter if the average dipole moment is 10^{-29}C · m.

Ans. 238 V/m; 5.89 nC/m^2; 5.89×10^{20} m^{-3}

6.2 BOUNDARY CONDITIONS FOR PERFECT DIELECTRIC MATERIALS

Illustrations

How do we attack a problem in which there are two different dielectrics, or a dielectric and a conductor? This is another example of a *boundary condition*, such as the condition at the surface of a conductor whereby the tangential fields are zero and the normal electric flux density is equal to the surface charge density on the conductor. Now we take the first step in solving a two-dielectric problem, or a dielectric-conductor problem, by determining the behavior of the fields at the dielectric interface.

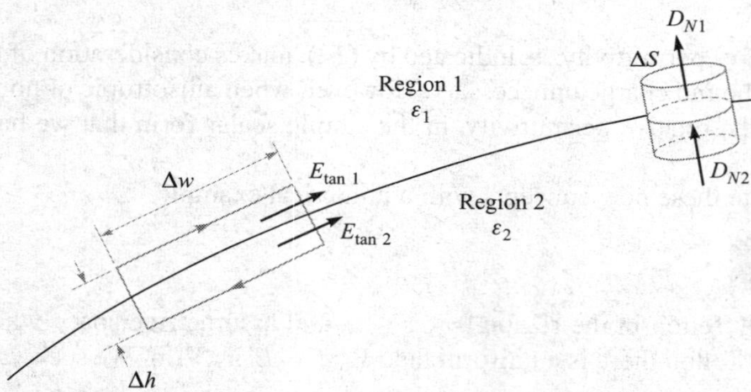

Fig. 6.2 The boundary between perfect dielectrics of permittivities ϵ_1 and ϵ_2. The continuity of D_N is shown by the gaussian surface on the right, and the continuity of $E_{\tan}$ is shown by the line integral about the closed path at the left.

Let us first consider the interface between two dielectrics having permittivities ϵ_1 and ϵ_2 and occupying regions 1 and 2, as shown in Figure 6.2. We first examine the tangential components by using

$$\oint \mathbf{E} \cdot d\mathbf{L} = 0$$

around the small closed path on the left, obtaining

$$E_{\tan 1}\, \Delta w - E_{\tan 2}\, \Delta w = 0$$

The small contribution to the line integral by the normal component of $\mathbf{E}$ along the sections of length Δh becomes negligible as Δh decreases and the closed path crowds the surface. Immediately, then,

$$E_{\tan 1} = E_{\tan 2} \tag{13}$$

and we might feel that Kirchhoff's voltage law is still applicable to this case. Certainly we have shown that the potential difference between any two points on the boundary that are separated by a distance Δw is the same immediately above or below the boundary.

If the tangential electric field intensity is continuous across the boundary, then tangential **D** is discontinuous, for

$$\frac{D_{\tan 1}}{\epsilon_1} = E_{\tan 1} = E_{\tan 2} = \frac{D_{\tan 2}}{\epsilon_2}$$

or

$$\frac{D_{\tan 1}}{D_{\tan 2}} = \frac{\epsilon_1}{\epsilon_2} \tag{14}$$

The boundary conditions on the normal components are found by applying Gauss's law to the small "pillbox" shown at the right in Figure 6.2. The sides are again very short, and the flux leaving the top and bottom surfaces is the difference

$$D_{N1}\Delta S - D_{N2}\Delta S = \Delta Q = \rho_S \Delta S$$

from which

$$\boxed{D_{N1} - D_{N2} = \rho_S} \tag{15}$$

What is this surface charge density? It cannot be a *bound* surface charge density, because we are taking the polarization of the dielectric into effect by using a dielectric constant different from unity; that is, instead of considering bound charges in free space, we are using an increased permittivity. Also, it is extremely unlikely that any *free* charge is on the interface, for no free charge is available in the perfect dielectrics we are considering. This charge must then have been placed there deliberately, thus unbalancing the total charge in and on this dielectric body. Except for this special case, then, we may assume ρ_S is zero on the interface and

$$\boxed{D_{N1} = D_{N2}} \tag{16}$$

or the normal component of **D** is continuous. It follows that

$$\epsilon_1 E_{N1} = \epsilon_2 E_{N2} \tag{17}$$

and normal **E** is discontinuous.

These conditions may be combined to show the change in the vectors **D** and **E** at the surface. Let $\mathbf{D}_1$ (and $\mathbf{E}_1$) make an angle θ_1 with a normal to the surface (Figure 6.3). Since the normal components of **D** are continuous,

$$D_{N1} = D_1 \cos\theta_1 = D_2 \cos\theta_2 = D_{N2} \tag{18}$$

The ratio of the tangential components is given by (14) as

$$\frac{D_{\tan 1}}{D_{\tan 2}} = \frac{D_1 \sin\theta_1}{D_2 \sin\theta_2} = \frac{\epsilon_1}{\epsilon_2}$$

or

$$\epsilon_2 D_1 \sin\theta_1 = \epsilon_1 D_2 \sin\theta_2 \tag{19}$$

and the division of this equation by (18) gives

$$\frac{\tan\theta_1}{\tan\theta_2} = \frac{\epsilon_1}{\epsilon_2} \tag{20}$$

In Figure 6.3 we have assumed that $\epsilon_1 > \epsilon_2$, and therefore $\theta_1 > \theta_2$.

The direction of **E** on each side of the boundary is identical with the direction of **D**, because $\mathbf{D} = \epsilon\mathbf{E}$.

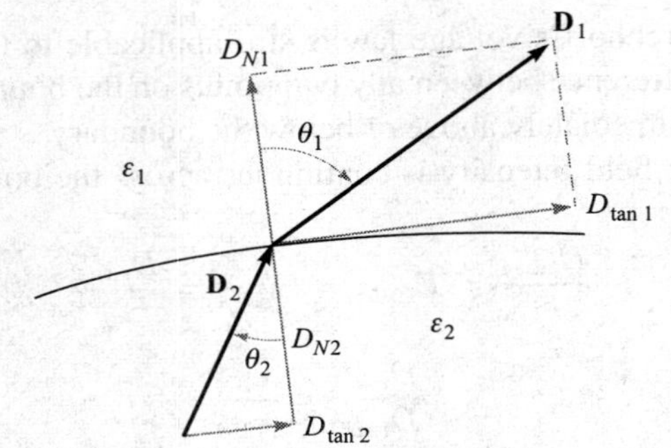

Fig. 6.3 The refraction of **D** at a dielectric interface. For the case shown, $\epsilon_1 > \epsilon_2$; $\mathbf{E}_1$ and $\mathbf{E}_2$ are directed along $\mathbf{D}_1$ and $\mathbf{D}_2$, with $D_1 > D_2$ and $E_1 < E_2$.

The magnitude of **D** in region 2 may be found from (18) and (19),

$$D_2 = D_1 \sqrt{\cos^2\theta_1 + \left(\frac{\epsilon_2}{\epsilon_1}\right)^2 \sin^2\theta_1} \tag{21}$$

and the magnitude of $\mathbf{E}_2$ is

$$E_2 = E_1 \sqrt{\sin^2\theta_1 + \left(\frac{\epsilon_1}{\epsilon_2}\right)^2 \cos^2\theta_1} \tag{22}$$

An inspection of these equations shows that D is larger in the region of larger permittivity (unless $\theta_1 = \theta_2 = 0^\circ$ where the magnitude is unchanged) and that E is larger in the region of smaller permittivity (unless $\theta_1 = \theta_2 = 90^\circ$, where its magnitude is unchanged).

These boundary conditions, (13), (14), (16), and (17), or the magnitude and direction relations derived from them, (20) to (22), allow us to find quickly the field on one side of a boundary *if we know the field on the other side*. In the example we began at the end of Section 6.1, this was the case. Now let's finish up that problem.

EXAMPLE 6.2

Complete Example 6.1 by finding the fields within the Teflon ($\epsilon_r = 2.1$), given the uniform external field $\mathbf{E}_{\text{out}} = E_0\mathbf{a}_x$ in free space.

Solution. We recall that we had a slab of Teflon extending from $x = 0$ to $x = a$, as shown in Figure 6.4, with free space on both sides of it and an external field $\mathbf{E}_{\text{out}} = E_0\mathbf{a}_x$. We also have $\mathbf{D}_{\text{out}} = \epsilon_0 E_0 \mathbf{a}_x$ and $\mathbf{P}_{\text{out}} = 0$.

Inside, the continuity of D_N at the boundary allows us to find that $\mathbf{D}_{\text{in}} = \mathbf{D}_{\text{out}} = \epsilon_0 E_0\mathbf{a}_x$. This gives us $\mathbf{E}_{\text{in}} = \mathbf{D}_{\text{in}}/\epsilon = \epsilon_0 E_0\mathbf{a}_x/(\epsilon_r\epsilon_0) = 0.476E_0\mathbf{a}_x$. To get the polarization field in the dielectric, we use $\mathbf{D} = \epsilon_0\mathbf{E} + \mathbf{P}$ and obtain

$$\mathbf{P}_{\text{in}} = \mathbf{D}_{\text{in}} - \epsilon_0\mathbf{E}_{\text{in}} = \epsilon_0 E_0\mathbf{a}_x - 0.476\epsilon_0 E_0\mathbf{a}_x = 0.524\epsilon_0 E_0\mathbf{a}_x$$

Summarizing then gives

$$\mathbf{D}_{\text{in}} = \epsilon_0 E_0\mathbf{a}_x \qquad (0 \le x \le a)$$
$$\mathbf{E}_{\text{in}} = 0.476E_0\mathbf{a}_x \qquad (0 \le x \le a)$$
$$\mathbf{P}_{\text{in}} = 0.524\epsilon_0 E_0\mathbf{a}_x \qquad (0 \le x \le a)$$

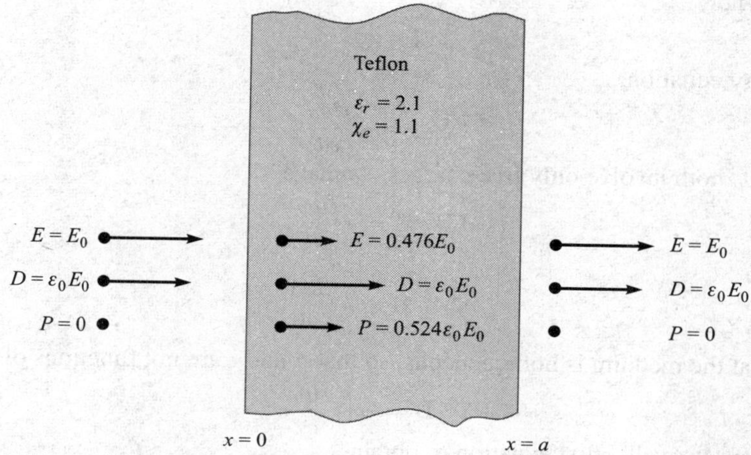

Fig. 6.4 A knowledge of the electric field external to the dielectric enables us to find the remaining external fields first and then to use the continuity of normal **D** to begin finding the internal fields.

A practical problem most often does not provide us with a direct knowledge of the field on either side of the boundary. The boundary conditions must be used to help us determine the fields on both sides of the boundary from the other information which is given. A simple problem of this type will be considered in Section 6.4.

The boundary conditions existing at the interface between a conductor and a dielectric are much simpler than those previously described. First, we know that **D** and **E** are both zero inside the conductor. Second, the tangential **E** and **D** field components must both be zero to satisfy

$$\oint \mathbf{E} \cdot d\mathbf{L} = 0$$

and

$$\mathbf{D} = \epsilon \mathbf{E}$$

Finally, the application of Gauss's law,

$$\oint_S \mathbf{D} \cdot d\mathbf{S} = Q$$

shows once more that both **D** and **E** are normal to the conductor surface and that $D_N = \rho_S$ and $E_N = \rho_S/\epsilon$. We see, then, that the boundary conditions we developed previously for the conductor–free space boundary are valid for the conductor-dielectric boundary if we replace ϵ_0 by ϵ. Thus

$$D_t = E_t = 0 \tag{23}$$

$$D_N = \epsilon E_N = \rho_S \tag{24}$$

It is interesting to spend a moment discovering how any change that is introduced *internally* within a conducting material arrives at the surface as a surface charge. We should understand that this is not a common occurrence, but it does give us some additional insight into the characteristics of a conductor.

Given Ohm's law,

$$\mathbf{J} = \sigma \mathbf{E}$$

and the continuity equation,

$$\nabla \cdot \mathbf{J} = -\frac{\partial \rho_\nu}{\partial t}$$

in which $\mathbf{J}$ and ρ_ν both involve only free charges, we have

$$\nabla \cdot \sigma \mathbf{E} = -\frac{\partial \rho_\nu}{\partial t}$$

or

$$\nabla \cdot \frac{\sigma}{\epsilon} \mathbf{D} = -\frac{\partial \rho_\nu}{\partial t}$$

If we assume that the medium is homogeneous, so that σ and ϵ are not functions of position,

$$\nabla \cdot \mathbf{D} = -\frac{\epsilon}{\sigma} \frac{\partial \rho_\nu}{\partial t}$$

Now we may use Maxwell's first equation to obtain

$$\rho_\nu = -\frac{\epsilon}{\sigma} \frac{\partial \rho_\nu}{\partial t}$$

Let us now make the simplifying assumption that σ is not a function of ρ_ν. This is probably not a very good assumption, for we found in Section 5.3, Eq. (9), that σ depended on both ρ_ν and the mobility, but it leads to an easy solution that at least permits us to compare different conductors. We simply rearrange and integrate directly, obtaining

$$\rho_\nu = \rho_0 e^{-(\sigma/\epsilon)t}$$

where $\rho_0 =$ charge density at $t = 0$. This shows an exponential decay of charge density at every point with a time constant of ϵ/σ. This time constant, often called the *relaxation time*, may be calculated for a relatively poor conductor, such as distilled water, from the data in Appendix C, giving

$$\frac{\epsilon}{\sigma} = \frac{80 \times 8.854 \times 10^{-12}}{2 \times 10^{-4}} = 3.54\ \mu\text{s}$$

In 3.54 μs any charge we place in the interior of a body of distilled water has dropped to about 37 percent of its initial value. This rapid decay is characteristic of good conductors and we see that, except for an extremely short transient period, we may safely consider the charge density to be zero within a good conductor.

With the physical materials with which we must work, no dielectric material is without some few free electrons; all have conductivities different from zero, and charge introduced internally in any of them will eventually reach the surface.

With the knowledge we now have of conducting materials, dielectric materials, and the necessary boundary conditions, we are ready to define and discuss capacitance.

D6.2 Let the region $z < 0$ be composed of a uniform dielectric material for which $\epsilon_r = 3.2$, while the region $z > 0$ is characterized by $\epsilon_r = 2$. Let $\mathbf{D}_1 = -30\mathbf{a}_x + 50\mathbf{a}_y + 70\mathbf{a}_z$ nC/m^2 and find: (*a*) D_{N1}; (*b*) $\mathbf{D}_{t1}$; (*c*) D_{t1}; (*d*) D_1; (*e*) θ_1; (*f*) $\mathbf{P}_1$.

Ans. 70 nC/m^2; $-30\mathbf{a}_x + 50\mathbf{a}_y$ nC/m^2; 58.3 nC/m^2; 91.1 nC/m^2; 39.8°; $-20.6\mathbf{a}_x + 34.4\mathbf{a}_y + 48.1\mathbf{a}_z$ nC/m^2

D6.3 Continue Problem D6.2 by finding: (*a*) $\mathbf{D}_{N2}$; (*b*) $\mathbf{D}_{t2}$; (*c*) $\mathbf{D}_2$; (*d*) $\mathbf{P}_2$; (*e*) θ_2

Ans. $70\mathbf{a}_z$ nC/m^2; $-18.75\mathbf{a}_x + 31.25\mathbf{a}_y$ nC/m^2; $-18.75\mathbf{a}_x + 31.25\mathbf{a}_y + 70\mathbf{a}_z$ nC/m^2; $-9.38\mathbf{a}_x$ $+15.63\mathbf{a}_y$ $+35\mathbf{a}_z$ nC/m^2; 27.5°

6.3 CAPACITANCE

Now let us consider two conductors embedded in a homogeneous dielectric (Figure 6.5). Conductor M_2 carries a total positive charge Q, and M_1 carries an equal negative charge. There are no other charges present, and the *total* charge of the system is zero.

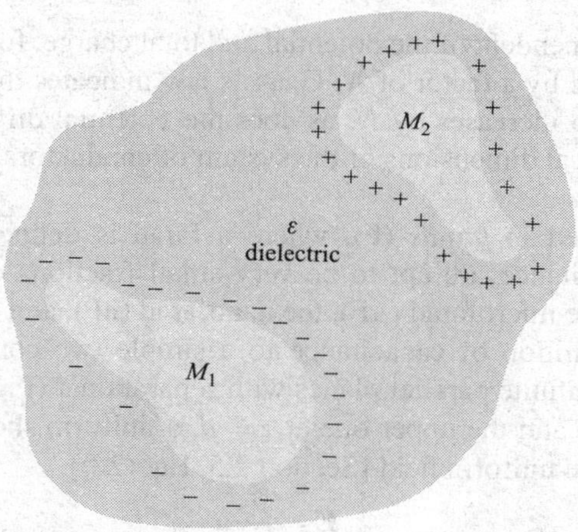

Fig. 6.5 Two oppositely charged conductors M_1 and M_2 surrounded by a uniform dielectric. The ratio of the magnitude of the charge on either conductor to the magnitude of the potential difference between them is the capacitance C.

We now know that the charge is carried on the surface as a surface charge density and also that the electric field is normal to the conductor surface. Each conductor is, moreover, an equipotential surface. Since M_2 carries the positive charge, the electric flux is directed from M_2 to M_1, and M_2 is at the more positive potential. In other words, work must be done to carry a positive charge from M_1 to M_2.

Let us designate the potential difference between M_2 and M_1 as V_0. We may now define the *capacitance* of this two-conductor system as the ratio of the magnitude of the total charge on either conductor to the magnitude of the potential difference between conductors,

$$C = \frac{Q}{V_0} \tag{25}$$

In general terms, we determine Q by a surface integral over the positive conductors, and we find V_0 by carrying a unit positive charge from the negative to the positive surface,

$$C = \frac{\oint_S \epsilon \mathbf{E} \cdot d\mathbf{S}}{-\int_-^+ \mathbf{E} \cdot d\mathbf{L}} \tag{26}$$

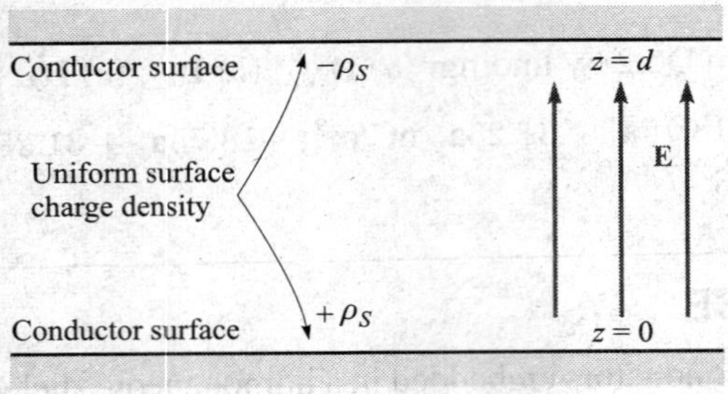

Fig. 6.6 The problem of the parallel-plate capacitor. The capacitance per square meter of surface area is ϵ/d.

The capacitance is independent of the potential and total charge, for their ratio is constant. If the charge density is increased by a factor of N, Gauss's law indicates that the electric flux density or electric field intensity also increases by N, as does the potential difference. The capacitance is a function only of the physical dimensions of the system of conductors and of the permittivity of the homogeneous dielectric.

Capacitance is measured in *farads* (F), where a farad is defined as one coulomb per volt. Common values of capacitance are apt to be very small fractions of a farad, and consequently more practical units are the microfarad (μF), the nanofarad (nF), and the picofarad (pF).

We can apply the definition of capacitance to a simple two-conductor system in which the conductors are identical, infinite parallel planes with separation d (Figure 6.6). Choosing the lower conducting plane at $z = 0$ and the upper one at $z = d$, a uniform sheet of surface charge $\pm\rho_S$ on each conductor leads to the uniform field [Section 2.5, Eq. (22)]

$$\mathbf{E} = \frac{\rho_S}{\epsilon}\mathbf{a}_z$$

where the permittivity of the homogeneous dielectric is ϵ, and

$$\mathbf{D} = \rho_S \mathbf{a}_z$$

The charge on the lower plane must then be positive, since **D** is directed upward, and the normal value of **D**,

$$D_N = D_z = \rho_S$$

is equal to the surface charge density there. On the upper plane,

$$D_N = -D_z$$

and the surface charge there is the negative of that on the lower plane.

The potential difference between lower and upper planes is

$$V_0 = -\int_{\text{upper}}^{\text{lower}} \mathbf{E} \cdot d\mathbf{L} = -\int_d^0 \frac{\rho_S}{\epsilon} dz = \frac{\rho_S}{\epsilon} d$$

Since the total charge on either plane is infinite, the capacitance is infinite. A more practical answer is obtained by considering planes, each of area S, whose linear dimensions are much greater than their separation d. The electric field and charge distribution are then almost uniform at all points not adjacent to the edges, and this latter region contributes only a small percentage of the total capacitance, allowing us to write the familiar result

$$Q = \rho_S S$$

$$V_0 = \frac{\rho_S}{\epsilon} d$$

$$C = \frac{Q}{V_0} = \frac{\epsilon S}{d} \tag{27}$$

More rigorously, we might consider (27) as the capacitance of a portion of the infinite-plane arrangement having a surface area S. Methods of calculating the effect of the unknown and nonuniform distribution near the edges must wait until we are able to solve more complicated potential problems.

EXAMPLE 6.3

Calculate the capacitance of a parallel-plate capacitor having a mica dielectric, $\epsilon_r = 6$, a plate area of 10 in.^{2}, and a separation of 0.01 in.

Solution. We may find that

$$S = 10 \times 0.0254^2 = 6.45 \times 10^{-3}\ \text{m}^2$$

$$d = 0.01 \times 0.0254 = 2.54 \times 10^{-4}\ \text{m}$$

and therefore

$$C = \frac{6 \times 8.854 \times 10^{-12} \times 6.45 \times 10^{-3}}{2.54 \times 10^{-4}} = 1.349\ \text{nF}$$

A large plate area is obtained in capacitors of small physical dimensions by stacking smaller plates in 50- or 100-decker sandwiches, or by rolling up foil plates separated by a flexible dielectric. Table C.1 in Appendix C also indicates that materials are available having dielectric constants greater than 1000.

If more than two conductors are involved, *partial capacitances* between each pair of conductors must be defined. This is interestingly discussed in Maxwell's works.[2]

Finally, the total energy stored in the capacitor is

$$W_E = \tfrac{1}{2}\int_{\text{vol}} \epsilon E^2\, dv = \tfrac{1}{2}\int_0^S \int_0^d \frac{\epsilon \rho_S^2}{\epsilon^2}\, dz\, dS = \tfrac{1}{2}\frac{\rho_S^2}{\epsilon} Sd = \tfrac{1}{2}\frac{\epsilon S}{d}\frac{\rho_S^2 d^2}{\epsilon^2}$$

or

$$W_E = \tfrac{1}{2} C V_0^2 = \tfrac{1}{2} Q V_0 = \tfrac{1}{2}\frac{Q^2}{C} \tag{28}$$

which are all familiar expressions. Equation (28) also indicates that the energy stored in a capacitor with a fixed potential difference across it increases as the dielectric constant of the medium increases.

[2] See the References at the end of Chapter 5.

D6.4 Find the relative permittivity of the dielectric material present in a parallel-plate capacitor if: (*a*) $S = 0.12\,\text{m}^2$, $d = 80\,\mu\text{m}$, $V_0 = 12$ V, and the capacitor contains $1\,\mu\text{J}$ of energy; (*b*) the stored energy density is 100 J/m^3, $V_0 = 200$ V, and $d = 45\,\mu\text{m}$; (*c*) $E = 200$ kV/m, $\rho_S = 20\,\mu\text{C/m}^2$, and $d = 100\,\mu\text{m}$.

Ans. 1.05; 1.14; 11.3

6.4 SEVERAL CAPACITANCE EXAMPLES

As a first brief example we choose a coaxial cable or coaxial capacitor of inner radius a, outer radius b, and length L. No great derivational struggle is required, because the potential difference is given as Eq. (11) in Section 4.3, and we find the capacitance very simply by dividing this by the total charge $\rho_L L$ in the length L. Thus,

$$C = \frac{2\pi\epsilon L}{\ln(b/a)} \tag{29}$$

Next we consider a spherical capacitor formed of two concentric spherical conducting shells of radius a and b, $b > a$. The expression for the electric field was obtained previously by Gauss's law,

$$E_r = \frac{Q}{4\pi\epsilon r^2}$$

where the region between the spheres is a dielectric with permittivity ϵ. The expression for potential difference was found from this by the line integral [Section 4.3, Eq. (12)]. Thus,

$$V_{ab} = \frac{Q}{4\pi\epsilon}\left(\frac{1}{a} - \frac{1}{b}\right)$$

Here Q represents the total charge on the inner sphere, and the capacitance becomes

$$C = \frac{Q}{V_{ab}} = \frac{4\pi\epsilon}{\dfrac{1}{a} - \dfrac{1}{b}} \tag{30}$$

If we allow the outer sphere to become infinitely large, we obtain the capacitance of an isolated spherical conductor,

$$C = 4\pi\epsilon a \tag{31}$$

For a diameter of 1 cm, or a sphere about the size of a marble,

$$C = 0.556 \text{ pF}$$

in free space.

Coating this sphere with a different dielectric layer, for which $\epsilon = \epsilon_1$, extending from $r = a$ to $r = r_1$,

$$D_r = \frac{Q}{4\pi r^2}$$

$$E_r = \frac{Q}{4\pi\epsilon_1 r^2} \qquad (a < r < r_1)$$
$$= \frac{Q}{4\pi\epsilon_0 r^2} \qquad (r_1 < r)$$

and the potential difference is

$$V_a - V_\infty = -\int_{r_1}^{a} \frac{Q\,dr}{4\pi\epsilon_1 r^2} - \int_{\infty}^{r_1} \frac{Q\,dr}{4\pi\epsilon_0 r^2}$$
$$= \frac{Q}{4\pi}\left[\frac{1}{\epsilon_1}\left(\frac{1}{a} - \frac{1}{r_1}\right) + \frac{1}{\epsilon_0 r_1}\right]$$

Therefore,

$$C = \frac{4\pi}{\frac{1}{\epsilon_1}\left(\frac{1}{a} - \frac{1}{r_1}\right) + \frac{1}{\epsilon_0 r_1}} \qquad (32)$$

In order to look at the problem of multiple dielectrics a little more thoroughly, let us consider a parallel-plate capacitor of area S and spacing d, with the usual assumption that d is small compared to the linear dimensions of the plates. The capacitance is $\epsilon_1 S/d$, using a dielectric of permittivity ϵ_1. Now let us replace a part of this dielectric by another of permittivity ϵ_2, placing the boundary between the two dielectrics parallel to the plates (Figure 6.7).

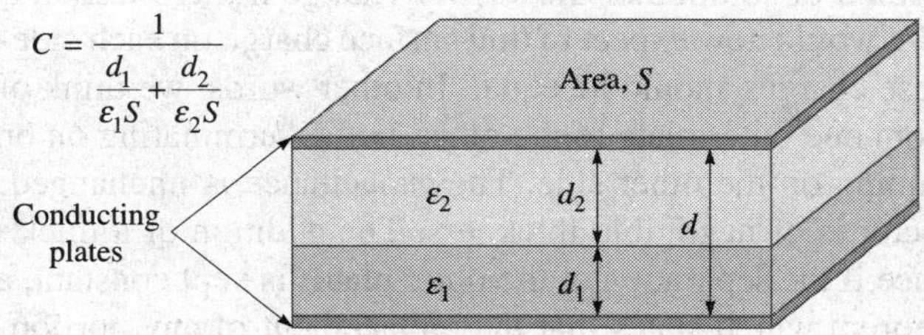

Fig. 6.7 A parallel-plate capacitor containing two dielectrics with the dielectric interface parallel to the conducting plates.

Some of us may immediately suspect that this combination is effectively two capacitors in series, yielding a total capacitance of

$$C = \frac{1}{\frac{1}{C_1} + \frac{1}{C_2}}$$

where $C_1 = \epsilon_1 S/d_1$ and $C_2 = \epsilon_2 S/d_2$. This is the correct result, but we can obtain it using less intuition and a more basic approach.

Since our capacitance definition, $C = Q/V$, involves a charge and a voltage, we may assume either and then find the other in terms of it. The capacitance is not a function of either, but only of the dielectrics and the geometry. Suppose we assume a potential difference V_0 between the plates. The electric field intensities in the two regions, E_2 and E_1, are both uniform, and $V_0 = E_1 d_1 + E_2 d_2$. At the dielectric interface, E is normal and $D_{N1} = D_{N2}$, or $\epsilon_1 E_1 = \epsilon_2 E_2$. Eliminating E_2 in our V_0

relation, we have

$$E_1 = \frac{V_0}{d_1 + d_2(\epsilon_1/\epsilon_2)}$$

and the surface charge density therefore has the magnitude

$$\rho_{S1} = D_1 = \epsilon_1 E_1 = \frac{V_0}{\dfrac{d_1}{\epsilon_1} + \dfrac{d_2}{\epsilon_2}}$$

Since $D_1 = D_2$, the magnitude of the surface charge is the same on each plate. The capacitance is then

$$C = \frac{Q}{V_0} = \frac{\rho_S S}{V_0} = \frac{1}{\dfrac{d_1}{\epsilon_1 S} + \dfrac{d_2}{\epsilon_2 S}} = \frac{1}{\dfrac{1}{C_1} + \dfrac{1}{C_2}}$$

As an alternate (and slightly simpler) solution, we might assume a charge Q on one plate, leading to a charge density Q/S and a value of D that is also Q/S. This is true in both regions, as $D_{N1} = D_{N2}$ and D is normal. Then $E_1 = D/\epsilon_1 = Q/(\epsilon_1 S)$, $E_2 = D/\epsilon_2 = Q/(\epsilon_2 S)$, and the potential differences across the regions are $V_1 = E_1 d_1 = Qd_1/(\epsilon_1 S)$, and $V_2 = E_2 d_2 = Qd_2/(\epsilon_2 S)$. The capacitance is

$$C = \frac{Q}{V} = \frac{Q}{V_1 + V_2} = \frac{1}{\dfrac{d_1}{\epsilon_1 S} + \dfrac{d_2}{\epsilon_2 S}} \tag{33}$$

How would the method of solution or the answer change if there were a third conducting plane along the interface? We would now expect to find surface charge on each side of this conductor, and the magnitudes of these charges should be equal. In other words, we think of the electric lines not as passing directly from one outer plate to the other, but as terminating on one side of this interior plane and then continuing on the other side. The capacitance is unchanged, provided, of course, that the added conductor is of negligible thickness. The addition of a thick conducting plate will increase the capacitance if the separation of the outer plates is kept constant, and this is an example of a more general theorem which states that the replacement of any portion of the dielectric by a conducting body will cause an increase in the capacitance.

If the dielectric boundary were placed *normal* to the two conducting plates and the dielectrics occupied areas of S_1 and S_2, then an assumed potential difference V_0 would produce field strengths $E_1 = E_2 = V_0/d$. These are tangential fields at the interface, and they must be equal. Then we may find in succession $D_1, D_2, \rho_{S1}, \rho_{S2}$, and Q, obtaining a capacitance

$$C = \frac{\epsilon_1 S_1 + \epsilon_2 S_2}{d} = C_1 + C_2 \tag{34}$$

as we should expect.

At this time we can do very little with a capacitor in which two dielectrics are used in such a way that the interface is not everywhere normal or parallel to the fields. Certainly we know the boundary conditions at each conductor and at the dielectric interface; however, we do not know the fields to which to apply the boundary conditions. Such a problem must be put aside until our knowledge of field theory has increased and we are willing and able to use more advanced mathematical techniques.

D6.5 Determine the capacitance of: (*a*) a 1 ft length of 35B/U coaxial cable, which has an inner conductor 0.1045 in. in diameter, a polyethylene dielectric ($\epsilon_r = 2.26$ from Table C.1), and an outer conductor which has an inner diameter of 0.680 in.; (*b*) a conducting sphere of radius 2.5 mm, covered with a polyethylene layer 2 mm thick, surrounded by a conducting sphere of radius 4.5 mm; (*c*) two rectangular conducting plates, 1 cm by 4 cm, with negligible thickness, between which are three sheets of dielectric, each 1 cm by 4 cm, and 0.1 mm thick, having dielectric constants of 1.5, 2.5, and 6.

Ans. 20.5 pF; 1.41 pF; 28.7 pF

6.5 CAPACITANCE OF A TWO-WIRE LINE

We next consider the problem of the two-wire line. The configuration will consist of two parallel conducting cylinders, each of circular cross section, and we shall be able to find complete information about the electric field intensity, the potential field, the surface-charge-density distribution, and the capacitance. This arrangement is an important type of transmission line, as is the coaxial cable we have discussed several times before.

We begin by investigating the potential field of two infinite line charges. Figure 6.8 shows a positive line charge in the xz plane at $x = a$ and a negative line charge at $x = -a$. The potential of a single line charge with zero reference at a radius of R_0 is

$$V = \frac{\rho_L}{2\pi\epsilon}\ln\frac{R_0}{R}$$

We now write the expression for the combined potential field in terms of the radial distances from the positive and negative lines, R_1 and R_2, respectively,

$$V = \frac{\rho_L}{2\pi\epsilon}\left(\ln\frac{R_{10}}{R_1} - \ln\frac{R_{20}}{R_2}\right) = \frac{\rho_L}{2\pi\epsilon}\ln\frac{R_{10}R_2}{R_{20}R_1}$$

We choose $R_{10} = R_{20}$, thus placing the zero reference at equal distances from each line. This surface is the $x = 0$ plane. Expressing R_1 and R_2 in terms of x and y,

$$V = \frac{\rho_L}{2\pi\epsilon}\ln\sqrt{\frac{(x+a)^2+y^2}{(x-a)^2+y^2}} = \frac{\rho_L}{4\pi\epsilon}\ln\frac{(x+a)^2+y^2}{(x-a)^2+y^2} \tag{35}$$

In order to recognize the equipotential surfaces and adequately understand the problem we are going to solve, some algebraic manipulations are necessary. Choosing an equipotential surface $V = V_1$, we define K_1 as a dimensionless parameter that is a function of the potential V_1,

$$K_1 = e^{4\pi\epsilon V_1/\rho_L} \tag{36}$$

so that

$$K_1 = \frac{(x+a)^2+y^2}{(x-a)^2+y^2}$$

After multiplying and collecting like powers, we obtain

$$x^2 - 2ax\frac{K_1+1}{K_1-1} + y^2 + a^2 = 0$$

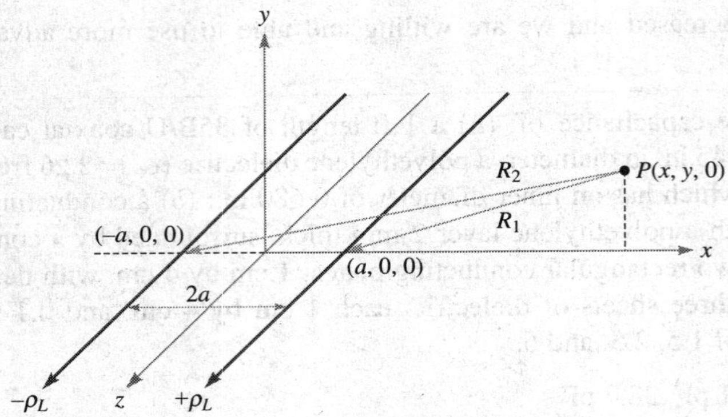

Fig. 6.8 Two parallel infinite line charges carrying opposite charge. The positive line is at $x = a$, $y = 0$, and the negative line is at $x = -a$, $y = 0$. A general point $P(x, y, 0)$ in the xy plane is radially distant R_1 and R_2 from the positive and negative lines, respectively. The equipotential surfaces are circular cylinders.

We next work through a couple of lines of algebra and complete the square,

$$\left(x - a\frac{K_1+1}{K_1-1}\right)^2 + y^2 = \left(\frac{2a\sqrt{K_1}}{K_1-1}\right)^2$$

This shows that the $V = V_1$ equipotential surface is independent of z (or is a cylinder) and intersects the xy plane in a circle of radius b,

$$b = \frac{2a\sqrt{K_1}}{K_1-1}$$

which is centered at $x = h, y = 0$, where

$$h = a\frac{K_1+1}{K_1-1}$$

Now let us attack a physical problem by considering a zero-potential conducting plane located at $x = 0$, and a conducting cylinder of radius b and potential V_0 with its axis located a distance h from the plane. We solve the last two equations for a and K_1 in terms of the dimensions b and h,

$$a = \sqrt{h^2 - b^2} \tag{37}$$

and

$$\sqrt{K_1} = \frac{h + \sqrt{h^2 - b^2}}{b} \tag{38}$$

But the potential of the cylinder is V_0, so (36) leads to

$$\sqrt{K_1} = e^{2\pi\epsilon V_0/\rho_L}$$

Therefore,

$$\rho_L = \frac{4\pi\epsilon V_0}{\ln K_1} \tag{39}$$

Thus, given h, b, and V_0, we may determine a, ρ_L, and the parameter K_1. The capacitance between the cylinder and plane is now available.

For a length L in the z direction, we have

$$C = \frac{\rho_L L}{V_0} = \frac{4\pi\epsilon L}{\ln K_1} = \frac{2\pi\epsilon L}{\ln\sqrt{K_1}}$$

or

$$C = \frac{2\pi\epsilon L}{\ln[h + \sqrt{h^2 - b^2}/b]} = \frac{2\pi\epsilon L}{\cosh^{-1}(h/b)} \tag{40}$$

The heavy black circle in Figure 6.9 shows the cross section of a cylinder of 5 m radius at a potential of 100 V in free space, with its axis 13 m distant from a plane at zero potential. Thus, $b = 5$, $h = 13$, $V_0 = 100$, and we rapidly find the location of the equivalent line charge from (37),

$$a = \sqrt{h^2 - b^2} = \sqrt{13^2 - 5^2} = 12\,\text{m}$$

the value of the potential parameter K_1 from (38),

$$\sqrt{K_1} = \frac{h + \sqrt{h^2 - b^2}}{b} = \frac{13 + 12}{5} = 5 \qquad K_1 = 25$$

the strength of the equivalent line charge from (39),

$$\rho_L = \frac{4\pi\epsilon V_0}{\ln K_1} = \frac{4\pi \times 8.854 \times 10^{-12} \times 100}{\ln 25} = 3.46\ \text{nC/m}$$

and the capacitance between cylinder and plane from (40),

$$C = \frac{2\pi\epsilon}{\cosh^{-1}(h/b)} = \frac{2\pi \times 8.854 \times 10^{-12}}{\cosh^{-1}(13/5)} = 34.6\ \text{pF/m}$$

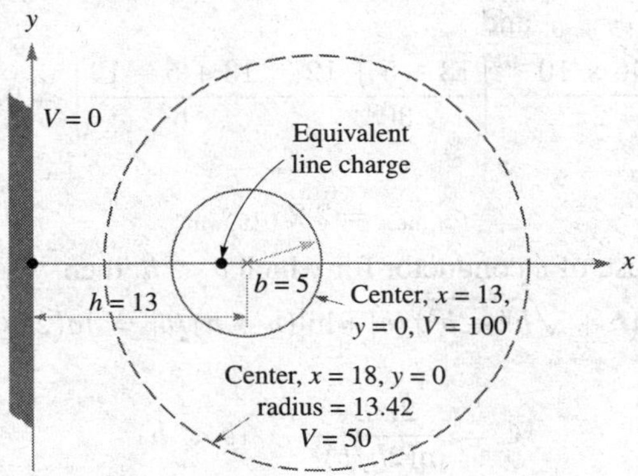

Fig. 6.9 A numerical example of the capacitance, linear charge density, position of an equivalent line charge, and characteristics of the mid-equipotential surface for a cylindrical conductor of 5 m radius at a potential of 100 V, parallel to and 13 m from a conducting plane at zero potential.

We may also identify the cylinder representing the 50 V equipotential surface by finding new values for K_1, h, and b. We first use (36) to obtain

$$K_1 = e^{4\pi\epsilon V_1/\rho_L} = e^{4\pi \times 8.854 \times 10^{-12} \times 50/3.46 \times 10^{-9}} = 5.00$$

Then the new radius is

$$b = \frac{2a\sqrt{K_1}}{K_1 - 1} = \frac{2 \times 12\sqrt{5}}{5 - 1} = 13.42\ \text{m}$$

and the corresponding value of h becomes

$$h = a\frac{K_1+1}{K_1-1} = 12\frac{5+1}{5-1} = 18 \text{ m}$$

This cylinder is shown in color in Figure 6.9.

The electric field intensity can be found by taking the gradient of the potential field, as given by (35),

$$\mathbf{E} = -\nabla\left[\frac{\rho_L}{4\pi\epsilon}\ln\frac{(x+a)^2+y^2}{(x-a)^2+y^2}\right]$$

Thus,

$$\mathbf{E} = -\frac{\rho_L}{4\pi\epsilon}\left[\frac{2(x+a)\mathbf{a}_x+2y\mathbf{a}_y}{(x+a)^2+y^2} - \frac{2(x-a)\mathbf{a}_x+2y\mathbf{a}_y}{(x-a)^2+y^2}\right]$$

and

$$\mathbf{D} = \epsilon\mathbf{E} = -\frac{\rho_L}{2\pi}\left[\frac{(x+a)\mathbf{a}_x+y\mathbf{a}_y}{(x+a)^2+y^2} - \frac{(x-a)\mathbf{a}_x+y\mathbf{a}_y}{(x-a)^2+y^2}\right].$$

If we evaluate D_x at $x = h - b, y = 0$, we may obtain $\rho_{S,\max}$

$$\rho_{S,\max} = -D_{x,x=h-b,y=0} = \frac{\rho_L}{2\pi}\left[\frac{h-b+a}{(h-b+a)^2} - \frac{h-b-a}{(h-b-a)^2}\right]$$

For our example,

$$\rho_{S,\max} = \frac{3.46\times10^{-9}}{2\pi}\left[\frac{13-5+12}{(13-5+12)^2} - \frac{13-5-12}{(13-5-12)^2}\right] = 0.165 \text{ nC/m}^2$$

Similarly, $\rho_{S,\min} = D_{x,x=h+b,y=0}$, and

$$\rho_{S,\min} = \frac{3.46\times10^{-9}}{2\pi}\left[\frac{13+5+12}{30^2} - \frac{13+5-12}{6^2}\right] = 0.073 \text{ nC/m}^2$$

Thus,

$$\rho_{S,\max} = 2.25\rho_{S,\min}$$

If we apply (40) to the case of a conductor for which $b \ll h$, then

$$\ln\left[\left(h+\sqrt{h^2-b^2}\right)/b\right] \doteq \ln[(h+h)/b] \doteq \ln(2h/b)$$

and

$$C = \frac{2\pi\epsilon L}{\ln(2h/b)} \qquad (b \ll h) \tag{41}$$

The capacitance between two circular conductors separated by a distance $2h$ is one-half the capacitance given by (40) or (41). This last answer is of interest because it gives us an expression for the capacitance of a section of two-wire transmission line, one of the types of transmission lines studied later in Chapter 14.

D6.6 A conducting cylinder with a radius of 1 cm and at a potential of 20 V is parallel to a conducting plane which is at zero potential. The plane is 5 cm distant from the cylinder axis. If the conductors are embedded in a perfect dielectric for which $\epsilon_r = 4.5$, find: (*a*) the capacitance per unit length between cylinder and plane; (*b*) $\rho_{S,\max}$ on the cylinder.

Ans. 109.2 pF/m; 42.6 nC/m

6.6 USING FIELD SKETCHES TO ESTIMATE CAPACITANCE IN TWO-DIMENSIONAL PROBLEMS

In capacitance problems in which the conductor configurations cannot be easily described using a single coordinate system, other analysis techniques are usually applied. Such methods typically involve a numerical determination of field or potential values over a grid within the region of interest, as will be briefly considered in Chapter 7. In this section, another approach is described that involves making sketches of field lines and equipotential surfaces in a manner that follows a few simple rules. This approach, while lacking the accuracy of more elegant methods, allows fairly quick estimates of capacitance while providing a useful physical picture of the field configuration.

The method requires only pencil and paper. Besides being economical, it is also capable of yielding good accuracy if used skillfully and patiently. Fair accuracy (5 to 10 percent on a capacitance determination) may be obtained by a beginner who does no more than follow the few rules and hints of the art. The method to be described is applicable only to fields in which no variation exists in the direction normal to the plane of the sketch. The procedure is based on several facts, that we have already demonstrated:

1. A conductor boundary is an equipotential surface.
2. The electric field intensity and electric flux density are both perpendicular to the equipotential surfaces.
3. **E** and **D** are therefore perpendicular to the conductor boundaries and possess zero tangential values.
4. The lines of electric flux, or streamlines, begin and terminate on charge and hence, in a charge-free, homogeneous dielectric, begin and terminate only on the conductor boundaries.

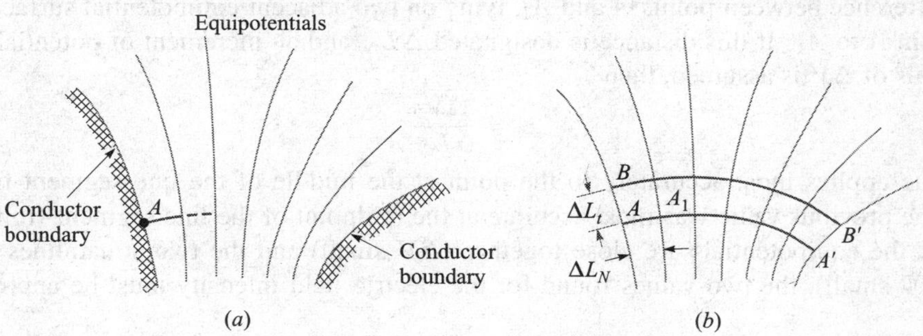

Fig. 6.10 (a) Sketch of the equipotential surfaces between two conductors. The increment of potential between each of the two adjacent equipotentials is the same. (b) One flux line has been drawn from A to A', and a second from B to B'.

Let us consider the implications of these statements by drawing the streamlines on a sketch which already shows the equipotential surfaces. In Figure 6.10a, two conductor boundaries are shown, and equipotentials are drawn with a constant potential difference between lines. We should remember that these lines are only the cross sections of the equipotential surfaces, which are cylinders (although not circular), since no variation in the direction normal to the surface of the paper is permitted. We arbitrarily choose to begin a streamline, or flux line, at A on the surface of the more positive conductor. It leaves the surface normally and must cross at right angles the undrawn but very real equipotential surfaces between the conductor and the first surface shown. The line is continued to the other conductor, obeying the single rule that the intersection with each equipotential must be square. Turning the paper from side to side as the line progresses enables us to maintain perpendicularity more accurately. The line has been completed in Figure 6.10*b*.

In a similar manner, we may start at B and sketch another streamline ending at B'. Before continuing, let us interpret the meaning of this pair of streamlines. The streamline, by definition, is everywhere tangent to the electric field intensity or to the electric flux density. Since the streamline is tangent to the electric flux density, the flux density is tangent to the streamline, and no electric flux may cross any streamline. In other words, if there is a charge of 5 μC on the surface between A and B (and extending 1 m into the paper), then 5 μC of flux begins in this region, and all must terminate between A' and B'. Such a pair of lines is sometimes called a flux *tube*, because it physically seems to carry flux from one conductor to another without losing any.

We now wish to construct a third streamline, and both the mathematical and visual interpretations we may make from the sketch will be greatly simplified if we draw this line starting from some point C chosen so that the same amount of flux is carried in the tube *BC* as is contained in *AB*. How do we choose the position of C?

The electric field intensity at the midpoint of the line joining A to B may be found approximately by assuming a value for the flux in the tube *AB*, say $\Delta\Psi$, which allows us to express the electric flux density by $\Delta\Psi/\Delta L_t$, where the depth of the tube into the paper is 1 m and ΔL_t is the length of the line joining A to B. The magnitude of E is then

$$E = \frac{1}{\epsilon}\frac{\Delta\Psi}{\Delta L_t}$$

However, we may also find the magnitude of the electric field intensity by dividing the potential difference between points A and A_1, lying on two adjacent equipotential surfaces, by the distance from A to A_1. If this distance is designated ΔL_N and an increment of potential between equipotentials of ΔV is assumed, then

$$E = \frac{\Delta V}{\Delta L_N}$$

This value applies most accurately to the point at the middle of the line segment from A to A_1, while the previous value was most accurate at the midpoint of the line segment from A to B. If, however, the equipotentials are close together (ΔV small) and the two streamlines are close together ($\Delta\Psi$ small), the two values found for the electric field intensity must be approximately equal,

$$\frac{1}{\epsilon}\frac{\Delta\Psi}{\Delta L_t} = \frac{\Delta V}{\Delta L_N} \tag{42}$$

Throughout our sketch we have assumed a homogeneous medium (ϵ constant), a constant increment of potential between equipotentials (ΔV constant), and a constant amount of flux per tube ($\Delta\Psi$ constant). To satisfy all these conditions, (42) shows that

$$\boxed{\frac{\Delta L_t}{\Delta L_N} = \text{constant} = \frac{1}{\epsilon}\frac{\Delta\Psi}{\Delta V}} \tag{43}$$

A similar argument might be made at any point in our sketch, and we are therefore led to the conclusion that a constant ratio must be maintained between the distance between streamlines as measured along an equipotential, and the distance between equipotentials as measured along a streamline. It is this *ratio* which must have the same value at every point, not the individual lengths. Each length must decrease in regions of greater field strength, because ΔV is constant.

The simplest ratio we can use is unity, and the streamline from B to B' shown in Figure 6.10b was started at a point for which $\Delta L_t = \Delta L_N$. Since the ratio of these distances is kept at unity, the streamlines and equipotentials divide the field-containing region into curvilinear squares, a term implying a planar geometric figure which differs from a true square in having slightly curved and slightly unequal sides but which approaches a square as its dimensions decrease. Those incremental surface elements in our three coordinate systems which are planar may also be drawn as curvilinear squares.

We may now rapidly sketch in the remainder of the streamlines by keeping each small box as square as possible. The complete sketch is shown in Figure 6.11.

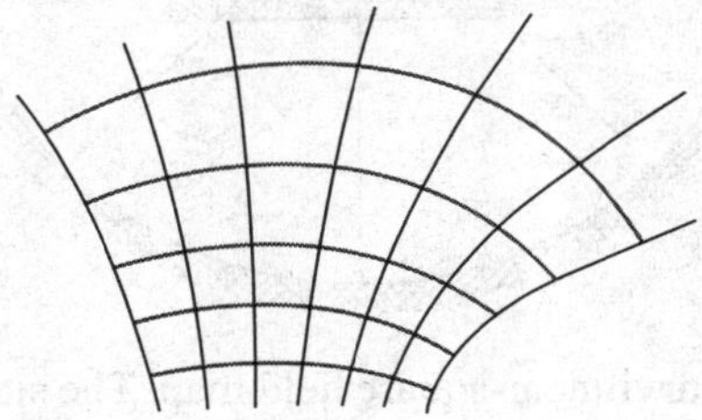

Fig. 6.11 The remainder of the streamlines have been added to Fig. 6.10b by beginning each new line normally to the conductor and maintaining curvilinear squares throughout the sketch.

The only difference between this example and the production of a field map using the method of curvilinear squares is that the intermediate potential surfaces are not given. The streamlines and equipotentials must both be drawn on an original sketch which shows only the conductor boundaries. Only one solution is possible, as we shall prove later by the uniqueness theorem for Laplace's equation, and the rules we have outlined are sufficient. One streamline is begun, an equipotential line is roughed in, another streamline is added, forming a curvilinear square, and the map is gradually extended throughout the desired region. Since none of us can ever expect to be perfect at this, we shall soon find that we can no longer make squares and also maintain right-angle corners. An error is accumulating in the drawing, and our present troubles should indicate the nature of the correction to make on some of the earlier work. It is usually best to start again on a fresh drawing, with the old one available as a guide.

The construction of a useful field map is an art; the science merely furnishes the rules. Proficiency in any art requires practice. A good problem for beginners is the coaxial cable or coaxial capacitor, since all the equipotentials are circles while the flux lines are straight lines. The next sketch attempted should be two parallel circular conductors, where the equipotentials are again circles but with different centers. Each of these is given as a problem at the end of the chapter, and the accuracy of the sketch may be checked by a capacitance calculation as outlined in the following paragraphs.

Figure 6.12 shows a completed map for a cable containing a square inner conductor surrounded by a circular conductor. The capacitance is found from $C = Q/V_0$ by replacing Q by $N_Q \Delta Q = N_Q \Delta \Psi$, where N_Q is the number of flux tubes joining the two conductors, and letting $V_0 = N_V \Delta V$, where N_V is the number of potential increments between conductors,

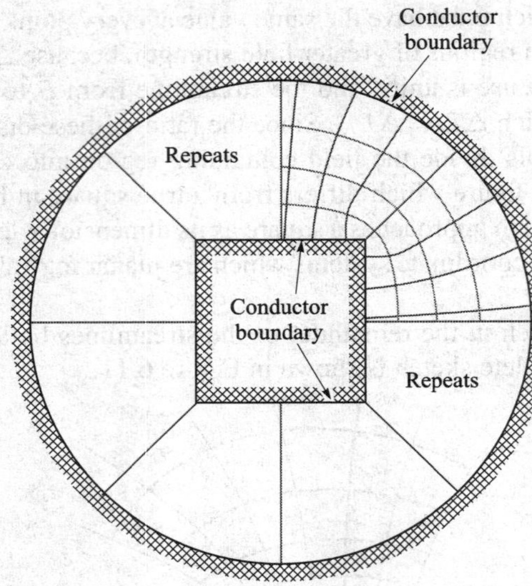

Fig. 6.12 An example of a curvilinear-square field map. The side of the square is two-thirds the radius of the circle. $N_V = 4$ and $N_Q = 8 \times 3.25 \times 26$, and therefore $C = \epsilon_0 N_Q/N_V = 57.6$ pF/m.

$$C = \frac{N_Q \Delta Q}{N_V \Delta V}$$

and then using (43),

$$C = \frac{N_Q}{N_V} \epsilon \frac{\Delta L_t}{\Delta L_N} = \epsilon \frac{N_Q}{N_V} \tag{44}$$

since $\Delta L_t/\Delta L_N = 1$. The determination of the capacitance from a flux plot merely consists of counting squares in two directions, between conductors and around either conductor. From Figure 6.12 we obtain

$$C = \epsilon_0 \frac{8 \times 3.25}{4} = 57.6 \text{ pF/m}$$

Ramo, Whinnery, and Van Duzer have an excellent discussion with examples of the construction of field maps by curvilinear squares. They offer the following suggestions:[3]

1. Plan on making a number of rough sketches, taking only a minute or so apiece, before starting any plot to be made with care. The use of transparent paper over the basic boundary will speed up this preliminary sketching.
2. Divide the known potential difference between electrodes into an equal number of divisions, say four or eight to begin with.
3. Begin the sketch of equipotentials in the region where the field is known best, as for example in some region where it approaches a uniform field. Extend the equipotentials according to your

[3] By permission from S. Ramo, J. R. Whinnery, and T. Van Duzer, pp. 51–52. See References at the end of this chapter. Curvilinear maps are discussed on pp. 50–52.

best guess throughout the plot. Note that they should tend to hug acute angles of the conducting boundary and be spread out in the vicinity of obtuse angles of the boundary.

4. Draw in the orthogonal set of field lines. As these are started, they should form curvilinear squares, but, as they are extended, the condition of orthogonality should be kept paramount, even though this will result in some rectangles with ratios other than unity.
5. Look at the regions with poor side ratios and try to see what was wrong with the first guess of equipotentials. Correct them and repeat the procedure until reasonable curvilinear squares exist throughout the plot.
6. In regions of low field intensity, there will be large figures, often of five or six sides. To judge the correctness of the plot in this region, these large units should be subdivided. The subdivisions should be started back away from the region needing subdivision, and each time a flux tube is divided in half, the potential divisions in this region must be divided by the same factor.

D6.7 Figure 6.13 shows the cross section of two circular cylinders at potentials of 0 and 60 V. The axes are parallel and the region between the cylinders is air-filled. Equipotentials at 20 V and 40 V are also shown. Prepare a curvilinear-square map on the figure and use it to establish suitable values for: (*a*) the capacitance per meter length; (*b*) E at the left side of the 60 V conductor if its true radius is 2 mm; (*c*) ρ_S at that point.

Ans. 69 pF/m; 60 kV/m; 550 nC/m^2

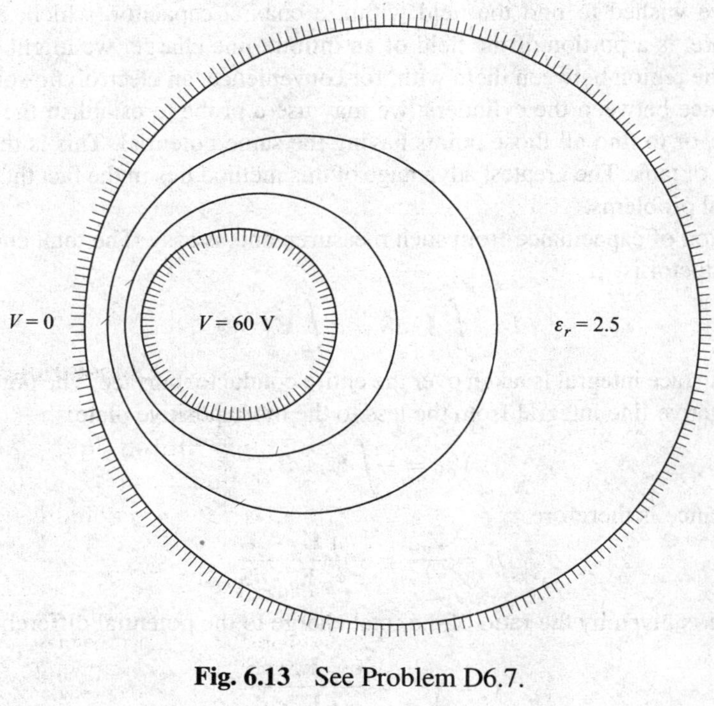

Fig. 6.13 See Problem D6.7.

6.7 CURRENT ANALOGIES

A useful analogy exists between current density in conducting media and electric flux density in dielectric media. The analogy is easily demonstrated, for in a conducting medium Ohm's law and the gradient relationship are, for direct currents only,

$$\mathbf{J} = \sigma \mathbf{E}_\sigma$$

$$\mathbf{E}_\sigma = -\nabla V_\sigma$$

whereas in a homogeneous dielectric

$$\mathbf{D} = \epsilon \mathbf{E}_\epsilon$$

$$\mathbf{E}_\epsilon = -\nabla V_\epsilon$$

The subscripts serve to identify the analogous problems. It is evident that the potentials V_σ and V_ϵ, the electric field intensities $\mathbf{E}_\sigma$ and $\mathbf{E}_\epsilon$, the conductivity and permittivity σ and ϵ, and the current density and electric flux density $\mathbf{J}$ and $\mathbf{D}$ are analogous in pairs. Referring to a curvilinear-square map, we should interpret flux tubes as current tubes, each tube now carrying an incremental current which cannot leave the tube.

Finally, we must look at the boundaries. What is analogous to a conducting boundary which terminates electric flux normally and is an equipotential surface? The analogy furnishes the answer, and we see that the surface must terminate current density normally and again be an equipotential surface. This is the surface of a *perfect* conductor, although in practice it is necessary only to use one whose conductivity is many times that of the conducting medium.

Therefore, if we wished to find the field within a coaxial capacitor, which, as we have seen several times before, is a portion of the field of an infinite line charge, we might take two copper cylinders and fill the region between them with, for convenience, an electrolytic solution. Applying a potential difference between the cylinders, we may use a probe to establish the potential at any intermediate point, or to find all those points having the same potential. This is the essence of the electrolytic trough or tank. The greatest advantage of this method lies in the fact that it is not limited to two-dimensional problems.

The determination of capacitance from such measurements is easy. The total current leaving the more positive conductor is

$$I = \oint_S \mathbf{J} \cdot d\mathbf{S} = \sigma \oint_S \mathbf{E}_\sigma \cdot d\mathbf{S}$$

where the closed surface integral is taken over the entire conductor surface. The potential difference is given by the negative line integral from the less to the more positive plate,

$$V_{\sigma 0} = -\int \mathbf{E}_\sigma \cdot d\mathbf{L}$$

and the total resistance is therefore

$$R = \frac{V_{\sigma 0}}{I} = \frac{-\int \mathbf{E}_\sigma \cdot d\mathbf{L}}{\sigma \oint_S \mathbf{E}_\sigma \cdot d\mathbf{S}}$$

The capacitance is given by the ratio of the total charge to the potential difference,

$$C = \frac{Q}{V_{\epsilon 0}} = \frac{\epsilon \oint_S \mathbf{E}_\epsilon \cdot d\mathbf{S}}{-\int \mathbf{E}_\epsilon \cdot d\mathbf{L}}$$

We now invoke the analogy by letting $V_{\epsilon 0} = V_{\sigma 0}$ and $\mathbf{E}_\epsilon = \mathbf{E}_\sigma$. The result is

$$RC = \frac{\epsilon}{\sigma} \tag{45}$$

Knowing the conductivity of the electrolyte and the permittivity of the dielectric, we may determine the capacitance by a simple resistance measurement.

Quizzes

Another obvious use of (45) is in quickly evaluating the resistance of a given two-conductor configuration, knowing its capacitance, or vice versa. We will find this useful when we determine expressions for transmission line parameters in Chapter 14. It should be noted, however, that (45) can be applied safely only to linear, homogeneous, and isotropic media. Specifically, permittivity and conductivity cannot depend on field strength, position within the medium, or field orientation.

References

1. Fano, R. M., L. J. Chu, and R. B. Adler. *Electromagnetic Fields, Energy, and Forces*. New York: John Wiley & Sons, 1960. Polarization in dielectrics is discussed in the first part of Chapter 5. This junior-level text presupposes a full-term physics course in electricity and magnetism, and it is therefore slightly more advanced in level. The introduction beginning on p. 1 should be read.
2. Fink, D. G., and H. W. Beaty. *Standard Handbook for Electrical Engineers*. 12th ed. New York: McGraw-Hill, 1987.
3. Matsch, L. W. *Capacitors, Magnetic Circuits, and Transformers*. Englewood Cliffs, N.J.: Prentice-Hall, 1964. Many of the practical aspects of capacitors are discussed in Chapter 2.
4. Ramo, S., J. R. Whinnery, and T. Van Duzer. *Fields and Waves in Communications Electronics*. 3rd ed. New York: John Wiley & Sons, 1994. This book is essentially the fifth edition of the senior authors' popular texts of 1944 and 1953. Although it is directed primarily toward beginning graduate students, it may be profitably read by anyone who is familiar with basic electromagnetic concepts. Anisotropic dielectric materials are discussed on pp. 699–712. Curvilinear square plotting is described on pp. 50–52.

CHAPTER 6 Problems

6.1 Atomic hydrogen contains 5.5×10^{25} atoms/m^{3}at a certain temperature and pressure. When an electric field of 4 kV/m is applied, each dipole formed by the electron and positive nucleus has an effective length of 7.1×10^{-19} m. (*a*) Find P. (*b*) Find ϵ_r.

6.2 Find the dielectric constant of a material in which the electric flux density is four times the polarization.

6.3 A coaxial conductor has radii $a = 0.8$ mm and $b = 3$ mm and a polystyrene dielectric for which $\epsilon_r = 2.56$. If $\mathbf{P} = (2/\rho)\mathbf{a}_\rho$ nC/m^2 in the dielectric, find: (*a*) **D** and **E** as functions of ρ; (*b*) V_{ab} and χ_e. (*c*) If there are 4×10^{19} molecules per cubic meter in the dielectric, find $\mathbf{p}(\rho)$.

6.4 Consider a composite material made up of two species, having number densities N_1 and N_2 molecules/m^3, respectively. The two materials are uniformly mixed, yielding a total number density of $N = N_1 + N_2$. The presence of an electric field **E** induces molecular dipole moments $\mathbf{p}_1$ and $\mathbf{p}_2$ within the individual species, whether mixed or not. Show that the dielectric constant of the composite material is given by $\epsilon_r = f\epsilon_{r1} + (1 - f)\epsilon_{r2}$, where f is the number fraction of species 1 dipoles in the composite, and where ϵ_{r1} and ϵ_{r2} are the dielectric constants that the unmixed species would have if each had number density N.

6.5 The surface $x = 0$ separates two perfect dielectrics. For $x > 0$ let $\epsilon_r = \epsilon_{r1} = 3$, while $\epsilon_{r2} = 5$ where $x < 0$. If $\mathbf{E}_1 = 80\mathbf{a}_x - 60\mathbf{a}_y - 30\mathbf{a}_z$ V/m, find: (*a*) E_{N1}; (*b*) $\mathbf{E}_{T1}$; (*c*) E_1; (*d*) the angle θ_1 between $\mathbf{E}_1$ and a normal to the surface; (*e*) D_{N2}; (*f*) D_{T2}; (*g*) $\mathbf{D}_2$; (*h*) $\mathbf{P}_2$; (*i*) the angle θ_2 between $\mathbf{E}_2$ and a normal to the surface.

6.6 The potential field in a slab of dielectric material for which $\epsilon_r = 1.6$ is given by $V = -5000x$. (*a*) Find **D**, **E**, and **P** in the material. (*b*) Evaluate ρ, ρ_b, and ρ_t in the material.

6.7 Two perfect dielectrics have relative permittivities $\epsilon_{r1} = 2$ and $\epsilon_{r2} = 8$. The planar interface between them is the surface $x - y + 2z = 5$. The origin lies in region 1. If $\mathbf{E}_1 = 100\mathbf{a}_x + 200\mathbf{a}_y - 50\mathbf{a}_z$ V/m, find $\mathbf{E}_2$.

6.8 Region 1 ($x \geq 0$) is a dielectric with $\epsilon_{r1} = 2$, while region $2(x < 0)$ has $\epsilon_{r2} = 5$. Let $\mathbf{E}_1 = 20\mathbf{a}_x - 10\mathbf{a}_y + 50\mathbf{a}_z$ V/m. (*a*) Find $\mathbf{D}_2$. (*b*) Find the energy density in both regions.

6.9 Let the cylindrical surfaces $\rho = 4$ cm and $\rho = 9$ cm enclose two wedges of perfect dielectrics, $\epsilon_{r1} = 2$ for $0 < \phi < \pi/2$ and $\epsilon_{r2} = 5$ for $\pi/2 < \phi < 2\pi$. If $\mathbf{E}_1 = (2000/\rho)\mathbf{a}_\rho$ V/m, find: (*a*) $\mathbf{E}_2$; (*b*) the total electrostatic energy stored in a 1 m length of each region.

6.10 Let $S = 100$ mm^2, $d = 3$ mm, and $\epsilon_r = 12$ for a parallel-plate capacitor. (*a*) Calculate the capacitance. (*b*) After connecting a 6 V battery across the capacitor, calculate E, D, Q, and the total stored electrostatic energy. (*c*) With the source still connected, the dielectric is carefully withdrawn from between the plates. With the dielectric gone, recalculate E, D, Q, and the energy stored in the capacitor. (*d*) If the charge and energy found in part (*c*) are less than the values found in part (*b*) (which you should have discovered), what became of the missing charge and energy?

6.11 Capacitors tend to be more expensive as their capacitance and maximum voltage, $V_{\max}$ increase. The voltage $V_{\max}$ is limited by the field strength at which the dielectric breaks down, E_{BD}. Which of these dielectrics will give the largest $CV_{\max}$ product for equal plate areas: (*a*) air: $\epsilon_r = 1$, $E_{BD} = 3$ MV/m; (*b*) barium titanate: $\epsilon_r = 1200$, $E_{BD} = 3$ MV/m; (*c*) silicon dioxide: $\epsilon_r = 3.78$, $E_{BD} = 16$ MV/m; (*d*) polyethylene: $\epsilon_r = 2.26$, $E_{BD} = 4.7$ MV/m?

6.12 An air-filled parallel-plate capacitor with plate separation d and plate area A is connected to a battery which applies a voltage V_0 between plates. With the battery left connected, the plates are moved apart to a distance of $10d$. Determine by what factor each of the following quantities changes: (*a*) V_0; (*b*) C; (*c*) E; (*d*) D; (*e*) Q; (*f*) ρ_S; (*g*) W_E.

6.13 A parallel-plate capacitor is filled with a nonuniform dielectric characterized by $\epsilon_r = 2 + 2 \times 10^6 x^2$, where x is the distance from one plate in meters. If $S = 0.02\,\text{m}^2$ and $d = 1$ mm, find C.

6.14 Repeat Problem 6.12, assuming the battery is disconnected before the plate separation is increased.

6.15 Let $\epsilon_{r1} = 2.5$ for $0 < y < 1$ mm, $\epsilon_{r2} = 4$ for $1 < y < 3$ mm, and ϵ_{r3} for $3 < y < 5$ mm (region 3). Conducting surfaces are present at $y = 0$ and $y = 5$ mm. Calculate the capacitance per square meter of surface area if: (*a*) region 3 is air; (*b*) $\epsilon_{r3} = \epsilon_{r1}$; (*c*) $\epsilon_{r3} = \epsilon_{r2}$; (*d*) region 3 is silver.

6.16 A parallel-plate capacitor is made using two circular plates of radius a, with the bottom plate on the xy plane, centered at the origin. The top plate is located at $z = d$, with its center on the z axis. Potential V_0 is on the top plate; the bottom plate is grounded. Dielectric having *radially dependent* permittivity fills the region between plates. The permittivity is given by $\epsilon(\rho) = \epsilon_0(1 + \rho/a)$. Find: (*a*) $\mathbf{E}$; (*b*) $\mathbf{D}$; (*c*) Q; (*d*) C.

6.17 Two coaxial conducting cylinders of radius 2 cm and 4 cm have a length of 1 m. The region between the cylinders contains a layer of dielectric from $\rho = c$ to $\rho = d$ with $\epsilon_r = 4$. Find the capacitance if: (*a*) $c = 2$ cm, $d = 3$ cm; (*b*) $d = 4$ cm, and the volume of the dielectric is the same as in part (*a*).

6.18 (*a*) If we could specify a material to be used as the dielectric in a coaxial capacitor for which the permittivity varied continuously with radius, what variation with ρ should be used in order to maintain a uniform value of the electric field intensity? (*b*) Under the conditions of part (*a*) how do the inner and outer radii appear in the expression for the capacitance per unit distance?

6.19 Two conducting spherical shells have radii $a = 3$ cm and $b = 6$ cm. The interior is a perfect dielectric for which $\epsilon_r = 8$. (*a*) Find C. (*b*) A portion of the dielectric is now removed so that $\epsilon_r = 1.0, 0 < \phi < \pi/2$, and $\epsilon_r = 8, \pi/2 < \phi < 2\pi$. Again find C.

6.20 Show that the capacitance per unit length of a cylinder of radius a is zero.

6.21 With reference to Figure 6.9, let $b = 6$ m, $h = 15$ m, and the conductor potential be 250 V. Take $\epsilon = \epsilon_0$. Find values for K_1, ρ_L, a, and C.

6.22 Two #16 copper conductor (1.29 mm diameter) are parallel with a separation d between axes. Determine d so that the capacitance between wires in air is 30 pF/m.

6.23 A 2 cm diameter conductors is suspended in air with its axis 5 cm from a conducting plane. Let the potential of the cylinder be 100 V and that of the plane be 0 V. Find the surface charge density on the (*a*) cylinder at a point nearest the plane; (*b*) plane at a point nearest the cylinder.

6.24 For the conductor configuration of Problem 6.23, determine the capacitance per unit length.

6.25 Construct a curvilinear-square map for a coaxial capacitor of 3 cm inner radius and 8 cm outer radius. These dimensions are suitable for the drawing. (*a*) Use your sketch to calculate the capacitance per meter length, assuming $\epsilon_r = 1$. (*b*) Calculate an exact value for the capacitance per unit length.

6.26 Construct a curvilinear-square map of the potential field about two parallel circular cylinders, each of 2.5 cm radius, separated by a center- to-center distance of 13 cm. These dimensions are suitable for the actual sketch if symmetry is considered. As a check, compute the capacitance per meter both from your sketch and from the exact formula. Assume $\epsilon_r = 1$.

6.27 Construct a curvilinear-square map of the potential field between two parallel circular cylinders, one of 4 cm radius inside another of 8 cm radius. The two axes are displaced by 2.5 cm. These dimensions are suitable for the drawing. As a check on the accuracy, compute the capacitance per meter from the sketch and from the exact expression:

$$C = \frac{2\pi\epsilon}{\cosh^{-1}\left[(a^2 + b^2 - D^2)/(2ab)\right]}$$

where a and b are the conductor radii and D is the axis separation.

6.28 A solid conducting cylinder of 4 cm radius is centered within a rectangular conducting cylinder with a 12 cm by 20 cm cross section. (*a*) Make a full-size sketch of one quadrant of

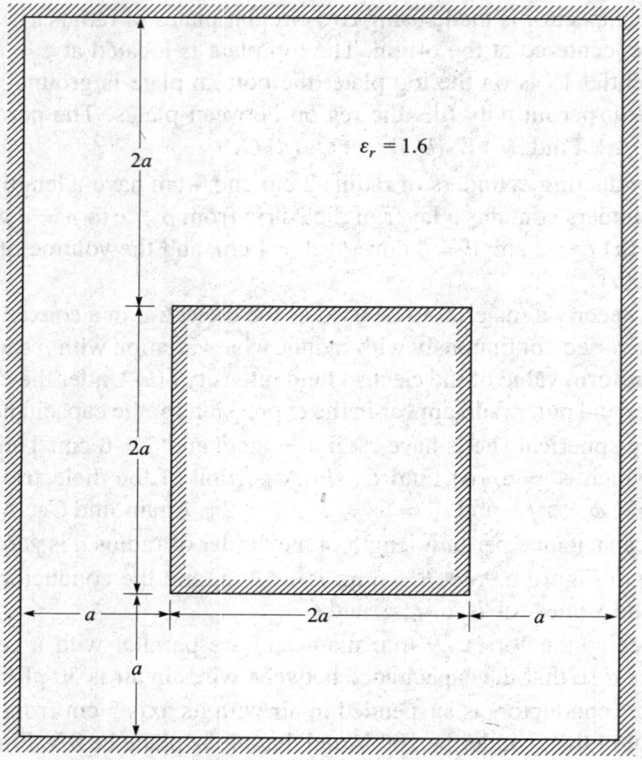

Fig. 6.14 See Problem 6.29.

this configuration and construct a curvilinear-square map for its interior. (b) Assume $\epsilon = \epsilon_0$ and estimate C per meter length.

6.29 The inner conductor of the transmission line shown in Figure 6.14 has a square cross section $2a \times 2a$, while the outer square is $4a \times 5a$. The axes are displaced as shown. (a) Construct a good-sized drawing of this transmission line, say with $a = 2.5$ cm, and then prepare a curvilinear-square plot of the electrostatic field between the conductors. (b) Use the map to calculate the capacitance per meter length if $\epsilon = 1.6\epsilon_0$. (c) How would your result to part (b) change if $a = 0.6$ cm?

6.30 For the coaxial capacitor of Problem 6.18, suppose that the dielectric is leaky, allowing current to flow between the inner and outer conductors, while the electric field is still uniform with radius. (a) What functional form must the dielectric conductivity assume? (b) What is the basic functional form of the resistance per unit distance, R? (c) What parameters remain in the product, RC, where the form of C, the capacitance per unit distance, has been determined in Problem 6.18?

6.31 A two-wire transmission line consists of two parallel perfectly conducting cylinders, each having a radius of 0.2 mm, separated by a center-to-center distance of 2 mm. The medium surrounding the wires has $\epsilon_r = 3$ and $\sigma = 1.5$ mS/m. A 100 V battery is connected between the wires. Calculate: (a) the magnitude of the charge per meter length on each wire; (b) the battery current.

7

Poisson's and Laplace's Equations

At this stage, we have learned a few methods of finding the electric field intensity, given a known charge distribution or a potential field. For a known charge having a spatial distribution of a general form, we may use Coulomb's law (as exemplified in Eq. [18], Chapter 2) to obtain the vector sum of the fields from all point charges that comprise the distribution. In cases involving symmetric charge distributions, Gauss's law (as represented in Eq. [6], Chapter 3) is sometimes useful to more rapidly obtain the field. We also found that the electric field intensity can be indirectly obtained by first evaluating the potential field associated with a charge distribution (as exemplified in Eq. [18], Chapter 4), and then taking the negative gradient of the result to obtain **E**.

In Chapter 6, we learned more about the close relationship between the potential and electric fields, as occurring within regions in which perfect conductors serve as the boundaries. On the boundaries, constant potentials exist and are often specified as a first step in evaluating capacitance or in constructing a curvilinear square map. Consequently, our studies of these configurations in Chapter 6 provided a first look at solving *boundary value problems*, in which potential and field are determined in a given region subject to stated values of potential or charge density on the boundary surfaces.

In this chapter, we proceed with the solution of such problems in a more direct way, by developing and using the mathematical tools for this specific purpose: Laplace's and Poisson's equations. These equations enable us to find potential fields within regions bounded by known potentials or charge densities. Problems involving one to three dimensions can be solved either analytically or numerically, as will be shown later in the chapter. Laplace's and Poisson's equations, when compared to our other methods, are probably the most widely useful because many problems in engineering practice involve devices in which applied potential differences are known, and in which constant potentials occur on the boundaries. Electric fields are, as before, readily obtained from the potential distributions through the negative gradient.

7.1 DERIVATION OF POISSON'S AND LAPLACE'S EQUATIONS

Obtaining Poisson's equation is exceedingly simple, for from the point form of Gauss's law,

$$\nabla \cdot \mathbf{D} = \rho_\nu \tag{1}$$

the definition of **D**,

$$\mathbf{D} = \epsilon \mathbf{E} \tag{2}$$

and the gradient relationship,

$$\mathbf{E} = -\nabla V \tag{3}$$

by substitution we have

$$\nabla \cdot \mathbf{D} = \nabla \cdot (\epsilon \mathbf{E}) = -\nabla \cdot (\epsilon \nabla V) = \rho_\nu$$

or

$$\nabla \cdot \nabla V = -\frac{\rho_\nu}{\epsilon} \tag{4}$$

for a homogeneous region in which ϵ is constant.

Equation (4) is *Poisson's equation*, but the "double ∇" operation must be interpreted and expanded, at least in rectangular coordinates, before the equation can be useful. In rectangular coordinates,

$$\nabla \cdot \mathbf{A} = \frac{\partial A_x}{\partial x} + \frac{\partial A_y}{\partial y} + \frac{\partial A_z}{\partial z}$$

$$\nabla V = \frac{\partial V}{\partial x}\mathbf{a}_x + \frac{\partial V}{\partial y}\mathbf{a}_y + \frac{\partial V}{\partial z}\mathbf{a}_z$$

and therefore

$$\nabla \cdot \nabla V = \frac{\partial}{\partial x}\left(\frac{\partial V}{\partial x}\right) + \frac{\partial}{\partial y}\left(\frac{\partial V}{\partial y}\right) + \frac{\partial}{\partial z}\left(\frac{\partial V}{\partial z}\right)$$

$$= \frac{\partial^2 V}{\partial x^2} + \frac{\partial^2 V}{\partial y^2} + \frac{\partial^2 V}{\partial z^2} \tag{5}$$

Usually the operation $\nabla \cdot \nabla$ is abbreviated ∇^2 (and pronounced "del squared"), a good reminder of the second-order partial derivatives appearing in (5), and we have

$$\nabla^2 V = \frac{\partial^2 V}{\partial x^2} + \frac{\partial^2 V}{\partial y^2} + \frac{\partial^2 V}{\partial z^2} = -\frac{\rho_\nu}{\epsilon} \tag{6}$$

in rectangular coordinates.

If $\rho_\nu = 0$, indicating zero *volume* charge density, but allowing point charges, line charge, and surface charge density to exist at singular locations as sources of the field, then

$$\nabla^2 V = 0 \tag{7}$$

which is *Laplace's equation*. The ∇^2 operation is called the *Laplacian of V*.

In rectangular coordinates Laplace's equation is

$$\nabla^2 V = \frac{\partial^2 V}{\partial x^2} + \frac{\partial^2 V}{\partial y^2} + \frac{\partial^2 V}{\partial z^2} = 0 \quad \text{(rectangular)} \tag{8}$$

and the form of $\nabla^2 V$ in cylindrical and spherical coordinates may be obtained by using the expressions for the divergence and gradient already obtained in those coordinate systems. For reference, the Laplacian in cylindrical coordinates is

$$\nabla^2 V = \frac{1}{\rho}\frac{\partial}{\partial \rho}\left(\rho \frac{\partial V}{\partial \rho}\right) + \frac{1}{\rho^2}\left(\frac{\partial^2 V}{\partial \phi^2}\right) + \frac{\partial^2 V}{\partial z^2} \quad \text{(cylindrical)} \tag{9}$$

and in spherical coordinates is

$$\nabla^2 V = \frac{1}{r^2}\frac{\partial}{\partial r}\left(r^2\frac{\partial V}{\partial r}\right) + \frac{1}{r^2 \sin\theta}\frac{\partial}{\partial\theta}\left(\sin\theta\frac{\partial V}{\partial\theta}\right) + \frac{1}{r^2\sin^2\theta}\frac{\partial^2 V}{\partial\phi^2} \quad \text{(spherical)} \tag{10}$$

These equations may be expanded by taking the indicated partial derivatives, but it is usually more helpful to have them in the forms just given; furthermore, it is much easier to expand them later if necessary than it is to put the broken pieces back together again.

Laplace's equation is all-embracing, for, applying as it does wherever volume charge density is zero, it states that every conceivable configuration of electrodes or conductors produces a field for which $\nabla^2 V = 0$. All these fields are different, with different potential values and different spatial rates of change, yet for each of them $\nabla^2 V = 0$. Since *every* field (if $\rho_\nu = 0$) satisfies Laplace's equation, how can we expect to reverse the procedure and use Laplace's equation to find one specific field in which we happen to have an interest? Obviously, more information is required, and we shall find that we must solve Laplace's equation subject to certain *boundary conditions*.

Every physical problem must contain at least one conducting boundary and usually contains two or more. The potentials on these boundaries are assigned values, perhaps V_0, $V_1, \ldots$, or perhaps numerical values. These definite equipotential surfaces will provide the boundary conditions for the type of problem to be solved in this chapter. In other types of problems, the boundary conditions take the form of specified values of E (alternatively, a surface charge density, ρ_S) on an enclosing surface, or a mixture of known values of V and E.

Before using Laplace's equation or Poisson's equation in several examples, we must pause to show that if our answer satisfies Laplace's equation and also satisfies the boundary conditions, then it is the only possible answer. It would be very distressing to work a problem by solving Laplace's equation with two different approved methods and then to obtain two different answers. We shall show that the two answers must be identical.

D7.1 Calculate numerical values for V and ρ_ν at point P in free space if: (*a*) $V = \dfrac{4yz}{x^2+1}$, at $P(1,2,3)$; (*b*) $V = 5\rho^2 \cos 2\phi$, at $P(\rho = 3, \phi = \dfrac{\pi}{3}, z = 2)$; (*c*) $V = \dfrac{2\cos\phi}{r^2}$, at $P(r = 0.5, \theta = 45°, \phi = 60°)$.

Ans. 12 V, −106.2 pC/m^3; 22.5 V, 0; 4 V, 0

7.2 UNIQUENESS THEOREM

Let us assume that we have two solutions of Laplace's equation, V_1 and V_2, both general functions of the coordinates used. Therefore

$$\nabla^2 V_1 = 0$$

and

$$\nabla^2 V_2 = 0$$

from which

$$\nabla^2 (V_1 - V_2) = 0$$

Each solution must also satisfy the boundary conditions, and if we represent the given potential values on the boundaries by V_b, then the value of V_1 on the boundary V_{1b} and the value of V_2 on the boundary V_{2b} must both be identical to V_b,

$$V_{1b} = V_{2b} = V_b$$

or

$$V_{1b} - V_{2b} = 0$$

In Section 4.8, Eq. (44), we made use of a vector identity,

$$\nabla\cdot(V\mathbf{D}) \equiv V(\nabla\cdot\mathbf{D}) + \mathbf{D}\cdot(\nabla V)$$

which holds for any scalar V and any vector $\mathbf{D}$. For the present application we shall select $V_1 - V_2$ as the scalar and $\nabla(V_1 - V_2)$ as the vector, giving

$$\nabla\cdot[(V_1 - V_2)\nabla(V_1 - V_2)] \equiv (V_1 - V_2)[\nabla\cdot\nabla(V_1 - V_2)] + \nabla(V_1 - V_2)\cdot\nabla(V_1 - V_2)$$

which we shall integrate throughout the volume *enclosed* by the boundary surfaces specified:

$$\int_{\text{vol}} \nabla\cdot[(V_1 - V_2)\nabla(V_1 - V_2)]\,dv \equiv \int_{\text{vol}} (V_1 - V_2)[\nabla\cdot\nabla(V_1 - V_2)]\,dv + \int_{\text{vol}} [\nabla(V_1 - V_2)]^2\,dv \qquad (11)$$

The divergence theorem allows us to replace the volume integral on the left side of the equation with the closed surface integral over the surface surrounding the volume. This surface consists of the boundaries already specified on which $V_{1b} = V_{2b}$, and therefore

$$\int_{\text{vol}} \nabla\cdot[(V_1 - V_2)\nabla(V_1 - V_2)]\,dv = \oint_S [(V_{1b} - V_{2b})\nabla(V_{1b} - V_{2b})]\cdot d\mathbf{S} = 0$$

One of the factors of the first integral on the right side of (11) is $\nabla\cdot\nabla(V_1 - V_2)$, or $\nabla^2(V_1 - V_2)$, which is zero by hypothesis, and therefore that integral is zero. Hence the remaining volume integral must be zero:

$$\int_{\text{vol}} [\nabla(V_1 - V_2)]^2\,dv = 0$$

There are two reasons why an integral may be zero: either the integrand (the quantity under the integral sign) is everywhere zero, or the integrand is positive in some regions and negative in others, and the contributions cancel algebraically. In this case the first reason must hold because $[\nabla(V_1 - V_2)]^2$ cannot be negative. Therefore

$$[\nabla(V_1 - V_2)]^2 = 0$$

and

$$\nabla(V_1 - V_2) = 0$$

Finally, if the gradient of $V_1 - V_2$ is everywhere zero, then $V_1 - V_2$ cannot change with any coordinates, and

$$V_1 - V_2 = \text{constant}$$

If we can show that this constant is zero, we shall have accomplished our proof. The constant is easily evaluated by considering a point on the boundary. Here $V_1 - V_2 = V_{1b} - V_{2b} = 0$, and we see that the constant is indeed zero, and therefore

$$V_1 = V_2$$

giving two identical solutions.

The uniqueness theorem also applies to Poisson's equation, for if $\nabla^2 V_1 = -\rho_\nu/\epsilon$ and $\nabla^2 V_2 = -\rho_\nu/\epsilon$, then $\nabla^2(V_1 - V_2) = 0$ as before. Boundary conditions still require that $V_{1b} - V_{2b} = 0$, and the proof is identical from this point.

This constitutes the proof of the uniqueness theorem. Viewed as the answer to a question, "How do two solutions of Laplace's or Poisson's equation compare if they both satisfy the same boundary conditions?" the uniqueness theorem should please us by its ensurance that the answers are identical. Once we can find any method of solving Laplace's or Poisson's equation subject to given boundary conditions, we have solved our problem once and for all. No other method can ever give a different answer.

D7.2 Consider the two potential fields $V_1 = y$ and $V_2 = y + e^x \sin y$. (*a*) Is $\nabla^2 V_1 = 0$? (*b*) Is $\nabla^2 V_2 = 0$? (*c*) Is $V_1 = 0$ at $y = 0$? (*d*) Is $V_2 = 0$ at $y = 0$? (*e*) Is $V_1 = \pi$ at $y = \pi$? (*f*) Is $V_2 = \pi$ at $y = \pi$? (*g*) Are V_1 and V_2 identical? (*h*) Why does the uniqueness theorem not apply?

Ans. Yes; yes; yes; yes; yes; yes; no; boundary conditions not given for a *closed* surface

7.3 EXAMPLES OF THE SOLUTION OF LAPLACE'S EQUATION

Several methods have been developed for solving the second-order partial differential equation known as Laplace's equation. The first and simplest method is that of direct integration. We shall use this technique to work several examples involving one-dimensional potential variation in various coordinate systems in this section. In Section 7.5, problems involving two-dimensional variations in potential are considered from an analytic point of view. Such problems are revisited in Section 7.6, in which a numerical iteration method for their solution is discussed. Additional methods requiring a more advanced mathematical knowledge are described in the references at the end of the chapter.

The method of direct integration is applicable only to problems which are "one-dimensional," or in which the potential field is a function of only one of the three coordinates. Since we are working with only three coordinate systems, it might seem, then, that there are nine problems to be solved, but a little reflection will show that a field which varies only with x is fundamentally the same as a field which varies only with y. Rotating the physical problem a quarter turn is no change. Actually, there are only five problems to be solved, one in rectangular coordinates, two in cylindrical, and two in spherical. We shall enjoy life to the fullest by solving them all.

EXAMPLE 7.1

Let us assume that V is a function only of x and worry later about which physical problem we are solving when we have a need for boundary conditions. Laplace's equation reduces to

$$\frac{\partial^2 V}{\partial x^2} = 0$$

and the partial derivative may be replaced by an ordinary derivative, since V is not a function of y or z,

$$\frac{d^2V}{dx^2} = 0$$

We integrate twice, obtaining

$$\frac{dV}{dx} = A$$

and

$$V = Ax + B \tag{12}$$

where A and B are constants of integration. Equation (12) contains two such constants, as we should expect for a second-order differential equation. These constants can be determined only from the boundary conditions.

What boundary conditions should we supply? They are our choice, since no physical problem has yet been specified, with the exception of the original hypothesis that the potential varied only with x. We should now attempt to visualize such a field. Most of us probably already have the answer, but it may be obtained by exact methods.

Since the field varies only with x and is not a function of y and z, then V is a constant if x is a constant or, in other words, the equipotential surfaces are described by setting x constant. These surfaces are parallel planes normal to the x axis. The field is thus that of a parallel-plate capacitor, and as soon as we specify the potential on any two planes, we may evaluate our constants of integration.

To be very general, let $V = V_1$ at $x = x_1$ and $V = V_2$ at $x = x_2$. These values are then substituted into (12), giving

$$V_1 = Ax_1 + B \qquad V_2 = Ax_2 + B$$

$$A = \frac{V_1 - V_2}{x_1 - x_2} \qquad B = \frac{V_2x_1 - V_1x_2}{x_1 - x_2}$$

and

$$V = \frac{V_1(x - x_2) - V_2(x - x_1)}{x_1 - x_2}$$

A simpler answer would have been obtained by choosing simpler boundary conditions. If we had fixed $V = 0$ at $x = 0$ and $V = V_0$ at $x = d$, then

$$A = \frac{V_0}{d} \qquad B = 0$$

and

$$V = \frac{V_0 x}{d} \tag{13}$$

Suppose our primary aim is to find the capacitance of a parallel-plate capacitor. We have solved Laplace's equation, obtaining (12) with the two constants A and B. Should they be evaluated or left alone? Presumably we are not interested in the potential field itself, but only in the capacitance, and we may continue successfully with A and B or we may simplify the algebra by a little foresight. Capacitance is given by the ratio of charge to potential difference, so we may choose now the potential *difference* as V_0, which is equivalent to one boundary condition, and then choose whatever second boundary condition seems to help the form of the equation the most. This is the essence of the second set of boundary conditions which produced (13). The potential difference was fixed as V_0 by making the potential of one plate zero and the other V_0; the location of these plates was made as simple as possible by letting $V = 0$ at $x = 0$.

Using (13), then, we still need the total charge on either plate before the capacitance can be found. We should remember that when we first solved this capacitor problem in Chapter 5, the sheet of charge provided our starting point. We did not have to work very hard to find the charge, for all the fields were expressed in terms of it. The work then was spent in finding potential difference. Now the problem is reversed (and simplified).

The necessary steps are these, after the choice of boundary conditions has been made:

1. Given V, use $\mathbf{E} = -\nabla V$ to find $\mathbf{E}$.
2. Use $\mathbf{D} = \epsilon\mathbf{E}$ to find $\mathbf{D}$.
3. Evaluate $\mathbf{D}$ at either capacitor plate, $\mathbf{D} = \mathbf{D}_S = D_N\mathbf{a}_N$.
4. Recognize that $\rho_S = D_N$.
5. Find Q by a surface integration over the capacitor plate, $Q = \int_S \rho_S\, dS$.

Here we have

$$V = V_0 \frac{x}{d}$$

$$\mathbf{E} = -\frac{V_0}{d}\mathbf{a}_x$$

$$\mathbf{D} = -\epsilon\frac{V_0}{d}\mathbf{a}_x$$

$$\mathbf{D}_S = \mathbf{D}\Big|_{x=0} = -\epsilon\frac{V_0}{d}\mathbf{a}_x$$

$$\mathbf{a}_N = \mathbf{a}_x$$

$$D_N = -\epsilon\frac{V_0}{d} = \rho_S$$

$$Q = \int_S \frac{-\epsilon V_0}{d}\, dS = -\epsilon\frac{V_0 S}{d}$$

and the capacitance is

$$C = \frac{|Q|}{V_0} = \frac{\epsilon S}{d} \tag{14}$$

We shall use this procedure several times in the examples to follow.

EXAMPLE 7.2

Since no new problems are solved by choosing fields which vary only with y or with z in rectangular coordinates, we pass on to cylindrical coordinates for our next example. Variations with respect to z are again nothing new, and we next assume variation with respect to ρ only. Laplace's equation becomes

$$\frac{1}{\rho}\frac{\partial}{\partial\rho}\left(\rho\frac{\partial V}{\partial\rho}\right)=0$$

or

$$\frac{1}{\rho}\frac{d}{d\rho}\left(\rho\frac{dV}{d\rho}\right)=0$$

Noting the ρ in the denominator, we exclude $\rho = 0$ from our solution and then multiply by ρ and integrate,

$$\rho\frac{dV}{d\rho}=A$$

rearrange, and integrate again,

$$V = A\ln\rho + B \tag{15}$$

The equipotential surfaces are given by $\rho =$ constant and are cylinders, and the problem is that of the coaxial capacitor or coaxial transmission line. We choose a potential difference of V_0 by letting $V = V_0$ at $\rho = a$, $V = 0$ at $\rho = b$, $b > a$, and obtain

$$V = V_0\frac{\ln(b/\rho)}{\ln(b/a)} \tag{16}$$

from which

$$\mathbf{E} = \frac{V_0}{\rho}\frac{1}{\ln(b/a)}\mathbf{a}_\rho$$

$$D_{N(\rho=a)} = \frac{\epsilon V_0}{a\ln(b/a)}$$

$$Q = \frac{\epsilon V_0 2\pi aL}{a\ln(b/a)}$$

$$C = \frac{2\pi\epsilon L}{\ln(b/a)} \tag{17}$$

which agrees with our results in Chapter 6.

EXAMPLE 7.3

Now let us assume that V is a function only of ϕ in cylindrical coordinates. We might look at the physical problem first for a change and see that equipotential surfaces are given by $\phi =$ constant. These are radial planes. Boundary conditions might be $V = 0$ at $\phi = 0$ and $V = V_0$ at $\phi = \alpha$, leading to the physical problem detailed in Figure 7.1.

Laplace's equation is now

$$\frac{1}{\rho^2}\frac{\partial^2 V}{\partial\phi^2}=0$$

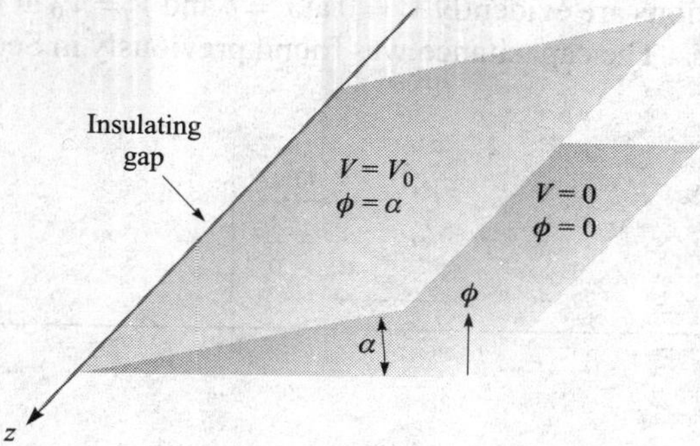

Fig. 7.1 Two infinite radial planes with an interior angle α. An infinitesimal insulating gap exists at $\rho = 0$. The potential field may be found by applying Laplace's equation in cylindrical coordinates.

We exclude $\rho = 0$ and have

$$\frac{d^2V}{d\phi^2} = 0$$

The solution is

$$V = A\phi + B$$

The boundary conditions determine A and B, and

$$V = V_0 \frac{\phi}{\alpha} \tag{18}$$

Taking the gradient of (18) produces the electric field intensity,

$$\mathbf{E} = -\frac{V_0 \mathbf{a}_\phi}{\alpha\rho} \tag{19}$$

and it is interesting to note that E is a function of ρ and not of ϕ. This does not contradict our original assumptions, which were restrictions only on the potential field. Note, however, that the *vector* field **E** is a function of ϕ.

A problem involving the capacitance of these two radial planes is included at the end of the chapter.

EXAMPLE 7.4

We now turn to spherical coordinates, dispose immediately of variations with respect to ϕ only as having just been solved, and treat first $V = V(r)$.

The details are left for a problem later, but the final potential field is given by

$$V = V_0 \frac{\frac{1}{r} - \frac{1}{b}}{\frac{1}{a} - \frac{1}{b}} \tag{20}$$

where the boundary conditions are evidently $V = 0$ at $r = b$ and $V = V_0$ at $r = a$, $b > a$. The problem is that of concentric spheres. The capacitance was found previously in Section 5.10 (by a somewhat different method) and is

$$C = \frac{4\pi\epsilon}{\dfrac{1}{a} - \dfrac{1}{b}} \tag{21}$$

EXAMPLE 7.5

In spherical coordinates we now restrict the potential function to $V = V(\theta)$, obtaining

$$\frac{1}{r^2 \sin\theta}\frac{d}{d\theta}\left(\sin\theta\frac{dV}{d\theta}\right) = 0$$

We exclude $r = 0$ and $\theta = 0$ or π and have

$$\sin\theta\frac{dV}{d\theta} = A$$

The second integral is then

$$V = \int \frac{A\,d\theta}{\sin\theta} + B$$

which is not as obvious as the previous ones. From integral tables (or a good memory) we have

$$V = A\ln\left(\tan\frac{\theta}{2}\right) + B$$

The equipotential surfaces are cones. Figure 7.2 illustrates the case where $V = 0$ at $\theta = \pi/2$ and $V = V_0$ at $\theta = \alpha$, $\alpha < \pi/2$. We obtain

$$V = V_0\frac{\ln\left(\tan\dfrac{\theta}{2}\right)}{\ln\left(\tan\dfrac{\alpha}{2}\right)} \tag{22}$$

In order to find the capacitance between a conducting cone with its vertex separated from a conducting plane by an infinitesimal insulating gap and its axis normal to the plane, let us first find the field strength:

$$\mathbf{E} = -\nabla V = \frac{-1}{r}\frac{\partial V}{\partial\theta}\mathbf{a}_\theta = -\frac{V_0}{r\sin\theta\ln\left(\tan\dfrac{\alpha}{2}\right)}\mathbf{a}_\theta$$

The surface charge density on the cone is then

$$\rho_S = \frac{-\epsilon V_0}{r\sin\alpha\ln\left(\tan\dfrac{\alpha}{2}\right)}$$

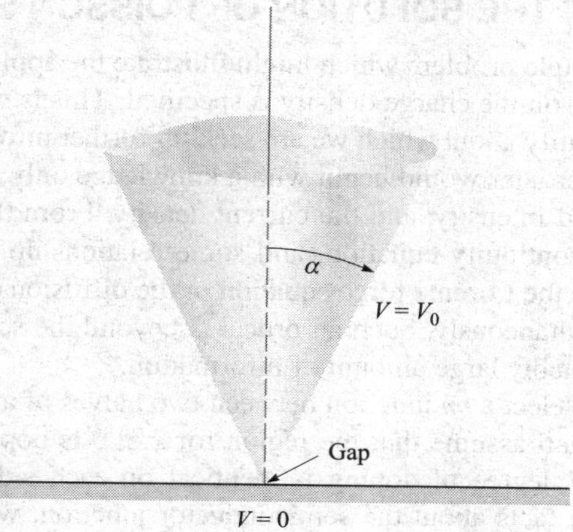

Fig. 7.2 For the cone $\theta = \alpha$ at V_0 and the plane $\theta = \pi/2$ at $V = 0$, the potential field is given by $V = V_0[\ln(\tan\theta/2)]/[\ln(\tan\alpha/2)]$.

producing a total charge Q,

$$Q = \frac{-\epsilon V_0}{\sin\alpha \ln\left(\tan\dfrac{\alpha}{2}\right)} \int_0^{\infty}\int_0^{2\pi} \frac{r \sin\alpha \, d\phi \, dr}{r}$$

$$= \frac{-2\pi\epsilon_0 V_0}{\ln\left(\tan\dfrac{\alpha}{2}\right)} \int_0^{\infty} dr$$

This leads to an infinite value of charge and capacitance, and it becomes necessary to consider a cone of finite size. Our answer will now be only an approximation because the theoretical equipotential surface is $\theta = \alpha$, a conical surface extending from $r = 0$ to $r = \infty$, whereas our physical conical surface extends only from $r = 0$ to, say, $r = r_1$. The approximate capacitance is

$$C \doteq \frac{2\pi\epsilon r_1}{\ln\left(\cot\dfrac{\alpha}{2}\right)} \tag{23}$$

If we desire a more accurate answer, we may make an estimate of the capacitance of the base of the cone to the zero-potential plane and add this amount to our answer. Fringing, or nonuniform, fields in this region have been neglected and introduce an additional source of error.

D7.3 Find $|\mathbf{E}|$ at $P(3, 1, 2)$ for the field of: (a) two coaxial conducting cylinders, $V = 50$ V at $\rho = 2$ m, and $V = 20$ V at $\rho = 3$ m; (b) two radial conducting planes, $V = 50$ V at $\phi = 10°$, and $V = 20$ V at $\phi = 30°$.

Ans. 23.4 V/m; 27.2 V/m

7.4 EXAMPLE OF THE SOLUTION OF POISSON'S EQUATION

To select a reasonably simple problem which might illustrate the application of Poisson's equation, we must assume that the volume charge density is specified. This is not usually the case, however; in fact, it is often the quantity about which we are seeking further information. The type of problem which we might encounter later would begin with a knowledge only of the boundary values of the potential, the electric field intensity, and the current density. From these we would have to apply Poisson's equation, the continuity equation, and some relationship expressing the forces on the charged particles, such as the Lorentz force equation or the diffusion equation, and solve the whole system of equations simultaneously. Such an ordeal is beyond the scope of this text, and we shall therefore assume a reasonably large amount of information.

As an example, let us select a *pn* junction between two halves of a semiconductor bar extending in the x direction. We shall assume that the region for $x < 0$ is doped *p* type and that the region for $x > 0$ is *n* type. The degree of doping is identical on each side of the junction. To review qualitatively some of the facts about the semiconductor junction, we note that initially there are excess holes to the left of the junction and excess electrons to the right. Each diffuses across the junction until an electric field is built up in such a direction that the diffusion current drops to zero. Thus, to prevent more holes from moving to the right, the electric field in the neighborhood of the junction must be directed to the left; E_x is negative there. This field must be produced by a net positive charge to the right of the junction and a net negative charge to the left. Note that the layer of positive charge consists of two parts—the holes which have crossed the junction and the positive donor ions from which the electrons have departed. The negative layer of charge is constituted in the opposite manner by electrons and negative acceptor ions.

The type of charge distribution which results is shown in Figure 7.3*a*, and the negative field which it produces is shown in Figure 7.3*b*. After looking at these two figures, one might profitably read the previous paragraph again.

A charge distribution of this form may be approximated by many different expressions. One of the simpler expressions is

$$\rho_\nu = 2\rho_{\nu 0}\,\mathrm{sech}\,\frac{x}{a}\,\tanh\,\frac{x}{a} \tag{24}$$

which has a maximum charge density $\rho_{v,max} = \rho_{v0}$ that occurs at $x = 0.881a$. The maximum charge density ρ_{v0} is related to the acceptor and donor concentrations N_a and N_d by noting that all the donor and acceptor ions in this region (the *depletion* layer) have been stripped of an electron or a hole, and thus

$$\rho_{v0} = eN_a = eN_d$$

Let us now solve Poisson's equation,

$$\nabla^2 V = -\frac{\rho_\nu}{\epsilon}$$

subject to the charge distribution assumed above,

$$\frac{d^2V}{dx^2} = -\frac{2\rho_{v0}}{\epsilon}\mathrm{sech}\frac{x}{a}\,\tanh\,\frac{x}{a}$$

in this one-dimensional problem in which variations with y and z are not present. We integrate once,

$$\frac{dV}{dx} = \frac{2\rho_{v0}a}{\epsilon}\mathrm{sech}\frac{x}{a} + C_1$$

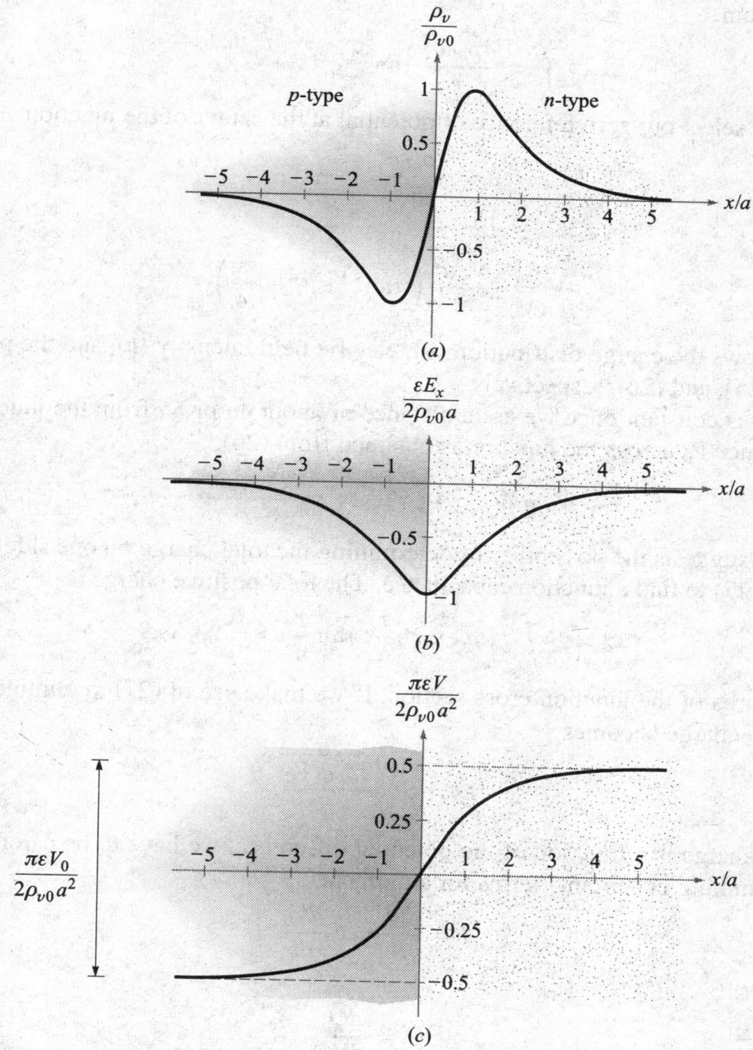

Fig. 7.3 (a) The charge density, (b) the electric field intensity, and (c) the potential are plotted for a pn junction as functions of distance from the center of the junction. The p-type material is on the left, and the n-type is on the right.

and obtain the electric field intensity,

$$E_x = -\frac{2\rho_{v0}a}{\epsilon}\text{sech}\frac{x}{a} - C_1$$

To evaluate the constant of integration C_1, we note that no net charge density and no fields can exist *far* from the junction. Thus, as $x \to \pm\infty$, E_x must approach zero. Therefore $C_1 = 0$, and

$$E_x = -\frac{2\rho_{v0}a}{\epsilon}\text{sech}\frac{x}{a} \tag{25}$$

Integrating again,

$$V = \frac{4\rho_{v0}a^2}{\epsilon} \tan^{-1} e^{x/a} + C_2$$

Let us arbitrarily select our zero reference of potential at the center of the junction, $x = 0$,

$$0 = \frac{4\rho_{v0}a^2}{\epsilon} \frac{\pi}{4} + C_2$$

and finally,

$$V = \frac{4\rho_{v0}a^2}{\epsilon}\left(\tan^{-1} e^{x/a} - \frac{\pi}{4}\right) \tag{26}$$

Figure 7.3 shows the charge distribution (a), electric field intensity (b), and the potential (c), as given by (24), (25), and (26), respectively.

The potential is constant once we are a distance of about $4a$ or $5a$ from the junction. The total potential difference V_0 across the junction is obtained from (26),

$$V_0 = V_{x\to\infty} - V_{x\to-\infty} = \frac{2\pi\rho_{v0}a^2}{\epsilon} \tag{27}$$

This expression suggests the possibility of determining the total charge on one side of the junction and then using (27) to find a junction capacitance. The total positive charge is

$$Q = S\int_0^\infty 2\rho_{\nu 0}\text{sech}\frac{x}{a}\tanh\frac{x}{a}\,dx = 2\rho_{\nu 0}aS$$

where S is the area of the junction cross section. If we make use of (27) to eliminate the distance parameter a, the charge becomes

$$Q = S\sqrt{\frac{2\rho_{\nu 0}\epsilon V_0}{\pi}} \tag{28}$$

Since the total charge is a function of the potential difference, we have to be careful in defining a capacitance. Thinking in "circuit" terms for a moment,

$$I = \frac{dQ}{dt} = C\frac{dV_0}{dt}$$

and thus

$$C = \frac{dQ}{dV_0}$$

By differentiating (28) we therefore have the capacitance,

$$C = \sqrt{\frac{\rho_{\nu 0}\epsilon}{2\pi V_0}}S = \frac{\epsilon S}{2\pi a} \tag{29}$$

The first form of (29) shows that the capacitance varies inversely as the square root of the voltage. That is, a higher voltage causes a greater separation of the charge layers and a smaller capacitance. The second form is interesting in that it indicates that we may think of the junction as a parallel-plate capacitor with a "plate" separation of $2\pi a$. In view of the dimensions of the region in which the charge is concentrated, this is a logical result.

Poisson's equation enters into any problem involving volume charge density. Besides semiconductor diode and transistor models, we find that vacuum tubes, magnetohydrodynamic energy conversion, and ion propulsion require its use in constructing satisfactory theories.

D7.4 In the neighborhood of a certain semiconductor junction the volume charge density is given by $\rho_\nu = 750 \text{ sech } 10^6 \pi x \tanh \pi x \text{ C/m}^3$. The dielectric constant of the semiconductor material is 10 and the junction area is $2 \times 10^{-7} \text{ m}^2$. Find: (*a*) V_0; (*b*) C; (*c*) E at the junction.

Ans. 2.70 V; 8.85 pF; 2.70 MV/m

D7.5 Given the volume charge density $\rho_\nu = -2 \times 10^7 \epsilon_0 \sqrt{x} \text{ C/m}^3$ in free space, let $V = 0$ at $x = 0$ and $V = 2$ V at $x = 2.5$ mm. At $x = 1$ mm, find: (*a*) V; (*b*) E_x.

Ans. 0.302 V; −555 V/m

7.5 PRODUCT SOLUTION OF LAPLACE'S EQUATION

Interactives

In this section we are confronted with the class of potential fields which vary with more than one of the three coordinates. Although our examples are taken in the rectangular coordinate system, the general method is applicable to the other coordinate systems. We shall avoid those applications, however, because the potential fields are given in terms of more advanced mathematical functions, such as Bessel functions and spherical and cylindrical harmonics, and our interest now does not lie with new mathematical functions but with the techniques and methods of solving electrostatic field problems.

We may give ourselves a general class of problems by specifying merely that the potential is a function of x and y alone, so that

$$\frac{\partial^2 V}{\partial x^2} + \frac{\partial^2 V}{\partial y^2} = 0 \tag{30}$$

We now assume that the potential is expressible as the *product* of a function of x alone and a function of y alone. It might seem that this prohibits too many solutions, such as $V = x + y$, or any sum of a function of x and a function of y, but we should realize that Laplace's equation is linear and the sum of any two solutions is also a solution. We could treat $V = x + y$ as the sum of $V_1 = x$ and $V_2 = y$, where each of these latter potentials is now a (trivial) product solution.

Representing the function of x by X and the function of y by Y, we have

$$V = XY \tag{31}$$

which is substituted into (30),

$$Y\frac{\partial^2 X}{\partial x^2} + X\frac{\partial^2 Y}{\partial y^2} = 0$$

Since X does not involve y and Y does not involve x, ordinary derivatives may be used,

$$Y\frac{d^2 X}{dx^2} + X\frac{d^2 Y}{dy^2} = 0 \tag{32}$$

Equation (32) may be solved by separating the variables through division by XY, giving

$$\frac{1}{X}\frac{d^2 X}{dx^2} + \frac{1}{Y}\frac{d^2 Y}{dy^2} = 0$$

or

$$\frac{1}{X}\frac{d^2X}{dx^2} = -\frac{1}{Y}\frac{d^2Y}{dy^2}$$

Now we need one of the cleverest arguments of mathematics: since $(1/X)d^2X/dx^2$ involves no y and $-(1/Y)d^2Y/dy^2$ involves no x, and since the two quantities are equal, then $(1/X)d^2X/dx^2$ cannot be a function of x either, and similarly, $-(1/Y)d^2Y/dy^2$ cannot be a function of y! In other words, we have shown that each of these terms must be a constant. For convenience, let us call this constant α^2,

$$\frac{1}{X}\frac{d^2X}{dx^2} = \alpha^2 \tag{33}$$

$$-\frac{1}{Y}\frac{d^2Y}{dy^2} = \alpha^2 \tag{34}$$

The constant α^2 is called the *separation constant* because its use results in separating one equation into two simpler equations.

Equation (33) may be written as

$$\frac{d^2X}{dx^2} = \alpha^2 X \tag{35}$$

and must now be solved. There are several methods by which a solution may be obtained. The first method is experience, or recognition, which becomes more powerful with practice. We are just beginning and can barely recognize Laplace's equation itself. The second method might be that of direct integration, when applicable, of course. Applying it here, we should write

$$d\left(\frac{dX}{dx}\right) = \alpha^2 X\,dx$$

$$\frac{dX}{dx} = \alpha^2 \int X\,dx$$

and then pass on to the next method, for X is some unknown function of x, and the method of integration is not applicable here. The third method we might describe as intuition, common sense, or inspection. It involves taking a good look at the equation, perhaps putting the operation into words. This method will work on (35) for some of us if we ask ourselves, "What function has a second derivative which has the same form as the function itself, except for multiplication by a constant?" The answer is the exponential function, of course, and we could go on from here to construct the solution. Instead, let us work with those of us whose intuition is suffering from exposure and apply a very powerful but long method, the infinite-power-series substitution.

We assume hopefully that X may be represented by

$$X = \sum_{a=0}^{\infty} a_n x^n$$

and substitute into (35), giving

$$\frac{d^2X}{dx^2} = \sum_{0}^{\infty} n(n-1)a_n x^{n-2} = \alpha^2 \sum_{0}^{\infty} a_n x^n$$

If these two different infinite series are to be equal for all x, they must be identical, and the coefficients of like powers of x may be equated term by term.

Thus,

$$2 \times 1 \times a_2 = \alpha^2 a_0$$
$$3 \times 2 \times a_3 = \alpha^2 a_1$$

and in general we have the recurrence relationship

$$(n+2)(n+1)a_{n+2} = \alpha^2 a_n$$

The even coefficients may be expressed in terms of a_0 as

$$a_2 = \frac{\alpha^2}{1 \times 2} a_0$$
$$a_4 = \frac{\alpha^2}{3 \times 4} a_2 = \frac{\alpha^4}{4!} a_0$$
$$a_6 = \frac{\alpha^6}{6!} a_0$$

and, in general, for n even, as

$$a_n = \frac{\alpha^n}{n!} a_0 \quad (n \text{ even})$$

For odd values of n, we have

$$a_3 = \frac{\alpha^2}{2 \times 3} a_1 = \frac{\alpha^3}{3!} \frac{a_1}{\alpha}$$
$$a_5 = \frac{\alpha^5}{5!} \frac{a_1}{\alpha}$$

and in general, for n odd,

$$a_n = \frac{\alpha^n}{n!} \frac{a_1}{\alpha} \quad (n \text{ odd})$$

Substituting back into the original power series for X, we obtain

$$X = a_0 \sum_{0,\text{even}}^{\infty} \frac{\alpha^n}{n!} x^n + \frac{a_1}{\alpha} \sum_{1,\text{odd}}^{\infty} \frac{\alpha^n}{n!} x^n$$

or

$$X = a_0 \sum_{0,\text{even}}^{\infty} \frac{(\alpha x)^n}{n!} + \frac{a_1}{\alpha} \sum_{1,\text{odd}}^{\infty} \frac{(\alpha x)^n}{n!}$$

Although the sum of these two infinite series is the solution of the differential equation in x, the form of the solution may be improved immeasurably by recognizing the first series as the hyperbolic cosine,

$$\cosh \alpha x = \sum_{0,\text{even}}^{\infty} \frac{(\alpha x)^n}{n!} = 1 + \frac{(\alpha x)^2}{2!} + \frac{(\alpha x)^4}{4!} + \ldots$$

and the second series as the hyperbolic sine,

$$\sinh \alpha x = \sum_{1,\text{odd}}^{\infty} \frac{(\alpha x)^n}{n!} = \alpha x + \frac{(\alpha x)^3}{3!} + \frac{(\alpha x)^5}{5!} + \ldots$$

The solution may therefore be written as

$$X = a_0 \cosh \alpha x + \frac{a_1}{\alpha} \sinh \alpha x$$

or

$$X = A\cosh \alpha x + B\sinh \alpha x$$

where the slightly simpler terms A and B have replaced a_0 and a_1/α, respectively, and are the two constants which must be evaluated in terms of the boundary conditions. The separation constant is not an arbitrary constant as far as the solution of (35) is concerned, for it appears in that equation.

An alternate form of the solution is obtained by expressing the hyperbolic functions in terms of exponentials, collecting terms, and selecting new arbitrary constants, A' and B',

$$X = A'e^{\alpha x} + B'e^{-\alpha x}$$

Turning our attention now to (34), we see the solution proceeds along similar lines, leading to two power series representing the sine and cosine, and we have

$$Y = C\cos \alpha y + D\sin \alpha y$$

from which the potential is

$$V = XY = (A\cosh \alpha x + B\sinh \alpha x)(C\cos \alpha y + D\sin \alpha y) \tag{36}$$

Before describing a physical problem and forcing the constants appearing in (36) to fit the boundary conditions prescribed, let us consider the physical nature of the potential field given by a simple choice of these constants. Letting $A = 0$, $C = 0$, and $BD = V_1$, we have

$$V = V_1 \sinh \alpha x \sin \alpha y \tag{37}$$

The sinh αx factor is zero at $x = 0$ and increases smoothly with x, soon becoming nearly exponential in form, since

$$\sinh \alpha x = \tfrac{1}{2}(e^{\alpha x} - e^{-\alpha x})$$

The sin αy term causes the potential to be zero at $y = 0$, $y = \pi/\alpha$, $y = 2\pi/\alpha$, and so forth. We therefore may place zero-potential conducting planes at $x = 0$, $y = 0$, and $y = \pi/\alpha$. Finally, we can describe the V_1 equipotential surface by setting $V = V_1$ in (37), obtaining

$$\sinh \alpha x \sin \alpha y = 1$$

or

$$\alpha y = \sin^{-1}\frac{1}{\sinh \alpha x}$$

This is not a familiar equation, but a hand calculator or a set of tables can furnish enough material values to allow us to plot αy as a function of αx. Such a curve is shown in Figure 7.4. Note that the curve is double-valued and symmetrical about the line $\alpha y = \pi/2$ when αy is restricted to the interval between 0 and π. The information of Figure 7.4 is transferred directly to the $V = 0$ and $V = V_1$ equipotential conducting surfaces in Figure 7.5. The surfaces are shown in cross section, since the potential is not a function of z.

It is very unlikely that we shall ever be asked to find the potential field of these peculiarly shaped electrodes, but we should bear in mind the possibility of combining a number of the fields having the form given by (36) or (37) and thus satisfying the boundary conditions of a more practical problem. We close this chapter with such an example.

The problem to be solved is that shown in Figure 7.6. The boundary conditions shown are $V = 0$ at $x = 0$, $y = 0$, and $y = b$, and $V = V_0$ at $x = d$ for all y between 0 and b. It is immediately apparent that the potential field given by (37) and outlined in Figure 7.5 satisfies two of the four

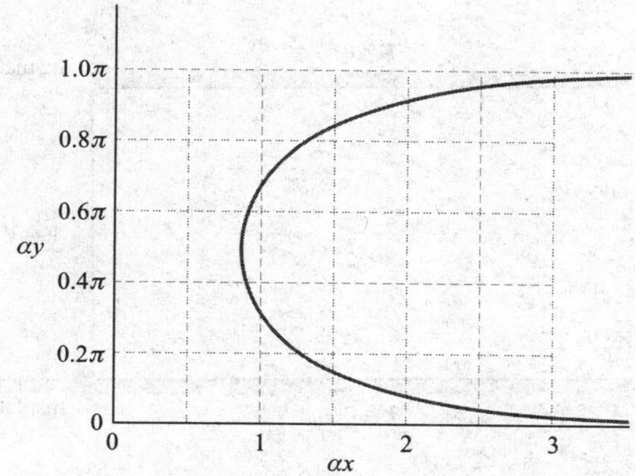

Fig. 7.4 A graph of the double-valued function $\alpha y = \sin^{-1}(1/\sinh \alpha x), 0 < \alpha y < \pi$.

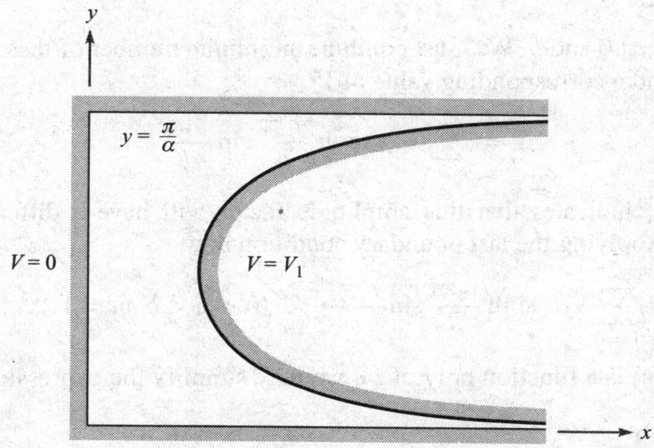

Fig. 7.5 Cross section of the $V = 0$ and $V = V_1$ equipotential surfaces for the potential field $V = V_1 \sinh \alpha x \sin \alpha y$.

boundary conditions. A third condition, $V = 0$ at $y = b$, may be satisfied by the choice of a, for the substitution of these values of (37) leads to the equation

$$0 = V_1 \sinh \alpha x \sin \alpha b$$

which may be satisfied by setting

$$\alpha b = m\pi \qquad (m = 1, 2, 3, \dots)$$

or

$$\alpha = \frac{m\pi}{b}$$

The potential function

$$V = V_1 \sinh \frac{m\pi x}{b} \sin \frac{m\pi y}{b} \tag{38}$$

thus produces the correct potential at $x = 0$, $y = 0$, and $y = b$, regardless of the choice of m or the value of V_1. It is impossible to choose m or V_1 in such a way that $V = V_0$ at $x = d$ for each and

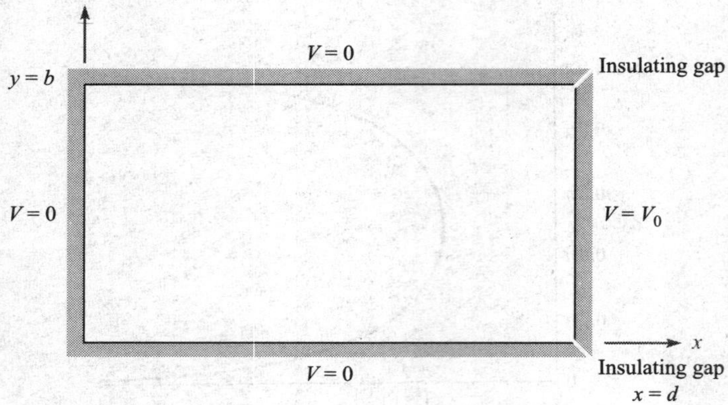

Fig. 7.6 Potential problem requiring an infinite summation of fields of the form $V = V_1 \sinh \alpha x \sin \alpha y$. A similar configuration was analyzed by the iteration method in Chapter 6.

every value of y between 0 and b. We must combine an infinite number of these fields, each with a different value of m and a corresponding value of V_1,

$$V = \sum_{m=0}^{\infty} V_{1m} \sinh \frac{m\pi x}{b} \sin \frac{m\pi y}{b}$$

The subscript on V_{1m} indicates that this amplitude factor will have a different value for each different value of m. Applying the last boundary condition now,

$$V_0 = \sum_{m=0}^{\infty} V_{1m} \sinh \frac{m\pi d}{d} \sin \frac{m\pi y}{b} \qquad (0 < y < b, m = 1, 2, \ldots)$$

Since $V_{1m} \sinh (m\pi d/b)$ is a function only of m, we may simplify the expression by replacing this factor by c_m,

$$V_0 = \sum_{m=0}^{\infty} c_m \sin \frac{m\pi y}{b} \qquad (0 < y < b, m = 1, 2, \ldots)$$

This is a Fourier sine series, and the c_m coefficients may be determined by the standard Fourier-series methods[1] if we can interpret V_0 as a periodic function of y. Since our physical problem is bounded by conducting planes at $y = 0$ and $y = b$, and our interest in the potential does not extend outside of this region, we may *define* the potential at $x = d$ for *y outside* of the range 0 to b in any manner we choose. Probably the simplest periodic expression is obtained by selecting the interval $0 < y < b$ as the half-period and choosing $V = -V_0$ in the adjacent half-period, or

$$V = V_0 \qquad (x = d, 0 < y < b)$$
$$V = -V_0 \qquad (x = d, b < y < 2b)$$

[1] Fourier series are discussed in almost every electrical engineering text on circuit theory. The authors are partial to the Hayt and Kemmerly reference given in the References at the end of the chapter.

The c_m coefficients are then

$$c_m = \frac{1}{b}\left[\int_0^b V_0 \sin\frac{m\pi y}{b}dy + \int_b^{2b} (-V_0)\sin\frac{m\pi y}{b}dy\right]$$

leading to

$$c_m = \frac{4V_0}{m\pi} \qquad (m \text{ odd})$$
$$= 0 \qquad (m \text{ even})$$

However, $c_m = V_{1m}\sinh(m\pi d/b)$, and therefore

$$V_{1m} = \frac{4V_0}{m\pi \sinh(m\pi d/b)} \qquad (m \text{ odd only})$$

which may be substituted into (38) to give the desired potential function,

$$V = \frac{4V_0}{\pi}\sum_{1,\text{odd}}^{\infty}\frac{1}{m}\frac{\sinh(m\pi x/b)}{\sinh(m\pi d/b)}\sin\frac{m\pi y}{b} \tag{39}$$

The map of this field may be obtained by evaluating (39) at a number of points and drawing equipotentials by interpolation between these points. If we let $b = d$ and $V_0 = 100$, the problem is identical with that used as the example in the discussion of the iteration method. Checking one of the grid points in that problem, we let $x = d/4 = b/4$, $y = b/2 = d/2$, and $V_0 = 100$ and obtain

$$V = \frac{400}{\pi}\sum_{1,\text{odd}}^{\infty}\frac{1}{m}\frac{\sinh(m\pi/4)}{\sinh m\pi}\sin\frac{m\pi}{2}$$
$$= \frac{400}{\pi}\left(\frac{\sinh(\pi/4)}{\sinh \pi} - \frac{1}{3}\frac{\sinh(3\pi/4)}{\sinh 3\pi} + \frac{1}{5}\frac{\sinh(5\pi/4)}{\sinh 5\pi} - \cdots\right)$$
$$= \frac{400}{\pi}\left(\frac{0.8687}{11.549} - \frac{5.228}{3 \times 6195.8} + \cdots\right)$$
$$= 9.577 - 0.036 + \cdots = 9.541 \text{ V}$$

The equipotentials are drawn for increments of 10 V in Figure 7.7, and flux lines have been added graphically to produce a curvilinear map.

The material covered in this discussion of the product solution was more difficult than much of the preceding work, and moreover, it presented three new ideas. The first new technique was the assumption that the potential might be expressed as the product of a function of x and a function of y, and the resultant separation of Laplace's equation into two simpler ordinary differential equations. The second new approach was employed when an infinite-power-series solution was assumed as the solution for one of the ordinary differential equations. Finally, we considered an example which required the combination of an infinite number of simpler product solutions, each having a different amplitude and a different variation in one of the coordinate directions. All these techniques are very powerful. They are useful in all coordinate systems, and they can be used in problems in which the potential varies with all three coordinates.

Interactives

We have merely introduced the subject here, and more information can be obtained from the end-of-chapter references, several of which devote hundreds of pages to the solution of Laplace's equation.

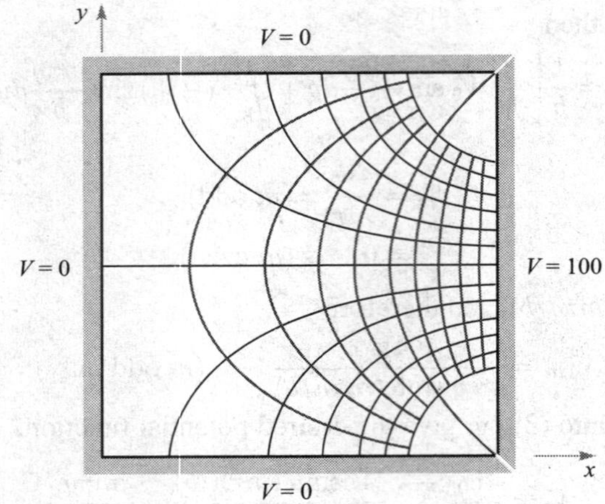

Fig. 7.7 The field map corresponding to $V = \frac{4V_0}{\pi} \sum_{1,\text{odd}}^{\infty} \frac{1}{m} \frac{\sinh(m\pi x/b)}{\sinh(m\pi d/b)} \sin \frac{m\pi y}{b}$ with $b = d$ and $V_0 = 100$ V.

7.6 SOLVING LAPLACE'S EQUATION THROUGH NUMERICAL ITERATION

In practice, many if not most problems that involve solving for a potential field are not readily amenable to analytic solution. In addition to other complications, the regions of interest may be irregularly shaped or configured in such a way that boundaries cannot be represented by constant-coordinate surfaces within a single coordinate system. We encountered such cases when studying curvilinear square mapping as a method of estimating capacitance in Chapter 6. In general, numerical methods must be employed in order to obtain results that are reasonably accurate. In one method, to be presented here, the region within which potential must be found is partitioned into a grid of appropriate shape, followed by the assignment of estimated values of potential at locations on the grid. Further refinements of these values are accomplished numerically such that the overall result is consistant with Laplace's equation. A great many numerical techniques similar to this one exist, in addition to many that are distinctly different. Commercial software is available that employs one or more of these. In this section we show an example of a basic iterative method, applicable to simple media. Broader discussions with more detail can be found in references 8 through 10 at the end of the chapter.

Let us assume a two-dimensional problem in which the potential does not vary with the z coordinate and divide the interior of a cross section of the region where the potential is desired into squares of length h on a side. A portion of this region is shown in Figure 7.8. The unknown values of the potential at five adjacent points are indicated as V_0, V_1, V_2, V_3, and V_4. The two-dimensional Laplace equation to be solved is

$$\frac{\partial^2 V}{\partial x^2} + \frac{\partial^2 V}{\partial y^2} = 0$$

Approximate values for the partial derivatives may be obtained in terms of the assumed potentials, or

$$\left.\frac{\partial V}{\partial x}\right|_a \doteq \frac{V_1 - V_0}{h}$$

and

$$\left.\frac{\partial V}{\partial x}\right|_c \doteq \frac{V_0 - V_3}{h}$$

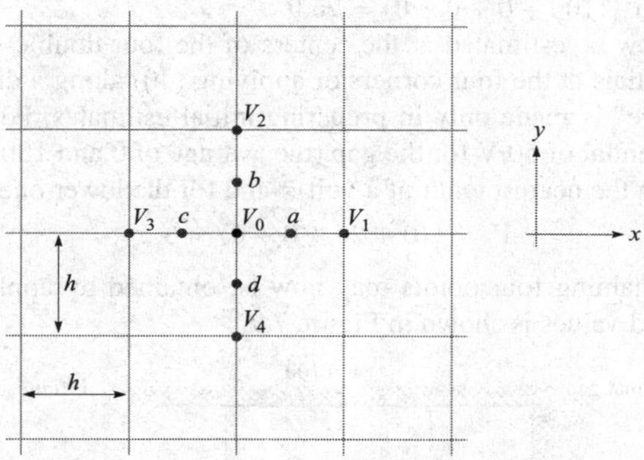

Fig. 7.8 A portion of a region containing a two-dimensional potential field, divided into squares of side h. The potential V_0 is approximately equal to the average of the potentials at the four neighboring points.

from which

$$\left.\frac{\partial^2 V}{\partial x^2}\right|_0 \doteq \frac{\left.\frac{\partial V}{\partial x}\right|_a - \left.\frac{\partial V}{\partial x}\right|_c}{h} \doteq \frac{V_1 - V_0 - V_0 + V_3}{h^2}$$

and similarly,

$$\left.\frac{\partial^2 V}{\partial y^2}\right|_0 \doteq \frac{V_2 - V_0 - V_0 + V_4}{h^2}$$

Combining, we have

$$\frac{\partial^2 V}{\partial x^2} + \frac{\partial^2 V}{\partial y^2} \doteq \frac{V_1 + V_2 + V_3 + V_4 - 4V_0}{h^2} = 0$$

or

$$V_0 \doteq \tfrac{1}{4}(V_1 + V_2 + V_3 + V_4) \tag{40}$$

The expression becomes exact as h approaches zero, and we shall write it without the approximation sign. It is intuitively correct, telling us that the potential is the average of the potential at the four neighboring points. The iterative method merely uses (40) to determine the potential at the corner of every square subdivision in turn, and then the process is repeated over the entire region as many times as is necessary until the values no longer change. The method is best shown in detail by an example.

For simplicity, consider a square region with conducting boundaries (Figure 7.9). The potential of the top is 100 V and that of the sides and bottom is zero. The problem is two-dimensional, and the sketch is a cross section of the physical configuration. The region is divided first into 16 squares, and some estimate of the potential must now be made at every corner before applying the iterative method. The better the estimate, the shorter the solution, although the final result is independent of these initial estimates. When the computer is used for iteration, the initial potentials are usually set equal to zero to simplify the program. Reasonably accurate values could be obtained from a rough curvilinear-square map, or we could apply (40) to the large squares. At the center of the figure the potential estimate is then $\frac{1}{4}(100 + 0 + 0 + 0) = 25.0$.

The potential may now be estimated at the centers of the four double-sized squares by taking the average of the potentials at the four corners or applying (40) along a diagonal set of axes. Use of this "diagonal average" is made only in preparing initial estimates. For the two upper double squares, we select a potential of 50 V for the gap (the average of 0 and 100), and then $V = \frac{1}{4}(50 + 100 + 25 + 0) = 43.8$ (to the nearest tenth of a volt),[2] and for the lower ones,

$$V = \tfrac{1}{4}(0 + 25 + 0 + 0) = 6.2$$

The potential at the remaining four points may now be obtained by applying (40) directly. The complete set of estimated values is shown in Figure 7.9.

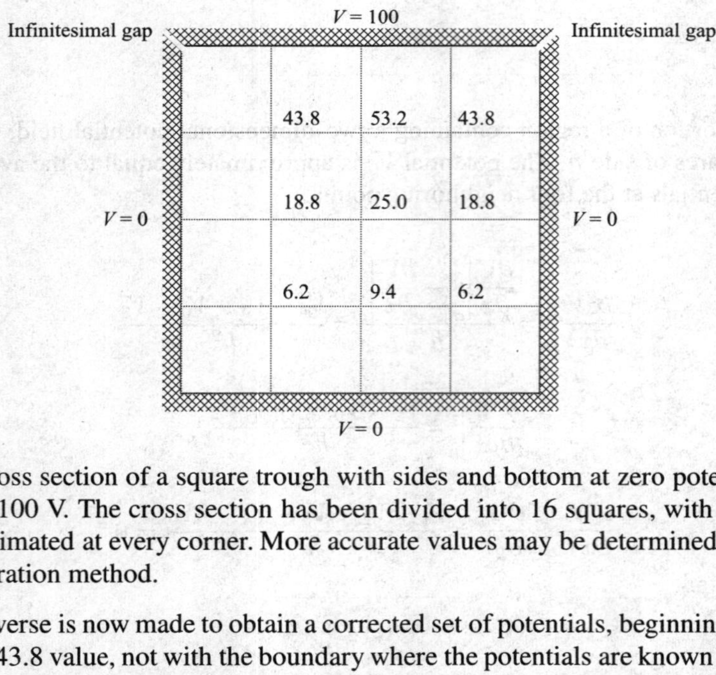

Fig. 7.9 Cross section of a square trough with sides and bottom at zero potential and top at 100 V. The cross section has been divided into 16 squares, with the potential estimated at every corner. More accurate values may be determined by using the iteration method.

The initial traverse is now made to obtain a corrected set of potentials, beginning in the upper left corner (with the 43.8 value, not with the boundary where the potentials are known and fixed), working across the row to the right, and then dropping down to the second row and proceeding from left to right again. Thus the 43.8 value changes to $\frac{1}{4}(100 + 53.2 + 18.8 + 0) = 43.0$. The best or newest

[2] When rounding off a decimal ending exactly with a five, the preceding digit should be made *even* (e.g., 42.75 becomes 42.8 and 6.25 becomes 6.2). This generally ensures a random process leading to better accuracy than would be obtained by always increasing the previous digit by 1.

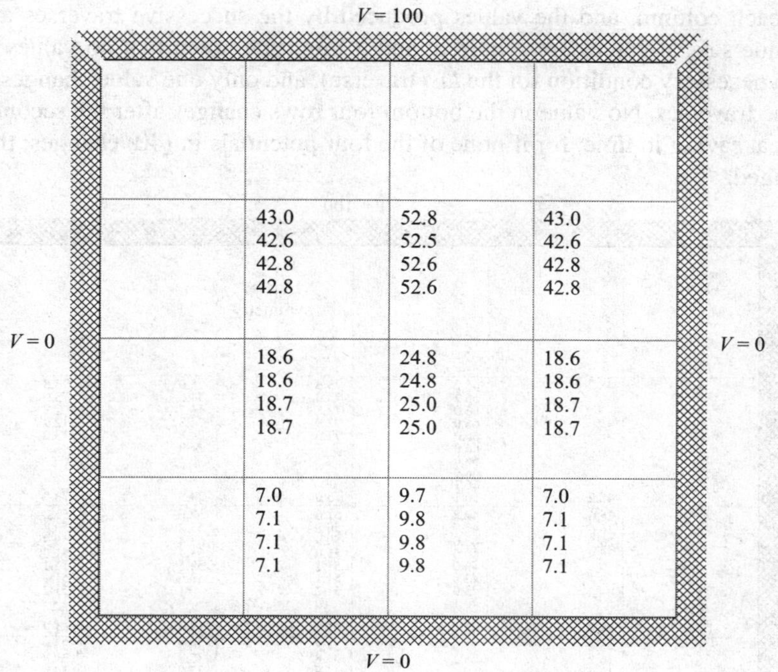

Fig. 7.10 The results of each of the four necessary traverses of the problem of Figure 7.8 are shown in order in the columns. The final values, unchanged in the last traverse, are at the bottom of each column.

potentials are always used when applying (40), so both points marked 43.8 are changed to 43.0, because of the evident symmetry, and the 53.2 value becomes $\frac{1}{4}(100 + 43.0 + 25.0 + 43.0) = 52.8$.

Because of the symmetry, little would be gained by continuing across the top line. Each point of this line has now been improved once. Dropping down to the next line, the 18.8 value becomes

$$V = \tfrac{1}{4}(43.0 + 25.0 + 6.2 + 0) = 18.6$$

and the traverse continues in this manner. The values at the end of this traverse are shown as the top numbers in each column of Figure 7.10. Additional traverses must now be made until the value at each corner shows no change. The values for the successive traverses are usually entered below each other in column form, as shown in Figure 7.10, and the final value is shown at the bottom of each column. Only four traverses are required in this example.

If each of the nine initial values were set equal to zero, it is interesting to note that ten traverses would be required. The cost of having a computer do these additional traverses is probably much less than the cost of the programming necessary to make decent initial estimates.

Since there is a large difference in potential from square to square, we should not expect our answers to be accurate to the tenth of a volt shown (and perhaps not to the nearest volt). Increased accuracy comes from dividing each square into four smaller squares, not from finding the potential to a larger number of significant figures at each corner.

In Figure 7.11, which shows only one of the symmetrical halves plus an additional column, this subdivision is accomplished, and the potential at the newly created corners is estimated by applying (40) directly where possible and diagonally when necessary. The set of estimated values appears at the top of each column, and the values produced by the successive traverses appear in order below. Here nine sets of values are required, and it might be noted that no values change on the last traverse (a necessary condition for the *last* traverse), and only one value changes on each of the preceding three traverses. No value in the bottom four rows changes after the second traverse; this results in a great saving in time, for if none of the four potentials in (40) changes, the answer is of course unchanged.

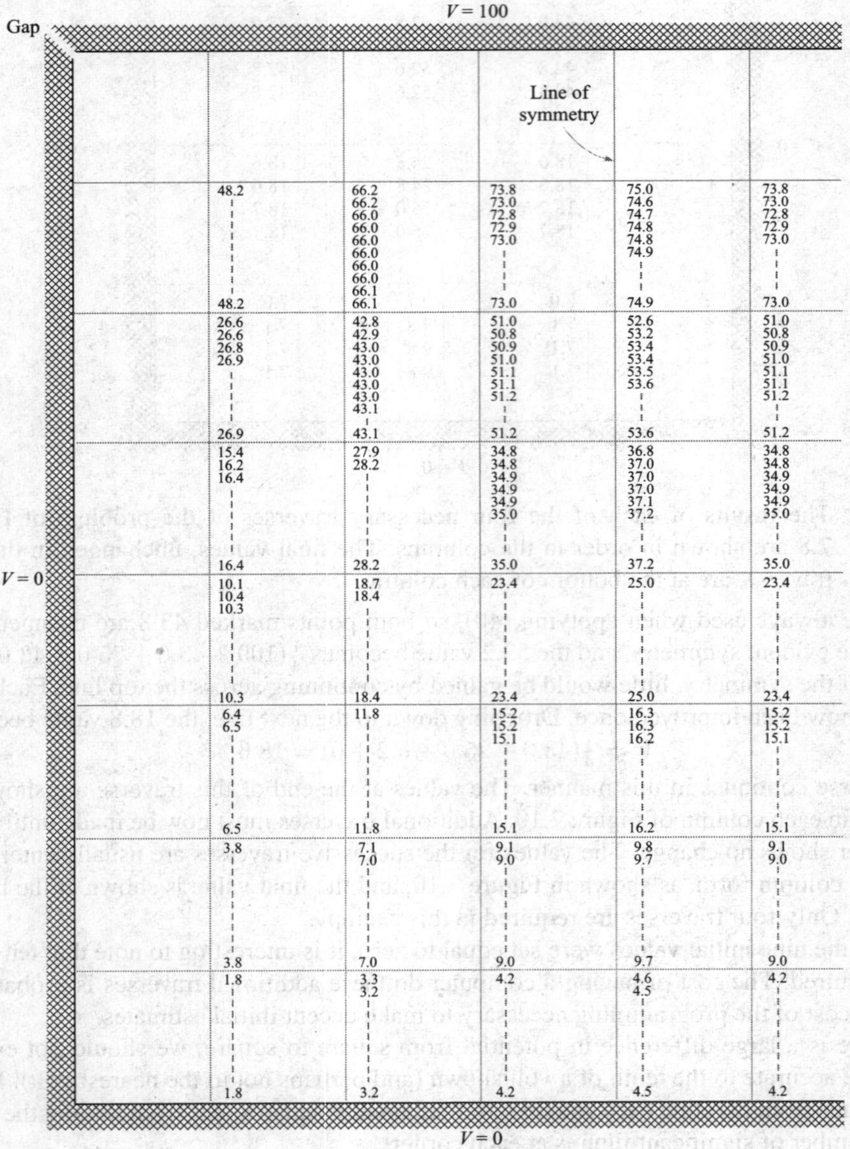

Fig. 7.11 The problem of Figures 7.9 and 7.10 is divided into smaller squares. Values obtained on the nine successive traverses are listed in order in the columns.

For this problem it is possible to compare our final values with the exact potentials, as found analytically through the product solution method. At the point for which the original estimate was 53.2, the final value found using the coarse grid was 52.6, the final value from the finer grid was 53.6, and the final value using a 16×16 grid is 53.93 V to two decimals. The exact potential obtained by the product solution method is 54.05 V to two decimals. Two other points are also compared, as shown in Table 7.1.

Table 7.1

Original estimate	53.2	25.0	9.4
4×4 grid	52.6	25.0	9.8
8×8 grid	53.6	25.0	9.7
16×16 grid	53.93	25.00	9.56
Exact	54.05	25.00	9.54

Very few electrode configurations have a square or rectangular cross section that can be neatly subdivided into a square grid. Curved boundaries, acute- or obtuse-angled corners, reentrant shapes, and other irregularities require slight modifications of the basic method. An important one of these is described in Problem 33 at the end of this chapter, and other irregular examples appear as Problems 30, 32, and 34.

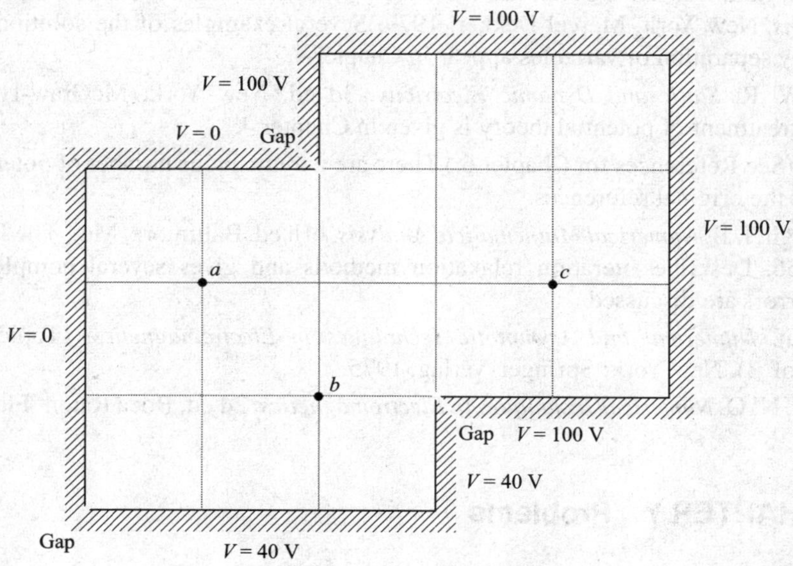

Fig. 7.12 See Problem D7.6.

A refinement of the iteration method is known as the *relaxation method*. In general it requires less work but more care in carrying out the arithmetical steps.[3]

[3] A detailed description appears in Scarborough. See References at the end of the chapter.

D7.6 In Figure 7.12, a square grid is shown within an irregular potential trough. Using the iteration method to find the potential to the nearest volt, determine the final value at: (a) point a; (b) point b; (c) point c.

Ans. 18 V; 46 V; 91 V

References

1. Dekker, A. J. See References for Chapter 5.
2. Hayt, W. H., Jr., and J. E. Kemmerly. *Engineering Circuit Analysis*. 5th ed. New York: McGraw-Hill, 1993.
3. Push, E. M., and E. W. Pugh. *Principles of Electricity and Magnetism*. 2d ed. Reading, Mass.: Addison-Wesley, 1970. This text provides the physicist's view of electricity and magnetism, but electrical engineering students should find it easy to read. The solution to Laplace's equation by a number of methods is discussed in Chapter 4.
4. Ramo, S., J. R. Whinnery, and T. Van Duzer. (See References for Chapter 6.) A more complete and advanced discussion of methods of solving Laplace's equation is given in Chapter 7.
5. Seeley, S., and A. D. Poularikas. *Electromagnetics: Classical and Modern Theory and Applications*. New York: Marcel Dekker, 1979. Several examples of the solution of Laplace's equation by separation of variables appear in Chapter 4.
6. Smythe, W. R. *Static and Dynamic Electricity*. 3d ed. New York: McGraw-Hill, 1968. An advanced treatment of potential theory is given in Chapter 4.
7. Weber, E. (See References for Chapter 6.) There are a tremendous number of potential solutions given with the original references.
8. Scarborough, J. B. *Numerical Mathematical Analysis*. 6th ed. Baltimore, Md.: The Johns Hopkins Press, 1966. Describes iteration relaxation methods and gives several complete examples. Inherent errors are discussed.
9. Mittra, Raj. *Numerical and Asymptotic Techniques in Electromagnetics*. (Topics in Applied Physics, vol. 3), New York: Springer-Verlag, 1975.
10. Sadiku, M. N. O. *Numerical Techniques in Electromagnetics*. 2d ed. Boca Raton, Fla.: CRC Press, 2000.

CHAPTER 7 Problems

Quizzes

7.1 Let $V = 2xy^2z^3$ and $\epsilon = \epsilon_0$. Given point $P(1, 2, -1)$, find: (a) V at P; (b) $\mathbf{E}$ at P; (c) ρ_ν at P; (d) the equation of the equipotential surface passing through P; (e) the equation of the streamline passing through P. (f) Does V satisfy Laplace's equation?

7.2 Given the spherically symmetric potential field in free space, $V = V_0 e^{-r/a}$, find: (a) ρ_ν at $r = a$; (b) the electric field at $r = a$; (c) the total charge.

7.3 Let $V(x,y) = 4e^{2x} + f(x) - 3y^2$ in a region of free space where $\rho_\nu = 0$. It is known that both E_x and V are zero at the origin. Find $f(x)$ and $V(x,y)$.

7.4 Given the potential field $V(\rho,\phi) = (V_0\rho/d)\cos\phi$: (a) Show that $V(\rho,\phi)$ satisfies Laplace's equation. (b) Describe the constant-potential surfaces. (c) Specifically describe the surfaces on which $V = V_0$ and $V = 0$.(d) Write the potential expression in rectangular coordinates.

7.5 Given the potential field $V = (A\rho^4 + B\rho^{-4})\sin 4\phi$: (a) Show that $\nabla^2 V = 0$. (b) Select A and B so that $V = 100$ V and $|\mathbf{E}| = 500$ V/m at $P(\rho = 1, \phi = 22.5°, z = 2)$.

7.6 A parallel-plate capacitor has plates located at $z = 0$ and $z = d$. The region between plates is filled with a material that contains volume charge of uniform density ρ_0 C/m^3 and has permittivity ϵ. Both plates are held at ground potential. (a) Determine the potential field between plates. (b) Determine the electric field intensity $\mathbf{E}$ between plates. (c) Repeat parts (a) and (b) for the case of the plate at $z = d$ raised to potential V_0, with the $z = 0$ plate grounded.

7.7 Let $V = (\cos 2\phi)/\rho$ in free space. (a) Find the volume charge density at point $A(0.5, 60°, 1)$. (b) Find the surface charge density on a conductor surface passing through the point $B(2, 30°, 1)$.

7.8 A uniform volume charge has constant density $\rho_\nu = \rho_0$ C/m^3 and fills the region $r < a$, in which permittivity ϵ is assumed. A conducting spherical shell is located at $r = a$ and is held at ground potential. Find: (a) the potential everywhere; (b) the electric field intensity, $\mathbf{E}$, everywhere.

7.9 The functions $V_1(\rho,\phi,z)$ and $V_2(\rho,\phi,z)$ both satisfy Laplace's equation in the region $a < \rho < b$, $0 \le \phi < 2\pi$, $-L < z < L$; each is zero on the surfaces $\rho = b$ for $-L < z < L$; $z = -L$ for $a < \rho < b$; and $z = L$ for $a < \rho < b$; and each is 100 V on the surface $\rho = a$ for $-L < z < L$. (a) In the region specified, is Laplace's equation satisfied by the functions $V_1 + V_2$, $V_1 - V_2$, $V_1 + 3$, and $V_1 V_2$? (b) On the boundary surfaces specified, are the potential values given in this problem obtained from the functions $V_1 + V_2$, $V_1 - V_2$, $V_1 + 3$, and $V_1 V_2$? (c) Are the functions $V_1 + V_2$, $V_1 - V_2$, $V_1 + 3$, and $V_1 V_2$ identical with V_1?

7.10 Consider the parallel-plate capacitor of Problem 7.6, but this time the charged dielectric exists only between $z = 0$ and $z = b$, where $b < d$. Free space fills the region $b < z < d$. Both plates are at ground potential. By solving Laplace's *and* Poisson's equations, find: (a) $V(z)$ for $0 < z < d$; (b) The electric field intensity for $0 < z < d$. No surface charge exists at $z = b$, so both V and $\mathbf{D}$ are continuous there.

7.11 The conducting planes $2x + 3y = 12$ and $2x + 3y = 18$ are at potentials of 100 V and 0, respectively. Let $\epsilon = \epsilon_0$ and find: (a) V at $P(5, 2, 6)$; (b) $\mathbf{E}$ at P.

7.12 The derivation of Laplace's and Poisson's equations assumed constant permittivity, but there are cases of spatially varying permittivity in which the equations will still apply. consider the vector identity, $\nabla \cdot (\psi \mathbf{G}) = \mathbf{G} \cdot \nabla \psi + \psi \nabla \cdot \mathbf{G}$, where ψ and $\mathbf{G}$ are scalar and vector functions, respectively. Determine a general rule on the allowed *directions* in which ϵ may vary with respect to the local electric field.

7.13 Coaxial conducting cylinders are located at $\rho = 0.5$ cm and $\rho = 1.2$ cm. The region between the cylinders is filled with a homogeneous perfect dielectric. If the inner cylinder is at 100 V and the outer at 0 V, find: (a) the location of the 20 V equipotential surface; (b) $E_{\rho\max}$; (c) ϵ_r if the charge per meter length on the inner cylinder is 20 nC/m.

7.14 Repeat Problem 7.13, but with the dielectric only partially filling the volume, within $0 < \phi < \pi$, and with free space in the remaining volume.

7.15 The two conducting planes illustrated in Figure 7.13 are defined by $0.001 < \rho < 0.120$ m, $0 < z < 0.1$ m, $\phi = 0.179$ and 0.188 rad. The medium surrounding the planes is air. For Region 1, $0.179 < \phi < 0.188$; neglect fringing and find: (*a*) $V(\phi)$; (*b*) $\mathbf{E}(\rho)$; (*c*) $\mathbf{D}(\rho)$; (*d*) ρ_s on the upper surface of the lower plane; (*e*) Q on the upper surface of the lower plane. (*f*) Repeat parts (*a*) through (*c*) for Region 2 by letting the location of the upper plane be $\phi = .188 - 2\pi$, and then find ρ_s and Q on the lower surface of the lower plane. (*g*) Find the total charge on the lower plane and the capacitance between the planes.

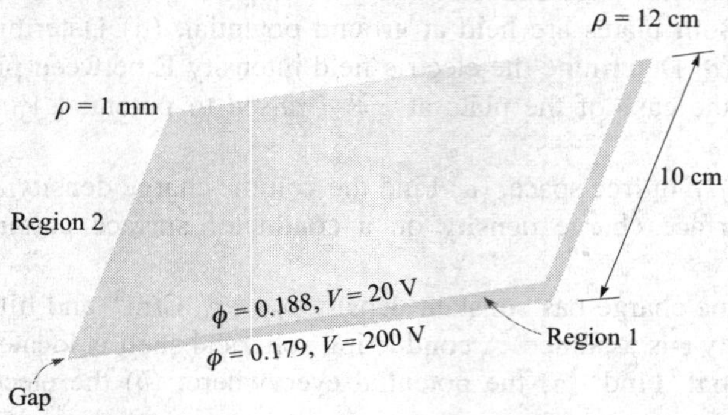

Fig. 7.13 See Problem 7.15.

7.16 A parallel-plate capacitor is made using two circular plates of radius a, with the bottom plate on the xy plane, centered at the origin. The top plate is located at $z = d$, with its center on the z axis. Potential V_0 is on the top plate; the bottom plate is grounded. Dielectric having *radially dependent* permittivity fills the region between plates. The permittivity is given by $\epsilon(\rho) = \epsilon_0(1 + \rho/a)$. Find: (*a*)$V(z)$; (*b*) $\mathbf{E}$; (*c*) Q; (*d*) C. This is a reprise of Problem 6.16, but it starts with Laplace's equation.

7.17 Concentric conducting spheres are located at $r = 5$ mm and $r = 20$ mm. The region between the spheres is filled with a perfect dielectric. If the inner sphere is at 100 V and the outer sphere is at 0 V: (*a*) Find the location of the 20 V equipotential surface. (*b*) Find $E_{r,\max}$. (*c*) Find ϵ_r if the surface charge density on the inner sphere is $1.0\ \mu\text{C/m}^2$.

7.18 The hemisphere $0 < r < a$, $0 < \theta < \pi/2$, is composed of homogeneous conducting material of conductivity σ. The flat side of the hemisphere rests on a perfectly conducting plane. Now, the material within the conical region $0 < \theta < \alpha$, $0 < r < a$ is drilled out and replaced with material that is perfectly conducting. An air gap is maintained between the $r = 0$ tip of this new material and the plane. What resistance is measured between the two perfect conductors? Neglect fringing fields.

7.19 Two coaxial conducting cones have their vertices at the origin and the z axis as their axis. Cone A has the point $A(1, 0, 2)$ on its surface, while cone B has the point $B(0, 3, 2)$ on its surface. Let $V_A = 100$ V and $V_B = 20$ V. Find: (*a*) α for each cone; (*b*) V at $P(1, 1, 1)$.

7.20 A potential field in free space is given as $V = 100 \ln \tan(\theta/2) + 50$ V. (*a*) Find the maximum value of $|\mathbf{E}_\theta|$ on the surface $\theta = 40°$ for $0.1 < r < 0.8$ m, $60° < \phi < 90°$. (*b*) Describe the surface $V = 80$ V.

7.21 In free space, let $\rho_\nu = 200\epsilon_0/r^{2.4}$. ($a$) Use Poisson's equation to find $V(r)$ if it is assumed that $r^2E_r \to 0$ when $r \to 0$, and also that $V \to 0$ as $r \to \infty$. (b) Now find $V(r)$ by using Gauss's law and a line integral.

7.22 By appropriate solution of Laplace's *and* Poisson's equations, determine the absolute potential at the center of a sphere of radius a, containing uniform volume charge of density ρ_0. Assume permittivity ϵ_0 everywhere. *Hint*: What must be true about the potential and the electric field at $r = 0$ and at $r = a$?

7.23 A rectangular trough is formed by four conducting planes located at $x = 0$ and 8 cm and $y = 0$ and 5 cm in air. The surface at $y = 5$ cm is at a potential of 100 V, the other three are at zero potential, and the necessary gaps are placed at two corners. Find the potential at $x = 3$ cm, $y = 4$ cm.

7.24 The four sides of a square trough are held at potentials of 0, 20, −30, and 60 V; the highest and lowest potentials are on opposite sides. Find the potential at the center of the trough.

7.25 In Figure 7.7, change the right side so that the potential varies linearly from 0 at the bottom of that side to 100 V at the top. Solve for the potential at the center of the trough.

7.26 If X is a function of x and $X'' + (x-1)X - 2X = 0$, assume a solution in the form of an infinite power series and determine numerical values for a_2 to a_8 if $a_0 = 1$ and $a_1 = -1$.

7.27 It is known that $V = XY$ is a solution of Laplace's equation, where X is a function of x alone and Y is a function of y alone. Determine which of the following potential functions are also solutions of Laplace's equation: (a) $V = 100X$; (b) $V = 50XY$; (c) $V = 2XY + x - 3y$; (d) $V = xXY$; (e) $V = X^2Y$.

7.28 Assume a product solution of Laplace's equation in cylindrical coordinates, $V = PF$, where V is not a function of z, P is a function only of ρ, and F is a function only of ϕ. (a) Obtain the two separated equations if the separation constant is n^2. (b) Show that $P = A\rho^n + B\rho^{-n}$ satisfies the ρ equation. (c) Construct the solution $V(\rho, \phi)$. Functions of this form are called *circular harmonics*.

7.29 Referring to Chapter 6, Figure 6.14, let the inner conductor of the transmission line be at a potential of 100 V, while the outer is at zero potential. Construct a grid, $0.5a$ on a side, and use iteration to find V at a point that is a units above the upper right corner of the inner conductor. Work to the nearest volt.

7.30 Use the iteration method to estimate the potentials at points x and y in the triangular trough of Figure 7.14. Work only to the nearest volt.

7.31 Use iteration methods to estimate the potential at point x in the trough shown in Figure 7.15. Working to the nearest volt is sufficient.

7.32 Using the grid in Figure 7.16, work to the nearest volt to estimate the potential at point A.

7.33 Conductors having boundaries that are curved or skewed usually do not permit every grid point to coincide with the actual boundary. Figure 7.17a illustrates the situation where the potential at V_0 is to be estimated in terms of V_1, V_2, V_3, and V_4 and the unequal distances

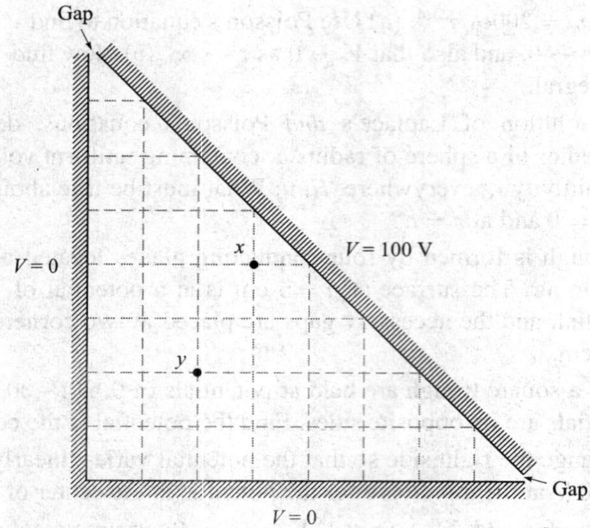

Fig. 7.14 See Problem 7.30.

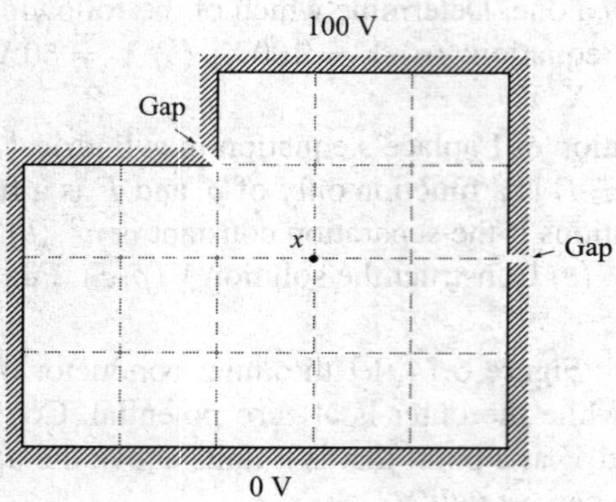

Fig. 7.15 See Problem 7.31.

h_1, h_2, h_3, and h_4. (a) Show that

$$V_0 \doteq \frac{V_1}{\left(1+\dfrac{h_1}{h_3}\right)\left(1+\dfrac{h_1 h_3}{h_4 h_2}\right)} + \frac{V_2}{\left(1+\dfrac{h_2}{h_4}\right)\left(1+\dfrac{h_2 h_4}{h_1 h_3}\right)}$$

$$+\frac{V_3}{\left(1+\dfrac{h_3}{h_1}\right)\left(1+\dfrac{h_1 h_3}{h_4 h_2}\right)} + \frac{V_4}{\left(1+\dfrac{h_4}{h_2}\right)\left(1+\dfrac{h_4 h_2}{h_3 h_1}\right)}$$

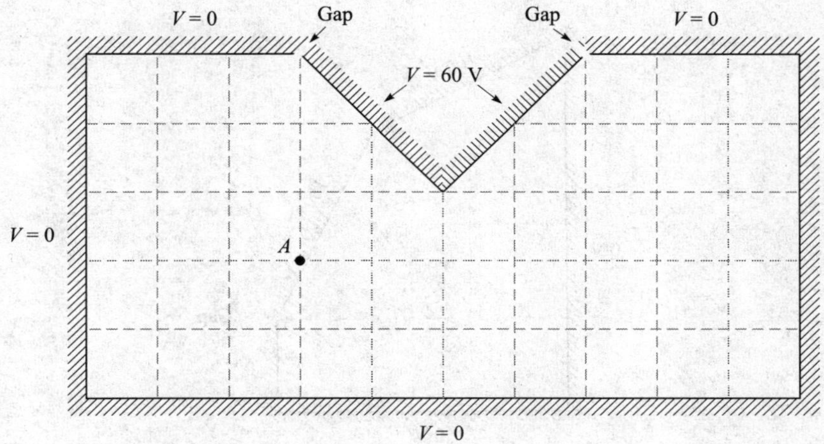

Fig. 7.16 See Problem 7.32.

(*b*) Determine V_0 in Figure 7.17*b*.

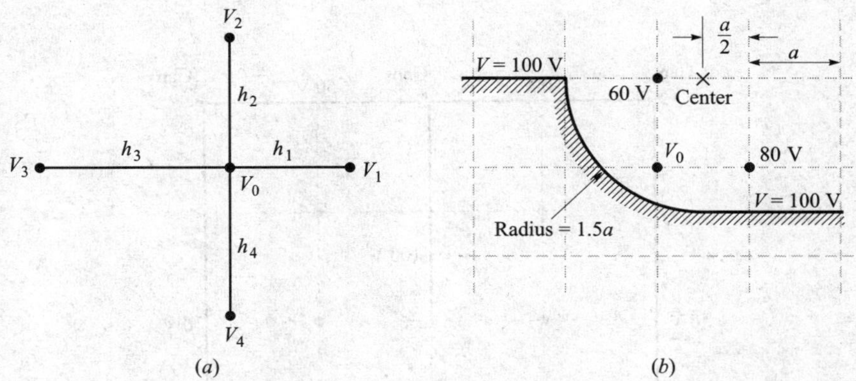

Fig. 7.17 See Problem 7.33.

7.34 Consider the configuration of conductors and potentials shown in Figure 7.18. Using the method described in Problem 7.33, estimate the potentials at points x, y, and z.

7.35 (*a*) After estimating potentials for the configuration of Figure 7.19, use the iteration method with a square grid 1 cm on a side to find better estimates at the seven grid points. Work to the nearest volt. (*b*) Construct a 0.5 cm grid, establish new rough estimates, and then use the iteration method on the 0.5 cm grid. Again, work to the nearest volt. (*c*) Use a computer to obtain values for a 0.25 cm grid. Work to the nearest 0.1 V.

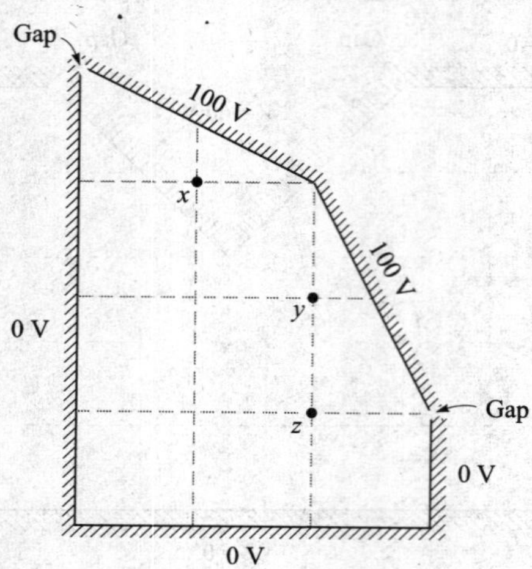

Fig. 7.18 See Problem 7.34.

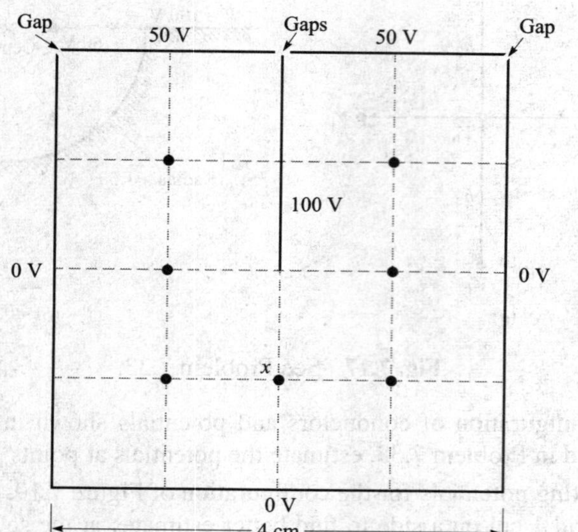

Fig. 7.19 See Problem 7.35.

8

The Steady Magnetic Field

At this point the concept of a field should be a familiar one. Since we first accepted the experimental law of forces existing between two point charges and defined electric field intensity as the force per unit charge on a test charge in the presence of a second charge, we have discussed numerous fields. These fields possess no real physical basis, for physical measurements must always be in terms of the forces on the charges in the detection equipment. Those charges which are the source cause measurable forces to be exerted on other charges, which we may think of as detector charges. The fact that we attribute a field to the source charges and then determine the effect of this field on the detector charges amounts merely to a division of the basic problem into two parts for convenience.

We shall begin our study of the magnetic field with a definition of the magnetic field itself and show how it arises from a current distribution. The effect of this field on other currents, or the second half of the physical problem, will be discussed in Chapter 9. As we did with the electric field, we shall confine our initial discussion to free-space conditions, and the effect of material media will also be saved for discussion in Chapter 9.

The relation of the steady magnetic field to its source is more complicated than is the relation of the electrostatic field to its source. We shall find it necessary to accept several laws temporarily on faith alone, relegating their proof to the (rather difficult) final section in this chapter. This section may well be omitted when studying magnetic fields for the first time. It is included to make acceptance of the laws a little easier; the proof of the laws does exist and is available for the disbelievers or the more advanced student.

8.1 BIOT-SAVART LAW

The source of the steady magnetic field may be a permanent magnet, an electric field changing linearly with time, or a direct current. We shall largely ignore the permanent magnet and save the time-varying electric field for a later discussion. Our present relationships will concern the magnetic field produced by a differential dc element in free space.

We may think of this differential current element as a vanishingly small section of a current-carrying filamentary conductor, where a filamentary conductor is the limiting case of a cylindrical conductor of circular cross section as the radius approaches zero. We assume a current I flowing in a differential vector length of the filament $d\mathbf{L}$. The law of Biot-Savart[1] then states that at any point P the magnitude of the magnetic field intensity produced by the differential element is proportional to the product of the current, the magnitude of the differential length, and the sine of the angle lying

[1] Biot and Savart were colleagues of Ampère, and all three were professors of physics at the Collège de France at one time or another. The Biot-Savart law was proposed in 1820.

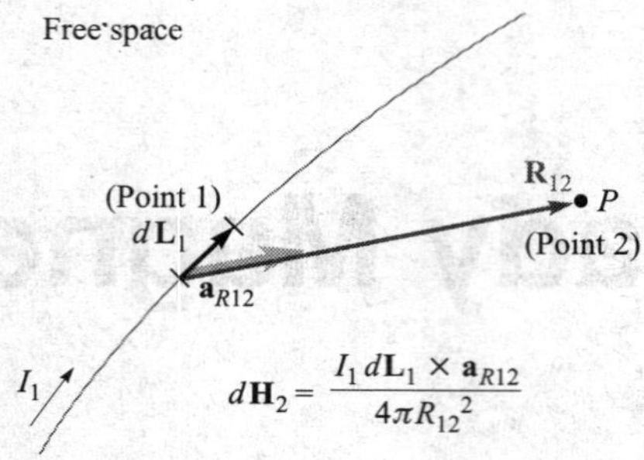

Fig. 8.1 The law of Biot-Savart expresses the magnetic field intensity $d\mathbf{H}_2$ produced by a differential current element $I_1 d\mathbf{L}_1$. The direction of $d\mathbf{H}_2$ is into the page.

between the filament and a line connecting the filament to the point P at which the field is desired; also, the magnitude of the magnetic field intensity is inversely proportional to the square of the distance from the differential element to the point P. The direction of the magnetic field intensity is normal to the plane containing the differential filament and the line drawn from the filament to the point P. Of the two possible normals, that one is to be chosen which is in the direction of progress of a right-handed screw turned from $d\mathbf{L}$ through the smaller angle to the line from the filament to P. Using rationalized mks units, the constant of proportionality is $1/4\pi$.

The *Biot-Savart law,* just described in some 150 words, may be written concisely using vector notation as

$$d\mathbf{H} = \frac{I d\mathbf{L} \times \mathbf{a}_R}{4\pi R^2} = \frac{I d\mathbf{L} \times \mathbf{R}}{4\pi R^3} \tag{1}$$

The units of the *magnetic field intensity* **H** are evidently amperes per meter (A/m). The geometry is illustrated in Figure 8.1. Subscripts may be used to indicate the point to which each of the quantities in (1) refers. If we locate the current element at point 1 and describe the point P at which the field is to be determined as point 2, then

$$d\mathbf{H}_2 = \frac{I_1 d\mathbf{L}_1 \times \mathbf{a}_{R12}}{4\pi R_{12}^2} \tag{2}$$

The law of Biot-Savart is sometimes called *Ampère's law for the current element,* but we shall retain the former name because of possible confusion with Ampère's circuital law, to be discussed later.

In some aspects, the Biot-Savart law is reminiscent of Coulomb's law when that law is written for a differential element of charge,

$$d\mathbf{E}_2 = \frac{dQ_1 \mathbf{a}_{R12}}{4\pi\epsilon_0 R_{12}^2}$$

Both show an inverse-square-law dependence on distance, and both show a linear relationship between source and field. The chief difference appears in the direction of the field.

It is impossible to check experimentally the law of Biot-Savart as expressed by (1) or (2) because the differential current element cannot be isolated. We have restricted our attention to direct currents only, so the charge density is not a function of time. The continuity equation in Section 5.2, Eq. (5),

$$\nabla \cdot \mathbf{J} = -\frac{\partial \rho_\nu}{\partial t}$$

therefore shows that

$$\mathbf{V} \cdot \mathbf{J} = 0$$

or upon applying the divergence theorem,

$$\oint_s \mathbf{J} \cdot d\mathbf{S} = 0$$

The total current crossing any closed surface is zero, and this condition may be satisfied only by assuming a current flow around a closed path. It is this current flowing in a closed circuit which must be our experimental source, not the differential element.

It follows that only the integral form of the Biot-Savart law can be verified experimentally,

$$\mathbf{H} = \oint \frac{I d\mathbf{L} \times \mathbf{a}_R}{4\pi R^2} \tag{3}$$

Equation (1) or (2), of course, leads directly to the integral form (3), but other differential expressions also yield the same integral formulation. Any term may be added to (1) whose integral around a closed path is zero. That is, any conservative field could be added to (1). The gradient of any scalar field always yields a conservative field, and we could therefore add a term ∇G to (1), where G is a general scalar field, without changing (3) in the slightest. This qualification on (1) or (2) is mentioned to show that if we later ask some foolish questions, not subject to any experimental check, concerning the force exerted by one *differential* current element on another, we should expect foolish answers.

The Biot-Savart law may also be expressed in terms of distributed sources, such as current density $\mathbf{J}$ and *surface current density* $\mathbf{K}$. Surface current flows in a sheet of vanishingly small thickness, and the current density $\mathbf{J}$, measured in amperes per square meter, is therefore infinite. Surface current density, however, is measured in amperes per meter width and designated by $\mathbf{K}$. If the surface current density is uniform, the total current I in any width b is

$$I = Kb$$

where we have assumed that the width b is measured perpendicularly to the direction in which the current is flowing. The geometry is illustrated by Figure 8.2. For a nonuniform surface current density, integration is necessary:

$$I = \int K dN \tag{4}$$

where dN is a differential element of the path *across* which the current is flowing. Thus the differential current element $I\,d\mathbf{L}$, where $d\mathbf{L}$ is in the direction of the current, may be expressed in terms of surface current density $\mathbf{K}$ or current density $\mathbf{J}$,

$$I\,d\mathbf{L} = \mathbf{K}\,dS = \mathbf{J}\,d\nu \tag{5}$$

and alternate forms of the Biot-Savart law obtained,

$$\mathbf{H} = \int_s \frac{\mathbf{K} \times \mathbf{a}_R dS}{4\pi R^2} \tag{6}$$

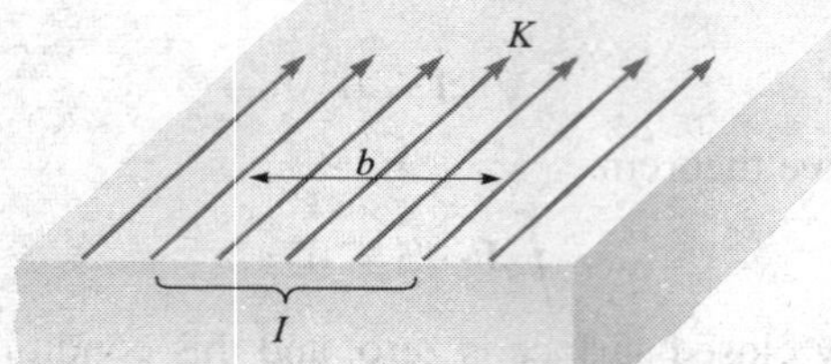

Fig. 8.2 The total current I within a transverse width b, in which there is a *uniform* surface current density K, is Kb.

and

$$\mathbf{H} = \int_{\text{vol}} \frac{\mathbf{J} \times \mathbf{a}_R d\nu}{4\pi R^2} \tag{7}$$

We may illustrate the application of the Biot-Savart law by considering an infinitely long straight filament. We shall apply (2) first and then integrate. This, of course, is the same as using the integral form (3) in the first place.[2]

Referring to Figure 8.3, we should recognize the symmetry of this field. No variation with z or with ϕ can exist. Point 2, at which we shall determine the field, is therefore chosen in the $z = 0$ plane. The field point $\mathbf{r}$ is therefore $r = \rho\mathbf{a}_\rho$. The source point $\mathbf{r}'$ is given by $\mathbf{r}' = z'\mathbf{a}_z$, and therefore

$$\mathbf{R}_{12} = \mathbf{r} - \mathbf{r}' = \rho\mathbf{a}_\rho - z'\mathbf{a}_z$$

so that

$$\mathbf{a}_{R12} = \frac{\rho\mathbf{a}_\rho - z'\mathbf{a}_z}{\sqrt{\rho^2 + z'^2}}$$

We take $d\mathbf{L} = dz'\mathbf{a}_z$ and (2) becomes

$$d\mathbf{H}_2 = \frac{I\,dz'\mathbf{a}_z \times (\rho\mathbf{a}_\rho - z'\mathbf{a}_z)}{4\pi(\rho^2 + z'^2)^{3/2}}$$

Since the current is directed toward increasing values of z', the limits are $-\infty$ and ∞ on the integral, and we have

$$\mathbf{H}_2 = \int_{-\infty}^{\infty} \frac{I\,dz'\mathbf{a}_z \times (\rho\mathbf{a}_\rho - z'\mathbf{a}_z)}{4\pi(\rho^2 + z'^2)^{3/2}} = \frac{I}{4\pi}\int_{-\infty}^{\infty} \frac{\rho dz'\mathbf{a}_\phi}{(\rho^2 + z'^2)^{3/2}}$$

Illustrations

At this point the unit vector $\mathbf{a}_\phi$ under the integral sign should be investigated, for it is not always a constant, as are the unit vectors of the rectangular coordinate system. A vector is constant when its magnitude and direction are both constant. The unit vector certainly has constant magnitude, but its direction may change. Here $\mathbf{a}_\phi$ changes with the coordinate ϕ but not with ρ or z. Fortunately, the integration here is with respect to z', and $\mathbf{a}_\phi$ is

[2] The closed path for the current may be considered to include a return filament parallel to the first filament and infinitely far removed. An outer coaxial conductor of infinite radius is another theoretical possibility. Practically, the problem is an impossible one, but we should realize that our answer will be quite accurate near a very long, straight wire having a distant return path for the current.

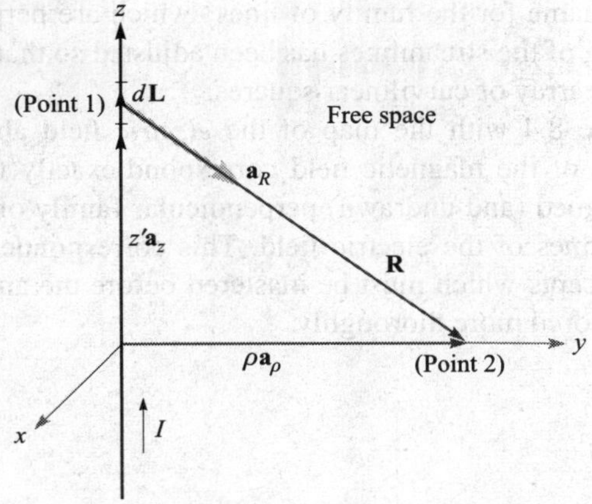

Fig. 8.3 An infinitely long straight filament carrying a direct current I. The field at point 2 is $\mathbf{H} = (I/2\pi\rho)\mathbf{a}_\phi$.

a constant and may be removed from under the integral sign,

$$\mathbf{H}_2 = \frac{I\rho\mathbf{a}_\phi}{4\pi}\int_{-\infty}^{\infty}\frac{dz'}{(\rho^2 + z'^2)^{3/2}} = \frac{I\rho\mathbf{a}_\phi}{4\pi}\,\frac{z'}{\rho^2\sqrt{\rho^2 + z'^2}}\bigg|_{-\infty}^{\infty}$$

and

$$\mathbf{H}_2 = \frac{I}{2\pi\rho}\mathbf{a}_\phi \tag{8}$$

The magnitude of the field is not a function of ϕ or z, and it varies inversely with the distance from the filament. The direction of the magnetic-field-intensity vector is circumferential. The streamlines are therefore circles about the filament, and the field may be mapped in cross section as in Figure 8.4.

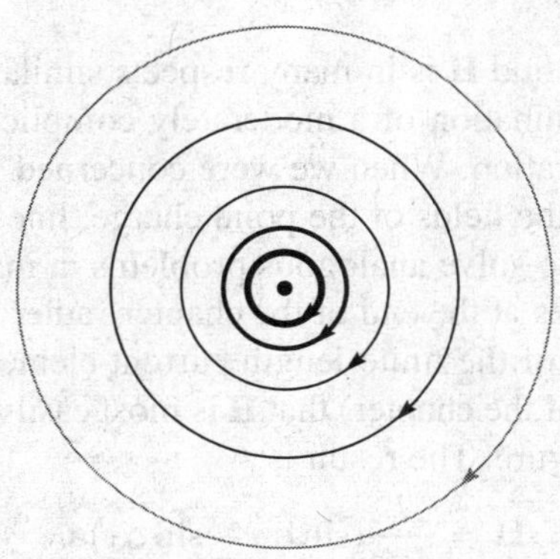

Fig. 8.4 The streamlines of the magnetic field intensity about an infinitely long straight filament carrying a direct current I. The direction of I is into the page.

The separation of the streamlines is proportional to the radius, or inversely proportional to the magnitude of **H**. To be specific, the streamlines have been drawn with curvilinear squares in mind. As yet we have no name for the family of lines[3] which are perpendicular to these circular streamlines, but the spacing of the streamlines has been adjusted so that the addition of this second set of lines will produce an array of curvilinear squares.

A comparison of Figure 8.4 with the map of the *electric* field about an infinite line *charge* shows that the streamlines of the magnetic field correspond exactly to the equipotentials of the electric field, and the unnamed (and undrawn) perpendicular family of lines in the magnetic field corresponds to the streamlines of the electric field. This correspondence is not an accident, but there are several other concepts which must be mastered before the analogy between electric and magnetic fields can be explored more thoroughly.

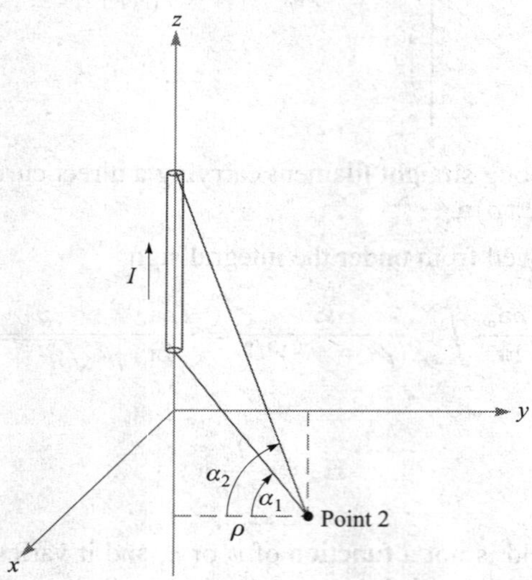

Fig. 8.5 The magnetic field intensity caused by a finite-length current filament on the z axis is $(I/4\pi\rho)(\sin\alpha_2 - \sin\alpha_1)\mathbf{a}_\phi$.

Using the Biot-Savart law to find **H** is in many respects similar to the use of Coulomb's law to find **E**. Each requires the determination of a moderately complicated integrand containing vector quantities, followed by an integration. When we were concerned with Coulomb's law we solved a number of examples, including the fields of the point charge, line charge, and sheet of charge. The law of Biot-Savart can be used to solve analogous problems in magnetic fields, and some of these problems now appear as exercises at the end of the chapter rather than as examples here.

One useful result is the field of the finite-length current element, shown in Figure 8.5. It turns out (see Problem 8.8 at the end of the chapter) that **H** is most easily expressed in terms of the angles α_1 and α_2, as identified in the figure. The result is

$$\mathbf{H} = \frac{I}{4\pi\rho}(\sin\alpha_2 - \sin\alpha_1)\mathbf{a}_\phi \tag{9}$$

If one or both ends are below point 2, then α_1, or both α_1 and α_2, are negative.

[3] If you can't wait, see Section 8.6.

Equation (9) may be used to find the magnetic field intensity caused by current filaments arranged as a sequence of straight-line segments.

EXAMPLE 8.1

As a numerical example illustrating the use of (9), let us determine **H** at $P_2(0.4, 0.3, 0)$ in the field of an 8 A filamentary current directed inward from infinity to the origin on the positive x axis, and then outward to infinity along the y axis. This arrangement is shown in Figure 8.6.

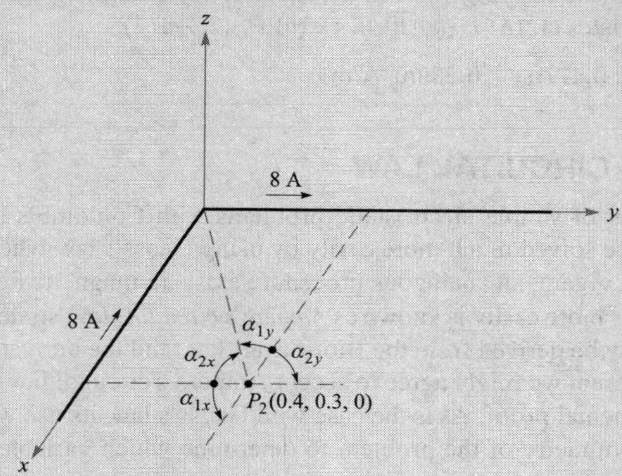

Fig. 8.6 The individual fields of two semi-infinite current segments are found by (9) and added to obtain $\mathbf{H}_2$ at P_2.

Solution. We first consider the semi-infinite current on the x axis, identifying the two angles, $\alpha_{1x} = -90°$ and $\alpha_{2x} = \tan^{-1}(0.4/0.3) = 53.1°$. The radial distance ρ is measured from the x axis, and we have $\rho_x = 0.3$. Thus, this contribution to $\mathbf{H}_2$ is

$$\mathbf{H}_{2(x)} = \frac{8}{4\pi(0.3)}(\sin 53.1° + 1)\mathbf{a}_\phi = \frac{2}{0.3\pi}(1.8)\mathbf{a}_\phi = \frac{12}{\pi}\mathbf{a}_\phi$$

The unit vector $\mathbf{a}_\phi$ must also be referred to the x axis. We see that it becomes $-\mathbf{a}_z$. Therefore,

$$\mathbf{H}_{2(x)} = -\frac{12}{\pi}\mathbf{a}_z \text{ A/m}$$

For the current on the y axis, we have $\alpha_{1y} = -\tan^{-1}(0.3/0.4) = -36.9°$, $\alpha_{2y} = 90°$, and $\rho_y = 0.4$. It follows that

$$\mathbf{H}_{2(y)} = \frac{8}{4\pi(0.4)}(1 + \sin 36.9°)(-\mathbf{a}_z) = -\frac{8}{\pi}\mathbf{a}_z \text{ A/m}$$

Adding these results, we have

$$\mathbf{H}_2 = \mathbf{H}_{2(x)} + \mathbf{H}_{2(y)} = -\frac{20}{\pi}\mathbf{a}_z = -6.37\mathbf{a}_z \text{ A/m}$$

D8.1 Given the following values for P_1, P_2, and $I_1\Delta L_1$, calculate $\Delta\mathbf{H}_2$: (*a*) $P_1(0,0,2)$, $P_2(4,2,0)$, $2\pi\mathbf{a}_z\mu\text{A}\cdot\text{m}$; (*b*) $P_1(0,2,0)$, $P_2(4,2,3)$, $2\pi\mathbf{a}_z\mu\text{A}\cdot\text{m}$; (*c*) $P_1(1,2,3)$, $P_2(-3,-1,2)$, $2\pi(-\mathbf{a}_x+\mathbf{a}_y+2\mathbf{a}_z)\mu\text{A}\cdot\text{m}$.

Ans. $-8.51\mathbf{a}_x + 17.01\mathbf{a}_y$ nA/m; $16\mathbf{a}_y$ nA/m; $18.9\mathbf{a}_x - 33.9\mathbf{a}_y + 26.4\mathbf{a}_z$ nA/m

D8.2 A current filament carrying 15 A in the $\mathbf{a}_z$ direction lies along the entire z axis. Find $\mathbf{H}$ in rectangular coordinates at: (*a*) $P_A(\sqrt{20},0,4)$; (*b*) $P_B(2,-4,4)$.

Ans. $0.534\mathbf{a}_y$ A/m; $0.477\mathbf{a}_x + 0.239\mathbf{a}_y$ A/m

8.2 AMPÈRE'S CIRCUITAL LAW

After solving a number of simple electrostatic problems with Coulomb's law, we found that the same problems could be solved much more easily by using Gauss's law whenever a high degree of symmetry was present. Again, an analogous procedure exists in magnetic fields. Here, the law that helps us solve problems more easily is known as *Ampère's circuital*[4] *law,* sometimes called Ampère's work law. This law may be derived from the Biot-Savart law, and the derivation is accomplished in Section 8.7. For the present we might agree to accept Ampère's circuital law temporarily as another law capable of experimental proof. As is the case with Gauss's law, its use will also require careful consideration of the symmetry of the problem to determine which variables and components are present.

Ampère's circuital law states that the line integral of **H** about any *closed* path is exactly equal to the direct current enclosed by that path,

$$\oint \mathbf{H}\cdot d\mathbf{L} = I \tag{10}$$

We define positive current as flowing in the direction of advance of a right-handed screw turned in the direction in which the closed path is traversed.

Referring to Figure 8.7, which shows a circular wire carrying a direct current *I*, the line integral of **H** about the closed paths lettered *a* and *b* results in an answer of *I*; the integral about the closed path *c* which passes through the conductor gives an answer less than *I* and is exactly that portion of the total current which is enclosed by the path *c*. Although paths *a* and *b* give the same answer, the integrands are, of course, different. The line integral directs us to multiply the component of **H** in the direction of the path by a small increment of path length at one point of the path, move along the path to the next incremental length, and repeat the process, continuing until the path is completely traversed. Since **H** will generally vary from point to point, and since paths *a* and *b* are not alike, the contributions to the integral made by, say, each micrometer of path length are quite different. Only the final answers are the same.

We should also consider exactly what is meant by the expression "current enclosed by the path." Suppose we solder a circuit together after passing the conductor once through a rubber band, which

[4] The preferred pronunciation puts the accent on "circ-."

we shall use to represent the closed path. Some strange and formidable paths can be constructed by twisting and knotting the rubber band, but if neither the rubber band nor the conducting circuit is broken, the current enclosed by the path is that carried by the conductor. Now let us replace the rubber band by a circular ring of spring steel across which is stretched a rubber sheet. The steel loop forms the closed path, and the current-carrying conductor must pierce the rubber sheet if the current is to be enclosed by the path. Again, we may twist the steel loop, and we may also deform the rubber sheet by pushing our fist into it or folding it in any way we wish. A single current-carrying conductor still pierces the sheet once, and this is the true measure of the current enclosed by the path. If we should thread the conductor once through the sheet from front to back and once from back to front, the total current enclosed by the path is the algebraic sum, which is zero.

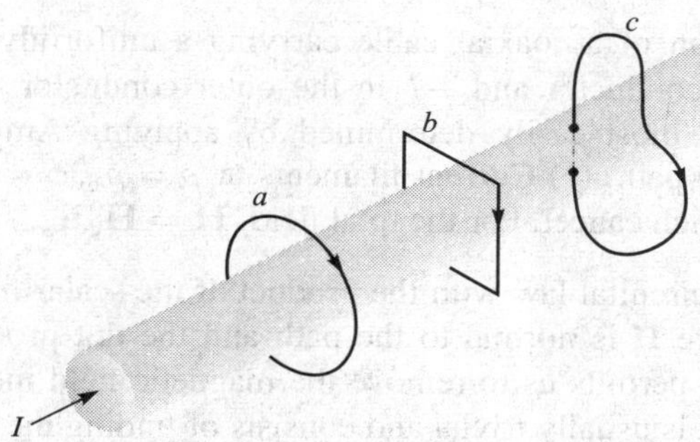

Fig. 8.7 A conductor has a total current I. The line integral of H about the closed paths a and b is equal to I, and the integral around path c is less than I, since the entire current is not enclosed by the path.

In more general language, given a closed path, we recognize this path as the perimeter of an infinite number of surfaces (not closed surfaces). Any current-carrying conductor enclosed by the path must pass through every one of these surfaces once. Certainly some of the surfaces may be chosen in such a way that the conductor pierces them twice in one direction and once in the other direction, but the algebraic total current is still the same.

We shall find that the nature of the closed path is usually extremely simple and can be drawn on a plane. The simplest surface is, then, that portion of the plane enclosed by the path. We need merely find the total current passing through this region of the plane.

The application of Gauss's law involves finding the total charge enclosed by a closed surface; the application of Ampère's circuital law involves finding the total current enclosed by a closed path.

Let us again find the magnetic field intensity produced by an infinitely long filament carrying a current I. The filament lies on the z axis in free space (as in Figure 8.3), and the current flows in the direction given by $\mathbf{a}_z$. Symmetry inspection comes first, showing that there is no variation with z or ϕ. Next we determine which components of $\mathbf{H}$ are present by using the Biot-Savart law. Without specifically using the cross product, we may say that the direction of $d\mathbf{H}$ is perpendicular to the plane conaining $d\mathbf{L}$ and $\mathbf{R}$ and therefore is in the direction of $\mathbf{a}_\phi$. Hence the only component of $\mathbf{H}$ is H_ϕ, and it is a function only of ρ.

We therefore choose a path to any section of which $\mathbf{H}$ is either perpendicular or tangential and along which H is constant. The first requirement (perpendicularity or tangency) allows us to replace

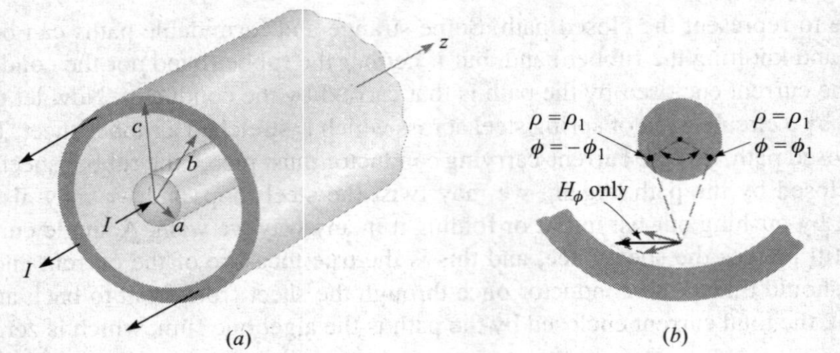

Fig. 8.8 (a) Cross section of a coaxial cable carrying a uniformly distributed current I in the inner conductor and $-I$ in the outer conductor. The magnetic field at any point is most easily determined by applying Ampère's circuital law about a circular path. (b) Current filaments at $\rho = \rho_1, \phi = \pm\phi_1$, produces $\mathbf{H}_\rho$ components which cancel. For the total field, $\mathbf{H} = \mathbf{H}_\phi \mathbf{a}_\phi$.

the dot product of Ampère's circuital law with the product of the scalar magnitudes, except along that portion of the path where **H** is normal to the path and the dot product is zero; the second requirement (constancy) then permits us to remove the magnetic field intensity from the integral sign. The integration required is usually trivial and consists of finding the length of that portion of the path to which **H** is parallel.

In our example the path must be a circle of radius ρ, and Ampère's circuital law becomes

$$\oint \mathbf{H} \cdot d\mathbf{L} = \int_0^{2\pi} H_\phi \rho d\phi = H_\phi \rho \int_0^{2\pi} d\phi = H_\phi 2\pi\rho = I$$

or

$$H_\phi = \frac{I}{2\pi\rho}$$

as before.

As a second example of the application of Ampère's circuital law, consider an infinitely long coaxial transmission line carrying a uniformly distributed total current I in the center conductor and $-I$ in the outer conductor. The line is shown in Figure 8.8a. Symmetry shows that H is not a function of ϕ or z. In order to determine the components present, we may use the results of the previous example by considering the solid conductors as being composed of a large number of filaments. No filament has a z component of **H**. Furthermore, the H_ρ component at $\phi = 0°$, produced by one filament located at $\rho = \rho_1, \phi = \phi_1$, is canceled by the H_ρ component produced by a symmetrically located filament at $\rho = \rho_1, \phi = -\phi_1$. This symmetry is illustrated by Figure 8.8b. Again we find only an H_ϕ component which varies with ρ.

A circular path of radius ρ, where ρ is larger than the radius of the inner conductor but less than the inner radius of the outer conductor, then leads immediately to

$$H_\phi = \frac{I}{2\pi\rho} \qquad (a < \rho < b)$$

Illustrations

If we choose ρ smaller than the radius of the inner conductor, the current enclosed is

$$I_{\text{encl}} = I\frac{\rho^2}{a^2}$$

and

$$2\pi\rho H_\phi = I\frac{\rho^2}{a^2}$$

or

$$H_\phi = \frac{I\rho}{2\pi a^2} \quad (\rho < a)$$

If the radius ρ is larger than the outer radius of the outer conductor, no current is enclosed and

$$H_\phi = 0 \quad (\rho > c)$$

Finally, if the path lies within the outer conductor, we have

$$2\pi\rho H_\phi = I - I\left(\frac{\rho^2 - b^2}{c^2 - b^2}\right)$$

$$H_\phi = \frac{I}{2\pi\rho}\frac{c^2 - \rho^2}{c^2 - b^2} \quad (b < \rho < c)$$

The magnetic-field-strength variation with radius is shown in Figure 8.9 for a coaxial cable in which $b = 3a$, $c = 4a$. It should be noted that the magnetic field intensity $\mathbf{H}$ is continuous at all the conductor boundaries. In other words, a slight increase in the radius of the closed path does not result in the enclosure of a tremendously different current. The value of H_ϕ shows no sudden jumps.

The external field is zero. This, we see, results from equal positive and negative currents enclosed by the path. Each produces an external field of magnitude $I/2\pi\rho$, but complete cancellation occurs. This is another example of "shielding"; such a coaxial cable carrying large currents would not produce any noticeable effect in an adjacent circuit.

As a final example, let us consider a sheet of current flowing in the positive y direction and located in the $z = 0$ plane. We may think of the return current as equally divided between two

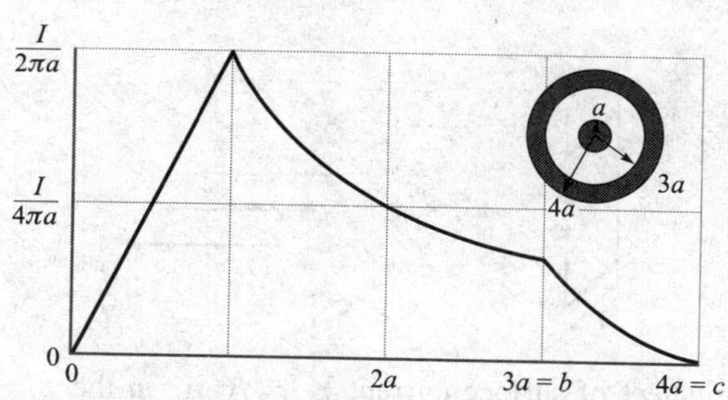

Fig. 8.9 The magnetic field intensity as a function of radius in an infinitely long coaxial transmission line with the dimensions shown.

distant sheets on either side of the sheet we are considering. A sheet of uniform surface current density $\mathbf{K} = K_y\mathbf{a}_y$ is shown in Figure 8.10. **H** cannot vary with x or y. If the sheet is subdivided into a number of filaments, it is evident that no filament can produce an H_y component. Moreover, the Biot-Savart law shows that the contributions to H_z produced by a symmetrically located pair of filaments cancel. Thus, H_z is zero also; only an H_x component is present. We therefore choose the path 1-1′-2′-2-1 composed of straight-line segments which are either parallel or perpendicular to H_x. Ampère's circuital law gives

$$H_{x1}L + H_{x2}(-L) = K_yL$$

or

$$H_{x1} - H_{x2} = K_y$$

If the path 3-3′-2′-2-3 is now chosen, the same current is enclosed, and

$$H_{x3} - H_{x2} = K_y$$

and therefore

$$H_{x3} = H_{x1}$$

It follows that H_x is the same for all positive z. Similarly, H_x is the same for all negative z. Because of the symmetry, then, the magnetic field intensity on one side of the current sheet is the negative of that on the other. Above the sheet,

$$H_x = \tfrac{1}{2}K_y \quad (z > 0)$$

while below it

$$H_x = -\tfrac{1}{2}K_y \quad (z < 0)$$

Letting $\mathbf{a}_N$ be a unit vector normal (outward) to the current sheet, the result may be written in a form correct for all z as

$$\mathbf{H} = \tfrac{1}{2}\mathbf{K} \times \mathbf{a}_N \tag{11}$$

If a second sheet of current flowing in the opposite direction, $\mathbf{K} = -K_y\mathbf{a}_y$, is placed at $z = h$, (11) shows that the field in the region between the current sheets is

$$\mathbf{H} = \mathbf{K} \times \mathbf{a}_N \quad (0 < z < h) \tag{12}$$

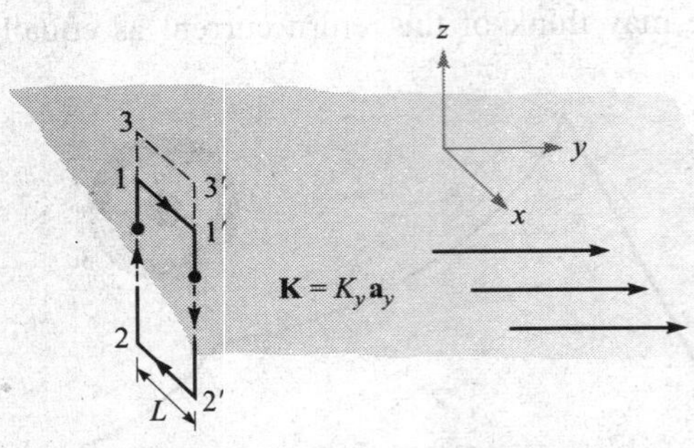

Fig. 8.10 A uniform sheet of surface current $\mathbf{K} = K_y\mathbf{a}_y$ in the $z = 0$ plane. H may be found by applying Ampère's circuital law about the paths 1-1′-2′-2-1 and 3-3′-2′-2-3.

and is zero elsewhere,

$$\mathbf{H} = 0 \quad (z < 0, z > h) \tag{13}$$

The most difficult part of the application of Ampère's circuital law is the determination of the components of the field which are present. The surest method is the logical application of the Biot-Savart law and a knowledge of the magnetic fields of simple form.

Problem 8.13 at the end of this chapter outlines the steps involved in applying Ampère's circuital law to an infinitely long solenoid of radius a and uniform current density $K_a\mathbf{a}_\phi$, as shown in Figure 8.11a. For reference, the result is

$$\mathbf{H} = K_a\mathbf{a}_z \quad (\rho < a) \tag{14a}$$

$$\mathbf{H} = 0 \quad (\rho > a) \tag{14b}$$

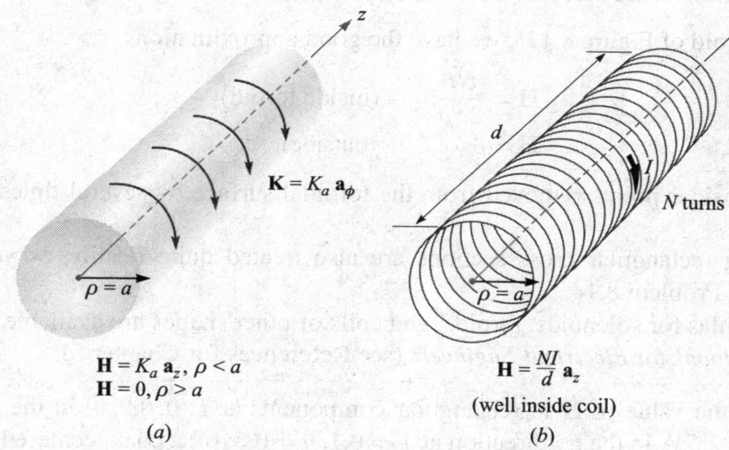

Fig. 8.11 (a) An ideal solenoid of infinite length with a circular current sheet $\mathbf{K} = K_a\mathbf{a}_\phi$. ($b$) An N-turn solenoid of finite length d.

If the solenoid has a finite length d and consists of N closely wound turns of a filament that carries a current I (Figure 8.11b), then the field at points well within the solenoid is given closely by

$$\mathbf{H} = \frac{NI}{d}\mathbf{a}_z \quad \text{(well within the solenoid)} \tag{15}$$

The approximation is useful it if is not applied closer than two radii to the open ends, nor closer to the solenoid surface than twice the separation between turns.

For the toroids shown in Figure 8.12, it can be shown that the magnetic field intensity for the ideal case, Figure 8.12a, is

$$\mathbf{H} = K_a\frac{\rho_0 - a}{\rho}\mathbf{a}_\phi \quad \text{(inside toroid)} \tag{16a}$$

$$\mathbf{H} = 0 \quad \text{(outside)} \tag{16b}$$

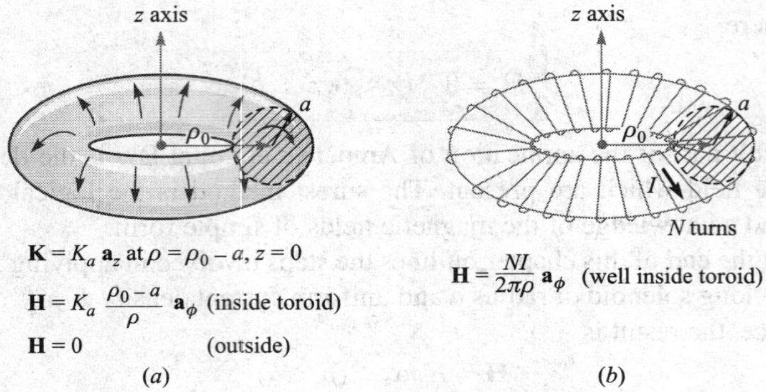

Fig. 8.12 (*a*) An ideal toroid carrying a surface current K in the direction shown. (*b*) An N-turn toroid carrying a filamentary current *I*.

For the N-turn toroid of Figure 8.12b, we have the good approximations,

$$\mathbf{H} = \frac{NI}{2\pi\rho}\mathbf{a}_\phi \quad \text{(inside toroid)} \tag{17a}$$

$$\mathbf{H} = 0 \quad \text{(outside)} \tag{17b}$$

as long as we consider points removed from the toroidal surface by several times the separation between turns.

Toroids having rectangular cross sections are also treated quite readily, as you can see for yourself by trying Problem 8.14.

Accurate formulas for solenoids, toroids, and coils of other shapes are available in Section 2 of the *Standard Handbook for Electrical Engineers* (see References for Chapter 5).

D8.3 Express the value of **H** in rectangular components at $P(0, 0.2, 0)$ in the field of: (*a*) a current filament, 2.5 A in the $\mathbf{a}_z$ direction at $x = 0.1, y = 0.3$; (*b*) a coax, centered on the z axis, with $a = 0.3, b = 0.5, c = 0.6, I = 2.5$ A in the $\mathbf{a}_z$ direction in the center conductor; (*c*) three current sheets, $2.7\mathbf{a}_x$ A/m at $y = 0.1$, $-1.4\mathbf{a}_x$ A/m at $y = 0.15$, and $-1.3\mathbf{a}_x$ A/m at $y = 0.25$.

Ans. $1.989\mathbf{a}_x - 1.989\mathbf{a}_y$ A/m; $-0.884\mathbf{a}_x$ A/m; $1.300\mathbf{a}_z$ A/m

8.3 CURL

We completed our study of Gauss's law by applying it to a differential volume element and were led to the concept of divergence. We now apply Ampère's circuital law to the perimeter of a differential surface element and discuss the third and last of the special derivatives of vector analysis, the curl. Our immediate objective is to obtain the point form of Ampère's circuital law.

Again we shall choose rectangular coordinates, and an incremental closed path of sides Δx and Δy is selected (Figure 8.13). We assume that some current, as yet unspecified, produces a reference value for **H** at the *center* of this small rectangle,

$$\mathbf{H}_0 = H_{x0}\mathbf{a}_x + H_{y0}\mathbf{a}_y + H_{z0}\mathbf{a}_z$$

The closed line integral of **H** about this path is then approximately the sum of the four values of $\mathbf{H} \cdot \Delta\mathbf{L}$ on each side. We choose the direction of traverse as 1-2-3-4-1, which corresponds to a

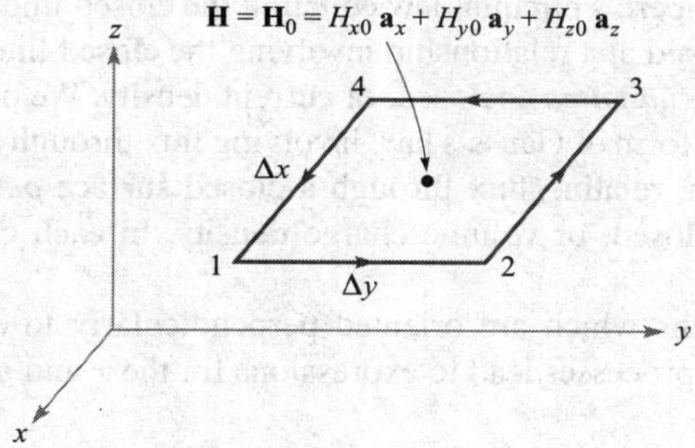

Fig. 8.13 An incremental closed path in rectangular coordinates is selected for the application of Ampère's circuital law to determine the spatial rate of change of H.

current in the $\mathbf{a}_z$ direction, and the first contribution is therefore

$$(\mathbf{H}\cdot\Delta\mathbf{L})_{1-2} = H_{y,1-2}\Delta y$$

The value of H_y *on this section* of the path may be given in terms of the reference value H_{y0} at the center of the rectangle, the rate of change of H_y with x, and the distance $\Delta x/2$ from the center to the midpoint of side 1–2:

$$H_{y,1-2} \doteq H_{y0} + \frac{\partial H_y}{\partial x}\left(\frac{1}{2}\Delta x\right)$$

Thus

$$(\mathbf{H}\cdot\Delta\mathbf{L})_{1-2} \doteq \left(H_{y0} + \frac{1}{2}\frac{\partial H_y}{\partial x}\Delta x\right)\Delta y$$

Along the next section of the path we have

$$(\mathbf{H}\cdot\Delta\mathbf{L})_{2-3} \doteq H_{x,2-3}(-\Delta x) \doteq -\left(H_{x0} + \frac{1}{2}\frac{\partial H_x}{\partial y}\Delta y\right)\Delta x$$

Continuing for the remaining two segments and adding the results,

$$\oint \mathbf{H}\cdot d\mathbf{L} \doteq \left(\frac{\partial H_y}{\partial x} - \frac{\partial H_x}{\partial y}\right)\Delta x\Delta y$$

By Ampère's circuital law, this result must be equal to the current enclosed by the path, or the current crossing any surface bounded by the path. If we assume a general current density **J**, the enclosed current is then $\Delta I \doteq J_z\Delta x\Delta y$, and

$$\oint \mathbf{H}\cdot d\mathbf{L} \doteq \left(\frac{\partial H_y}{\partial x} - \frac{\partial H_x}{\partial y}\right)\Delta x\Delta y \doteq J_z\Delta x\Delta y$$

or

$$\frac{\oint \mathbf{H}\cdot d\mathbf{L}}{\Delta x\Delta y} \doteq \frac{\partial H_y}{\partial x} - \frac{\partial H_x}{\partial y} \doteq J_z$$

As we cause the closed path to shrink, the preceding expression becomes more nearly exact, and in the limit we have the equality

$$\lim_{\Delta x,\Delta y\to 0}\frac{\oint \mathbf{H}\cdot d\mathbf{L}}{\Delta x\Delta y} = \frac{\partial H_y}{\partial x} - \frac{\partial H_x}{\partial y} = J_z \qquad (18)$$

After beginning with Ampère's circuital law equating the closed line integral of **H** to the current enclosed, we have now arrived at a relationship involving the closed line integral of **H** *per unit area* enclosed and the current *per unit area* enclosed, or current density. We performed a similar analysis in passing from the integral form of Gauss's law, involving flux through a closed surface and charge enclosed, to the point form, relating flux through a closed surface *per unit volume* enclosed and charge *per unit volume* enclosed, or volume charge density. In each case a limit is necessary to produce an equality.

If we choose closed paths which are oriented perpendicularly to each of the remaining two coordinate axes, analogous processes lead to expressions for the x and y components of the current density,

$$\lim_{\Delta y,\Delta z\to 0} \frac{\oint \mathbf{H}\cdot d\mathbf{L}}{\Delta y \Delta z} = \frac{\partial H_z}{\partial y} - \frac{\partial H_y}{\partial z} = J_x \tag{19}$$

and

$$\lim_{\Delta z,\Delta x\to 0} \frac{\oint \mathbf{H}\cdot d\mathbf{L}}{\Delta z \Delta x} = \frac{\partial H_x}{\partial z} - \frac{\partial H_z}{\partial x} = J_y \tag{20}$$

Comparing (18), (19), and (20), we see that a component of the current density is given by the limit of the quotient of the closed line integral of **H** about a small path in a plane normal to that component and of the area enclosed as the path shrinks to zero. This limit has its counterpart in other fields of science and long ago received the name of *curl*. The curl of any vector is a vector, and any component of the curl is given by the limit of the quotient of the closed line integral of the vector about a small path in a plane normal to that component desired and the area enclosed, as the path shrinks to zero. It should be noted that this definition of curl does not refer specifically to a particular coordinate system. The mathematical form of the definition is

$$(\text{curl } \mathbf{H})_N = \lim_{\Delta S_N\to 0} \frac{\oint \mathbf{H}\cdot d\mathbf{L}}{\Delta S_N} \tag{21}$$

where ΔS_N is the planar area enclosed by the closed line integral. The N subscript indicates that the component of the curl is that component which is *normal* to the surface enclosed by the closed path. It may represent any component in any coordinate system.

In rectangular coordinates the definition (21) shows that the x, y, and z components of the curl **H** are given by (18), (19), and (20), and therefore

$$\text{curl } \mathbf{H} = \left(\frac{\partial H_z}{\partial y} - \frac{\partial H_y}{\partial z}\right)\mathbf{a}_x + \left(\frac{\partial H_x}{\partial z} - \frac{\partial H_z}{\partial x}\right)\mathbf{a}_y + \left(\frac{\partial H_y}{\partial x} - \frac{\partial H_x}{\partial y}\right)\mathbf{a}_z \tag{22}$$

This result may be written in the form of a determinant,

$$\text{curl } \mathbf{H} = \begin{vmatrix} \mathbf{a}_x & \mathbf{a}_y & \mathbf{a}_z \\ \dfrac{\partial}{\partial x} & \dfrac{\partial}{\partial y} & \dfrac{\partial}{\partial z} \\ H_x & H_y & H_z \end{vmatrix} \tag{23}$$

and may also be written in terms of the vector operator,

$$\text{curl } \mathbf{H} = \nabla \times \mathbf{H} \tag{24}$$

Equation (22) is the result of applying the definition (21) to the rectangular coordinate system. We obtained the z component of this expression by evaluating Ampère's circuital law about an incremental path of sides Δx and Δy, and we could have obtained the other two components just as easily by choosing the appropriate paths. Equation (23) is a neat method of storing the rectangular coordinate expression for curl; the form is symmetrical and easily remembered. Equation (24) is even more concise and leads to (22) upon applying the definitions of the cross product and vector operator.

The expressions for curl **H** in cylindrical and spherical coordinates are derived in Appendix A by applying the definition (21). Although they may be written in determinant form, as explained there, the determinants do not have one row of unit vectors on top and one row of components on the bottom, and they are not easily memorized. For this reason, the curl expansions in cylindrical and spherical coordinates that follow here and appear inside the back cover are usually referred to whenever necessary.

$$\nabla \times \mathbf{H} = \left(\frac{1}{\rho}\frac{\partial H_z}{\partial \phi} - \frac{\partial H_\phi}{\partial z}\right)\mathbf{a}_\rho + \left(\frac{\partial H_\rho}{\partial z} - \frac{\partial H_z}{\partial \rho}\right)\mathbf{a}_\phi + \left(\frac{1}{\rho}\frac{\partial(\rho H_\phi)}{\partial \rho} - \frac{1}{\rho}\frac{\partial H_\rho}{\partial \phi}\right)\mathbf{a}_z \quad \text{(cylindrical)} \tag{25}$$

$$\nabla \times \mathbf{H} = \frac{1}{r\sin\theta}\left(\frac{\partial(H_\phi \sin\theta)}{\partial \theta} - \frac{\partial H_\theta}{\partial \phi}\right)\mathbf{a}_r + \frac{1}{r}\left(\frac{1}{\sin\theta}\frac{\partial H_r}{\partial \phi} - \frac{\partial(rH_\phi)}{\partial r}\right)\mathbf{a}_\theta + \frac{1}{r}\left(\frac{\partial(rH_\theta)}{\partial r} - \frac{\partial H_r}{\partial \theta}\right)\mathbf{a}_\phi \quad \text{(spherical)} \tag{26}$$

Although we have described curl as a line integral per unit area, this does not provide everyone with a satisfactory physical picture of the nature of the curl operation, for the closed line integral itself requires physical interpretation. This integral was first met in the electrostatic field, where we saw that $\oint \mathbf{E} \cdot d\mathbf{L} = 0$. Inasmuch as the integral was zero, we did not belabor the physical picture. More recently we have discussed the closed line integral of **H**, $\oint \mathbf{H} \cdot d\mathbf{L} = I$. Either of these closed line integrals is also known by the name of *circulation,* a term obviously borrowed from the field of fluid dynamics.

The circulation of **H**, or $\oint \mathbf{H} \cdot d\mathbf{L}$, is obtained by multiplying the component of **H** parallel to the specified closed path at each point along it by the differential path length and summing the results as the differential lengths approach zero and as their number becomes infinite. We do not require a vanishingly small path. Ampère's circuital law tells us that if **H** does possess circulation about a given path, then current passes through this path. In electrostatics we see that the circulation of **E** is zero about every path, a direct consequence of the fact that zero work is required to carry a charge around a closed path.

We may now describe curl as *circulation per unit area*. The closed path is vanishingly small, and curl is defined at a point. The curl of **E** must be zero, for the circulation is zero. The curl of **H** is not zero, however; the circulation of **H** per unit area is the current density by Ampère's circuital law [or (18), (19), and (20)].

Skilling[5] suggests the use of a very small paddle wheel as a "curl meter." Our vector quantity, then, must be thought of as capable of applying a force to each blade of the paddle wheel, the force being proportional to the component of the field normal to the surface of that blade. To test a field for curl we dip our paddle wheel into the field, with the axis of the paddle wheel lined up with the direction of the component of curl desired, and note the action of the field on the paddle. No rotation means no curl; larger angular velocities mean greater values of the curl; a reversal in the direction of spin means a reversal in the sign of the curl. To find the direction of the vector curl and not merely to establish the presence of any particular component, we should place our paddle wheel in the field and hunt around for the orientation which produces the greatest torque. The direction of the curl is then along the axis of the paddle wheel, as given by the right-hand rule.

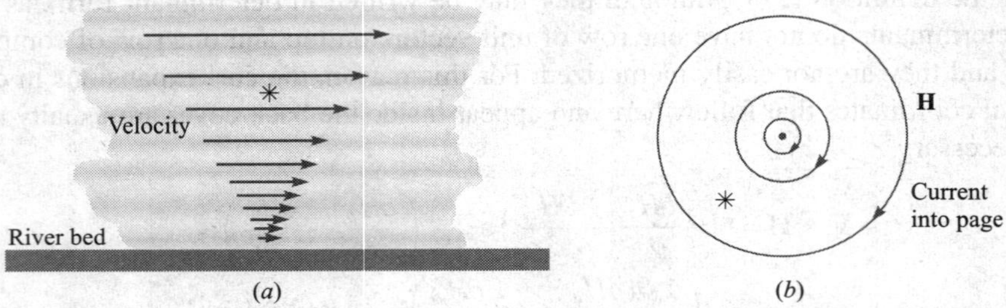

Fig. 8.14 (*a*) The curl meter shows a component of the curl of the water velocity into the page. (*b*) The curl of the magnetic field intensity about an infinitely long filament is shown.

As an example, consider the flow of water in a river. Figure 8.14*a* shows the longitudinal section of a wide river taken at the middle of the river. The water velocity is zero at the bottom and increases linearly as the surface is approached. A paddle wheel placed in the position shown, with its axis perpendicular to the paper, will turn in a clockwise direction, showing the presence of a component of curl in the direction of an inward normal to the surface of the page. If the velocity of water does not change as we go up- or downstream and also shows no variation as we go across the river (or even if it decreases in the same fashion toward either bank), then this component is the only component present at the center of the stream, and the curl of the water velocity has a direction into the page.

In Figure 8.14*b* the streamlines of the magnetic field intensity about an infinitely long filamentary conductor are shown. The curl meter placed in this field of curved lines shows that a larger number of blades have a clockwise force exerted on them but that this force is in general smaller than the counterclockwise force exerted on the smaller number of blades closer to the wire. It seems possible that if the curvature of the streamlines is correct and also if the variation of the field strength is just right, the net torque on the paddle wheel may be zero. Actually, the paddle wheel does not rotate in this case, for since $\mathbf{H} = (I/2\pi\rho)\mathbf{a}_\phi$, we may substitute into (25) obtaining

$$\text{curl } \mathbf{H} = -\frac{\partial H_\phi}{\partial z}\mathbf{a}_\rho + \frac{1}{\rho}\frac{\partial(\rho H_\phi)}{\partial \rho}\mathbf{a}_z = 0$$

[5] See the References at the end of the chapter.

EXAMPLE 8.2

As an example of the evaluation of curl **H** from the definition and of the evaluation of another line integral, let us suppose that $\mathbf{H} = 0.2z^2\mathbf{a}_x$ for $z > 0$, and $\mathbf{H} = 0$ elsewhere, as shown in Figure 8.15. Calculate $\oint \mathbf{H} \cdot d\mathbf{L}$ about a square path with side d, centered at $(0, 0, z_1)$ in the $y = 0$ plane where $z_1 > d/2$.

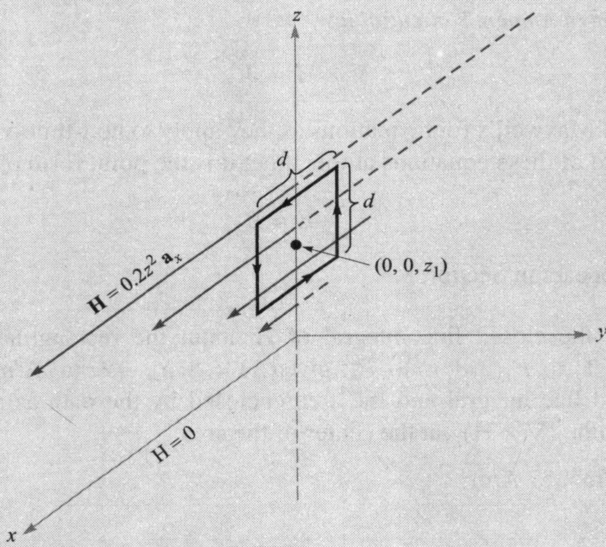

Fig. 8.15 A square path of side d with its center on the z axis at $z = z_1$ is used to evaluate $\oint \mathbf{H} \cdot d\mathbf{L}$ and find curl H.

Solution. We evaluate the line integral of **H** along the four segments, beginning at the top:

$$\oint \mathbf{H} \cdot d\mathbf{L} = 0.2\left(z_1 + \tfrac{1}{2}d\right)^2 d + 0 - 0.2\left(z_1 - \tfrac{1}{2}d\right)^2 d + 0 = 0.4z_1 d^2$$

In the limit as the area approaches zero, we find

$$(\nabla \times \mathbf{H})_y = \lim_{d \to 0} \frac{\oint \mathbf{H} \cdot d\mathbf{L}}{d^2} = \lim_{d \to 0} \frac{0.4z_1 d^2}{d^2} = 0.4z_1$$

The other components are zero, so $\nabla \times \mathbf{H} = 0.4z_1\mathbf{a}_y$.

To evaluate the curl without trying to illustrate the definition or the evaluation of a line integral, we simply take the partial derivative indicated by (23):

$$\nabla \times \mathbf{H} = \begin{vmatrix} \mathbf{a}_x & \mathbf{a}_y & \mathbf{a}_z \\ \dfrac{\partial}{\partial x} & \dfrac{\partial}{\partial y} & \dfrac{\partial}{\partial z} \\ 0.2z^2 & 0 & 0 \end{vmatrix} = \frac{\partial}{\partial z}(0.2z^2)\mathbf{a}_y = 0.4z\mathbf{a}_y$$

which checks with the preceding result when $z = z_1$.

Returning now to complete our original examination of the application of Ampère's circuital law to a differential-sized path, we may combine (18), (19), (20), (22), and (24),

$$\text{curl } \mathbf{H} = \nabla \times \mathbf{H} = \left(\frac{\partial H_z}{\partial y} - \frac{\partial H_y}{\partial z}\right)\mathbf{a}_x + \left(\frac{\partial H_x}{\partial z} - \frac{\partial H_z}{\partial x}\right)\mathbf{a}_y + \left(\frac{\partial H_y}{\partial x} - \frac{\partial H_x}{\partial y}\right)\mathbf{a}_z = \mathbf{J} \qquad (27)$$

and write the *point form of Ampère's circuital law,*

$$\nabla \times \mathbf{H} = \mathbf{J} \qquad (28)$$

This is the second of Maxwell's four equations as they apply to non-time-varying conditions. We may also write the third of these equations at this time; it is the point form of $\oint \mathbf{E} \cdot d\mathbf{L} = 0$, or

$$\nabla \times \mathbf{E} = 0 \qquad (29)$$

The fourth equation appears in Section 8.5.

D8.4 (*a*) Evaluate the closed line integral of **H** about the rectangular path $P_1(2, 3, 4)$ to $P_2(4, 3, 4)$ to $P_3(4, 3, 1)$ to $P_4(2, 3, 1)$ to P_1, given $\mathbf{H} = 3z\mathbf{a}_x - 2x^3\mathbf{a}_z$ A/m. (*b*) Determine the quotient of the closed line integral and the area enclosed by the path as an approximation to $(\nabla \times \mathbf{H})_y$. (*c*) Determine $(\nabla \times \mathbf{H})_y$ at the center of the area.

Ans. 354 A; 59 A/m^2; 57 A/m^2

D8.5 Calculate the value of the vector current density: (*a*) in rectangular coordinates at $P_A(2, 3, 4)$ if $\mathbf{H} = x^2z\mathbf{a}_y - y^2x\mathbf{a}_z$; (*b*) in cylindrical coordinates at $P_B(1.5, 90°, 0.5)$ if $\mathbf{H} = \dfrac{2}{\rho}(\cos 0.2\phi)\mathbf{a}_\rho$; (*c*) in spherical coordinates at $P_C(2, 30°, 20°)$ if $\mathbf{H} = \dfrac{1}{\sin\theta}\mathbf{a}_\theta$.

Ans. $-16\mathbf{a}_x + 9\mathbf{a}_y + 16\mathbf{a}_z$ A/m^2; $0.055\mathbf{a}_z$ A/m^2; $\mathbf{a}_\phi$ A/m^2

8.4 STOKES' THEOREM

Although Section 8.3 was devoted primarily to a discussion of the curl operation, the contribution to the subject of magnetic fields should not be overlooked. From Ampère's circuital law we derived one of Maxwell's equations, $\nabla \times \mathbf{H} = \mathbf{J}$. This latter equation should be considered the point form of Ampère's circuital law and applies on a "per-unit-area" basis. In this section we shall again devote a major share of the material to the mathematical theorem known as Stokes' theorem, but in the process we shall show that we may obtain Ampère's circuital law from $\nabla \times \mathbf{H} = \mathbf{J}$. In other words, we are then prepared to obtain the integral form from the point form or to obtain the point form from the integral form.

Consider the surface S of Figure 8.16, which is broken up into incremental surfaces of area ΔS. If we apply the definition of the curl to one of these incremental surfaces, then

$$\frac{\oint \mathbf{H} \cdot d\mathbf{L}_{\Delta S}}{\Delta S} \doteq (\nabla \times \mathbf{H})_N$$

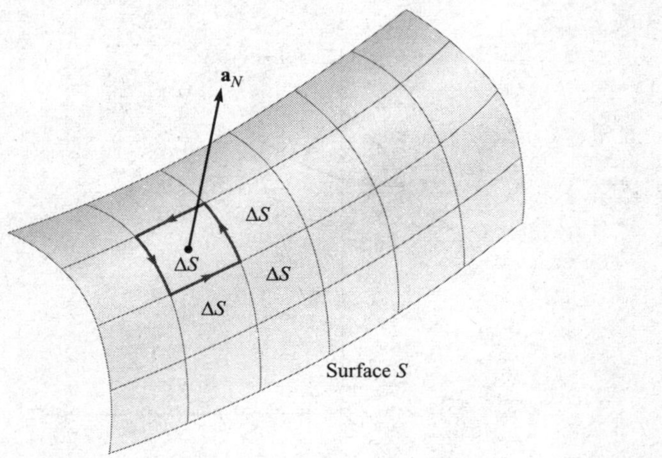

Fig. 8.16 The sum of the closed line integrals about the perimeter of every ΔS is the same as the closed line integral about the perimeter of S because of cancellation on every interior path.

where the N subscript again indicates the right-hand normal to the surface. The subscript on $d\mathbf{L}_{\Delta S}$ indicates that the closed path is the perimeter of an incremental area ΔS. This result may also be written

$$\frac{\oint \mathbf{H} \cdot d\mathbf{L}_{\Delta S}}{\Delta S} \doteq (\nabla \times \mathbf{H}) \cdot \mathbf{a}_N$$

or

$$\oint \mathbf{H} \cdot d\mathbf{L}_{\Delta S} \doteq (\nabla \times \mathbf{H}) \cdot \mathbf{a}_N \Delta S = (\nabla \times \mathbf{H}) \cdot \Delta \mathbf{S}$$

where $\mathbf{a}_N$ is a unit vector in the direction of the right-hand normal to ΔS.

Now let us determine this circulation for every ΔS comprising S and sum the results. As we evaluate the closed line integral for each ΔS, some cancellation will occur because every *interior* wall is covered once in each direction. The only boundaries on which cancellation cannot occur form the outside boundary, the path enclosing S. Therefore we have

$$\oint \mathbf{H} \cdot d\mathbf{L} \equiv \int_S (\nabla \times \mathbf{H}) \cdot d\mathbf{S} \tag{30}$$

where $d\mathbf{L}$ is taken only on the perimeter of S.

Equation (30) is an identity, holding for any vector field, and is known as *Stokes' theorem*.

EXAMPLE 8.3

A numerical example may help to illustrate the geometry involved in Stokes' theorem. Consider the portion of a sphere shown in Figure 8.17. The surface is specified by $r = 4$, $0 \le \theta \le 0.1\pi$, $0 \le \phi \le 0.3\pi$, and the closed path forming its perimeter is composed of three circular arcs. We are given the field $\mathbf{H} = 6r\sin\phi\mathbf{a}_r + 18r\sin\theta\cos\phi\mathbf{a}_\phi$ and are asked to evaluate each side of Stokes' theorem.

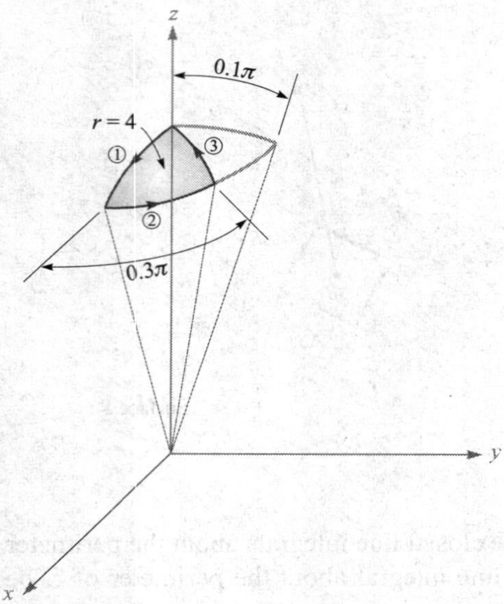

Fig. 8.17 A portion of a spherical cap is used as a surface and a closed path to illustrate Stokes' theorem.

Solution. The first path segment is described in spherical coordinates by $r = 4, 0 \le \theta \le 0.1\pi, \phi = 0$; the second one by $r = 4, \theta = 0.1\pi, 0 \le \phi \le 0.3\pi$; and the third by $r = 4, 0 \le \theta \le 0.1\pi, \phi = 0.3\pi$. The differential path element $d\mathbf{L}$ is the vector sum of the three differential lengths of the spherical coordinate system first discussed in Section 1.9,

$$d\mathbf{L} = dr\,\mathbf{a}_r + r\,d\theta\,\mathbf{a}_\theta + r\sin\theta\,d\phi\,\mathbf{a}_\phi$$

The first term is zero on all three segments of the path since $r = 4$ and $dr = 0$, the second is zero on segment 2 since θ is constant, and the third term is zero on both segments 1 and 3. Thus

$$\oint \mathbf{H}\cdot d\mathbf{L} = \int_1 H_\theta r\,d\theta + \int_2 H_\phi r\sin\theta\,d\phi + \int_3 H_\theta r\,d\theta$$

Since $H_\theta = 0$, we have only the second integral to evaluate,

$$\oint \mathbf{H}\cdot d\mathbf{L} = \int_0^{0.3\pi} [18(4)\sin 0.1\pi\cos\phi]4\sin 0.1\pi d\phi = 288\sin^2 0.1\pi\sin 0.3\pi = 22.2\text{ A}$$

We next attack the surface integral. First, we use (26) to find

$$\nabla\times\mathbf{H} = \frac{1}{r\sin\theta}(36r\sin\theta\cos\theta\cos\phi)\mathbf{a}_r + \frac{1}{r}\left(\frac{1}{\sin\theta}6r\cos\phi - 36r\sin\theta\cos\phi\right)\mathbf{a}_\theta$$

Since $d\mathbf{S} = r^2\sin\theta\,d\theta\,d\phi\,\mathbf{a}_r$, the integral is

$$\int_S (\nabla\times\mathbf{H})\cdot d\mathbf{S} = \int_0^{0.3\pi}\int_0^{0.1\pi}(36\cos\theta\cos\phi)16\sin\theta\,d\theta\,d\phi$$

$$= \int_0^{0.3\pi} 576\left(\tfrac{1}{2}\sin^2\theta\right)\Big|_0^{0.1\pi}\cos\phi\,d\phi$$

$$= 288\sin^2 0.1\pi\sin 0.3\pi = 22.2\text{ A}$$

Thus, the results check Stokes' theorem, and we note in passing that a current of 22.2 A is flowing upward through this section of a spherical cap.

Next, let us see how easy it is to obtain Ampère's circuital law from $\nabla \times \mathbf{H} = \mathbf{J}$. We merely have to dot each side by $d\mathbf{S}$, integrate each side over the same (open) surface S, and apply Stokes' theorem:

$$\int_S (\nabla \times \mathbf{H}) \cdot d\mathbf{S} = \int_S \mathbf{J} \cdot d\mathbf{S} = \oint \mathbf{H} \cdot d\mathbf{L}$$

The integral of the current density over the surface S is the total current I passing through the surface, and therefore

$$\oint \mathbf{H} \cdot d\mathbf{L} = I$$

This short derivation shows clearly that the current I, described as being "enclosed by the closed path," is also the current passing through any of the infinite number of surfaces which have the closed path as a perimeter.

Stokes' theorem relates a surface integral to a closed line integral. It should be recalled that the divergence theorem relates a volume integral to a closed surface integral. Both theorems find their greatest use in general vector proofs. As an example, let us find another expression for $\nabla \cdot \nabla \times \mathbf{A}$, where $\mathbf{A}$ represents any vector field. The result must be a scalar (why?), and we may let this scalar be T, or

$$\nabla \cdot \nabla \times \mathbf{A} = T$$

Multiplying by $d\nu$ and integrating throughout any volume ν,

$$\int_{\text{vol}} (\nabla \cdot \nabla \times \mathbf{A})\, d\nu = \int_{\text{vol}} T\, d\nu$$

we first apply the divergence theorem to the left side, obtaining

$$\oint_S (\nabla \times \mathbf{A}) \cdot d\mathbf{S} = \int_{\text{vol}} T\, d\nu$$

The left side is the surface integral of the curl of $\mathbf{A}$ over the *closed* surface surrounding the volume ν. Stokes' theorem relates the surface integral of the curl of $\mathbf{A}$ over the *open* surface enclosed by a given closed path. If we think of the path as the opening of a laundry bag and the open surface as the surface of the bag itself, we see that as we gradually approach a closed surface by pulling on the drawstrings, the closed path becomes smaller and smaller and finally disappears as the surface becomes closed. Hence the application of Stokes' theorem to a *closed* surface produces a zero result, and we have

$$\int_{\text{vol}} T\, d\nu = 0$$

Since this is true for any volume, it is true for the differential volume $d\nu$,

$$T\, d\nu = 0$$

and therefore

$$T = 0$$

or

$$\nabla \cdot \nabla \times \mathbf{A} \equiv 0 \tag{31}$$

Equation (31) is a useful identity of vector calculus.[6] Of course, it may also be proven easily by direct expansion in rectangular coordinates.

Let us apply the identity to the non-time-varying magnetic field for which

$$\nabla \times \mathbf{H} = \mathbf{J}$$

This shows quickly that

$$\nabla \cdot \mathbf{J} = 0$$

which is the same result we obtained earlier in the chapter by using the continuity equation.

Before introducing several new magnetic field quantities in the following section, we may review our accomplishments at this point. We initially accepted the Biot-Savart law as an experimental result,

$$\mathbf{H} = \oint \frac{I\, d\mathbf{L} \times \mathbf{a}_R}{4\pi R^2}$$

and tentatively accepted Ampère's circuital law, subject to later proof,

$$\oint \mathbf{H} \cdot d\mathbf{L} = I$$

From Ampère's circuital law the definition of curl led to the point form of this same law,

$$\nabla \times \mathbf{H} = \mathbf{J}$$

We now see that Stokes' theorem enables us to obtain the integral form of Ampère's circuital law from the point of form.

D8.6 Evaluate both sides of Stokes' theorem for the field $\mathbf{H} = 6xy\mathbf{a}_x - 3y^2\mathbf{a}_y$ A/m and the rectangular path around the region, $2 \leq x \leq 5, -1 \leq y \leq 1, z = 0$. Let the positive direction of $d\mathbf{S}$ be $\mathbf{a}_z$.

Ans. -126 A; -126 A

8.5 MAGNETIC FLUX AND MAGNETIC FLUX DENSITY

In free space, let us define the *magnetic flux density* **B** as

$$\mathbf{B} = \mu_0 \mathbf{H} \quad \text{(free space only)} \tag{32}$$

where **B** is measured in webers per square meter (Wb/m^2) or in a newer unit adopted in the International System of Units, tesla (T). An older unit that is often used for magnetic flux density is the gauss (G), where 1 T or 1Wb/m^2 is the same as 10 000 G. The constant μ_0 is not dimensionless

[6] This and other vector identities are tabulated in Appendix A.3.

and has the *defined value* for free space, in henrys per meter (H/m), of

$$\mu_0 = 4\pi \times 10^{-7} \text{ H/m} \tag{33}$$

The name given to μ_0 is the *permeability* of free space.

We should note that since **H** is measured in amperes per meter, the weber is dimensionally equal to the product of henrys and amperes. Considering the henry as a new unit, the weber is merely a convenient abbreviation for the product of henrys and amperes. When time-varying fields are introduced, it will be shown that a weber is also equivalent to the product of volts and seconds.

The magnetic-flux-density vector **B**, as the name weber per square meter implies, is a member of the flux-density family of vector fields. One of the possible analogies between electric and magnetic fields[7] compares the laws of Biot-Savart and Coulomb, thus establishing an analogy between **H** and **E**. The relations $\mathbf{B} = \mu_0\mathbf{H}$ and $\mathbf{D} = \epsilon_0\mathbf{E}$ then lead to an analogy between **B** and **D**. If **B** is measured in teslas or webers per square meter, then magnetic flux should be measured in webers. Let us represent magnetic flux by Φ and define Φ as the flux passing through any designated area,

$$\Phi = \int_S \mathbf{B} \cdot d\mathbf{S} \text{ Wb} \tag{34}$$

Our analogy should now remind us of the electric flux Ψ, measured in coulombs, and of Gauss's law, which states that the total flux passing through any closed surface is equal to the charge enclosed,

$$\Psi = \oint_S \mathbf{D} \cdot d\mathbf{S} = Q$$

The charge Q is the source of the lines of electric flux and these lines begin and terminate on positive and negative charge, respectively.

No such source has ever been discovered for the lines of magnetic flux. In the example of the infinitely long straight filament carrying a direct current I, the **H** field formed concentric circles about the filament. Since $\mathbf{B} = \mu_0\mathbf{H}$, the **B** field is of the same form. The magnetic flux lines are closed and do not terminate on a "magnetic charge." For this reason Gauss's law for the magnetic field is

$$\oint_S \mathbf{B} \cdot d\mathbf{S} = 0 \tag{35}$$

and application of the divergence theorem shows us that

$$\nabla \cdot \mathbf{B} = 0 \tag{36}$$

We have not proved (35) or (36) but have only suggested the truth of these statements by considering the single field of the infinite filament. It is possible to show that (35) or (36) follows from the Biot-Savart law and the definition of **B**, $\mathbf{B} = \mu_0\mathbf{H}$, but this is another proof which we shall postpone to Section 8.7.

---

[7] An alternate analogy is presented in Section 10.2.

Equation (36) is the last of Maxwell's four equations as they apply to static electric fields and steady magnetic fields. Collecting these equations, we then have for static electric fields and steady magnetic fields

$$\begin{aligned} \nabla \cdot \mathbf{D} &= \rho_\nu \\ \nabla \times \mathbf{E} &= 0 \\ \nabla \times \mathbf{H} &= \mathbf{J} \\ \nabla \cdot \mathbf{B} &= 0 \end{aligned} \tag{37}$$

To these equations we may add the two expressions relating **D** to **E** and **B** to **H** in free space,

$$\mathbf{D} = \epsilon_0 \mathbf{E} \tag{38}$$

$$\mathbf{B} = \mu_0 \mathbf{H} \tag{39}$$

We have also found it helpful to define an electrostatic potential,

$$\mathbf{E} = -\nabla V \tag{40}$$

and we shall discuss a potential for the steady magnetic field in the following section. In addition, we have extended our coverage of electric fields to include conducting materials and dielectrics, and we have introduced the polarization **P**. A similar treatment will be applied to magnetic fields in the next chapter.

Returning to (37), it may be noted that these four equations specify the divergence and curl of an electric and a magnetic field. The corresponding set of four integral equations that apply to static electric fields and steady magnetic fields is

$$\begin{aligned} &\oint_S \mathbf{D} \cdot d\mathbf{S} = Q = \int_{\text{vol}} \rho_\nu \, d\nu \\ &\oint \mathbf{E} \cdot d\mathbf{L} = 0 \\ &\oint \mathbf{H} \cdot d\mathbf{L} = I = \int_S \mathbf{J} \cdot d\mathbf{S} \\ &\oint_S \mathbf{B} \cdot d\mathbf{S} = 0 \end{aligned} \tag{41}$$

Our study of electric and magnetic fields would have been much simpler if we could have begun with either set of equations, (37) or (41). With a good knowledge of vector analysis, such as we should now have, either set may be readily obtained from the other by applying the divergence theorem or Stokes' theorem. The various experimental laws can be obtained easily from these equations.

As an example of the use of flux and flux density in magnetic fields, let us find the flux between the conductors of the coaxial line of Figure 8.8a. The magnetic field intensity was found to be

$$H_\phi = \frac{I}{2\pi\rho} \quad (a < \rho < b)$$

and therefore

$$\mathbf{B} = \mu_0 \mathbf{H} = \frac{\mu_0 I}{2\pi\rho} \mathbf{a}_\phi$$

The magnetic flux contained between the conductors in a length d is the flux crossing any radial plane extending from $\rho = a$ to $\rho = b$ and from, say, $z = 0$ to $z = d$

$$\Phi = \int_S \mathbf{B} \cdot d\mathbf{S} = \int_0^d \int_a^b \frac{\mu_0 I}{2\pi\rho} \mathbf{a}_\phi \cdot d\rho\, dz\, \mathbf{a}_\phi$$

or

$$\Phi = \frac{\mu_0 I d}{2\pi} \ln \frac{b}{a} \tag{42}$$

This expression will be used later to obtain the inductance of the coaxial transmission line.

D8.7 A solid conductor of circular cross section is made of a homogeneous nonmagnetic material. If the radius $a = 1$ mm, the conductor axis lies on the z axis, and the total current in the $\mathbf{a}_z$ direction is 20 A, find: (*a*) H_ϕ at $\rho = 0.5$ mm; (*b*) B_ϕ at $\rho = 0.8$ mm; (*c*) the total magnetic flux per unit length inside the conductor; (*d*) the total flux for $\rho < 0.5$ mm; (*e*) the total magnetic flux outside the conductor.

Ans. 1592 A/m; 3.2 mT; 2 μWb/m; 0.5 μWb; ∞

8.6 THE SCALAR AND VECTOR MAGNETIC POTENTIALS

The solution of electrostatic field problems is greatly simplified by the use of the scalar electrostatic potential V. Although this potential possesses a very real physical significance for us, it is mathematically no more than a stepping-stone which allows us to solve a problem by several smaller steps. Given a charge configuration, we may first find the potential and then from it the electric field intensity.

We should question whether or not such assistance is available in magnetic fields. Can we define a potential function which may be found from the current distribution and from which the magnetic fields may be easily determined? Can a scalar magnetic potential be defined, similar to the scalar electrostatic potential? We shall show in the next few pages that the answer to the first question is yes, but the second must be answered "sometimes." Let us attack the second question first by assuming the existence of a scalar magnetic potential, which we designate V_m, whose negative gradient gives the magnetic field intensity

$$\mathbf{H} = -\nabla V_m$$

The selection of the negative gradient will provide us with a closer analogy to the electric potential and to problems which we have already solved.

This definition must not conflict with our previous results for the magnetic field, and therefore

$$\nabla \times \mathbf{H} = \mathbf{J} = \nabla \times (-\nabla V_m)$$

However, the curl of the gradient of any scalar is identically zero, a vector identity the proof of which is left for a leisure moment. Therefore we see that if $\mathbf{H}$ is to be defined as the gradient of a scalar magnetic potential, then current density must be zero throughout the region in which the scalar magnetic potential is so defined. We then have

$$\mathbf{H} = -\nabla V_m \quad (\mathbf{J} = 0) \tag{43}$$

Since many magnetic problems involve geometries in which the current-carrying conductors occupy a relatively small fraction of the total region of interest, it is evident that a scalar magnetic potential can be useful. The scalar magnetic potential is also applicable in the case of permanent magnets. The dimensions of V_m are obviously amperes.

This scalar potential also satisfies Laplace's equation. In free space,

$$\nabla \cdot \mathbf{B} = \mu_0 \nabla \cdot \mathbf{H} = 0$$

and hence

$$\mu_0 \nabla \cdot (-\nabla V_m) = 0$$

or

$$\nabla^2 V_m = 0 \quad (\mathbf{J} = 0) \tag{44}$$

We shall see later that V_m continues to satisfy Laplace's equation in homogeneous magnetic materials; it is not defined in any region in which current density is present.

Although we shall consider the scalar magnetic potential to a much greater extent in Chapter 9, when we introduce magnetic materials and discuss the magnetic circuit, one difference between V and V_m should be pointed out now: V_m is not a single-valued function of position. The electric potential V is single-valued; once a zero reference is assigned, there is only one value of V associated with each point in space. Such is not the case with V_m. Consider the cross section of the coaxial line shown in Figure 8.18. In the region $a < \rho < b$, $\mathbf{J} = 0$, and we may establish a scalar magnetic potential. The value of $\mathbf{H}$ is

$$\mathbf{H} = \frac{I}{2\pi\rho}\mathbf{a}_\phi$$

where I is the total current flowing in the $\mathbf{a}_z$ direction in the inner conductor. Let us find V_m by integrating the appropriate component of the gradient. Applying (43),

$$\frac{I}{2\pi\rho} = -\nabla V_m \Big|_\phi = -\frac{1}{\rho}\frac{\partial V_m}{\partial \phi}$$

or

$$\frac{\partial V_m}{\partial \phi} = -\frac{I}{2\pi}$$

Thus

$$V_m = -\frac{I}{2\pi}\phi$$

where the constant of integration has been set equal to zero. What value of potential do we associate with point P, where $\phi = \pi/4$? If we let V_m be zero at $\phi = 0$ and proceed counterclockwise around the circle, the magnetic potential goes negative linearly. When we have made one circuit, the potential is $-I$, but that was the point at which we said the potential was zero a moment ago. At P, then, $\phi = \pi/4, 9\pi/4, 17\pi/4, \ldots$, or $-7\pi/4, -15\pi/4, -23\pi/4, \ldots$, or

$$V_{mP} = \frac{I}{2\pi}\left(2n - \tfrac{1}{4}\right)\pi \quad (n = 0, \pm1, \pm2, \ldots)$$

or

$$V_{mP} = I\left(n - \tfrac{1}{8}\right) \quad (n = 0, \pm1, \pm2, \ldots)$$

The reason for this multivaluedness may be shown by a comparison with the electrostatic case. There, we know that

$$\nabla \times \mathbf{E} = 0$$

$$\oint \mathbf{E} \cdot d\mathbf{L} = 0$$

and therefore the line integral

$$V_{ab} = -\int_b^a \mathbf{E} \cdot d\mathbf{L}$$

is independent of the path. In the magnetostatic case, however,

$$\nabla \times \mathbf{H} = 0 \quad (\text{wherever } \mathbf{J} = 0)$$

but

$$\oint \mathbf{H} \cdot d\mathbf{L} = I$$

even if **J** is zero along the path of integration. Every time we make another complete lap around the current, the result of the integration increases by I. If no current I is enclosed by the path, then a single-valued potential function may be defined. In general, however,

$$V_{m,ab} = -\int_b^a \mathbf{H} \cdot d\mathbf{L} \quad (\text{specified path}) \tag{45}$$

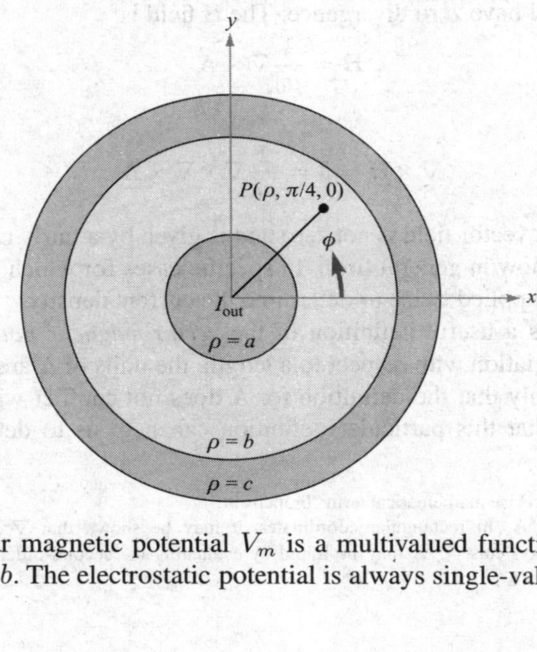

Fig. 8.18 The scalar magnetic potential V_m is a multivalued function of ϕ in the region $a < \rho < b$. The electrostatic potential is always single-valued.

where a specific path or type of path must be selected. We should remember that the electrostatic potential V is a conservative field; the magnetic scalar potential V_m is not a conservative field. In our coaxial problem let us erect a barrier[8] at $\phi = \pi$; we agree not to select a path which crosses this plane. Therefore we cannot encircle I, and a single-valued potential is possible. The result is seen to be

$$V_m = -\frac{I}{2\pi}\phi \quad (-\pi < \phi < \pi)$$

and

$$V_{mP} = -\frac{I}{8} \quad \left(\phi = \frac{\pi}{4}\right)$$

The scalar magnetic potential is evidently the quantity whose equipotential surfaces will form curvilinear squares with the streamlines of **H** in Figure 8.4. This is one more facet of the analogy between electric and magnetic fields about which we will have more to say in the next chapter.

Let us temporarily leave the scalar magnetic potential now and investigate a vector magnetic potential. This vector field is one which is extremely useful in studying radiation from antennas and from apertures, and radiation leakage from transmission lines, waveguides, and microwave ovens. The vector magnetic potential may be used in regions where the current density is zero or nonzero, and we shall also be able to extend it to the time-varying case later.

Our choice of a vector magnetic potential is indicated by noting that

$$\nabla \cdot \mathbf{B} = 0$$

Next, a vector identity which we proved in Section 8.4 shows that the divergence of the curl of any vector field is zero. Therefore we select

$$\mathbf{B} = \nabla \times \mathbf{A} \tag{46}$$

where A signifies a *vector magnetic potential*, and we automatically satisfy the condition that the magnetic flux density shall have zero divergence. The **H** field is

$$\mathbf{H} = \frac{1}{\mu_0}\nabla \times \mathbf{A}$$

and

$$\nabla \times \mathbf{H} = \mathbf{J} = \frac{1}{\mu_0}\nabla \times \nabla \times \mathbf{A}$$

The curl of the curl of a vector field is not zero and is given by a fairly complicated expression,[9] which we need not know now in general form. In specific cases for which the form of **A** is known, the curl operation may be applied twice to determine the current density.

Equation (46) serves as a useful definition of the *vector magnetic potential* **A**. Since the curl operation implies differentiation with respect to a length, the units of **A** are webers per meter.

As yet we have seen only that the definition for **A** does not conflict with any previous results. It still remains to show that this particular definition can help us to determine magnetic fields

[8] This corresponds to the more precise mathematical term "branch cut."

[9] $\nabla \times \nabla \times \mathrm{A} \equiv \nabla(\nabla \cdot \mathbf{A}) - \nabla^2\mathrm{A}$. In rectangular coordinates, it may be shown that $\nabla^2\mathrm{A} \equiv \nabla^2\mathrm{A}_x\mathbf{a}_x + \nabla^2\mathrm{A}_y\mathbf{a}_y + \nabla^2\mathrm{A}_z\mathbf{a}_z$. In other coordinate systems, $\nabla^2\mathrm{A}$ may be found by evaluating the second-order partial derivatives in $\nabla^2\mathrm{A} = \nabla(\nabla \cdot \mathbf{A}) - \nabla \times \nabla \times \mathrm{A}$.

more easily. We certainly cannot identify **A** with any easily measured quantity or history-making experiment.

We shall show in Section 8.7 that, given the Biot-Savart law, the definition of **B**, and the definition of **A**, **A** may be determined from the differential current elements by

$$\mathbf{A} = \oint \frac{\mu_0 I \, d\mathbf{L}}{4\pi R} \tag{47}$$

The significance of the terms in (47) is the same as in the Biot-Savart law; a direct current I flows along a filamentary conductor of which any differential length $d\mathbf{L}$ is distant R from the point at which **A** is to be found. Since we have defined **A** only through specification of its curl, it is possible to add the gradient of any scalar field to (47) without changing **B** or **H**, for the curl of the gradient is identically zero. In steady magnetic fields, it is customary to set this possible added term equal to zero.

The fact that **A** is a vector magnetic *potential* is more apparent when (47) is compared with the similar expression for the electrostatic potential,

$$V = \int \frac{\rho_L dL}{4\pi\epsilon_0 R}$$

Each expression is the integral along a line source, in one case line charge and in the other case line current; each integrand is inversely proportional to the distance from the source to the point of interest; and each involves a characteristic of the medium (here free space), the permeability or the permittivity.

Equation (47) may be written in differential form,

$$d\mathbf{A} = \frac{\mu_0 I \, d\mathbf{L}}{4\pi R} \tag{48}$$

if we again agree not to attribute any physical significance to any magnetic fields we obtain from (48) until the *entire closed path in which the current flows is considered.*

With this reservation, let us go right ahead and consider the vector magnetic potential field about a differential filament. We locate the filament at the origin in free space, as shown in Figure 8.19, and allow it to extend in the positive z direction so that $d\mathbf{L} = dz\,\mathbf{a}_z$. We use cylindrical coordinates to find $d\mathbf{A}$ at the point (ρ, ϕ, z):

$$d\mathbf{A} = \frac{\mu_0 I \, dz\,\mathbf{a}_z}{4\pi\sqrt{\rho^2 + z^2}}$$

or

$$d\mathbf{A}_z = \frac{\mu_0 I \, dz}{4\pi\sqrt{\rho^2 + z^2}} \qquad dA_\phi = 0 \quad dA_\rho = 0 \tag{49}$$

Illustrations

We note that the direction of $d\mathbf{A}$ is the same as that of $I\,d\mathbf{L}$. Each small section of a current-carrying conductor produces a contribution to the total vector magnetic potential which is in the same direction as the current flow in the conductor. The magnitude of the vector magnetic potential varies inversely with the distance to the current element, being strongest in the neighborhood of the current and gradually

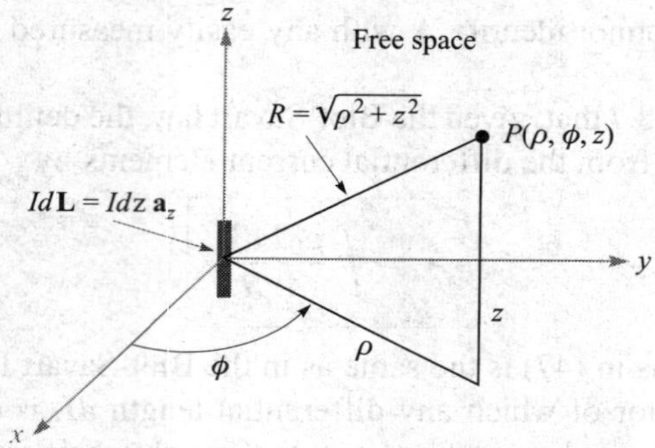

Fig. 8.19 The differential current element $I\,dz\mathbf{a}_z$ at the origin establishes the differential vector magnetic potential field, $d\mathbf{A} = \dfrac{\mu_0 I\,dz\mathbf{a}_z}{4\pi\sqrt{\rho^2+z^2}}$ at $P(\rho, \phi, z)$.

falling off to zero at distant points. Skilling[10] describes the vector magnetic potential field as "like the current distribution but fuzzy around the edges, or like a picture of the current out of focus."

In order to find the magnetic field intensity, we must take the curl of (49) in cylindrical coordinates, leading to

$$d\mathbf{H} = \frac{1}{\mu_0}\nabla \times d\mathbf{A} = \frac{1}{\mu_0}\left(-\frac{\partial dA_z}{\partial \rho}\right)\mathbf{a}_\phi$$

or

$$d\mathbf{H} = \frac{I\,dz}{4\pi}\frac{\rho}{(\rho^2+z^2)^{3/2}}\mathbf{a}_\phi$$

which is easily shown to be the same as the value given by the Biot-Savart law.

Expressions for the vector magnetic potential **A** can also be obtained for a current source which is distributed. For a current sheet **K**, the differential current element becomes

$$I\,d\mathbf{L} = \mathbf{K}\,dS$$

In the case of current flow throughout a volume with a density **J**, we have

$$I\,d\mathbf{L} = \mathbf{J}\,d\nu$$

In each of these two expressions the vector character is given to the current. For the filamentary element it is customary, although not necessary, to use $I\,d\mathbf{L}$ instead of $\mathbf{I}\,dL$. Since the magnitude of the filamentary element is constant, we have chosen the form which allows us to remove one quantity from the integral. The alternative expressions for **A** are then

$$\mathbf{A} = \int_S \frac{\mu_0 \mathbf{K}\,dS}{4\pi R} \tag{50}$$

and

$$\mathbf{A} = \int_{\text{vol}} \frac{\mu_0 \mathbf{J}\,d\nu}{4\pi R} \tag{51}$$

[10] See the References at the end of the chapter.

Equations (47), (50), and (51) express the vector magnetic potential as an integration over all of its sources. From a comparison of the form of these integrals with those which yield the electrostatic potential, it is evident that once again the zero reference for **A** is at infinity, for no finite current element can produce any contribution as $R \to \infty$. We should remember that we very seldom used the similar expressions for V; too often our theoretical problems included charge distributions which extended to infinity, and the result would be an infinite potential everywhere. Actually, we calculated very few potential fields until the differential form of the potential equation was obtained, $\nabla^2 V = -\rho_\nu/\epsilon$, or better yet, $\nabla^2 V = 0$. We were then at liberty to select our own zero reference.

The analogous expressions for **A** will be derived in the next section, and an example of the calculation of a vector magnetic potential field will be completed.

D8.8 A current sheet, $\mathbf{K} = 2.4\mathbf{a}_z$ A/m, is present at the surface $\rho = 1.2$ in free space. (*a*) Find **H** for $\rho > 1.2$. Find V_m at $P(\rho = 1.5, \phi = 0.6\pi, z = 1)$ if: (*b*) $V_m = 0$ at $\phi = 0$ and there is a barrier at $\phi = \pi$; (*c*) $V_m = 0$ at $\phi = 0$ and there is a barrier at $\phi = \pi/2$; (*d*) $V_m = 0$ at $\phi = \pi$ and there is a barrier at $\phi = 0$; (*e*) $V_m = 5$ V at $\phi = \pi$ and there is a barrier at $\phi = 0.8\pi$.

Ans. $\frac{2.88}{\rho}\mathbf{a}_\phi$; -5.43 V; 12.7 V; 3.62 V; -9.48 V

D8.9 The value of **A** within a solid nonmagnetic conductor of radius a carrying a total current I in the $\mathbf{a}_z$ direction may be found easily. Using the known value of **H** or **B** for $\rho < a$, then (46) may be solved for **A**. Select $A = (\mu_0 I \ln 5)/2\pi$ at $\rho = a$ (to correspond with an example in the next section) and find **A** at $\rho =$: (*a*) 0; (*b*) $0.25a$; (*c*) $0.75a$; (*d*) a.

Ans. $0.422I\mathbf{a}_z$ μWb/m; $0.416I\mathbf{a}_z$ μWb/m; $0.366I\mathbf{a}_z$ μWb/m; $0.322I\mathbf{a}_z$ μWb/m

8.7 DERIVATION OF THE STEADY-MAGNETIC-FIELD LAWS

We shall now carry out our threat to supply the promised proofs of the several relationships between the magnetic field quantities. All these relationships may be obtained from the definitions of **H**,

$$\mathbf{H} = \oint \frac{I\,d\mathbf{L} \times \mathbf{a}_R}{4\pi R^2} \tag{3}$$

of **B** (in free space),

$$\mathbf{B} = \mu_0 \mathbf{H} \tag{32}$$

and of **A**,

$$\mathbf{B} = \nabla \times \mathbf{A} \tag{46}$$

Let us first assume that we may express **A** by the last equation of Section 8.6,

$$\mathbf{A} = \int_{\text{vol}} \frac{\mu_0 \mathbf{J}\,d\nu}{4\pi R} \tag{51}$$

and then demonstrate the correctness of (51) by showing that (3) follows. First we should add subscripts to indicate the point at which the current element is located (x_1, y_1, z_1) and the point at which **A** is given (x_2, y_2, z_2). The differential volume element $d\nu$ is then written $d\nu_1$ and in rectangular coordinates would be $dx_1\,dy_1\,dz_1$. The variables of integration are x_1, y_1, and z_1.

Using these subscripts, then,

$$\mathbf{A}_2 = \int_{\text{vol}} \frac{\mu_0 \mathbf{J}_1 dv_1}{4\pi R_{12}} \tag{52}$$

From (32) and (46), we have

$$\mathbf{H} = \frac{\mathbf{B}}{\mu_0} = \frac{\nabla \times \mathbf{A}}{\mu_0} \tag{53}$$

To show that (3) follows from (52), it is necessary to substitute (52) into (53). This step involves taking the curl of $\mathbf{A}_2$, a quantity expressed in terms of the variables x_2, y_2, and z_2, and the curl therefore involves partial derivatives with respect to x_2, y_2, and z_2. We do this, placing a subscript on the del operator to remind us of the variables involved in the partial differentiation process,

$$\mathbf{H}_2 = \frac{\nabla_2 \times \mathbf{A}_2}{\mu_0} = \frac{1}{\mu_0} \nabla_2 \times \int_{\text{vol}} \frac{\mu_0 \mathbf{J}_1 dv_1}{4\pi R_{12}}$$

The order of partial differentiation and integration is immaterial, and $\mu_0/4\pi$ is constant, allowing us to write

$$\mathbf{H}_2 = \frac{1}{4\pi} \int_{\text{vol}} \nabla_2 \times \frac{\mathbf{J}_1 dv_1}{R_{12}}$$

The curl operation within the integrand represents partial differentiation with respect to x_2, y_2, and z_2. The differential volume element dv_1 is a scalar and a function only of x_1, y_1, and z_1. Consequently, it may be factored out of the curl operation as any other constant, leaving

$$\mathbf{H}_2 = \frac{1}{4\pi} \int_{\text{vol}} \left(\nabla_2 \times \frac{\mathbf{J}_1}{R_{12}} \right) dv_1 \tag{54}$$

The curl of the product of a scalar and a vector is given by an identity which may be checked by expansion in rectangular coordinates or gratefully accepted from Appendix A.3,

$$\nabla \times (S\mathbf{V}) \equiv (\nabla S) \times \mathbf{V} + S(\nabla \times \mathbf{V}) \tag{55}$$

This identity is used to expand the integrand of (54),

$$\mathbf{H}_2 = \frac{1}{4\pi} \int_{\text{vol}} \left[\left(\nabla_2 \frac{1}{R_{12}} \right) \times \mathbf{J}_1 + \frac{1}{R_{12}} (\nabla_2 \times \mathbf{J}_1) \right] dv_1 \tag{56}$$

The second term of this integrand is zero because $\nabla_2 \times \mathbf{J}_1$ indicates partial derivatives of a function of x_1, y_1, and z_1, taken with respect to the variables x_2, y_2, and z_2; the first set of variables is not a function of the second set, and all partial derivatives are zero.

The first term of the integrand may be determined by expressing R_{12} in terms of the coordinate values,

$$R_{12} = \sqrt{(x_2 - x_1)^2 + (y_2 - y_1)^2 + (z_2 - z_1)^2}$$

and taking the gradient of its reciprocal. Problem 8.42 shows that the result is

$$\nabla_2 \frac{1}{R_{12}} = -\frac{\mathbf{R}_{12}}{R_{12}^3} = -\frac{\mathbf{a}_{R12}}{R_{12}^2}$$

Substituting this result into (56), we have

$$\mathbf{H}_2 = -\frac{1}{4\pi} \int_{\text{vol}} \frac{\mathbf{a}_{R12} \times \mathbf{J}_1}{R_{12}^2} dv_1$$

or

$$\mathbf{H}_2 = \int_{\text{vol}} \frac{\mathbf{J}_1 \times \mathbf{a}_{R12}}{4\pi R_{12}^2} d\nu_1$$

which is the equivalent of (3) in terms of current density. Replacing $\mathbf{J}_1\, d\nu_1$ by $I_1\, d\mathbf{L}_1$, we may rewrite the volume integral as a closed line integral,

$$\mathbf{H}_2 = \oint \frac{I_1 d\mathbf{L}_1 \times \mathbf{a}_{R12}}{4\pi R_{12}^2}$$

Equation (51) is therefore correct and agrees with the three definitions (3), (32), and (46).

Next we shall continue with our mathematical orgy and prove Ampère's circuital law in point form,

$$\nabla \times \mathbf{H} = \mathbf{J} \tag{28}$$

Combining (28), (32), and (46), we obtain

$$\nabla \times \mathbf{H} = \nabla \times \frac{\mathbf{B}}{\mu_0} = \frac{1}{\mu_0} \nabla \times \nabla \times \mathbf{A} \tag{57}$$

We now need the expansion in rectangular coordinates for $\nabla \times \nabla \times \mathbf{A}$. Performing the indicated partial differentiations and collecting the resulting terms, we may write the result as

$$\nabla \times \nabla \times \mathbf{A} \equiv \nabla(\nabla \cdot \mathbf{A}) - \nabla^2 \mathbf{A} \tag{58}$$

where

$$\nabla^2 \mathbf{A} \equiv \nabla^2 A_x \mathbf{a}_x + \nabla^2 A_y \mathbf{a}_y + \nabla^2 A_z \mathbf{a}_z \tag{59}$$

Equation (59) is the definition (in rectangular coordinates) of the *Laplacian of a vector*.

Substituting (58) into (57), we have

$$\nabla \times \mathbf{H} = \frac{1}{\mu_0}[\nabla(\nabla \cdot \mathbf{A}) - \nabla^2 \mathbf{A}] \tag{60}$$

and now require expressions for the divergence and the Laplacian of **A**.

We may find the divergence of **A** by applying the divergence operation to (52),

$$\nabla_2 \cdot \mathbf{A}_2 = \frac{\mu_0}{4\pi} \int_{\text{vol}} \nabla_2 \cdot \frac{\mathbf{J}_1}{R_{12}} d\nu_1 \tag{61}$$

and using the vector identity (44) of Section 4.8,

$$\nabla \cdot (S\mathbf{V}) \equiv \mathbf{V} \cdot (\nabla S) + S(\nabla \cdot \mathbf{V})$$

Thus,

$$\nabla_2 \cdot \mathbf{A}_2 = \frac{\mu_0}{4\pi} \int_{\text{vol}} \left[\mathbf{J}_1 \cdot \left(\nabla_2 \frac{1}{R_{12}} \right) + \frac{1}{R_{12}} (\nabla_2 \cdot \mathbf{J}_1) \right] d\nu_1 \tag{62}$$

The second part of the integrand is zero because $\mathbf{J}_1$ is not a function of x_2, y_2, and z_2.

We have already used the result that $\nabla_2(1/R_{12}) = -\mathbf{R}_{12}/R_{12}^3$, and it is just as easily shown that

$$\nabla_1 \frac{1}{R_{12}} = \frac{\mathbf{R}_{12}}{R_{12}^3}$$

or that

$$\nabla_1 \frac{1}{R_{12}} = -\nabla_2 \frac{1}{R_{12}}$$

Equation (62) can therefore be written as

$$\nabla_2 \cdot \mathbf{A}_2 = \frac{\mu_0}{4\pi} \int_{\text{vol}} \left[-\mathbf{J}_1 \cdot \left(\nabla_1 \frac{1}{R_{12}} \right) \right] d\nu_1$$

and the vector identity applied again,

$$\nabla_2 \cdot \mathbf{A}_2 = \frac{\mu_0}{4\pi} \int_{\text{vol}} \left[\frac{1}{R_{12}} (\nabla_1 \cdot \mathbf{J}_1) - \nabla_1 \cdot \left(\frac{\mathbf{J}_1}{R_{12}} \right) \right] d\nu_1 \tag{63}$$

Since we are concerned only with steady magnetic fields, the continuity equation shows that the first term of (63) is zero. Application of the divergence theorem to the second term gives

$$\nabla_2 \cdot \mathbf{A}_2 = -\frac{\mu_0}{4\pi} \oint_{S_1} \frac{\mathbf{J}_1}{R_{12}} \cdot d\mathbf{S}_1$$

where the surface S_1 encloses the volume throughout which we are integrating. This volume must include all the current, for the original integral expression for **A** was an integration such as to include the effect of all the current. Since there is no current outside this volume (otherwise we should have had to increase the volume to include it), we may integrate over a slightly larger volume or a slightly larger enclosing surface without changing **A.** On this larger surface the current density $\mathbf{J}_1$ must be zero, and therefore the closed surface integral is zero, since the integrand is zero. Hence the divergence of **A** is zero.

In order to find the Laplacian of the vector **A**, let us compare the x component of (51) with the similar expression for electrostatic potential,

$$A_x = \int_{\text{vol}} \frac{\mu_0 J_x d\nu}{4\pi R} \qquad V = \int_{\text{vol}} \frac{\rho_\nu \, d\nu}{4\pi\epsilon_0 R}$$

We note that one expression can be obtained from the other by a straightforward change of variable, J_x for ρ_ν, μ_0 for $1/\epsilon_0$, and A_x for V. However, we have derived some additional information about the electrostatic potential which we shall not have to repeat now for the x component of the vector magnetic potential. This takes the form of Poisson's equation,

$$\nabla^2 V = -\frac{\rho_\nu}{\epsilon_0}$$

which becomes, after the change of variables,

$$\nabla^2 A_x = -\mu_0 J_x$$

Similarly, we have

$$\nabla^2 A_y = -\mu_0 J_y$$

and

$$\nabla^2 A_z = -\mu_0 J_z$$

or

$$\boxed{\nabla^2 \mathbf{A} = -\mu_0 \mathbf{J}} \tag{64}$$

Returning to (60), we can now substitute for the divergence and Laplacian of **A** and obtain the desired answer,

$$\nabla \times \mathbf{H} = \mathbf{J} \tag{28}$$

We have already shown the use of Stokes' theorem in obtaining the integral form of Ampère's circuital law from (28) and need not repeat that labor here.

We thus have succeeded in showing that every result we have essentially pulled from thin air[11] for magnetic fields follows from the basic definitions of **H**, **B**, and **A**. The derivations are not simple, but they should be understandable on a step-by-step basis. It is hoped that the procedure need never be committed to memory.

Finally, let us return to (64) and make use of this formidable second-order vector partial differential equation to find the vector magnetic potential in one simple example. We select the field between conductors of a coaxial cable, with radii of a and b as usual, and current I in the $\mathbf{a}_z$ direction in the inner conductor. Between the conductors, $\mathbf{J} = 0$, and therefore

$$\nabla^2 \mathbf{A} = 0$$

We have already been told (and Problem 8.44 gives us the opportunity to check the results for ourselves) that the vector Laplacian may be expanded as the vector sum of the scalar Laplacians of the three components in rectangular coordinates,

$$\nabla^2 \mathbf{A} = \nabla^2 A_x \mathbf{a}_x + \nabla^2 A_y \mathbf{a}_y + \nabla^2 A_z \mathbf{a}_z$$

but such a relatively simple result is not possible in other coordinate systems. That is, in cylindrical coordinates, for example,

$$\nabla^2 \mathbf{A} \neq \nabla^2 A_\rho \mathbf{a}_\rho + \nabla^2 A_\phi \mathbf{a}_\phi + \nabla^2 A_z \mathbf{a}_z$$

However, it is not difficult to show for cylindrical coordinates that the z component of the vector Laplacian is the scalar Laplacian of the z component of **A**, or

$$\nabla^2 \mathbf{A}\Big|_z = \nabla^2 A_z \tag{65}$$

and since the current is entirely in the z direction in this problem, **A** has only a z component. Therefore

$$\nabla^2 A_z = 0$$

or

$$\frac{1}{\rho}\frac{\partial}{\partial \rho}\left(\rho \frac{\partial A_z}{\partial \rho}\right) + \frac{1}{\rho^2}\frac{\partial^2 A_z}{\partial \phi^2} + \frac{\partial^2 A_z}{\partial z^2} = 0$$

Thinking symmetrical thoughts about (51) shows us that A_z is a function only of ρ, and thus

$$\frac{1}{\rho}\frac{d}{d\rho}\left(\rho \frac{dA_z}{d\rho}\right) = 0$$

We have solved this equation before, and the result is

$$A_z = C_1 \ln \rho + C_2$$

If we choose a zero reference at $\rho = b$, then

$$A_z = C_1 \ln \frac{\rho}{b}$$

[11] Free space.

In order to relate C_1 to the sources in our problem, we may take the curl of **A**,

$$\nabla \times \mathbf{A} = -\frac{\partial A_z}{\partial \rho}\mathbf{a}_\phi = -\frac{C_1}{\rho}\mathbf{a}_\phi = \mathbf{B}$$

obtain **H**,

$$\mathbf{H} = -\frac{C_1}{\mu_0 \rho}\mathbf{a}_\phi$$

and evaluate the line integral,

$$\oint \mathbf{H} \cdot d\mathbf{L} = I = \int_0^{2\pi} -\frac{C_1}{\mu_0 \rho}\mathbf{a}_\phi \cdot \rho\, d\phi\, \mathbf{a}_\phi = -\frac{2\pi C_1}{\mu_0}$$

Thus

$$C_1 = -\frac{\mu_0 I}{2\pi}$$

or

$$A_z = \frac{\mu_0 I}{2\pi} \ln \frac{b}{\rho} \tag{66}$$

and

$$H_\phi = \frac{I}{2\pi\rho}$$

as before. A plot of A_z versus ρ for $b = 5a$ is shown in Figure 8.20; the decrease of $|\mathbf{A}|$ with distance from the concentrated current source which the inner conductor represents is evident. The results of Problem D8.9 have also been added to Figure 8.20. The extension of the curve into the outer conductor is left as Problem 8.43.

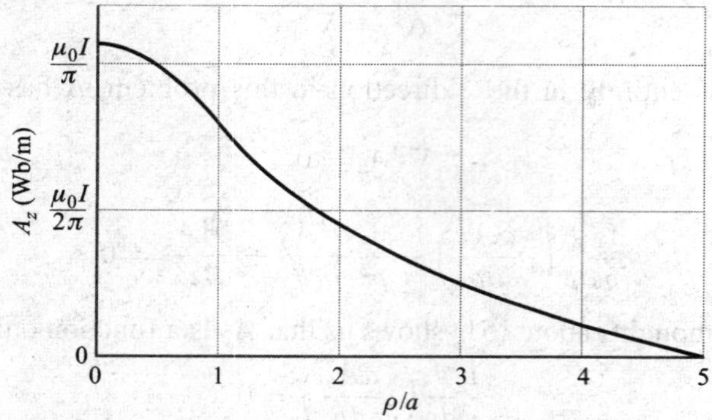

Fig. 8.20 The vector magnetic potential is shown within the inner conductor and in the region between conductors for a coaxial cable with $b = 5a$ carrying I in the $\mathbf{a}_z$ direction. $A_z = 0$ is arbitrarily selected at $\rho = b$.

It is also possible to find A_z between conductors by applying a process some of us informally call "uncurling." That is, we know **H** or **B** for the coax, and we may therefore select the ϕ component of $\nabla \times \mathbf{A} = \mathbf{B}$ and integrate to obtain A_z. Try it, you'll like it!

D8.10 Equation (66) is obviously also applicable to the exterior of any conductor of circular cross section carrying a current I in the $\mathbf{a}_z$ direction in free space. The zero reference is arbitrarily set at $\rho = b$. Now consider two conductors, each of 1 cm radius, parallel to the z axis with their axes lying in the $x = 0$ plane. One conductor whose axis is at $(0, 4\,\text{cm}, z)$ carries 12 A in the $\mathbf{a}_z$ direction; the other axis is at $(0, -4\ \text{cm}, z)$ and carries 12 A in the $-\mathbf{a}_z$ direction. Each current has its zero reference for $\mathbf{A}$ located 4 cm from its axis. Find the total $\mathbf{A}$ field at: (*a*) $(0, 0, z)$; (*b*) $(0, 8\,\text{cm}, z)$; (*c*) $(4\,\text{cm},\ 4\,\text{cm}, z)$; (*d*) $(2\,\text{cm}, 4\,\text{cm}, z)$.

Ans. 0; 2.64 μWb/m; 1.93 μWb/m; 3.40 μWb/m

References

1. Boast, W. B. (See References for Chapter 2.) The scalar magnetic potential is defined on p. 220, and its use in mapping magnetic fields is discussed on p. 444.
2. Jordan, E. C., and K. G. Balmain. *Electromagnetic Waves and Radiating Systems.* 2d ed. Englewood Cliffs, N.J.: Prentice-Hall, 1968. Vector magnetic potential is discussed on pp. 90–96.
3. Paul, C. R., K. W. Whites, and S. Y. Nasar. *Introduction to Electromagnetic Fields*. 3d ed. New York: McGraw-Hill, 1998. The vector magnetic potential is presented on pp. 216–20.
4. Skilling, H. H. (See References for Chapter 3.) The "paddle wheel" is introduced on pp. 23–25.

CHAPTER 8 Problems

Quizzes

8.1 (*a*) Find $\mathbf{H}$ in rectangular components at $P(2, 3, 4)$ if there is a current filament on the z axis carrying 8 mA in the $\mathbf{a}_z$ direction. (*b*) Repeat if the filament is located at $x = -1$, $y = 2$. (*c*) Find $\mathbf{H}$ if both filaments are present.

8.2 A filamentary conductor is formed into an equilateral triangle with sides of length ℓ carrying current I. Find the magnetic field intensity at the center of the triangle.

8.3 Two semi-infinite filaments on the z axis lie in the regions $-\infty < z < -a$ and $a < z < \infty$. Each carries a current I in the $\mathbf{a}_z$ direction. (*a*) Calculate $\mathbf{H}$ as a function of ρ and ϕ at $z = 0$. (*b*) What value of a will cause the magnitude of $\mathbf{H}$ at $\rho = 1$, $z = 0$, to be one-half the value obtained for an infinite filament?

8.4 (*a*) A filament is formed into a circle of radius a, centered at the origin in the plane $z = 0$. It carries a current I in the $\mathbf{a}_\phi$ direction. Find $\mathbf{H}$ at the origin. (*b*) A second filament is shaped into a square in the $z = 0$ plane. The sides are parallel to the coordinate axes and a current I flows in the general $\mathbf{a}_\phi$ direction. Determine the side length b (in terms of a), such that $\mathbf{H}$ at the origin is the same magnitude as that of the circular loop of part (*a*).

8.5 The parallel filamentary conductors shown in Figure 8.21 lie in free space. Plot $|\mathbf{H}|$ versus y, $-4 < y < 4$, along the line $x = 0$, $z = 2$.

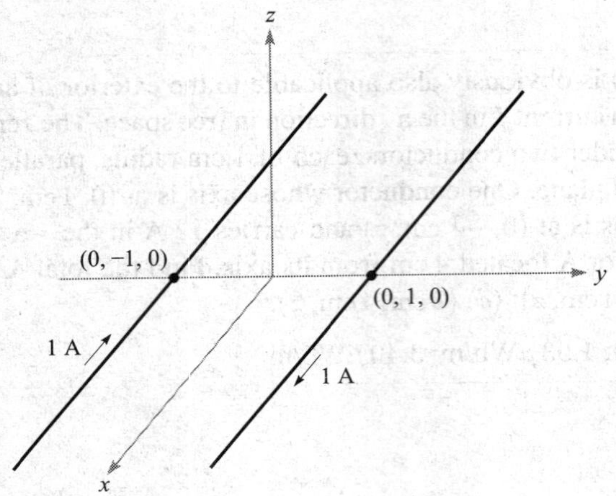

Fig. 8.21 See Problem 8.5.

8.6 A disk of radius a lies in the xy plane, with the z axis through its center. Surface charge of uniform density ρ_s lies on the disk, which rotates about the z axis at angular velocity Ω rad/s. Find **H** at any point on the z axis.

8.7 Given points $C(5, -2, 3)$ and $P(4, -1, 2)$, a current element $Id\mathbf{L} = 10^{-4}(4, -3, 1)$ A · m at C produces a field $d\mathbf{H}$ at P. (*a*) Specify the direction of $d\mathbf{H}$ by a unit vector $\mathbf{a}_H$. (*b*) Find $|d\mathbf{H}|$. (*c*) What direction $\mathbf{a}_l$ should $Id\mathbf{L}$ have at C so that $d\mathbf{H} = 0$?

8.8 For the finite-length current element on the z axis, as shown in Figure 8.5, use the Biot-Savart law to derive Eq. (9) of Section 8.1.

8.9 A current sheet $\mathbf{K} = 8\mathbf{a}_x$ A/m flows in the region $-2 < y < 2$ in the plane $z = 0$. Calculate H at $P(0, 0, 3)$.

8.10 A hollow spherical conducting shell of radius a has filamentary connections made at the top $(r = a, \theta = 0)$ and bottom $(r = a, \theta = \pi)$. A direct current I flows down the upper filament, down the spherical surface, and out the lower filament. Find **H** in spherical coordinates (*a*) inside and (*b*) outside the sphere.

8.11 An infinite filament on the z axis carries 20π mA in the $\mathbf{a}_z$ direction. Three uniform cylindrical current sheets are also present: 400 mA/m at $\rho = 1$ cm, -250 mA/m at $\rho = 2$ cm, and -300 mA/m at $\rho = 3$ cm. Calculate H_ϕ at $\rho = 0.5, 1.5, 2.5$, and 3.5 cm.

8.12 In Figure 8.22, let the regions $0 < z < 0.3$ m and $0.7 < z < 1.0$ m be conducting slabs carrying uniform current densities of 10 A/m^2 in opposite directions as shown. Find **H** at $z =$: (*a*) -0.2; (*b*) 0.2; (*c*) 0.4; (*d*) 0.75; (*e*) 1.2 m.

8.13 A hollow cylindrical shell of radius a is centered on the z axis and carries a uniform surface current density of $K_a\mathbf{a}_\phi$. (*a*) Show that H is not a function of ϕ or z. (*b*) Show that H_ϕ and H_ρ are everywhere zero. (*c*) Show that $H_z = 0$ for $\rho > a$. (*d*) Show that $H_z = K_a$ for $\rho < a$. (*e*) A second shell, $\rho = b$, carries a current $K_b\mathbf{a}_\phi$. Find **H** everywhere.

8.14 A toroid having a cross section of rectangular shape is defined by the following surfaces: the cylinders $\rho = 2$ and $\rho = 3$ cm, and the planes $z = 1$ and $z = 2.5$ cm. The toroid carries a surface current density of $-50\mathbf{a}_z$ A/m on the surface $\rho = 3$ cm. Find **H** at the point

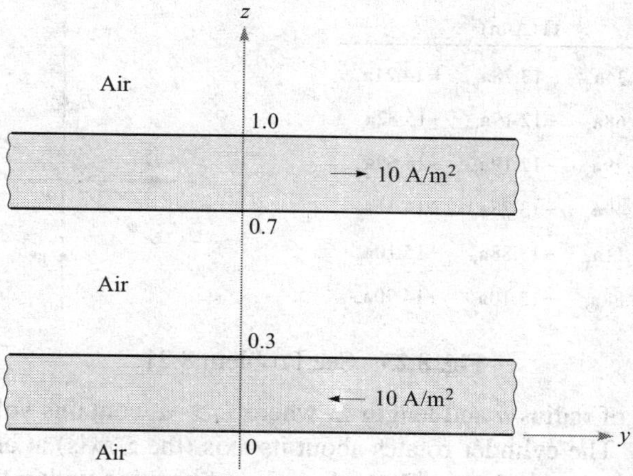

Fig. 8.22 See Problem 8.12.

$P(\rho, \phi, z)$: (*a*) $P_A(1.5\,\text{cm}, 0, 2\,\text{cm})$; (*b*) $P_B(2.1\,\text{cm}, 0, 2\,\text{cm})$; (*c*) $P_C(2.7\,\text{cm}, \pi/2, 2\,\text{cm})$; (*d*) $P_D(3.5\,\text{cm}, \pi/2, 2\,\text{cm})$.

8.15 Assume that there is a region with cylindrical symmetry in which the conductivity is given by $\sigma = 1.5e^{-150\rho}$kS/m. An electric field of $30\mathbf{a}_z$ V/m is present. (*a*) Find **J**. (*b*) Find the total current crossing the surface $\rho < \rho_0$, $z = 0$, all ϕ. (*c*) Make use of Ampère's circuital law to find **H**.

8.16 A balanced coaxial cable contains three coaxial conductors of negligible resistance. Assume a solid inner conductor of radius a, an intermediate conductor of inner radius b_i, outer radius b_o, and an outer conductor having inner and outer radii c_i and c_o, respectively. The intermediate conductor carries current I in the positive $\mathbf{a}_z$ direction and is at potential V_0. The inner and outer conductors are both at zero potential and carry currents $I/2$ (in each) in the negative $\mathbf{a}_z$ direction. Assuming that the current distribution in each conductor is uniform, find: (*a*) **J** in each conductor; (*b*) **H** everywhere; (*c*) **E** everywhere.

8.17 A current filament on the z axis carries a current of 7 mA in the $\mathbf{a}_z$ direction, and current sheets of 0.5 $\mathbf{a}_z$ A/m and -0.2 $\mathbf{a}_z$ A/m are located at $\rho = 1$ cm and $\rho = 0.5$ cm, respectively. Calculate **H** at: (*a*) $\rho = 0.5$ cm; (*b*) $\rho = 1.5$ cm; (*c*) $\rho = 4$ cm. (*d*) What current sheet should be located at $\rho = 4$ cm so that $\mathbf{H} = 0$ for all $\rho > 4$ cm?

8.18 A wire of 3 mm radius is made up of an inner material ($0 < \rho < 2$ mm) for which $\sigma = 10^7$ S/m, and an outer material (2 mm $< \rho <$ 3 mm) for which $\sigma = 4 \times 10^7$ S/m. If the wire carries a total current of 100 mA dc, determine **H** everywhere as a function of ρ.

8.19 Calculate $\nabla \times [\nabla(\nabla \cdot \mathbf{G})]$ if $\mathbf{G} = 2x^2yz\,\mathbf{a}_x - 20y\,\mathbf{a}_y + (x^2 - z^2)\,\mathbf{a}_z$.

8.20 A solid conductor of circular cross section with a radius of 5 mm has a conductivity that varies with radius. The conductor is 20 m long, and there is a potential difference of 0.1 V dc between its two ends. Within the conductor, $\mathbf{H} = 10^5\rho^2\mathbf{a}_\phi$ A/m. (*a*) Find σ as a function of ρ. (*b*) What is the resistance between the two ends?

8.21 Points A, B, C, D, E, and F are each 2 mm from the origin on the coordinate axes indicated in Figure 8.23. The value of **H** at each point is given. Calculate an approximate value for $\nabla \times \mathbf{H}$ at the origin.

Point	H (A/m)		
A	$11.34\mathbf{a}_x$	$-13.78\mathbf{a}_y$	$+14.21\mathbf{a}_z$
B	$10.68\mathbf{a}_x$	$-12.19\mathbf{a}_y$	$+15.82\mathbf{a}_z$
C	$10.49\mathbf{a}_x$	$-12.19\mathbf{a}_y$	$+15.69\mathbf{a}_z$
D	$11.49\mathbf{a}_x$	$-13.78\mathbf{a}_y$	$+14.35\mathbf{a}_z$
E	$11.11\mathbf{a}_x$	$-13.88\mathbf{a}_y$	$+15.10\mathbf{a}_z$
F	$10.88\mathbf{a}_x$	$-13.10\mathbf{a}_y$	$+14.90\mathbf{a}_z$

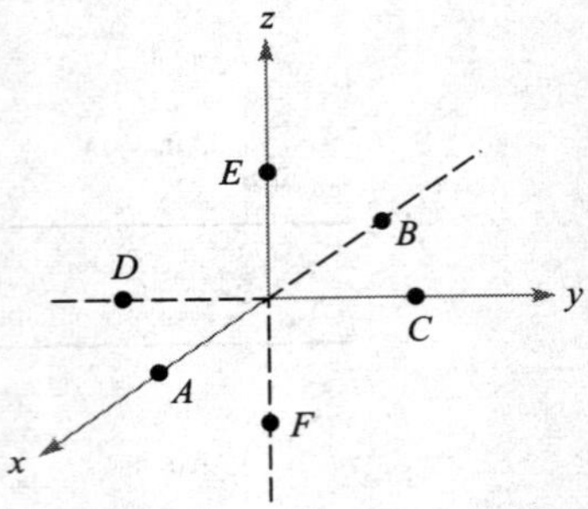

Fig. 8.23 See Problem 8.21.

8.22 A solid cylinder of radius a and length L, where $L \gg a$, contains volume charge of uniform density ρ_0 C/m^3. The cylinder rotates about its axis (the z axis) at angular velocity Ω rad/s. (a) Determine the current density $\mathbf{J}$ as a function of position within the rotating cylinder. (b) Determine $\mathbf{H}$ on-axis by applying the results of Problem 8.6. (c) Determine the magnetic field intensity $\mathbf{H}$ inside and outside. (d) Check your result of part (c) by taking the curl of $\mathbf{H}$.

8.23 Given the field $\mathbf{H} = 20\rho^2\mathbf{a}_\phi$ A/m: (a) Determine the current density $\mathbf{J}$. (b) Integrate $\mathbf{J}$ over the circular surface $\rho = 1, 0 < \phi < 2\pi, z = 0$, to determine the total current passing through that surface in the $\mathbf{a}_z$ direction. (c) Find the total current once more, this time by a line integral around the circular path $\rho = 1, 0 < \phi < 2\pi, z = 0$.

8.24 Evaluate both sides of Stokes' theorem for the field $\mathbf{G} = 10 \sin\theta\, \mathbf{a}_\phi$ and the surface $r = 3, 0 \leq \theta \leq 90°, 0 \leq \phi \leq 90°$. Let the surface have the $\mathbf{a}_r$ direction.

8.25 When x, y, and z are positive and less than 5, a certain magnetic field intensity may be expressed as $\mathbf{H} = [x^2yz/(y+1)]\mathbf{a}_x + 3x^2z^2\mathbf{a}_y - [xyz^2/(y+1)]\mathbf{a}_z$. Find the total current in the $\mathbf{a}_x$ direction that crosses the strip $x = 2, 1 \leq y \leq 4, 3 \leq z \leq 4$, by a method utilizing: (a) a surface integral; (b) a closed line integral.

8.26 Let $\mathbf{G} = 15r\mathbf{a}_\phi$. ($a$) Determine $\oint \mathbf{G} \cdot d\mathbf{L}$ for the circular path $r = 5, \theta = 25°, 0 \leq \phi \leq 2\pi$. ($b$) Evaluate $\int_S (\nabla \times \mathbf{G}) \cdot d\mathbf{S}$ over the spherical cap $r = 5, 0 \leq \theta \leq 25°, 0 \leq \phi \leq 2\pi$.

8.27 The magnetic field intensity is given in a certain region of space as $\mathbf{H} = [(x+2y)/z^2]\mathbf{a}_y + (2/z)\mathbf{a}_z$ A/m. (a) Find $\nabla \times \mathbf{H}$. (b) Find $\mathbf{J}$. (c) Use $\mathbf{J}$ to find the total current passing through the surface $z = 4, 1 \leq x \leq 2, 3 \leq z \leq 5$, in the $\mathbf{a}_z$ direction. (d) Show that the same result is obtained using the other side of Stokes' theorem.

8.28 Given $\mathbf{H} = (3r^2/\sin\theta)\mathbf{a}_\theta + 54r\cos\theta\mathbf{a}_\phi$ A/m in free space: (a) Find the total current in the $\mathbf{a}_\theta$ direction through the conical surface $\theta = 20°, 0 \leq \phi \leq 2\pi, 0 \leq r \leq 5$, by whatever side of Stokes' theorem you like the best. (b) Check the result by using the other side of Stokes' theorem.

8.29 A long, straight, nonmagnetic conductor of 0.2 mm radius carries a uniformly distributed current of 2 A dc. (a) Find $\mathbf{J}$ within the conductor. (b) Use Ampère's circuital law to find $\mathbf{H}$ and $\mathbf{B}$ within the conductor. (c) Show that $\nabla \times \mathbf{H} = \mathbf{J}$ within the conductor. (d) Find $\mathbf{H}$ and $\mathbf{B}$ outside the conductor. (e) Show that $\nabla \times \mathbf{H} = \mathbf{J}$ outside the conductor.

8.30 (An inversion of Problem 8.20). A solid, nonmagnetic conductor of circular cross section has a radius of 2 mm. The conductor is inhomogeneous, with $\sigma = 10^6(1 + 10^6\rho^2)$ S/m. If the conductor is 1 m in length and has a voltage of 1 mV between its ends, find: (a) $\mathbf{H}$ inside; (b) the total magnetic flux inside the conductor.

8.31 The cylindrical shell defined by 1 cm $< \rho <$ 1.4 cm consists of a nonmagnetic conducting material and carries a total current of 50 A in the $\mathbf{a}_z$ direction. Find the total magnetic flux crossing the plane $\phi = 0, 0 < z < 1$: (*a*) $0 < \rho <$ 1.2 cm; (*b*) 1.0 cm $< \rho <$ 1.4 cm; (*c*) 1.4 cm $< \rho <$ 20 cm.

8.32 The free space region defined by $1 < z < 4$ cm and $2 < \rho < 3$ cm is a toroid of rectangular cross section. Let the surface at $\rho = 3$ cm carry a surface current $\mathbf{K} = 2\mathbf{a}_z$ kA/m. (*a*) Specify the current densities on the surfaces at $\rho = 2$ cm, $z = 1$ cm, and $z = 4$ cm. (*b*) Find $\mathbf{H}$ everywhere. (*c*) Calculate the total flux within the toroid.

8.33 Use an expansion in rectangular coordinates to show that the curl of the gradient of any scalar field G is identically equal to zero.

8.34 A filamentary conductor on the z axis carries a current of 16 A in the $\mathbf{a}_z$ direction, a conducting shell at $\rho = 6$ carries a total current of 12 A in the $-\mathbf{a}_z$ direction, and another shell at $\rho = 10$ carries a total current of 4 A in the $-\mathbf{a}_z$ direction. (*a*) Find $\mathbf{H}$ for $0 < \rho < 12$. (*b*) Plot H_ϕ versus ρ. (*c*) Find the total flux Φ crossing the surface $1 < \rho < 7, 0 < z < 1$.

8.35 A current sheet, $\mathbf{K} = 20\,\mathbf{a}_z$ A/m, is located at $\rho = 2$, and a second sheet, $\mathbf{K} = -10\mathbf{a}_z$ A/m, is located at $\rho = 4$. (*a*) Let $V_m = 0$ at $P(\rho = 3, \phi = 0, z = 5)$ and place a barrier at $\phi = \pi$. Find $V_m(\rho, \phi, z)$ for $-\pi < \phi < \pi$. (*b*) Let $\mathbf{A} = 0$ at P and find $\mathbf{A}(\rho, \phi, z)$ for $2 < \rho < 4$.

8.36 Let $\mathbf{A} = (3y - z)\mathbf{a}_x + 2xz\mathbf{a}_y$ Wb/m in a certain region of free space. (*a*) Show that $\nabla \cdot \mathbf{A} = 0$. (*b*) At $P(2, -1, 3)$, find $\mathbf{A}$, $\mathbf{B}$, $\mathbf{H}$, and $\mathbf{J}$.

8.37 Let $N = 1000$, $I = 0.8$ A, $\rho_0 = 2$ cm, and $a = 0.8$ cm for the toroid shown in Figure 8.12*b*. Find V_m in the interior of the toroid if $V_m = 0$ at $\rho = 2.5$ cm, $\phi = 0.3\pi$. Keep ϕ within the range $0 < \phi < 2\pi$.

8.38 Assume a direct current I amps flowing in the $\mathbf{a}_z$ direction in a filament extending between $-L < z < L$ on the z axis. (*a*) Using cylindrical coordinates, find $\mathbf{A}$ at any general point $P(\rho, 0°, z)$. HINT: use the $\sinh^{-1}$ form of the integral evaluation. (*b*) From part (*a*) find $\mathbf{B}$ and $\mathbf{H}$. (*c*) Let $L \to \infty$ and show that the expression for $\mathbf{H}$ reduces to the known one for an infinite filament.

8.39 Planar current sheets of $\mathbf{K} = 30\mathbf{a}_z$ A/m and $-30\mathbf{a}_z$ A/m are located in free space at $x = 0.2$ and $x = -0.2$, respectively. For the region $-0.2 < x < 0.2$: (*a*) find $\mathbf{H}$; (*b*) obtain an expression for V_m if $V_m = 0$ at $P(0.1, 0.2, 0.3)$; (*c*) find $\mathbf{B}$; (*d*) obtain an expression for $\mathbf{A}$ if $\mathbf{A} = 0$ at P.

8.40 Show that the line integral of the vector potential $\mathbf{A}$ about any closed path is equal to the magnetic flux enclosed by the path, or $\oint \mathbf{A} \cdot d\mathbf{L} = \int \mathbf{B} \cdot d\mathbf{S}$.

8.41 Assume that $\mathbf{A} = 50\rho^2\mathbf{a}_z$ Wb/m in a certain region of free space. (*a*) Find $\mathbf{H}$ and $\mathbf{B}$. (*b*) Find $\mathbf{J}$. (*c*) Use $\mathbf{J}$ to find the total current crossing the surface $0 \le \rho \le 1$, $0 \le \phi < 2\pi$, $z = 0$. (*d*) Use the value of H_ϕ at $\rho = 1$ to calculate $\oint \mathbf{H} \cdot d\mathbf{L}$ for $\rho = 1$, $z = 0$.

8.42 Show that $\nabla_2(1/R_{12}) = -\nabla_1(1/R_{12}) = \mathbf{R}_{21}/R_{12}^3$.

8.43 Compute the vector magnetic potential within the outer conductor for the coaxial line whose vector magnetic potential is shown in Figure 8.20 if the outer radius of the outer conductor is $7a$. Select the proper zero reference and sketch the results on the figure.

8.44 By expanding Eq. (58), Section 8.7 in rectangular coordinates, show that (59) is correct.

9

Magnetic Forces, Materials, and Inductance

The magnetic field quantities **H**, **B**, Φ, V_m, and **A** introduced in Chapter 8 were not given much physical significance. Each of these quantities is merely defined in terms of the distribution of current sources throughout space. If the current distribution is known, we should feel that **H**, **B**, and **A** are determined at every point in space, even though we may not be able to evaluate the defining integrals because of mathematical complexity.

We are now ready to undertake the second half of the magnetic field problem, that of determining the forces and torques exerted by the magnetic field on other charges. The electric field causes a force to be exerted on a charge which may be either stationary or in motion; we shall see that the steady magnetic field is capable of exerting a force only on a *moving* charge. This result appears reasonable; a magnetic field may be produced by moving charges and may exert forces on moving charges; a magnetic field cannot arise from stationary charges and cannot exert any force on a stationary charge.

This chapter initially considers the forces and torques on current-carrying conductors which may either be of a filamentary nature or possess a finite cross section with a known current density distribution. The problems associated with the motion of particles in a vacuum are largely avoided.

With an understanding of the fundamental effects produced by the magnetic field, we may then consider the varied types of magnetic materials, the analysis of elementary magnetic circuits, the forces on magnetic materials, and finally, the important electrical circuit concepts of self-inductance and mutual inductance.

9.1 FORCE ON A MOVING CHARGE

In an electric field the definition of the electric field intensity shows us that the force on a charged particle is

$$\mathbf{F} = Q\mathbf{E} \tag{1}$$

The force is in the same direction as the electric field intensity (for a positive charge) and is directly proportional to both **E** and Q. If the charge is in motion, the force at any point in its trajectory is then given by (1).

A charged particle in motion in a magnetic field of flux density **B** is found experimentally to experience a force whose magnitude is proportional to the product of the magnitudes of the charge Q, its velocity **v**, and the flux density **B**, and to the sine of the angle between the vectors **v** and

B. The direction of the force is perpendicular to both **v** and **B** and is given by a unit vector in the direction of $\mathbf{v} \times \mathbf{B}$. The force may therefore be expressed as

$$\mathbf{F} = Q\mathbf{v} \times \mathbf{B} \tag{2}$$

A fundamental difference in the effect of the electric and magnetic fields on charged particles is now apparent, for a force which is always applied in a direction at right angles to the direction in which the particle is proceeding can never change the magnitude of the particle velocity. In other words, the *acceleration* vector is always normal to the velocity vector. The kinetic energy of the particle remains unchanged, and it follows that the steady magnetic field is incapable of transferring energy to the moving charge. The electric field, on the other hand, exerts a force on the particle which is independent of the direction in which the particle is progressing and therefore effects an energy transfer between field and particle in general.

The first two problems at the end of this chapter illustrate the different effects of electric and magnetic fields on the kinetic energy of a charged particle moving in free space.

The force on a moving particle arising from combined electric and magnetic fields is obtained easily by superposition,

$$\mathbf{F} = Q(\mathbf{E} + \mathbf{v} \times \mathbf{B}) \tag{3}$$

This equation is known as the *Lorentz force equation*, and its solution is required in determining electron orbits in the magnetron, proton paths in the cyclotron, plasma characteristics in a magnetohydrodynamic (MHD) generator, or, in general, charged-particle motion in combined electric and magnetic fields.

D9.1 The point charge $Q = 18$ nC has a velocity of 5×10^6 m/s in the direction $\mathbf{a}_v = 0.60\mathbf{a}_x + 0.75\mathbf{a}_y + 0.30\mathbf{a}_z$. Calculate the magnitude of the force exerted on the charge by the field: (*a*) $\mathbf{B} = -3\mathbf{a}_x + 4\mathbf{a}_y + 6\mathbf{a}_z$ mT; (*b*) $\mathbf{E} = -3\mathbf{a}_x + 4\mathbf{a}_y + 6\mathbf{a}_z$ kV/m; (*c*) B and E acting together.

Ans. 660 μN; 140 μN; 670 μN

9.2 FORCE ON A DIFFERENTIAL CURRENT ELEMENT

The force on a charged particle moving through a steady magnetic field may be written as the differential force exerted on a differential element of charge,

$$d\mathbf{F} = dQ\,\mathbf{v} \times \mathbf{B} \tag{4}$$

Physically, the differential element of charge consists of a large number of very small, discrete charges occupying a volume which, although small, is much larger than the average separation between the charges. The differential force expressed by (4) is thus merely the sum of the forces on the individual charges. This sum, or resultant force, is not a force applied to a single object. In an analogous way, we might consider the differential gravitational force experienced by a small volume taken in a shower of falling sand. The small volume contains a large number of sand grains, and the differential force is the sum of the forces on the individual grains within the small volume.

If our charges are electrons in motion in a conductor, however, we can show that the force is transferred to the conductor and that the sum of this extremely large number of extremely small forces is of practical importance. Within the conductor, electrons are in motion throughout a region of immobile positive ions which form a crystalline array, giving the conductor its solid properties. A magnetic field which exerts forces on the electrons tends to cause them to shift position slightly and produces a small displacement between the centers of "gravity" of the positive and negative charges. The Coulomb forces between electrons and positive ions, however, tend to resist such a displacement. Any attempt to move the electrons, therefore, results in an attractive force between electrons and the positive ions of the crystalline lattice. The magnetic force is thus transferred to the crystalline lattice, or to the conductor itself. The Coulomb forces are so much greater than the magnetic forces in good conductors that the actual displacement of the electrons is almost immeasurable. The charge separation that does result, however, is disclosed by the presence of a slight potential difference across the conductor sample in a direction perpendicular to both the magnetic field and the velocity of the charges. The voltage is known as the *Hall voltage,* and the effect itself is called the *Hall effect*.

Figure 9.1 illustrates the direction of the Hall voltage for both positive and negative charges in motion. In Figure 9.1a, $\mathbf{v}$ is in the $-\mathbf{a}_x$ direction, $\mathbf{v} \times \mathbf{B}$ is in the $\mathbf{a}_y$ direction, and Q is positive, causing $\mathbf{F}_Q$ to be in the $\mathbf{a}_y$ direction; thus, the positive charges move to the right. In Figure 9.1b, $\mathbf{v}$ is now in the $+\mathbf{a}_x$ direction, $\mathbf{B}$ is still in the $\mathbf{a}_z$ direction, $\mathbf{v} \times \mathbf{B}$ is in the $-\mathbf{a}_y$ direction, and Q is negative; thus $\mathbf{F}_Q$ is again in the $\mathbf{a}_y$ direction. Hence, the negative charges end up at the right edge. Equal currents provided by holes and electrons in semiconductors can therefore be differentiated by their Hall voltages. This is one method of determining whether a given semiconductor is n-type or p-type.

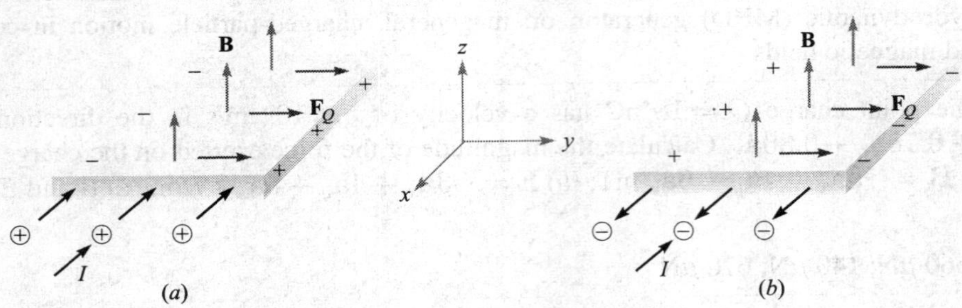

Fig. 9.1 Equal currents directed into the material are provided by positive charges moving inward in (a) and negative charges moving outward in (b). The two cases can be distinguished by oppositely directed Hall voltages, as shown.

Devices employ the Hall effect to measure the magnetic flux density and, in some applications where the current through the device can be made proportional to the magnetic field across it, to serve as electronic wattmeters, squaring elements, and so forth.

Returning to (4), we may therefore say that if we are considering an element of moving charge in an electron beam, the force is merely the sum of the forces on the individual electrons in that small volume element, but if we are considering an element of moving charge within a conductor, the total force is applied to the solid conductor itself. We shall now limit our attention to the forces on current-carrying conductors.

In Chapter 5 we defined convection current density in terms of the velocity of the volume charge density,

$$\mathbf{J} = \rho_\nu \mathbf{v}$$

The differential element of charge in (4) may also be expressed in terms of volume charge density,[1]

$$dQ = \rho_\nu d\nu$$

Thus

$$d\mathbf{F} = \rho_\nu d\nu\, \mathbf{v} \times \mathbf{B}$$

or

$$d\mathbf{F} = \mathbf{J} \times \mathbf{B}\, d\nu \tag{5}$$

We saw in Chapter 8 that $\mathbf{J}\, d\nu$ may be interpreted as a differential current element; that is,

$$\mathbf{J}\, d\nu = \mathbf{K}\, dS = I\, d\mathbf{L}$$

and thus the Lorentz force equation may be applied to surface current density,

$$d\mathbf{F} = \mathbf{K} \times \mathbf{B}\, dS \tag{6}$$

or to a differential current filament,

$$d\mathbf{F} = I\, d\mathbf{L} \times \mathbf{B} \tag{7}$$

Integrating (5), (6), or (7) over a volume, a surface which may be either open or closed (why?), or a closed path, respectively, leads to the integral formulations

$$\mathbf{F} = \int_{\text{vol}} \mathbf{J} \times \mathbf{B}\, d\nu \tag{8}$$

$$\mathbf{F} = \int_S \mathbf{K} \times \mathbf{B}\, dS \tag{9}$$

and

$$\mathbf{F} = \oint I\, d\mathbf{L} \times \mathbf{B} = -I \oint \mathbf{B} \times d\mathbf{L} \tag{10}$$

One simple result is obtained by applying (7) or (10) to a straight conductor in a uniform magnetic field,

$$\mathbf{F} = I\mathbf{L} \times \mathbf{B} \tag{11}$$

The magnitude of the force is given by the familiar equation

$$F = BIL \sin\theta \tag{12}$$

[1] Remember that $d\nu$ is a differential volume element and not a differential increase in velocity.

where θ is the angle between the vectors representing the direction of the current flow and the direction of the magnetic flux density. Equation (11) or (12) applies only to a portion of the closed circuit, and the remainder of the circuit must be considered in any practical problem.

EXAMPLE 9.1

As a numerical example of these equations, consider Figure 9.2. We have a square loop of wire in the $z = 0$ plane carrying 2 mA in the field of an infinite filament on the y axis, as shown. We desire the total force on the loop.

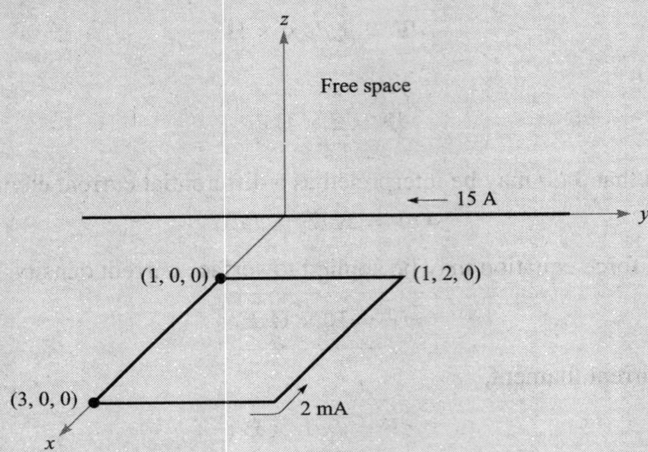

Fig. 9.2 A square loop of wire in the xy plane carrying 2 mA is subjected to a nonuniform B field.

Solution. The field produced in the plane of the loop by the straight filament is

$$\mathbf{H} = \frac{I}{2\pi x}\mathbf{a}_z = \frac{15}{2\pi x}\mathbf{a}_z \text{ A/m}$$

Therefore,

$$\mathbf{B} = \mu_0\mathbf{H} = 4\pi \times 10^{-7}\mathbf{H} = \frac{3 \times 10^{-6}}{x}\mathbf{a}_z \text{ T}$$

We use the integral form (10),

$$\mathbf{F} = -I\oint \mathbf{B} \times d\mathbf{L}$$

Let us assume a rigid loop so that the total force is the sum of the forces on the four sides. Beginning with the left side:

$$\mathbf{F} = -2 \times 10^{-3} \times 3 \times 10^{-6}\left[\int_{x=1}^{3} \frac{\mathbf{a}_z}{x} \times dx\,\mathbf{a}_x + \int_{y=0}^{2} \frac{\mathbf{a}_z}{3} \times dy\,\mathbf{a}_y\right.$$

$$\left. + \int_{x=3}^{1} \frac{\mathbf{a}_z}{x} \times dx\,\mathbf{a}_x + \int_{y=2}^{0} \frac{\mathbf{a}_z}{1} \times dy\,\mathbf{a}_y\right]$$

$$= -6 \times 10^{-9} \left[\ln x \Big|_1^3 \mathbf{a}_y + \frac{1}{3} y \Big|_0^2 (-\mathbf{a}_x) + \ln x \Big|_3^1 \mathbf{a}_y + y \Big|_2^0 (-\mathbf{a}_x) \right]$$

$$= -6 \times 10^{-9} \left[(\ln 3)\mathbf{a}_y - \frac{2}{3}\mathbf{a}_x + \left(\ln \frac{1}{3} \right) \mathbf{a}_y + 2\mathbf{a}_x \right] = -8\mathbf{a}_x \text{ nN}$$

Thus, the net force on the loop is in the $-\mathbf{a}_x$ direction.

D9.2 The field $\mathbf{B} = -2\mathbf{a}_x + 3\mathbf{a}_y + 4\mathbf{a}_z$ mT is present in free space. Find the vector force exerted on a straight wire carrying 12 A in the $\mathbf{a}_{AB}$ direction, given $A(1, 1, 1)$ and: (*a*) $B(2, 1, 1)$; (*b*) $B(3, 5, 6)$.

Ans. $-48\mathbf{a}_y + 36\mathbf{a}_z$ mN; $12\mathbf{a}_x - 216\mathbf{a}_y + 168\mathbf{a}_z$ mN

D9.3 The semiconductor sample shown in Figure 9.1 is n-type silicon, having a rectangular cross section of 0.9 mm by 1.1 cm and a length of 1.3 cm. Assume the electron and hole mobilities are 0.13 and 0.03 $\text{m}^2/\text{V} \cdot \text{s}$, respectively, at the operating temperature. Let $B = 0.07$ T and the electric field intensity in the direction of the current flow be 800 V/m. Find the magnitude of: (*a*) the voltage across the sample length; (*b*) the drift velocity; (*c*) the transverse force per coulomb of moving charge caused by B; (*d*) the transverse electric field intensity; (*e*) the Hall voltage.

Ans. 10.40 V; 104.0 m/s; 7.28 N/C; 7.28 V/m; 80.1 mV

9.3 FORCE BETWEEN DIFFERENTIAL CURRENT ELEMENTS

The concept of the magnetic field was introduced to break into two parts the problem of finding the interaction of one current distribution on a second current distribution. It is possible to express the force on one current element directly in terms of a second current element without finding the magnetic field. Since we claimed that the magnetic-field concept simplifies our work, it then behooves us to show that avoidance of this intermediate step leads to more complicated expressions.

The magnetic field at point 2 due to a current element at point 1 was found to be

$$d\mathbf{H}_2 = \frac{I_1 d\mathbf{L}_1 \times \mathbf{a}_{R12}}{4\pi R_{12}^2}$$

Now, the differential force on a differential current element is

$$d\mathbf{F} = I\, d\mathbf{L} \times \mathbf{B}$$

and we apply this to our problem by letting $\mathbf{B}$ be $d\mathbf{B}_2$ (the differential flux density at point 2 caused by current element 1), by identifying $I\, d\mathbf{L}$ as $I_2 d\mathbf{L}_2$, and by symbolizing the differential amount of our differential force on element 2 as $d(d\mathbf{F}_2)$:

$$d(d\mathbf{F}_2) = I_2 d\mathbf{L}_2 \times d\mathbf{B}_2$$

Since $d\mathbf{B}_2 = \mu_0 d\mathbf{H}_2$, we obtain the force between two differential current elements,

$$d(d\mathbf{F}_2) = \mu_0 \frac{I_1 I_2}{4\pi R_{12}^2} d\mathbf{L}_2 \times (d\mathbf{L}_1 \times \mathbf{a}_{R12}) \tag{13}$$

EXAMPLE 9.2

As an example that illustrates the use (and misuse) of these results, consider the two differential current elements shown in Figure 9.3. We seek the differential force on $d\mathbf{L}_2$.

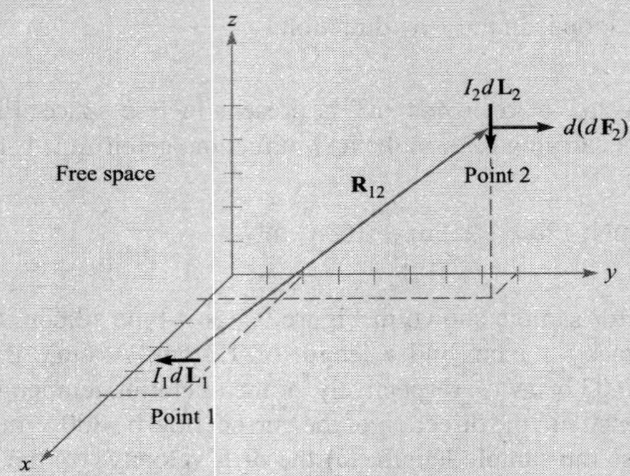

Fig. 9.3 Given $P_1(5, 2, 1)$, $P_2(1, 8, 5)$, $I_1\, d\mathbf{L}_1 = -3\mathbf{a}_y$ A · m, and $I_2\, d\mathbf{L}_2 = -4\mathbf{a}_z$ A · m, the force on $I_2\, d\mathbf{L}_2$ is 8.56 nN in the $\mathbf{a}_y$ direction.

Solution. We have $I_1 d\mathbf{L}_1 = -3\mathbf{a}_y \text{A}\cdot\text{m}$ at $P_1(5, 2, 1)$, and $I_2 d\mathbf{L}_2 = -4\mathbf{a}_z \text{A}\cdot\text{m}$ at $P_2(1, 8, 5)$. Thus, $\mathbf{R}_{12} = -4\mathbf{a}_x + 6\mathbf{a}_y + 4\mathbf{a}_z$, and we may substitute these data into (13),

$$d(d\mathbf{F}_2) = \frac{4\pi 10^{-7}}{4\pi} \frac{(-4\mathbf{a}_z) \times [(-3\mathbf{a}_y) \times (-4\mathbf{a}_x + 6\mathbf{a}_y + 4\mathbf{a}_z)]}{(16 + 36 + 16)^{1.5}} = 8.56\mathbf{a}_y \text{ nN}$$

Many chapters ago when we discussed the force exerted by one point charge on another point charge, we found that the force on the first charge was the negative of that on the second. That is, the total force on the system was zero. This is not the case with the differential current elements, and $d(d\mathbf{F}_1) = -12.84\mathbf{a}_z$ nN in Example 9.2. The reason for this different behavior lies with the nonphysical nature of the current element. Whereas point charges may be approximated quite well by small charges, the continuity of current demands that a complete circuit be considered. This we shall now do.

The total force between two filamentary circuits is obtained by integrating twice:

$$\begin{aligned}\mathbf{F}_2 &= \mu_0 \frac{I_1 I_2}{4\pi} \oint \left[d\mathbf{L}_2 \times \oint \frac{d\mathbf{L}_1 \times \mathbf{a}_{R12}}{R_{12}^2} \right] \\ &= \mu_0 \frac{I_1 I_2}{4\pi} \oint \left[\oint \frac{\mathbf{a}_{R12} \times d\mathbf{L}_1}{R_{12}^2} \right] \times d\mathbf{L}_2\end{aligned} \tag{14}$$

Equation (14) is quite formidable, but the familiarity gained in Chapter 8 with the magnetic field should enable us to recognize the inner integral as the integral necessary to find the magnetic field at point 2 due to the current element at point 1.

Although we shall only give the result, it is not very difficult to make use of (14) to find the force of repulsion between two infinitely long, straight, parallel, filamentary conductors with separation d, and carrying equal but opposite currents I, as shown in Figure 9.4. The integrations are simple, and most errors are made in determining suitable expressions for $\mathbf{a}_{R12}$, $d\mathbf{L}_1$, and $d\mathbf{L}_2$. However, since the magnetic field intensity at either wire caused by the other is already known to be $I/(2\pi d)$, it is readily apparent that the answer is a force of $\mu_0 I^2/(2\pi d)$ newtons per meter length.

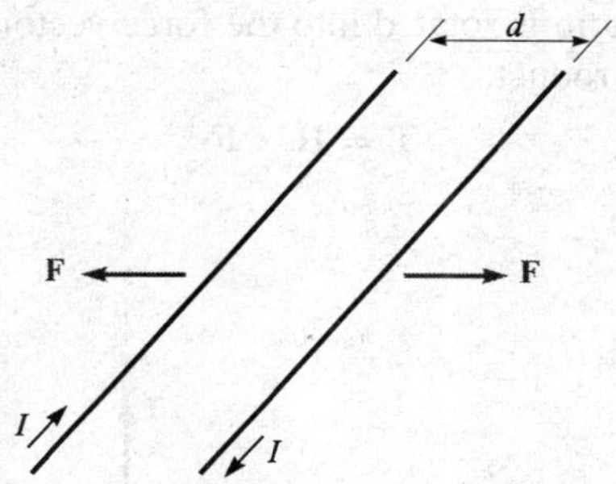

Fig. 9.4 Two infinite parallel filaments with separation d and equal but opposite currents I experience a repulsive force of $\mu_0 I^2/(2\pi d)$ N/m.

D9.4 Two differential current elements, $I_1\Delta\mathbf{L}_1 = 3\times10^{-6}\mathbf{a}_y$ A · m at $P_1(1,0,0)$ and $I_2\Delta\mathbf{L}_2 = 3\times10^{-6}(-0.5\mathbf{a}_x + 0.4\mathbf{a}_y + 0.3\mathbf{a}_z)$ A · m at $P_2(2,2,2)$, are located in free space. Find the vector force exerted on: (*a*) $I_2\Delta\mathbf{L}_2$ by $I_1\Delta\mathbf{L}_1$; (*b*) $I_1\Delta\mathbf{L}_1$ by $I_2\Delta\mathbf{L}_2$.

Ans. $(-1.333\mathbf{a}_x + 0.333\mathbf{a}_y - 2.67\mathbf{a}_z)10^{-20}$ N; $(4.67\mathbf{a}_x + 0.667\mathbf{a}_z)10^{-20}$ N

9.4 FORCE AND TORQUE ON A CLOSED CIRCUIT

We have already obtained general expressions for the forces exerted on current systems. One special case is easily disposed of, for if we take our relationship for the force on a filamentary closed circuit, as given by Eq. (10), Section 9.2,

$$\mathbf{F} = -I\oint \mathbf{B}\times d\mathbf{L}$$

and assume a *uniform* magnetic flux density, then $\mathbf{B}$ may be removed from the integral:

$$\mathbf{F} = -I\mathbf{B}\times\oint d\mathbf{L}$$

However, we discovered during our investigation of closed line integrals in an electrostatic potential field that $\oint d\mathbf{L} = 0$, and therefore the force on a closed filamentary circuit in a uniform magnetic field is zero.

If the field is not uniform, the total force need not be zero.

This result for uniform fields does not have to be restricted to filamentary circuits only. The circuit may contain surface currents or volume current density as well. If the total current is divided into filaments, the force on each one is zero, as we have shown, and the total force is again zero. Therefore any real closed circuit carrying direct currents experiences a total vector force of zero in a uniform magnetic field.

Although the force is zero, the torque is generally not equal to zero.

In defining the *torque,* or *moment,* of a force, it is necessary to consider both an origin at or about which the torque is to be calculated, and the point at which the force is applied. In Figure 9.5a, we apply a force $\mathbf{F}$ at point P, and we establish an origin at O with a rigid lever arm $\mathbf{R}$ extending from O to P. The torque about point O is a vector whose magnitude is the product of the magnitudes of $\mathbf{R}$, of $\mathbf{F}$, and of the sine of the angle between these two vectors. The direction of the vector torque $\mathbf{T}$ is normal to both the force $\mathbf{F}$ and the lever arm $\mathbf{R}$ and is in the direction of progress of a right-handed screw as the lever arm is rotated into the force vector through the smaller angle. The torque is expressible as a cross product,

$$\mathbf{T} = \mathbf{R} \times \mathbf{F}$$

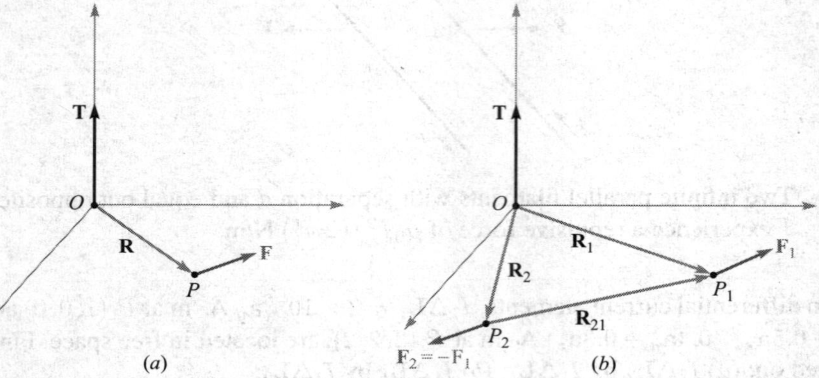

Fig. 9.5 (a) Given a lever arm R extending from an origin O to a point P where force $\mathbf{F}$ is applied, the torque about O is $\mathbf{T} = \mathbf{R} \times \mathbf{F}$. ($b$) If $\mathbf{F}_2 = -\mathbf{F}_1$, then the torque $\mathbf{T} = \mathbf{R}_{21} \times \mathbf{F}_1$ is independent of the choice of origin for $\mathbf{R}_1$ and $\mathbf{R}_2$.

Now let us assume that two forces, $\mathbf{F}_1$ at P_1 and $\mathbf{F}_2$ at P_2, having lever arms $\mathbf{R}_1$ and $\mathbf{R}_2$ extending from a common origin O, as shown in Figure 9.5b, are applied to an object of fixed shape and that the object does not undergo any translation. Then the torque about the origin is

$$\mathbf{T} = \mathbf{R}_1 \times \mathbf{F}_1 + \mathbf{R}_2 \times \mathbf{F}_2$$

where

$$\mathbf{F}_1 + \mathbf{F}_2 = 0$$

and therefore

$$\mathbf{T} = (\mathbf{R}_1 - \mathbf{R}_2) \times \mathbf{F}_1 = \mathbf{R}_{21} \times \mathbf{F}_1$$

The vector $\mathbf{R}_{21} = \mathbf{R}_1 - \mathbf{R}_2$ joins the point of application of $\mathbf{F}_2$ to that of $\mathbf{F}_1$ and is independent of the choice of origin for the two vectors $\mathbf{R}_1$ and $\mathbf{R}_2$. Therefore, the torque is also independent of the choice of origin, provided that the total force is zero. This may be extended to any number of forces.

Consider the application of a vertically upward force at the end of a horizontal crank handle on an elderly automobile. This cannot be the only applied force, for if it were, the entire handle would be accelerated in an upward direction. A second force, equal in magnitude to that exerted at the end of the handle, is applied in a downward direction by the bearing surface at the axis of rotation.

For a 40 N force on a crank handle 0.3 m in length, the torque is 12 N · m. This figure is obtained regardless of whether the origin is considered to be on the axis of rotation (leading to 12 N · m plus 0 N · m), at the midpoint of the handle (leading to 6 N · m plus 6 N · m), or at some point not even on the handle or an extension of the handle.

We may therefore choose the most convenient origin, and this is usually on the axis of rotation and in the plane containing the applied forces if the several forces are coplanar.

With this introduction to the concept of torque, let us now consider the torque on a differential current loop in a magnetic field **B**. The loop lies in the xy plane (Figure 9.6); the sides of the loop are parallel to the x and y axes and are of length dx and dy. The value of the magnetic field at the center of the loop is taken as $\mathbf{B}_0$. Since the loop is of differential size, the value of **B** at all points on the loop may be taken as $\mathbf{B}_0$. (Why was this not possible in the discussion of curl and divergence?) The total force on the loop is therefore zero, and we are free to choose the origin for the torque at the center of the loop.

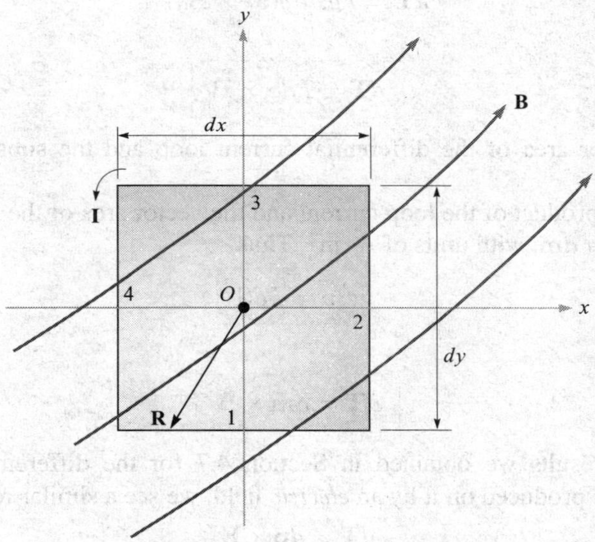

Fig. 9.6 A differential current loop in a magnetic field B. The torque on the loop is $d\mathbf{T} = I(dx\, dy\mathbf{a}_z) \times \mathbf{B}_0 = I d\mathbf{S} \times \mathbf{B}$.

The vector force on side 1 is

$$d\mathbf{F}_1 = I\, dx\, \mathbf{a}_x \times \mathbf{B}_0$$

or

$$d\mathbf{F}_1 = I\, dx(B_{0y}\mathbf{a}_z - B_{0z}\mathbf{a}_y)$$

For this side of the loop the lever arm **R** extends from the origin to the midpoint of the side, $\mathbf{R}_1 = -\frac{1}{2}dy\, \mathbf{a}_y$, and the contribution to the total torque is

$$\begin{aligned} d\mathbf{T}_1 &= \mathbf{R}_1 \times d\mathbf{F}_1 \\ &= -\tfrac{1}{2}dy\, \mathbf{a}_y \times I\, dx(B_{0y}\mathbf{a}_z - B_{0z}\mathbf{a}_y) = -\tfrac{1}{2}dx\, dy\, I B_{0y}\mathbf{a}_x \end{aligned}$$

The torque contribution on side 3 is found to be the same,

$$d\mathbf{T}_3 = \mathbf{R}_3 \times d\mathbf{F}_3 = \tfrac{1}{2}dy\,\mathbf{a}_y \times (-I\,dx\,\mathbf{a}_x \times \mathbf{B}_0)$$
$$= -\tfrac{1}{2}dx\,dy\,IB_{0y}\mathbf{a}_x = d\mathbf{T}_1$$

and

$$d\mathbf{T}_1 + d\mathbf{T}_3 = -dx\,dy\,IB_{0y}\mathbf{a}_x$$

Evaluating the torque on sides 2 and 4, we find

$$d\mathbf{T}_2 + d\mathbf{T}_4 = dx\,dy\,IB_{0x}\mathbf{a}_y$$

and the total torque is then

$$d\mathbf{T} = I\,dx\,dy(B_{0x}\mathbf{a}_y - B_{0y}\mathbf{a}_x)$$

The quantity within the parentheses may be represented by a cross product,

$$d\mathbf{T} = I\,dx\,dy(\mathbf{a}_z \times \mathbf{B}_0)$$

or

$$d\mathbf{T} = I\,d\mathbf{S} \times \mathbf{B} \tag{15}$$

where $d\mathbf{S}$ is the vector area of the differential current loop and the subscript on $\mathbf{B}_0$ has been dropped.

We now define the product of the loop current and the vector area of the loop as the differential *magnetic dipole moment* $d\mathbf{m}$, with units of $\text{A} \cdot \text{m}^2$. Thus

$$d\mathbf{m} = I\,d\mathbf{S} \tag{16}$$

and

$$d\mathbf{T} = d\mathbf{m} \times \mathbf{B} \tag{17}$$

If we extend the results we obtained in Section 4.7 for the differential *electric* dipole by determining the torque produced on it by an *electric* field, we see a similar result,

$$d\mathbf{T} = d\mathbf{p} \times \mathbf{E}$$

Equations (15) and (17) are general results which hold for differential loops of any shape, not just rectangular ones. The torque on a circular or triangular loop is also given in terms of the vector surface or the moment by (15) or (17).

Since we selected a differential current loop so that we might assume $\mathbf{B}$ was constant throughout it, it follows that the torque on a *planar* loop of any size or shape in a *uniform* magnetic field is given by the same expression,

$$\mathbf{T} = I\mathbf{S} \times \mathbf{B} = \mathbf{m} \times \mathbf{B} \tag{18}$$

We should note that the torque on the current loop always tends to turn the loop so as to align the magnetic field produced by the loop with the applied magnetic field that is causing the torque. This is perhaps the easiest way to determine the direction of the torque.

EXAMPLE 9.3

To illustrate some force and torque calculations, consider the rectangular loop shown in Figure 9.7. Calculate the torque by using $\mathbf{T} = I\mathbf{S} \times \mathbf{B}$.

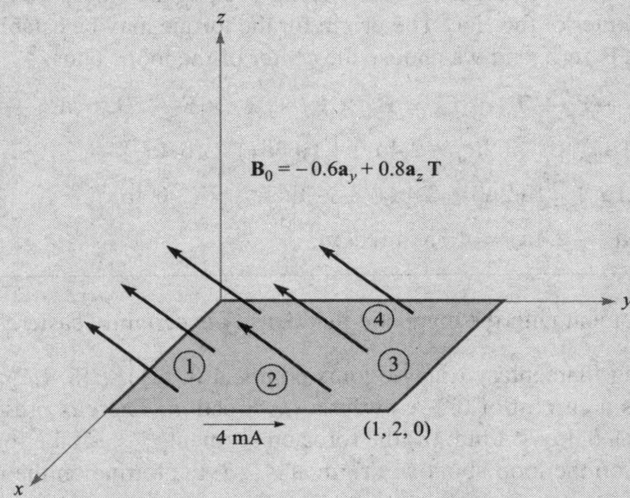

Fig. 9.7 A rectangular loop is located in a uniform magnetic flux density $\mathbf{B}_0$.

Solution. The loop has dimensions of 1 m by 2 m and lies in the uniform field $\mathbf{B}_0 = -0.6\mathbf{a}_y + 0.8\mathbf{a}_z$T. The loop current is 4 mA, a value that is sufficiently small to avoid causing any magnetic field that might affect $\mathbf{B}_0$.

We have

$$\mathbf{T} = 4 \times 10^{-3}[(1)(2)\mathbf{a}_z] \times (-0.6\mathbf{a}_y + 0.8\mathbf{a}_z) = 4.8\mathbf{a}_x \text{ mN} \cdot \text{m}$$

Thus, the loop tends to rotate about an axis parallel to the positive x axis. The small magnetic field produced by the 4 mA loop current tends to line up with $\mathbf{B}_0$.

EXAMPLE 9.4

Now let us find the torque once more, this time by calculating the total force and torque contribution for each side.

Solution. On side 1 we have

$$\mathbf{F}_1 = I\mathbf{L}_1 \times \mathbf{B}_0 = 4 \times 10^{-3}(1\mathbf{a}_x) \times (-0.6\mathbf{a}_y + 0.8\mathbf{a}_z)$$
$$= -3.2\mathbf{a}_y - 2.4\mathbf{a}_z \text{ mN}$$

On side 3 we obtain the negative of this result,

$$\mathbf{F}_3 = 3.2\mathbf{a}_y + 2.4\mathbf{a}_z \text{ mN}$$

Next we attack side 2:

$$\mathbf{F}_2 = I\mathbf{L}_2 \times \mathbf{B}_0 = 4 \times 10^{-3}(2\mathbf{a}_y) \times (-0.6\mathbf{a}_y + 0.8\mathbf{a}_z) = 6.4\mathbf{a}_x \text{ mN}$$

with side 4 again providing the negative of this result,

$$\mathbf{F}_4 = -6.4\mathbf{a}_x \text{ mN}$$

Since these forces are distributed uniformly along each of the sides, we treat each force as if it were applied at the center of the side. The origin for the torque may be established anywhere since the sum of the forces is zero, and we choose the center of the loop. Thus,

$$\begin{aligned}\mathbf{T} &= \mathbf{T}_1 + \mathbf{T}_2 + \mathbf{T}_3 + \mathbf{T}_4 = \mathbf{R}_1 \times \mathbf{F}_1 + \mathbf{R}_2 \times \mathbf{F}_2 + \mathbf{R}_3 \times \mathbf{F}_3 + \mathbf{R}_4 \times \mathbf{F}_4 \\ &= (-1\mathbf{a}_y) \times (-3.2\mathbf{a}_y - 2.4\mathbf{a}_z) + (0.5\mathbf{a}_x) \times (6.4\mathbf{a}_x) \\ &\quad + (1\mathbf{a}_y) \times (3.2\mathbf{a}_y + 2.4\mathbf{a}_z) + (-0.5\mathbf{a}_x) \times (-6.4\mathbf{a}_x) \\ &= 2.4\mathbf{a}_x + 2.4\mathbf{a}_x = 4.8\mathbf{a}_x \text{ mN} \cdot \text{m}\end{aligned}$$

Crossing the loop moment with the magnetic flux density is certainly easier.

D9.5 A conducting filamentary triangle joins points $A(3,1,1)$, $B(5,4,2)$, and $C(1,2,4)$. The segment AB carries a current of 0.2 A in the $\mathbf{a}_{AB}$ direction. There is present a magnetic field $\mathbf{B} = 0.2\mathbf{a}_x - 0.1\mathbf{a}_y + 0.3\mathbf{a}_z$ T. Find: (*a*) the force on segment BC; (*b*) the force on the triangular loop; (*c*) the torque on the loop about an origin at A; (*d*) the torque on the loop about an origin at C.

Ans. $-0.08\mathbf{a}_x + 0.32\mathbf{a}_y + 0.16\mathbf{a}_z$ N; 0; $-0.16\mathbf{a}_x - 0.08\mathbf{a}_y + 0.08\mathbf{a}_z$ N·m; $-0.16\mathbf{a}_x - 0.08\mathbf{a}_y + 0.08\mathbf{a}_z$ N·m

9.5 THE NATURE OF MAGNETIC MATERIALS

We are now in a position to combine our knowledge of the action of a magnetic field on a current loop with a simple model of an atom and obtain some appreciation of the difference in behavior of various types of materials in magnetic fields.

Although accurate quantitative results can only be predicted through the use of quantum theory, the simple atomic model which assumes that there is a central positive nucleus surrounded by electrons in various circular orbits yields reasonable quantitative results and provides a satisfactory qualitative theory. An electron in an orbit is analogous to a small current loop (in which the current is directed oppositely to the direction of electron travel) and as such experiences a torque in an external magnetic field, the torque tending to align the magnetic field produced by the orbiting electron with the external magnetic field. If there were no other magnetic moments to consider, we would then conclude that all the orbiting electrons in the material would shift in such a way as to add their magnetic fields to the applied field, and thus that the resultant magnetic field at any point in the material would be greater than it would be at that point if the material were not present.

A second moment, however, is attributed to *electron spin*. Although it is tempting to model this phenomenon by considering the electron as spinning about its own axis and thus generating a magnetic dipole moment, satisfactory quantitative results are not obtained from such a theory. Instead, it is necessary to digest the mathematics of relativistic quantum theory to show that an electron may have a spin magnetic moment of about $\pm 9 \times 10^{-24}$ A · m^2; the plus and minus signs indicate that alignment aiding or opposing an external magnetic field is possible. In an atom with many electrons present, only the spins of those electrons in shells which are not completely filled will contribute to a magnetic moment for the atom.

A third contribution to the moment of an atom is caused by *nuclear spin*. Although this factor provides a negligible effect on the overall magnetic properties of materials, it is the basis of the nuclear magnetic resonance imaging (MRI) procedure provided by many of the larger hospitals.

Thus each atom contains many different component moments, and their combination determines the magnetic characteristics of the material and provides its general magnetic classification. We shall describe briefly six different types of material: diamagnetic, paramagnetic, ferromagnetic, antiferromagnetic, ferrimagnetic, and superparamagnetic.

Let us first consider those atoms in which the small magnetic fields produced by the motion of the electrons in their orbits and those produced by the electron spin combine to produce a net field of zero. Note that we are considering here the fields produced by the electron motion itself in the absence of any external magnetic field; we might also describe this material as one in which the permanent magnetic moment $\mathbf{m}_0$ of each atom is zero. Such a material is termed *diamagnetic*. It would seem, therefore, that an external magnetic field would produce no torque on the atom, no realignment of the dipole fields, and consequently an internal magnetic field that is the same as the applied field. With an error that only amounts to about one part in a hundred thousand, this is correct.

Let us select an orbiting electron whose moment $\mathbf{m}$ is in the same direction as the applied field $\mathbf{B}_0$ (Figure 9.8). The magnetic field produces an outward force on the orbiting electron. Since the orbital radius is quantized and cannot change, the inward Coulomb force of attraction is also unchanged. The force unbalance created by the outward magnetic force must therefore be compensated for by a reduced orbital velocity. Hence, the orbital moment decreases, and a smaller internal field results.

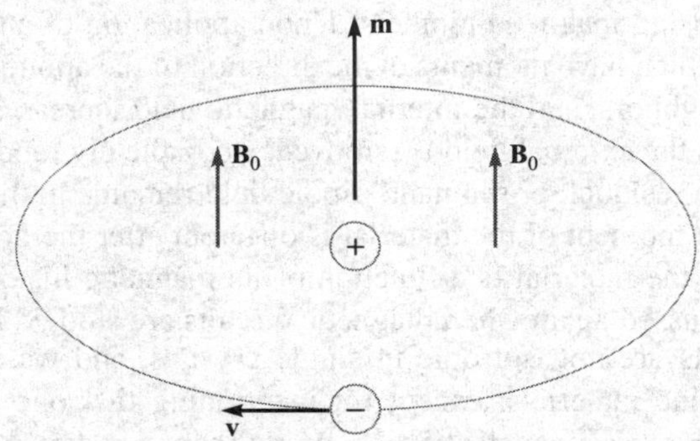

Fig. 9.8 An orbiting electron is shown having a magnetic moment m in the same direction as an applied field $\mathbf{B}_0$.

If we had selected an atom for which $\mathbf{m}$ and $\mathbf{B}_0$ were opposed, the magnetic force would be inward, the velocity would increase, the orbital moment would increase, and greater cancellation of $\mathbf{B}_0$ would occur. Again a smaller internal field would result.

Metallic bismuth shows a greater diamagnetic effect than most other diamagnetic materials, among which are hydrogen, helium, the other "inert" gases, sodium chloride, copper, gold, silicon, germanium, graphite, and sulfur. We should also realize that the diamagnetic effect is present in all materials, because it arises from an interaction of the external magnetic field with every orbiting electron; however, it is overshadowed by other effects in the materials we shall consider next.

Now let us discuss an atom in which the effects of the electron spin and orbital motion do not quite cancel. The atom as a whole has a small magnetic moment, but the random orientation of the atoms in a larger sample produces an *average* magnetic moment of zero. The material shows no magnetic effects in the absence of an external field. When an external field is applied, however, there is a small torque on each atomic moment, and these moments tend to become aligned with the external field. This alignment acts to increase the value of **B** within the material over the external value. However, the diamagnetic effect is still operating on the orbiting electrons and may counteract the increase. If the net result is a decrease in **B**, the material is still called diamagnetic. However, if there is an increase in **B**, the material is termed *paramagnetic*. Potassium, oxygen, tungsten, and the rare earth elements and many of their salts, such as erbium chloride, neodymium oxide, and yttrium oxide, one of the materials used in masers, are examples of paramagnetic substances.

The remaining four classes of material, ferromagnetic, antiferromagnetic, ferrimagnetic, and superparamagnetic, all have strong atomic moments. Moreover, the interaction of adjacent atoms causes an alignment of the magnetic moments of the atoms in either an aiding or exactly opposing manner.

Animations

In *ferromagnetic* materials each atom has a relatively large dipole moment, caused primarily by uncompensated electron spin moments. Interatomic forces cause these moments to line up in a parallel fashion over regions containing a large number of atoms. These regions are called *domains,* and they may have a variety of shapes and sizes ranging from one micrometer to several centimeters, depending on the size, shape, material, and magnetic history of the sample. Virgin ferromagnetic materials will have domains which each have a strong magnetic moment; the domain moments, however, vary in direction from domain to domain. The overall effect is therefore one of cancellation, and the material as a whole has no magnetic moment. Upon application of an external magnetic field, however, those domains which have moments in the direction of the applied field increase their size at the expense of their neighbors, and the internal magnetic field increases greatly over that of the external field alone. When the external field is removed, a completely random domain alignment is not usually attained, and a residual, or remnant, dipole field remains in the macroscopic structure. The fact that the magnetic moment of the material is different after the field has been removed, or that the magnetic state of the material is a function of its magnetic history, is called *hysteresis,* a subject which will be discussed again when magnetic circuits are studied in Section 9.8.

Ferromagnetic materials are not isotropic in single crystals, and we shall therefore limit our discussion to polycrystalline materials, except for mentioning that one of the characteristics of anisotropic magnetic materials is magnetostriction, or the change in dimensions of the crystal when a magnetic field is impressed on it.

The only elements that are ferromagnetic at room temperature are iron, nickel, and cobalt, and they lose all their ferromagnetic characteristics above a temperature called the Curie temperature, which is 1043 K (770°C) for iron. Some alloys of these metals with each other and with other metals are also ferromagnetic, as for example alnico, an aluminum-nickel-cobalt alloy with a small amount of copper. At lower temperatures some of the rare earth elements, such as gadolinium and dysprosium, are ferromagnetic. It is also interesting that some alloys of nonferromagnetic metals are ferromagnetic, such as bismuth-manganese and copper-manganese-tin.

In *antiferromagnetic* materials, the forces between adjacent atoms cause the atomic moments to line up in an antiparallel fashion. The net magnetic moment is zero, and antiferromagnetic materials are affected only slightly by the presence of an external magnetic field. This effect was first discovered in manganese oxide, but several hundred antiferromagnetic materials have been identified since then. Many oxides, sulfides, and chlorides are included, such as nickel oxide (NiO), ferrous sulfide (FeS), and cobalt chloride ($CoCl_2$). Antiferromagnetism is only present at relatively low temperatures, often well below room temperature. The effect is not of engineering importance at present.

The *ferrimagnetic* substances also show an antiparallel alignment of adjacent atomic moments, but the moments are not equal. A large response to an external magnetic field therefore occurs, although not as large as that in ferromagnetic materials. The most important group of ferrimagnetic materials are the *ferrites*, in which the conductivity is low, several orders of magnitude less than that of semiconductors. The fact that these substances have greater resistance than the ferromagnetic materials results in much smaller induced currents in the material when alternating fields are applied, as for example in transformer cores which operate at the higher frequencies. The reduced currents (eddy currents) lead to lower ohmic losses in the transformer core. The iron oxide magnetite (Fe_3O_4), a nickel-zinc ferrite ($Ni_{1/2}Zn_{1/2}Fe_2O_4$), and a nickel ferrite ($NiFe_2O_4$) are examples of this class of materials. Ferrimagnetism also disappears above the Curie temperature.

Table 9.1 Characteristics of magnetic materials

Classification	Magnetic Moments	**B** Values	Comments
Diamagnetic	$\mathbf{m}_{\text{orb}} + \mathbf{m}_{\text{spin}} = 0$	$B_{\text{int}} < B_{\text{appl}}$	$B_{\text{int}} \doteq B_{\text{appl}}$
Paramagnetic	$\mathbf{m}_{\text{orb}} + \mathbf{m}_{\text{spin}} = \text{small}$	$B_{\text{int}} > B_{\text{appl}}$	$B_{\text{int}} \doteq B_{\text{appl}}$
Ferromagnetic	$\lvert\mathbf{m}_{\text{spin}}\rvert \gg \lvert\mathbf{m}_{\text{orb}}\rvert$	$B_{\text{int}} \gg B_{\text{appl}}$	Domains
Antiferromagnetic	$\lvert\mathbf{m}_{\text{spin}}\rvert \gg \lvert\mathbf{m}_{\text{orb}}\rvert$	$B_{\text{int}} \doteq B_{\text{appl}}$	Adjacent moments oppose
Ferrimagnetic	$\lvert\mathbf{m}_{\text{spin}}\rvert \gg \lvert\mathbf{m}_{\text{orb}}\rvert$	$B_{\text{int}} > B_{\text{appl}}$	Unequal adjacent moments oppose; low σ
Superparamagnetic	$\lvert\mathbf{m}_{\text{spin}}\rvert \gg \lvert\mathbf{m}_{\text{orb}}\rvert$	$B_{\text{int}} > B_{\text{appl}}$	Nonmagnetic matrix; recording tapes

Superparamagnetic materials are composed of an assembly of ferromagnetic particles in a nonferromagnetic matrix. Although domains exist within the individual particles, the domain walls cannot penetrate the intervening matrix material to the adjacent particle. An important example is the magnetic tape used in audiotape or videotape recorders.

Table 9.1 summarizes the characteristics of the six types of magnetic materials we have discussed.

9.6 MAGNETIZATION AND PERMEABILITY

To place our description of magnetic materials on a more quantitative basis, we shall now devote a page or so to showing how the magnetic dipoles act as a distributed source for the magnetic field. Our result will be an equation that looks very much like Ampère's circuital law, $\oint \mathbf{H} \cdot d\mathbf{L} = I$.

The current, however, will be the movement of *bound* charges (orbital electrons, electron spin, and nuclear spin), and the field, which has the dimensions of **H**, will be called the magnetization **M**. The current produced by the bound charges is called a *bound current* or *Amperian current*.

Let us begin by defining the magnetization **M** in terms of the magnetic dipole moment **m**. The bound current I_b circulates about a path enclosing a differential area $d\mathbf{S}$, establishing a dipole moment (A · m^2),

$$\mathbf{m} = I_b d\mathbf{S}$$

If there are n magnetic dipoles per unit volume and we consider a volume $\Delta\nu$, then the total magnetic dipole moment is found by the vector sum

$$\mathbf{m}_{\text{total}} = \sum_{i=1}^{n\Delta\nu} \mathbf{m}_i \tag{19}$$

Each of the $\mathbf{m}_i$ may be different. Next, we define the *magnetization* **M** as the *magnetic dipole moment per unit volume,*

$$\mathbf{M} = \lim_{\Delta\nu \to 0} \frac{1}{\Delta\nu} \sum_{i=1}^{n\Delta\nu} \mathbf{m}_i$$

and see that its units must be the same as for **H**, amperes per meter.

Now let us consider the effect of some alignment of the magnetic dipoles as the result of the application of a magnetic field. We shall investigate this alignment along a closed path, a short portion of which is shown in Figure 9.9. The figure shows several magnetic moments **m** that make an angle θ with the element of path $d\mathbf{L}$; each moment consists of a bound current I_b circulating about an area $d\mathbf{S}$. We are therefore considering a small volume, $dS \cos\theta dL$, or $d\mathbf{S} \cdot d\mathbf{L}$, within which there are $n d\mathbf{S} \cdot d\mathbf{L}$ magnetic dipoles. In changing from a random orientation to this partial alignment, the bound current crossing the surface enclosed by the path (to our left as we travel in the $\mathbf{a}_L$ direction in Figure 9.9) has increased by I_b for each of the $n d\mathbf{S} \cdot d\mathbf{L}$ dipoles. Thus

$$dI_b = nI_b d\mathbf{S} \cdot d\mathbf{L} = \mathbf{M} \cdot d\mathbf{L} \tag{20}$$

and within an entire closed contour,

$$I_b = \oint \mathbf{M} \cdot d\mathbf{L} \tag{21}$$

Equation (21) merely says that if we go around a closed path and find dipole moments going our way more often than not, there will be a corresponding current composed of, for example, orbiting electrons crossing the interior surface.

This last expression has some resemblance to Ampère's circuital law, and we may now generalize the relationship between **B** and **H** so that it applies to media other than free space. Our present discussion is based on the forces and torques on differential current loops in a **B** field, and we therefore take **B** as our fundamental quantity and seek an improved definition of **H**. We thus write Ampère's circuital law in terms of the *total* current, bound plus free,

$$\oint \frac{\mathbf{B}}{\mu_0} \cdot d\mathbf{L} = I_T \tag{22}$$

where

$$I_T = I_b + I$$

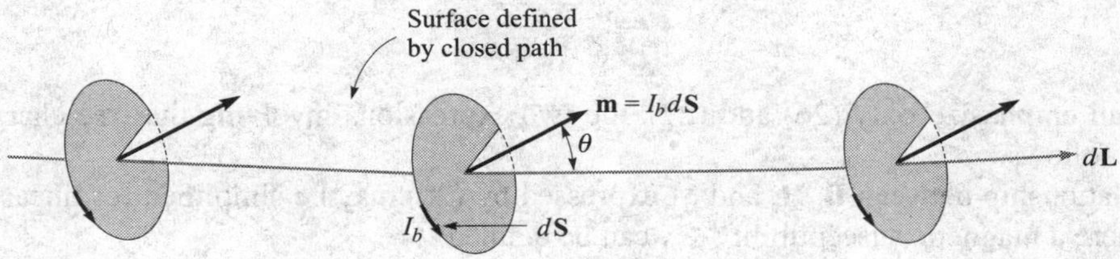

Fig. 9.9 A section $d\mathbf{L}$ of a closed path along which magnetic dipoles have been partially aligned by some external magnetic field. The alignment has caused the bound current crossing the surface defined by the closed path to increase by $nI_b d\mathbf{S} \cdot d\mathbf{L}$ amperes.

and I is the total *free* current enclosed by the closed path. Note that the free current appears without subscript since it is the most important type of current and will be the only current appearing in Maxwell's equations.

Combining these last three equations, we obtain an expression for the free current enclosed,

$$I = I_T - I_b = \oint \left(\frac{\mathbf{B}}{\mu_0} - \mathbf{M} \right) \cdot d\mathbf{L} \tag{23}$$

We may now define $\mathbf{H}$ in terms of $\mathbf{B}$ and $\mathbf{M}$,

$$\mathbf{H} = \frac{\mathbf{B}}{\mu_0} - \mathbf{M} \tag{24}$$

and we see that $\mathbf{B} = \mu_0 \mathbf{H}$ in free space where the magnetization is zero. This relationship is usually written in a form that avoids fractions and minus signs:

$$\mathbf{B} = \mu_0 (\mathbf{H} + \mathbf{M}) \tag{25}$$

We may now use our newly defined $\mathbf{H}$ field in (23),

$$I = \oint \mathbf{H} \cdot d\mathbf{L} \tag{26}$$

obtaining Ampère's circuital law in terms of the free currents.

Utilizing the several current densities, we have

$$I_b = \int_S \mathbf{J}_b \cdot d\mathbf{S}$$

$$I_T = \int_S \mathbf{J}_T \cdot d\mathbf{S}$$

$$I = \int_S \mathbf{J} \cdot d\mathbf{S}$$

With the help of Stokes' theorem, we may therefore transform (21), (26), and (22) into the equivalent curl relationships:

$$\nabla \times \mathbf{M} = \mathbf{J}_b$$

$$\nabla \times \frac{\mathbf{B}}{\mu_0} = \mathbf{J}_T$$

$$\nabla \times \mathbf{H} = \mathbf{J} \tag{27}$$

We shall emphasize only (26) and (27), the two expressions involving the free charge, in the work that follows.

The relationship between **B**, **H**, and **M** expressed by (25) may be simplified for linear isotropic media where a magnetic susceptibility χ_m can be defined:

$$\mathbf{M} = \chi_m \mathbf{H} \tag{28}$$

Thus we have

$$\mathbf{B} = \mu_0(\mathbf{H} + \chi_m \mathbf{H}) = \mu_0 \mu_r \mathbf{H}$$

where

$$\mu_r = 1 + \chi_m \tag{29}$$

is defined as the *relative permeability* μ_r. We next define the *permeability* μ:

$$\mu = \mu_0 \mu_r \tag{30}$$

and this enables us to write the simple relationship between **B** and **H**,

$$\mathbf{B} = \mu \mathbf{H} \tag{31}$$

EXAMPLE 9.5

Given a ferrite material which we shall specify to be operating in a linear mode with $B = 0.05$ T, let us assume $\mu_r = 50$, and calculate values for χ_m, M, and H.

Solution. Since $\mu_r = 1 + \chi_m$, we have

$$\chi_m = \mu_r - 1 = 49$$

Also,

$$B = \mu_r \mu_0 H$$

and

$$H = \frac{0.05}{50 \times 4\pi \times 10^{-7}} = 796 \text{ A/m}$$

The magnetization is $\chi_m H$, or 390 00 A/m. The alternate ways of relating B and H are, first,

$$B = \mu_0(H + M)$$

or

$$0.05 = 4\pi \times 10^{-7}(796 + 39\,000)$$

showing that Amperian currents produce 49 times the magnetic field intensity that the free charges do; and second,

$$B = \mu_r \mu_0 H$$

or

$$0.05 = 50 \times 4\pi \times 10^{-7} \times 796$$

where we utilize a relative permeability of 50 and let this quantity account completely for the notion of the bound charges. We shall emphasize the latter interpretation in the chapters that follow.

The first two laws that we investigated for magnetic fields were the Biot-Savart law and Ampère's circuital law. Both were restricted to free space in their application. We may now extend their use to any homogeneous, linear, isotropic magnetic material that may be described in terms of a relative permeability μ_r.

Just as we found for anisotropic dielectric materials, the permeability of an anisotropic magnetic material must be given as a 3×3 matrix, while **B** and **H** are both 3×1 matrices. We have

$$B_x = \mu_{xx} H_x + \mu_{xy} H_y + \mu_{xz} H_z$$
$$B_y = \mu_{yx} H_x + \mu_{yy} H_y + \mu_{yz} H_z$$
$$B_z = \mu_{zx} H_x + \mu_{zy} H_y + \mu_{zz} H_z$$

For anisotropic materials, then, $\mathbf{B} = \mu\mathbf{H}$ is a matrix equation; however $\mathbf{B} = \mu_0(\mathbf{H} + \mathbf{M})$ remains valid, although **B**, **H**, and **M** are no longer parallel in general. The most common anisotropic magnetic material is a single ferromagnetic crystal, although thin magnetic films also exhibit anisotropy. Most applications of ferromagnetic materials, however, involve polycrystalline arrays that are much easier to make.

Our definitions of susceptibility and permeability also depend on the assumption of linearity. Unfortunately, this is true only in the less interesting paramagnetic and diamagnetic materials for which the relative permeability rarely differs from unity by more than one part in a thousand. Some typical values of the susceptibility for diamagnetic materials are hydrogen, -2×10^{-5}; copper, -0.9×10^{-5}; germanium, -0.8×10^{-5}; silicon, -0.3×10^{-5}; and graphite, -12×10^{-5}. Several representative paramagnetic susceptibilities are oxygen, 2×10^{-6}; tungsten, 6.8×10^{-5}; ferric oxide (Fe_2O_3), 1.4×10^{-3}; and yttrium oxide (Y_2O_3), 0.53×10^{-6}. If we simply take the ratio of B to $\mu_0 H$ as the relative permeability of a ferromagnetic material, typical values of μ_r would range from 10 to 100 000. Diamagnetic, paramagnetic, and antiferromagnetic materials are commonly said to be nonmagnetic.

D9.6 Find the magnetization in a magnetic material where: (*a*) $\mu = 1.8 \times 10^{-5}$ H/m and $H = 120$ A/m; (*b*) $\mu_r = 22$, there are 8.3×10^{28} atoms/m^3, and each atom has a dipole moment of 4.5×10^{-27} A $\cdot$ m^2; (*c*) $B = 300\,\mu$T and $\chi_m = 15$.

Ans. 1599 A/m; 374 A/m; 224 A/m

D9.7 The magnetization in a magnetic material for which $\chi_m = 8$ is given in a certain region as $150z^2\mathbf{a}_x$ A/m. At $z = 4$ cm, find the magnitude of: (*a*) $\mathbf{J}_T$; (*b*) J; (*c*) $\mathbf{J}_b$.

Ans. 13.5 A/m^2; 1.5 A/m^2; 12 A/m^2

9.7 MAGNETIC BOUNDARY CONDITIONS

We should have no difficulty in arriving at the proper boundary conditions to apply to **B**, **H**, and **M** at the interface between two different magnetic materials, for we have solved similar problems for both conducting materials and dielectrics. We need no new techniques.

Figure 9.10 shows a boundary between two isotropic homogeneous linear materials with permeabilities μ_1 and μ_2. The boundary condition on the normal components is determined by allowing the surface to cut a small cylindrical gaussian surface. Applying Gauss's law for the magnetic field from Section 8.5,

$$\oint_S \mathbf{B} \cdot d\mathbf{S} = 0$$

we find that

$$B_{N1}\Delta S - B_{N2}\Delta S = 0$$

or

$$B_{N2} = B_{N1} \tag{32}$$

Thus

$$H_{N2} = \frac{\mu_1}{\mu_2} H_{N1} \tag{33}$$

The normal component of **B** is continuous, but the normal component of **H** is discontinuous by the ratio μ_1/μ_2.

The relationship between the normal components of **M**, of course, is fixed once the relationship between the normal components of **H** is known. For linear magnetic materials, the result is written simply as

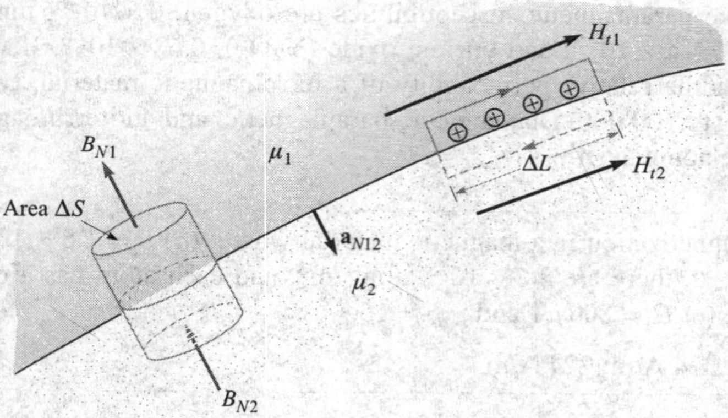

Fig. 9.10 A gaussian surface and a closed path are constructed at the boundary between media 1 and 2, having permeabilities of μ_1 and μ_2, respectively. From this we determine the boundary conditions $B_{N1} = B_{N2}$ and $H_{t1} - H_{t2} = K$, the component of the surface current density directed into the page.

$$M_{N2} = \chi_{m2}\frac{\mu_1}{\mu_2}H_{N1} = \frac{\chi_{m2}\mu_1}{\chi_{m1}\mu_2}M_{N1} \tag{34}$$

Next, Ampère's circuital law

$$\oint \mathbf{H}\cdot d\mathbf{L} = I$$

is applied about a small closed path in a plane normal to the boundary surface, as shown to the right in Figure 9.10. Taking a clockwise trip around the path, we find that

$$H_{t1}\Delta L - H_{t2}\Delta L = K\Delta L$$

where we assume that the boundary may carry a surface current **K** whose component normal to the plane of the closed path is K. Thus

$$H_{t1} - H_{t2} = K \tag{35}$$

The directions are specified more exactly by using the cross product to identify the tangential components,

$$(\mathbf{H}_1 - \mathbf{H}_2) \times \mathbf{a}_{N12} = \mathbf{K}$$

where $\mathbf{a}_{N12}$ is the unit normal at the boundary directed from region 1 to region 2. An equivalent formulation in terms of the vector tangential components may be more convenient for **H**:

$$\mathbf{H}_{t1} - \mathbf{H}_{t2} = \mathbf{a}_{N12} \times \mathbf{K}$$

For tangential **B**, we have

$$\frac{B_{t1}}{\mu_1} - \frac{B_{t2}}{\mu_2} = K \tag{36}$$

The boundary condition on the tangential component of the magnetization for linear materials is therefore

$$M_{t2} = \frac{\chi_{m2}}{\chi_{m1}}M_{t1} - \chi_{m2}K \tag{37}$$

The last three boundary conditions on the tangential components are much simpler, of course, if the surface current density is zero. This is a free current density, and it must be zero if neither material is a conductor.

EXAMPLE 9.6

To illustrate these relationships with an example, let us assume that $\mu = \mu_1 = 4\,\mu$H/m in region 1 where $z > 0$, while $\mu_2 = 7\,\mu$H/m in region 2 wherever $z < 0$. Moreover, let $\mathbf{K} = 80\mathbf{a}_x$ A/m on the surface $z = 0$. We establish a field, $\mathbf{B}_1 = 2\mathbf{a}_x - 3\mathbf{a}_y + \mathbf{a}_z$ mT, in region 1 and seek the value of $\mathbf{B}_2$.

Solution. The normal component of $\mathbf{B}_1$ is

$$\mathbf{B}_{N1} = (\mathbf{B}_1 \cdot \mathbf{a}_{N12})\mathbf{a}_{N12} = [(2\mathbf{a}_x - 3\mathbf{a}_y + \mathbf{a}_z)\cdot(-\mathbf{a}_z)](-\mathbf{a}_z) = \mathbf{a}_z \text{ mT}$$

Thus,

$$\mathbf{B}_{N2} = \mathbf{B}_{N1} = \mathbf{a}_z \text{ mT}$$

We next determine the tangential components:

$$\mathbf{B}_{t1} = \mathbf{B}_1 - \mathbf{B}_{N1} = 2\mathbf{a}_x - 3ay \text{ mT}$$

and

$$\mathbf{H}_{t1} = \frac{\mathbf{B}_{t1}}{\mu_1} = \frac{(2\mathbf{a}_x - 3\mathbf{a}_y)10^{-3}}{4 \times 10^{-6}} = 500\mathbf{a}_x - 750\mathbf{a}_y \text{ A/m}$$

Thus,

$$\mathbf{H}_{t2} = \mathbf{H}_{t1} - \mathbf{a}_{N12} \times \mathbf{K} = 500\mathbf{a}_x - 750\mathbf{a}_y - (-\mathbf{a}_z) \times 80\mathbf{a}_x$$
$$= 500\mathbf{a}_x - 750\mathbf{a}_y + 80\mathbf{a}_y = 500\mathbf{a}_x - 670\mathbf{a}_y \text{ A/m}$$

and

$$\mathbf{B}_{t2} = \mu_2 \mathbf{H}_{t2} = 7 \times 10^{-6}(500\mathbf{a}_x - 670\mathbf{a}_y) = 3.5\mathbf{a}_x - 4.69\mathbf{a}_y \text{ mT}$$

Therefore,

$$\mathbf{B}_2 = \mathbf{B}_{N2} + \mathbf{B}_{t2} = 3.5\mathbf{a}_x - 4.69\mathbf{a}_y + \mathbf{a}_z \text{ mT}$$

D9.8 Let the permittivity be $5\,\mu$H/m in region A where $x < 0$, and $20\,\mu$H/m in region B where $x > 0$. If there is a surface current density $\mathbf{K} = 150\mathbf{a}_y - 200\mathbf{a}_z$ A/m at $x = 0$, and if $H_A = 300\mathbf{a}_x - 400\mathbf{a}_y + 500\mathbf{a}_z$ A/m, find: (*a*) $|\mathbf{H}_{tA}|$; (*b*) $|\mathbf{H}_{NA}|$; (*c*) $|\mathbf{H}_{tB}|$; (*d*) $|\mathbf{H}_{NB}|$.

Ans. 640 A/m; 300 A/m; 695 A/m; 75 A/m

9.8 THE MAGNETIC CIRCUIT

In this section we shall digress briefly to discuss the fundamental techniques involved in solving a class of magnetic problems known as magnetic circuits. As we shall see shortly, the name arises from the great similarity to the dc-resistive-circuit analysis with which it is assumed we are all familiar. The only important difference lies in the nonlinear nature of the ferromagnetic portions of the magnetic circuit; the methods which must be adopted are similar to those required in nonlinear electric circuits which contain diodes, thermistors, incandescent filaments, and other nonlinear elements.

As a convenient starting point, let us identify those field equations upon which resistive circuit analysis is based. At the same time we shall point out or derive the analogous equations for the magnetic circuit. We begin with the electrostatic potential and its relationship to electric field intensity,

$$\mathbf{E} = -\nabla V \tag{38a}$$

The scalar magnetic potential has already been defined, and its analogous relation to the magnetic field intensity is

$$\mathbf{H} = -\nabla V_m \tag{38b}$$

In dealing with magnetic circuits, it is convenient to call V_m the *magnetomotive force*, or mmf, and we shall acknowledge the analogy to the electromotive force, or emf, by doing so. The units of the mmf are, of course, amperes, but it is customary to recognize that coils with many turns are often employed by using the term "ampere-turns." Remember that no current may flow in any region in which V_m is defined.

The electric potential difference between points A and B may be written as

$$V_{AB} = \int_A^B \mathbf{E} \cdot d\mathbf{L} \tag{39a}$$

and the corresponding relationship between the mmf and the magnetic field intensity,

$$V_{mAB} = \int_A^B \mathbf{H} \cdot d\mathbf{L} \tag{39b}$$

was developed in Chapter 8, where we learned that the path selected must not cross the chosen barrier surface.

Ohm's law for the electric circuit has the point form

$$\mathbf{J} = \sigma \mathbf{E} \tag{40a}$$

and we see that the magnetic flux density will be the analog of the current density,

$$\mathbf{B} = \mu \mathbf{H} \tag{40b}$$

To find the total current, we must integrate:

$$I = \int_S \mathbf{J} \cdot d\mathbf{S} \tag{41a}$$

A corresponding operation is necessary to determine the total magnetic flux flowing through the cross section of a magnetic circuit:

$$\Phi = \int_S \mathbf{B} \cdot d\mathbf{S} \tag{41b}$$

We then defined resistance as the ratio of potential difference and current, or

$$V = IR \tag{42a}$$

and we shall now define *reluctance* as the ratio of the magnetomotive force to the total flux; thus

$$V_m = \Phi \Re \tag{42b}$$

where reluctance is measured in ampere-turns per weber (A · t/Wb). In resistors which are made of a linear isotropic homogeneous material of conductivity σ and have a uniform cross section of area S and length d, the total resistance is

$$R = \frac{d}{\sigma S} \tag{43a}$$

If we are fortunate enough to have such a linear isotropic homogeneous magnetic material of length d and uniform cross section S, then the total reluctance is

$$\Re = \frac{d}{\mu S} \tag{43b}$$

The only such material to which we shall commonly apply this relationship is air.

Finally, let us consider the analog of the source voltage in an electric circuit. We know that the closed line integral of **E** is zero,

$$\oint \mathbf{E} \cdot d\mathbf{L} = 0$$

In other words, Kirchhoff's voltage law states that the rise in potential through the source is exactly equal to the fall in potential through the load. The expression for magnetic phenomena takes on a slightly different form,

$$\oint \mathbf{H} \cdot d\mathbf{L} = I_{\text{total}}$$

for the closed line integral is not zero. Since the total current linked by the path is usually obtained by allowing a current I to flow through an N-turn coil, we may express this result as

$$\oint \mathbf{H} \cdot d\mathbf{L} = NI \tag{44}$$

In an electric circuit the voltage source is a part of the closed path; in the magnetic circuit the current-carrying coil will surround or link the magnetic circuit. In tracing a magnetic circuit, we shall not be able to identify a pair of terminals at which the magnetomotive force is applied. The analogy is closer here to a pair of coupled circuits in which induced voltages exist (and in which we shall see in Chapter 10 that the closed line integral of **E** is also not zero).

Let us try out some of these ideas on a simple magnetic circuit. In order to avoid the complications of ferromagnetic materials at this time, we shall assume that we have an air-core toroid with 500 turns, a cross-sectional area of 6 cm^2, a mean radius of 15 cm, and a coil current of 4 A. As we already know, the magnetic field is confined to the interior of the toroid, and if we consider the closed path of our magnetic circuit along the mean radius, we link 2000 A · t,

$$V_{m,\text{ source}} = 2000 \text{ A} \cdot \text{t}$$

Although the field in the toroid is not quite uniform, we may assume that it is for all practical purposes and calculate the total reluctance of the circuit as

$$\mathfrak{R} = \frac{d}{\mu S} = \frac{2\pi(0.15)}{4\pi 10^{-7} \times 6 \times 10^{-4}} = 1.25 \times 10^9 \text{ A·t/Wb}$$

Thus,

$$\Phi = \frac{V_{m,S}}{\mathfrak{R}} = \frac{2000}{1.25 \times 10^9} = 1.6 \times 10^{-6} \text{ Wb}$$

This value of the total flux is in error by less than $\frac{1}{4}$ percent, in comparison with the value obtained when the exact distribution of flux over the cross section is used.

Hence,

$$B = \frac{\Phi}{S} = \frac{1.6 \times 10^{-6}}{6 \times 10^{-4}} = 2.67 \times 10^{-3} \text{ T}$$

and finally,

$$H = \frac{B}{\mu} = \frac{2.67 \times 10^{-3}}{4\pi 10^{-7}} = 2120 \text{ A·t/m}$$

As a check, we may apply Ampère's circuital law directly in this symmetrical problem,

$$H_\phi 2\pi r = NI$$

and obtain

$$H_\phi = \frac{NI}{2\pi r} = \frac{500 \times 4}{6.28 \times 0.15} = 2120 \text{ A/m}$$

at the mean radius.

Our magnetic circuit in this example does not give us any opportunity to find the mmf across different elements in the circuit, for there is only one type of material. The analogous electric circuit is, of course, a single source and a single resistor. We could make it look just as long as the preceding analysis, however, if we found the current density, the electric field intensity, the total current, the resistance, and the source voltage.

More interesting and more practical problems arise when ferromagnetic materials are present in the circuit. Let us begin by considering the relationship between B and H in such a material. We may assume that we are establishing a curve of B versus H for a sample of ferromagnetic material which is completely demagnetized; both B and H are zero. As we begin to apply an mmf, the flux density also rises, but not linearly, as the experimental data of Figure 9.11 show near the origin. After H reaches a value of about 100 A · t/m, the flux density rises more slowly and begins to saturate when H is several hundred A · t/m. Having reached partial saturation, let us now turn to Figure 9.12, where we may continue our experiment at point x by reducing H. As we do so, the effects of hysteresis begin to show, and we do not retrace our original curve. Even after H is zero, $B = B_r$, the remnant flux density. As H is reversed, then brought back to zero, and the complete cycle traced several times, the hysteresis loop of Figure 9.12 is obtained. The mmf required to reduce the flux density to zero is identified as H_c, the coercive "force." For smaller maximum values of H, smaller hysteresis loops are obtained, and the locus of the tips is about the same as the virgin magnetization curve of Figure 9.11.

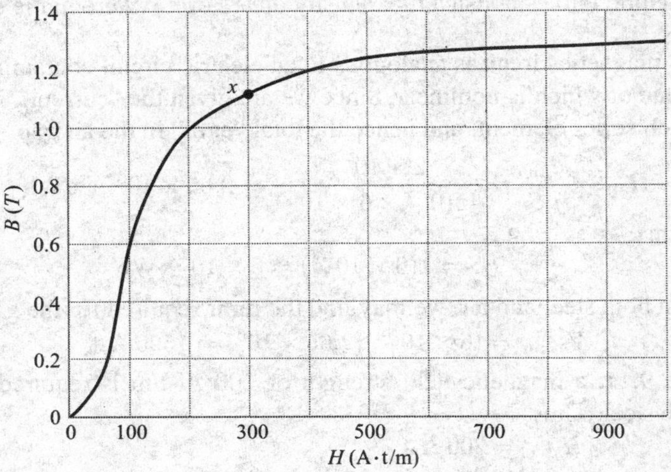

Fig. 9.11 Magnetization curve of a sample of silicon sheet steel.

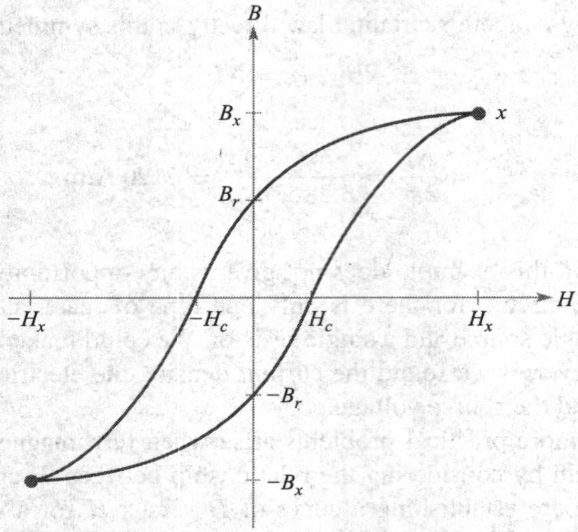

Fig. 9.12 A hysteresis loop for silicon steel. The coercive force H_c and remnant flux density B_r are indicated.

EXAMPLE 9.7

Let us make use of the magnetization curve for silicon steel to solve a magnetic circuit problem that is slightly different from our previous example. We shall use a steel core in the toroid, except for an air gap of 2 mm. Magnetic circuits with air gaps occur because gaps are deliberately introduced in some devices, such as inductors, which must carry large direct currents, because they are unavoidable in other devices such as rotating machines, or because of unavoidable problems in assembly. There are still 500 turns about the toroid, and we ask what current is required to establish a flux density of 1 T everywhere in the core.

Solution. This magnetic circuit is analogous to an electric circuit containing a voltage source and two resistors, one of which is nonlinear. Since we are given the "current," it is easy to find the "voltage" across each series element, and hence the total "emf." In the air gap,

$$\Re_{\text{air}} = \frac{d_{\text{air}}}{\mu S} = \frac{2 \times 10^{-3}}{4\pi 10^{-7} \times 6 \times 10^{-4}} = 2.65 \times 10^6 \text{ A·t/Wb}$$

Knowing the total flux,

$$\Phi = BS = 1(6 \times 10^{-4}) = 6 \times 10^{-4} \text{ Wb}$$

which is the same in both steel and air, we may find the mmf required for the gap,

$$V_{m,\text{air}} = (6 \times 10^{-4})(2.65 \times 10^6) = 1590 \text{ A·t}$$

Referring to Figure 9.11, a magnetic field strength of 200 A · t/m is required to produce a flux density of 1 T in the steel. Thus

$$H_{\text{steel}} = 200 \text{ A·t}$$
$$V_{m,\text{steel}} = H_{\text{steel}} d_{\text{steel}} = 200 \times 0.30\pi = 188 \text{ A·t}$$

The total mmf is therefore 1778 A·t, and a coil current of 3.56 A is required.

We should realize that we have made several approximations in obtaining this answer. We have already mentioned the lack of a completely uniform cross section, or cylindrical symmetry; the path of every flux line is not of the same length. The choice of a "mean" path length can help compensate for this error in problems in which it may be more important than it is in our example. Fringing flux in the air gap is another source of error, and formulas are available by which we may calculate an effective length and cross-sectional area for the gap which will yield more accurate results. There is also a leakage flux between the turns of wire, and in devices containing coils concentrated on one section of the core, a few flux lines bridge the interior of the toroid. Fringing and leakage are problems which seldom arise in the electric circuit because the ratio of the conductivities of air and the conductive or resistive materials used is so high. In contrast, the magnetization curve for silicon steel shows that the ratio of H to B in the steel is about 200 up to the "knee" of the magnetization curve; this compares with a ratio in air of about 800 000. Thus, although flux prefers steel to air by the commanding ratio of 4000 to 1, this is not very close to the ratio of conductivities of, say, 10^{15} for a good conductor and a fair insulator.

EXAMPLE 9.8

As a last example, let us consider the reverse problem. Given a coil current of 4 A in the magnetic circuit of Example 9.7, what will the flux density be?

Solution. First let us try to linearize the magnetization curve by a straight line from the origin to $B = 1$, $H = 200$. We then have $B = H/200$ in steel and $B = \mu_0 H$ in air. The two reluctances are found to be 0.314×10^6 for the steel path and 2.65×10^6 for the air gap, or $2.96 \times 10^6 \text{A}\cdot\text{t/Wb}$ total. Since V_m is 2000 A · t, the flux is 6.76×10^{-4} Wb, and $B = 1.13$ T. A more accurate solution may be obtained by assuming several values of B and calculating the necessary mmf. Plotting the results enables us to determine the true value of B by interpolation. With this method we obtain $B = 1.10$ T. The good accuracy of the linear model results from the fact that the reluctance of the air gap in a magnetic circuit is often much greater than the reluctance of the ferromagnetic portion of the circuit. A relatively poor approximation for the iron or steel can thus be tolerated.

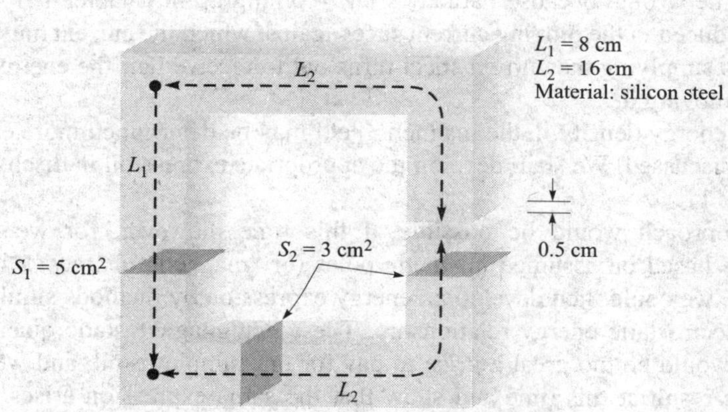

Fig. 9.13 See Problem D9.9.

D9.9 Given the magnetic circuit of Figure 9.13, assume $B = 0.6$ T at the midpoint of the left leg and find: (*a*) $V_{m,\text{air}}$; (*b*) $V_{m,\text{steel}}$; (*c*) the current required in a 1300-turn coil linking the left leg.

Ans. 3980 A · t; 72 A · t; 3.12 A

D9.10 The magnetization curve for material X under normal operating conditions may be approximated by the expression $B = (H/160)(0.25 + e^{-H/320})$, where H is in A/m and B is in T. If a magnetic circuit contains a 12 cm length of material X, as well as a 0.25 mm air gap, assume a uniform cross section of 2.5 cm^2 and find the total mmf required to produce a flux of: (*a*) 10 μWb; (*b*) 100 μWb.

Ans. 8.58 A · t; 86.7 A · t

9.9 POTENTIAL ENERGY AND FORCES ON MAGNETIC MATERIALS

In the electrostatic field we first introduced the point charge and the experimental law of force between point charges. After defining electric field intensity, electric flux density, and electric potential, we were able to find an expression for the energy in an electrostatic field by establishing the work necessary to bring the prerequisite point charges from infinity to their final resting places. The general expression for energy is

$$W_E = \frac{1}{2}\int_{\text{vol}} \mathbf{D}\cdot\mathbf{E}\, d\nu \tag{45}$$

where a linear relationship between **D** and **E** is assumed.

This is not as easily done for the steady magnetic field. It would seem that we might assume two simple sources, perhaps two current sheets, find the force on one due to the other, move the sheet a differential distance against this force, and equate the necessary work to the change in energy. If we did, we would be wrong, because Faraday's law (coming up in Chapter 10) shows that there will be a voltage induced in the moving current sheet against which the current must be maintained. Whatever source is supplying the current sheet turns out to receive half the energy we are putting into the circuit by moving it.

In other words, energy density in the magnetic field may be determined more easily after time-varying fields are discussed. We shall develop the appropriate expression in discussing Poynting's theorem in Chapter 11.

An alternate approach would be possible at this time, however, for we might define a magnetostatic field based on assumed magnetic poles (or "magnetic charges"). Using the scalar magnetic potential, we could then develop an energy expression by methods similar to those used in obtaining the electrostatic energy relationship. These new magnetostatic quantities we would have to introduce would be too great a price to pay for one simple result, and we shall therefore merely present the result at this time and show that the same expression arises in the Poynting theorem later. The total energy stored in a steady magnetic field in which **B** is linearly related to **H** is

$$W_H = \frac{1}{2}\int_{\text{vol}} \mathbf{B}\cdot\mathbf{H}\, d\nu \tag{46}$$

Letting $\mathbf{B} = \mu\mathbf{H}$, we have the equivalent formulations

$$W_H = \frac{1}{2}\int_{\text{vol}} \mu H^2 d\nu \tag{47}$$

or

$$W_H = \frac{1}{2}\int_{\text{vol}} \frac{B^2}{\mu} d\nu \tag{48}$$

It is again convenient to think of this energy as being distributed throughout the volume with an energy density of $\frac{1}{2}\mathbf{B}\cdot\mathbf{H}\,\text{J/m}^3$, although we have no mathematical justification for such a statement.

In spite of the fact that these results are valid only for linear media, we may use them to calculate the forces on nonlinear magnetic materials if we focus our attention on the linear media (usually air) which may surround them. For example, suppose that we have a long solenoid with a silicon-steel core. A coil containing n turns/m with a current I surrounds it. The magnetic field intensity in the core is therefore $nI\,\text{A}\cdot\text{t/m}$, and the magnetic flux density can be obtained from the magnetization curve for silicon steel. Let us call this value B_{st}. Suppose that the core is composed of two semi-infinite cylinders[2] which are just touching. We now apply a mechanical force to separate these two sections of the core while keeping the flux density constant. We apply a force F over a distance dL, thus doing work $F\,dL$. Faraday's law does not apply here, for the fields in the core have not changed, and we can therefore use the principle of virtual work to determine that the work we have done in moving one core appears as stored energy in the air gap we have created. By (48), this increase is

$$dW_H = F\,dL = \frac{1}{2}\frac{B_{\text{st}}^2}{\mu_0} S\,dL$$

where S is the core cross-sectional area. Thus

$$F = \frac{B_{\text{st}}^2 S}{2\mu_0}$$

If, for example, the magnetic field intensity is sufficient to produce saturation in the steel, approximately 1.4 T, the force is

$$F = 7.80 \times 10^5 S \quad \text{N}$$

or about $113\,\text{lb}_f/\text{in}^2$.

D9.11 (*a*) What force is being exerted on the pole faces of the circuit described in Problem D9.9 and Figure 9.13? (*b*) Is the force trying to open or close the air gap?

Ans. 1194 N; as Wilhelm Eduard Weber would put it, "schliessen"

[2] A semi-infinite cylinder is a cylinder of infinite length having one end located in finite space.

9.10 INDUCTANCE AND MUTUAL INDUCTANCE

Inductance is the last of the three familiar parameters from circuit theory which we are defining in more general terms. Resistance was defined in Chapter 5 as the ratio of the potential difference between two equipotential surfaces of a conducting material to the total current crossing either equipotential surface. The resistance is a function of conductor geometry and conductivity only. Capacitance was defined in the same chapter as the ratio of the total charge on either of two equipotential conducting surfaces to the potential difference between the surfaces. Capacitance is a function only of the geometry of the two conducting surfaces and the permittivity of the dielectric medium between or surrounding them.

Illustrations

As a prelude to defining inductance, we first need to introduce the concept of flux linkage. Let us consider a toroid of N turns in which a current I produces a total flux Φ. We shall assume first that this flux links or encircles each of the N turns, and we also see that each of the N turns links the total flux Φ. The *flux linkage* $N\Phi$ is defined as the product of the number of turns N and the flux Φ linking each of them.[3] For a coil having a single turn, the flux linkage is equal to the total flux.

We now define *inductance* (or self-inductance) as the ratio of the total flux linkages to the current which they link,

$$L = \frac{N\Phi}{I} \tag{49}$$

The current I flowing in the N-turn coil produces the total flux Φ and $N\Phi$ flux linkages, where we assume for the moment that the flux Φ links each turn. This definition is applicable only to magnetic media which are linear, so that the flux is proportional to the current. If ferromagnetic materials are present, there is no single definition of inductance which is useful in all cases, and we shall restrict our attention to linear materials.

The unit of inductance is the henry (H), equivalent to one weber-turn per ampere.

Let us apply (49) in a straightforward way to calculate the inductance per meter length of a coaxial cable of inner radius a and outer radius b. We may take the expression for total flux developed as Eq. (42) in Chapter 8,

$$\Phi = \frac{\mu_0 Id}{2\pi} \ln \frac{b}{a}$$

and obtain the inductance rapidly for a length d,

$$L = \frac{\mu_0 d}{2\pi} \ln \frac{b}{a} \quad \text{H}$$

or, on a per-meter basis,

$$L = \frac{\mu_0}{2\pi} \ln \frac{b}{a} \quad \text{H/m} \tag{50}$$

In this case, $N = 1$ turn, and all the flux links all the current.

---

[3] The symbol λ is commonly used for flux linkages. We shall only occasionally make use of this concept, however, and we will continue to write it as $N\Phi$.

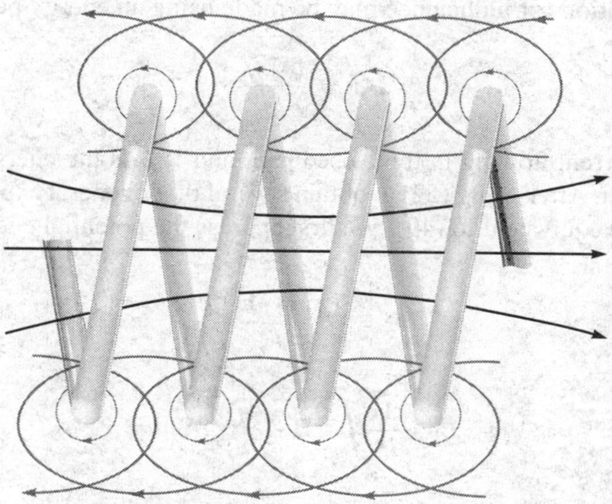

Fig. 9.14 A portion of a coil showing partial flux linkages. The total flux linkages are obtained by adding the fluxes linking each turn.

In the problem of a toroidal coil of N turns and a current I, as shown in Figure 8.12b, we have

$$B_\phi = \frac{\mu_0 NI}{2\pi\rho}$$

If the dimensions of the cross section are small compared with the mean radius of the toroid ρ_0, then the total flux is

$$\Phi = \frac{\mu_0 NIS}{2\pi\rho_0}$$

where S is the cross-sectional area. Multiplying the total flux by N, we have the flux linkages, and dividing by I, we have the inductance

$$L = \frac{\mu_0 N^2 S}{2\pi\rho_0} \tag{51}$$

Once again we have assumed that all the flux links all the turns, and this is a good assumption for a toroidal coil of many turns packed closely together. Suppose, however, that our toroid has an appreciable spacing between turns, a short part of which might look like Figure 9.14. The flux linkages are no longer the product of the flux at the mean radius times the total number of turns. In order to obtain the total flux linkages we must look at the coil on a turn-by-turn basis.

$$(N\Phi)_{\text{total}} = \Phi_1 + \Phi_2 + \cdots + \Phi_i + \cdots + \Phi_N$$
$$= \sum_{i=1}^{N} \Phi_i$$

where Φ_i is the flux linking the ith turn. Rather than doing this, we usually rely on experience and empirical quantities called winding factors and pitch factors to adjust the basic formula to apply to the real physical world.

An equivalent definition for inductance may be made using an energy point of view,

$$L = \frac{2W_H}{I^2} \tag{52}$$

where I is the total current flowing in the closed path and W_H is the energy in the magnetic field produced by the current. After using (52) to obtain several other general expressions for inductance, we shall show that it is equivalent to (49). We first express the potential energy W_H in terms of the magnetic fields,

$$L = \frac{\int_{\text{vol}} \mathbf{B} \cdot \mathbf{H}\, d\nu}{I^2} \tag{53}$$

and then replace $\mathbf{B}$ by $\nabla \times \mathbf{A}$,

$$L = \frac{1}{I^2} \int_{\text{vol}} \mathbf{H} \cdot (\nabla \times \mathbf{A}) d\nu$$

The vector identity

$$\nabla \cdot (\mathbf{A} \times \mathbf{H}) \equiv \mathbf{H} \cdot (\nabla \times \mathbf{A}) - \mathbf{A} \cdot (\nabla \times \mathbf{H}) \tag{54}$$

may be proved by expansion in rectangular coordinates. The inductance is then

$$L = \frac{1}{I^2} \left[\int_{\text{vol}} \nabla \cdot (\mathbf{A} \times \mathbf{H})\, d\nu + \int_{\text{vol}} \mathbf{A} \cdot (\nabla \times \mathbf{H})\, d\nu \right] \tag{55}$$

After applying the divergence theorem to the first integral and letting $\nabla \times \mathbf{H} = \mathbf{J}$ in the second integral, we have

$$L = \frac{1}{I^2} \left[\oint_S (\mathbf{A} \times \mathbf{H}) \cdot d\mathbf{S} + \int_{\text{vol}} \mathbf{A} \cdot \mathbf{J}\, d\nu \right]$$

The surface integral is zero, since the surface encloses the volume containing all the magnetic energy, and this requires that $\mathbf{A}$ and $\mathbf{H}$ be zero on the bounding surface. The inductance may therefore be written as

$$L = \frac{1}{I^2} \int_{\text{vol}} \mathbf{A} \cdot \mathbf{J}\, d\nu \tag{56}$$

Equation (56) expresses the inductance in terms of an integral of the values of $\mathbf{A}$ and $\mathbf{J}$ at every point. Since current density exists only within the conductor, the integrand is zero at all points *outside* the conductor, and the vector magnetic potential need not be determined there. The vector potential is that which arises from the current $\mathbf{J}$, and any other current source contributing a vector potential field in the region of the original current density is to be ignored for the present. Later we shall see that this leads to a *mutual inductance*.

The vector magnetic potential $\mathbf{A}$ due to $\mathbf{J}$ is given by Eq. (51), Chapter 8,

$$\mathbf{A} = \int_{\text{vol}} \frac{\mu \mathbf{J}}{4\pi R} d\nu$$

and the inductance may therefore be expressed more basically as a rather formidable double volume integral,

$$L = \frac{1}{I^2} \int_{\text{vol}} \left(\int_{\text{vol}} \frac{\mu \mathbf{J}}{4\pi R} d\nu \right) \cdot \mathbf{J}\, d\nu \tag{57}$$

A slightly simpler integral expression is obtained by restricting our attention to current filaments of small cross section for which $\mathbf{J}\,d\nu$ may be replaced by $I\,d\mathbf{L}$ and the volume integral by a closed line integral along the axis of the filament,

$$\begin{aligned} L &= \frac{1}{I^2}\oint\left(\oint\frac{\mu I\,d\mathbf{L}}{4\pi R}\right)\cdot I\,d\mathbf{L} \\ &= \frac{\mu}{4\pi}\oint\left(\oint\frac{d\mathbf{L}}{R}\right)\cdot d\mathbf{L} \end{aligned} \tag{58}$$

Our only present interest in Eqs. (57) and (58) lies in their implication that the inductance is a function of the distribution of the current in space or the geometry of the conductor configuration.

To obtain our original definition of inductance (49) let us hypothesize a uniform current distribution in a filamentary conductor of small cross section so that $\mathbf{J}\,d\nu$ in (56) becomes $I\,d\mathbf{L}$,

$$L = \frac{1}{I}\oint\mathbf{A}\cdot d\mathbf{L} \tag{59}$$

For a small cross section, $d\mathbf{L}$ may be taken along the center of the filament. We now apply Stokes' theorem and obtain

$$L = \frac{1}{I}\int_S(\nabla\times\mathbf{A})\cdot d\mathbf{S}$$

or

$$L = \frac{1}{I}\int_S\mathbf{B}\cdot d\mathbf{S}$$

or

$$L = \frac{\Phi}{I} \tag{60}$$

Retracing the steps by which (60) is obtained, we should see that the flux Φ is that portion of the total flux which passes through any and every open surface whose perimeter is the filamentary current path.

If we now let the filament make N identical turns about the total flux, an idealization which may be closely realized in some types of inductors, the closed line integral must consist of N laps about this common path, and (60) becomes

$$L = \frac{N\Phi}{I} \tag{61}$$

The flux Φ is now the flux crossing any surface whose perimeter is the path occupied by any *one* of the N turns. The inductance of an N-turn coil may still be obtained from (60), however, if we realize that the flux is that which crosses the complicated surface[4] whose perimeter consists of all N turns.

Use of any of the inductance expressions for a true filamentary conductor (having zero radius) leads to an infinite value of inductance, regardless of the configuration of the filament. Near the conductor, Ampère's circuital law shows that the magnetic field intensity varies inversely with the distance from the conductor, and a simple integration soon shows that an infinite amount of energy and an infinite amount of flux are contained within any finite cylinder about the filament. This difficulty is eliminated by specifying a small but finite filamentary radius.

[4] Somewhat like a spiral ramp.

The interior of any conductor also contains magnetic flux, and this flux links a variable fraction of the total current, depending on its location. These flux linkages lead to an *internal inductance,* which must be combined with the external inductance to obtain the total inductance. The internal inductance of a long, straight wire of circular cross section, radius a, and uniform current distribution is

$$L_{a,\text{int}} = \frac{\mu}{8\pi} \quad \text{H/m} \tag{62}$$

a result requested in Problem 9.43 at the end of this chapter.

In Chapter 12 it will be seen that the current distribution in a conductor at high frequencies tends to be concentrated near the surface. The internal flux is reduced, and it is usually sufficient to consider only the external inductance. At lower frequencies, however, internal inductance may become an appreciable part of the total inductance.

We conclude by defining the *mutual inductance* between circuits 1 and 2, M_{12}, in terms of mutual flux linkages,

$$M_{12} = \frac{N_2 \Phi_{12}}{I_1} \tag{63}$$

where Φ_{12} signifies the flux produced by I_1 which links the path of the filamentary current I_2, and N_2 is the number of turns in circuit 2. The mutual inductance, therefore, depends on the magnetic interaction between two currents. With either current alone, the total energy stored in the magnetic field can be found in terms of a single inductance, or self-inductance; with both currents having nonzero values, the total energy is a function of the two self-inductances and the mutual inductance. In terms of a mutual energy, it can be shown that (63) is equivalent to

$$M_{12} = \frac{1}{I_1 I_2} \int_{\text{vol}} (\mathbf{B}_1 \cdot \mathbf{H}_2) d\nu \tag{64}$$

or

$$M_{12} = \frac{1}{I_1 I_2} \int_{\text{vol}} (\mu \mathbf{H}_1 \cdot \mathbf{H}_2) d\nu \tag{65}$$

where $\mathbf{B}_1$ is the field resulting from I_1 (with $I_2 = 0$) and $\mathbf{H}_2$ is the field arising from I_2 (with $I_1 = 0$). Interchange of the subscripts does not change the right-hand side of (65), and therefore

$$M_{12} = M_{21} \tag{66}$$

Mutual inductance is also measured in henrys, and we rely on the context to allow us to differentiate it from magnetization, also represented by M.

EXAMPLE 9.9

Calculate the self-inductances of and the mutual inductances between two coaxial solenoids of radius R_1 and R_2, $R_2 > R_1$, carrying currents I_1 and I_2 with n_1 and n_2 turns/m, respectively.

Solution. We first attack the mutual inductances. From Eq. (15), Chapter 8, we let $n_1 = N/d$, and obtain

$$\mathbf{H}_1 = n_1 I_1 \mathbf{a}_z \quad (0 < \rho < R_1)$$
$$= 0 \quad (\rho > R_1)$$

and

$$\mathbf{H}_2 = n_2 I_2 \mathbf{a}_z \quad (0 < \rho < R_2)$$
$$= 0 \quad (\rho > R_2)$$

Thus, for this uniform field

$$\Phi_{12} = \mu_0 n_1 I_1 \pi R_1^2$$

and

$$M_{12} = \mu_0 n_1 n_2 \pi R_1^2$$

Similarly,

$$\Phi_{21} = \mu_0 n_2 I_2 \pi R_1^2$$
$$M_{21} = \mu_0 n_1 n_2 \pi R_1^2 = M_{12}$$

If $n_1 = 50$ turns/cm, $n_2 = 80$ turns/cm, $R_1 = 2$ cm, and $R_2 = 3$ cm, then

$$M_{12} = M_{21} = 4\pi \times 10^{-7}(5000)(8000)\pi(0.02^2) = 63.2 \text{ mH/m}$$

The self-inductances are easily found. The flux produced in coil 1 by I_1 is

$$\Phi_{11} = \mu_0 n_1 I_1 \pi R_1^2$$

and thus

$$L_1 = \mu_0 n_1^2 S_1 d \quad \text{H}$$

The inductance per unit length is therefore

$$L_1 = \mu_0 n_1^2 S_1 \quad \text{H/m}$$

or

$$L_1 = 39.5 \quad \text{mH/m}$$

Similarly,

$$L_2 = \mu_0 n_2^2 S_2 = 22.7 \quad \text{mH/m}$$

We see, therefore, that there are many methods available for the calculation of self-inductance and mutual inductance. Unfortunately, even problems possessing a high degree of symmetry present very challenging integrals for evaluation, and only a few problems are available for us to try our skill on.

Inductance will be discussed in circuit terms in Chapter 11.

D9.12 Calculate the self-inductance of: (*a*) 3.5 m of coaxial cable with $a = 0.8$ mm and $b = 4$ mm, filled with a material for which $\mu_r = 50$; (*b*) a toroidal coil of 500 turns, wound on a fiberglass form having a 2.5×2.5 cm square cross section and an inner radius of 2 cm; (*c*) a solenoid having 500 turns about a cylindrical core of 2 cm radius in which $\mu_r = 50$ for $0 < \rho < 0.5$ cm and $\mu_r = 1$ for $0.5 < \rho < 2$ cm; the length of the solenoid is 50 cm.

Ans. 56.3 μH; 1.01 mH; 3.2 mH

D9.13 A solenoid is 50 cm long, 2 cm in diameter, and contains 1500 turns. The cylindrical core has a diameter of 2 cm and a relative permeability of 75. This coil is coaxial with a second solenoid, also 50 cm long, but with a 3 cm diameter and 1200 turns. Calculate: (*a*) L for the inner solenoid; (*b*) L for the outer solenoid; (*c*) M between the two solenoids.

Ans. 133.2 mH; 86.7 mH; 106.6 mH

References

1. Kraus, J. D., and D. A. Fleisch. (See References for Chapter 3.) Examples of the calculation of inductance are given on pp. 99–108.
2. Matsch, L. W. (See References for Chapter 6.) Chapter 3 is devoted to magnetic circuits and ferromagnetic materials.
3. Paul, C. R., K. W. Whites, and S. Y. Nasar. (See References for Chapter 8.) Magnetic circuits, including those with permanent magnets, are discussed on pp. 263–70.

CHAPTER 9 Problems

Quizzes

9.1 A point charge, $Q = -0.3\,\mu$C and $m = 3 \times 10^{-16}$ kg, is moving through the field $\mathbf{E} = 30\mathbf{a}_z$ V/m. Use Eq. (1) and Newton's laws to develop the appropriate differential equations and solve them, subject to the initial conditions at $t = 0, \mathbf{v} = 3 \times 10^5\mathbf{a}_x$ m/s at the origin. At $t = 3\,\mu$s, find: (*a*) the position $P(x, y, z)$ of the charge; (*b*) the velocity $\mathbf{v}$; (*c*) the kinetic energy of the charge.

9.2 A point charge, $Q = -0.3\,\mu$C and $m = 3 \times 10^{-16}$ kg, is moving through the field $\mathbf{B} = 30\mathbf{a}_z$ mT. Make use of Eq. (2) and Newton's laws to develop the appropriate differential equations and solve them, subject to the initial condition at $t = 0, \mathbf{v} = 3 \times 10^5\mathbf{a}_x$ m/s at the origin. Solve these equations (perhaps with the help of an example given in Section 7.5) to evaluate at $t = 3\,\mu$s: (*a*) the position $P(x, y, z)$ of the charge; (*b*) its velocity; (*c*) and its kinetic energy.

9.3 A point charge for which $Q = 2 \times 10^{-16}$ C and $m = 5 \times 10^{-26}$ kg is moving in the combined fields $\mathbf{E} = 100\mathbf{a}_x - 200\mathbf{a}_y + 300\mathbf{a}_z$ V/m and $\mathbf{B} = -3\mathbf{a}_x + 2\mathbf{a}_y - \mathbf{a}_z$ mT. If the charge velocity at $t = 0$ is $\mathbf{v}(0) = (2\mathbf{a}_x - 3\mathbf{a}_y - 4\mathbf{a}_z)10^5$ m/s: (*a*) give the unit vector showing the direction in which the charge is accelerating at $t = 0$; (*b*) find the kinetic energy of the charge at $t = 0$.

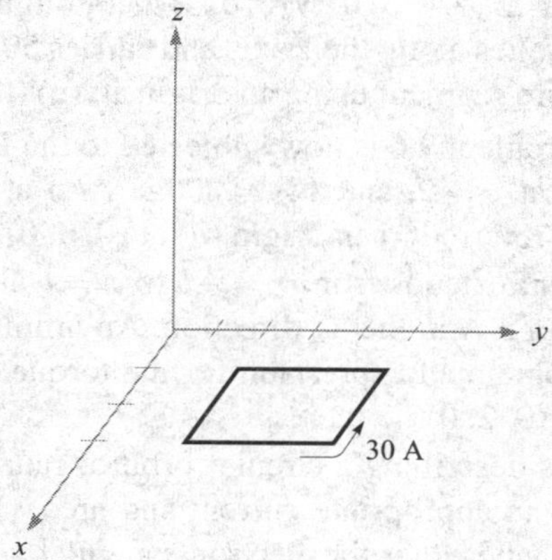

Fig. 9.15 See Problem 9.6.

9.4 Show that a charged particle in a uniform magnetic field describes a circular orbit with an orbital period that is independent of the radius. Find the relationship between the angular velocity and magnetic flux density for an electron (the *cyclotron frequency*).

9.5 A rectangular loop of wire in free space joins points $A(1, 0, 1)$ to $B(3, 0, 1)$ to $C(3, 0, 4)$ to $D(1, 0, 4)$ to A. The wire carries a current of 6 mA, flowing in the $\mathbf{a}_z$ direction from B to C. A filamentary current of 15 A flows along the entire z axis in the $\mathbf{a}_z$ direction. (*a*) Find F on side BC. (*b*) Find **F** on side AB. (*c*) Find $\mathbf{F}_{\text{total}}$ on the loop.

9.6 The magnetic flux density in a region of free space is given as $\mathbf{B} = -3x\mathbf{a}_x + 5y\mathbf{a}_y - 2z\mathbf{a}_z$ T. Find the total force on the rectangular loop shown in Figure 9.15 if it lies in the plane $z = 0$ and is bounded by $x = 1$, $x = 3$, $y = 2$, and $y = 5$, all dimensions in cm.

9.7 Uniform current sheets are located in free space as follows: $8\mathbf{a}_z$ A/m at $y = 0$, $-4\mathbf{a}_z$ A/m at $y = 1$, and $-4\mathbf{a}_z$ A/m at $y = -1$. Find the vector force per meter length exerted on a current filament carrying 7 mA in the $\mathbf{a}_L$ direction if the filament is located at: (*a*) $x = 0$, $y = 0.5$, and $\mathbf{a}_L = \mathbf{a}_z$; (*b*) $y = 0.5$, $z = 0$, and $\mathbf{a}_L = \mathbf{a}_x$; (*c*) $x = 0$, $y = 1.5$, and $\mathbf{a}_L = \mathbf{a}_z$.

9.8 Filamentary currents of $-25\mathbf{a}_z$ and $25\mathbf{a}_z$ A are located in the $x = 0$ plane in free space at $y = -1$ and $y = 1$ m, respectively. A third filamentary current of $10^{-3}\mathbf{a}_z$ A is located at $x = k$, $y = 0$. Find the vector force on a 1 m length of the 1 mA filament and plot $|\mathbf{F}|$ versus k.

9.9 A current of $-100\mathbf{a}_z$ A/m flows on the conducting cylinder $\rho = 5$ mm, and $+500\mathbf{a}_z$ A/m is present on the conducting cylinder $\rho = 1$ mm. Find the magnitude of the total force per meter length that is acting to split the outer cylinder apart along its length.

9.10 A planar transmission line consists of two conducting planes of width b separated d m in air, carrying equal and opposite currents of I A. If $b \gg d$, find the force of repulsion per meter of length between the two conductors.

9.11 (*a*) Use Eq. (14), Section 9.3, to show that the force of attraction per unit length between two filamentary conductors in free space with currents $I_1\mathbf{a}_z$ at $x = 0$, $y = d/2$, and $I_2\mathbf{a}_z$ at $x = 0$, $y = -d/2$, is $\mu_0 I_1 I_2/(2\pi d)$. (*b*) Show how a simpler method can be used to check your result.

9.12 A conducting current strip carrying $\mathbf{K} = 12\mathbf{a}_z$ A/m lies in the $x = 0$ plane between $y = 0.5$ and $y = 1.5$ m. There is also a current filament of $I = 5$ A in the $\mathbf{a}_z$ direction on the z axis. Find the force exerted on the (*a*) filament by the current strip; (*b*) strip by the filament.

9.13 A current of 6 A flows from $M(2, 0, 5)$ to $N(5, 0, 5)$ in a straight, solid conductor in free space. An infinite current filament lies along the z axis and carries 50 A in the $\mathbf{a}_z$ direction. Compute the vector torque on the wire segment using an origin at: (*a*) (0, 0, 5); (*b*) (0, 0, 0); (*c*) (3, 0, 0).

9.14 The rectangular loop of Problem 9.6 is now subjected to the **B** field produced by two current sheets, $\mathbf{K}_1 = 400\mathbf{a}_y$ A/m at $z = 2$, and $\mathbf{K}_2 = 300\mathbf{a}_z$ A/m at $y = 0$, in free space. Find the vector torque on the loop, referred to an origin (*a*) at (0, 0, 0); (*b*) at the center of the loop.

9.15 A solid conducting filament extends from $x = -b$ to $x = b$ along the line $y = 2$, $z = 0$. This filament carries a current of 3 A in the $\mathbf{a}_x$ direction. An infinite filament on the z axis carries 5 A in the $\mathbf{a}_z$ direction. Obtain an expression for the torque exerted on the finite conductor about an origin located at (0, 2, 0).

9.16 Assume that an electron is describing a circular orbit of radius a about a positively charged nucleus. (*a*) By selecting an appropriate current and area, show that the equivalent orbital dipole moment is $ea^2\omega/2$, where ω is the electron's angular velocity. (*b*) Show that the torque produced by a magnetic field parallel to the plane of the orbit is $ea^2\omega B/2$. (*c*) By equating the Coulomb and centrifugal forces, show that ω is $(4\pi\epsilon_0 m_e a^3/e^2)^{-1/2}$, where m_e is the electron mass. (*d*) Find values for the angular velocity, torque, and the orbital magnetic moment for a hydrogen atom, where a is about 6×10^{-11} m; let $B = 0.5$ T.

9.17 The hydrogen atom described in Problem 9.16 is now subjected to a magnetic field having the same direction as that of the atom. Show that the forces caused by B result in a decrease of the angular velocity by $eB/(2m_e)$ and a decrease in the orbital moment by $e^2a^2B/(4m_e)$. What are these decreases for the hydrogen atom in parts per million for an external magnetic flux density of 0.5 T?

9.18 Calculate the vector torque on the square loop shown in Figure 9.16 about an origin at A in the field B, given: (*a*) $A(0, 0, 0)$ and $\mathbf{B} = 100\mathbf{a}_y$ mT; (*b*) $A(0, 0, 0)$ and $\mathbf{B} = 200\mathbf{a}_x + 100\mathbf{a}_y$ mT; (*c*) $A(1, 2, 3)$ and $\mathbf{B} = 200\mathbf{a}_x + 100\mathbf{a}_y - 300\mathbf{a}_z$ mT; (*d*) $A(1, 2, 3)$ and $\mathbf{B} = 200\mathbf{a}_x + 100\mathbf{a}_y - 300\mathbf{a}_z$ mT for $x \geq 2$ and $\mathbf{B} = 0$ elsewhere.

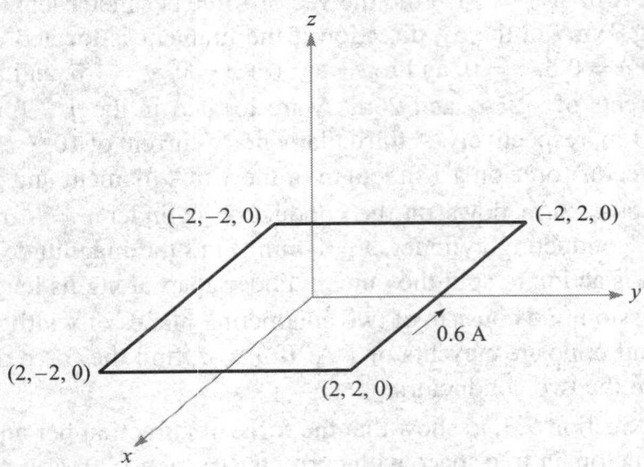

Fig. 9.16 See Problem 9.18.

9.19 Given a material for which $\chi_m = 3.1$ and within which $\mathbf{B} = 0.4y\mathbf{a}_z$ T, find: (*a*)$\mathbf{H}$; (*b*) μ; (*c*) μ_r; (*d*) $\mathbf{M}$; (*e*) $\mathbf{J}$; (*f*) $\mathbf{J}_b$; (*g*) $\mathbf{J}_T$.

9.20 Find **H** in a material where (*a*) $\mu_r = 4.2$, there are 2.7×10^{29} atoms/m^3, and each atom has a dipole moment of $2.6 \times 10^{-30}\mathbf{a}_y$ A $\cdot$ m^2; (*b*) $\mathbf{M} = 270\mathbf{a}_z$ A/m and $\mu = 2\mu$ H/m; (*c*) $\chi_m = 0.7$ and $\mathbf{B} = 2\mathbf{a}_z$ T. (*d*) Find **M** in a material where bound surface current densities of $12\mathbf{a}_z$ A/m and $-9\mathbf{a}_z$ A/m exist at $\rho = 0.3$ m and 0.4 m, respectively.

9.21 Find the magnitude of the magnetization in a material for which: (*a*) the magnetic flux density is 0.02 Wb/m^2; (*b*) the magnetic field intensity is 1200 A/m and the relative permeability is 1.005; (*c*) there are 7.2×10^{28} atoms per cubic meter, each having a dipole moment of 4×10^{-30} A$\cdot$m^2 in the same direction, and the magnetic susceptibility is 0.003.

9.22 Under some conditions, it is possible to approximate the effects of ferromagnetic materials by assuming linearity in the relationship of **B** and **H**. Let $\mu_r = 1000$ for a certain material of which a cylindrical wire of radius 1 mm is made. If $I = 1$ A and the current distribution is uniform, find (*a*) **B**, (*b*) **H**, (*c*) **M**, (*d*) **J**, and (*e*) $\mathbf{J}_b$ within the wire.

9.23 Calculate values for H_ϕ, B_ϕ, and M_ϕ at $\rho = c$ for a coaxial cable with $a = 2.5$ mm and $b = 6$ mm if it carries a current $I = 12$ A in the center conductor, and $\mu = 3\mu$H/m for $2.5\,\text{mm} < \rho < 3.5\,\text{mm}$, $\mu = 5\,\mu$H/m for $3.5\,\text{mm} < \rho < 4.5\,\text{mm}$, and $\mu = 10\ \mu$H/m for $4.5\,\text{mm} < \rho < 6$ mm. Use $c =$: (*a*) 3 mm; (*b*) 4 mm; (*c*) 5 mm.

9.24 A coaxial transmission line has $a = 5$ mm and $b = 20$ mm. Let its center lie on the z axis and let a dc current I flow in the $\mathbf{a}_z$ direction in the center conductor. The volume between the conductors contains a magnetic material for which $\mu_r = 2.5$, as well as air. Find **H**, **B**, and **M** everywhere between conductors if $H_\phi = \dfrac{600}{\pi}$ A/m at $\rho = 10$ mm, $\phi = \dfrac{\pi}{2}$, and the magnetic material is located where: (*a*) $a < \rho < 3a$; (*b*) $0 < \phi < \pi$.

9.25 A conducting filament at $z = 0$ carries 12 A in the $\mathbf{a}_z$ direction. Let $\mu_r = 1$ for $\rho < 1$ cm, $\mu_r = 6$ for $1 < \rho < 2$ cm, and $\mu_r = 1$ for $\rho > 2$ cm. Find: (*a*) **H** everywhere; (*b*) **B** everywhere.

9.26 Two current sheets, $K_0\mathbf{a}_y$ A/m at $z = 0$ and $-K_0\mathbf{a}_y$ A/m at $z = d$, are separated by two slabs of magnetic material, μ_{r1} for $0 < z < a$, and μ_{r2} for $a < z < d$. If $\mu_{r2} = 3\mu_{r1}$, find the ratio a/d such that 10 percent of the total magnetic flux is in the region $0 < z < a$.

9.27 Let $\mu_{r1} = 2$ in region 1, defined by $2x + 3y - 4z > 1$, while $\mu_{r2} = 5$ in region 2 where $2x + 3y - 4z < 1$. In region 1, $\mathbf{H}_1 = 50\mathbf{a}_x - 30\mathbf{a}_y + 20\mathbf{a}_z$ A/m. Find: (*a*) $\mathbf{H}_{N1}$; (*b*) $\mathbf{H}_{t1}$; (*c*) $\mathbf{H}_{t2}$; (*d*) $\mathbf{H}_{N2}$; (*e*) θ_1, the angle between $\mathbf{H}_1$ and $\mathbf{a}_{N21}$; (*f*) θ_2, the angle between $\mathbf{H}_2$ and $\mathbf{a}_{N21}$.

9.28 For values of B below the knee on the magnetization curve for silicon steel, approximate the curve by a straight line with $\mu = 5$ mH/m. The core shown in Figure 9.17 has areas of 1.6 cm^2 and lengths of 10 cm in each outer leg, and an area of 2.5 cm^2 and a length of 3 cm in the central leg. A coil of 1200 turns carrying 12 mA is placed around the central leg. Find B in the: (*a*) center leg; (*b*) center leg if a 0.3 mm air gap is present in the center leg.

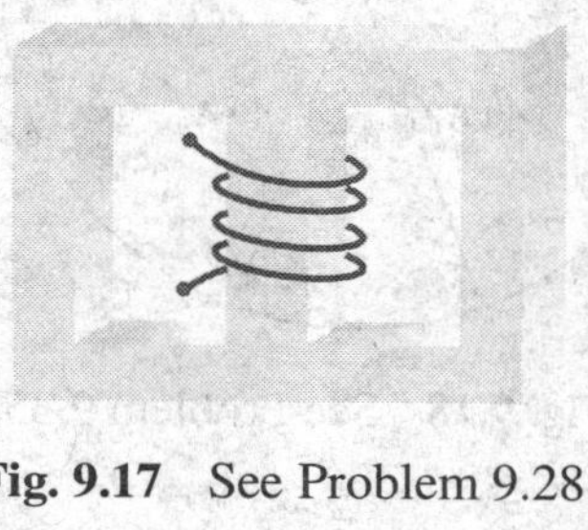

Fig. 9.17 See Problem 9.28.

9.29 In Problem 9.28, the linear approximation suggested in the statement of the problem leads to flux density of 0.666 T in the central leg. Using this value of B and the magnetization curve for silicon steel, what current is required in the 1200-turn coil?

9.30 A toroidal core has a circular cross section of 4 cm^2 area. The mean radius of the toroid is 6 cm. The core is composed of two semicircular segments, one of silicon steel and the other of a linear material with $\mu_r = 200$. There is a 0.4 mm air gap at each of the two joints, and the core is wrapped by a 4000-turn coil carrying a dc current I_1. (*a*) Find I_1 if the flux density in the core is 1.2 T. (*b*) Find the flux density in the core if $I_1 = -0.3$ A.

9.31 A toroid is constructed of a magnetic material having a cross-sectional area of 2.5 cm^2 and an effective length of 8 cm. There is also a short air gap of 0.25 mm length and an effective area of 2.8 cm^2. An mmf of 200 A $\cdot$ t is applied to the magnetic circuit. Calculate the total flux in the toroid if the magnetic material: (*a*) is assumed to have infinite permeability; (*b*) is assumed to be linear with $\mu_r = 1000$; (*c*) is silicon steel.

9.32 Determine the total energy stored in a spherical region 1 cm in radius, centered at the origin in free space, in the uniform field: (*a*) $\mathbf{H}_1 = -600\mathbf{a}_y$ A/m; (*b*) $\mathbf{H}_2 = 600\mathbf{a}_x + 1200\mathbf{a}_y$ A/m; (*c*) $\mathbf{H}_3 = -600\mathbf{a}_x + 1200\mathbf{a}_y$ A/m; (*d*) $\mathbf{H}_4 = \mathbf{H}_2 + \mathbf{H}_3$; (*e*) $1000\mathbf{a}_x$A/m $+ 0.001\mathbf{a}_x$ T.

9.33 A toroidal core has a square cross section, $2.5\,\text{cm} < \rho < 3.5\,\text{cm}$, $-0.5\,\text{cm} < z < 0.5\,\text{cm}$. The upper half of the toroid, $0 < z < 0.5\,\text{cm}$, is constructed of a linear material for which $\mu_r = 10$, while the lower half, $-0.5\,\text{cm} < z < 0$, has $\mu_r = 20$. An mmf of 150 A $\cdot$ t establishes a flux in the $\mathbf{a}_\phi$ direction. For $z > 0$, find: (*a*) $H_\phi(\rho)$; (*b*) $B_\phi(\rho)$; (*c*) $\Phi_{z>0}$. (*d*) Repeat for $z > 0$. (*e*) Find Φ_{total}.

9.34 Determine the energy stored per unit length in the internal magnetic field of an infinitely long, straight wire of radius a, carrying uniform current I.

9.35 The cones $\theta = 21°$ and $\theta = 159°$ are conducting surfaces and carry total currents of 40 A, as shown in Figure 9.18. The currents return on a spherical conducting surface of 0.25 m radius. (*a*) Find $\mathbf{H}$ in the region $0 < r < 0.25$, $21° < \theta < 159°$, $0 < \phi < 2\pi$. (*b*) How much energy is stored in this region?

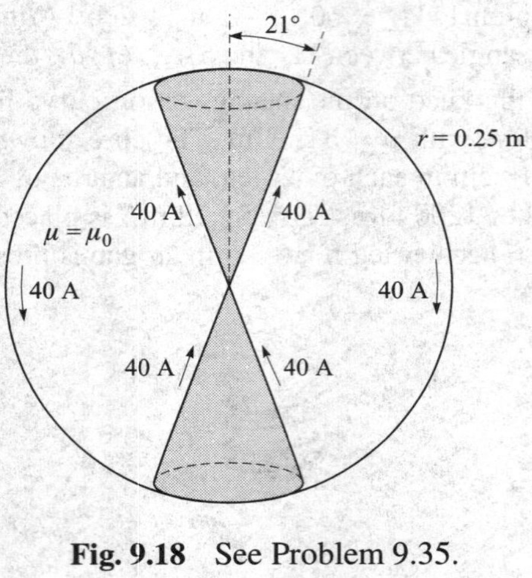

Fig. 9.18 See Problem 9.35.

9.36 The dimensions of the outer conductor of a coaxial cable are b and c, where $c > b$. Assuming $\mu = \mu_0$, find the magnetic energy stored per unit length in the region $b < \rho < c$ for a uniformly distributed total current I flowing in opposite directions in the inner and outer conductors.

9.37 Find the inductance of the cone-sphere configuration described in Problem 9.35 and Figure 9.18. The inductance is that offered at the origin between the vertices of the cone.

9.38 A toroidal core has a rectangular cross section defined by the surfaces $\rho = 2$ cm, $\rho = 3$ cm, $z = 4$ cm, and $z = 4.5$ cm. The core material has a relative permeability of 80. If the core is wound with a coil containing 8000 turns of wire, find its inductance.

9.39 Conducting planes in air at $z = 0$ and $z = d$ carry surface currents of $\pm K_0 \mathbf{a}_x$ A/m. (*a*) Find the energy stored in the magnetic field per unit length $(0 < x < 1)$ in a width $w(0 < y < w)$. (*b*) Calculate the inductance per unit length of this transmission line from $W_H = \frac{1}{2}LI^2$, where I is the total current in a width w in either conductor. (*c*) Calculate the total flux passing through the rectangle $0 < x < 1$, $0 < z < d$, in the plane $y = 0$, and from this result again find the inductance per unit length.

9.40 A coaxial cable has conductor dimensions of 1 and 5 mm. The region between conductors is air for $0 < \phi < \dfrac{\pi}{2}$ and $\pi < \phi < \dfrac{3\pi}{2}$, and a nonconducting material having $\mu_r = 8$ for $\dfrac{\pi}{2} < \phi < \pi$ and $\dfrac{3\pi}{2} < \phi < 2\pi$. Find the inductance per meter length.

9.41 A rectangular coil is composed of 150 turns of a filamentary conductor. Find the mutual inductance in free space between this coil and an infinite straight filament on the z axis if the four corners of the coil are located at: (*a*) (0, 1, 0), (0, 3, 0), (0, 3, 1), and (0, 1, 1); (*b*) (1, 1, 0), (1, 3, 0), (1, 3, 1), and (1, 1, 1).

9.42 Find the mutual inductance between two filaments forming circular rings of radii a and Δa, where $\Delta a \ll a$. The field should be determined by approximate methods. The rings are coplanar and concentric.

9.43 (*a*) Use energy relationships to show that the internal inductance of a nonmagnetic cylindrical wire of radius a carrying a uniformly distributed current I is $\mu_0/(8\pi)$ H/m. (*b*) Find the internal inductance if the portion of the conductor for which $\rho < c < a$ is removed.

10

Time-Varying Fields and Maxwell's Equations

The basic relationships of the electrostatic and the steady magnetic field were obtained in the previous nine chapters, and we are now ready to discuss time-varying fields. The discussion will be short, for vector analysis and vector calculus should now be more familiar tools; some of the relationships are unchanged, and most of the relationships are changed only slightly.

Two new concepts will be introduced: the electric field produced by a changing magnetic field and the magnetic field produced by a changing electric field. The first of these concepts resulted from experimental research by Michael Faraday and the second from the theoretical efforts of James Clerk Maxwell.

Maxwell actually was inspired by Faraday's experimental work and by the mental picture provided through the "lines of force" that Faraday introduced in developing his theory of electricity and magnetism. He was 40 years younger than Faraday, but they knew each other during the five years Maxwell spent in London as a young professor, a few years after Faraday had retired. Maxwell's theory was developed subsequent to his holding this university position, while he was working alone at his home in Scotland. It occupied him for five years between the ages of 35 and 40.

The four basic equations of electromagnetic theory presented in this chapter bear his name.

10.1 FARADAY'S LAW

After Oersted[1] demonstrated in 1820 that an electric current affected a compass needle, Faraday professed his belief that if a current could produce a magnetic field, then a magnetic field should be able to produce a current. The concept of the "field" was not available at that time, and Faraday's goal was to show that a current could be produced by "magnetism."

Animations

He worked on this problem intermittently over a period of 10 years, until he was finally successful in 1831.[2] He wound two separate windings on an iron toroid and placed a galvanometer in one circuit and a battery in the other. Upon closing the battery circuit, he noted a momentary deflection of the galvanometer; a similar deflection in the opposite direction occurred when the battery was disconnected. This, of course, was the first experiment he made involving a *changing*

[1] Hans Christian Oersted was professor of physics at the University of Copenhagen in Denmark.

[2] Joseph Henry produced similar results at Albany Academy in New York at about the same time.

magnetic field, and he followed it with a demonstration that either a *moving* magnetic field or a moving coil could also produce a galvanometer deflection.

In terms of fields, we now say that a time-varying magnetic field produces an *electromotive force* (emf) which may establish a current in a suitable closed circuit. An electromotive force is merely a voltage that arises from conductors moving in a magnetic field or from changing magnetic fields, and we shall define it in this section. Faraday's law is customarily stated as

$$\text{emf} = -\frac{d\Phi}{dt} \text{ V} \tag{1}$$

Equation (1) implies a closed path, although not necessarily a closed conducting path; the closed path, for example, might include a capacitor, or it might be a purely imaginary line in space. The magnetic flux is that flux which passes through any and every surface whose perimeter is the closed path, and $d\Phi/dt$ is the time rate of change of this flux.

A nonzero value of $d\Phi/dt$ may result from any of the following situations:

1. A time-changing flux linking a stationary closed path
2. Relative motion between a steady flux and a closed path
3. A combination of the two

The minus sign is an indication that the emf is in such a direction as to produce a current whose flux, if added to the original flux, would reduce the magnitude of the emf. This statement that the induced voltage acts to produce an opposing flux is known as *Lenz's law*.[3]

If the closed path is that taken by an N-turn filamentary conductor, it is often sufficiently accurate to consider the turns as coincident and let

$$\text{emf} = -N\frac{d\Phi}{dt} \tag{2}$$

where Φ is now interpreted as the flux passing through any one of N coincident paths.

We need to define emf as used in (1) or (2). The emf is obviously a scalar, and (perhaps not so obviously) a dimensional check shows that it is measured in volts. We define the emf as

$$\text{emf} = \oint \mathbf{E} \cdot d\mathbf{L} \tag{3}$$

and note that it is the voltage about a specific *closed path*. If any part of the path is changed, generally the emf changes. The departure from static results is clearly shown by (3), for an electric field intensity resulting from a static charge distribution must lead to zero potential difference about a closed path. In electrostatics, the line integral leads to a potential difference; with time-varying fields, the result is an emf or a voltage.

Replacing Φ in (1) with the surface integral of $\mathbf{B}$, we have

$$\text{emf} = \oint \mathbf{E} \cdot d\mathbf{L} = -\frac{d}{dt}\int_S \mathbf{B} \cdot d\mathbf{S} \tag{4}$$

[3] Henri Frederic Emile Lenz was born in Germany but worked in Russia. He published his law in 1834.

where the fingers of our right hand indicate the direction of the closed path, and our thumb indicates the direction of $d\mathbf{S}$. A flux density $\mathbf{B}$ in the direction of $d\mathbf{S}$ and increasing with time thus produces an average value of $\mathbf{E}$ which is *opposite* to the positive direction about the closed path. The right-handed relationship between the surface integral and the closed line integral in (4) should always be kept in mind during flux integrations and emf determinations.

Let us divide our investigation into two parts by first finding the contribution to the total emf made by a changing field within a stationary path (transformer emf), and then we will consider a moving path within a constant (motional, or generator, emf).

We first consider a stationary path. The magnetic flux is the only time-varying quantity on the right side of (4), and a partial derivative may be taken under the integral sign,

$$\text{emf} = \oint \mathbf{E}\cdot d\mathbf{L} = -\int_S \frac{\partial \mathbf{B}}{\partial t}\cdot d\mathbf{S} \tag{5}$$

Before we apply this simple result to an example, let us obtain the point form of this integral equation. Applying Stokes' theorem to the closed line integral, we have

$$\int_S (\nabla \times \mathbf{E})\cdot d\mathbf{S} = -\int_S \frac{\partial \mathbf{B}}{\partial t}\cdot d\mathbf{S}$$

where the surface integrals may be taken over identical surfaces. The surfaces are perfectly general and may be chosen as differentials,

$$(\nabla \times \mathbf{E})\cdot d\mathbf{S} = -\frac{\partial \mathbf{B}}{\partial t}\cdot d\mathbf{S}$$

and

$$\boxed{\nabla \times \mathbf{E} = -\frac{\partial \mathbf{B}}{\partial t}} \tag{6}$$

This is one of Maxwell's four equations as written in differential, or point, form, the form in which they are most generally used. Equation (5) is the integral form of this equation and is equivalent to Faraday's law as applied to a fixed path. If $\mathbf{B}$ is not a function of time, (5) and (6) evidently reduce to the electrostatic equations

$$\oint \mathbf{E}\cdot d\mathbf{L} = 0 \quad \text{(electrostatics)}$$

and

$$\nabla \times \mathbf{E} = 0 \quad \text{(electrostatics)}$$

As an example of the interpretation of (5) and (6), let us assume a simple magnetic field which increases exponentially with time within the cylindrical region $\rho < b$,

$$\mathbf{B} = B_0 e^{kt}\mathbf{a}_z \tag{7}$$

where B_0 = constant. Choosing the circular path $\rho = a$, $a < b$, in the $z = 0$ plane, along which E_ϕ must be constant by symmetry, we then have from (5)

$$\text{emf} = 2\pi a E_\phi = -kB_0 e^{kt}\pi a^2$$

The emf around this closed path is $-kB_0e^{kt}\pi a^2$. It is proportional to a^2 because the magnetic flux density is uniform and the flux passing through the surface at any instant is proportional to the area.

If we now replace a with ρ, $\rho < b$, the electric field intensity at any point is

$$\mathbf{E} = -\tfrac{1}{2}kB_0e^{kt}\rho\mathbf{a}_\phi \tag{8}$$

Let us now attempt to obtain the same answer from (6), which becomes

$$(\nabla \times \mathbf{E})_z = -kB_0e^{kt} = \frac{1}{\rho}\frac{\partial(\rho E_\phi)}{\partial \rho}$$

Multiplying by ρ and integrating from 0 to ρ (treating t as a constant, since the derivative is a partial derivative),

$$-\tfrac{1}{2}kB_0e^{kt}\rho^2 = \rho E_\phi$$

or

$$\mathbf{E} = -\tfrac{1}{2}kB_0e^{kt}\rho\mathbf{a}_\phi$$

once again.

If B_0 is considered positive, a filamentary conductor of resistance R would have a current flowing in the negative $\mathbf{a}_\phi$ direction, and this current would establish a flux within the circular loop in the negative $\mathbf{a}_z$ direction. Since E_ϕ increases exponentially with time, the current and flux do also, and thus they tend to reduce the time rate of increase of the applied flux and the resultant emf in accordance with Lenz's law.

Before leaving this example, it is well to point out that the given field **B** does not satisfy all of Maxwell's equations. Such fields are often assumed (*always* in ac-circuit problems) and cause no difficulty when they are interpreted properly. They occasionally cause surprise, however. This particular field is discussed further in Problem 10.19 at the end of the chapter.

Now let us consider the case of a time-constant flux and a moving closed path. Before we derive any special results from Faraday's law (1), let us use the basic law to analyze the specific problem outlined in Figure 10.1. The closed circuit consists of two parallel conductors which are connected at one end by a high-resistance voltmeter of negligible dimensions and at the other end by a sliding bar moving at a velocity **v**. The magnetic flux density **B** is constant (in space and time) and is normal to the plane containing the closed path.

Let the position of the shorting bar be given by y; the flux passing through the surface within the closed path at any time t is then

$$\Phi = Byd$$

From (1), we obtain

$$\text{emf} = -\frac{d\Phi}{dt} = -B\frac{dy}{dt}d = -B\upsilon d \tag{9}$$

The emf is defined as $\oint \mathbf{E}\cdot d\mathbf{L}$ and we have a conducting path, so we may actually determine **E** at every point along the closed path. We found in electrostatics that the tangential component of **E** is zero at the surface of a conductor, and we shall show in Section 10.4 that the tangential component is zero at the surface of a *perfect* conductor ($\sigma = \infty$) for all time-varying conditions. This is equivalent to saying that a perfect conductor is a "short circuit." The entire closed path in Figure 10.1 may be considered a perfect conductor, with the exception of the voltmeter. The actual computation of $\oint \mathbf{E}\cdot d\mathbf{L}$ then must involve no contribution along the entire moving bar, both rails, and the voltmeter leads. Since we are integrating in a counterclockwise direction (keeping the interior of the positive side of the surface on our left as usual), the contribution $E\,\Delta L$ across

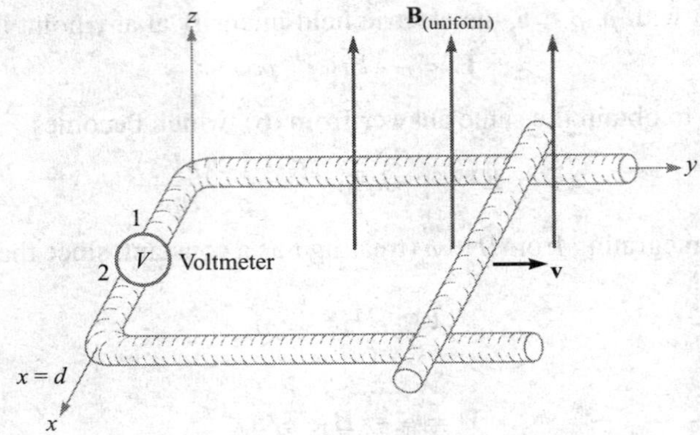

Fig. 10.1 An example illustrating the application of Faraday's law to the case of a constant magnetic flux density $\mathbf{B}$ and a moving path. The shorting bar moves to the right with a velocity $\mathbf{v}$, and the circuit is completed through the two rails and an extremely small high-resistance voltmeter. The voltmeter reading is $V_{12} = -B\nu d$.

the voltmeter must be $-B\nu d$, showing that the electric field intensity in the instrument is directed from terminal 2 to terminal 1. For an up-scale reading, the positive terminal of the voltmeter should therefore be terminal 2.

The direction of the resultant small current flow may be confirmed by noting that the enclosed flux is reduced by a clockwise current in accordance with Lenz's law. The voltmeter terminal 2 is again seen to be the positive terminal.

Let us now consider this example using the concept of *motional emf*. The force on a charge Q moving at a velocity $\mathbf{v}$ in a magnetic field $\mathbf{B}$ is

$$\mathbf{F} = Q\mathbf{v} \times \mathbf{B}$$

or

$$\frac{\mathbf{F}}{Q} = \mathbf{v} \times \mathbf{B} \tag{10}$$

The sliding conducting bar is composed of positive and negative charges, and each experiences this force. The force per unit charge, as given by (10), is called the *motional* electric field intensity $\mathbf{E}_m$,

$$\mathbf{E}_m = \mathbf{v} \times \mathbf{B} \tag{11}$$

If the moving conductor were lifted off the rails, this electric field intensity would force electrons to one end of the bar (the far end) until the *static field* due to these charges just balanced the field induced by the motion of the bar. The resultant tangential electric field intensity would then be zero along the length of the bar.

The motional emf produced by the moving conductor is then

$$\text{emf} = \oint \mathbf{E}_m \cdot d\mathbf{L} = \oint (\mathbf{v} \times \mathbf{B}) \cdot d\mathbf{L} \tag{12}$$

where the last integral may have a nonzero value only along that portion of the path which is in motion, or along which $\mathbf{v}$ has some nonzero value.

Evaluating the right side of (12), we obtain

$$\oint(\mathbf{v}\times\mathbf{B})\cdot d\mathbf{L}=\int_d^0 \nu B\,dx=-B\nu d$$

as before. This is the total emf, since **B** is not a function of time.

In the case of a conductor moving in a uniform constant magnetic field, we may therefore ascribe a motional electric field intensity $\mathbf{E}_m=\mathbf{v}\times\mathbf{B}$ to every portion of the moving conductor and evaluate the resultant emf by

$$\text{emf}=\oint\mathbf{E}\cdot d\mathbf{L}=\oint\mathbf{E}_m\cdot d\mathbf{L}=\oint(\mathbf{v}\times\mathbf{B})\cdot d\mathbf{L} \tag{13}$$

If the magnetic flux density is also changing with time, then we must include both contributions, the transformer emf (5) and the motional emf (12),

$$\text{emf}=\oint\mathbf{E}\cdot d\mathbf{L}=-\int_S\frac{\partial\mathbf{B}}{\partial t}\cdot d\mathbf{S}+\oint(\mathbf{v}\times\mathbf{B})\cdot d\mathbf{L} \tag{14}$$

This expression is equivalent to the simple statement

$$\text{emf}=-\frac{d\Phi}{dt} \tag{1}$$

and either can be used to determine these induced voltages.

Although (1) appears simple, there are a few contrived examples in which its proper application is quite difficult. These usually involve sliding contacts or switches; they always involve the substitution of one part of a circuit by a new part.[4] As an example, consider the simple circuit of Figure 10.2, which contains several perfectly conducting wires, an ideal voltmeter, a uniform constant field **B**, and a switch. When the switch is opened, there is obviously more flux enclosed in the voltmeter circuit; however, it continues to read zero. The change in flux has not been produced by either a time-changing **B** [first term of (14)] or a conductor moving through a magnetic field [second part of (14)]. Instead, a new circuit has been substituted for the old. Thus it is necessary to use care in evaluating the change in flux linkages.

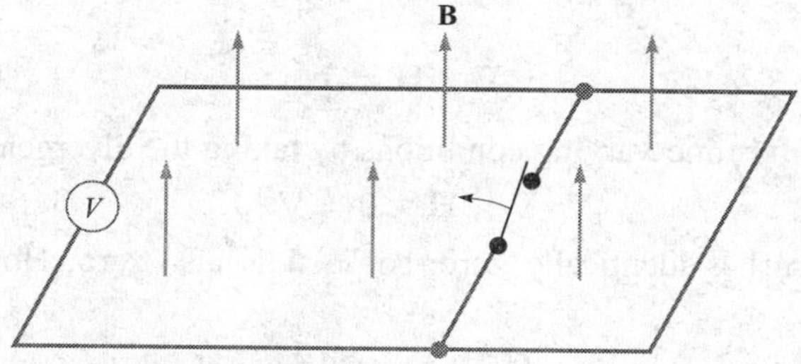

Fig. 10.2 An apparent increase in flux linkages does not lead to an induced voltage when one part of a circuit is simply substituted for another by opening the switch. No indication will be observed on the voltmeter.

The separation of the emf into the two parts indicated by (14), one due to the time rate of change of **B** and the other to the motion of the circuit, is somewhat arbitrary in that it depends on the relative velocity of the *observer* and the system. A field that is changing with both time and space

[4] See Bewley, in References at the end of the chapter, particularly pp. 12–19.

may look constant to an observer moving with the field. This line of reasoning is developed more fully in applying the special theory of relativity to electromagnetic theory.[5]

D10.1 Within a certain region, $\epsilon = 10^{-11}$ F/m and $\mu = 10^{-5}$ H/m. If $B_x = 2 \times 10^{-4} \cos 10^5 t \sin 10^{-3} y$ T: (*a*) use $\nabla \times \mathbf{H} = \epsilon \frac{\partial \mathbf{E}}{\partial t}$ to find **E**; (*b*) find the total magnetic flux passing through the surface $x = 0$, $0 < y < 40$ m, $0 < z < 2$ m, at $t = 1$ μs; (*c*) find the value of the closed line integral of **E** around the perimeter of the given surface.

Ans. $-20\,000 \sin 10^5 t \cos 10^{-3} y \mathbf{a}_z$ V/m; 0.318 mWb; −3.19 V

D10.2 With reference to the sliding bar shown in Figure 10.1, let $d = 7$ cm, $\mathbf{B} = 0.3\mathbf{a}_z$ T, and $\mathbf{v} = 0.1\mathbf{a}_y e^{20y}$ m/s. Let $y = 0$ at $t = 0$. Find: (*a*) $\nu(t = 0)$; (*b*) $y(t = 0.1)$; (*c*) $\nu(t = 0.1)$; (*d*) V_{12} at $t = 0.1$.

Ans. 0.1 m/s; 1.12 cm; 0.125 m/s; −2.63 mV

10.2 DISPLACEMENT CURRENT

Faraday's experimental law has been used to obtain one of Maxwell's equations in differential form,

$$\nabla \times \mathbf{E} = -\frac{\partial \mathbf{B}}{\partial t} \tag{15}$$

which shows us that a time-changing magnetic field produces an electric field. Remembering the definition of curl, we see that this electric field has the special property of circulation; its line integral about a general closed path is not zero. Now let us turn our attention to the time-changing electric field.

We should first look at the point form of Ampère's circuital law as it applies to steady magnetic fields,

$$\nabla \times \mathbf{H} = \mathbf{J} \tag{16}$$

and show its inadequacy for time-varying conditions by taking the divergence of each side,

$$\nabla \cdot \nabla \times \mathbf{H} \equiv 0 = \nabla \cdot \mathbf{J}$$

The divergence of the curl is identically zero, so $\nabla \cdot \mathbf{J}$ is also zero. However, the equation of continuity,

$$\nabla \cdot \mathbf{J} = -\frac{\partial \rho_\nu}{\partial t}$$

then shows us that (16) can be true only if $\partial \rho_\nu / \partial t = 0$. This is an unrealistic limitation, and (16) must be amended before we can accept it for time-varying fields. Suppose we add an unknown term **G** to (16),

$$\nabla \times \mathbf{H} = \mathbf{J} + \mathbf{G}$$

[5] This is discussed in several of the references listed in the References at the end of the chapter. See Panofsky and Phillips, pp. 142–51; Owen, pp. 231–45; and Harman in several places.

Again taking the divergence, we have

$$0 = \nabla \cdot \mathbf{J} + \nabla \cdot \mathbf{G}$$

Thus

$$\nabla \cdot \mathbf{G} = \frac{\partial \rho_\nu}{\partial t}$$

Replacing ρ_ν with $\nabla \cdot \mathbf{D}$,

$$\nabla \cdot \mathbf{G} = \frac{\partial}{\partial t}(\nabla \cdot \mathbf{D}) = \nabla \cdot \frac{\partial \mathbf{D}}{\partial t}$$

from which we obtain the simplest solution for $\mathbf{G}$,

$$\mathbf{G} = \frac{\partial \mathbf{D}}{\partial t}$$

Ampère's circuital law in point form therefore becomes

$$\nabla \times \mathbf{H} = \mathbf{J} + \frac{\partial \mathbf{D}}{\partial t} \tag{17}$$

Equation (17) has not been derived. It is merely a form we have obtained which does not disagree with the continuity equation. It is also consistent with all our other results, and we accept it as we did each experimental law and the equations derived from it. We are building a theory, and we have every right to our equations *until they are proved wrong*. This has not yet been done.

We now have a second one of Maxwell's equations and shall investigate its significance. The additional term $\partial \mathbf{D}/\partial t$ has the dimensions of current density, amperes per square meter. Since it results from a time-varying electric flux density (or displacement density), Maxwell termed it a *displacement current density*. We sometimes denote it by $\mathbf{J}_d$:

$$\nabla \times \mathbf{H} = \mathbf{J} + \mathbf{J}_d$$

$$\mathbf{J}_d = \frac{\partial \mathbf{D}}{\partial t}$$

This is the third type of current density we have met. Conduction current density,

$$\mathbf{J} = \sigma \mathbf{E}$$

is the motion of charge (usually electrons) in a region of zero net charge density, and convection current density,

$$\mathbf{J} = \rho_\nu \mathbf{v}$$

is the motion of volume charge density. Both are represented by $\mathbf{J}$ in (17). Bound current density is, of course, included in $\mathbf{H}$. In a nonconducting medium in which no volume charge density is present, $\mathbf{J} = 0$, and then

$$\nabla \times \mathbf{H} = \frac{\partial \mathbf{D}}{\partial t} \quad (\text{if } \mathbf{J} = 0) \tag{18}$$

Notice the symmetry between (18) and (15):

$$\nabla \times \mathbf{E} = -\frac{\partial \mathbf{B}}{\partial t} \tag{15}$$

Again the analogy between the intensity vectors $\mathbf{E}$ and $\mathbf{H}$ and the flux density vectors $\mathbf{D}$ and $\mathbf{B}$ is apparent. We cannot place too much faith in this analogy, however, for it fails when we investigate forces on particles. The force on a charge is related to **E** and to **B**, and some good arguments may be presented showing an analogy between **E** and **B** and between **D** and **H**. We shall omit them, however, and merely say that the concept of displacement current was probably suggested to Maxwell by the symmetry first mentioned in this paragraph.[6]

The total displacement current crossing any given surface is expressed by the surface integral,

$$I_d = \int_S \mathbf{J}_d \cdot d\mathbf{S} = \int_S \frac{\partial \mathbf{D}}{\partial t} \cdot d\mathbf{S}$$

and we may obtain the time-varying version of Ampère's circuital law by integrating (17) over the surface S,

$$\int_S (\nabla \times \mathbf{H}) \cdot d\mathbf{S} = \int_S \mathbf{J} \cdot d\mathbf{S} + \int_S \frac{\partial \mathbf{D}}{\partial t} \cdot d\mathbf{S}$$

and applying Stokes' theorem,

$$\oint \mathbf{H} \cdot d\mathbf{L} = I + I_d = I + \int_S \frac{\partial \mathbf{D}}{\partial t} \cdot d\mathbf{S} \qquad (19)$$

What is the nature of displacement current density? Let us study the simple circuit of Figure 10.3, which contains a filamentary loop and a parallel-plate capacitor. Within the loop, a magnetic field varying sinusoidally with time is applied to produce an emf about the closed path (the filament plus the dashed portion between the capacitor plates), which we shall take as

$$\text{emf} = V_0 \cos \omega t$$

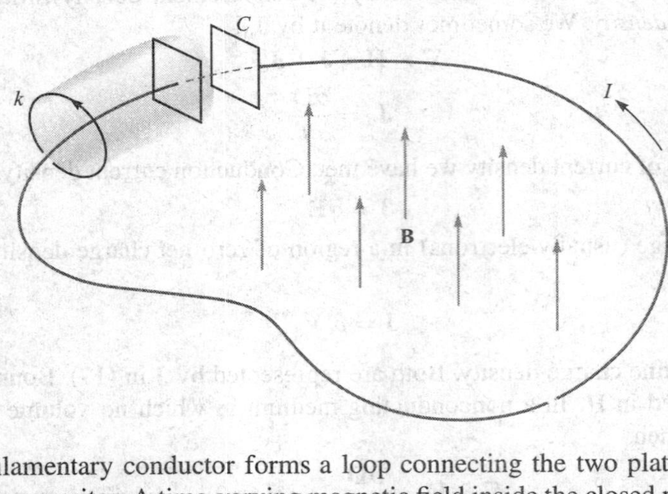

Fig. 10.3 A filamentary conductor forms a loop connecting the two plates of a parallel-plate capacitor. A time-varying magnetic field inside the closed path produces an emf of $V_0 \cos \omega t$ around the closed path. The conduction current I is equal to the displacement current between the capacitor plates.

[6] The analogy that relates **B** to **D** and **H** to **E** is strongly advocated by Fano, Chu, and Adler (see References for Chapter 6) on pp.159–60 and 179; the case for comparing **B** to **E** and **D** to **H** is presented in Halliday and Resnick (see References for this chapter) on pp. 665–68 and 832–36.

Using elementary circuit theory and assuming that the loop has negligible resistance and inductance, we may obtain the current in the loop as

$$I = -\omega C V_0 \sin \omega t$$
$$= -\omega \frac{\epsilon S}{d} V_0 \sin \omega t$$

where the quantities ϵ, S, and d pertain to the capacitor. Let us apply Ampère's circuital law about the smaller closed circular path k and neglect displacement current for the moment:

$$\oint_k \mathbf{H} \cdot d\mathbf{L} = I_k$$

The path and the value of **H** along the path are both definite quantities (although difficult to determine), and $\oint_k \mathbf{H} \cdot d\mathbf{L}$ is a definite quantity. The current I_k is that current through every surface whose perimeter is the path k. If we choose a simple surface punctured by the filament, such as the plane circular surface defined by the circular path k, the current is evidently the conduction current. Suppose now we consider the closed path k as the mouth of a paper bag whose bottom passes between the capacitor plates. The bag is not pierced by the filament, and the conductor current is zero. Now we need to consider displacement current, for within the capacitor

$$D = \epsilon E = \epsilon \left(\frac{V_0}{d} \cos \omega t \right)$$

and therefore

$$I_d = \frac{\partial D}{\partial t} S = -\omega \frac{\epsilon S}{d} V_0 \sin \omega t$$

This is the same value as that of the conduction current in the filamentary loop. Therefore the application of Ampère's circuital law, including displacement current to the path k, leads to a definite value for the line integral of **H**. This value must be equal to the total current crossing the chosen surface. For some surfaces the current is almost entirely conduction current, but for those surfaces passing between the capacitor plates, the conduction current is zero, and it is the displacement current which is now equal to the closed line integral of **H**.

Physically, we should note that a capacitor stores charge and that the electric field between the capacitor plates is much greater than the small leakage fields outside. We therefore introduce little error when we neglect displacement current on all those surfaces which do not pass between the plates.

Displacement current is associated with time-varying electric fields and therefore exists in all imperfect conductors carrying a time-varying conduction current. The last part of the following drill problem indicates the reason why this additional current was never discovered experimentally.

D10.3 Find the amplitude of the displacement current density: (*a*) adjacent to an automobile antenna where the magnetic field intensity of an FM signal is $H_x = 0.15 \cos[3.12(3 \times 10^8 t - y)]$ A/m; (*b*) in the air space at a point within a large power distribution transformer where $\mathbf{B} = 0.8 \cos[1.257 \times 10^{-6}(3 \times 10^8 t - x)]\mathbf{a}_y$ T; (*c*) within a large, oil-filled power capacitor where $\epsilon_r = 5$ and $\mathbf{E} = 0.9 \cos[1.257 \times 10^{-6}(3 \times 10^8 t - z\sqrt{5})]\mathbf{a}_x$ MV/m; (*d*) in a metallic conductor at 60 Hz, if $\epsilon = \epsilon_0$, $\mu = \mu_0$, $\sigma = 5.8 \times 10^7$ S/m, and $\mathbf{J} = \sin(377t - 117.1z)\mathbf{a}_x$ MA/m^2.

Ans. 0.468 A/m^2; 0.800 A/m^2; 0.0150 A/m^2; 57.6 pA/m^2

10.3 MAXWELL'S EQUATIONS IN POINT FORM

We have already obtained two of Maxwell's equations for time-varying fields,

$$\nabla \times \mathbf{E} = -\frac{\partial \mathbf{B}}{\partial t} \tag{20}$$

and

$$\nabla \times \mathbf{H} = \mathbf{J} + \frac{\partial \mathbf{D}}{\partial t} \tag{21}$$

The remaining two equations are unchanged from their non–time-varying form:

$$\nabla \cdot \mathbf{D} = \rho_\nu \tag{22}$$

$$\nabla \cdot \mathbf{B} = 0 \tag{23}$$

Equation (22) essentially states that charge density is a source (or sink) of electric flux lines. Note that we can no longer say that *all* electric flux begins and terminates on charge, because the point form of Faraday's law (20) shows that **E**, and hence **D**, may have circulation if a changing magnetic field is present. Thus the lines of electric flux may form closed loops. However, the converse is still true, and every coulomb of charge must have one coulomb of electric flux diverging from it.

Equation (23) again acknowledges the fact that "magnetic charges," or poles, are not known to exist. Magnetic flux is always found in closed loops and never diverges from a point source.

These four equations form the basis of all electromagnetic theory. They are partial differential equations and relate the electric and magnetic fields to each other and to their sources, charge and current density. The auxiliary equations relating **D** and **E**,

$$\mathbf{D} = \epsilon \mathbf{E} \tag{24}$$

relating **B** and **H**,

$$\mathbf{B} = \mu \mathbf{H} \tag{25}$$

defining conduction current density,

$$\mathbf{J} = \sigma \mathbf{E} \tag{26}$$

and defining convection current density in terms of the volume charge density ρ_ν,

$$\mathbf{J} = \rho_\nu \mathbf{v} \tag{27}$$

are also required to define and relate the quantities appearing in Maxwell's equations.

The potentials V and **A** have not been included because they are not strictly necessary, although they are extremely useful. They will be discussed at the end of this chapter.

If we do not have "nice" materials to work with, then we should replace (24) and (25) with the relationships involving the polarization and magnetization fields,

$$\mathbf{D} = \epsilon_0 \mathbf{E} + \mathbf{P} \tag{28}$$

$$\mathbf{B} = \mu_0(\mathbf{H} + \mathbf{M}) \tag{29}$$

For linear materials we may relate **P** to **E**

$$\mathbf{P} = \chi_e \epsilon_0 \mathbf{E} \tag{30}$$

and **M** to **H**

$$\mathbf{M} = \chi_m \mathbf{H} \tag{31}$$

Finally, because of its fundamental importance we should include the Lorentz force equation, written in point form as the force per unit volume,

$$\mathbf{f} = \rho_\nu(\mathbf{E} + \mathbf{v} \times \mathbf{B}) \tag{32}$$

The following chapters are devoted to the application of Maxwell's equations to several simple problems.

D10.4 Let $\mu = 10^{-5}$ H/m, $\epsilon = 4 \times 10^{-9}$ F/m, $\sigma = 0$, and $\rho_\nu = 0$. Find k (including units) so that each of the following pairs of fields satisfies Maxwell's equations: (*a*) $\mathbf{D} = 6\mathbf{a}_x - 2y\mathbf{a}_y + 2z\mathbf{a}_z$ nC/m^2, $\mathbf{H} = kx\mathbf{a}_x + 10y\mathbf{a}_y - 25z\mathbf{a}_z$ A/m; (*b*) $\mathbf{E} = (20y - kt)\mathbf{a}_x$ V/m, $\mathbf{H} = (y + 2 \times 10^6 t)\mathbf{a}_z$ A/m.

Ans. 15 A/m^2; -2.5×10^8 V/(m · s)

10.4 MAXWELL'S EQUATIONS IN INTEGRAL FORM

The integral forms of Maxwell's equations are usually easier to recognize in terms of the experimental laws from which they have been obtained by a generalization process. Experiments must treat physical macroscopic quantities, and their results therefore are expressed in terms of integral relationships. A differential equation always represents a theory. Let us now collect the integral forms of Maxwell's equations from Section 10.3.

Integrating (20) over a surface and applying Stokes' theorem, we obtain Faraday's law,

$$\oint \mathbf{E} \cdot d\mathbf{L} = -\int_S \frac{\partial \mathbf{B}}{\partial t} \cdot d\mathbf{S} \tag{33}$$

and the same process applied to (21) yields Ampère's circuital law,

$$\oint \mathbf{H} \cdot d\mathbf{L} = I + \int_S \frac{\partial \mathbf{D}}{\partial t} \cdot d\mathbf{S} \tag{34}$$

Gauss's laws for the electric and magnetic fields are obtained by integrating (22) and (23) throughout a volume and using the divergence theorem:

$$\oint_S \mathbf{D} \cdot d\mathbf{S} = \int_{\text{vol}} \rho_L \, dv \tag{35}$$

$$\oint_S \mathbf{B} \cdot d\mathbf{S} = 0 \tag{36}$$

These four integral equations enable us to find the boundary conditions on **B**, **D**, **H**, and **E** which are necessary to evaluate the constants obtained in solving Maxwell's equations in partial differential form. These boundary conditions are in general unchanged from their forms for static or steady fields, and the same methods may be used to obtain them. Between any two real physical media (where **K** must be zero on the boundary surface), (33) enables us to relate the tangential **E**-field components,

$$E_{t1} = E_{t2} \tag{37}$$

and from (34),

$$H_{t1} = H_{t2} \tag{38}$$

The surface integrals produce the boundary conditions on the normal components,

$$D_{N1} - D_{N2} = \rho_S \tag{39}$$

and

$$B_{N1} = B_{N2} \tag{40}$$

It is often desirable to idealize a physical problem by assuming a perfect conductor for which σ is infinite but **J** is finite. From Ohm's law, then, in a perfect conductor,

$$\mathbf{E} = 0$$

and it follows from the point form of Faraday's law that

$$\mathbf{H} = 0$$

for time-varying fields. The point form of Ampère's circuital law then shows that the finite value of **J** is

$$\mathbf{J} = 0$$

and current must be carried on the conductor surface as a surface current **K**. Thus, if region 2 is a perfect conductor, (37) to (40) become, respectively,

$$E_{t1} = 0 \tag{41}$$

$$H_{t1} = K \quad (\mathbf{H}_{t1} = \mathbf{K} \times \mathbf{a}_N) \tag{42}$$

$$D_{N1} = \rho_s \tag{43}$$

$$B_{N1} = 0 \tag{44}$$

where $\mathbf{a}_N$ is an outward normal at the conductor surface.

Note that surface charge density is considered a physical possibility for either dielectrics, perfect conductors, or imperfect conductors, but that surface *current* density is assumed only in conjunction with perfect conductors.

The preceding boundary conditions are a very necessary part of Maxwell's equations. All real physical problems have boundaries and require the solution of Maxwell's equations in two or more regions and the matching of these solutions at the boundaries. In the case of perfect conductors, the solution of the equations within the conductor is trivial (all time-varying fields are zero), but the application of the boundary conditions (41) to (44) may be very difficult.

Certain fundamental properties of wave propagation are evident when Maxwell's equations are solved for an *unbounded* region. This problem is treated in Chapter 12. It represents the simplest application of Maxwell's equations because it is the only problem which does not require the application of any boundary conditions.

D10.5 The unit vector $0.64\mathbf{a}_x + 0.6\mathbf{a}_y - 0.48\mathbf{a}_z$ is directed from region 2 ($\epsilon_r = 2, \mu_r = 3, \sigma_2 = 0$) toward region 1 ($\epsilon_{r1} = 4, \mu_{r1} = 2, \sigma_1 = 0$). If $\mathbf{B}_1 = (\mathbf{a}_x - 2\mathbf{a}_y + 3\mathbf{a}_z)\sin 300t$ T at point P in region 1 adjacent to the boundary, find the amplitude at P of: (*a*) $\mathbf{B}_{N1}$; (*b*) $\mathbf{B}_{t1}$; (*c*) $\mathbf{B}_{N2}$; (*d*) $\mathbf{B}_2$.

Ans. 2.00 T; 3.16 T; 2.00 T; 5.15 T

D10.6 The surface $y = 0$ is a perfectly conducting plane, while the region $y > 0$ has $\epsilon_r = 5$, $\mu_r = 3$, and $\sigma = 0$. Let $\mathbf{E} = 20\cos(2 \times 10^8 t - 2.58z)\mathbf{a}_y$ V/m for $y > 0$, and find at $t = 6$ ns; (*a*) ρ_S at $P(2, 0, 0.3)$; (*b*) $\mathbf{H}$ at P; (*c*) $\mathbf{K}$ at P.

Ans. 0.81 nC/m^2; $-62.3\mathbf{a}_x$ mA/m; $-62.3\mathbf{a}_z$ mA/m

10.5 THE RETARDED POTENTIALS

Animations

The time-varying potentials, usually called *retarded* potentials for a reason which we shall see shortly, find their greatest application in radiation problems in which the distribution of the source is known approximately. We should remember that the scalar electric potential V may be expressed in terms of a static charge distribution,

$$V = \int_{\text{vol}} \frac{\rho_\nu d\nu}{4\pi\epsilon R} \quad \text{(static)} \tag{45}$$

and the vector magnetic potential may be found from a current distribution which is constant with time,

$$\mathbf{A} = \int_{\text{vol}} \frac{\mu \mathbf{J}\, d\nu}{4\pi R} \quad \text{(dc)} \tag{46}$$

The differential equations satisfied by V,

$$\nabla^2 V = -\frac{\rho_\nu}{\epsilon} \quad \text{(static)} \tag{47}$$

and $\mathbf{A}$,

$$\nabla^2 \mathbf{A} = -\mu \mathbf{J} \quad \text{(dc)} \tag{48}$$

may be regarded as the point forms of the integral equations (45) and (46), respectively.

Having found V and $\mathbf{A}$, the fundamental fields are then simply obtained by using the gradient,

$$\mathbf{E} = -\nabla V \quad \text{(static)} \tag{49}$$

or the curl,

$$\mathbf{B} = \nabla \times \mathbf{A} \quad \text{(dc)} \tag{50}$$

We now wish to define suitable time-varying potentials which are consistent with the preceding expressions when only static charges and direct currents are involved.

Equation (50) apparently is still consistent with Maxwell's equations. These equations state that $\nabla \cdot \mathbf{B} = 0$, and the divergence of (50) leads to the divergence of the curl which is identically zero. Let us therefore tentatively accept (50) as satisfactory for time-varying fields and turn our attention to (49).

The inadequacy of (49) is obvious because application of the curl operation to each side and recognition of the curl of the gradient as being identically zero confront us with $\nabla \times \mathbf{E} = 0$. However, the point form of Faraday's law states that $\nabla \times \mathbf{E}$ is not generally zero, so let us try to effect an improvement by adding an unknown term to (49),

$$\mathbf{E} = -\nabla V + \mathbf{N}$$

taking the curl,

$$\nabla \times \mathbf{E} = 0 + \nabla \times \mathbf{N}$$

using the point form of Faraday's law,

$$\nabla \times \mathbf{N} = -\frac{\partial \mathbf{B}}{\partial t}$$

and using (50), giving us

$$\nabla \times \mathbf{N} = -\frac{\partial}{\partial t}(\nabla \times \mathbf{A})$$

or

$$\nabla \times \mathbf{N} = -\nabla \times \frac{\partial \mathbf{A}}{\partial t}$$

The simplest solution of this equation is

$$\mathbf{N} = -\frac{\partial \mathbf{A}}{\partial t}$$

and this leads to

$$\boxed{\mathbf{E} = -\nabla V - \frac{\partial \mathbf{A}}{\partial t}} \tag{51}$$

We still must check (50) and (51) by substituting them into the remaining two of Maxwell's equations:

$$\nabla \times \mathbf{H} = \mathbf{J} + \frac{\partial \mathbf{D}}{\partial t}$$

$$\nabla \cdot \mathbf{D} = \rho_\nu$$

Doing this, we obtain the more complicated expressions:

$$\frac{1}{\mu}\nabla \times \nabla \times \mathbf{A} = \mathbf{J} + \epsilon \left(-\nabla \frac{\partial V}{\partial t} - \frac{\partial^2 \mathbf{A}}{\partial t^2} \right)$$

and

$$\epsilon \left(-\nabla \cdot \nabla V - \frac{\partial}{\partial t} \nabla \cdot \mathbf{A} \right) = \rho_\nu$$

or

$$\nabla(\nabla \cdot \mathbf{A}) - \nabla^2 \mathbf{A} = \mu \mathbf{J} - \mu\epsilon \left(\nabla \frac{\partial V}{\partial t} + \frac{\partial^2 \mathbf{A}}{\partial t^2} \right) \tag{52}$$

and

$$\nabla^2 V + \frac{\partial}{\partial t}(\nabla \cdot \mathbf{A}) = -\frac{\rho_\nu}{\epsilon} \tag{53}$$

There is no apparent inconsistency in (52) and (53). Under static or dc conditions $\nabla \cdot \mathbf{A} = 0$, and (52) and (53) reduce to (48) and (47), respectively. We shall therefore assume that the time-varying potentials may be defined in such a way that **B** and **E** may be obtained from them through (50) and (51). These latter two equations do not serve, however, to define **A** and V *completely*. They represent necessary, but not sufficient, conditions. Our initial assumption was merely that $B = \nabla \times \mathbf{A}$, and a vector cannot be defined by giving its curl alone. Suppose, for example, that we have a very simple vector potential field in which A_y and A_z are zero. Expansion of (50) leads to

$$B_x = 0$$
$$B_y = \frac{\partial A_x}{\partial z}$$
$$B_z = -\frac{\partial A_x}{\partial y}$$

and we see that no information is available about the manner in which A_x varies with x. This information could be found if we also knew the value of the divergence of **A**, for in our example

$$\nabla \cdot \mathbf{A} = \frac{\partial A_x}{\partial x}$$

Finally, we should note that our information about **A** is given only as partial derivatives and that a space-constant term might be added. In all physical problems in which the region of the solution extends to infinity, this constant term must be zero, for there can be no fields at infinity.

Generalizing from this simple example, we may say that a vector field is defined completely when both its curl and divergence are given and when its value is known at any one point (including infinity). We are therefore at liberty to specify the divergence of **A**, and we do so with an eye on (52) and (53), seeking the simplest expressions. We define

$$\nabla \cdot \mathbf{A} = -\mu\epsilon \frac{\partial V}{\partial t} \tag{54}$$

and (52) and (53) become

$$\nabla^2 \mathbf{A} = -\mu \mathbf{J} + \mu\epsilon \frac{\partial^2 \mathbf{A}}{\partial t^2} \tag{55}$$

and

$$\nabla^2 V = -\frac{\rho_\nu}{\epsilon} + \mu\epsilon \frac{\partial^2 V}{\partial t^2} \tag{56}$$

These equations are related to the wave equation, which will be discussed in Chapters 11 and 12. They show considerable symmetry, and we should be highly pleased with our definitions of V and $\mathbf{A}$,

$$\mathbf{B} = \nabla \times \mathbf{A} \tag{50}$$

$$\nabla \cdot \mathbf{A} = -\mu\epsilon \frac{\partial V}{\partial t} \tag{54}$$

$$\mathbf{E} = -\nabla V - \frac{\partial \mathbf{A}}{\partial t} \tag{51}$$

The integral equivalents of (45) and (46) for the time-varying potentials follow from the definitions (50), (51), and (54), but we shall merely present the final results and indicate their general nature. In Chapter 12, we will find that any electromagnetic disturbance will travel at a velocity

$$\nu = \frac{1}{\sqrt{\mu\epsilon}}$$

through any homogeneous medium described by μ and ϵ. In the case of free space this velocity turns out to be the velocity of light, approximately 3×10^8 m/s. It is logical, then, to suspect that the potential at any point is due not to the value of the charge density at some distant point at the same instant, but to its value at some previous time, because the effect propagates at a finite velocity. Thus (45) becomes

$$V = \int_{\text{vol}} \frac{[\rho_\nu]}{4\pi\epsilon R} d\nu \tag{57}$$

where $[\rho_\nu]$ indicates that every t appearing in the expression for ρ_ν has been replaced by a *retarded* time,

$$t' = t - \frac{R}{\nu}$$

Thus, if the charge density throughout space were given by

$$\rho_\nu = e^{-r} \cos \omega t$$

then

$$[\rho_\nu] = e^{-r} \cos \left[\omega \left(t - \frac{R}{\nu} \right) \right]$$

where R is the distance between the differential element of charge being considered and the point at which the potential is to be determined.

The retarded vector magnetic potential is given by

$$\mathbf{A} = \int_{\text{vol}} \frac{\mu[\mathbf{J}]}{4\pi R} d\nu \tag{58}$$

The use of a retarded time has resulted in the time-varying potentials being given the name of retarded potentials. In Chapter 14 we shall apply (58) to the simple situation of a differential current element in which I is a sinusoidal function of time. Other simple applications of (58) are considered in several problems at the end of this chapter.

We may summarize the use of the potentials by stating that a knowledge of the distribution of ρ_ν and $\mathbf{J}$ throughout space theoretically enables us to determine V and $\mathbf{A}$ from (57) and (58). The electric and magnetic fields are then obtained by applying (50) and (51). If the charge and current distributions are unknown, or reasonable approximations cannot be made for them, these potentials usually offer no easier path toward the solution than does the direct application of Maxwell's equations.

D10.7 A point charge of $4\cos 10^8\pi t\ \mu$C is located at $P_+(0, 0, 1.5)$, while $-4\cos 10^8\pi t\ \mu$C is at $P_-(0, 0, -1.5)$, both in free space. Find V at $P(r = 450, \theta, \phi = 0)$ at $t = 15$ ns for $\theta =$: (*a*) 0°; (*b*) 90°; (*c*) 45°.

Ans. 159.8 V; 0; 143 V

References

1. Bewley, L. V. *Flux Linkages and Electromagnetic Induction*. New York: Macmillan, 1952. This little book discusses many of the paradoxical examples involving induced (?) voltages.
2. Faraday, M. *Experimental Researches in Electricity*. London: B. Quaritch, 1839, 1855. Very interesting reading of early scientific research. A more recent and available source is *Great Books of the Western World,* vol. 45, Encyclopaedia Britannica, Inc., Chicago, 1952.
3. Halliday, D., R. Resnick, and J. Walker. *Fundamentals of Physics*. 5th ed. New York: John Wiley & Sons, 1997. This text is widely used in the first university-level course in physics.
4. Harman, W. W. *Fundamentals of Electronic Motion*. New York: McGraw-Hill, 1953. Relativistic effects are discussed in a clear and interesting manner.
5. Nussbaum, A. *Electromagnetic Theory for Engineers and Scientists*. Englewood Cliffs, N.J.: Prentice-Hall, 1965. See the rocket-generator example beginning on p. 211.
6. Owen, G. E. *Electromagnetic Theory*. Boston: Allyn and Bacon, 1963. Faraday's law is discussed in terms of the frame of reference in Chapter 8.
7. Panofsky, W. K. H., and M. Phillips. *Classical Electricity and Magnetism*. 2d ed. Reading, Mass.: Addison-Wesley, 1962. Relativity is treated at a moderately advanced level in Chapter 15.

CHAPTER 10 Problems

Quizzes

10.1 In Figure 10.4, let $B = 0.2\cos 120\pi t$ T, and assume that the conductor joining the two ends of the resistor is perfect. It may be assumed that the magnetic field produced by $I(t)$ is negligible. Find: (*a*) $V_{ab}(t)$; (*b*) $I(t)$.

10.2 In Figure 10.1, replace the voltmeter with a resistance, R. (*a*) Find the current I that flows as a result of the motion of the sliding bar. (*b*) The bar current results in a force exerted on the bar as it moves. Determine this force. (*c*) Determine the mechanical power required to maintain a constant velocity $\mathbf{v}$ and show that this power is equal to the power absorbed by R.

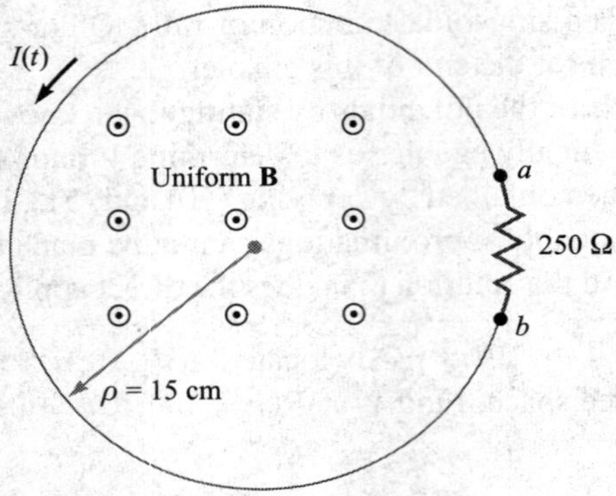

Fig. 10.4 See Problem 10.1.

10.3 Given $\mathbf{H} = 300\mathbf{a}_z \cos(3 \times 10^8 t - y)$ A/m in free space, find the emf developed in the general $\mathbf{a}_\phi$ direction about the closed path having corners at: (*a*) (0, 0, 0), (1, 0, 0), (1, 1, 0), and (0, 1, 0); (*b*) (0, 0, 0) $(2\pi, 0, 0)$, $(2\pi, 2\pi, 0)$, $(0, 2\pi, 0)$.

10.4 Conductor surfaces are located at $\rho = 1$ cm and $\rho = 2$ cm in free space. The volume 1 cm $< \rho < 2$ cm contains the fields $H_\phi = \dfrac{2}{\rho}\cos(6 \times 10^8 \pi t - 2\pi z)$ A/m and $E_\rho = \dfrac{240\pi}{\rho}\cos(6 \times 10^8 \pi t - 2\pi z)$ V/m. (*a*) Show that these two fields satisfy Eq. (6), Section 10.1. (*b*) Evaluate both integrals in Eq. (4) for the planar surface defined by $\phi = 0$, 1 cm $< \rho <$ 2 cm, $0 < z < 0.1$ m, and its perimeter, and show that the same results are obtained.

10.5 The location of the sliding bar in Figure 10.5 is given by $x = 5t + 2t^3$, and the separation of the two rails is 20 cm. Let $\mathbf{B} = 0.8x^2\mathbf{a}_z$ T. Find the voltmeter reading at: (*a*) $t = 0.4$ s; (*b*) $x = 0.6$ m.

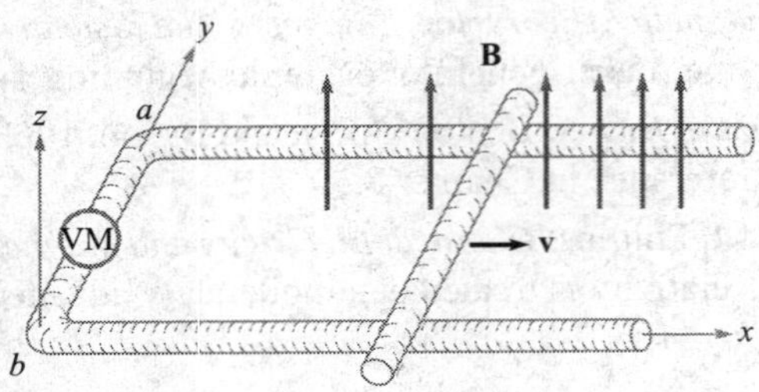

Fig. 10.5 See Problem 10.5.

10.6 A perfectly conducting filament containing a small 500 Ω resistor is formed into a square, as illustrated by Figure 10.6. Find $I(t)$ if $\mathbf{B} =$:(*a*) $0.3\cos(120\pi t - 30°)\mathbf{a}_z$ T; (*b*) $0.4\cos[\pi(ct - y)]\mathbf{a}_z$ μT, where $c = 3 \times 10^8$ m/s.

10.7 The rails in Figure 10.7 each have a resistance of 2.2 Ω/m. The bar moves to the right at a constant speed of 9 m/s in a uniform magnetic field of 0.8 T. Find $I(t)$, $0 < t < 1$ s, if the bar is at $x = 2$ m at $t = 0$ and: (*a*) a 0.3 Ω resistor is present across the left end with the right end open-circuited; (*b*) a 0.3 Ω resistor is present across each end.

10.8 Figure 10.1 is modified to show that the rail separation is larger when y is larger. Specifically, let the separation $d = 0.2 + 0.02y$. Given a uniform velocity $v_y = 8$ m/s and a uniform

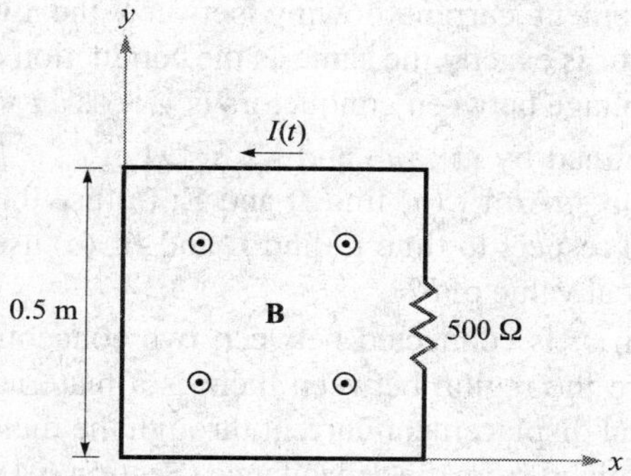

Fig. 10.6 See Problem 10.6.

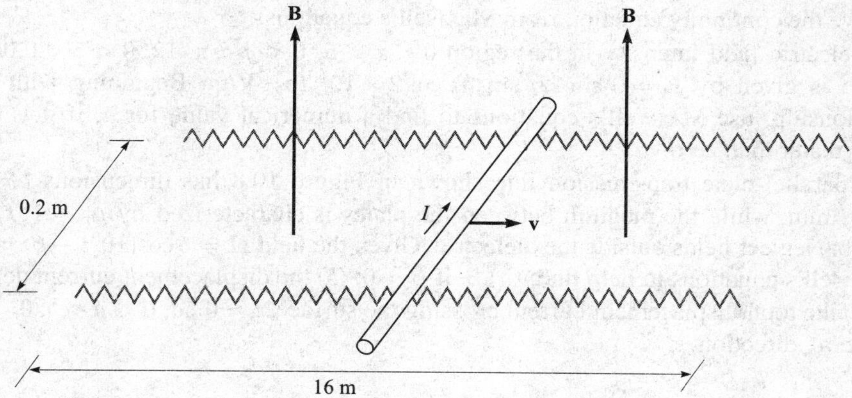

Fig. 10.7 See Problem 10.7.

magnetic flux density $B_z = 1.1$ T, find V_{12} as a function of time if the bar is located at $y = 0$ at $t = 0$.

10.9 A square filamentary loop of wire is 25 cm on a side and has a resistance of 125 Ω per meter length. The loop lies in the $z = 0$ plane with its corners at (0, 0, 0), (0.25, 0, 0), (0.25, 0.25, 0), and (0, 0.25, 0) at $t = 0$. The loop is moving with a velocity $v_y = 50$ m/s in the field $B_z = 8\cos(1.5 \times 10^8 t - 0.5x)\ \mu$T. Develop a function of time which expresses the ohmic power being delivered to the loop.

10.10 (*a*) Show that the ratio of the amplitudes of the conduction current density and the displacement current density is $\sigma/\omega\epsilon$ for the applied field $E = E_m \cos \omega t$. Assume $\mu = \mu_0$. (*b*) What is the amplitude ratio if the applied field is $E = E_m e^{-t/\tau}$, where τ is real?

10.11 Let the internal dimensions of a coaxial capacitor be $a = 1.2$ cm, $b = 4$ cm, and $l =$ 40 cm. The homogeneous material inside the capacitor has the parameters $\epsilon = 10^{-11}$ F/m, $\mu = 10^{-5}$ H/m, and $\sigma = 10^{-5}$ S/m. If the electric field intensity is $\mathbf{E} = (10^6/\rho)\cos 10^5 t\mathbf{a}_\rho$ V/m, find: (*a*) $\mathbf{J}$; (*b*) the total conduction current I_c through the capacitor; (*c*) the total displacement current I_d through the capacitor; (*d*) the ratio of the amplitude of I_d to that of I_c, the quality factor of the capacitor.

10.12 Show that the displacement current flowing between the two conducting cylinders in a lossless coaxial capacitor is exactly the same as the conduction current flowing in the external circuit if the applied voltage between conductors is $V_0 \cos \omega t$ volts.

10.13 Consider the region defined by $|x|$, $|y|$, and $|z| < 1$. Let $\epsilon_r = 5$, $\mu_r = 4$, and $\sigma = 0$. If $J_d = 20 \cos(1.5 \times 10^8 t - bx)\mathbf{a}_y$ μA/m^2: (*a*) find **D** and **E**; (*b*) use the point form of Faraday's law and an integration with respect to time to find **B** and **H**; (*c*) use $\nabla \times \mathbf{H} = \mathbf{J}_d + \mathbf{J}$ to find $\mathbf{J}_d$. (*d*) What is the numerical value of b?

10.14 A voltage source $V_0 \sin \omega t$ is connected between two concentric conducting spheres, $r = a$ and $r = b$, $b > a$, where the region between them is a material for which $\epsilon = \epsilon_r \epsilon_0$, $\mu = \mu_0$, and $\sigma = 0$. Find the total displacement current through the dielectric and compare it with the source current as determined from the capacitance (Section 6.4) and circuit-analysis methods.

10.15 Let $\mu = 3 \times 10^{-5}$ H/m, $\epsilon = 1.2 \times 10^{-10}$ F/m, and $\sigma = 0$ everywhere. If $\mathbf{H} = 2\cos(10^{10} t - \beta x)\mathbf{a}_z$ A/m, use Maxwell's equations to obtain expressions for **B**, **D**, **E**, and β.

10.16 Derive the continuity equation from Maxwell's equations.

10.17 The electric field intensity in the region $0 < x < 5$, $0 < y < \pi/12$, $0 < z < 0.06$ m in free space is given by $\mathbf{E} = C \sin 12y \sin\ az \cos 2 \times 10^{10} t\mathbf{a}_x$ V/m. Beginning with the $\nabla \times \mathbf{E}$ relationship, use Maxwell's equations to find a numerical value for a, if it is known that a is greater than zero.

10.18 The parallel-plate transmission line shown in Figure 10.8 has dimensions $b = 4$ cm and $d = 8$ mm, while the medium between the plates is characterized by $\mu_r = 1$, $\epsilon_r = 20$, and $\sigma = 0$. Neglect fields outside the dielectric. Given the field $\mathbf{H} = 5\cos(10^9 t - \beta z)\mathbf{a}_y$ A/m, use Maxwell's equations to help find: (*a*) β, if $\beta > 0$; (*b*) the displacement current density at $z = 0$; (*c*) the total displacement current crossing the surface $x = 0.5d$, $0 < y < b$, $0 < z < 0.1$ m in the $\mathbf{a}_x$ direction.

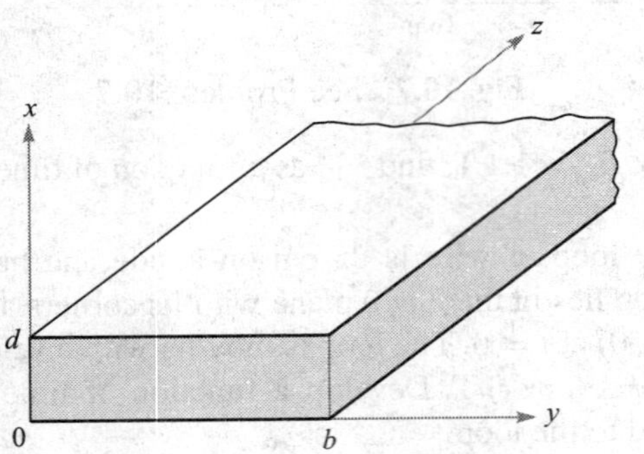

Fig. 10.8 See Problem 10.18.

10.19 In Section 10.1, Faraday's law was used to show that the field $\mathbf{E} = -\frac{1}{2}kB_0 e^{kt}\rho\mathbf{a}_\phi$ results from the changing magnetic field $\mathbf{B} = B_0 e^{kt}\mathbf{a}_z$. (*a*) Show that these fields do not satisfy Maxwell's other curl equation. (*b*) If we let $B_0 = 1$ T and $k = 10^6\ s^{-1}$, we are establishing a fairly large magnetic flux density in 1 μs. Use the $\nabla \times \mathbf{H}$ equation to show that the rate at which B_z should (but does not) change with ρ is only about 5×10^{-6} T per meter in free space at $t = 0$.

10.20 Point $C(-0.1, -0.2, 0.3)$ lies on the surface of a perfect conductor. The electric field intensity at C is $(500\mathbf{a}_x - 300\mathbf{a}_y + 600\mathbf{a}_z)\cos 10^7 t$ V/m, and the medium surrounding the conductor is characterized by $\mu_r = 5$, $\epsilon_r = 10$, and $\sigma = 0$. (*a*) Find a unit vector normal to the conductor surface at C, if the origin lies within the conductor. (*b*) Find the surface charge density at C.

10.21 (*a*) Show that under static field conditions, Eq. (55) reduces to Ampère's circuital law. (*b*) Verify that Eq. (51) becomes Faraday's law when we take the curl.

10.22 In a sourceless medium in which $\mathbf{J} = 0$ and $\rho_v = 0$, assume a rectangular coordinate system in which $\mathbf{E}$ and $\mathbf{H}$ are functions only of z and t. The medium has permittivity ϵ and permeability μ. (*a*) If $\mathbf{E} = E_x\mathbf{a}_x$ and $\mathbf{H} = H_y\mathbf{a}_y$, begin with Maxwell's equations and determine the second-order partial differential equation that E_x must satisfy. (*b*) Show that $E_x = E_0\cos(\omega t - \beta z)$ is a solution of that equation for a particular value of β. (*c*) Find β as a function of given parameters.

10.23 In region 1, $z < 0$, $\epsilon_1 = 2 \times 10^{-11}$ F/m, $\mu_1 = 2 \times 10^{-6}$ H/m, and $\sigma_1 = 4 \times 10^{-3}$ S/m; in region 2, $z > 0$, $\epsilon_2 = \epsilon_1/2$, $\mu_2 = 2\mu_1$, and $\sigma_2 = \sigma_1/4$. It is known that $\mathbf{E}_1 = (30\mathbf{a}_x + 20\mathbf{a}_y + 10\mathbf{a}_z)\cos 10^9 t$ V/m at $P(0, 0, 0^-)$. (*a*) Find $\mathbf{E}_{N1}$, $\mathbf{E}_{t1}$, $\mathbf{D}_{N1}$, and $\mathbf{D}_{t1}$ at P_1. (*b*) Find $\mathbf{J}_{N1}$ and $\mathbf{J}_{t1}$ at P_1. (c) Find $\mathbf{E}_{t2}$, $\mathbf{D}_{t2}$, and $\mathbf{J}_{t2}$ at $P_2(0, 0, 0^+)$. (*d*) (Harder) Use the continuity equation to help show that $J_{N1} - J_{N2} = \partial D_{N2}/\partial t - \partial D_{N1}/\partial t$, and then determine $\mathbf{D}_{N2}$, $\mathbf{J}_{N2}$, and $\mathbf{E}_{N2}$.

10.24 In a medium in which $\rho_v = 0$, but in which the permittivity is a function of position, determine the conditions on the permittivity variation such that (*a*) $\nabla \cdot \mathbf{E} = 0$; (*b*) $\nabla \cdot \mathbf{E} \doteq 0$.

10.25 In a region where $\mu_r = \epsilon_r = 1$ and $\sigma = 0$, the retarded potentials are given by $V = x(z - ct)$ V and $\mathbf{A} = x\left(\dfrac{z}{c} - t\right)\mathbf{a}_z$ Wb/m, where $c = 1\sqrt{\mu_0\epsilon_0}$. (*a*) Show that $\nabla \cdot \mathbf{A} = -\mu\epsilon\dfrac{\partial V}{\partial t}$. (*b*) Find $\mathbf{B}$, $\mathbf{H}$, $\mathbf{E}$, and $\mathbf{D}$. (*c*) Show that these results satisfy Maxwell's equations if $\mathbf{J}$ and ρ_v are zero.

10.26 Let the current $I = 80t$ A be present in the $\mathbf{a}_z$ direction on the z axis in free space within the interval $-0.1 < z < 0.1$ m. (*a*) Find A_z at $P(0, 2, 0)$ and (*b*) sketch A versus t over the time interval $-0.1 < t < 0.1\ \mu$s.

11 Transmission Lines

Transmission lines are used to transmit electric energy and signals from one point to another, specifically from a source to a load. Examples include the connection between a transmitter and an antenna, connections between computers in a network, or connections between a hydroelectric generating plant and a substation several hundred miles away. Other familiar examples include the interconnects between components of a stereo system and the connection between a cable service provider and your television set. Examples that are less familiar include the connections between devices on a circuit board that are designed to operate at high frequencies.

What all of these examples have in common is that the devices to be connected are separated by distances on the order of a wavelength or much larger, whereas in basic circuit analysis methods, connections between elements are assumed to have negligible length. The latter condition enabled us, for example, to take for granted that the voltage across a resistor on one side of a circuit was exactly in phase with the voltage source on the other side, or, more generally, that the time measured at the source location is precisely the same time as measured at all other points in the circuit. When distances are sufficiently large between source and receiver, time delay effects become appreciable, leading to delay-induced phase differences. In short, we deal with *wave phenomena* on transmission lines, in the same manner that we will find with point-to-point energy propagation in free space or in dielectrics.

The basic elements in a circuit, such as resistors, capacitors, inductors, and the connections between them, are considered *lumped* elements if the time delay in traversing the elements is negligible. On the other hand, if the elements or interconnections are large enough, it may be necessary to consider them as *distributed* elements. This means that their resistive, capacitive, and inductive characteristics must be evaluated on a per-unit-distance basis. Transmission lines have this property in general, and thus they become circuit elements in themselves, possessing impedances that contribute to the circuit problem. The basic rule is that one must consider elements as distributed if the propagation delay across the element dimension is of the order of the shortest time interval of interest. In the time-harmonic case, this condition would lead to a measurable phase difference between each end of the device in question.

In this chapter, we investigate wave phenomena in transmission lines. Our objectives include (1) to understand how to treat transmission lines as circuit elements possessing complex impedances that are functions of line length and frequency, (2) to understand wave propagation on lines, including cases in which losses may occur, (3) to learn methods of combining different transmission lines to accomplish a desired objective, and (4) to understand transient phenomena on lines.

11.1 PHYSICAL DESCRIPTION OF TRANSMISSION LINE PROPAGATION

To obtain a feel for the manner in which waves propagate on transmission lines, the following demonstration may be helpful. Consider a *lossless* line, as shown in Figure 11.1. By lossless, we mean that all power that is launched into the line at the input end eventually arrives at the output end. A battery having voltage V_0 is connected to the input by closing switch S_1 at time $t = 0$. When the switch is closed, the effect is to launch voltage, $V^+ = V_0$. This voltage does not instantaneously appear everywhere on the line, but rather begins to travel from the battery toward the load resistor, R, at a certain velocity. The *wavefront*, represented by the vertical dashed line in Figure 11.1, represents the instantaneous boundary between the section of the line that has been charged to V_0 and the remaining section that is yet to be charged. It also represents the boundary between the section of the line that carries the charging current, I^+, and the remaining section that carries no current. Both current and voltage are discontinuous across the wavefront.

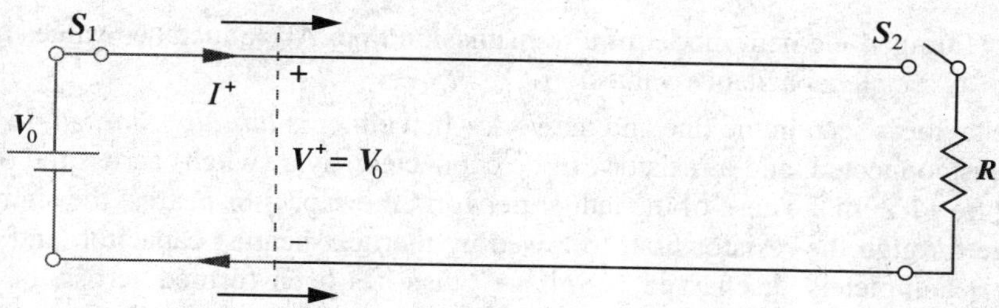

Fig. 11.1 Basic transmission line circuit, showing voltage and current waves initiated by closing switch S_1.

As the line charges, the wavefront moves from left to right at velocity ν, which is to be determined. On reaching the far end, all or a fraction of the wave voltage and current will reflect, depending on what the line is attached to. For example, if the resistor at the far end is left disconnected (switch S_2 is open), then all of the wavefront voltage will be reflected. If the resistor is connected, then some fraction of the incident voltage will reflect. The details of this will be treated in Section 11.9. Of interest at the moment are the factors that determine the wave velocity. The key to understanding and quantifying this is to note that the conducting transmission line will possess capacitance and inductance that are expressed on a per-unit-length basis. We have already derived expressions for these and evaluated them in Chapters 6 and 9 for certain transmission line geometries. Knowing these line characteristics, we can construct a model for the transmission line using lumped capacitors and inductors, as shown in Figure 11.2. The ladder network thus formed is referred to as a *pulse-forming network*, for reasons that will soon become clear.[1]

Consider now what happens when we connect the same switched voltage source to the network. Referring to Figure 11.2, on closing the switch at the battery location, current begins to increase in L_1, allowing C_1 to charge. As C_1 approaches full charge, current in L_2 begins to increase, allowing C_2 to charge next. This progressive charging process continues down the network, until all three capacitors are fully charged. In the network, a "wavefront" location can be identified as the point

[1] Designs and applications of pulse-forming networks are discussed in Reference 1.

between two adjacent capacitors that exhibit the most difference between their charge levels. As the charging process continues, the wavefront moves from left to right. Its speed depends on how fast each inductor can reach its full-current state, and simultaneously by how fast each capacitor is able to charge to full voltage. The wave is faster if the values of L_i and C_i are lower. We therefore expect the wave velocity to be inversely proportional to a function involving the product of inductance and capacitance. In the lossless transmission line, it turns out (as will be shown) that the wave velocity is given by $\nu = 1/\sqrt{LC}$, where L and C are specified per unit length.

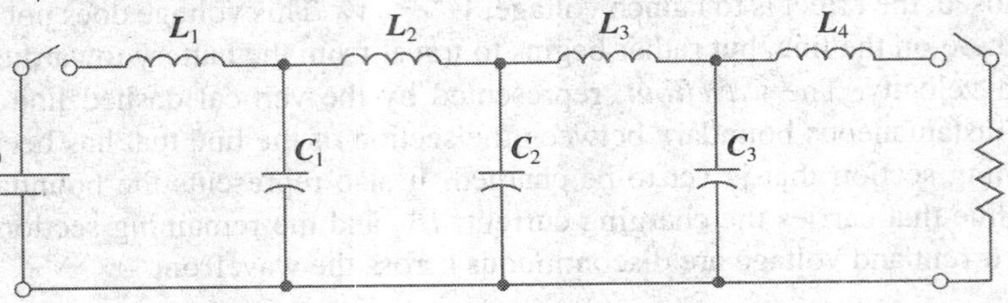

Fig. 11.2 Lumped-element model of a transmission line. All inductance values are equal, as are all capacitance values.

Similar behavior is seen in the line and network when either is *initially charged*. In this case, the battery remains connected, and a resistor can be connected (by a switch) across the output end, as shown in Figure 11.2. In the case of the ladder network, the capacitor nearest the shunted end (C_3) will discharge through the resistor first, followed by the next-nearest capacitor, and so on. When the network is completely discharged, a voltage pulse has been formed across the resistor, and so we see why this ladder configuration is called a pulse-forming network. Essentially identical behavior is seen in a charged transmission line when connecting a resistor between conductors at the output end. The switched voltage exercises, as used in these discussions, are examples of transient problems on transmission lines. Transients will be treated in detail in Section 11.14. In the beginning, line responses to sinusoidal signals are emphasized.

Finally, we surmise that the existence of voltage and current across and within the transmission line conductors implies the existence of electric and magnetic fields in the space around the conductors, and which are associated with the voltage and current. Consequently, we have two possible approaches to the analysis of transmission lines: (1) We can solve Maxwell's equations subject to the line configuration to obtain the fields, and with these find general expressions for the wave power, velocity, and other parameters of interest. (2) Or we can (for now) avoid the fields and solve for the voltage and current using an appropriate circuit model. It is the latter approach that we use in this chapter; the contribution of field theory is solely in the prior (and assumed) evaluation of the inductance and capacitance parameters. We will find, however, that circuit models become inconvenient or useless when losses in transmission lines are to be fully characterized, or when analyzing more complicated wave behavior (i.e., moding) which may occur as frequencies get high. The loss issues will be taken up in Section 11.5. Moding phenomena will be considered in Chapter 14.

11.2 THE TRANSMISSION LINE EQUATIONS

Our first goal is to obtain the differential equations, known as the *wave equations,* which the voltage or current must satisfy on a uniform transmission line. To do this, we construct a circuit model for an incremental length of line, write two circuit equations, and use these to obtain the wave equations.

Our circuit model contains the *primary constants* of the transmission line. These include the inductance, L, and capacitance, C, as well as the shunt conductance, G, and series resistance, R—all of which have values that are specified *per unit length*. The shunt conductance is used to model leakage current through the dielectric that may occur throughout the line length; the assumption is that the dielectric may possess conductivity, σ_d, in addition to a dielectric constant, ϵ_r where the latter affects the capacitance. The series resistance is associated with any finite conductivity, σ_c, in the conductors. Either one of the latter parameters, R and G, will be responsible for power loss in transmission. In general, both are functions of frequency. Knowing the frequency and the dimensions, we can determine the values of R, G, L, and C by using formulas developed in earlier chapters.

We assume propagation in the $\mathbf{a}_z$ direction. Our model consists of a line section of length Δz containing resistance $R\Delta z$, inductance $L\Delta z$, conductance $G\Delta z$, and capacitance $C\Delta z$, as indicated in Figure 11.3. Since the section of the line looks the same from either end, we divide the series elements in half to produce a symmetrical network. We could equally well have placed half the conductance and half the capacitance at each end.

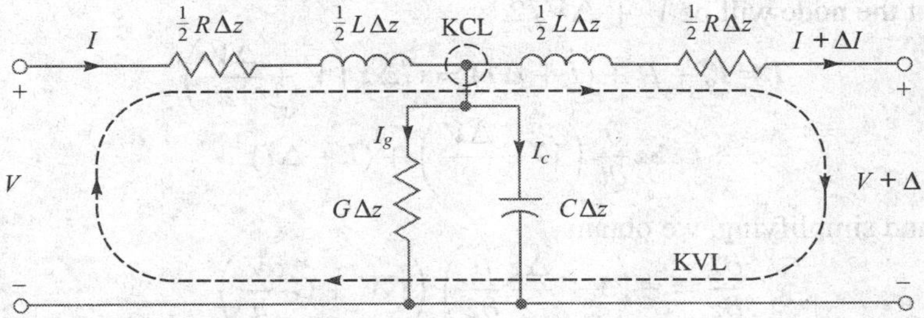

Fig. 11.3 Lumped-element model of a short transmission line section with losses. The length of the section is Δz. Analysis involves applying Kirchoff's voltage and current laws (KVL and KCL) to the indicated loop and node respectively.

Our objective is to determine the manner and extent to which the output voltage and current are changed from their input values in the limit as the length approaches a very small value. We will consequently obtain a pair of differential equations that describe the rates of change of voltage and current with respect to z. In Figure 11.3, the input and output voltages and currents differ respectively by quantities ΔV and ΔI, which are to be determined. The two equations are obtained by successive applications of Kirchoff's voltage law (KVL) and Kirchoff's current law (KCL).

First, KVL is applied to the loop that encompasses the entire section length, as shown in Figure 11.3:

$$V = \frac{1}{2}RI\Delta z + \frac{1}{2}L\frac{\partial I}{\partial t}\Delta z + \frac{1}{2}L\left(\frac{\partial I}{\partial t} + \frac{\partial \Delta I}{\partial t}\right)\Delta z$$
$$+ \frac{1}{2}R(I + \Delta I)\Delta z + (V + \Delta V) \tag{1}$$

We can solve Eq. (1) for the ratio, $\Delta V/\Delta z$, obtaining

$$\frac{\Delta V}{\Delta z} = -\left(RI + L\frac{\partial I}{\partial t} + \frac{1}{2}L\frac{\partial \Delta I}{\partial t} + \frac{1}{2}R\Delta I\right) \tag{2}$$

Next, we write

$$\Delta I = \frac{\partial I}{\partial z}\Delta z \qquad \text{and} \qquad \Delta V = \frac{\partial V}{\partial z}\Delta z \tag{3}$$

which are then substituted into (2) to result in

$$\frac{\partial V}{\partial z} = -\left(1 + \frac{\Delta z}{2}\frac{\partial}{\partial z}\right)\left(RI + L\frac{\partial I}{\partial t}\right) \tag{4}$$

Now, in the limit as Δz approaches zero (or a value small enough to be negligible), (4) simplifies to the final form:

$$\frac{\partial V}{\partial z} = -\left(RI + L\frac{\partial I}{\partial t}\right) \tag{5}$$

Equation (5) is the first of the two equations that we are looking for. To find the second equation, we apply KCL to the upper central node in the circuit of Figure 11.3, noting from the symmetry that the voltage at the node will be $V + \Delta V/2$:

$$I = I_g + I_c + (I + \Delta I) = G\Delta z\left(V + \frac{\Delta V}{2}\right) + C\Delta z\frac{\partial}{\partial t}\left(V + \frac{\Delta V}{2}\right) + (I + \Delta I) \tag{6}$$

Then, using (3) and simplifying, we obtain

$$\frac{\partial I}{\partial z} = -\left(1 + \frac{\Delta z}{2}\frac{\partial}{\partial z}\right)\left(GV + C\frac{\partial V}{\partial t}\right) \tag{7}$$

Again, we obtain the final form by allowing Δz to be reduced to a negligible magnitude. The result is

$$\frac{\partial I}{\partial z} = -\left(GV + C\frac{\partial V}{\partial t}\right) \tag{8}$$

The coupled differential equations, (5) and (8), describe the evolution of current and voltage in any transmission line. Historically, they have been referred to as the *telegraphist's equations*. Their solution leads to the wave equation for the transmission line, which we now undertake. We begin by differentiating Eq. (5) with respect to z and Eq. (8) with respect to t, obtaining

$$\frac{\partial^2 V}{\partial z^2} = -R\frac{\partial I}{\partial z} - L\frac{\partial^2 I}{\partial t\partial z} \tag{9}$$

and

$$\frac{\partial I}{\partial z\partial t} = -G\frac{\partial V}{\partial t} - C\frac{\partial^2 V}{\partial t^2} \tag{10}$$

Next, Eqs. (8) and (10) are substituted into (9). After rearranging terms, the result is

$$\frac{\partial^2 V}{\partial z^2} = LC\frac{\partial^2 V}{\partial t^2} + (LG + RC)\frac{\partial V}{\partial t} + RGV \tag{11}$$

An analogous procedure involves differentiating Eq. (5) with respect to t and Eq. (8) with respect to z. Then, Eq. (5) and its derivative are substituted into the derivative of (8) to obtain an equation for the current that is in identical form to that of (11):

$$\frac{\partial^2 I}{\partial z^2} = LC\frac{\partial^2 I}{\partial t^2} + (LG + RC)\frac{\partial I}{\partial t} + RGI \tag{12}$$

Equations (11) and (12) are the *general wave equations* for the transmission line. Their solutions under various conditions form a major part of our study.

11.3 LOSSLESS PROPAGATION

Lossless propagation means that power is not dissipated or otherwise deviated as the wave travels down the transmission line; all power at the input end eventually reaches the output end. More realistically, any mechanisms that would cause losses to occur have negligible effect. In our model, lossless propagation occurs when $R = G = 0$. Under this condition, only the first term on the right-hand side of either Eq. (11) or Eq. (12) survives. Equation (11), for example, becomes

$$\frac{\partial^2 V}{\partial z^2} = LC\frac{\partial^2 V}{\partial t^2} \tag{13}$$

In considering the voltage function that will satisfy (13), it is most expedient to simply state the solution, and then show that it is correct. The solution of (13) is of the form:

$$V(z,t) = f_1\left(t - \frac{z}{\nu}\right) + f_2\left(t + \frac{z}{\nu}\right) = V^+ + V^- \tag{14}$$

where ν, the *wave velocity*, is a constant. The expressions $(t \pm z/\nu)$ are the arguments of functions f_1 and f_2. The identities of the functions themselves are not critical to the solution of (13). Therefore f_1 and f_2 can be *any* function.

Animations

The arguments of f_1 and f_2 indicate, respectively, travel of the functions in the forward and backward z directions. We assign the symbols V^+ and V^- to identify the forward and backward voltage wave components. To understand the behavior, consider for example the value of f_1 (whatever this might be) at the zero value of its argument, occurring when $z = t = 0$. Now, as time increases to positive values (as it must), and if we are to keep track of $f_1(0)$, then the value of z must also increase to keep the argument $(t - z/\nu)$ equal to zero. The function f_1 therefore moves (or propagates) in the positive z direction. Using similar reasoning, the function f_2 will propagate in the *negative* z direction, as z in the argument $(t + z/\nu)$ must *decrease* to offset the increase in t. Therefore we associate the argument $(t - z/\nu)$ with *forward* z propagation, and the argument $(t + z/\nu)$ with *backward* z travel. This behavior occurs irrespective of what f_1 and f_2 are. As is evident in the argument forms, the propagation velocity is ν in both cases.

We next verify that functions having the argument forms expressed in (14) are solutions to (13). First, we take partial derivatives of f_1, for example with respect to z and t. Using the chain rule, the z partial derivative is

$$\frac{\partial f_1}{\partial z} = \frac{\partial f_1}{\partial(t - z/\nu)} \frac{\partial(t - z/\nu)}{\partial z} = -\frac{1}{\nu} f_1' \tag{15}$$

where it is apparent that the primed function, f_1', denotes the derivative of f_1 with respect to its argument. The partial derivative with respect to time is

$$\frac{\partial f_1}{\partial t} = \frac{\partial f_1}{\partial(t - z/\nu)} \frac{\partial(t - z/\nu)}{\partial t} = f_1' \tag{16}$$

Next, the second partial derivatives with respect to z and t can be taken using similar reasoning:

$$\frac{\partial^2 f_1}{\partial z^2} = \frac{1}{\nu^2} f_1'' \quad \text{and} \quad \frac{\partial^2 f_1}{\partial t^2} = f_1'' \tag{17}$$

where f_1'' is the second derivative of f_1 with respect to its argument. The results in (17) can now be substituted into (13), obtaining

$$\frac{1}{\nu^2} f_1'' = LC f_1'' \tag{18}$$

We now identify the wave velocity for lossless propagation, which is the condition for equality in (18):

$$\boxed{\nu = \frac{1}{\sqrt{LC}}} \tag{19}$$

Performing the same procedure using f_2 (and its argument) leads to the same expression for ν.

The form of ν as expressed in Eq. (19) confirms our original expectation that the wave velocity would be in some inverse proportion to L and C. The same result will be true for current, since Eq. (12) under lossless conditions would lead to a solution of the form identical to that of (14), with velocity given by (19). What is not known yet, however, is the relation *between* voltage and current.

We have already found that voltage and current are related through the telegraphist's equations, (5) and (8). These, under lossless conditions $(R = G = 0)$ become

$$\boxed{\frac{\partial V}{\partial z} = -L\frac{\partial I}{\partial t}} \tag{20}$$

$$\boxed{\frac{\partial I}{\partial z} = -C\frac{\partial V}{\partial t}} \tag{21}$$

Using the voltage function, we can substitute (14) into (20) and use the methods demonstrated in (15) to write

$$\frac{\partial I}{\partial t} = -\frac{1}{L}\frac{\partial V}{\partial z} = \frac{1}{L\nu}(f_1' - f_2') \tag{22}$$

We next integrate (22) over time, obtaining the current in terms of its forward and backward propagating components:

$$\boxed{I(z,t) = \frac{1}{L\nu}\left[f_1\left(t - \frac{z}{\nu}\right) - f_2\left(t + \frac{z}{\nu}\right)\right] = I^+ + I^-} \tag{23}$$

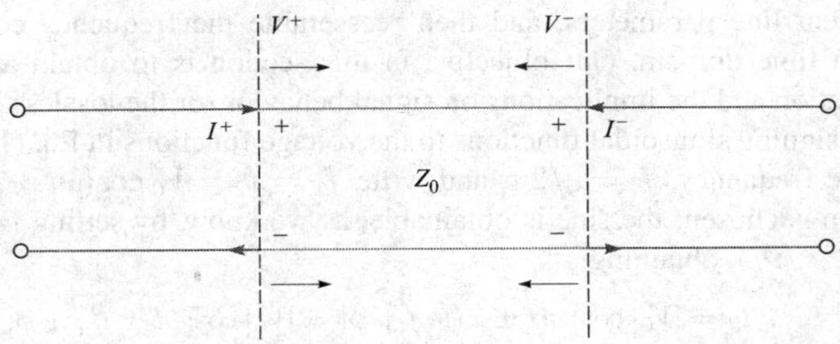

Fig. 11.4 Current directions in waves having positive voltage polarity.

In performing this integration, all integration constants are set to zero. The reason for this, as demonstrated by (20) and (21), is that a time-varying voltage must lead to a time-varying current, with the reverse also true. The factor $1/L\nu$ appearing in (23) multiplies voltage to obtain current, and so we identify the product $L\nu$ as the *characteristic impedance*, Z_0, of the lossless line. Z_0 is defined as the ratio of the voltage to the current in a single propagating wave. Using (19), we write the characteristic impedance as

$$Z_0 = L\nu = \sqrt{\frac{L}{C}} \tag{24}$$

By inspecting (14) and (23), we now note that

$$V^+ = Z_0 I^+ \tag{25a}$$

and

$$V^- = -Z_0 I^- \tag{25b}$$

The significance of the preceding relations can be seen in Figure 11.4. The figure shows forward- and backward-propagating voltage waves, V^+ and V^-, both of which have positive polarity. The currents that are associated with these voltages will flow in opposite directions. We define *positive current* as having a *clockwise* flow in the line, and *negative current* as having a *counterclockwise* flow. The minus sign in (25*b*) thus assures that negative current will be associated with a backward-propagating wave that has positive polarity. This is a general convention, applying to lines with losses also. Propagation with losses is studied by solving (11) under the assumption that either R or G (or both) are not zero. We will do this in Section 11.7 under the special case of sinusoidal voltages and currents. Sinusoids in lossless transmission lines are considered in Section 11.4.

11.4 LOSSLESS PROPAGATION OF SINUSOIDAL VOLTAGES

An understanding of sinusoidal waves on transmission lines is important because any signal that is transmitted in practice can be decomposed into a discrete or continuous summation of sinusoids. This is the basis of *frequency domain* analysis of signals on lines. In such studies, the effect of the transmission line on any signal can be determined by noting the effects on the frequency components. This means that one can effectively propagate the spectrum of a given signal, using frequency-dependent line parameters, and then reassemble the frequency components into the resultant signal in time domain. Our objective in this section is to obtain an understanding of sinusoidal propagation and the implications on signal behavior for the lossless line case.

We begin by assigning sinusoidal functions to the voltage functions in Eq. (14). Specifically, we consider a specific frequency, $f = \omega/2\pi$, and write $f_1 = f_2 = V_0 \cos(\omega t + \phi)$. By convention, the cosine function is chosen; the sine is obtainable, as we know, by setting $\phi = -\pi/2$. We next replace t with $(t \pm z/\nu_p)$, obtaining

$$\mathcal{V}(z,t) = |V_0| \cos[\omega(t \pm z/\nu_p) + \phi] = |V_0| \cos[\omega t \pm \beta z + \phi] \tag{26}$$

where we have assigned a new notation to the velocity, which is now called the *phase velocity*, ν_p. This is applicable to a pure sinusoid (having a single frequency) and will be found to depend on frequency in some cases. Choosing, for the moment, $\phi = 0$, we obtain the two possibilities of forward or backward z travel by choosing the minus or plus sign in (26). The two cases are

$$\mathcal{V}_f(z,t) = |V_0| \cos(\omega t - \beta z) \qquad \text{(forward } z \text{ propagation)} \tag{27a}$$

and

$$\mathcal{V}_b(z,t) = |V_0| \cos(\omega t + \beta z) \qquad \text{(backward } z \text{ propagation)} \tag{27b}$$

where the magnitude factor, $|V_0|$, is the value of $\mathcal{V}$ at $z = 0, t = 0$. We define the *phase constant* β, obtained from (26), as

$$\beta \equiv \frac{\omega}{\nu_p} \tag{28}$$

We refer to the solutions expressed in (27*a*) and (27*b*) as the *real instantaneous* forms of the transmission-line voltage. They are the mathematical representations of what one would experimentally measure. The terms ωt and βz, appearing in these equations, have units of angle and are usually expressed in radians. We know that ω is the radian time frequency, measuring phase shift *per unit time,* and it has units of rad/s. In a similar way, we see that β will be interpreted as a *spatial* frequency, which in the present case measures the phase shift *per unit distance* along the z direction. Its units are rad/m. If we were to fix the time at $t = 0$, Eqs. (27*a*) and (27*b*) would become

$$\mathcal{V}_f(z,0) = \mathcal{V}_b(z,0) = |V_0| \cos(\beta z) \tag{29}$$

which we identify as a simple periodic function that repeats every incremental distance λ, known as the *wavelength*. The requirement is that $\beta\lambda = 2\pi$, and so

$$\lambda = \frac{2\pi}{\beta} = \frac{\nu_p}{f} \tag{30}$$

We next consider a point (such as a wave crest) on the cosine function of Eq. (27*a*), the occurrence of which requires the argument of the cosine to be an integer multiple of 2π.

Considering the mth crest of the wave, the condition at $t = 0$ becomes

$$\beta z = 2m\pi$$

To keep track of this point on the wave, we require that the entire cosine argument be the same multiple of 2π for all time. From (27a) the condition becomes

$$\omega t - \beta z = \omega(t - z/\nu_p) = 2m\pi \tag{31}$$

Again, with increasing time, the position z must also increase in order to satisfy (31). Consequently the wave crest (and the entire wave) travels in the positive z direction at velocity ν_p. Equation (27b), having cosine argument $(\omega t + \beta z)$, describes a wave that travels in the *negative* z direction, since as time increases, z must now *decrease* to keep the argument constant. Similar behavior is found for the wave current, but complications arise from line-dependent phase differences that occur between current and voltage. These issues are best addressed once we are familiar with complex analysis of sinusoidal signals.

11.5 COMPLEX ANALYSIS OF SINUSOIDAL WAVES

Expressing sinusoidal waves as complex functions is useful (and essentially indispensable) because it greatly eases the evaluation and visualization of phase that will be found to accumulate by way of many mechanisms. In addition, we will find many cases in which two or more sinusoidal waves must be combined to form a resultant wave—a task made much easier if complex analysis is used.

Expressing sinusoidal functions in complex form is based on the Euler identity:

$$e^{\pm jx} = \cos(x) \pm j\sin(x) \tag{32}$$

from which we may write the cosine and sine, respectively, as the real and imaginary parts of the complex exponent:

$$\cos(x) = \text{Re}[e^{\pm jx}] = \frac{1}{2}(e^{jx} + e^{-jx}) = \frac{1}{2}e^{jx} + c.c. \tag{33a}$$

$$\sin(x) = \pm\text{Im}[e^{\pm jx}] = \frac{1}{2j}(e^{jx} - e^{-jx}) = \frac{1}{2j}e^{jx} + c.c. \tag{33b}$$

where $j \equiv \sqrt{-1}$, and where $c.c.$ denotes the complex conjugate of the preceding term. The conjugate is formed by changing the sign of j wherever it appears in the complex expression.

We may next apply (33a) to our voltage wave function, Eq. (26):

$$\mathcal{V}(z,t) = |V_0|\cos[\omega t \pm \beta z + \phi] = \frac{1}{2}\underbrace{(|V_0|e^{j\phi})}_{V_0}e^{\pm j\beta z}e^{j\omega t} + c.c. \tag{34}$$

Note that we have arranged the phases in (34) such that we identify the *complex amplitude* of the wave as $V_0 = (|V_0|e^{j\phi})$. In future usage, a single symbol (V_0 in the present example) will usually be used for the voltage or current amplitudes, with the understanding that these will generally be complex (having magnitude and phase).

Two additional definitions follow from Eq. (34). First, we define the *complex instantaneous* voltage as

$$V_c(z,t) = V_0 e^{\pm j\beta z}e^{j\omega t} \tag{35}$$

The *phasor* voltage is then formed by dropping the $e^{j\omega t}$ factor from the complex instantaneous form:

$$V_s(z) = V_0 e^{\pm j\beta z} \tag{36}$$

The phasor voltage can be defined provided we have *sinusoidal steady-state* conditions—meaning that V_0 is independent of time. This has in fact been our assumption all along, because a time-varying amplitude would imply the existence of other frequency components in our signal. Again, we are treating only a single-frequency wave. The significance of the phasor voltage is that we are effectively letting time stand still and observing the stationary wave in space at $t = 0$. The processes of evaluating relative phases between various line positions and of combining multiple waves is made much simpler in phasor form. Again, this works only if all waves under consideration have the same frequency. With the definitions in (35) and (36), the real instantaneous voltage can be constructed using (34):

$$\mathcal{V}(z,t) = |V_0|\cos[\omega t \pm \beta z + \phi] = \text{Re}[V_c(z,t)] = \frac{1}{2}V_c + c.c. \tag{37a}$$

Or, in terms of the phasor voltage:

$$\mathcal{V}(z,t) = |V_0|\cos[\omega t \pm \beta z + \phi] = \text{Re}[V_s(z)e^{j\omega t}] = \frac{1}{2}V_s(z)e^{j\omega t} + c.c. \tag{37b}$$

In words, we may obtain our real sinusoidal voltage wave by multiplying the phasor voltage by $e^{j\omega t}$ (reincorporating the time dependence) and then taking the real part of the resulting expression. It is imperative that one becomes familiar with these relations and their meaning before proceeding further.

EXAMPLE 11.1

Two voltage waves having equal frequencies and amplitudes propagate in opposite directions on a lossless transmission line. Determine the total voltage as a function of time and position.

Solution. Since the waves have the same frequency, we can write their combination using their phasor forms. Assuming phase constant, β, and real amplitude, V_0, the two wave voltages combine in this way:

$$V_{sT}(z) = V_0 e^{-j\beta z} + V_0 e^{+j\beta z} = 2V_0\cos(\beta z)$$

In real instantaneous form, this becomes

$$\mathcal{V}(z,t) = \text{Re}[2V_0\cos(\beta z)e^{j\omega t}] = 2V_0\cos(\beta z)\cos(\omega t)$$

We recognize this as a *standing wave,* in which the amplitude varies as $\cos(\beta z)$, and oscillates in time as $\cos(\omega t)$. Zeros in the amplitude (nulls) occur at fixed locations, $z_n = (m\pi)/(2\beta)$ where m is an odd integer. We extend the concept in Section 11.10, where we explore the *voltage standing wave ratio* as a measurement technique.

11.6 TRANSMISSION LINE EQUATIONS AND THEIR SOLUTIONS IN PHASOR FORM

We now apply our results of the previous section to the transmission line equations, beginning with the general wave equation, (11). This is rewritten as follows, for the real instantaneous voltage, $\mathcal{V}(z,t)$:

$$\frac{\partial^2 \mathcal{V}}{\partial z^2} = LC\frac{\partial^2 \mathcal{V}}{\partial t^2} + (LG + RC)\frac{\partial \mathcal{V}}{\partial t} + RG\mathcal{V} \tag{38}$$

We next substitute $\mathcal{V}(z,t)$ as given by the far right-hand side of (37*b*), noting that the complex conjugate term (*c.c.*) will form a separate redundant equation. We also use the fact that the operator $\partial/\partial t$, when applied to the complex form, is equivalent to multiplying by a factor of $j\omega$. After substitution, and after all time derivatives are taken, the factor $e^{j\omega t}$ divides out. We are left with the wave equation in terms of the phasor voltage:

$$\frac{d^2 V_s}{dz^2} = -\omega^2 LCV_s + j\omega(LG + RC)V_s + RGV_s \tag{39}$$

Rearranging terms leads to the simplified form:

$$\frac{d^2 V_s}{dz^2} = \underbrace{(R + j\omega L)}_{Z}\underbrace{(G + j\omega C)}_{Y} V_s = \gamma^2 V_s \tag{40}$$

where Z and Y, as indicated, are respectively the *net series impedance* and the *net shunt admittance* in the transmission line—both as per-unit-distance measures. The *propagation constant* in the line is defined as

$$\gamma = \sqrt{(R + j\omega L)(G + j\omega C)} = \sqrt{ZY} = \alpha + j\beta \tag{41}$$

The significance of the term will be explained in Section 11.7. For our immediate purposes, the solution of (40) will be

$$V_s(z) = V_0^+ e^{-\gamma z} + V_0^- e^{+\gamma z} \tag{42a}$$

The wave equation for current will be identical in form to (40). We therefore expect the phasor current to be in the form:

$$I_s(z) = I_0^+ e^{-\gamma z} + I_0^- e^{\gamma z} \tag{42b}$$

The relation between the current and voltage waves is now found, as before, through the telegraphist's equations, (5) and (8). In a manner consistent with Eq. (37*b*), we write the sinusoidal current as

$$\mathcal{I}(z,t) = |I_0|\cos(\omega t \pm \beta z + \xi) = \frac{1}{2}\underbrace{(|I_0|e^{j\xi})}_{I_0} e^{\pm j\beta z} e^{j\omega t} + c.c. = \frac{1}{2} I_s(z) e^{j\omega t} + c.c. \tag{43}$$

Substituting the far right-hand sides of (37*b*) and (43) into (5) and (8) transforms the latter equations as follows:

$$\frac{\partial \mathcal{V}}{\partial z} = -\left(R\mathcal{I} + L\frac{\partial \mathcal{I}}{\partial t}\right) \quad \Rightarrow \quad \boxed{\frac{dV_s}{dz} = -(R + j\omega L)I_s = -ZI_s} \tag{44a}$$

and

$$\frac{\partial \mathcal{I}}{\partial z} = -\left(G\mathcal{V} + C\frac{\partial \mathcal{V}}{\partial t}\right) \quad \Rightarrow \quad \boxed{\frac{dI_s}{dz} = -(G + j\omega C)V_s = -YV_s} \tag{44b}$$

We can now substitute (42*a*) and (42*b*) into either (44*a*) or (44*b*) [we will use (44a)] to find:

$$-\gamma V_0^+ e^{-\gamma z} + \gamma V_0^- e^{\gamma z} = -Z(I_0^+ e^{-\gamma z} + I_0^- e^{\gamma z}) \tag{45}$$

Next, equating coefficients of $e^{-\gamma z}$ and $e^{\gamma z}$, we find the general expression for the line characteristic impedance:

$$Z_0 = \frac{V_0^+}{I_0^+} = -\frac{V_0^-}{I_0^-} = \frac{Z}{\gamma} = \frac{Z}{\sqrt{ZY}} = \sqrt{\frac{Z}{Y}} \tag{46}$$

Incorporating the expressions for Z and Y, we find the characteristic impedance in terms of our known line parameters:

$$\boxed{Z_0 = \sqrt{\frac{R + j\omega L}{G + j\omega C}} = |Z_0|e^{j\theta}} \tag{47}$$

Note that with the voltage and current as given in (37*b*) and (43), we would identify the phase of the characteristic impedance, $\theta = \phi - \xi$.

EXAMPLE 11.2

A lossless transmission line is 80 cm long and operates at a frequency of 600 MHz. The line parameters are $L = 0.25$ μH/m and $C = 100$ pF/m. Find the characteristic impedance, the phase constant, and the phase velocity.

Solution. Since the line is lossless, both R and G are zero. The characteristic impedance is

$$Z_0 = \sqrt{\frac{L}{C}} = \sqrt{\frac{0.25 \times 10^{-6}}{100 \times 10^{-12}}} = 50\ \Omega$$

Since $\gamma = \alpha + j\beta = \sqrt{(R + j\omega L)(G + j\omega C)} = j\omega\sqrt{LC}$, we see that

$$\beta = \omega\sqrt{LC} = 2\pi(600 \times 10^6)\sqrt{(0.25 \times 10^{-6})(100 \times 10^{-12})} = 18.85 \text{ rad/m}$$

Also,

$$\nu_p = \frac{\omega}{\beta} = \frac{2\pi(600 \times 10^6)}{18.85} = 2 \times 10^8 \text{ m/s}$$

11.7 LOSSLESS AND LOW-LOSS PROPAGATION

Having obtained the phasor forms of voltage and current in a general transmission line [Eqs. (42a) and (42b)], we can now look more closely at the significance of these results. First we incorporate (41) into (42a) to obtain

$$V_s(z) = V_0^+ e^{-\alpha z} e^{-j\beta z} + V_0^- e^{\alpha z} e^{j\beta z} \tag{48}$$

Next, multiplying (48) by $e^{j\omega t}$ and taking the real part gives the real instantaneous voltage:

$$\mathcal{V}(z,t) = V_0^+ e^{-\alpha z}\cos(\omega t - \beta z) + V_0^- e^{\alpha z}\cos(\omega t + \beta z) \tag{49}$$

In this exercise, we have assigned V_0^+ and V_0^- to be real. Eq. (49) is recognized as describing forward- and backward-propagating waves that diminish in amplitude with distance according to $e^{-\alpha z}$ for the forward wave, and $e^{\alpha z}$ for the backward wave. Both waves are said to *attenuate* with propagation distance at a rate determined by the *attenuation coefficient*, α, expressed in units of nepers/m [Np/m].[2]

The phase constant, β, found by taking the imaginary part of (41), is likely to be a somewhat complicated function, and will in general depend on R and G. Nevertheless, β is still defined as the ratio ω/ν_p, and the wavelength is still defined as the distance that provides a phase shift of 2π rad, so that $\lambda = 2\pi/\beta$. By inspecting (41), we observe that losses in propagation are avoided (or $\alpha = 0$) only when $R = G = 0$. In that case, (41) gives $\gamma = j\beta = j\omega\sqrt{LC}$, and so $\nu_p = 1/\sqrt{LC}$, as we found before.

Expressions for α and β when losses are small can be readily obtained from (41). In the *low-loss approximation,* we require $R \ll \omega L$ and $G \ll \omega C$, a condition that is often true in practice. Before we apply these conditions, Eq. (41) can be written in the form:

$$\begin{aligned}\gamma &= \alpha + j\beta = [(R + j\omega L)(G + j\omega C)]^{1/2} \\ &= j\omega\sqrt{LC}\left[\left(1 + \frac{R}{j\omega L}\right)^{1/2}\left(1 + \frac{G}{j\omega C}\right)^{1/2}\right]\end{aligned} \tag{50}$$

The low-loss approximation then allows us to use the first three terms in the binomial series:

$$\sqrt{1+x} \doteq 1 + \frac{x}{2} - \frac{x^2}{8} \quad (x \ll 1) \tag{51}$$

We use (51) to expand the terms in large parentheses in (50), obtaining

$$\gamma \doteq j\omega\sqrt{LC}\left[\left(1 + \frac{R}{j2\omega L} + \frac{R^2}{8\omega^2 L^2}\right)\left(1 + \frac{G}{j2\omega C} + \frac{G^2}{8\omega^2 C^2}\right)\right] \tag{52}$$

All products in (52) are then carried out, neglecting the terms involving RG^2, R^2G, and R^2G^2, as these will be negligible compared to all others. The result is

$$\gamma = \alpha + j\beta \doteq j\omega\sqrt{LC}\left[1 + \frac{1}{j2\omega}\left(\frac{R}{L} + \frac{G}{C}\right) + \frac{1}{8\omega^2}\left(\frac{R^2}{L^2} - \frac{2RG}{LC} + \frac{G^2}{C^2}\right)\right] \tag{53}$$

[2] The term *neper* was selected (by some poor speller) to honor John Napier, a Scottish mathematician who first proposed the use of logarithms.

Now, separating real and imaginary parts of (53) yields α and β:

$$\alpha \doteq \frac{1}{2}\left(R\sqrt{\frac{C}{L}} + G\sqrt{\frac{L}{C}}\right) \tag{54a}$$

and

$$\beta \doteq \omega\sqrt{LC}\left[1 + \frac{1}{8}\left(\frac{G}{\omega C} - \frac{R}{\omega L}\right)^2\right] \tag{54b}$$

We note that α scales in direct proportion to R and G, as would be expected. We also note that the terms in (54*b*) that involve R and G lead to a phase velocity, $\nu_p = \omega/\beta$, that is frequency-dependent. Moreover, the *group velocity*, $\nu_g = d\omega/d\beta$, will also depend on frequency, and will lead to signal distortion, as we will explore in Chapter 13. Note that with nonzero R and G, phase and group velocities that are constant with frequency can be obtained when $R/L = G/C$, known as *Heaviside's condition*. In this case, (54*b*) becomes $\beta \doteq \omega\sqrt{LC}$, and the line is said to be *distortionless*. Further complications occur when accounting for possible frequency dependencies within R, G, L, and C. Consequently, conditions of low-loss or distortion-free propagation will usually occur over limited frequency ranges. As a rule, loss increases with increasing frequency, mostly because of the increase in R with frequency. The nature of this latter effect, known as *skin effect* loss, requires field theory to understand and quantify. We will study this in Chapter 12, and we will apply it to transmission line structures in Chapter 14.

Finally, we can apply the low-loss approximation to the characteristic impedance, Eq. (47). Using (51), we find

$$Z_0 = \sqrt{\frac{R + j\omega L}{G + j\omega C}} = \sqrt{\frac{j\omega L\left(1 + \frac{R}{j\omega L}\right)}{j\omega C\left(1 + \frac{G}{j\omega C}\right)}} \doteq \sqrt{\frac{L}{C}}\left[\frac{\left(1 + \frac{R}{j2\omega L} + \frac{R^2}{8\omega^2 L^2}\right)}{\left(1 + \frac{G}{j2\omega C} + \frac{G^2}{8\omega^2 C^2}\right)}\right] \tag{55}$$

Next, we multiply (55) by a factor of 1, in the form of the complex conjugate of the denominator of (55) divided by itself. The resulting expression is simplified by neglecting all terms on the order of R^2G, G^2R, and higher. Additionally, the approximation, $1/(1+x) \doteq 1 - x$, where $x \ll 1$ is used. The result is

$$Z_0 \doteq \sqrt{\frac{L}{C}}\left\{1 + \frac{1}{2\omega^2}\left[\frac{1}{4}\left(\frac{R}{L} + \frac{G}{C}\right)^2 - \frac{G^2}{C^2}\right] + \frac{j}{2\omega}\left(\frac{G}{C} - \frac{R}{L}\right)\right\} \tag{56}$$

Note that when Heaviside's condition (again, $R/L = G/C$) holds, Z_0 simplifies to just $\sqrt{L/C}$, as is true when both R and G are zero.

EXAMPLE 11.3

Suppose in a certain transmission line $G = 0$, but R is finite-valued and satisfies the low-loss requirement, $R \ll \omega L$. Use Eq. (56) to write the approximate magnitude and phase of Z_0.

Solution. With $G = 0$, the imaginary part of (56) is much greater than the second term in the real part [proportional to $(R/\omega L)^2$].

Therefore, the characteristic impedance becomes

$$Z_0(G=0) \doteq \sqrt{\frac{L}{C}}\left(1 - j\frac{R}{2\omega L}\right) = |Z_0|e^{j\theta}$$

where $|Z_0| \doteq \sqrt{L/C}$, and $\theta = \tan^{-1}(-R/2\omega L)$.

D11.1 At an operating radian frequency of 500 Mrad/s, typical circuit values for a certain transmission line are: $R = 0.2\ \Omega$/m, $L = 0.25\ \mu$H/m, $G = 10\ \mu$S/m, and $C = 100$ pF/m. Find: (*a*) α; (*b*) β; (*c*) λ; (*d*) ν_p; (*e*) Z_0.

Ans. 2.25 mNp/m; 2.50 rad/m; 2.51 m; 2×10^8 m/sec; $50.0 - j0.0350\ \Omega$

11.8 POWER TRANSMISSION AND LOSS CHARACTERIZATION

Having found the sinusoidal voltage and current in a lossy transmission line, we next evaluate the power transmitted over a specified distance as a function of voltage and current amplitudes. We start with the *instantaneous* power, given simply as the product of the real voltage and current. Consider the forward-propagating term in (49), where again, the amplitude, $V_0^+ = |V_0|$, is taken to be real. The current waveform will be similar, but will generally be shifted in phase. Both current and voltage attenuate according to the factor $e^{-\alpha z}$. The instantaneous power therefore becomes

$$\mathcal{P}(z,t) = \mathcal{V}(z,t)\mathcal{I}(z,t) = |V_0||I_0|e^{-2\alpha z}\cos(\omega t - \beta z)\cos(\omega t - \beta z + \theta) \tag{57}$$

Usually, the *time-averaged* power, $\langle\mathcal{P}\rangle$, is of interest. We find this through:

$$\langle\mathcal{P}\rangle = \frac{1}{T}\int_0^T |V_0||I_0|e^{-2\alpha z}\cos(\omega t - \beta z)\cos(\omega t - \beta z + \theta)dt \tag{58}$$

where $T = 2\pi/\omega$ is the time period for one oscillation cycle. Using a trigonometric identity, the product of cosines in the integrand can be written as the sum of individual cosines at the sum and difference frequencies:

$$\langle\mathcal{P}\rangle = \frac{1}{T}\int_0^T \frac{1}{2}|V_0||I_0|[\cos(2\omega t - 2\beta z + \theta) + \cos(\theta)]\,dt \tag{59}$$

The first cosine term integrates to zero, leaving the $\cos\theta$ term. The remaining integral easily evaluates as

$$\langle\mathcal{P}\rangle = \frac{1}{2}|V_0||I_0|e^{-2\alpha z}\cos\theta = \frac{1}{2}\frac{|V_0|^2}{|Z_0|}e^{-2\alpha z}\cos\theta\ [\text{W}] \tag{60}$$

The same result can be obtained directly from the phasor voltage and current. We begin with these, expressed as

$$V_s(z) = V_0e^{-\alpha z}e^{-j\beta z} \tag{61}$$

and

$$I_s(z) = I_0e^{-\alpha z}e^{-j\beta z} = \frac{V_0}{Z_0}e^{-\alpha z}e^{-j\beta z} \tag{62}$$

where $Z_0 = |Z_0|e^{j\theta}$. We now note that the time-averaged power as expressed in (60) can be obtained from the phasor forms through

$$\langle \mathcal{P} \rangle = \frac{1}{2}\text{Re}\{V_s I_s^*\} \tag{63}$$

where again, the asterisk (*) denotes the complex conjugate (applied in this case to the current phasor only). Using (61) and (62) in (63), it is found that

$$\begin{aligned}\langle \mathcal{P} \rangle &= \frac{1}{2}\text{Re}\left\{V_0 e^{-\alpha z} e^{-j\beta z} \frac{V_0^*}{|Z_0|e^{-j\theta}} e^{-\alpha z} e^{+j\beta z}\right\} \\ &= \frac{1}{2}\text{Re}\left\{\frac{V_0 V_0^*}{|Z_0|} e^{-2\alpha z} e^{j\theta}\right\} = \frac{1}{2}\frac{|V_0|^2}{|Z_0|} e^{-2\alpha z} \cos\theta\end{aligned} \tag{64}$$

which we note is identical to the time-integrated result in (60). Equation (63) applies to any single-frequency wave.

An important result of the preceding exercise is that power attenuates as $e^{-2\alpha z}$, or

$$\langle \mathcal{P}(z) \rangle = \langle \mathcal{P}(0) \rangle e^{-2\alpha z} \tag{65}$$

Power drops at twice the exponential rate with distance as either voltage or current.

A convenient measure of power loss is in *decibel* units. This is based on expressing the power decrease as a power of 10. Specifically, we write

$$\frac{\langle \mathcal{P}(z) \rangle}{\langle \mathcal{P}(0) \rangle} = e^{-2\alpha z} = 10^{-\kappa\alpha z} \tag{66}$$

where the constant, κ, is to be determined. Setting $\alpha z = 1$, we find

$$e^{-2} = 10^{-\kappa} \quad \Rightarrow \quad \kappa = \log_{10}(e^2) = 0.869 \tag{67}$$

Now, by definition, the power loss in decibels (dB) is

$$\text{Power Loss (dB)} = 10\log_{10}\left[\frac{\langle \mathcal{P}(0) \rangle}{\langle \mathcal{P}(z) \rangle}\right] = 8.69\alpha z \tag{68}$$

where we note that inverting the power ratio in the argument of the log function [as compared to the ratio in (66)] yields a positive number for the dB loss. Also, noting that $\langle \mathcal{P} \rangle \propto |V_0|^2$, we may write, equivalently:

$$\text{Power Loss (dB)} = 10\log_{10}\left[\frac{\langle \mathcal{P}(0) \rangle}{\langle \mathcal{P}(z) \rangle}\right] = 20\log_{10}\left[\frac{|V_0(0)|}{|V_0(z)|}\right] \tag{69}$$

where $|V_0(z)| = |V_0(0)|e^{-\alpha z}$.

EXAMPLE 11.4

A piece of coaxial line having characteristic impedance of 50 Ω is shown in the following Figure 11.5. The spacing between the inner and the outer conductor of this line is filled with a dielectric, which can be considered lossless for all practical purposes. However, the conductors are not perfect resulting into the attenuation along the line at the rate of 20 dB/km at the frequency of 500 MHz. The line inductance is given as $L = 238.1$ nH m^{-1}. Determine (a) the resistance R of this line per length; (b) the capacitance C of the line per unit length; (c) the velocity of the propagation along the line, and (d) the dielectric constant of the material which is filled between the two conductors given in the Figure 11.5.

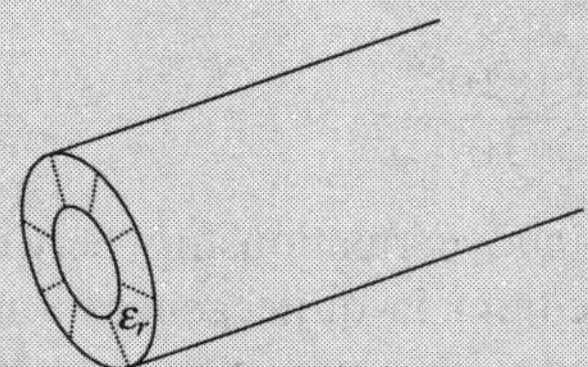

Fig. 11.5 A piece of coaxial line.

Solution.

First of all, the attenuation constant is to be converted into Np/m from dB/km. This can be done with the help of the following equation:

Attenuation in Np ≡ Attenuation in dB/8.69

Hence, the attenuation constant in Np/m is computed as follows:

$$\alpha = 20\,\text{dB/km} \equiv \frac{20}{1000}\,\text{dB/m} \equiv \frac{20}{8.69 \times 1000}\,\text{Np/m} \equiv 2.302 \times 10^{-3}[\text{Np/m}]$$

(a) It is given that the dielectric loss can be considered zero for all practical purposes. The dielectric loss in Figure 11.3 is represented by the shunt conductance G, and hence in the present situation, $G = 0$. The series resistance R in the present case can be computed using the formula for the attenuation constant of low-loss line given by (54a)

$$\alpha = \frac{1}{2}R\sqrt{\frac{C}{L}} \cong \frac{R}{2\,Z_0} = 2.302 \times 10^{-3}\ \text{Np/m}$$

where $G = 0$ is substituted, and the expression for the characteristic impedance $Z_0 \cong \sqrt{L/C}$ is used, which is valid for the low-loss line as per Eq. (56) under the condition that $G = 0$ and $R \ll \omega L$. Now, the characteristic impedance of the line $Z_0 = 50\ \Omega$ is given, which can be substituted in the above expression to obtain the series resistance of the line

$$R = 2\alpha Z_0 = 2 \times 2.302 \times 10^{-3}\ \text{Np/m} \times 50 = 0.2302\,\Omega\,\text{m}^{-1}$$

It is to be noted here that at 500 MHz, $\omega L \equiv 2 \times \pi \times 500 \times 10^6 \times 238.1 \times 10^{-9} = 748.01 \gg 0.2302$, which verifies our assumption that $R \ll \omega L$, and it is also the main reason to define the characteristic impedance as a purely real quantity in most of the practical situations at RF and microwave frequencies.

(*b*) The capacitance C of the line per unit length can be calculated as

$$Z_0 \cong \sqrt{\frac{L}{C}} \quad \Rightarrow \; C \cong \frac{L}{Z_0^2} = \frac{238.1 \times 10^{-9}}{(50)^2} = 95.2 \text{ pF m}^{-1}$$

(*c*) The phase velocity can be calculated with the help of (28) and (54b) by applying the condition that $G = 0$ and $R \ll \omega L$, which results into

$$v_p = \frac{\omega}{\beta};\; \beta \cong \omega\sqrt{LC} \quad \Rightarrow \; v_p \cong \frac{1}{\sqrt{LC}} = \frac{1}{\sqrt{238.1 \times 10^{-9} \times 95.2 \times 10^{-12}}} = 2.1 \times 10^8 \text{ m/sec}$$

(*d*) The dielectric constant of the material filled between the conductors can be computed by comparing the actual velocity along the line with the velocity in the free space or the air filled medium. This ratio is usually called the velocity factor, and for the non-magnetic dielectrics, can be written as

$$\frac{v_p}{c} = \frac{2.1 \times 10^8}{3 \times 10^8} = 0.7 \equiv \frac{1}{\sqrt{\epsilon_r}} \quad \Rightarrow \epsilon_r = 2.04$$

You should remember that the above relationship between the velocity in free space and the velocity in the dielectric medium is quite general, and is usually valid for all types of transmission lines which are filled with a non-magnetic dielectric medium other than air. The basis of the above relationship is that the capacitance per unit length between the two conductors is directly proportional to the dielectric constant of the material which is filled inside the region between these conductors.

EXAMPLE 11.5

A 20 m length of transmission line is known to produce a 2.0 dB drop in power from end to end. (*a*) What fraction of the input power reaches the output? (*b*) What fraction of the input power reaches the midpoint of the line? (*c*) What exponential attenuation coefficient, α, does this represent?

Solution.

(*a*) The power fraction will be

$$\frac{\langle \mathcal{P}(20) \rangle}{\langle \mathcal{P}(0) \rangle} = 10^{-0.2} = 0.63$$

(*b*) 2 dB in 20 m implies a loss rating of 0.1 dB/m. So, over a 10-metre span, the loss is 1.0 dB. This represents the power fraction, $10^{-0.1} = 0.79$.

(*c*) The exponential attenuation coefficient is found through

$$\alpha = \frac{2.0 \text{ dB}}{(8.69 \text{ dB/Np})(20 \text{ m})} = 0.012 \text{ [Np/m]}$$

A final point addresses the question: Why use decibels? The most compelling reason is that when evaluating the accumulated loss for several lines and devices that are all end-to-end connected, the net loss in dB for the entire span is just the sum of the dB losses of the individual elements.

D11.2 Two transmission lines are to be joined end-to-end. Line 1 is 30 m long and is rated at 0.1 dB/m. Line 2 is 45 m long and is rated at 0.15 dB/m. The joint is not done well and imparts a 3 dB loss. What percentage of the input power reaches the output of the combination?

Ans. 5.3%

EXAMPLE 11.6

A distortionless transmission line operating at 250 MHz has $R = 30\ \Omega/\text{m}$, $L = 200$ nH/m, and $C = 80$ pF/m.

(*a*) Determine the characteristic impedance Z_0, the propagation constant γ, and the velocity of propagation v_p along the line.

(*b*) After how many metres traveling along the line, will the voltage wave get reduced to 30% of its initial value?

(*c*) How far should a voltage wave travel along this line in order to undergo a phase shift of 20°?

Solution. For a distortionless line, $\dfrac{R}{L} = \dfrac{G}{C}$ (i)

(*a*) If the above condition is applied to Eq. (56), it results into

$$Z_0 = \sqrt{\frac{L}{C}} = \sqrt{\frac{200 \times 10^{-9}}{80 \times 10^{-12}}} = 50\,\Omega \tag{ii}$$

The propagation constant is given by

$$\gamma \equiv \alpha + j\beta = \sqrt{(R + j\omega L)(G + j\omega C)} \equiv \sqrt{(R + j\omega L)\left(\frac{RC}{L} + j\omega C\right)}$$

$$\equiv \sqrt{(R + j\omega L)\frac{C}{L}(R + j\omega L)} = \left(R\sqrt{\frac{C}{L}} + j\omega\sqrt{LC}\right) \tag{iii}$$

$$\equiv \left(30\sqrt{\frac{80 \times 10^{-12}}{200 \times 10^{-9}}} + j \times 2 \times \pi \times 250 \times 10^{6} \times \sqrt{200 \times 10^{-9} \times 80 \times 10^{-12}}\right)$$

$$\equiv (0.60 + j\,6.2832)\ \text{m}^{-1} \tag{iv}$$

where, we have made use of the relationship given by (i). The above equation can be separated into real and imaginary parts to define the value of attenuation constant and the phase constant as follows

$$\alpha = 0.6\,\text{Np/m}; \quad \beta = 6.2832\,\text{rad/m} \tag{v}$$

The velocity of propagation along the line is given by

$$\Rightarrow \quad v_p = \frac{\omega}{\beta} \equiv \frac{1}{\sqrt{LC}} = \frac{1}{\sqrt{200 \times 10^{-9} \times 80 \times 10^{-12}}} = 2.5 \times 10^{8}\ \text{m/sec}$$

where (iii) is used to define the propagation constant β.

(*b*) Let us assume that the wave is travelling along the z-direction, and let the magnitude of the voltage wave be V_0 at $z = 0$. The magnitude of this voltage wave at any arbitrary position along the line would then be given by $V_0 \exp(-\alpha z)$. Now, we have to find the position '$z = d$' at which the magnitude of this wave gets reduced to 30% of its initial value, i.e.,

$$V_0 \exp(-\alpha\, d) \equiv 0.3\, V_0 \Rightarrow \exp(\alpha\, d) \equiv \frac{1}{0.3}$$

$$\Rightarrow d \equiv \frac{1}{\alpha} \ln\left(\frac{1}{0.3}\right) = \frac{1}{0.6} \ln\left(\frac{1}{0.3}\right) = 2 \text{ m}$$

(*c*) As the wave travels along the line, it would continuously undergo a phase shift. If it is assumed that the wave is travelling along the +ve z-direction and at the initial position this wave is expressed by $V_0 \exp(-j0)$ then this wave can be expressed at any arbitrary position along the line by $V_0 \exp(-j\beta\, z)$. Hence, the phase shift after traversing a distance 'l' would be represented by βl. Now, it is given that

$$\beta l = 20 \text{ deg} \equiv 20 \times \frac{\pi}{180} \text{ rad}$$

$$\Rightarrow l \equiv \frac{20}{\beta} \times \frac{\pi}{180}\, m = \frac{20}{6.2832} \times \frac{\pi}{180}\, m = 5.56 \text{ cm}$$

where the value of the phase constant β is substituted from (v).

11.9 WAVE REFLECTION AT DISCONTINUITIES

The concept of wave reflection was introduced in Section 11.1. As implied there, the need for a reflected wave originates from the necessity to satisfy all voltage and current boundary conditions at the ends of transmission lines and at locations at which two dissimilar lines are connected to each other. The consequences of reflected waves are usually less than desirable, in that some of the power that was intended to be transmitted to a load, for example, reflects and propagates back to the source. Conditions for achieving *no* reflected waves are therefore important to understand.

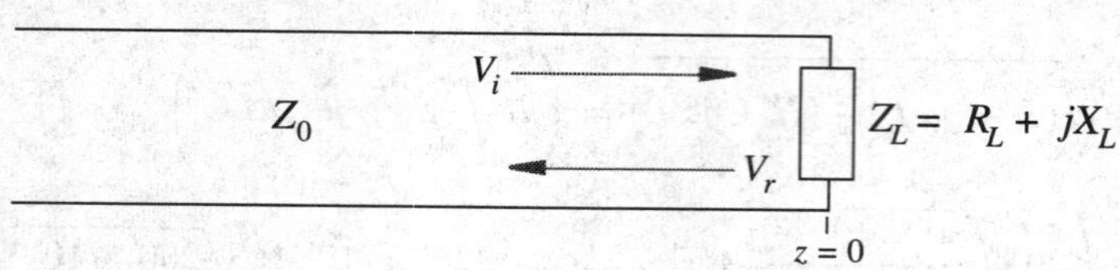

Fig. 11.6 Voltage wave reflection from a complex load impedance.

The basic reflection problem is illustrated in Figure 11.6. In it, a transmission line of characteristic impedance Z_0 is terminated by a load having complex impedance, $Z_L = R_L + jX_L$. If the line is lossy, then we know that Z_0 will also be complex. For convenience, we assign coordinates such that the load is at location $z = 0$. Therefore, the line occupies the region $z < 0$. A voltage wave is presumed to be incident on the load, and is expressed in phasor form for all z:

$$V_i(z) = V_{0i}e^{-\alpha z}e^{-j\beta z} \tag{70a}$$

When the wave reaches the load, a reflected wave is generated that back-propagates:

$$V_r(z) = V_{0r}e^{+\alpha z}e^{+j\beta z} \tag{70b}$$

The phasor voltage at the load is now the sum of the incident and reflected voltage phasors, evaluated at $z = 0$:

$$V_L = V_{0i} + V_{0r} \tag{71}$$

Additionally, the current through the load is the sum of the incident and reflected currents, also at $z = 0$:

$$I_L = I_{0i} + I_{0r} = \frac{1}{Z_0}[V_{0i} - V_{0r}] = \frac{V_L}{Z_L} = \frac{1}{Z_L}[V_{0i} + V_{0r}] \tag{72}$$

We can now solve for the ratio of the reflected voltage amplitude to the incident voltage amplitude, defined as the *reflection coefficient*, Γ:

$$\Gamma \equiv \frac{V_{0r}}{V_{0i}} = \frac{Z_L - Z_0}{Z_L + Z_0} = |\Gamma| e^{j\phi_r} \tag{73}$$

where we emphasize the complex nature of Γ—meaning that in general, a reflected wave will experience a reduction in amplitude and a phase shift, relative to the incident wave.

Now, using (71) with (73), we may write

$$V_L = V_{0i} + \Gamma V_{0i} \tag{74}$$

from which we find the *transmission coefficient,* defined as the ratio of the load voltage amplitude to the incident voltage amplitude:

$$\tau \equiv \frac{V_L}{V_{0i}} = 1 + \Gamma = \frac{2Z_L}{Z_0 + Z_L} = |\tau| e^{j\phi_t} \tag{75}$$

A point that may at first cause some alarm is that if Γ is a positive real number, then $\tau > 1$; the voltage amplitude at the load is thus greater than the incident voltage. Although this would seem counterintuitive, it is not a problem because the load current will be lower than that in the incident wave. We will find that this always results in an average *power* at the load that is less than or equal to that in the incident wave. An additional point concerns the possibility of loss in the line. The incident wave amplitude that is used in (73) and (75) is always the amplitude that occurs *at the load*—after loss has occurred in propagating from the input.

Usually, the main objective in transmitting power to a load is to configure the line/load combination such that there is no reflection. The load therefore receives all the transmitted power. The condition for this is $\Gamma = 0$, which means that the load impedance must be equal to the line impedance. In such cases the load is said to be *matched* to the line (or vice versa). Various impedance-matching methods exist, many of which will be explored later in this chapter.

Finally, the fractions of the incident wave *power* that are reflected and dissipated by the load need to be determined. The incident power is found from (64), where this time we position the load at $z = L$, with the line input at $z = 0$.

$$\langle \mathcal{P}_i \rangle = \frac{1}{2}\text{Re}\left\{ \frac{V_0 V_0^*}{|Z_0|} e^{-2\alpha L} e^{j\theta} \right\} = \frac{1}{2}\frac{|V_0|^2}{|Z_0|} e^{-2\alpha L} \cos\theta \tag{76a}$$

The reflected power is then found by substituting the reflected wave voltage into (76a), where the latter is obtained by multiplying the incident voltage by Γ:

$$\langle \mathcal{P}_r \rangle = \frac{1}{2}\mathrm{Re}\left\{ \frac{(\Gamma V_0)(\Gamma^* V_0^*)}{|Z_0|} e^{-2\alpha L} e^{j\theta} \right\} = \frac{1}{2}\frac{|\Gamma|^2 |V_0|^2}{|Z_0|} e^{-2\alpha L} \cos\theta \qquad (76b)$$

The reflected power fraction at the load is now determined by the ratio of (76b) to (76a):

$$\frac{\langle \mathcal{P}_r \rangle}{\langle \mathcal{P}_i \rangle} = \Gamma\Gamma^* = |\Gamma|^2 \qquad (77a)$$

The fraction of the incident power that is transmitted into the load (or dissipated by it) is therefore

$$\frac{\langle \mathcal{P}_t \rangle}{\langle \mathcal{P}_i \rangle} = 1 - |\Gamma|^2 \qquad (77b)$$

The reader should be aware that the transmitted power fraction is *not* $|\tau|^2$, as one might be tempted to conclude.

Illustrations

In situations involving the connection of two semi-infinite transmission lines having different characteristic impedances, reflections will occur at the junction, with the second line being treated as the load. For a wave incident from line 1 (Z_{01}) to line 2 (Z_{02}), we find

$$\Gamma = \frac{Z_{02} - Z_{01}}{Z_{02} + Z_{01}} \qquad (78)$$

The fraction of the power that propagates into the second line is then $1 - |\Gamma|^2$.

EXAMPLE 11.7

A 50 Ω lossless transmission line is terminated by a load impedance, $Z_L = 50 - j75\ \Omega$. If the incident power is 100 mW, find the power dissipated by the load.

Solution. The reflection coefficient is

$$\Gamma = \frac{Z_L - Z_0}{Z_L + Z_0} = \frac{50 - j75 - 50}{50 - j75 + 50} = 0.36 - j0.48 = 0.60e^{-j.93}$$

Then

$$\langle \mathcal{P}_t \rangle = (1 - |\Gamma|^2)\langle \mathcal{P}_i \rangle = [1 - (0.60)^2](100) = 64 \text{ mW}$$

EXAMPLE 11.8

Two lossy lines are to be joined end-to-end. The first line is 10 m long and has a loss rating of 0.20 dB/m. The second line is 15 m long and has a loss rating of 0.10 dB/m. The reflection coefficient at the junction (line 1 to line 2) is $\Gamma = 0.30$. The input power (to line 1) is 100 mW. (a) Determine the total loss of the combination in dB. (b) Determine the power transmitted to the output end of line 2.

Solution. (*a*) The dB loss of the joint is

$$L_j(\text{dB}) = 10\log_{10}\left(\frac{1}{1-|\Gamma|^2}\right) = 10\log_{10}\left(\frac{1}{1-0.09}\right) = 0.41 \text{ dB}$$

The total loss of the link in dB is now

$$L_t(\text{dB}) = (0.20)(10) + 0.41 + (0.10)(15) = 3.91 \text{ dB}$$

(*b*) The output power will be $P_{\text{out}} = 100 \times 10^{-0.391} = 41$ mW.

11.10 VOLTAGE STANDING WAVE RATIO

In many instances, transmission line performance characteristics are amenable to measurement. Included in these are measurements of unknown load impedances, or input impedances of lines that are terminated by known or unknown load impedances. Such techniques rely on the ability to measure voltage amplitudes that occur as functions of position within a line, usually designed for this purpose. A typical apparatus consists of a *slotted line,* which is a lossless coaxial transmission line having a longitudinal gap in the outer conductor along its entire length. The line is positioned between the sinusoidal voltage source and the impedance that is to be measured. Through the gap in the slotted line, a voltage probe may be inserted to measure the voltage amplitude between the inner and outer conductors. As the probe is moved along the length of the line, the maximum and minimum voltage amplitudes are noted, and their ratio, known as the *voltage standing wave ratio,* or VSWR, is determined. The significance of this measurement and its utility form the subject of this section.

To understand the meaning of the voltage measurements, we consider a few special cases. First, if the slotted line is terminated by a matched impedance, then no reflected wave occurs; the probe will indicate the same voltage amplitude at every point. Of course, the instantaneous voltage which the probe samples will differ in phase by $\beta(z_2 - z_1)$ rad as the probe is moved from $z = z_1$ to $z = z_2$, but the system is insensitive to the phase of the field. The equal-amplitude voltages are characteristic of an unattenuated travelling wave.

Second, if the slotted line is terminated by an open or short circuit (or in general a purely imaginary load impedance), the total voltage in the line is a standing wave and, as was shown in Example 11.1, the voltage probe provides no output when it is located at the nodes; these occur periodically with half-wavelength spacing. As the probe position is changed, its output varies as $|\cos(\beta z + \phi)|$, where z is the distance from the load, and where the phase, ϕ, depends on the load impedance. For example, if the load is a short circuit, the requirement of zero voltage at the short leads to a null occurring there, and so the voltage in the line will vary as $|\sin(\beta z)|$ (where $\phi = \pm\pi/2$).

A more complicated situation arises when the reflected voltage is neither 0 nor 100 percent of the incident voltage. Some energy is absorbed by the load and some is reflected. The slotted line therefore supports a voltage that is composed of both a travelling wave and a standing wave. It is customary to describe this voltage as a standing wave, even though a travelling wave is also present. We shall see that the voltage does not have zero amplitude at any point for all time, and the degree to which the voltage is divided between a travelling wave and a true standing wave is expressed by the ratio of the maximum amplitude found by the probe to the minimum amplitude (VSWR). This information, along with the positions of the voltage minima or maxima with respect to that of the load, enable one to determine the load impedance. The VSWR also provides a measure of the quality of the termination. Specifically, a perfectly matched load yields a VSWR of exactly 1. A totally reflecting load produces an infinite VSWR.

To derive the specific form of the total voltage, we begin with the forward and backward-propagating waves that occur within the slotted line. The load is positioned at $z = 0$, and so all positions within the slotted line occur at negative values of z. Taking the input wave amplitude as V_0, the total phasor voltage is

$$V_{sT}(z) = V_0 e^{-j\beta z} + \Gamma V_0 e^{j\beta z} \tag{79}$$

The line, being lossless, has real characteristic impedance, Z_0. The load impedance, Z_L, is in general complex, which leads to a complex reflection coefficient:

$$\Gamma = \frac{Z_L - Z_0}{Z_L + Z_0} = |\Gamma| e^{j\phi} \tag{80}$$

If the load is a short circuit ($Z_L = 0$), ϕ is equal to π; if Z_L is real and less than Z_0, ϕ is also equal to π; and if Z_L is real and greater than Z_0, ϕ is zero. Using (80), we may rewrite (79) in the form:

$$V_{sT}(z) = V_0 \left(e^{-j\beta z} + |\Gamma| e^{j(\beta z + \phi)} \right) = V_0 e^{j\phi/2} \left(e^{-j\beta z} e^{-j\phi/2} + |\Gamma| e^{j\beta z} e^{j\phi/2} \right) \tag{81}$$

To express (81) in a more useful form, we can apply the algebraic trick of adding and subtracting the term $V_0(1 - |\Gamma|)e^{-j\beta z}$:

$$V_{sT}(z) = V_0(1 - |\Gamma|)e^{-j\beta z} + V_0 |\Gamma| e^{j\phi/2} \left(e^{-j\beta z} e^{-j\phi/2} + e^{j\beta z} e^{j\phi/2} \right) \tag{82}$$

The last term in parentheses in (82) becomes a cosine, and we write

$$V_{sT}(z) = V_0(1 - |\Gamma|)e^{-j\beta z} + 2V_0 |\Gamma| e^{j\phi/2} \cos(\beta z + \phi/2) \tag{83}$$

The important characteristics of this result are most easily seen by converting it to real instantaneous form:

$$\mathcal{V}(z,t) = \mathrm{Re}[V_{sT}(z)e^{j\omega t}] = \underbrace{V_0(1 - |\Gamma|)\cos(\omega t - \beta z)}_{\textit{traveling wave}} + \underbrace{2|\Gamma| V_0 \cos(\beta z + \phi/2)\cos(\omega t + \phi/2)}_{\textit{standing wave}} \tag{84}$$

Equation (84) is recognized as the sum of a traveling wave of amplitude $(1 - |\Gamma|)V_0$ and a standing wave having amplitude $2|\Gamma|V_0$. We can visualize events as follows: The portion of the incident wave that reflects and back-propagates in the slotted line interferes with an equivalent portion of the incident wave to form a standing wave. The rest of the incident wave (which does not interfere) is the traveling wave part of (84). The maximum amplitude observed in the line is found where the amplitudes of the two terms in (84) add directly to give $(1 + |\Gamma|)V_0$. The minimum amplitude is found where the standing wave achieves a null, leaving only the traveling wave amplitude of $(1 - |\Gamma|)V_0$. The fact that the two terms in (84) combine in this way with the proper phasing is not immediately apparent, but the following arguments will show that this does occur.

To obtain the minimum and maximum voltage amplitudes, we may revisit the first part of Eq. (81):

$$V_{sT}(z) = V_0\left(e^{-j\beta z} + |\Gamma| e^{j(\beta z+\phi)}\right) \tag{85}$$

First, the minimum voltage amplitude is obtained when the two terms in (85) subtract directly (having a phase difference of π). This occurs at locations

$$z_{\min} = -\frac{1}{2\beta}(\phi + (2m+1)\pi) \quad (m = 0, 1, 2, \ldots) \tag{86}$$

Note again that all positions within the slotted line occur at negative values of z. Substituting (86) into (85) leads to the minimum amplitude:

$$V_{sT}(z_{\min}) = V_0(1 - |\Gamma|) \tag{87}$$

The same result is obtained by substituting (86) into the real voltage, (84). This produces a null in the standing wave part, and we obtain

$$\mathcal{V}(z_{\min}, t) = \pm V_0(1 - |\Gamma|)\sin(\omega t + \phi/2) \tag{88}$$

The voltage oscillates (through zero) in time, with amplitude $V_0(1 - |\Gamma|)$. The plus and minus signs in (88) apply to even and odd values of m in (86), respectively.

Next, the maximum voltage amplitude is obtained when the two terms in (85) add in-phase. This will occur at locations given by

$$z_{\max} = -\frac{1}{2\beta}(\phi + 2m\pi) \quad (m = 0, 1, 2, \ldots) \tag{89}$$

On substitution of (89) into (85), we obtain

$$V_{sT}(z_{\max}) = V_0(1 + |\Gamma|) \tag{90}$$

As before, we may substitute (89) into the real instantaneous voltage (84). The effect is to produce a maximum in the standing wave part, which then adds in-phase to the running wave. The result is

$$\mathcal{V}(z_{\max}, t) = \pm V_0(1 + |\Gamma|)\cos(\omega t + \phi/2) \tag{91}$$

where the plus and minus signs apply to positive and negative values of m in (89), respectively. Again, the voltage oscillates through zero in time, with amplitude $V_0(1 + |\Gamma|)$.

Note that a voltage maximum is located at the load ($z = 0$) if $\phi = 0$; moreover, $\phi = 0$ when Γ is real and positive. This occurs for real Z_L when $Z_L > Z_0$. Thus there is a voltage maximum at the load when the load impedance is greater than Z_0 and both impedances are real. With $\phi = 0$, maxima also occur at $z_{\max} = -m\pi/\beta = -m\lambda/2$. For a zero load impedance, $\phi = \pi$, and the maxima are found at $z_{\max} = -\pi/(2\beta), -3\pi/(2\beta)$, or $z_{\max} = -\lambda/4, -3\lambda/4$, and so forth.

The minima are separated by multiples of one half-wavelength (as are the maxima), and for a zero load impedance, the first minimum occurs when $-\beta z = 0$, or at the load. In general, a voltage minimum is found at $z = 0$ whenever $\phi = \pi$; this occurs if $Z_L < Z_0$ where Z_L is real. The general results are illustrated in Figure 11.7.

Finally, the voltage standing wave ratio is defined as

$$s \equiv \frac{V_{sT}(z_{\max})}{V_{sT}(z_{\min})} = \frac{1 + |\Gamma|}{1 - |\Gamma|} \tag{92}$$

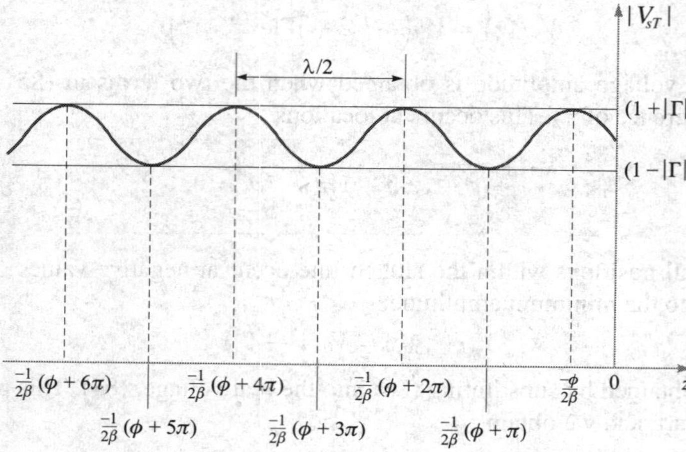

Fig. 11.7 Plot of the magnitude of V_{sT} as found from Eq. (85) as a function of position, z, in front of the load (at $z = 0$). The reflection coefficient phase is ϕ, which leads to the indicated locations of maximum and minimum voltage amplitude, as found from Eqs. (86) and (89).

Illustrations

Since the absolute voltage amplitudes have divided out, our measured VSWR permits the immediate evaluation of $|\Gamma|$. The phase of Γ is then found by measuring the location of the first maximum or minimum with respect to the load, and then using (86) or (89) as appropriate. Once Γ is known, the load impedance can be found, assuming Z_0 is known.

D11.3 What voltage standing wave ratio results when $\Gamma = \pm 1/2$?

Ans. 3

EXAMPLE 11.9

Slotted line measurements yield a VSWR of 5, a 15 cm spacing between successive voltage maxima, and the first maximum at a distance of 7.5 cm in front of the load. Determine the load impedance, assuming a 50 Ω impedance for the slotted line.

Solution. The 15 cm spacing between maxima is $\lambda/2$, implying a wavelength of 30 cm. Since the slotted line is air-filled, the frequency is $f = c/\lambda = 1$ GHz. The first maximum at 7.5 cm is thus at a distance of $\lambda/4$ from the load, which means that a voltage minimum occurs at the load. Thus Γ will be real and negative. We use (92) to write

$$|\Gamma| = \frac{s-1}{s+1} = \frac{5-1}{5+1} = \frac{2}{3}$$

So

$$\Gamma = -\frac{2}{3} = \frac{Z_L - Z_0}{Z_L + Z_0}$$

which we solve for Z_L to obtain

$$Z_L = \frac{1}{5} Z_0 = \frac{50}{5} = 10\ \Omega$$

11.11 TRANSMISSION LINES OF FINITE LENGTH

A new type of problem emerges when considering the propagation of sinusoidal voltages on finite-length lines which have loads that are not impedance-matched. In such cases, numerous reflections occur at the load and at the generator, setting up a multiwave bidirectional voltage distribution in the line. As always, the objective is to determine the net power transferred to the load in steady state, but we must now include the effect of the numerous forward and backward reflected waves.

Figure 11.8 shows the basic problem. The line, assumed to be lossless, has characteristic impedance Z_0 and is of length l. The sinusoidal voltage source at frequency ω provides phasor voltage V_s. Associated with the souce is a complex internal impedance, Z_g, as shown. The load impedance, Z_L, is also assumed to be complex and is located at $z = 0$. The line thus exists along the negative z axis. The easiest method of approaching the problem is not to attempt to analyze every reflection individually, but rather to recognize that in steady state, there will exist one net forward wave and one net backward wave, representing the superposition of all waves that are incident on the load and all waves that are reflected from it. We may thus write the total voltage in the line as

$$V_{sT}(z) = V_0^+ e^{-j\beta z} + V_0^- e^{j\beta z} \tag{93}$$

in which V_0^+ and V_0^- are complex amplitudes, composed respectively of the sum of all individual forward and backward wave amplitudes and phases. In a similar way, we may write the total current in the line:

$$I_{sT}(z) = I_0^+ e^{-j\beta z} + I_0^- e^{j\beta z} \tag{94}$$

We now define the *wave impedance*, $Z_w(z)$, as the ratio of the total phasor voltage to the total phasor current. Using (93) and (94), this becomes

$$Z_w(z) \equiv \frac{V_{sT}(z)}{I_{sT}(z)} = \frac{V_0^+ e^{-j\beta z} + V_0^- e^{j\beta z}}{I_0^+ e^{-j\beta z} + I_0^- e^{j\beta z}} \tag{95}$$

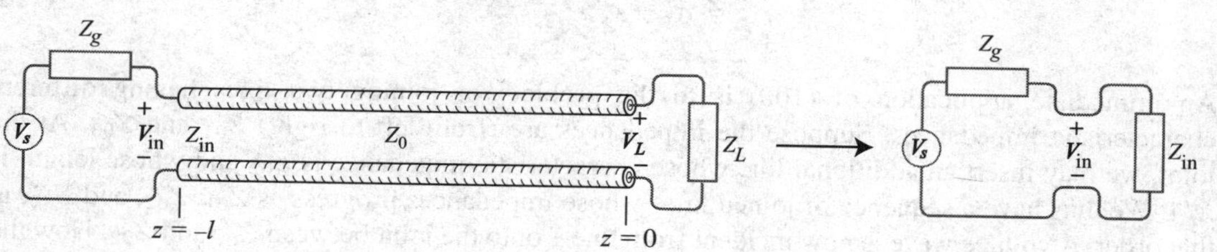

Fig. 11.8 Finite-length transmission line configuration and its equivalent circuit.

We next use the relations $V_0^- = \Gamma V_0^+$, $I_0^+ = V_0^+/Z_0$, and $I_0^- = -V_0^-/Z_0$.

Equation (95) simplifies to

$$Z_w(z) = Z_0 \left[\frac{e^{-j\beta z} + \Gamma e^{j\beta z}}{e^{-j\beta z} - \Gamma e^{j\beta z}} \right] \tag{96}$$

Now, using the Euler identity, (32), and substituting $\Gamma = (Z_L - Z_0)/(Z_L + Z_0)$, Eq. (96) becomes

$$Z_w(z) = Z_0 \left[\frac{Z_L \cos(\beta z) - jZ_0 \sin(\beta z)}{Z_0 \cos(\beta z) - jZ_L \sin(\beta z)} \right] \tag{97}$$

The wave impedance at the line input is now found by evaluating (97) at $z = -l$, obtaining

$$Z_{\text{in}} = Z_0 \left[\frac{Z_L \cos(\beta l) + jZ_0 \sin(\beta l)}{Z_0 \cos(\beta l) + jZ_L \sin(\beta l)} \right] \tag{98}$$

This is the quantity that we need in order to create the equivalent circuit in Figure 11.8.

One special case is that in which the line length is a half-wavelength, or an integer multiple thereof. In that case,

$$\beta l = \frac{2\pi}{\lambda} \frac{m\lambda}{2} = m\pi \quad (m = 0, 1, 2, \ldots)$$

Using this result in (98), we find

$$Z_{\text{in}}(l = m\lambda/2) = Z_L \tag{99}$$

For a half-wave line, the equivalent circuit can be constructed simply by removing the line completely and placing the load impedance at the input. This simplification works, of course, provided the line length is indeed an integer multiple of a half-wavelength. Once the frequency begins to vary, the condition is no longer satisfied, and (98) must be used in its general form to find Z_{in}.

Another important special case is that in which the line length is an odd multiple of a quarter wavelength:

$$\beta l = \frac{2\pi}{\lambda}(2m+1)\frac{\lambda}{4} = (2m+1)\frac{\pi}{2} \quad (m = 0, 1, 2, \ldots)$$

Using this result in (98) leads to

$$Z_{\text{in}}(l = \lambda/4) = \frac{Z_0^2}{Z_L} \tag{100}$$

An immediate application of (100) is to the problem of joining two lines having different characteristic impedances. Suppose the impedances are (from left to right) Z_{01} and Z_{03}. At the joint, we may insert an additional line whose characteristic impedance is Z_{02} and whose length is $\lambda/4$. We thus have a sequence of joined lines whose impedances progress as Z_{01}, Z_{02}, and Z_{03}, in that order. A voltage wave is now incident from line 1 onto the joint between Z_{01} and Z_{02}. Now the effective load at the far end of line 2 is Z_{03}. The input impedance to line 2 at any frequency is now

$$Z_{\text{in}} = Z_{02} \frac{Z_{03} \cos \beta_2 l + jZ_{02} \sin \beta_2 l}{Z_{02} \cos \beta_2 l + jZ_{03} \sin \beta_2 l} \tag{101}$$

Then, since the length of line 2 is $\lambda/4$,

$$Z_{\text{in}}(\text{line } 2) = \frac{Z_{02}^2}{Z_{03}} \tag{102}$$

Reflections at the Z_{01}–Z_{02} interface will not occur if $Z_{\text{in}} = Z_{01}$. Therefore, we can match the junction (allowing complete transmission through the three-line sequence) if Z_{02} is chosen so that

$$Z_{02} = \sqrt{Z_{01} Z_{03}} \tag{103}$$

This technique is called *quarter-wave matching,* and again is limited to the frequency (or narrow band of frequencies) such that $l \doteq (2m+1)\lambda/4$. We will encounter more examples of these techniques when we explore electromagnetic wave reflection in Chapter 13. Meanwhile, further examples that involve the use of the input impedance and the VSWR are presented in Section 11.12.

EXAMPLE 11.10

The transmission lines in the UHF and microwave frequency range can also be used as circuit elements showing very useful properties. The basic idea here is to consider a piece of lossless transmission line terminated in either *open* or *short* at its end. The input impedance of this special line can then exhibit either inductive or capacitive behavior depending upon the length and the operating frequency. Consider a piece of lossless transmission line of length 'l' filled with Teflon ($\epsilon_r = 2.1$), and characteristic impedance Z_0 as shown in Figure 11.9. The line is shorted at its end.

(*a*) Determine the expression for the input impedance of this line starting from Eq. (98).

(*b*) Now, suppose, the shorted 50 Ω transmission line shown in Figure 11.9 is to be designed to provide a capacitive reactance with an equivalent capacitance of 5 pF at the input for a device operating at 2 GHz. Find the shortest possible length l of the line.

(*c*) Explain whether there is any change in the behavior of the input impedance of the shorted line if the frequency is changed to 3 GHz.

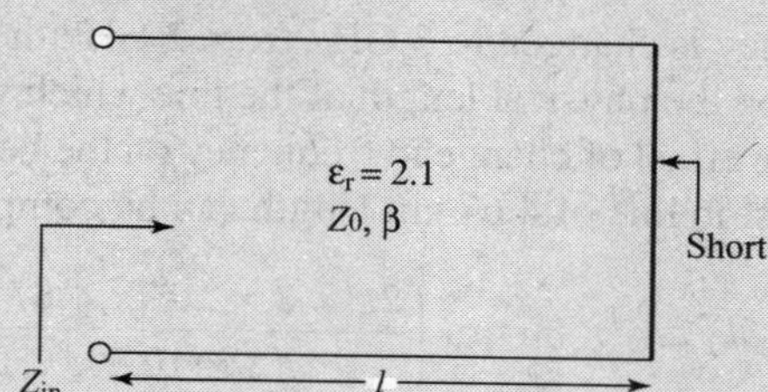

Fig. 11.9 A piece of short-circuited transmission line.

Solution.

(*a*) The input impedance of a transmission line of length 'l', whose characteristic impedance and the propagation constant are defined by 'Z_0' and 'β', respectively, is given by the following expression as per Eq. (98):

$$Z_{\text{in}} = Z_0 \left[\frac{Z_L \cos(\beta l) + j\, Z_0 \sin(\beta l)}{Z_0 \cos(\beta l) + j\, Z_L \sin(\beta l)} \right] \tag{i}$$

where Z_L is the terminated load impedance. Now, for the short-circuited transmission line shown in Figure 11.9, the load impedance would be zero, i.e., $Z_L = 0$, which basically modifies the basic equation for the input impedance as

$$Z_{\text{in}}^{\text{sc}} = Z_0 \left[\frac{0 \times \cos(\beta l) + j Z_0 \sin(\beta l)}{Z_0 \cos(\beta l) + j\, 0 \times \sin(\beta l)}\right] \equiv j Z_0 \tan(\beta l) \qquad \text{(ii)}$$

From the above equation, it can be postulated that a piece of short-circuited transmission line can provide either inductive or capacitive reactance at the input of the line depending upon the fact that whether the sign of the tangent function is positive or negative.

(*b*) For realizing a capacitive reactance at the input of the shorted transmission line shown in Figure 11.9, the argument of tangent function in the above equation should be selected in such a way that the value of $\tan(\beta l)$ becomes negative at the operating frequency. Under this condition, the above equation can be written as

$$Z_{\text{in}}^{\text{sc}} = jZ_0 \tan(\beta l) \equiv -\frac{j}{\omega C_{\text{eq}}}$$

$$\Rightarrow \tan(\beta l) \equiv -\frac{1}{\omega Z_0 C_{\text{eq}}} = -\frac{1}{2 \times \pi \times 2 \times 10^9 \times 50 \times 5 \times 10^{-12}} = -0.3183$$

Now, the tangent function would be negative when its argument $\theta \equiv \beta l$ lies either in the second or the fourth quadrant. It is, however, given that the length of the line 'l' should be minimum. Hence, the argument of the tangent function should lie in the second quadrant, i.e.,

$$\frac{\pi}{2} < \beta l < \pi$$

The value of the shortest possible length 'l_s' can then be computed as

$$\tan(\beta l_s) = -0.3183 \equiv \tan(0.9019\,\pi)$$

$$\Rightarrow l_s = \frac{0.9019\,\pi}{\beta} \equiv \frac{0.9019\,\pi}{2\pi/\lambda} \equiv \frac{0.9019\,v_p}{2f} = \frac{0.9019\,c}{2f\sqrt{\epsilon_r}} = \frac{0.9019 \times 3 \times 10^8}{2 \times 2 \times 10^9 \times \sqrt{2.1}} = 4.67 \text{ cm}$$

(*c*) If the operating frequency is changed to 3 GHz from 2 GHz then the phase constant β would change without changing the physical length of the line which would be kept fixed at 4.67 cm as computed above. The effect of change in frequency on the behavior of the input impedance of the shorted transmission line of 4.67 cm length can be computed as follows

$$Z_{in}^{sc} = jZ_0 \tan(\beta l) \equiv j \times 50 \tan\left[\frac{2\pi \times 3 \times 10^9 \times 4.67 \times 10^{-2}}{(3 \times 10^8/\sqrt{2.1})}\right] = j100.85$$

Hence, at 3 GHz, the input impedance of the shorted 50 Ω transmission line shown in Figure 11.9 behaves like an inductive reactance. The equivalent inductance of this line at 3 GHz can be computed using the following expression

$$Z_{\text{in}}^{sc} = j100.85 \equiv j\omega L_{\text{eq}}$$

$$\Rightarrow L_{\text{eq}} = \frac{100.85}{2 \times \pi \times 3 \times 10^9} = 5.35 \text{ nH}$$

11.12 SOME TRANSMISSION LINE EXAMPLES

In this section we shall apply many of the results that we obtained in the previous sections to several typical transmission line problems. We shall simplify our work by restricting our attention to the lossless line.

Let us begin by assuming a two-wire 300 Ω line ($Z_0 = 300\ \Omega$), such as the lead-in wire from the antenna to a television or FM receiver. The circuit is shown in Figure 11.10. The line is 2 m long, and the values of L and C are such that the velocity on the line is 2.5×10^8 m/s. We shall terminate the line with a receiver having an input resistance of 300 Ω and represent the antenna by its Thevenin equivalent $Z = 300\,\Omega$ in series with $V_s = 60$ V at 100 MHz. This antenna voltage is larger by a factor of about 10^5 than it would be in a practical case, but it also provides simpler values to work with; in order to think practical thoughts, divide currents or voltages by 10^5, divide powers by 10^{10}, and leave impedances alone.

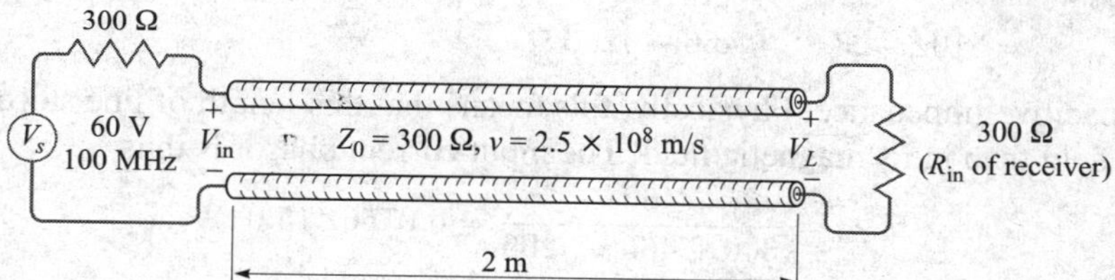

Fig. 11.10 A transmission line that is matched at both ends produces no reflections, and thus delivers maximum power to the load.

Since the load impedance is equal to the characteristic impedance, the line is matched; the reflection coefficient is zero, and the standing wave ratio is unity. For the given velocity and frequency, the wavelength on the line is $v/f = 2.5$ m, and the phase constant is $2\pi/\lambda = 0.8\pi$ rad/m; the attenuation constant is zero. The electrical length of the line is $\beta l = (0.8\pi)2$, or 1.6π rad. This length may also be expressed as 288°, or 0.8 wavelength.

The input impedance offered to the voltage source is 300 Ω, and since the internal impedance of the source is 300 Ω, the voltage at the input to the line is half of 60 V, or 30 V. The source is matched to the line and delivers the maximum available power to the line. Since there is no reflection and no attenuation, the voltage at the load is 30 V, but it is delayed in phase by 1.6π rad. Thus

$$V_{\text{in}} = 30\cos(2\pi 10^8 t)\ \text{V}$$

whereas

$$V_L = 30\cos(2\pi 10^8 t - 1.6\pi)\ \text{V}$$

The input current is

$$I_{\text{in}} = \frac{V_{\text{in}}}{300} = 0.1\cos(2\pi 10^8 t)\ \text{A}$$

while the load current is

$$I_L = 0.1\cos(2\pi 10^8 t - 1.6\pi)\ \text{A}$$

The average power delivered to the input of the line by the source must all be delivered to the load by the line,

$$P_{\text{in}} = P_L = \frac{1}{2} \times 30 \times 0.1 = 1.5 \text{ W}$$

Now let us connect a second receiver, also having an input resistance of 300 Ω, across the line in parallel with the first receiver. The load impedance is now 150 Ω, the reflection coefficient is

$$\Gamma = \frac{150 - 300}{150 + 300} = -\frac{1}{3}$$

and the standing wave ratio on the line is

$$s = \frac{1 + \frac{1}{3}}{1 - \frac{1}{3}} = 2$$

The input impedance is no longer 300 Ω, but is now

$$Z_{\text{in}} = Z_0 \frac{Z_L \cos \beta l + jZ_0 \sin \beta l}{Z_0 \cos \beta l + jZ_L \sin \beta l} = 300 \frac{150 \cos 288^\circ + j300 \sin 288^\circ}{300 \cos 288^\circ + j150 \sin 288^\circ}$$

$$= 510\angle{-23.8^\circ} = 466 - j206 \ \Omega$$

which is a capacitive impedance. Physically, this means that this length of line stores more energy in its electric field than in its magnetic field. The input current phasor is thus

$$I_{s,\text{in}} = \frac{60}{300 + 466 - j206} = 0.0756\angle 15.0^\circ \text{ A}$$

and the power supplied to the line by the source is

$$P_{\text{in}} = \frac{1}{2} \times (0.0756)^2 \times 466 = 1.333 \text{ W}$$

Since there are no losses in the line, 1.333 W must also be delivered to the load. Note that this is less than the 1.50 W which we were able to deliver to a matched load; moreover, this power must divide equally between two receivers, and thus each receiver now receives only 0.667 W. Since the input impedance of each receiver is 300 Ω, the voltage across the receiver is easily found as

$$0.667 = \frac{1}{2} \frac{|V_{s,L}|^2}{300}$$

$$|V_{s,L}| = 20 \text{ V}$$

in comparison with the 30 V obtained across the single load.

Before we leave this example, let us ask ourselves several questions about the voltages on the transmission line. Where is the voltage a maximum and a minimum, and what are these values? Does the phase of the load voltage still differ from the input voltage by 288°? Presumably, if we can answer these questions for the voltage, we could do the same for the current.

Equation (89) serves to locate the voltage maxima at

$$z_{\text{max}} = -\frac{1}{2\beta}(\phi + 2m\pi) \quad (m = 0, 1, 2, \ldots)$$

where $\Gamma = |\Gamma|e^{j\phi}$. Thus, with $\beta = 0.8\pi$ and $\phi = \pi$, we find

$$z_{\text{max}} = -0.625 \quad \text{and} \quad -1.875 \text{ m}$$

while the minima are $\lambda/4$ distant from the maxima;

$$z_{\text{min}} = 0 \quad \text{and} \quad -1.25 \text{ m}$$

and we find that the load voltage (at $z = 0$) is a voltage minimum. This, of course, verifies the general conclusion we reached earlier: a voltage minimum occurs at the load if $Z_L < Z_0$, and a voltage maximum occurs if $Z_L > Z_0$, where both impedances are pure resistances.

The minimum voltage on the line is thus the load voltage, 20 V; the maximum voltage must be 40 V, since the standing wave ratio is 2. The voltage at the input end of the line is

$$V_{s,\text{in}} = I_{s,\text{in}} Z_{\text{in}} = (0.0756\angle 15.0^\circ)(510\angle -23.8^\circ) = 38.5\angle -8.8^\circ$$

The input voltage is almost as large as the maximum voltage anywhere on the line because the line is about three-quarters of a wavelength long, a length which would place the voltage maximum at the input when $Z_L < Z_0$.

Finally, it is of interest to determine the load voltage in magnitude *and phase*. We begin with the total voltage in the line, using (93).

$$V_{sT} = \left(e^{-j\beta z} + \Gamma e^{j\beta z}\right) V_0^+ \tag{104}$$

We may use this expression to determine the voltage at any point on the line in terms of the voltage at any other point. Since we know the voltage at the input to the line, we let $z = -l$,

$$V_{s,\text{in}} = \left(e^{j\beta l} + \Gamma e^{-j\beta l}\right) V_0^+ \tag{105}$$

and solve for V_0^+,

$$V_0^+ = \frac{V_{s,\text{in}}}{e^{j\beta l} + \Gamma e^{-j\beta l}} = \frac{38.5\angle -8.8^\circ}{e^{j1.6\pi} - \frac{1}{3}e^{-j1.6\pi}} = 30.0\angle 72.0^\circ \text{ V}$$

We may now let $z = 0$ in (104) to find the load voltage,

$$V_{s,L} = (1 + \Gamma)V_0^+ = 20\angle 72^\circ = 20\angle -288^\circ$$

The amplitude agrees with our previous value. The presence of the reflected wave causes $V_{s,\text{in}}$ and $V_{s,L}$ to differ in phase by about -279° instead of -288°.

EXAMPLE 11.11

In order to provide a slightly more complicated example, let us now place a purely capacitive impedance of $-j300\,\Omega$ in parallel with the two 300 Ω receivers. We are to find the input impedance and the power delivered to each receiver.

Solution. The load impedance is now 150 Ω in parallel with $-j300\,\Omega$, or

$$Z_L = \frac{150(-j300)}{150 - j300} = \frac{-j300}{1 - j2} = 120 - j60\,\Omega$$

We first calculate the reflection coefficient and the VSWR:

$$\Gamma = \frac{120 - j60 - 300}{120 - j60 + 300} = \frac{-180 - j60}{420 - j60} = 0.447\angle -153.4^\circ$$

$$s = \frac{1 + 0.447}{1 - 0.447} = 2.62$$

Thus, the VSWR is higher and the mismatch is therefore worse. Let us next calculate the input impedance. The electrical length of the line is still 288°, so that

$$Z_{\text{in}} = 300\,\frac{(120 - j60)\cos 288^\circ + j300 \sin 288^\circ}{300 \cos 288^\circ + j(120 - j60)\sin 288^\circ} = 755 - j138.5\ \Omega$$

This leads to a source current of

$$I_{s,\text{in}} = \frac{V_{Th}}{Z_{Th} + Z_{\text{in}}} = \frac{60}{300 + 755 - j138.5} = 0.0564\angle 7.47^\circ\ \text{A}$$

Therefore, the average power delivered to the input of the line is $P_{\text{in}} = \frac{1}{2}(0.0564)^2(755) = 1.200$ W. Since the line is lossless, it follows that $P_L = 1.200$ W, and each receiver gets only 0.6 W.

EXAMPLE 11.12

A local transmitter represented by an equivalent voltage source of 50 V in series with an internal impedance of $R_g = 50\ \Omega$ is operating at 500 MHz. The transmitter is connected to an antenna being represented by a complex load of $Z_L = 200 - j50\ \Omega$ through a 4 m long, 50 Ω Teflon filled ($\epsilon_r = 2.10$) lossless transmission line as shown in Figure 11.11.

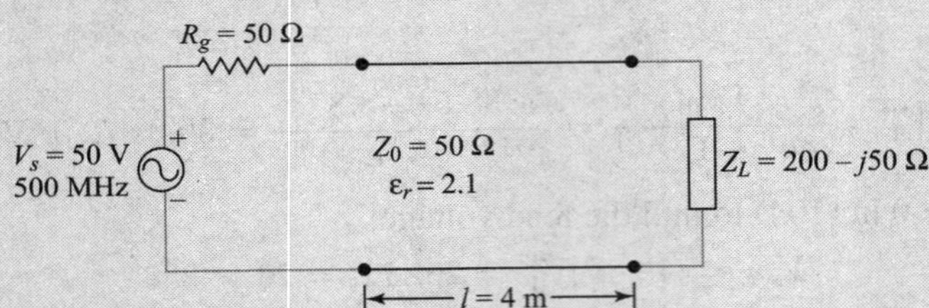

Fig. 11.11 Lossless transmission line.

Determine

(*a*) the reflection coefficient and the VSWR at the load
(*b*) the input impedance seen by the transmitter
(*c*) the time-averaged power delivered to the transmission line by the transmitter
(*d*) the time-averaged power delivered to the antenna
(*e*) the time-averaged power dissipated in the internal impedance R_g
(*f*) the total power supplied by the transmitter

Solution.

(*a*) The reflection coefficient at the load is given by

$$\Gamma = \frac{Z_L - Z_0}{Z_L - Z_0} = \frac{(200 - j50) - 50}{(200 - j50) + 50} = 0.6154 - j0.0769 \equiv 0.6202\angle - 0.1244$$

The voltage standing-wave ratio at the load is determined as

$$s \equiv \frac{1 + |\Gamma|}{1 - |\Gamma|} = \frac{1 + 0.6202}{1 - 0.6202} = 4.27$$

(*b*) For determining the input impedance, let us first calculate the phase constant β for the transmission line shown in Figure 11.11.

$$\beta = \frac{2\pi}{\lambda} \equiv \frac{2\pi f}{v_p} \equiv \frac{2\pi \times 500 \times 10^6}{(3 \times 10^8/\sqrt{2.1})} = 15.18 \text{ rad/m}$$

The input impedance seen by the transmitter is then calculated as follows

$$Z_{\text{in}} = Z_0 \frac{Z_L \cos\beta l + jZ_0 \sin\beta l}{Z_0 \cos\beta l + jZ_L \sin\beta l} = 50 \frac{(200 - j50)\cos(15.18 \times 4) + j50 \sin(15.18 \times 4)}{50 \cos(15.18 \times 4) + j\,(200 - j50)\sin(15.18 \times 4)}$$

$$= (14.664 - j24.17)\,\Omega$$

(*c*) For calculating the time-averaged power delivered to the transmission line by the source, the configuration shown in Figure 11.11 can be replaced by its equivalent circuit at the generator end as shown below in Figure 11.12

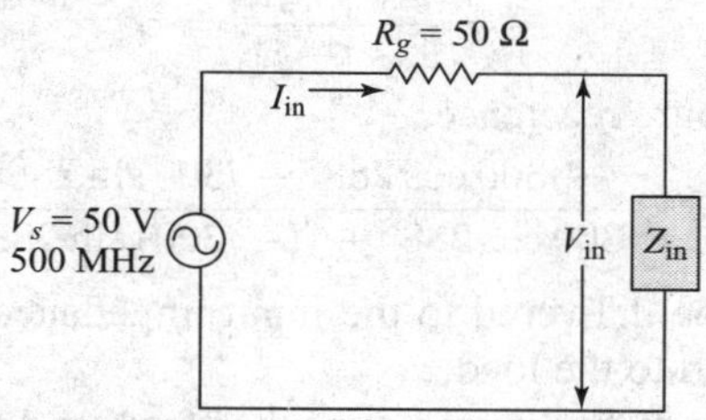

Fig. 11.12 Equivalent circuit.

The terminal voltage V_{in} in Figure 11.12 can be determined as

$$V_{\text{in}} = V_s \frac{Z_{\text{in}}}{Z_{\text{in}} + R_g} = 50 \frac{14.664 - j24.17}{14.664 - j24.17 + 50} = 20.48\angle -38.3^\circ \text{ V}$$

Similarly, the input terminal current can be calculated as

$$I_{\text{in}} = \frac{V_s}{R_g + Z_{\text{in}}} = \frac{50}{14.664 - j24.17 + 50} = 0.7243\angle 20.5^\circ \text{ A}$$

The time-averaged power delivered to the transmission line by the transmitter can finally be determined as

$$P_{\text{in}} = \frac{1}{2}(I_{\text{in}})^2\,\Re[Z_{\text{in}}] = \frac{1}{2} \times (0.7243)^2 \times 14.664 = 3.846 \text{ W}$$

(*d*) As the transmission line is lossless, hence the total power delivered to the transmission line by the source would be absorbed by the load, which represents the antenna in Figure 11.11. Hence, the time-averaged power delivered to the antenna is 3.846 W.

(*e*) The time-averaged power dissipated in the internal impedance R_g can be determined using the equivalent circuit of Figure 11.12, which provides following expression:

$$P_g = \frac{1}{2}(I_{\text{in}})^2 R_g = \frac{1}{2} \times (0.7243)^2 \times 50 = 13.1 \text{ W}$$

(*f*) The total power supplied by the transmitter would be given by

$$P_{\text{Total}} = P_{\text{in}} + P_g = (3.846 + 13.1) = 16.946 \text{ W}$$

EXAMPLE 11.13

As a final example, let us terminate the line shown in Figure 11.10 with a purely capacitive impedance, $Z_L = -j300\,\Omega$. We seek the reflection coefficient, the VSWR, and the power delivered to the load.

Solution. Obviously, we cannot deliver any average power to the load since it is a pure reactance. As a consequence, the reflection coefficient is

$$\Gamma = \frac{-j300 - 300}{-j300 + 300} = -j1 = 1\angle -90^\circ$$

and the reflected wave is equal in amplitude to the incident wave. Hence it should not surprise us to see that the VSWR is

$$s = \frac{1 + |-j1|}{1 - |-j1|} = \infty$$

and the input impedance is a pure reactance,

$$Z_{\text{in}} = 300\,\frac{-j300\cos 288^\circ + j300\sin 288^\circ}{300\cos 288^\circ + j(-j300)\sin 288^\circ} = j589$$

Thus, no average power can be delivered to the input impedance by the source, and therefore no average power can be delivered to the load.

Although we could continue to find numerous other facts and figures for these examples, much of the work may be done more easily for problems of this type by using graphical techniques. We shall encounter these in Section 11.13.

D11.4 A 50 W lossless line has a length of 0.4λ. The operating frequency is 300 MHz. A load $Z_L = 40 + j30\ \Omega$ is connected at $z = 0$, and the Thevenin-equivalent source at $z = -l$ is $12\angle 0^\circ$ V in series with $Z_{Th} = 50 + j0\ \Omega$. Find: (*a*) Γ; (*b*) s; (*c*) Z_{in}.

Ans. $0.333\angle 90^\circ$; 2.00; $25.5 + j5.90\ \Omega$

D11.5 For the transmission line of Problem D11.4, also find: (*a*) the phasor voltage at $z = -l$; (*b*) the phasor voltage at $z = 0$; (*c*) the average power delivered to Z_L.

Ans. $4.14\angle 8.58^\circ$ V; $6.32\angle -125.6^\circ$ V; 0.320 W

11.13 GRAPHICAL METHODS

Animations

Transmission line problems often involve manipulations with complex numbers, making the time and effort required for a solution several times greater than are needed for a similar sequence of operations on real numbers. One means of reducing the labor without seriously affecting the accuracy is by using transmission-line charts. Probably the most widely used one is the Smith chart.[3]

[3] P. H. Smith, "Transmission Line Calculator," *Electronics*, vol. 12, pp. 29–31, January 1939.

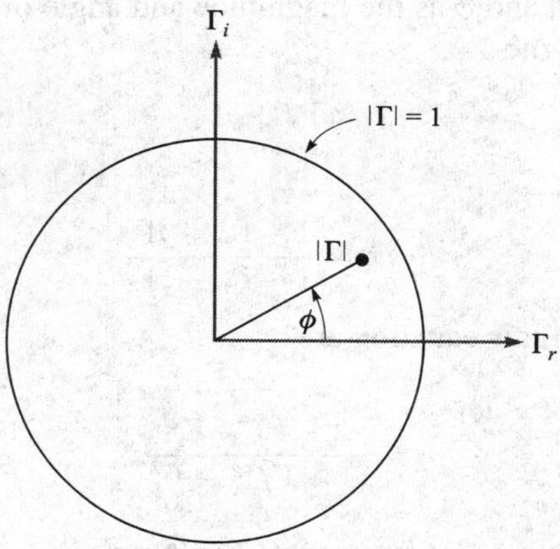

Fig. 11.13 The polar coordinates of the Smith chart are the magnitude and phase angle of the reflection coefficient; the rectangular coordinates are the real and imaginary parts of the reflection coefficient. The entire chart lies within the circle $|\Gamma| = 1$.

Basically, this diagram shows curves of constant resistance and constant reactance; these may represent either an input impedance or a load impedance. The latter, of course, is the input impedance of a zero-length line. An indication of location along the line is also provided, usually in terms of the fraction of a wavelength from a voltage maximum or minimum. Although they are not specifically shown on the chart, the standing-wave ratio and the magnitude and angle of the reflection coefficient are very quickly determined. As a matter of fact, the diagram is constructed within a circle of unit radius, using polar coordinates, with radius variable $|\Gamma|$ and counterclockwise angle variable ϕ, where $\Gamma = |\Gamma|e^{j\phi}$. Figure 11.13 shows this circle. Since $|\Gamma| < 1$, all our information must lie on or within the unit circle. Peculiarly enough, the reflection coefficient itself will not be plotted on the final chart, for these additional contours would make the chart very difficult to read.

The basic relationship upon which the chart is constructed is

$$\Gamma = \frac{Z_L - Z_0}{Z_L + Z_0} \tag{106}$$

The impedances which we plot on the chart will be *normalized* with respect to the characteristic impedance. Let us identify the normalized load impedance as z_L,

$$z_L = r + jx = \frac{Z_L}{Z_0} = \frac{R_L + jX_L}{Z_0}$$

and thus

$$\Gamma = \frac{z_L - 1}{z_L + 1}$$

or

$$z_L = \frac{1 + \Gamma}{1 - \Gamma} \tag{107}$$

In polar form, we have used $|\Gamma|$ and ϕ as the magnitude and angle of Γ. With Γ_r and Γ_i as the real and imaginary parts of Γ, we write

$$\Gamma = \Gamma_r + j\Gamma_i \tag{108}$$

Thus

$$r + jx = \frac{1 + \Gamma_r + j\Gamma_i}{1 - \Gamma_r - j\Gamma_i} \tag{109}$$

The real and imaginary parts of this equation are

$$r = \frac{1 - \Gamma_r^2 - \Gamma_i^2}{(1 - \Gamma_r)^2 + \Gamma_i^2} \tag{110}$$

$$x = \frac{2\Gamma_i}{(1 - \Gamma_r)^2 + \Gamma_i^2} \tag{111}$$

After several lines of elementary algebra, we may write (110) and (111) in forms which readily display the nature of the curves on Γ_r, Γ_i axes,

$$\left(\Gamma_r - \frac{r}{1+r}\right)^2 + \Gamma_i^2 = \left(\frac{1}{1+r}\right)^2 \tag{112}$$

$$(\Gamma_r - 1)^2 + \left(\Gamma_i - \frac{1}{x}\right)^2 = \left(\frac{1}{x}\right)^2 \tag{113}$$

The first equation describes a family of circles, where each circle is associated with a specific value of resistance r. For example, if $r = 0$ the radius of this zero-resistance circle is seen to be unity, and it is centered at the origin ($\Gamma_r = 0, \Gamma_i = 0$). This, for a pure reactance termination leads to a reflection coefficient of unity magnitude. On the other hand, if $r = \infty$, then $z_L = \infty$ and we have $\Gamma = 1 + j0$. The circle described by (112) is centered at $\Gamma_r = 1$, $\Gamma_i = 0$ and has zero radius. It is therefore the point $\Gamma = 1 + j0$, as we decided it should be. As another example, the circle for $r = 1$ is centered at $\Gamma_r = 0.5$, $\Gamma_i = 0$ and has a radius of 0.5. This circle is shown in Figure 11.14, along with circles for $r = 0.5$ and $r = 2$. All circles are centered on the Γ_r axis and pass through the point $\Gamma = 1 + j0$.

Equation (113) also represents a family of circles, but each of these circles is defined by a particular value of x, rather than r. If $x = \infty$, then $z_L = \infty$, and $\Gamma = 1 + j0$ again. The circle described by (113) is centered at $\Gamma = 1 + j0$ and has zero radius; it is therefore the point $\Gamma = 1 + j0$. If $x = +1$, then the circle is centered at $\Gamma = 1 + j1$ and has unit radius. Only one-quarter of this circle lies within the boundary curve $|\Gamma| = 1$, as shown in Figure 11.15. A similar quarter-circle appears below the Γ_r axis for $x = -1$. The portions of other circles for $x = 0.5$, -0.5, 2, and -2 are also shown. The "circle" representing $x = 0$ is the Γ_r axis; this is also labeled in Figure 11.15.

The two families of circles both appear on the Smith chart, as shown in Figure 11.16. It is now evident that if we are given Z_L, we may divide by Z_0 to obtain z_L, locate the appropriate r and x circles (interpolating as necessary), and determine Γ by the intersection of the two circles. Since the chart does not have concentric circles showing the values of $|\Gamma|$, it is necessary to measure the radial

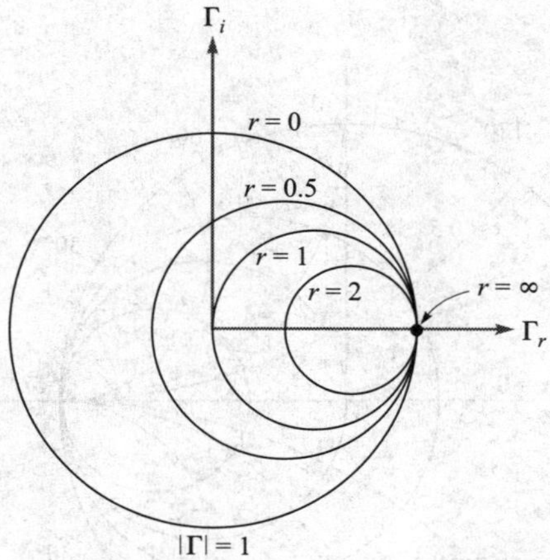

Fig. 11.14 Constant-r circles are shown on the Γ_r, Γ_i plane. The radius of any circle is $1/(1+r)$.

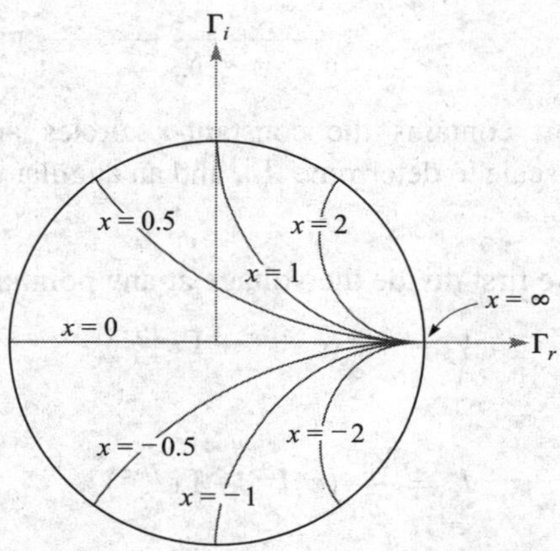

Fig. 11.15 The portions of the circles of constant x lying within $|\Gamma| = 1$ are shown on the Γ_r, Γ_i axes. The radius of a given circle is $1/|x|$.

distance from the origin to the intersection with dividers or compass and use an auxiliary scale to find $|\Gamma|$. The graduated line segment below the chart in Figure 11.16 serves this purpose. The angle of Γ is ϕ, and it is the counterclockwise angle from the Γ_r axis. Again, radial lines showing the angle would clutter up the chart badly, so the angle is indicated on the circumference of the circle. A straight line from the origin through the intersection may be extended to the perimeter of the chart. As an example, if $Z_L = 25 + j50\,\Omega$ on a 50 Ω line, $z_L = 0.5 + j1$, and point A on Figure 11.16 shows the intersection of the $r = 0.5$ and $x = 1$ circles. The reflection coefficient is approximately 0.62 at an angle ϕ of 83°.

The Smith chart is completed by adding a second scale on the circumference by which distance along the line may be computed. This scale is in wavelength units, but the values placed on it are

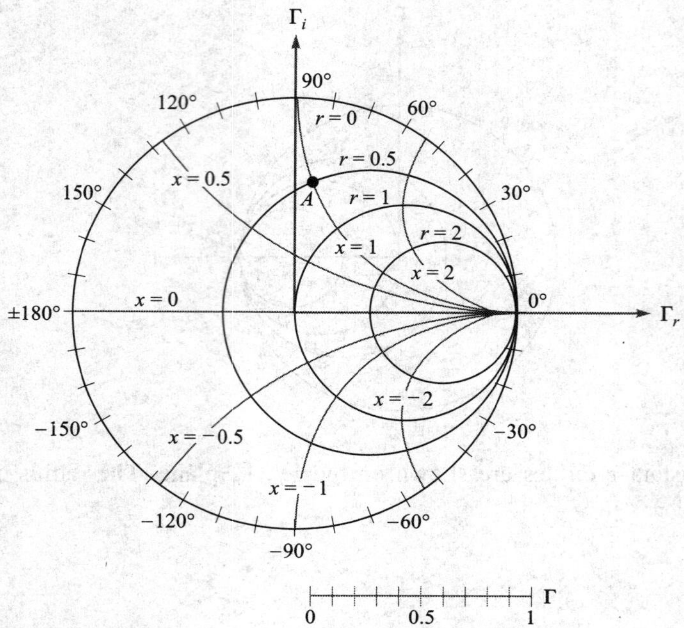

Fig. 11.16 The Smith chart contains the constant-r circles and constant-x circles, an auxiliary radial scale to determine $|\Gamma|$, and an angular scale on the circumference for measuring ϕ.

not obvious. To obtain them, we first divide the voltage at any point along the line,

$$V_s = V_0^+(e^{-j\beta z} + \Gamma e^{j\beta z})$$

by the current

$$I_s = \frac{V_0^+}{Z_0}(e^{-j\beta z} - \Gamma e^{j\beta z})$$

obtaining the normalized input impedance

$$z_{\text{in}} = \frac{V_s}{Z_0 I_s} = \frac{e^{-j\beta z} + \Gamma e^{j\beta z}}{e^{-j\beta z} - \Gamma e^{j\beta z}}$$

Replacing z with $-l$ and dividing numerator and denominator by $e^{j\beta l}$, we have the general equation relating normalized input impedance, reflection coefficient, and line length,

$$z_{\text{in}} = \frac{1 + \Gamma e^{-j2\beta l}}{1 - \Gamma e^{-j2\beta l}} = \frac{1 + |\Gamma| e^{j(\phi - 2\beta l)}}{1 - |\Gamma| e^{j(\phi - 2\beta l)}} \tag{114}$$

Note that when $l = 0$, we are located at the load, and $z_{\text{in}} = (1 + \Gamma)/(l - \Gamma) = z_L$, as shown by (107).

Equation (114) shows that the input impedance at any point $z = -l$ can be obtained by replacing Γ, the reflection coefficient of the load, by $\Gamma e^{-j2\beta l}$. That is, we decrease the angle of Γ by $2\beta l$

radians as we move from the load to the line input. Only the angle of Γ is changed; the magnitude remains constant.

Thus, as we proceed from the load z_L to the input impedance z_{in}, we move *toward* the generator a distance l on the transmission line, but we move through a *clockwise* angle of $2\beta l$ on the Smith chart. Since the magnitude of Γ stays constant, the movement toward the source is made along a constant-radius circle. One lap around the chart is accomplished whenever βl changes by π rad, or when l changes by one-half wavelength. This agrees with our earlier discovery that the input impedance of a half-wavelength lossless line is equal to the load impedance.

The Smith chart is thus completed by the addition of a scale showing a change of 0.5λ for one circumnavigation of the unit circle. For convenience, two scales are usually given, one showing

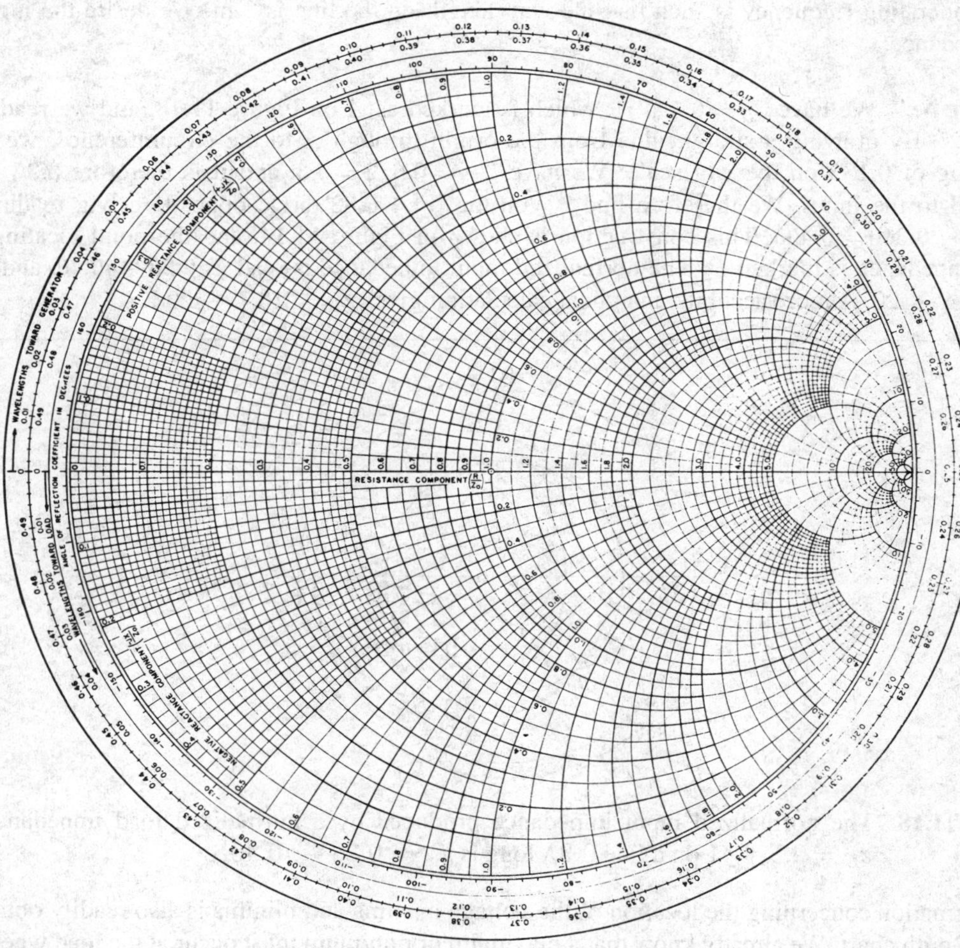

Fig. 11.17 A photographic reduction of one version of a useful Smith Chart (*courtesy of the Emeloid Company, Hillside, N.J.*). For accurate work, larger charts are available wherever fine technical books are sold.

an increase in distance for clockwise movement and the other an increase for counterclockwise travel. These two scales are shown in Figure 11.17. Note that the one marked "wavelengths toward generator" (wtg) shows increasing values of l/λ for clockwise travel, as described previously. The zero point of the wtg scale is rather arbitrarily located to the left. This corresponds to input impedances having phase angles of 0° and $R_L < Z_0$. We have also seen that voltage minima are always located here.

EXAMPLE 11.14

The use of the transmission line chart is best shown by example. Let us again consider a load impedance, $Z_L = 25 + j50\ \Omega$, terminating a 50 Ω line. The line length is 60 cm and the operating frequency is such that the wavelength on the line is 2 m. We desire the input impedance.

Solution. We have $z_L = 0.5 + j1$, which is marked as A on Figure 11.18, and we read $\Gamma = 0.62\angle 82°$. By drawing a straight line from the origin through A to the circumference, we note a reading of 0.135 on the wtg scale. We have $l/\lambda = 0.6/2 = 0.3$, and it is therefore 0.3λ from the load to the input. We therefore find z_{in} on the $|\Gamma| = 0.62$ circle opposite a wtg reading of $0.135 + 0.300 = 0.435$. This construction is shown in Figure 11.14, and the point locating the input impedance is marked B. The normalized input impedance is read as $0.28 - j0.40$, and thus $Z_{\text{in}} = 14 - j20$. A more accurate analytical calculation gives $Z_{\text{in}} = 13.7 - j20.2$.

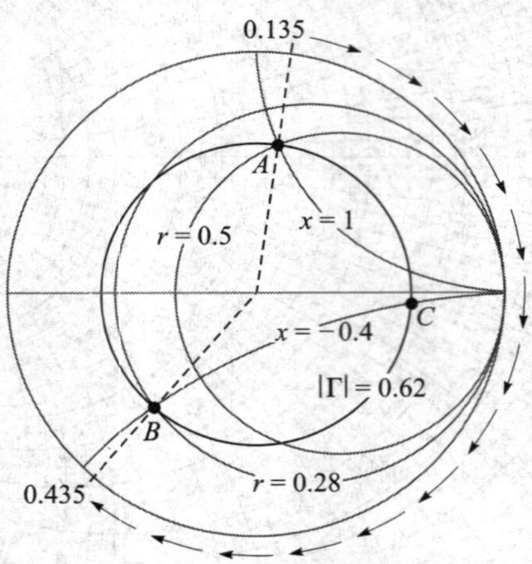

Fig. 11.18 The normalized input impedance produced by a normalized load impedance $z_L = 0.5 + j1$ on a line 0.3λ long is $z_{\text{in}} = 0.28 - j0.40$.

Information concerning the location of the voltage maxima and minima is also readily obtained on the Smith chart. We already know that a maximum or minimum must occur at the load when Z_L is a pure resistance; if $R_L > Z_0$ there is a maximum at the load, and if $R_L < Z_0$ there is a minimum. We may extend this result now by noting that we could cut off the load end of a transmission line at a point where the input impedance is a pure resistance and replace that section with a resistance R_{in}; there would be no changes on the generator portion of the line. It follows, then, that the location of voltage maxima and minima must be at those points where Z_{in} is a pure resistance. Purely resistive input impedances must occur on the $x = 0$ line (the Γ_r axis) of the Smith chart. Voltage maxima or current minima occur when $r > 1$, or at wtg $= 0.25$, and voltage minima or current maxima occur when $r < 1$, or at wtg $= 0$. In Example 11.14, then, the maximum at wtg $= 0.250$ must occur $0.250 - 0.135 = 0.115$ wavelengths toward the generator from the load. This is a distance of 0.115×200, or 23 cm from the load.

We should also note that since the standing wave ratio produced by a resistive load R_L is either R_L/R_0 or R_0/R_L, whichever is greater than unity, the value of s may be read directly as the value of r at the intersection of the $|\Gamma|$ circle and the r axis, $r > 1$. In our example this intersection is marked point C, and $r = 4.2$; thus, $s = 4.2$.

Transmission line charts may also be used for normalized admittances, although there are several slight differences in such use. We let $y_L = Y_L/Y_0 = g + jb$ and use the r circles as g circles and the x circles as b circles. The two differences are: first, the line segment where $g > 1$ and $b = 0$ corresponds to a voltage minimum; and second, 180° must be added to the angle of Γ as read from the perimeter of the chart. We shall use the Smith chart in this way in Section 11.14.

Special charts are also available for non-normalized lines, particularly 50 Ω charts and 20 mS charts.

EXAMPLE 11.15

A lossless transmission line having characteristic impedance Z_0 is terminated into a load $Z_L = R_L + jX_L$. Find out with the help of Smith chart theory, the points on the line where the impedance becomes purely real, and determine the value of the real impedance at these points. Show how the standing wave ratio along the line can be obtained by plotting the constant $|\Gamma|$ circle on the Smith chart.

Solution. The conventional Smith chart with a typical load $Z_L = R_L + jX_L$ represented by the point L is shown in Figure 11.19. As you know, you can draw a circle with the origin as O and the radius as 'OL' on the chart, which would cut the real axis at two points denoted by the point M on the right-hand side, and the point m on the left-hand side. This circle is usually called the constant $|\Gamma|$ circle because of the fact that if you move along the perimeter of this circle then the magnitude of the reflection coefficient remains constant. Now, since the voltage standing-wave ratio s depends exclusively on the magnitude of the reflection coefficient, hence by this virtue, the parameter s also remains constant if you move on this circle. The points M and m represent the points where the impedance becomes purely real. These points would always be located on the real axis on the right-hand side and the left-hand side of the origin O, respectively, and the exact location of these two points would depend upon the value of the load impedance Z_L.

Now, we have to find out the actual value of normalized impedance at two points denoted by M and m. Let us first consider the point M on the right-hand side of this Smith chart, and let us assume that the point M intersects the constant r (normalized resistance) circle. It can easily be seen from the Smith chart that at any point on the right-hand side of the origin O, the value of the normalized resistance would be greater than unity, i.e., $r > 1$. With this background, let us try to determine the value of the standing wave ratio s at the point M which is given by the following expression

$$s \equiv \frac{1 + |\Gamma|}{1 - |\Gamma|} \tag{i}$$

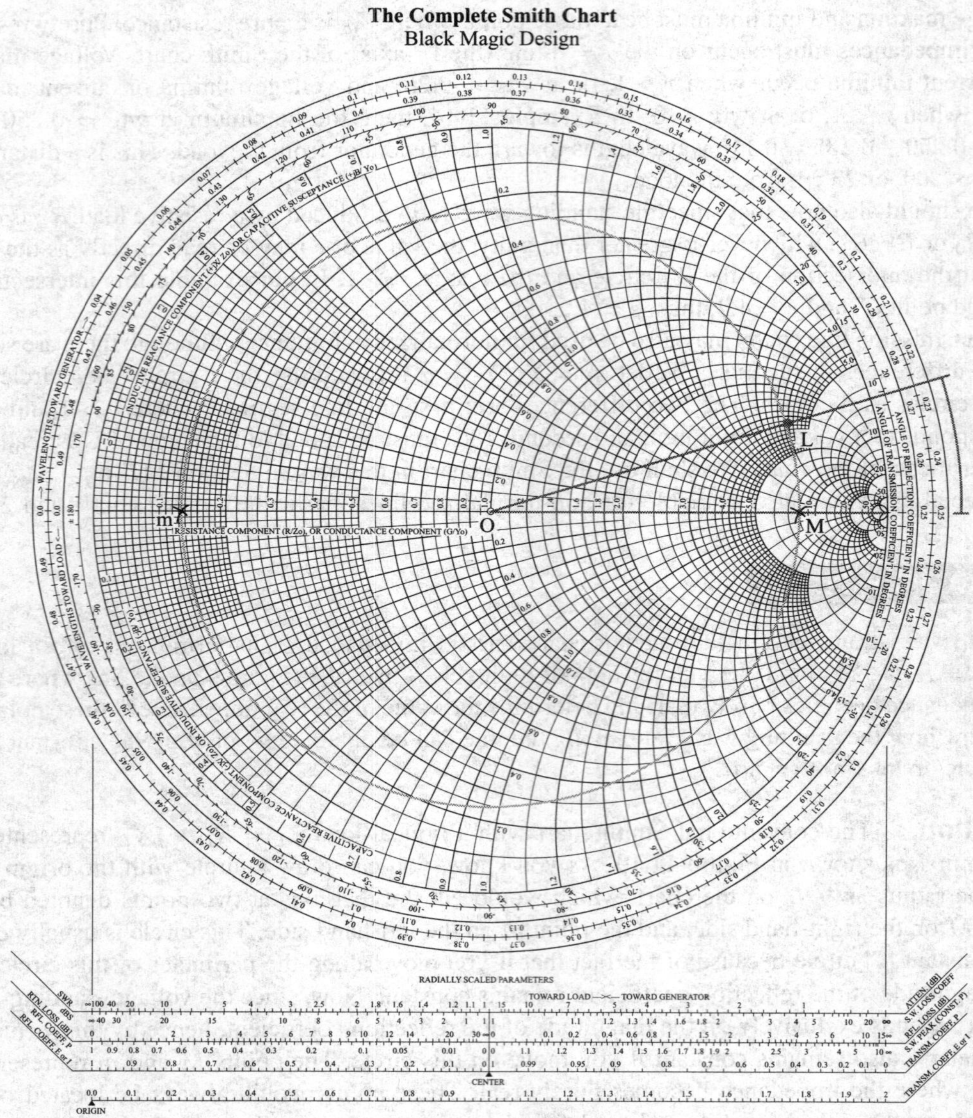

Fig. 11.19 The Smith chart representing the real impedance points.

where, the reflection coefficient is given by

$$\Gamma = \frac{R_L - Z_0}{R_L + Z_0} \equiv \frac{r-1}{r+1} \tag{ii}$$

In the above expression, the characteristic impedance of the line Z_0 is always a real quantity, and the load impedance R_L is considered purely real or resistive in the present situation because of the fact that at the point M, the imaginary part of the impedance is zero. It means that the normalized impedance $r \equiv R_L/Z_0$ would also be purely real in Eq. (ii). Now, since r is a purely real quantity in (ii), and its value is greater than unity, (ii) can also be written as

$$|\Gamma| \equiv \frac{r-1}{r+1} \tag{iii}$$

Finally, the value of $|\Gamma|$ can be substituted from (iii) into (i) to obtain the following expression for the voltage standing wave ratio s:

$$s = \frac{1+|\Gamma|}{1-|\Gamma|} \equiv \frac{1+\frac{r-1}{r+1}}{1-\frac{r-1}{r+1}} \equiv r \tag{iv}$$

From the above equation, we can say that the value of the normalized impedance at the point M, where the constant $|\Gamma|$ circle cuts the real axis, is purely real, and is equal to the standing-wave ratio.

In the second step, let us consider the point m on the left-hand side of the Smith chart intersecting the constant r circle. It can be observed from the Smith chart that at any point on the left-hand side of the origin O, the value of the normalized resistance would be less than unity, i.e., $r < 1$. Hence, at the point m, the normalized impedance is purely resistive, and the value of the normalized resistance r is less than unity, which then modifies Eq. (iii) as follows:

$$|\Gamma| \equiv \frac{1-r}{1+r} \tag{v}$$

The expression given by Eq. (5) is based on the simple fact that the magnitude of the reflection coefficient, i.e., the value of $|\Gamma|$ should always be a positive number which is not possible using the form given by (iii) if the value of 'r' is assumed to be less than unity. Once the value of $|\Gamma|$ is defined using (v), the voltage standing-wave ratio 's' can be computed using (i), i.e.,

$$s = \frac{1+|\Gamma|}{1-|\Gamma|} \equiv \frac{1+\frac{1-r}{1+r}}{1-\frac{1-r}{1+r}} \equiv \frac{1}{r} \tag{vi}$$

Hence, the value of the normalized impedance at the point 'm', where the constant $|\Gamma|$ circle cuts the real axis on the left-hand side of the origin 'O', is purely real, and is equal to the inverse of the standing wave ratio.

From Eq. (iv), it can also be postulated that the standing-wave ratio along the line can be obtained by plotting the constant $|\Gamma|$ circle on the Smith chart, and noting the point of intersection of this circle with the positive real axis represented by the point M in Figure 11.19. The value of the normalized resistance r circle, which passes through the point M, then provides the value of the voltage standing wave ratio s on the line.

EXAMPLE 11.16

What are the limitations of a quarter-wave transformer? Explain whether it is possible to match a reactive load having $Z_L = (100 + j150)\Omega$ with a lossless transmission line having $Z_0 = 50\,\Omega$ using a quarter-wave transformer. If yes, find out the characteristic impedance of the transformer and other relevant parameters required to achieve this matching.

Solution. The quarter-wave transformer is a piece of transmission line, which is used either for matching two different impedances or for matching a load with a transmission line of different characteristic impedance. It has two major limitations when used for matching purposes. The first limitation is that it can only be used to match the *real* impedances. The second limitation is that the matching is achieved in the very narrow frequency band with the condition that the ideal matching is achieved only at those frequencies where the length of the line is an odd multiple of quarter wavelength, i.e., $l \equiv (2m+1)\lambda/4$.

Now, if one wants to match a reactive load with a transmission line having a *real* value of the characteristic impedance then it can only be achieved if the reactive load is first converted into a *resistive* load. In other words, a piece of transmission line of appropriate length should be connected at the load so that the input impedance becomes *resistive* at the new position. Hence, you have to locate the position on the line shown in Figure 11.20 such that if a transmission line of length l is connected at the load, then the impedance Z_B seen from the new position towards the load becomes purely resistive. Now, you know from the Smith chart theory that the impedance can be purely resistive either at the position of the voltage maximum or at the voltage minimum. In principle, the quarter-wave section can be inserted at any of these two points. However, from the practical point of view, one tries to find a position which is closer to the load in order to minimize the length l of the extra line. The whole procedure is explained below with the help of the Smith chart shown in Figure 11.21.

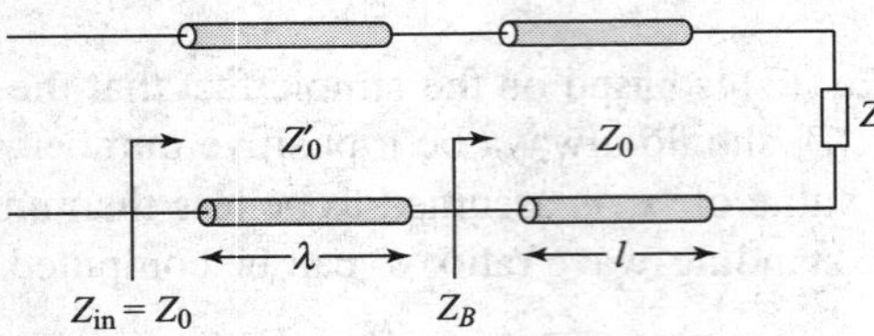

Fig. 11.20 Load line.

Major steps

1. First of all, connect a line of length l shown in Figure 11.20 so that the impedance seen from this point towards the load $Z_L = (100 + j150)\ \Omega$ becomes purely resistive. The characteristic impedance of this line is assumed to be same as that of the actual transmission line, i.e., $Z_0 = 50\ \Omega$.
2. Normalize the load impedance $Z_L = (100 + j150)\ \Omega$ with respect to the characteristic impedance of the line, i.e., $Z_L = \frac{(100+j150)}{50} = (2 + j3)$.
3. Plot $Z_L = (2 + j3)$ point on the Smith chart shown by L in Figure 11.21. Draw a constant $|\Gamma|$ circle on the Smith chart towards the generator (anti-clockwise direction) from the point L with the origin as O and the radius as OL as shown in this figure. This circle intersects the real axis on the right-hand side of the chart at the point M, where the voltage is maximum and the impedance is purely real.
4. The value of the normalized resistance at the point M may be read as $r = 7$. The distance between the load position L and the voltage maximum position M may be determined from the WTG scale provided on the Smith chart, which is given by $l = (0.25 - 0.214)\lambda \equiv 0.036\lambda$. It means that if you move a distance of 0.036 λ from the load towards the generator, then you would arrive at a position where the impedance becomes purely resistive, and this impedance can now be matched with the quarter-wave transformer.

5. The actual value of the resistive impedance at a distance $l = .036\lambda$ from the load in Figure 11.20 is then given by $Z_B = 50 \times 7 = 350\ \Omega$.
6. This impedance can now be matched with a lossless transmission line having characteristic impedance of $Z_0 = 50\ \Omega$ with the help of a quarter-wave transformer. The characteristic impedance of the quarter-wave section in this case would be given by

$$Z_0' = \sqrt{Z_B \times Z_0} = \sqrt{350 \times 50} = 132.29\ \Omega.$$

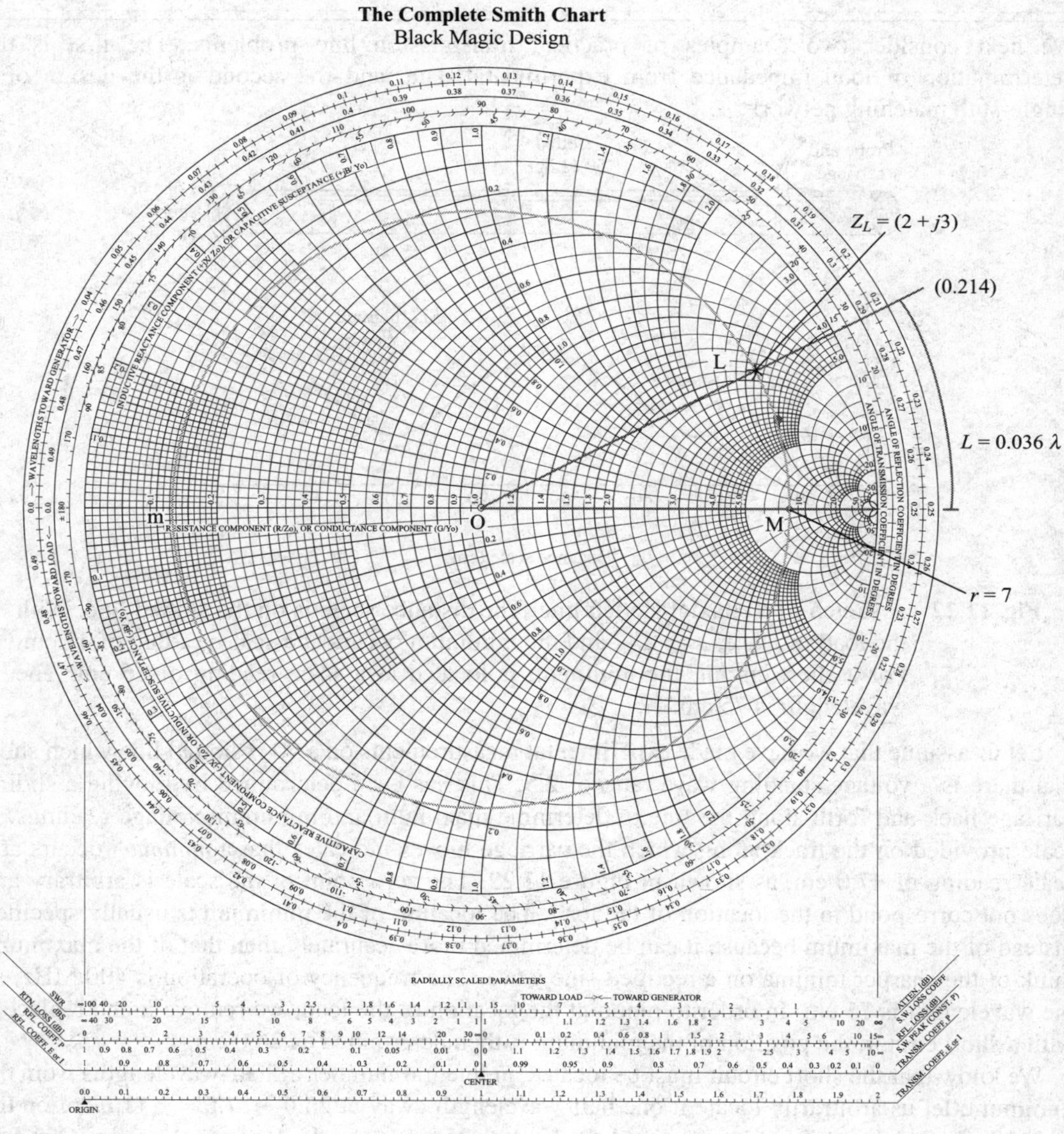

Fig. 11.21 Smith chart.

D11.6 A load $Z_L = 80 - j100\ \Omega$ is located at $z = 0$ on a lossless 50 Ω line. The operating frequency is 200 MHz and the wavelength on the line is 2 m. (*a*) If the line is 0.8 m in length, use the Smith chart to find the input impedance. (*b*) What is s? (*c*) What is the distance from the load to the nearest voltage maximum? (*d*) What is the distance from the input to the nearest point at which the remainder of the line could be replaced by a pure resistance?

Ans. $79 + j99\ \Omega$: 4.50; 0.0397 m; 0.760 m

We next consider two examples of practical transmission line problems. The first is the determination of load impedance from experimental data, and the second is the design of a single-stub matching network.

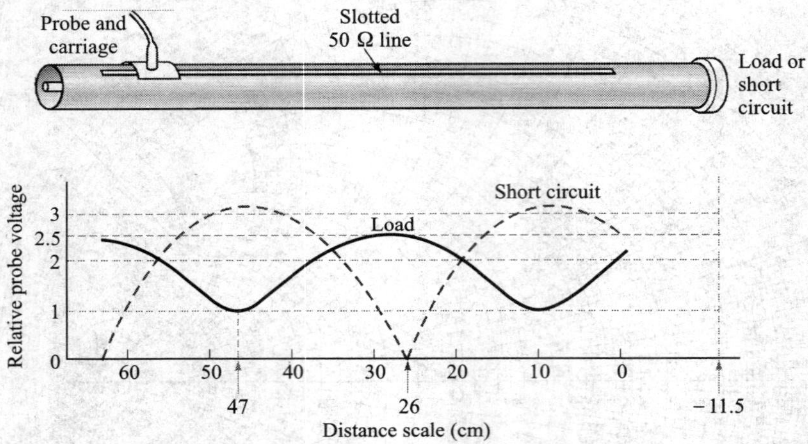

Fig. 11.22 A sketch of a coaxial slotted line. The distance scale is on the slotted line. With the load in place, $s = 2.5$, and the minimum occurs at a scale reading of 47 cm. For a short circuit, the minimum is located at a scale reading of 26 cm. The wavelength is 75 cm.

Let us assume that we have made experimental measurements on a 50 Ω slotted line which show that there is a voltage standing wave ratio of 2.5. This has been determined by moving a sliding carriage back and forth along the line to determine maximum and minimum voltage readings. A scale provided on the track along which the carriage moves indicates that a *minimum* occurs at a scale reading of 47.0 cm, as shown in Figure 11.22. The zero point of the scale is arbitrary and does not correspond to the location of the load. The location of the minimum is usually specified instead of the maximum because it can be determined more accurately than that of the maximum; think of the sharper minima on a rectified sine wave. The frequency of operation is 400 MHz, so the wavelength is 75 cm. In order to pinpoint the location of the load, we remove it and replace it with a short circuit; the position of the minimum is then determined as 26.0 cm.

We know that the short circuit must be located an integral number of half-wavelengths from the minimum; let us arbitrarily locate it one half-wavelength away at $26.0 - 37.5 = -11.5$ cm on the scale. Since the short circuit has replaced the load, the load is also located at -11.5 cm. Our data thus show that the minimum is $47.0 - (-11.5) = 58.5$ cm from the load, or subtracting one-half

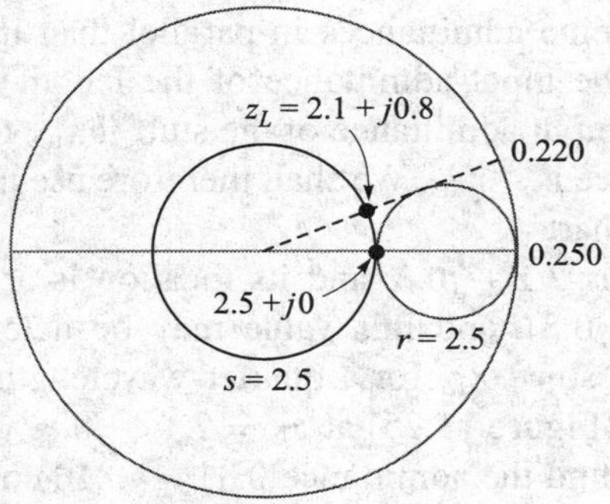

Fig. 11.23 If $z_{\text{in}} = 2.5 + j0$ on a line 0.3 wavelengths long, then $z_L = 2.1 + j0.8$.

wavelength, a minimum is 21.0 cm from the load. The voltage *maximum* is thus $21.0 - (37.5/2) = 2.25$ cm from the load, or $2.25/75 = 0.030$ wavelength from the load.

With this information, we can now turn to the Smith chart. At a voltage maximum the input impedance is a pure resistance equal to sR_0; on a normalized basis, $z_{\text{in}} = 2.5$. We therefore enter the chart at $z_{\text{in}} = 2.5$ and read 0.250 on the wtg scale. Subtracting 0.030 wavelength to reach the load, we find that the intersection of the $s = 2.5$ (or $|\Gamma| = 0.429$) circle and the radial line to 0.220 wavelength is at $z_L = 2.1 + j0.8$. The construction is sketched on the Smith chart of Figure 11.23. Thus $Z_L = 105 + j40\,\Omega$, a value which assumes its location at a scale reading of -11.5 cm, or an integral number of half-wavelengths from that position. Of course, we may select the "location" of our load at will by placing the short circuit at that point which we wish to consider the load location. Since load locations are not well defined, it is important to specify the point (or plane) at which the load impedance is determined.

As a final example, let us try to match this load to the 50 Ω line by placing a short-circuited stub of length d_1 a distance d from the load (see Figure 11.24). The stub line has the same characteristic impedance as the main line. The lengths d and d_1 are to be determined.

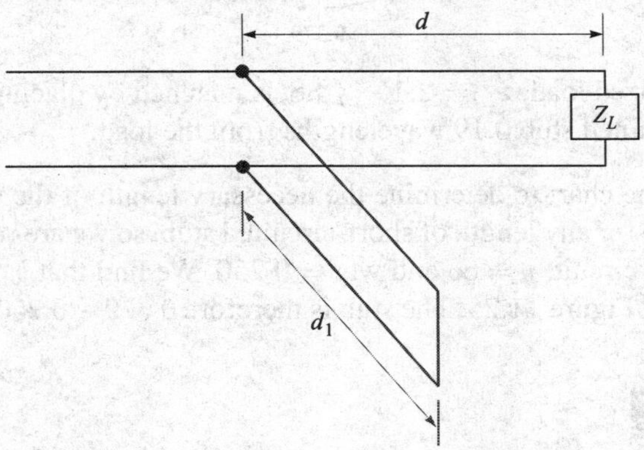

Fig. 11.24 A short-circuited stub of length d_1, located at a distance d from a load Z_L, is used to provide a matched load to the left of the stub.

The input impedance to the stub is a pure reactance; when combined in parallel with the input impedance of the length d containing the load, the resultant input impedance must be $1 + j0$.

Since it is much easier to combine admittances in parallel than impedances, let us rephrase our goal in admittance language: the input admittance of the length d containing the load must be $1 + jb_{\text{in}}$ for the addition of the input admittance of the stub jb_{stub} to produce a total admittance of $1 + j0$. Hence the stub admittance is $-jb_{\text{in}}$. We shall therefore use the Smith chart as an admittance chart instead of an impedance chart.

The impedance of the load is $2.1 + j0.8$, and its location is at -11.5 cm. The admittance of the load is therefore $1/(2.1 + j0.8)$, and this value may be determined by adding one-quarter wavelength on the Smith chart, since Z_{in} for a quarter-wavelength line is R_0^2/Z_L, or $z_{\text{in}} = 1/z_L$, or $y_{\text{in}} = z_L$. Entering the chart (Figure 11.25) at $z_L = 2.1 + j0.8$, we read 0.220 on the wtg scale; we add (or subtract) 0.250 and find the admittance $0.41 - j0.16$ corresponding to this impedance. This point is still located on the $s = 2.5$ circle. Now, at what point or points on this circle is the real part of the admittance equal to unity? There are two answers, $1 + j0.95$ at wtg $= 0.16$, and $1 - j0.95$ at wtg $= 0.34$, as shown in Figure 11.25. Let us select the former value since this leads to the shorter stub. Hence $y_{\text{stub}} = -j0.95$, and the stub location corresponds to wtg $= 0.16$. Since the load admittance was found at wtg $= 0.470$, then we must move $(0.5 - 0.47) + 0.16 = 0.19$ wavelength to get to the stub location.

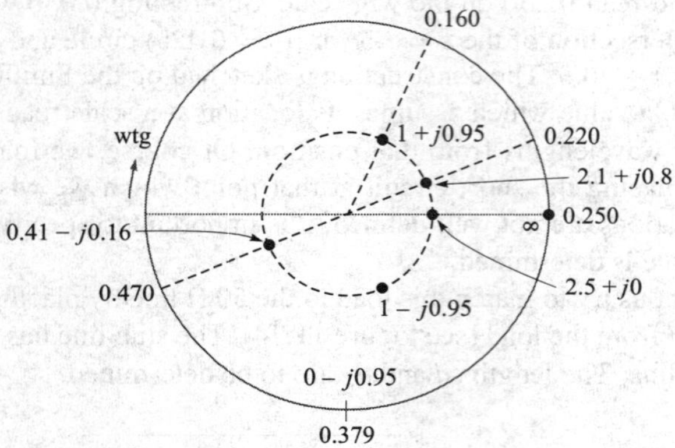

Fig. 11.25 A normalized load, $z_L = 2.1 + j0.8$, is matched by placing a 0.129-wavelength short-circuited stub 0.19 wavelengths from the load.

Finally, we may use the chart to determine the necessary length of the short-circuited stub. The input conductance is zero for any length of short-circuited stub, so we are restricted to the perimeter of the chart. At the short circuit, $y = \infty$ and wtg $= 0.250$. We find that $b_{\text{in}} = -0.95$ is achieved at wtg $= 0.379$, as shown in Figure 11.25. The stub is therefore $0.379 - 0.250 = 0.129$ wavelength, or 9.67 cm long.

EXAMPLE 11.17

It is desired to measure the impedance of an unknown load using the 50 Ω coaxial slotted line set-up shown in Figure 11.22. The experiment is performed by terminating the slotted line in the unknown impedance. The voltage standing-wave ratio measured using this set-up is found to be 3.0, and the voltage minimum is recorded at the scale reading of 50 cm.

Afterwards, the load is removed and a short is placed at the load position. The two consecutive voltage minima are now observed at scale readings of 20 cm and 45 cm, respectively. Using the Smith chart, determine (a) the wavelength, (b) the load impedance, and (c) the reflection coefficient of the load. (d) What would be the location of the first voltage minimum if the actual load is replaced by an open circuit?

Solution.

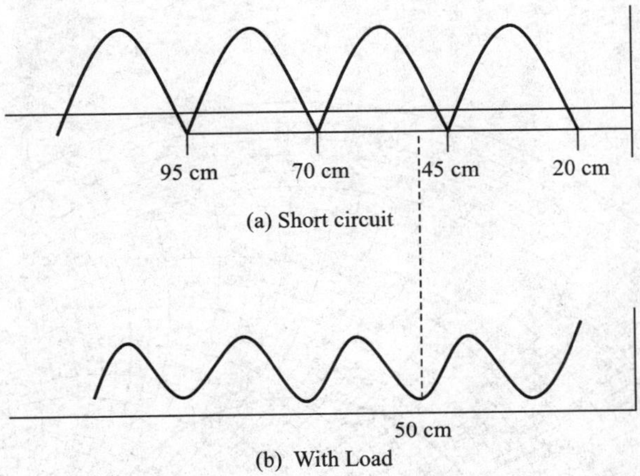

Fig. 11.26 Coaxial slotted line measurement.

The positions of the minimum under the load and the short conditions is shown in Figure 11.26. The detailed procedure is explained below.

(*a*) The consecutive voltage minima under the short conditions are at observed at 45 cm and 20 cm as seen in Figure 11.26(a). As these minima are spaced half wavelength apart, hence the wavelength may be calculated as

$$\lambda = (45 - 20) \times 2 = 50 \text{ cm}$$

(*b*) For finding the load impedance, first draw the constant VSWR circle with $s = 3$ as shown on the Smith chart. Mark the point A, where the $s = 3$ intersects the left-hand side of the real axis. This point A basically represents the voltage minimum point corresponding to the load. In other words, you can consider the point $z = 50$ cm in Figure 11.26 (b) mapped to the point A on the Smith chart.

Now, you can assume the load to be located at any of the minima point corresponding to the short, i.e., the load position can be either at $z = 20$ cm or 45 cm as shown in Figure 11.26 (a). It is convenient to consider the point which is nearest to the load minimum. Hence, the load can be assumed to be positioned at $z = 45$ cm.

It basically means that if you start from the load minima position 'A' on the Smith chart ($z = 50$ cm), and move a distance of $d = 5$ cm $= 0.1\ \lambda$ towards load then you would arrive at the load position ($z = 45$ cm).

On the Smith chart, move on the constant $s = 3$ circle from the point 'A' towards the load until you reach 0.1λ marked on the 'Wavelength towards Load' scale. Draw a line from the centre of the chart to this point intersecting the constant $s = 3$ circle at the point B shown in the chart. The constant r and constant x circles passing through the point 'B' provide the load point, which may be read as

$$Z_L = (0.5 - j0.6)$$

The actual value of the load impedance may then be calculated as

$$Z_L = 50 \times (0.5 - j0.6) = (25 - j30)\ \Omega$$

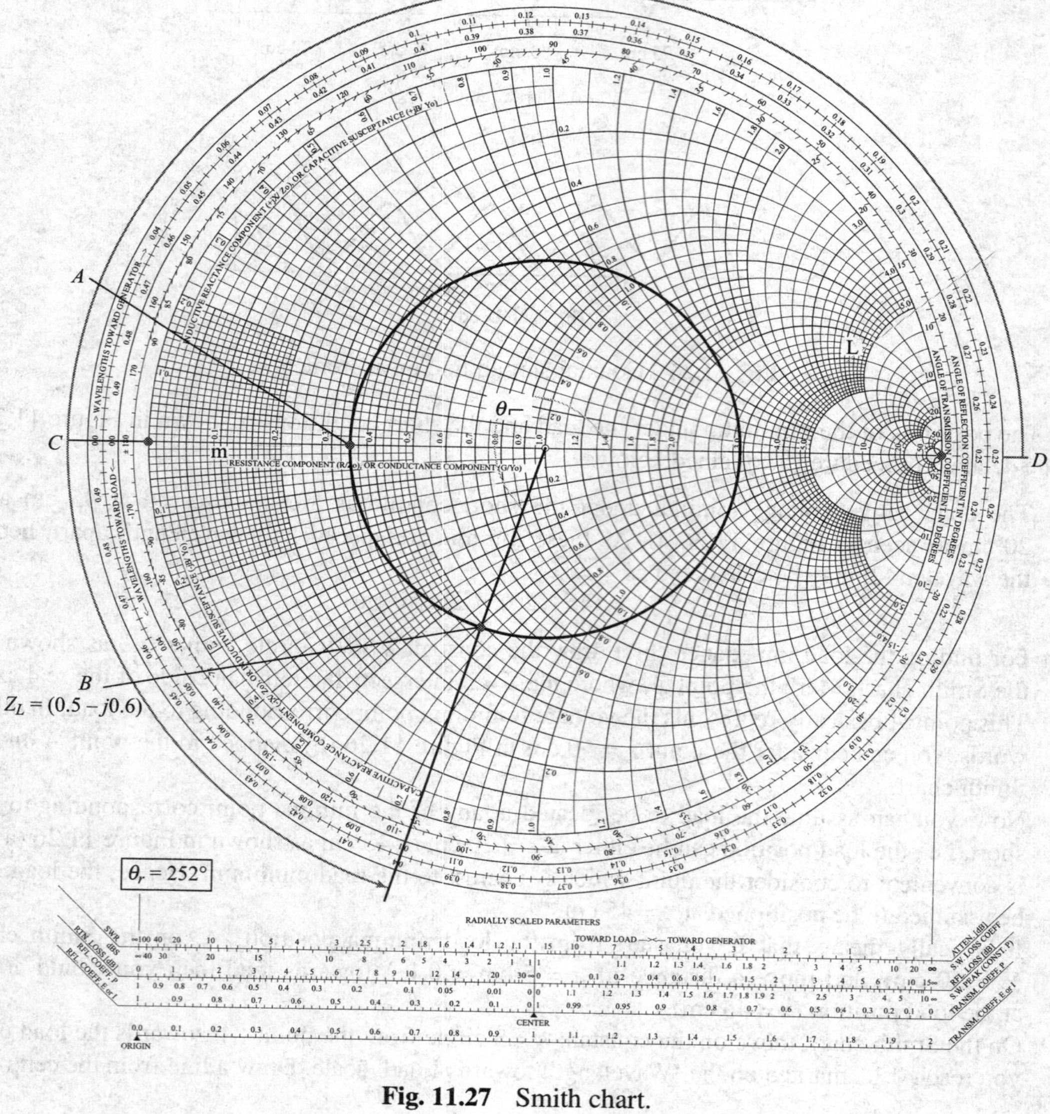

Fig. 11.27 Smith chart.

(*c*) The reflection coefficient of the load can be found out by measuring $|\Gamma|$and the angle θ_Γ from the Smith chart, which are given as

$$|\Gamma| = 0.5 \text{ and } \theta_\Gamma = 252^\circ$$

$$\text{Hence, } \Gamma = |\Gamma|\, e^{j\theta_r} = 0.5e^{j252^\circ} = 0.5e^{j4.398} = -0.1545 - j0.4755$$

$|\Gamma|$ can also be found out analytically using $|\Gamma| = \dfrac{\rho - 1}{\rho + 1} = \dfrac{3-1}{3+1} = 0.5$

(*d*) When the actual load is replaced with an *open* circuit, the first voltage minimum would be shifted by 0.25λ away from the load. If the load position is kept fixed at 45.0 cm as shown in Figure 11.26 (a) then the first voltage minima would be located at 45.0 cm + 0.25λ = 57.5 cm on the scale reading. On the Smith chart, the load position would now be on the extreme right-hand side denoted by the point D, and the voltage minimum position would be at the extreme left-hand side denoted by the point C. It can also be interpreted from the Smith chart that if you start from the voltage minimum position (point C) and move a distance of 0.25λ on the outer circle of chart towards load then you would arrive at the point D representing an *open* load.

D11.7 Standing wave measurements on a lossless 75 Ω line show maxima of 18 V and minima of 5 V. One minimum is located at a scale reading of 30 cm. With the load replaced by a short circuit, two adjacent minima are found at scale readings of 17 and 37 cm. Find: (*a*) s; (*b*) λ; (*c*) f; (*d*) Γ_L; (*e*) Z_L.

Ans. 3.60; 0.400 m; 750 MHz; $0.704\angle{-33.0}$; $77.9 + j104.7\,\Omega$

EXAMPLE 11.18

A load impedance given by $Z_L = 100 + j100\,\Omega$ is to be matched to a transmission line having characteristic impedance of 150 Ω. Design a matching network consisting of a short-circuited shunt stub having characteristic impedance of 50 Ω, which is connected at some distance d_{stub} away from the load. The stub should be located at shortest possible distance from the load, and its length should also be minimum. Please use the Smith chart and explain clearly all the steps.

Solution. First, we have to find out the location of the stub d_{stub} with reference to the load which is based on the criterion that the real part of the impedance at this point should be equal to the characteristic impedance of the transmission line. A short-circuited stub is then to be connected in parallel at this point, and the length of this stub L_{stub} should be determined such that it cancels the imaginary part of the load impedance. The overall arrangement of the shunt stub with the load and the transmission line is shown in Figure 11.28. It is also to be noted that for shunt stubs, it becomes more convenient to work with admittances, rather than impedances.

The overall procedure of matching the given load with the transmission line using the Smith chart is given below:

(*a*) Calculate the normalized load impedance and determine the corresponding location marked by the point 'L' on the Smith chart.

$$Z_L = \frac{100 + j100}{150} = (0.667 + j0.667)$$

(*b*) Draw a constant $|\Gamma|$ circle on the Smith chart with the origin as 'O' and the radius as OL as shown Figure 11.29.

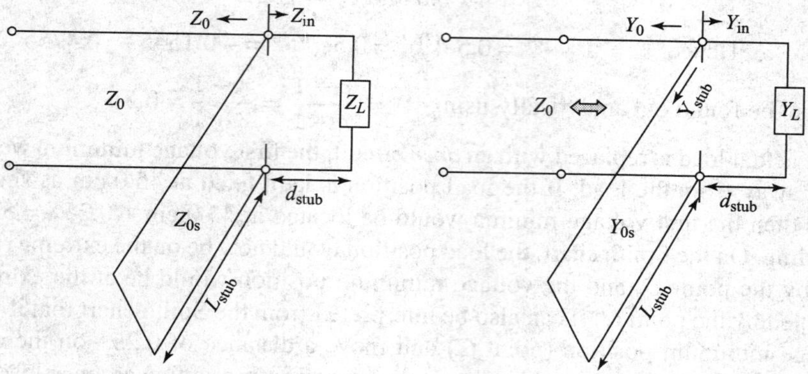

Fig. 11.28 Shunt stub with load and transmission line.

(*c*) Determine the normalized load admittance by drawing a straight line passing through the load point L and the origin O as shown on the chart. Locate the point Y on this line, which is situated at 180° opposite to the impedance point L intersecting the constant $|\Gamma|$ circle. This point represents the normalized admittance, and the value of this admittance read on the chart is given by

$$y_L = (0.75 - j0.75)$$

From now onwards, the Smith chart would be treated as an admittance chart.

(*d*) Start moving from the load admittance point Y towards the generator on the constant $|\Gamma|$ circle until you reach the g = 1 circle. The points where the constant $|\Gamma|$ circle intersects with the g = 1 circle, as denoted by P and Q in the chart, correspond to possible locations for the stub insertion.

(*e*) Now, it is given that the stub should be located nearest to the load. Hence, the point P which is closer to the load would be selected. The normalized admittance at this point as read from the chart is given by

$$y_{\text{in}} = (1 + j0.9)$$

The location of this stub with reference to the load can be determined from the WTG scale shown on the outer circle of the Smith chart, which can be read as

$$d_{\text{stub}} = (0.50 - 0.368)\lambda + 0.158\lambda \equiv 0.29\lambda$$

(*f*) The next part of the design is to determine the admittance and the length of the above shorted stub, which is connected in parallel with the main transmission line at a distance d_{stub} from the load as shown in Figure 11.28. As the main aim of this stub is to cancel the imaginary part of the admittance y_{in}, hence the stub should provide a normalized admittance of

$$y_{\text{stub}} = -j0.9$$

The above value of the normalized admittance is computed based on the assumption that the characteristic impedance of the stub is same as that of the transmission line, i.e., $Z_{0s} = Z_0$. However, in the present situation, it is given that the $Z_0 = 150\ \Omega$ while $Z_{0s} = 50\ \Omega$. Hence, we have to first determine the actual value of the admittance based on the characteristic impedance of the line, i.e.,

$$Y_{\text{stub}} \equiv \frac{y_{\text{stub}}}{Z_0} \equiv \frac{-j0.9}{150}\ \mho$$

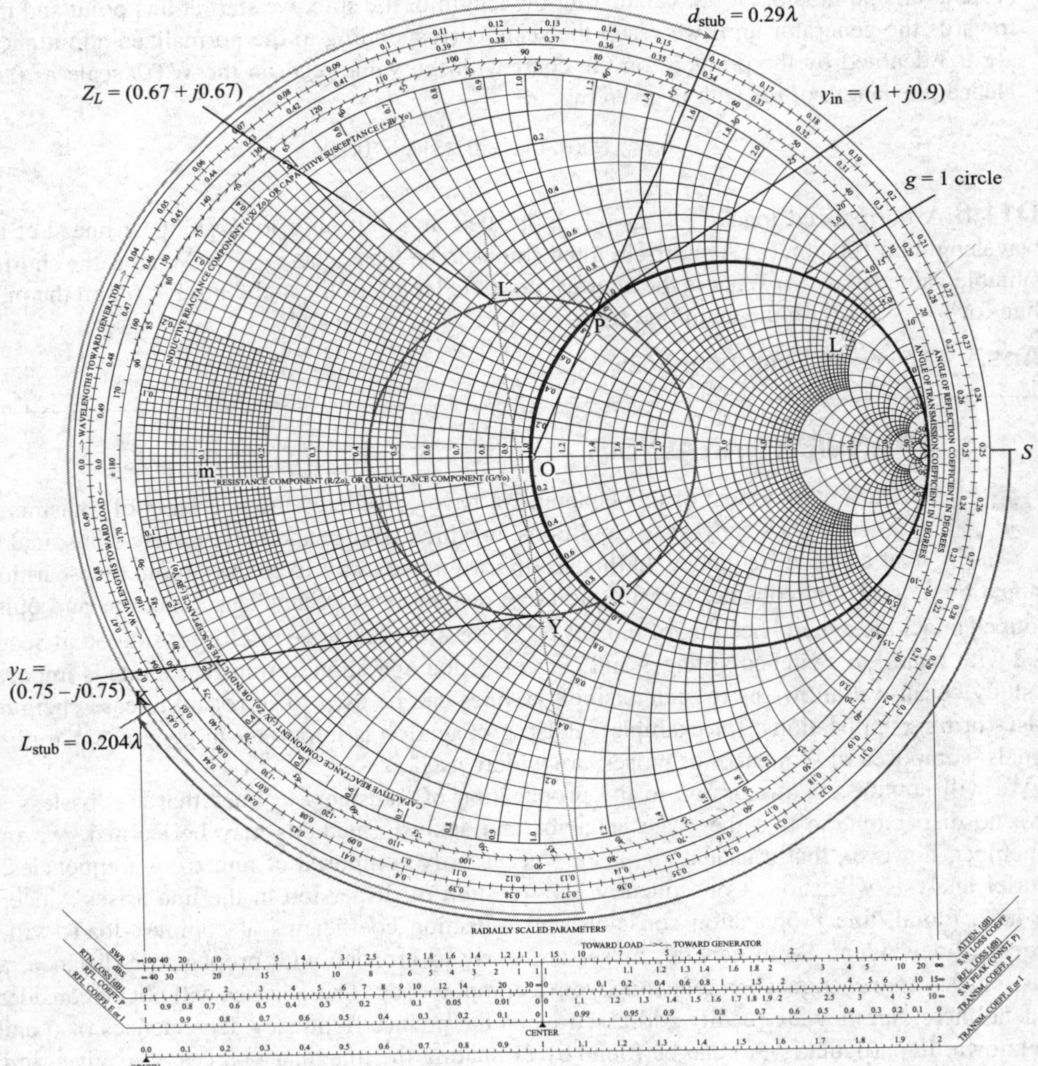

Fig. 11.29 Smith chart.

The above admittance is then normalized with respect to the characteristic impedance of the stub, i.e.,

$$\tilde{y}_{\text{stub}} \equiv Y_{\text{stub}} \times Z_{0s} \equiv \frac{-j0.9}{150} \times 50 = -j0.3$$

(*g*) Finally, we have to determine the length of the short-circuited stub having characteristic impedance of $50\,\Omega$, which provides the normalized admittance of $\tilde{y}_{\text{stub}} = -j0.3$. Now, since the Smith chart is used as the admittance chart, hence the short circuit point is represented by the extreme right-hand side of the chart denoted by the point S, and the value read on the WTG scale at this point is 0.25. For calculating the length of the stub, we start at this point and move towards the generator until we reach the scale corresponding to the normalized admittance of $-j\,0.3$ denoted by the point K on the chart with the value read on the WTG scale as 0.454. Hence, the length of the stub is given by

$$L_{\text{stub}} \equiv (0.454 - 0.25)\lambda = 0.204\lambda$$

D11.8 A normalized load, $z_L = 2 - j1$, is located at $z = 0$ on a lossless $50\ \Omega$ line. Let the wavelength be 100 cm. (*a*) A short-circuited stub is to be located at $z = -d$. What is the shortest suitable value for d? (*b*) What is the shortest possible length of the stub? Find s: (*c*) on the main line for $z < -d$; (*d*) on the main line for $-d < z < 0$; (*e*) on the stub.

Ans. 12.5 cm; 12.5 cm; 1.00; 2.62; ∞

11.14 TRANSIENT ANALYSIS

Animations

Throughout most of this chapter, we have considered the operation of transmission lines under steady-state conditions, in which voltage and current were sinusoidal and at a single frequency. In this section we move away from the simple time-harmonic case and consider transmission line responses to voltage step functions and pulses, grouped under the general heading of *transients*. These situations were briefly considered in Section 11.2 with regard to switched voltages and currents. Line operation in transient mode is important to study because it allows us to understand how lines can be used to store and release energy (in pulse-forming applications, for example). Pulse propagation is important in general since digital signals, composed of sequences of pulses, are widely used.

We will confine our discussion to the propagation of transients in lines that are lossless and have no dispersion, so that the basic behavior and analysis methods may be learned. We must remember, however, that transient signals are necessarily composed of numerous frequencies, as Fourier analysis will show. Consequently, the question of dispersion in the line arises, since, as we have found, line propagation constants and reflection coefficients at complex loads will be frequency-dependent. So in general, pulses are likely to broaden with propagation distance, and pulse shapes may change when reflecting from a complex load. These issues will not be considered in detail here, but they are readily addressed when the precise frequency dependences of β and Γ are known. In particular, $\beta(\omega)$ can be found by evaluating the imaginary part of γ, as given in Eq. (41), which would in general include the frequency dependences of R, C, G, and L arising from various mechanisms. For example, the skin effect (which affects both the conductor resistance and

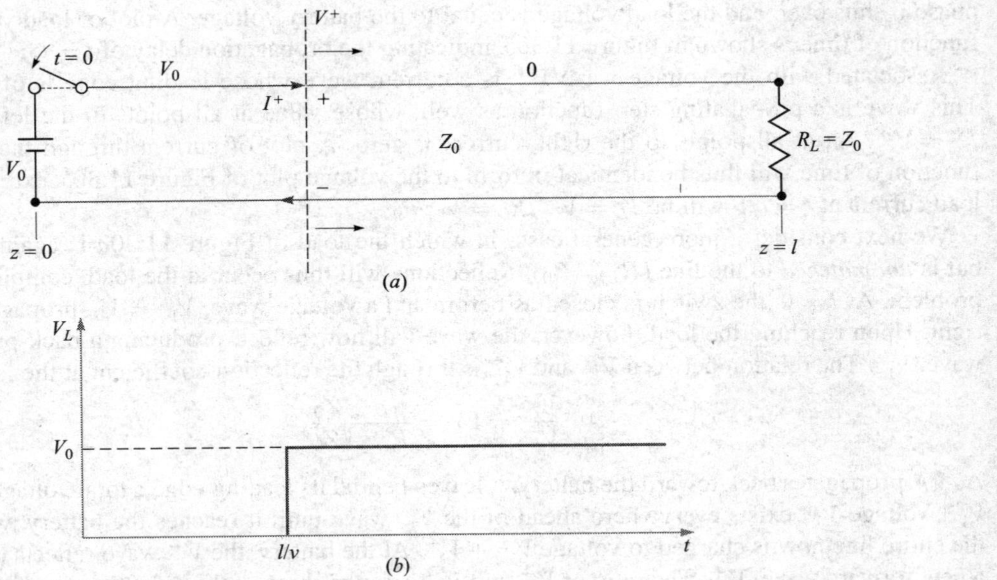

Fig. 11.30 (a) Closing the switch at time $t = 0$ initiates voltage and current waves V^+ and I^+. The leading edge of both waves is indicated by the dashed line, which propagates in the lossless line toward the load at velocity ν. In this case, $V^+ = V_0$; the line voltage is V^+ everywhere to the left of the leading edge, where current is $I^+ = V^+/Z_0$. To the right of the leading edge, voltage and current are both zero. Clockwise current, indicated here, is treated as positive and will occur when V^+ is positive. (b) Voltage across the load resistor as a function of time, showing the one-way transit time delay, l/ν.

the internal inductance) will result in frequency-dependent R and L. Once $\beta(\omega)$ is known, pulse broadening can be evaluated using the methods to be presented in Chapter 13.

We begin our basic discussion of transients by considering a lossless transmission line of length l terminated by a matched load, $R_L = Z_0$, as shown in Figure 11.30a. At the front end of the line is a battery of voltage V_0, which is connected to the line by closing a switch. At time $t = 0$, the switch is closed, and the line voltage at $z = 0$ becomes equal to the battery voltage. This voltage, however, does not appear across the load until adequate time has elapsed for the propagation delay. Specifically, at $t = 0$, a voltage wave is initiated in the line at the battery end, which then propagates toward the load. The leading edge of the wave, labeled V^+ in Figure 11.30, is of value $V^+ = V_0$. It can be thought of as a propagating step function, since at all points to the left of V^+, the line voltage is V_0; at all points to the right (not yet reached by the leading edge), the line voltage is zero. The wave propagates at velocity ν, which in general is the group velocity in the line.[4] The wave reaches the load at time $t = l/\nu$ and then does not reflect, since the load is matched. The transient

[4] Since we have a step function (composed of many frequencies) as opposed to a sinusoid at a single frequency, the wave will propagate at the group velocity. In a lossless line with no dispersion as considered in this section, $\beta = \omega\sqrt{LC}$, where L and C are constant with frequency. In this case we would find that the group and phase velocities are equal; i.e., $d\omega/d\beta = \omega/\beta = \nu = 1/\sqrt{LC}$. We will thus write the velocity as ν, knowing it to be both ν_p and ν_g.

phase is thus over, and the load voltage is equal to the battery voltage. A plot of load voltage as a function of time is shown in Figure 11.30*b*, indicating the propagation delay of $t = l/\nu$.

Associated with the voltage wave V^+ is a current wave whose leading edge is of value I^+. This wave is a propagating step function as well, whose value at all points to the left of V^+ is $I^+ = V^+/Z_0$; at all points to the right, current is zero. A plot of current through the load as a function of time will thus be identical in form to the voltage plot of Figure 11.30*b*, except that the load current at $t = l/\nu$ will be $I_L = V^+/Z_0 = V_0/R_L$.

We next consider a more general case, in which the load of Figure 11.30*a* is again a resistor but is *not matched* to the line ($R_L \neq Z_0$). Reflections will thus occur at the load, complicating the problem. At $t = 0$, the switch is closed as before and a voltage wave, $V_1^+ = V_0$, propagates to the right. Upon reaching the load, however, the wave will now reflect, producing a back-propagating wave, V_1^-. The relation between V_1^- and V_1^+ is through the reflection coefficient at the load

$$\frac{V_1^-}{V_1^+} = \Gamma_L = \frac{R_L - Z_0}{R_L + Z_0} \tag{115}$$

As V_1^- propagates back toward the battery, it leaves behind its leading edge a total voltage of $V_1^+ + V_1^-$. Voltage V_1^+ exists everywhere ahead of the V_1^- wave until it reaches the battery, whereupon the entire line now is charged to voltage $V_1^+ + V_1^-$. At the battery, the V_1^- wave reflects to produce a new forward wave, V_2^+. The ratio of V_2^+ and V_1^- is found through the reflection coefficient at the battery:

$$\frac{V_2^+}{V_1^-} = \Gamma_g = \frac{Z_g - Z_0}{Z_g + Z_0} = \frac{0 - Z_0}{0 + Z_0} = -1 \tag{116}$$

where the impedance at the generator end, Z_g, is that of the battery, or zero.

V_2^+ (equal to $-V_1^-$) now propagates to the load, where it reflects to produce backward wave $V_2^- = \Gamma_L V_2^+$. This wave then returns to the battery, where it reflects with $\Gamma_g = -1$, and the process repeats. Note that with each round trip the wave voltage is reduced in magnitude since $|\Gamma_L| < 1$. Because of this the propagating wave voltages will eventually approach zero, and steady state is reached.

The voltage across the load resistor can be found at any given time by summing the voltage waves that have reached the load and have reflected from it up to that time. After many round trips, the load voltage will be in general:

$$\begin{aligned} V_L &= V_1^+ + V_1^- + V_2^+ + V_2^- + V_3^+ + V_3^- + \cdots \\ &= V_1^+\left(1 + \Gamma_L + \Gamma_g\Gamma_L + \Gamma_g\Gamma_L^2 + \Gamma_g^2\Gamma_L^2 + \Gamma_g^2\Gamma_L^3 + \cdots\right) \end{aligned}$$

With a simple factoring operation, the preceding equation becomes

$$V_L = V_1^+\left(1 + \Gamma_L\right)\left(1 + \Gamma_g\Gamma_L + \Gamma_g^2\Gamma_L^2 + \cdots\right) \tag{117}$$

Allowing time to approach infinity, the second term in parentheses in (117) becomes the power series expansion for the expression $1/(1 - \Gamma_g\Gamma_L)$. Thus, in steady state we obtain

$$V_L = V_1^+\left(\frac{1 + \Gamma_L}{1 - \Gamma_g\Gamma_L}\right) \tag{118}$$

In our present example, $V_1^+ = V_0$ and $\Gamma_g = -1$. Substituting these into (118), we find the expected result in steady state: $V_L = V_0$.

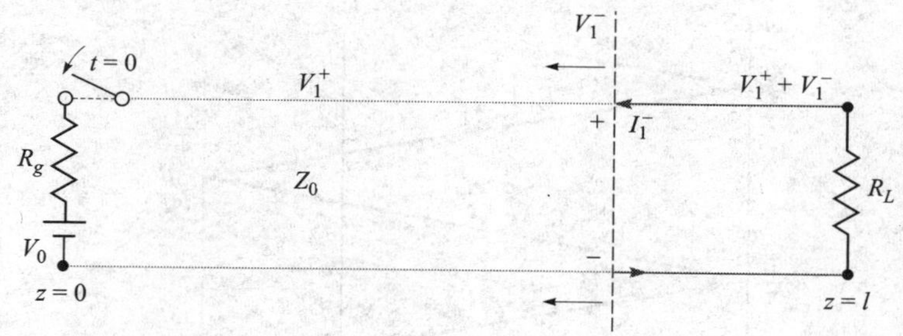

Fig. 11.31 With series resistance at the battery location, voltage division occurs when the switch is closed, such that $V_0 = V_{rg} + V_1^+$. Shown is the first reflected wave, which leaves voltage $V_1^+ + V_1^-$ behind its leading edge. Associated with the wave is current I_1^-, which is $-V_1^-/Z_0$. Counter-clockwise current is treated as negative and will occur when V_1^- is positive.

A more general situation would involve a nonzero impedance at the battery location, as shown in Figure 11.31. In this case, a resistor of value R_g is positioned in series with the battery. When the switch is closed, the battery voltage appears across the series combination of R_g and the line characteristic impedance, Z_0. The value of the initial voltage wave, V_1^+, is thus found through simple voltage division, or

$$V_1^+ = \frac{V_0 Z_0}{R_g + Z_0} \tag{119}$$

With this initial value, the sequence of reflections and the development of the voltage across the load occurs in the same manner as determined by (117), with the steady-state value determined by (118). The value of the reflection coefficient at the generator end, determined by (116), is $\Gamma_g = (R_g - Z_0)/(R_g + Z_0)$.

A useful way of keeping track of the voltage at any point in the line is through a *voltage reflection diagram*. Such a diagram for the line of Figure 11.31 is shown in Figure 11.32*a*. It is a two-dimensional plot in which position on the line, z, is shown on the horizontal axis. Time is plotted on the vertical axis and is conveniently expressed as it relates to position and velocity through $t = z/\nu$. A vertical line, located at $z = l$, is drawn which, together with the ordinate, defines the z axis boundaries of the transmission line. With the switch located at the battery position, the initial voltage wave, V_1^+, starts at the origin, or lower left corner of the diagram ($z = t = 0$). The location of the leading edge of V_1^+ as a function of time is shown as the diagonal line that joins the origin to the point along the right-hand vertical line that corresponds to time $t = l/\nu$ (the one-way transit time). From there (the load location), the position of the leading edge of the reflected wave, V_1^-, is shown as a "reflected" line which joins the $t = l/\nu$ point on the right boundary to the $t = 2l/\nu$ point on the ordinate. From there (at the battery location), the wave reflects again, forming V_2^+, shown as a line parallel to that for V_1^+. Subsequent reflected waves are shown, and their values are labeled.

The voltage as a function of time at a given position in the line can now be determined by adding the voltages in the waves as they intersect a vertical line drawn at the desired location. This addition is performed starting at the bottom of the diagram ($t = 0$) and progressing upward (in time). Whenever a voltage wave crosses the vertical line, its value is added to the total at that

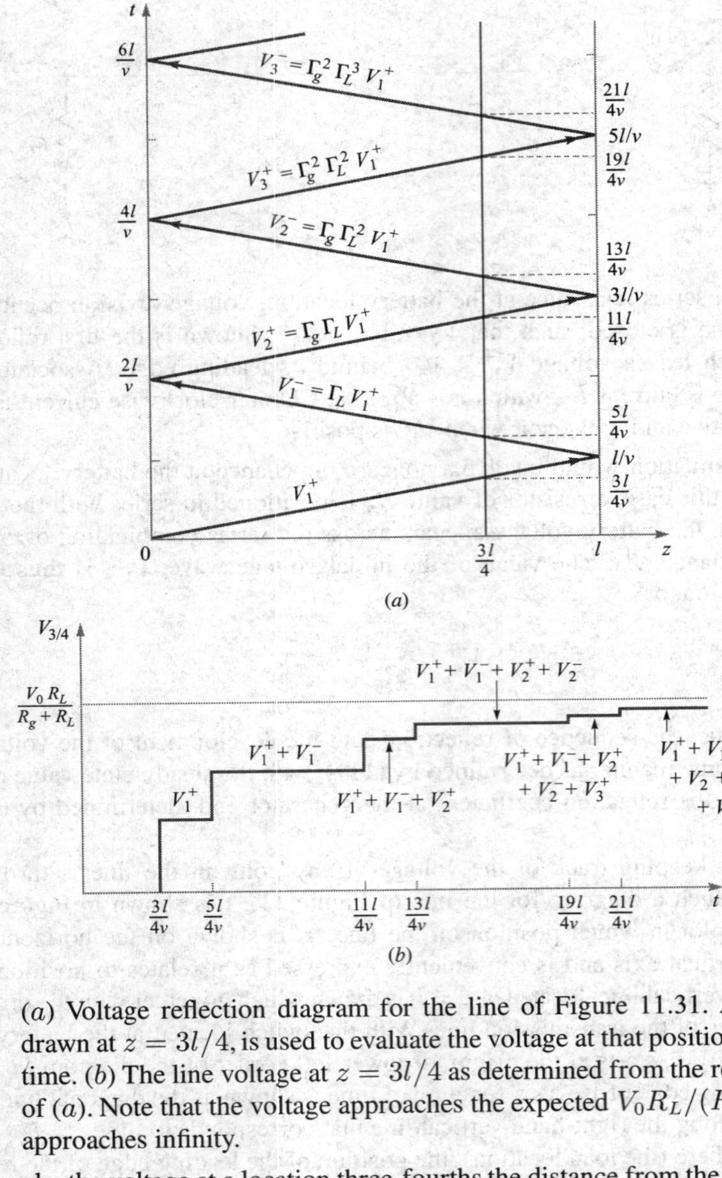

Fig. 11.32 (*a*) Voltage reflection diagram for the line of Figure 11.31. A reference line, drawn at $z = 3l/4$, is used to evaluate the voltage at that position as a function of time. (*b*) The line voltage at $z = 3l/4$ as determined from the reflection diagram of (*a*). Note that the voltage approaches the expected $V_0 R_L/(R_g + R_L)$ as time approaches infinity.

time. For example, the voltage at a location three-fourths the distance from the battery to the load is plotted in Figure 11.32*b*. To obtain this plot, the line $z = (3/4)l$ is drawn on the diagram. Whenever a wave crosses this line, the voltage in the wave is added to the voltage that has accumulated at $z = (3/4)l$ over all earlier times. This general procedure enables one to easily determine the voltage at any specific time and location. In doing so, the terms in (117) that have occurred up to the chosen time are being added, but with information on the time at which each term appears.

Line current can be found in a similar way through a *current reflection diagram*. It is easiest to construct the current diagram directly from the voltage diagram by determining a value for current that is associated with each voltage wave. In dealing with current, it is important to keep track of the *sign* of the current because it relates to the voltage waves and their polarities. Referring to Figures 11.30a and 11.31, we use the convention in which current associated with a *forward-z* traveling voltage wave of positive polarity is positive. This would result in current that flows in the clockwise direction, as shown in Figure 11.30a. Current associated with a *backward-z* traveling voltage wave of positive polarity (thus flowing counterclockwise) is negative. Such a case is illustrated in Figure 11.31. In our two-dimensional transmission-line drawings, we assign positive polarity to voltage waves propagating in *either* direction if the upper conductor carries a positive charge and the lower conductor a negative charge. In Figures 11.30a and 11.31, both voltage waves are of positive polarity, so their associated currents will be net positive for the forward wave and net negative for the backward wave. In general, we write

$$I^+ = \frac{V^+}{Z_0} \tag{120}$$

and

$$I^- = -\frac{V^-}{Z_0} \tag{121}$$

Finding the current associated with a backward-propagating voltage wave immediately requires a minus sign, as (121) indicates.

Figure 11.33a shows the current reflection diagram that is derived from the voltage diagram of Figure 11.32a. Note that the current values are labeled in terms of the voltage values, with the appropriate sign added as per (120) and (121). Once the current diagram is constructed, current at a given location and time can be found in exactly the same manner as voltage is found using the voltage diagram. Figure 11.33b shows the current as a function of time at the $z = (3/4)l$ position, determined by summing the current wave values as they cross the vertical line drawn at that location.

EXAMPLE 11.19

In Figure 11.31, $R_g = Z_0 = 50\,\Omega$, $R_L = 25\,\Omega$, and the battery voltage is $V_0 = 10$ V. The switch is closed at time $t = 0$. Determine the voltage at the load resistor and the current in the battery as functions of time.

Solution. Voltage and current reflection diagrams are shown in Figure 11.34a and b. At the moment the switch is closed, half the battery voltage appears across the 50-ohm resistor, with the other half comprising the initial voltage wave. Thus $V_1^+ = (1/2)V_0 = 5$ V. The wave reaches the 25-ohm load, where it reflects with reflection coefficient

$$\Gamma_L = \frac{25 - 50}{25 + 50} = -\frac{1}{3}$$

So $V_1^- = -(1/3)V_1^+ = -5/3$ V. This wave returns to the battery, where it encounters reflection coefficient $\Gamma_g = 0$. Thus, no further waves appear; steady state is reached.

Once the voltage wave values are known, the current reflection diagram can be constructed. The values for the two current waves are

$$I_1^+ = \frac{V_1^+}{Z_0} = \frac{5}{50} = \frac{1}{10}\text{ A}$$

and

$$I_1^- = -\frac{V_1^-}{Z_0} = -\left(-\frac{5}{3}\right)\left(\frac{1}{50}\right) = \frac{1}{30}\ \text{A}$$

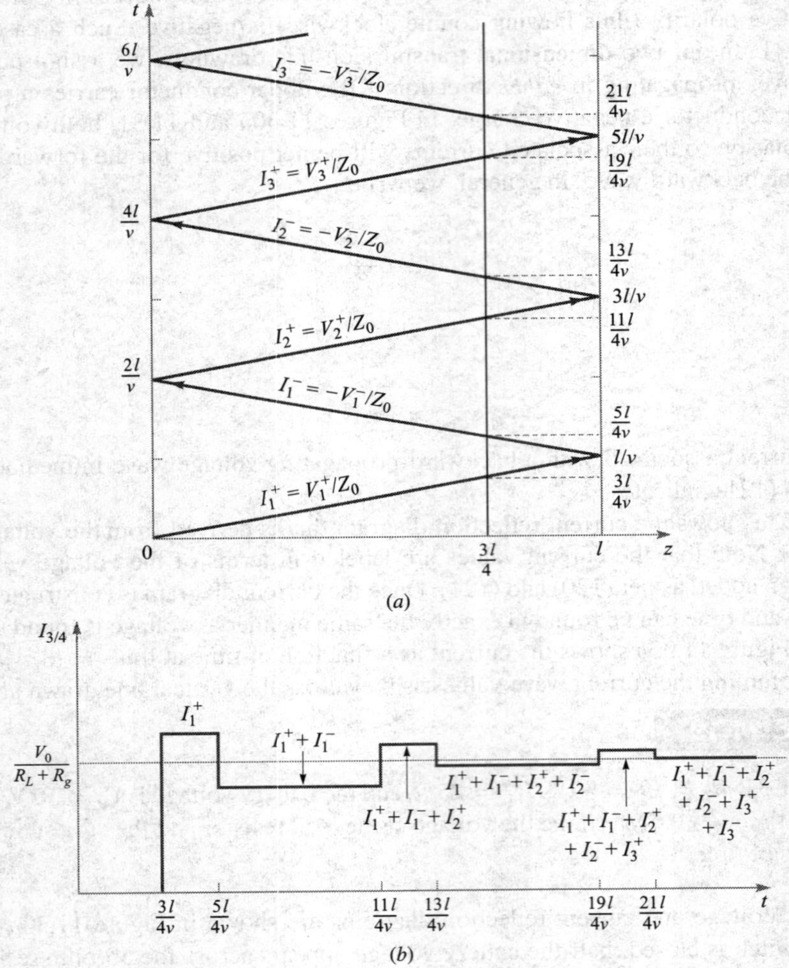

Fig. 11.33 (*a*) Current reflection diagram for the line of Figure 11.31 as obtained from the voltage diagram of Figure 11.32*a*. (*b*) Current at the $z = 3l/4$ position as determined from the current reflection diagram, showing the expected steady-state value of $V_0/(R_L + R_g)$.

Note that no attempt is made here to derive I_1^- from I_1^+. They are both obtained independently from their respective voltages.

The voltage at the load as a function of time is now found by summing the voltages along the vertical line at the load position. The resulting plot is shown Figure 11.35*a*. Current in the battery

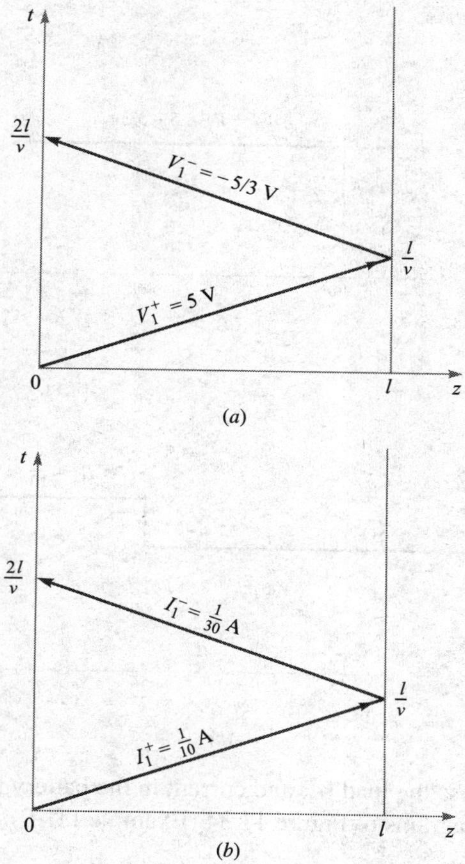

Fig. 11.34 Voltage (a) and current (b) reflection diagrams for Example 11.19.

is found by summing the currents along the vertical axis, with the resulting plot shown as Figure 11.35*b*. Note that in steady state, we treat the circuit as lumped, with the battery in series with the 50- and 25-ohm resistors. Therefore, we expect to see a steady-state current through the battery (and everywhere else) of

$$I_B(\text{steady state}) = \frac{10}{50+25} = \frac{1}{7.5}\text{ A}$$

This value is also found from the current reflection diagram for $t > 2l/\nu$. Similarly, the steady-state load voltage should be

$$V_L(\text{steady state}) = V_0\frac{R_L}{R_g+R_L} = \frac{(10)(25)}{50+25} = \frac{10}{3}\text{ V}$$

which is found also from the voltage reflection diagram for $t > l/\nu$.

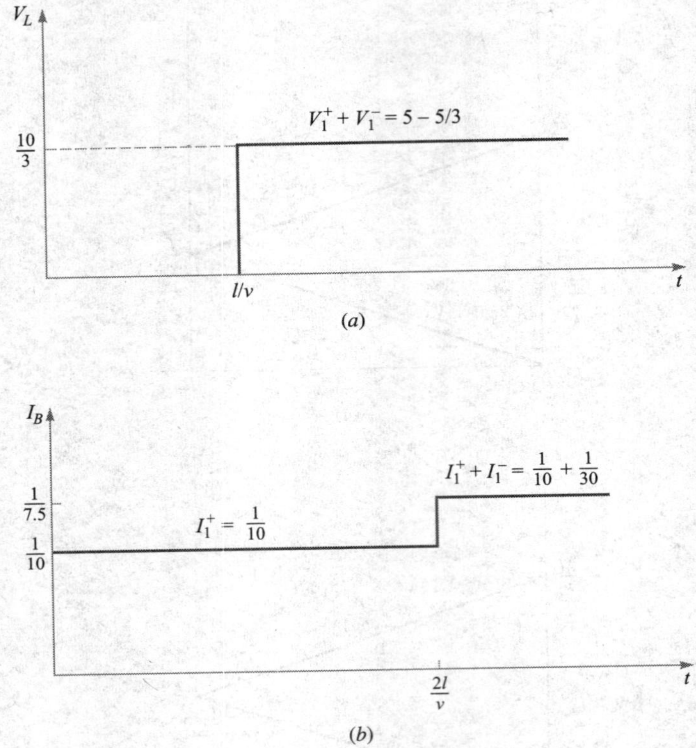

Fig. 11.35 Voltage across the load (a) and current in the battery (b) as determined from the reflection diagrams of Figure 11.34 (Example 11.19).

EXAMPLE 11.20

A quarter-wavelength-long transmission line, often known as the quarter-wave transformer, is quite often being used for matching a load with a line having different characteristic impedance. However, this concept can also be used to determine the length of a long line with the help of a variable-frequency voltage oscillator and an oscilloscope. Suppose, you are given a transmission line of length l, which is open-circuited at its far end. You apply a sinusoidal voltage of fixed amplitude, and measure the voltage observed at the input of the line by changing the frequency of the oscillator as shown in Table 11.1. You can assume the line to be filled with Teflon ($\epsilon_r = 2.1$). Determine the length of the line 'l'. Explain all the intermediate steps.

Solution. If the characteristic impedance of transmission line is assumed to be Z_0 then the input impedance of this quarter-wave line terminated in the load Z_L is given by the following expression:

$$Z_{\text{in}} = \frac{Z_0^2}{Z_L} \tag{i}$$

Table 11.1 Frequency v/s voltage at the source end

Frequency(kHz)	Voltage(m V)
1600	636
1699	428
1801	236
1850	176
1880	144
1900	162
1927	184
1952	212
2050	384
2186	672

Now, it is given that the transmission line is terminated in *open* circuit ($Z_L \equiv \infty$), and the voltage is measured at the input of the line. From Eq. (i), it can be postulated that for the open circuited line, the input impedance or the input voltage would be minimum (ideally it would be zero) at the center frequency when the length of this line becomes equal to the quarter-wavelength. From the above table, it can be observed that at 1.88 MHz, the voltage becomes minimum for the open circuited line. Hence, we can say that at 1.88 MHz, the given transmission line acts like a quarter-wave transformer, which gives:

$$l = \frac{\lambda_0}{4 \times \sqrt{\epsilon_r}} \equiv \frac{3 \times 10^8}{1.88 \times 10^6 \times 4 \times \sqrt{2.1}} = 27.529 \text{ m} \tag{ii}$$

where, λ_0 is the free space wavelength at the frequency of 1.88 MHz, and it is given that $\epsilon_r = 2.1$.

EXAMPLE 11.21

A time-domain reflectometer (TDR) is an instrument, which is based on transient analysis. It is mainly used to locate faults and damages on long transmission lines by sending a step voltage down the line, and observing the voltage at any specific position along the line as a function of time. Suppose a TDR having source impedance equal to the characteristic impedance of the line is connected to the input of a 100 Ω matched transmission line filled with Polystyrene ($\epsilon_r = 2.25$) as shown in Figure 11.36(a). The resistance R_d in this figure represents the additional shunt resistance, which might occur at this position due to some damage of the line at a distance l from the source point.

If the voltage waveform shown in Figure 11.36(*b*) was observed at a distance d from the source end of the TDR, determine (a) the position of the observation point d, (b) the source voltage V_0, (c) the position l at which the damage occurs on the line, and (d) the shunt resistance R_d resulting from the damage of the line. (e) Draw the waveform seen at the observation point of the TDR if no fault or damage is detected on the line.

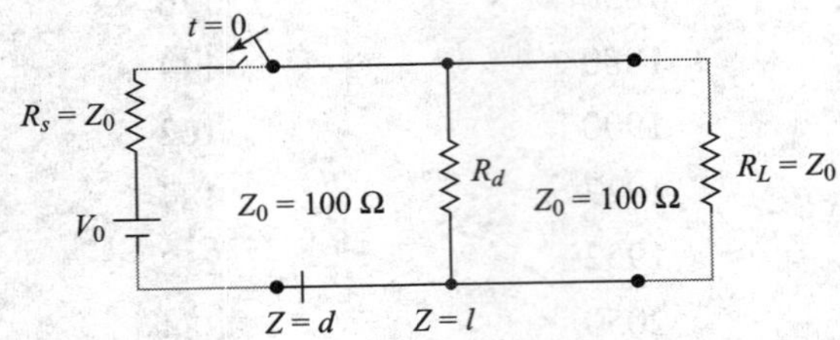

Fig. 11.36(*a*) Transmission line

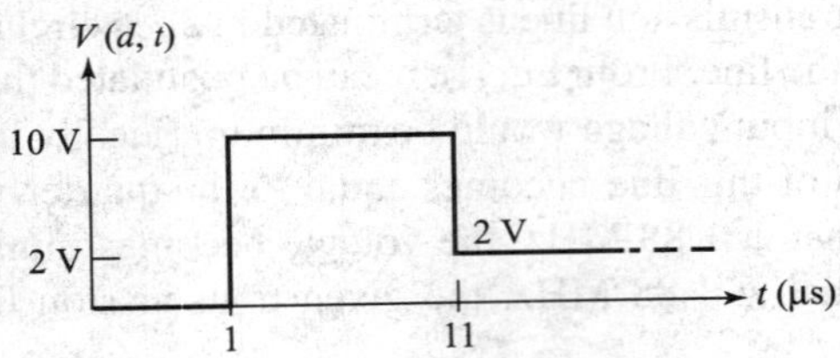

Fig. 11.36(*b*) Voltage waveform

Solution.

(*a*) From Figure 11.36(*b*), it seen that the voltage recorded at the position d is zero till t = 1 μs. It means that the voltage wave starting from the source end (d = 0) would take t = 1 μs to reach the observation point. The location of the observation point 'd' can hence be calculated using the following expression:

$$d = v \times t_1 \equiv \frac{t_1 \times c}{\sqrt{\epsilon_r}} = \frac{10^{-6} \times 3 \times 10^8}{\sqrt{2.25}} = 200 \text{ m} \qquad \text{(i)}$$

where, v is the velocity along the transmission line filled with the Polystyrene material t_1 = 1 μs is time taken by the voltage wave to travel the distance d from source to the observation point, and c is the free-space velocity.

(*b*) It is given that the line is fully matched at both the source and the load end when there is no fault or damage along the line. It can also be observed from Figure 11.36(*b*) that the damage on the line occurs only at t = 11μs when seen at the observation point. Hence, at t_1 = 1 μs, the transmission line at the observation point would appear to be fully matched and the voltage at this location would be simply the initial voltage wave V_1^+, which is given as 10 V as seen in Figure 11.36(*b*). Meanwhile, the initial voltage wave can be related to the source voltage V_0

using the following expression:

$$V_1^+ = V_0\left(\frac{Z_0}{R_s + Z_0}\right) \equiv= V_0\left(\frac{Z_0}{Z_0 + Z_0}\right) = \frac{V_0}{2} \equiv 10 \text{ V} \quad \text{(ii)}$$

From above equation, it is deduced that the source voltage V_0 = 20 V.

(*c*) When there is some damage or fault on the transmission line, then it would not appear to be matched as some additional shunt resistance appears on the line due to this damage. From Figure 11.36(*b*), it is seen that the damage on the line changes the level of the voltage at the observation point at t = 11 μs. Now, if the damage on the line is assumed to be at position l, then the voltage wave would travel a total distance of $2(l - d)$ so that the effect of this damage appears on the observation point $z = d$. Hence, the position of the disturbance on the transmission line can be determined using the following relationship:

$$\Delta t \equiv (t_2 - t_1) = \frac{2(l-d)}{v} \equiv \frac{2(l-d)}{c/\sqrt{\epsilon_r}} \equiv \frac{2(l-d) \times \sqrt{2.25}}{3 \times 10^8} \equiv (11 - 1) \times 10^{-6} \quad \text{(iii)}$$

From above equation, one obtains $(l - d) \equiv \dfrac{10 \times 10^{-6} \times 3 \times 10^8}{2 \times 1.5} = 1000$ m

Hence, the position l at which the damage occurs along the line is given by

$$l = 1000 + 200 = 1200 \text{ m}$$

(*d*) The shunt resistance R_d resulting from the damage of the line can be obtained by taking note of the fact that the total voltage recorded at the observation point after the occurrence of damage $t \geq 11\mu s$ is the sum of the initial voltage wave, and the reflected wave which appears due to the line mismatch at $t = 11\mu s$ in Figure 11.37 i.e.,

$$V(d, t = 11\mu s) \equiv V_1^+ + \Gamma_{Ld}V_1^+ \equiv 10[1 + \Gamma_{Ld}] = 2\ V \quad \text{(iv)}$$

where Γ_{Ld} is the reflection coefficient occurring due to mismatch at $z = l = 1200$ m shown in Figure 11.36(*a*). The value of this reflection coefficient from the above equation may be determined as

$$\Gamma_{Ld} = \frac{1}{5} - 1 \equiv -\frac{4}{5} = -0.8 \quad \text{(v)}$$

The reflection coefficient Γ_{Ld} may be related to the characteristic impedance of the line Z_0 and the resultant load impedance occurring due to damage at $z = l$ in Figure 11.37(*a*) using the following expression:

$$\Gamma_{Ld} = \left(\frac{Z_{Ld} - Z_0}{Z_{Ld} + Z_0}\right) \equiv \left(\frac{Z_{Ld} - 100}{Z_{Ld} + 100}\right) = -0.8 \quad \Rightarrow \quad Z_{Ld} = \frac{100}{9}\Omega \quad \text{(vi)}$$

Now, it is to be noted that the impedance Z_{Ld} defined in the above equation represents the parallel combination of the shunt resistance R_d and the characteristic impedance Z_0 as can be seen from Figure 11.36(*a*), i.e.,

$$Z_{Ld} \equiv \frac{R_d Z_0}{R_d + Z_0} \equiv \frac{R_d \times 100}{R_d + 100} = \frac{100}{9}\Omega \quad \text{(vii)}$$

The above equation can be solved to obtain the shunt resistance occurring due to damage, i.e.,

$$R_d = 12.5\ \Omega$$

(*e*) If no damage were detected on the transmission line then the shunt resistance R_d would not be present indicating that its value would be given by $R_d = \infty$. Under this situation, the line would be fully matched, and the waveform seen at the observation point of the TDR would like as shown in Figure 11.36(*c*). We can infer from Figure 11.36(*c*) that when no damage is detected on the transmission line then no reflection is present anywhere on the line.

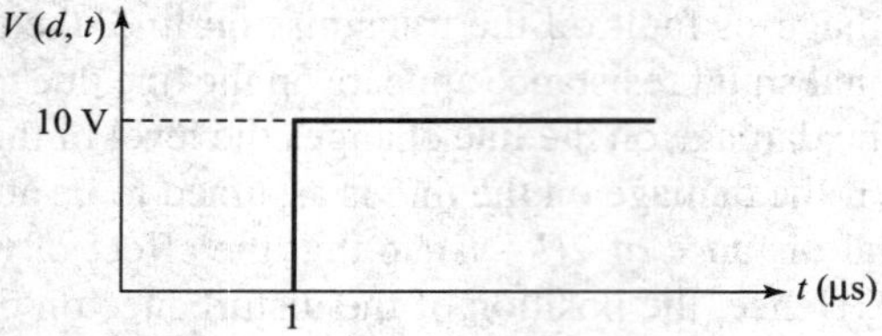

Fig. 11.36(*c*) Waveform.

Another type of transient problem involves lines that are *initially charged*. In these cases, the manner in which the line discharges through a load is of interest. Consider the situation shown in Figure 11.37, in which a charged line of characteristic impedance Z_0 is discharged through a resistor of value R_g when a switch at the resistor location is closed.[5] We consider the resistor at the $z = 0$ location; the other end of the line is open (as would be necessary) and is located at $z = l$.

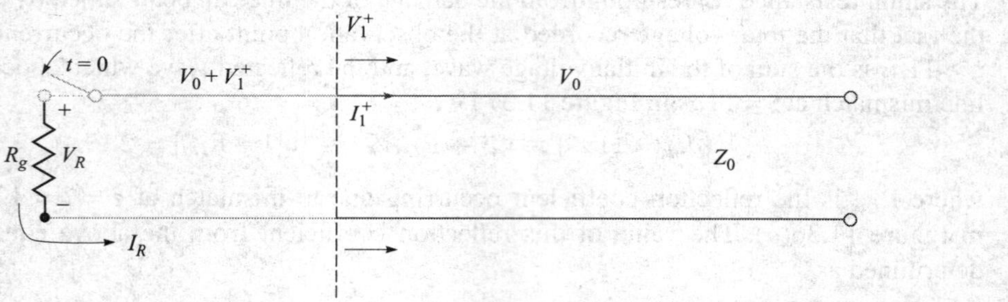

Fig. 11.37 In an initially charged line, closing the switch as shown initiates a voltage wave of opposite polarity to that of the initial voltage. The wave thus depletes the line voltage and will fully discharge the line in one round trip if $R_g = Z_0$.

When the switch is closed, current I_R begins to flow through the resistor, and the line discharge process begins. This current does not immediately flow everywhere in the transmission line but begins at the resistor and establishes its presence at more distant parts of the line as time progresses. By analogy, consider a long line of automobiles at a red light. When the light turns green, the cars at the front move through the intersection first, followed successively by those further toward the rear. The point which divides cars in motion and those standing still is in fact a wave which propagates toward the back of the line. In the transmission line, the flow of charge progresses in a similar way. A voltage wave, V_1^+, is initiated and propagates to the right. To the left of its leading edge, charge is in motion; to the right of the leading edge, charge is stationary and carries its original

[5] Even though this is a load resistor, we will call it R_g because it is located at the front (generator) end of the line.

density. Accompanying the charge in motion to the left of V_1^+ is a drop in the charge density as the discharge process occurs, and so the line voltage to the left of V_1^+ is partially reduced. This voltage will be given by the sum of the initial voltage, V_0, and V_1^+, which means that V_1^+ must in fact be negative (or of opposite sign to V_0). The line discharge process is analyzed by keeping track of V_1^+ as it propagates and undergoes multiple reflections at the two ends. Voltage and current reflection diagrams are used for this purpose in much the same way as before.

Referring to Figure 11.37, we see that for positive V_0 the current flowing through the resistor will be counterclockwise and hence negative. We also know that continuity requires that the resistor current be equal to the current associated with the voltage wave, or

$$I_R = -I_1^+ = -\frac{V_1^+}{Z_0}$$

Now the resistor voltage will be

$$V_R = V_0 + V_1^+ = I_R R_g = -I_1^+ R_g = -\frac{V_1^+}{Z_0} R_g$$

We solve for V_1^+ to obtain

$$V_1^+ = \frac{-V_0 Z_0}{Z_0 + R_g} \tag{122}$$

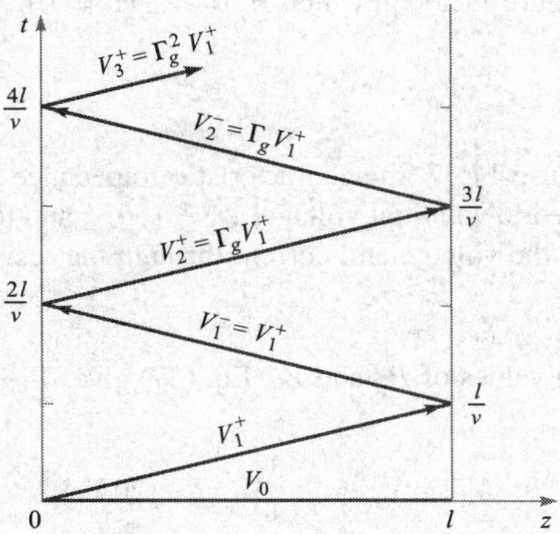

Fig. 11.38 Voltage reflection diagram for the charged line of Figure 11.37, showing the initial condition of V_0 everywhere on the line at $t = 0$.

Having found V_1^+, we can set up the voltage and current reflection diagrams. That for voltage is shown in Figure 11.38. Note that the initial condition of voltage V_0 everywhere on the line is accounted for by assigning voltage V_0 to the horizontal axis of the voltage diagram. The diagram is otherwise drawn as before, but with $\Gamma_L = 1$ (at the open-circuited load end). Variations in how the line discharges thus depend on the resistor value at the switch end, R_g, which determines the reflection coefficient, Γ_g, at that location. The current reflection diagram is derived from the voltage diagram in the usual way. There is no initial current to consider.

A special case of practical importance is that in which the resistor is matched to the line, or $R_g = Z_0$. In this case, Eq. (122) gives $V_1^+ = -V_0/2$. The line fully discharges in one round trip of V_1^+ and produces a voltage across the resistor of value $V_R = V_0/2$, which persists for time $T = 2l/\nu$. The resistor voltage as a function of time is shown in Figure 11.39. The transmission line in this application is known as a *pulse-forming line*; pulses that are generated in this way are well formed and of low noise, provided the switch is sufficiently fast. Commercial units are available that are capable of generating high-voltage pulses of widths on the order of a few nanoseconds, using thyratron-based switches.

When the resistor is not matched to the line, full discharge still occurs, but does so over several reflections, leading to a complicated pulse shape.

Fig. 11.39 Voltage across the resistor as a function of time, as determined from the reflection diagram of Figure 11.38, in which $R_g = Z_0$ $(\Gamma = 0)$.

EXAMPLE 11.22

In the charged line of Figure 11.37, the characteristic impedance is $Z_0 = 100\ \Omega$, and $R_g = 100/3\ \Omega$. The line is charged to an initial voltage, $V_0 = 160$ V, and the switch is closed at time $t = 0$. Determine and plot the voltage and current through the resistor for time $0 < t < 8l/\nu$ (four round trips).

Solution. With the given values of R_g and Z_0, Eq. (47) gives $\Gamma_g = -1/2$. Then, with $\Gamma_L = 1$, and using (122), we find

$$V_1^+ = V_1^- = -\frac{3}{4}V_0 = -120\text{ V}$$

$$V_2^+ = V_2^- = \Gamma_g V_1^- = +60\text{ V}$$

$$V_3^+ = V_3^- = \Gamma_g V_2^- = -30\text{ V}$$

$$V_4^+ = V_4^- = \Gamma_g V_3^- = +15\text{ V}$$

Using these values on the voltage reflection diagram, we evaluate the voltage in time at the resistor location by moving up the left-hand vertical axis, adding voltages as we progress, and beginning with $V_0 + V_1^+$ at $t = 0$. Note that when we add voltages along the vertical axis, we are encountering the intersection points between incident and reflected waves, which occur (in time) at each integer multiple of $2l/\nu$. So, when moving up the axis, we add the voltages of *both* waves to our total at each occurrence. The voltage within each time interval is thus:

$$
\begin{aligned}
V_R &= V_0 + V_1^+ = 40\,\text{V} && (0 < t < 2l/\nu)\\
&= V_0 + V_1^+ + V_1^- + V_2^+ = -20\,\text{V} && (2l/\nu < t < 4l/\nu)\\
&= V_0 + V_1^+ + V_1^- + V_2^+ + V_2^- + V_3^+ = 10\,\text{V} && (4l/\nu < t < 6l/\nu)\\
&= V_0 + V_1^+ + V_1^- + V_2^+ + V_2^- + V_3^+ + V_3^- + V_4^+ = -5\,\text{V} && (6l/\nu < t < 8l/\nu)
\end{aligned}
$$

The resulting voltage plot over the desired time range is shown in Figure 11.40a.

The current through the resistor is most easily obtained by dividing the voltages in Figure 11.40a by $-R_g$. As a demonstration, we can also use the current diagram of Figure 11.33a to obtain this

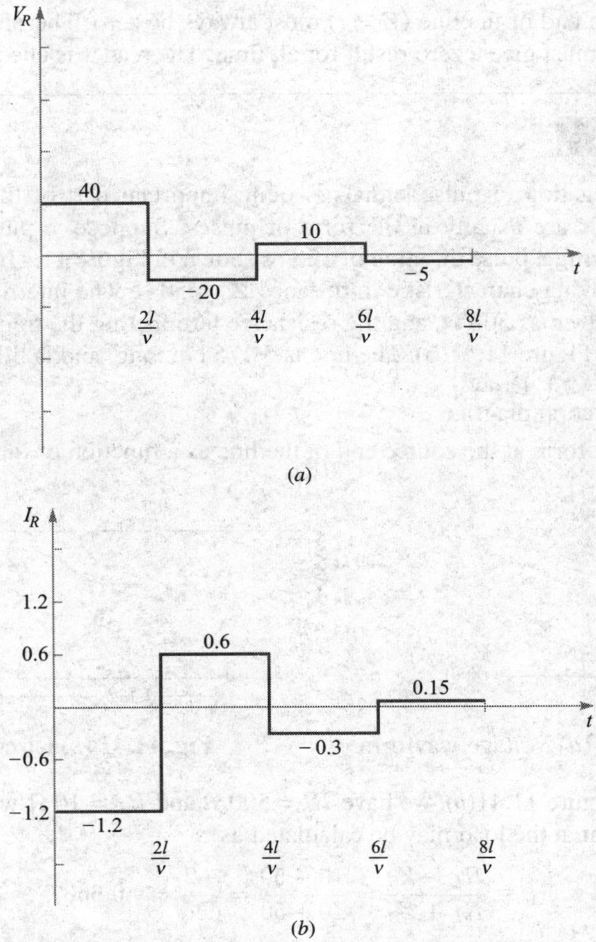

Fig. 11.40 Resistor voltage (a) and current (b) as functions of time for the line of Figure 11.37, with values as specified in Example 11.33.

result. Using (120) and (121), we evaluate the current waves as follows:

$$I_1^+ = V_1^+/Z_0 = -1.2\text{ A}$$
$$I_1^- = -V_1^-/Z_0 = +1.2\text{ A}$$
$$I_2^+ = -I_2^- = V_2^+/Z_0 = +0.6\text{ A}$$
$$I_3^+ = -I_3^- = V_3^+/Z_0 = -0.30\text{ A}$$
$$I_4^+ = -I_4^- = V_4^+/Z_0 = +0.15\text{ A}$$

Using these values on the current reflection diagram, Figure 11.33*a*, we add up currents in the resistor in time by moving up the left-hand axis, as we did with the voltage diagram. The result is shown in Figure 11.40*b*. As a further check to the correctness of our diagram construction, we note that current at the open end of the line ($Z = l$) must always be zero. Therefore, summing currents up the right-hand axis must give a zero result for all time. The reader is encouraged to verify this.

EXAMPLE 11.23

The study of propagation of pulse signals is quite important part of the transient analysis as the digital signals are usually in the form of pulses. Suppose, a pulse signal source of amplitude 0.5 V having a pulse duration of 125 ps shown in Figure 11.41(*a*) is connected to a transmission line having characteristic impedance $Z_0 = 50\ \Omega$. The internal impedance of the pulse generator is given as 500 Ω, and the resistance terminating the transmission line is $R_L = 10\ \Omega$ as shown in Figure 11.41(b). The line is 5.175 cm long, and is filled with a dielectric material having $\epsilon_r = 2.1$. Draw

(*a*) The voltage reflection diagram

(*b*) The voltage waveform at the source end of the line as a function of time up to 1 ns

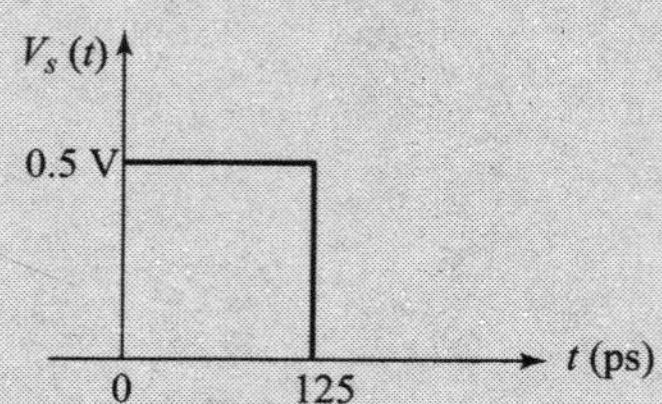

Fig. 11.41 (*a*) Voltage wavform

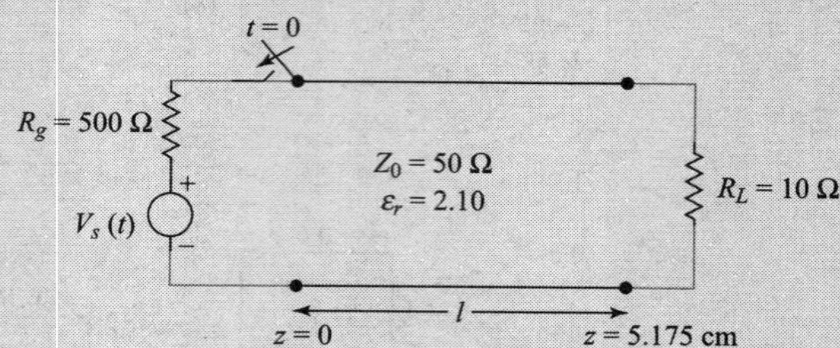

Fig. 11.41 (*b*) Circuit

Solution. From Figure 11.41(*b*), we have $R_g = 500\ \Omega$ and $R_L = 10\ \Omega$ with $Z_0 = 50\ \Omega$. Hence, the reflection coefficient at the load may be calculated as

$$\Gamma_L \equiv \frac{R_L - Z_0}{R_L + Z_0} = \frac{10 - 50}{10 + 50} = \frac{-2}{3} = -0.6667 \tag{i}$$

Similarly, the reflection coefficient at the source end may be calculated as

$$\Gamma_g \equiv \frac{R_g - Z_0}{R_g + Z_0} = \frac{500 - 50}{500 + 50} = \frac{45}{55} = 0.818 \tag{ii}$$

The velocity of pulse along the line would be given by

$$v = \frac{c}{\sqrt{\epsilon_r}} \equiv \frac{3 \times 10^8}{\sqrt{2.1}} = 2.07 \times 10^8 \text{ m/s} \quad \text{(iii)}$$

The time taken by the wave to travel from the source to the load end along the line would be given by

$$T = \frac{l}{v} \equiv \frac{5.175 \times 10^{-2}}{2.07 \times 10^8} = 0.25 \text{ ns} \quad \text{(iv)}$$

Now, the pulse signal shown in Figure 11.41(a) can be expressed as the summation of two step functions. The first step function of 0.5 V is appearing at t = 0 ps, while the second step function of –0.5 V is appearing at t = 125 ps. The step-signal source shown in Figure 11.41(a) can then be expressed as

$$V_s(t) \equiv 0.5[u(t) - u(t - 125 \times 10^{-12})] \text{ V} \quad \text{(v)}$$

The initial voltage wave in the present situation can be calculated as:

$$V_1^+ = \frac{0.5\, Z_0}{Z_0 + R_g} \equiv \frac{0.5 \times 50}{50 + 500} = \frac{1}{22} \text{ V} \quad \text{(vi)}$$

(a) Now, for drawing the voltage reflection diagram corresponding to the line of Figure 11.41(b) excited by the pulse function shown in Figure 11.41(a), we have to note that two sets of diagrams would have to be constructed. The first diagram would correspond to the excitation of 0.5 V applied at t = 0, while the second set would correspond to the excitation of –0.5 V applied at t = 125 ps. The voltage reflection diagram under the present situation is shown in Figure 11.41(c), where the sets of lines corresponding to both 0.5 V and –0.5 V are plotted on the same diagram. The solid lines in this figure are for +0.5 V step source applied at t = 0, while the dashed lines represent the response due to the step signal of –0.5 V applied at t = 0.125 ns.

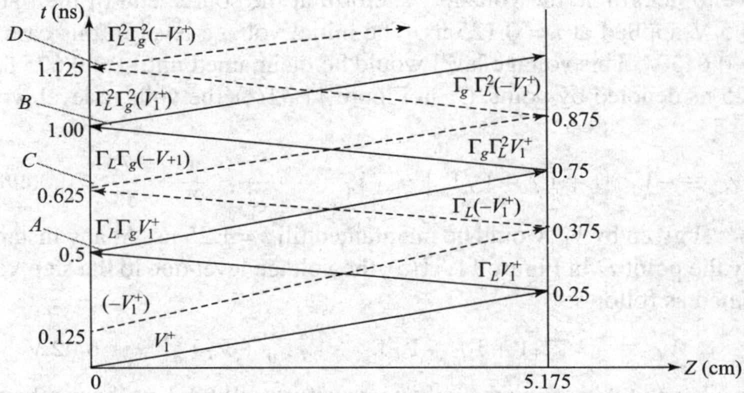

Fig. 11.41(c) Voltage reflection diagram.

The value of the voltage wave is shown along each line. For the solid line corresponding to the step source of 0.5 V, it can be observed that in the beginning, the initial voltage wave is just equal to V_1^+, which gets multiplied by the factor of Γ_L at the load end, and by the factor of Γ_g

at the source end, and this process repeats whenever the voltage wave reaches these two ends as shown in the figure. Similarly, it can be observed from this figure that for the step source of –0.5 V applied at $t = 0.125$ ns, the initial voltage wave is equal to $-V_1^+$, which keeps on multiplying by the factor of Γ_L and Γ_g at the load and the source ends, respectively.

(*b*) To obtain the voltage waveform at the source end of the line due to applied pulse of Figure 11.41(*a*), we can take help of the voltage reflection diagram shown in Figure 11.41(*c*). Our approach would be to plot the voltage waveform at the source end of the line due to two-step voltages of +0.5 V and –0.5 V separately, and then obtain the overall plot due to the pulse signal by adding these two plots.

It should be noted here that for the step signal of 0.5 V applied at $t = 0$ ns, the change in the signal level at the input side is at $t = 0.5$ ns, and $t = 1$ ns denoted by points A and B respectively. However, for the step signal of –0.5 V applied at $t = 0.125$ ns, the change in the signal level at the input side is at $t = 0.675$ ns, and $t = 1.125$ ns denoted by points 'C' and 'D' respectively. The exact value of the voltage levels at these points can now be calculated as follows:

For a step signal of +0.5 V applied at $t = 0$ ns, in the beginning, the voltage level would be given by $V_1^+ = \frac{1}{22}V \equiv 0.045$ V. This voltage level would be maintained until $t = 0.5$ ns. However, at any instant after $t = 0.5$ ns denoted by point 'A' in Figure 11.41(*c*), the voltage level would be modified as follows:

$$V_A \equiv V_1^+ + \Gamma_L V_1^+ + \Gamma_L \Gamma_g V_1^+ = V_1^+ \times -\frac{7}{33} = -\frac{1}{22} \times \frac{7}{33} = -0.00964 \text{ V}$$

Similarly, the voltage level at any instant after $t = 1$ ns denoted by the point B in Figure 11.41(*c*) due to the step voltage of +0.5 V can be computed as follows:

$$V_B \equiv V_1^+ \left[1 + \Gamma_L + \Gamma_L\Gamma_g + \Gamma_L^2\Gamma_g^+\Gamma_L^2\Gamma_g^2\right] \equiv V_A + V_1^+\Gamma_L^2\Gamma_g[1 + \Gamma_g] = V_A + 0.03 = 0.02 \text{ V}$$

The voltage waveform at the source end of the line up to 1 ns due to the applied step voltage of +0.5 V is shown in Figure 11.41(*d*).

Now, we have to determine the voltage waveform at the source end of the line due to the step signal of –0.5 V applied at $t = 0.125$ ns. The initial voltage level in this case would be given by $-V_1^+ \equiv -0.045$ V. This voltage level would be maintained until $t = 0.625$ ns. At any instant after $t = 0.625$ ns denoted by point 'C' in Figure 11.41(*c*), the voltage level would be modified as follows:

$$V_C \equiv -V_1^+ [1 + \Gamma_L + \Gamma_L\Gamma_g] = -V_1^+ \times -\frac{7}{33} = \frac{1}{22} \times \frac{7}{33} = 0.00964 \text{ V}$$

The voltage level given by V_C would be maintained till $t = 1.25$ ns. At any instant after $t = 1.125$ ns denoted by the point D in Figure 11.41(*c*), the voltage level due to the step voltage of –0.5 V can be computed as follows:

$$V_D \equiv -V_1^+ \left[1 + \Gamma_L + \Gamma_L\Gamma_g + \Gamma_L^2\Gamma_g + \Gamma_L^2\Gamma_g^2\right] = -0.02 \text{ V}$$

In our case, we have to determine the voltage waveform till $t = 1$ ns, hence the voltage level V_D would appear after this time. The voltage waveform at the source end of the line up to 1 ns due to the applied step voltage of –0.5 V is shown in Figure 11.41(*e*).

Once the voltage waveforms at the source end due to the step voltages of +0.5 V and –0.5 V have been obtained, the total voltage at the source end due to the pulse signal can be determined by adding these two individual waveforms shown in Figs. 11.41(*d*) and 11.41(*e*). The final result

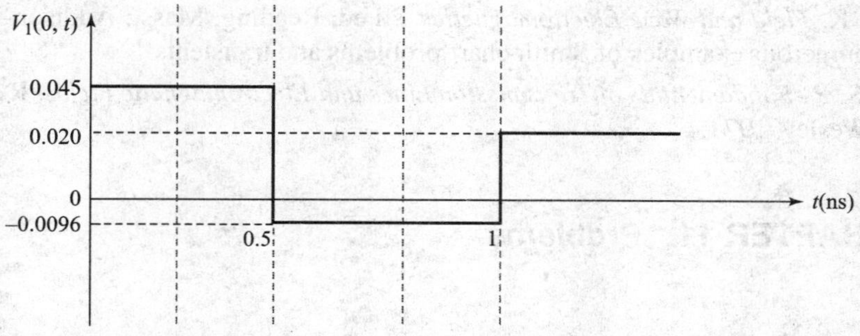

Fig. 11.41(*d*) The voltage (in V) at the source point due to the step signal of $+0.5$ V.

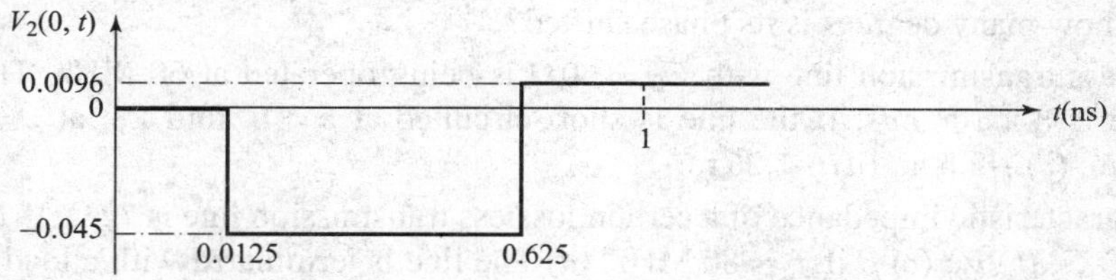

Fig. 11.41(*e*) The voltage (in V) at the source point due to the step signal of -0.5 V.

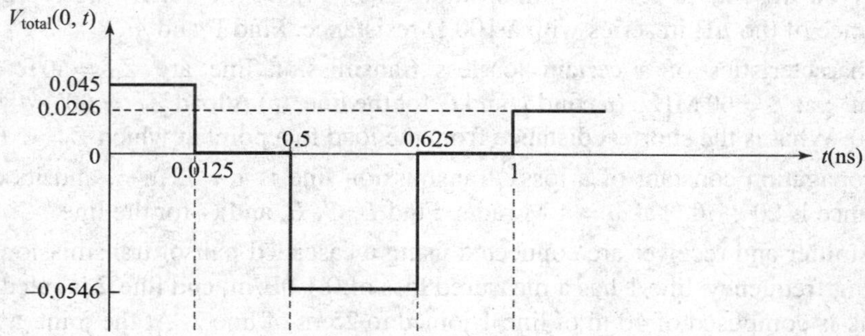

Fig. 11.41(*f*) The total voltage (in V) at the source point due to the pulse signal.

is shown in Figure 11.41(f), which represents the voltage waveform at the source end of the line as a function of time up to 1 ns due to the applied pulse signal shown in Figure 11.41(a).

References

1. White, H. J., P. R. Gillette, and J. V. Lebacqz. "The Pulse-Forming Network." Chapter 6 in *Pulse Generators,* edited by G. N, Glasoe and J. V. Lebacqz. New York: Dover, 1965.
2. Brown, R. G., R. A. Sharpe, W. L. Hughes, and R. E. Post. *Lines, Waves, and Antennas*. 2d ed. New York: The Ronald Press Company, 1973. Transmission lines are covered in the first six chapters, with numerous examples.

3. Cheng, D. K. *Field and Wave Electromagnetics*. 2d ed. Reading, Mass.: Addison-Wesley, 1989. Provides numerous examples of Smith chart problems and transients.

4. Seshadri, S. R. *Fundamentals of Transmission Lines and Electromagnetic Fields.* Reading, Mass.: Addison-Wesley, 1971.

CHAPTER 11 Problems

Quizzes

11.1 The parameters of a certain transmission line operating at 6×10^8 rad/s are $L = 0.4\,\mu$H/m, $C = 40$ pF/m, $G = 80\,\mu$S/m, and $R = 20\ \Omega$/m. (*a*) Find γ, α, β, λ, and Z_0. (*b*) If a voltage wave travels 20 m down the line, what percentage of the original wave amplitude remains, and by how many degrees is its phase shifted?

11.2 A lossless transmission line with $Z_0 = 60\,\Omega$ is being operated at 60 MHz. The velocity on the line is 3×10^8 m/s. If the line is short-circuited at $z = 0$, find Z_{in} at $z =$: (*a*) -1 m; (*b*) -2 m; (*c*) -2.5 m; (*d*) -1.25 m.

11.3 The characteristic impedance of a certain lossless transmission line is $72\,\Omega$. If $L = 0.5\,\mu$H/m, find: (*a*) C; (*b*) ν_p; (*c*) β if $f = 80$ MHz. (*d*) The line is terminated with a load of 60 Ω. Find Γ and s.

11.4 A lossless transmission line having $Z_0 = 120\ \Omega$ is operating at $\omega = 5 \times 10^8$ rad/s. If the velocity on the line is 2.4×10^8 m/s, find: (*a*) L; (*b*) C. (*c*) Let Z_L be represented by an inductance of 0.6 μH in series with a 100 Ω resistance. Find Γ and s.

11.5 Two characteristics of a certain lossless transmission line are $Z_0 = 50\,\Omega$ and $\gamma = 0 + j0.2\pi\,\text{m}^{-1}$ at $f = 60$ MHz: (*a*) find L and C for the line. (*b*) A load $Z_L = 60 + j80\,\Omega$ is located at $z = 0$. What is the shortest distance from the load to a point at which $Z_{\text{in}} = R_{\text{in}} + j0$?

11.6 The propagation constant of a lossy transmission line is $1 + j2\,\text{m}^{-1}$, and its characteristic impedance is $20 + j0\ \Omega$ at $\omega = 1$ M rad/s. Find L, C, R, and G for the line.

11.7 A transmitter and receiver are connected using a cascaded pair of transmission lines. At the operating frequency, line 1 has a measured loss of 0.1 dB/m, and line 2 is rated at 0.2 dB/m. The link is composed of 40 m of line 1 joined to 25 m of line 2. At the joint, a splice loss of 2 dB is measured. If the transmitted power is 100 mW, what is the received power?

11.8 A measure of absolute power is the dBm scale, in which power is specified in decibels relative to 1 milliwatt. Specifically, $P(\text{dBm}) = 10\log_{10}[P(\text{mW})/1\text{ mW}]$. Suppose a receiver is rated as having a *sensitivity* of -5 dBm—indicating the *minimum* power that it must receive in order to adequately interpret the transmitted data. Consider a transmitter having an output of 100 mW connected to this receiver through a length of transmission line whose loss is 0.1 dB/m. What is the maximum length of line that can be used?

11.9 A sinusoidal voltage source drives the series combination of an impedance, $Z_g = 50 - j50\ \Omega$, and a lossless transmission line of length L, shorted at the load end. The line characteristic impedance is 50 Ω, and wavelength λ is measured on the line. (*a*) Determine, in terms of wavelength, the shortest line length that will result in the voltage source driving a total impedance of 50 Ω. (*b*) Will other line lengths meet the requirements of part (*a*)? If so, what are they?

11.10 A 100 MHz voltage source drives the series combination of an impedance, $Z_g = 25 + j25\ \Omega$ and a lossless transmission line of length $\lambda/4$, terminated by a load impedance, Z_L. The line characteristic impedance is 50 Ω. (*a*) Determine the load impedance value required to achieve a net impedance (seen by the voltage source) of 50 Ω. (*b*) If the inductance of the line is $L = 1\ \mu$H/m, determine the line length in metres.

11.11 A transmission line having primary constants $L, C, R,$ and G has length ℓ and is terminated by a load having complex impedance $R_L + jX_L$. At the input end of the line, a dc voltage source, V_0, is connected. Assuming all parameters are known at zero frequency, find the steady-state power dissipated by the load if (*a*) $R = G = 0$; (*b*) $R \neq 0, G = 0$; (*c*) $R = 0, G \neq 0$; (*d*) $R \neq 0$, $G \neq 0$.

11.12 In a circuit in which a sinusoidal voltage source drives its internal impedance in series with a load impedance, it is known that maximum power transfer to the load occurs when the source and load impedances form a complex conjugate pair. Suppose the source (with its internal impedance) now drives a complex load of impedance $Z_L = R_L + jX_L$ that has been moved to the end of a lossless transmission line of length ℓ having characteristic impedance Z_0. If the source impedance is $Z_g = R_g + jX_g$, write an equation that can be solved for the required line length, ℓ, such that the displaced load will receive the maximum power.

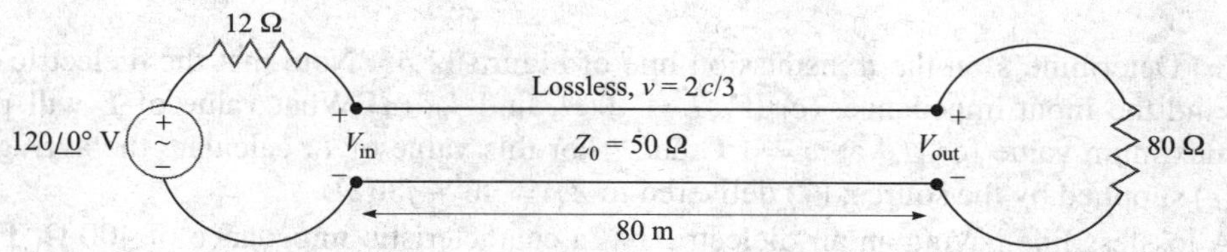

Fig. 11.42 See Problem 11.15.

11.13 The incident voltage wave on a certain lossless transmission line for which $Z_0 = 50\,\Omega$ and $\nu_p = 2 \times 10^8$ m/s is $V^+(z,t) = 200\cos(\omega t - \pi z)$ V. (*a*) Find ω. (*b*) Find $I^+(z,t)$. The section of line for which $z > 0$ is replaced by a load $Z_L = 50 + j30\ \Omega$ at $z = 0$. Find: (*c*) Γ_L; (*d*) $V_s^-(z)$; (*e*) V_s at $z = -2.2$ m.

11.14 A 50 Ω lossless line is terminated with 60 and 30 Ω resistors in parallel. The voltage at the input to the line is $\mathcal{V}(t) = 100\cos(5 \times 10^9 t)$ and the line is three-eighths of a wavelength long. What average power is delivered to each load resistor?

11.15 For the transmission line represented in Figure 11.42, find $V_{s,\text{out}}$ if $f =$: (*a*) 60 Hz; (*b*) 500 kHz.

11.16 A 300 Ω transmission line is 0.8 m long and terminated with a short circuit. The line is operating in air with a wavelength of 0.3 m and is lossless. (*a*) If the input voltage amplitude is 10 V, what is the maximum voltage amplitude at any point on the line? (*b*) What is the current amplitude in the short circuit?

11.17 Determine the average power absorbed by each resistor in Figure 11.43.

11.18 The line shown in Figure 11.44 is lossless. Find s on both sections 1 and 2.

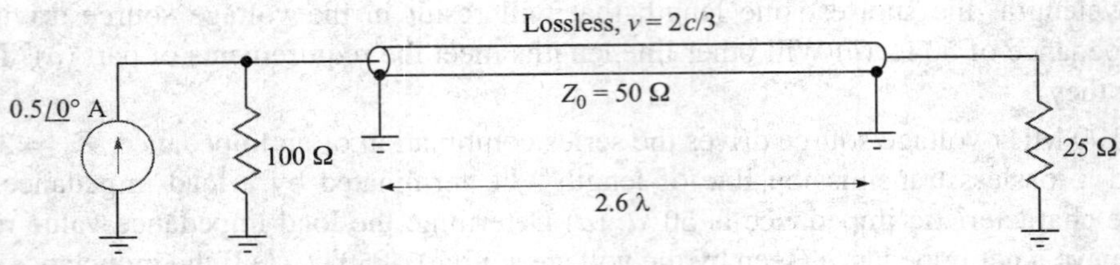

Fig. 11.43 See Problem 11.17.

11.19 A lossless transmission line is 50 cm in length and operating at a frequency of 100 MHz. The line parameters are $L = 0.2\,\mu$H/m and $C = 80$ pF/m. The line is terminated in a short circuit at $z = 0$, and there is a load $Z_L = 50 + j20\,\Omega$ across the line at location $z = -20$ cm. What average power is delivered to Z_L if the input voltage is $100\angle 0°$ V?

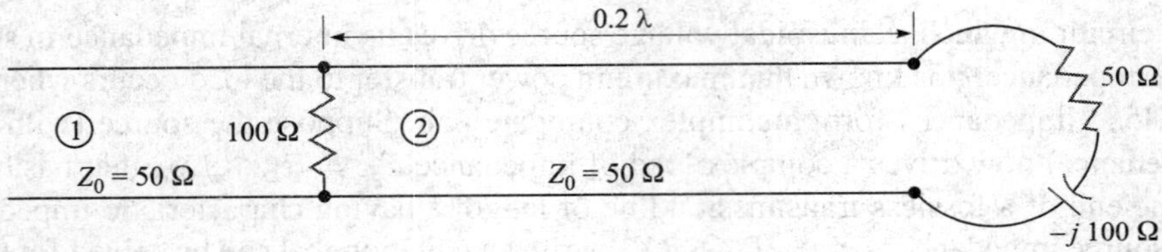

Fig. 11.44 See Problem 11.18.

11.20 (*a*) Determine s on the transmission line of Figure 11.45. Note that the dielectric is air. (*b*) Find the input impedance. (*c*) If $\omega L = 10\ \Omega$, find I_s. (*d*) What value of L will produce a maximum value for $|I_s|$ at $\omega = 1$ Grad/s? For this value of L, calculate the average power: (*e*) supplied by the source; (*f*) delivered to $Z_L = 40 + j30\ \Omega$.

11.21 A lossless line having an air dielectric has a characteristic impedance of 400 Ω. The line is operating at 200 MHz and $Z_{\text{in}} = 200 - j200\ \Omega$. Use analytic methods or the Smith chart (or both) to find: (*a*) s; (*b*) Z_L, if the line is 1 m long; (*c*) the distance from the load to the nearest voltage maximum.

11.22 A lossless two-wire line has a characteristic impedance of 300 Ω and a capacitance of 15 pF/m. The load at $z = 0$ consists of a 600 Ω resistor in parallel with a 10 pF capacitor. If $\omega = 10^8$ rad/s and the line is 20 m long, use the Smith chart to find: (*a*) $|\Gamma_L|$; (*b*) s; (*c*) Z_{in}.

11.23 The normalized load on a lossless transmission line is $2 + j1$. Let $\lambda = 20$ m and make use of the Smith chart to find: (*a*) the shortest distance from the load to a point at which $z_{\text{in}} = r_{\text{in}} + j0$, where $r_{\text{in}} > 0$; (*b*) z_{in} at this point. (*c*) The line is cut at this point and the portion containing z_L is thrown away. A resistor $r = r_{\text{in}}$ of part (*a*) is connected across the line. What

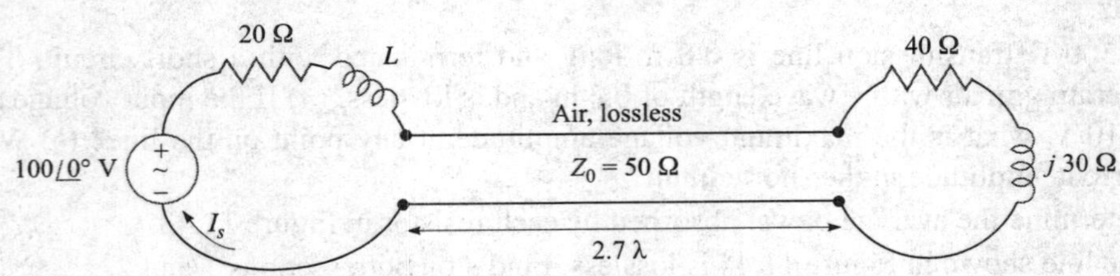

Fig. 11.45 See Problem 11.20.

is s on the remainder of the line? (d) What is the shortest distance from this resistor to a point at which $z_{\text{in}} = 2 + j1$?

11.24 With the aid of the Smith chart, plot a curve of $|Z_{\text{in}}|$ versus l for the transmission line shown in Figure 11.46. Cover the range $0 < l/\lambda < 0.25$.

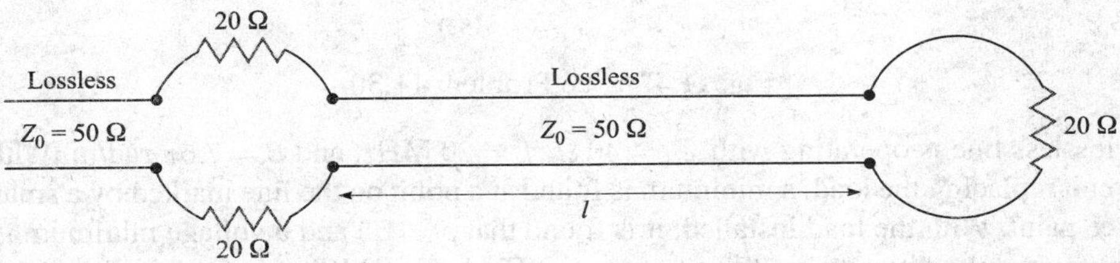

Fig. 11.46 See Problem 11.24.

11.25 A 300 Ω transmission line is short-circuited at $z = 0$. A voltage maximum, $|V|_{\text{max}} = 10$ V, is found at $z = -25$ cm, and the minimum voltage, $|V|_{\text{min}} = 0$, is at $z = -50$ cm. Use the Smith chart to find Z_L (with the short circuit replaced by the load) if the voltage readings are: (a) $|V|_{\text{max}} = 12$ V at $z = -5$ cm, and $|V|_{\text{min}} = 5$ V; (b) $|V|_{\text{max}} = 17$ V at $z = -20$ cm, and $|V|_{\text{min}} = 0$.

11.26 A lossless 50 Ω transmission line operates with a velocity that is 3/4 c. A load $Z_L = 60 + j30$ Ω is located at $z = 0$. Use the Smith chart to find: (a) s; (b) the distance from the load to the nearest voltage minimum if $f = 300$ MHz; (c) the input impedance if $f = 200$ MHz and the input is at $z = -110$ cm.

11.27 The characteristic admittance ($Y_0 = 1/Z_0$) of a lossless transmission line is 20 mS. The line is terminated in a load $Y_L = 40 - j20$ mS. Make use of the Smith chart to find: (a) s; (b) Y_{in} if $l = 0.15\lambda$; (c) the distance in wavelengths from Y_L to the nearest voltage maximum.

11.28 The wavelength on a certain lossless line is 10 cm. If the normalized input impedance is $z_{\text{in}} = 1 + j2$, use the Smith chart to determine: (a) s; (b) z_L, if the length of the line is 12 cm; (c) x_L, if $z_L = 2 + jx_L$ where $x_L > 0$.

11.29 A standing wave ratio of 2.5 exists on a lossless 60 Ω line. Probe measurements locate a voltage minimum on the line whose location is marked by a small scratch on the line. When the load is replaced by a short circuit, the minima are 25 cm apart, and one minimum is located at a point 7 cm toward the source from the scratch. Find Z_L.

11.30 A 2-wire line constructed of lossless wire of circular cross section is gradually flared into a coupling loop that looks like an egg beater. At the point X, indicated by the arrow in Figure 11.47, a short circuit is placed across the line. A probe is moved along the line and indicates that the first voltage minimum to the left of X is 16 cm from X. With the short circuit removed, a voltage minimum is found 5 cm to the left of X, and a voltage maximum is located that is 3 times the voltage of the minimum. Use the Smith chart to determine: (a) f; (b) s; (c) the normalized input impedance of the egg beater as seen looking to the right at point X.

11.31 In order to compare the relative sharpness of the maxima and minima of a standing wave, assume a load $z_L = 4 + j0$ is located at $z = 0$. Let $|V|_{\text{min}} = 1$ and $\lambda = 1$ m. Determine the width of the: (a) minimum where $|V| < 1.1$; (b) maximum where $|V| > 4/1.1$.

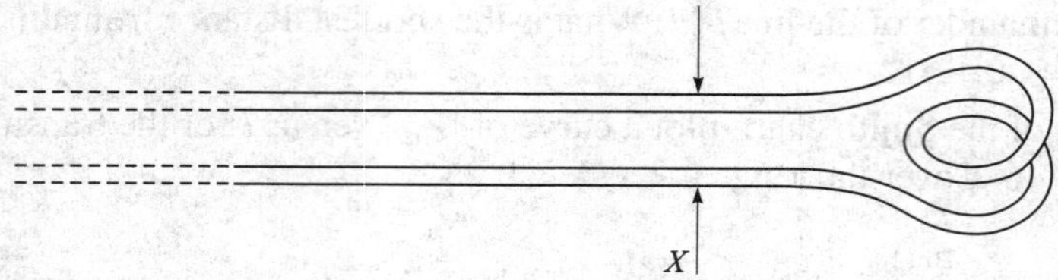

Fig. 11.47 See Problem 11.30.

11.32 A lossless line is operating with $Z_0 = 40\ \Omega$, $f = 20$ MHz, and $\beta = 7.5\pi$ rad/m. With a short circuit replacing the load, a minimum is found at a point on the line marked by a small spot of puce paint. With the load installed, it is found that $s = 1.5$ and a voltage minimum is located 1 m toward the source from the puce dot. (*a*) Find Z_L. (*b*) What load would produce $s = 1.5$ with $|V|_{\text{max}}$ at the paint spot?

11.33 In Figure 11.17, let $Z_L = 40 - j10\ \Omega$, $Z_0 = 50\ \Omega$, $f = 800$ MHz, and $v = c$. (*a*) Find the shortest length d_1 of a short-circuited stub, and the shortest distance d that it may be located from the load to provide a perfect match on the main line to the left of the stub. (*b*) Repeat for an open-circuited stub.

11.34 The lossless line shown in Figure 11.48 is operating with $\lambda = 100$ cm. If $d_1 = 10$ cm, $d = 25$ cm, and the line is matched to the left of the stub, what is Z_L?

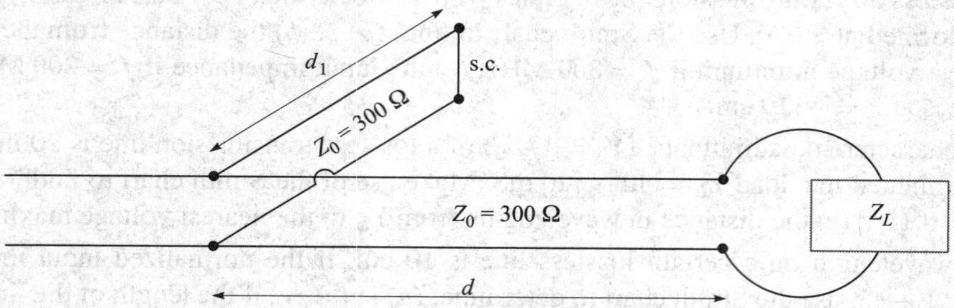

Fig. 11.48 See Problem 11.34.

11.35 A load, $Z_L = 25 + j75\ \Omega$, is located at $z = 0$ on a lossless two-wire line for which $Z_0 = 50\ \Omega$ and $v = c$. (*a*) If $f = 300$ MHz, find the shortest distance d $(z = -d)$ at which the input admittance has a real part equal to $1/Z_0$ and a negative imaginary part. (*b*) What value of capacitance C should be connected across the line at that point to provide unity standing wave ratio on the remaining portion of the line?

11.36 The two-wire lines shown in Figure 11.49 are all lossless and have $Z_0 = 200\ \Omega$. Find d and the shortest possible value for d_1 to provide a matched load if $\lambda = 100$ cm.

11.37 In the transmission line of Figure 11.31, $R_L = Z_0 = 50\ \Omega$, and $R_g = 25\ \Omega$. Determine and plot the voltage at the load resistor and the current in the battery as functions of time by constructing appropriate voltage and current reflection diagrams.

11.38 Repeat Problem 11.37, with $Z_0 = 50\ \Omega$, and $R_L = R_g = 25\ \Omega$. Carry out the analysis for the time period $0 < t < 8l/\nu$.

11.39 In the transmission line of Figure 11.31, $Z_0 = 50\ \Omega$, and $R_L = R_g = 25\ \Omega$. The switch is closed at $t = 0$ and *is opened again* at time $t = l/4\nu$, thus creating a rectangular voltage *pulse*

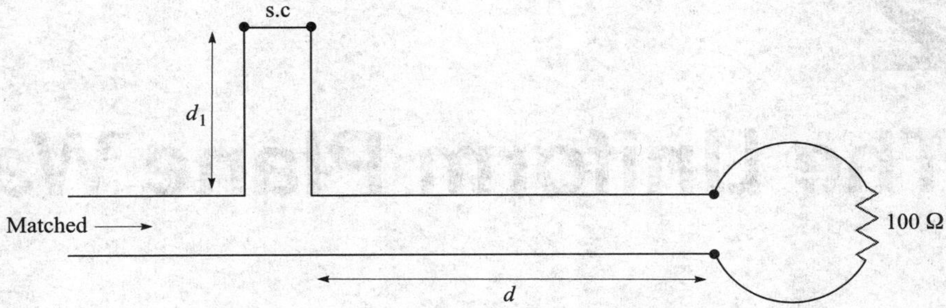

Fig. 11.49 See Problem 11.36.

in the line. Construct an appropriate voltage reflection diagram for this case and use it to make a plot of the voltage at the load resistor as a function of time for $0 < t < 8l/\nu$ (note that the effect of opening the switch is to initiate a second voltage wave, whose value is such that it leaves a net current of zero in its wake).

11.40 In the charged line of Figure 11.37, the characteristic impedance is $Z_0 = 100\ \Omega$, and $R_g = 300\ \Omega$. The line is charged to initial voltage, $V_0 = 160$ V, and the switch is closed at $t = 0$. Determine and plot the voltage and current through the resistor for time $0 < t < 8l/\nu$ (four round-trips). This problem accompanies Example 11.22 as the other special case of the basic charged-line problem, in which now $R_g > Z_0$.

11.41 In the transmission line of Figure 11.50, the switch is located *midway* down the line and is closed at $t = 0$. Construct a voltage reflection diagram for this case, where $R_L = Z_0$. Plot the load resistor voltage as a function of time.

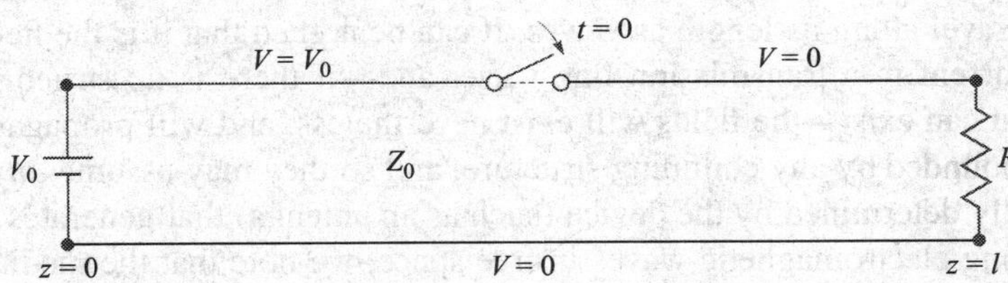

Fig. 11.50 See Problem 11.41.

11.42 A simple *frozen wave generator* is shown in Figure 11.51. Both switches are closed simultaneously at $t = 0$. Construct an appropriate voltage reflection diagram for the case in which $R_L = Z_0$. Determine and plot the load resistor voltage as a function of time.

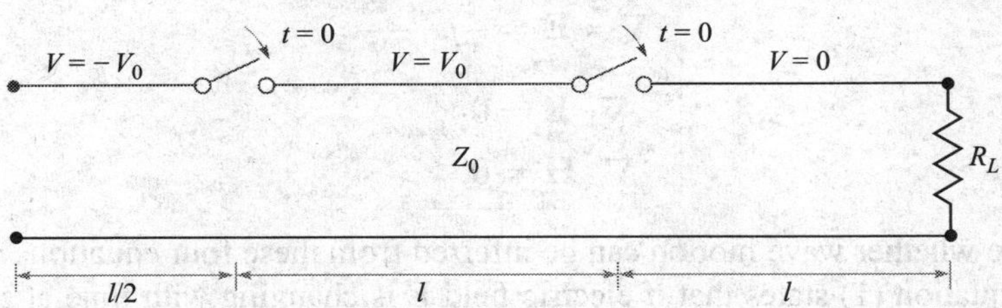

Fig. 11.51 See Problem 11.42.

12

The Uniform Plane Wave

This chapter is concerned with the application of Maxwell's equations to the problem of electromagnetic wave propagation. The uniform plane wave represents the simplest case, and while it is appropriate for an introduction, it is of great practical importance. Waves encountered in practice can often be assumed to be of this form. In this study, we will explore the basic principles of electromagnetic wave propagation, and we will come to understand the physical processes that determine the speed of propagation and the extent to which attenuation may occur. We will derive and use the Poynting theorem to find the power carried by a wave. Finally, we will learn how to describe wave polarization.

12.1 WAVE PROPAGATION IN FREE SPACE

We begin with a quick study of Maxwell's equations, in which we look for clues of wave phenomena. In Chapter 11, we saw how voltages and currents propagate as waves in transmission lines, and we know that the existence of voltages and currents implies the existence of electric and magnetic fields. So we can identify a transmission line as a structure that confines the fields while enabling them to travel along its length as waves. It can be argued that it is the fields that generate the voltage and current in a transmission line wave, and—if there is no structure on which the voltage and current can exist—the fields will exist nevertheless, and will propagate. In free space, the fields are not bounded by any confining structure, and so they may assume *any* magnitude and direction, as initially determined by the device (such as an antenna) that generates them.

When considering electromagnetic waves in free space, we note that the medium is *sourceless* ($\rho_\nu = \mathbf{J} = 0$). Under these conditions, Maxwell's equations may be written in terms of $\mathbf{E}$ and $\mathbf{H}$ only as

$$\nabla \times \mathbf{H} = \epsilon_0 \frac{\partial \mathbf{E}}{\partial t} \tag{1}$$

$$\nabla \times \mathbf{E} = -\mu_0 \frac{\partial \mathbf{H}}{\partial t} \tag{2}$$

$$\nabla \cdot \mathbf{E} = 0 \tag{3}$$

$$\nabla \cdot \mathbf{H} = 0 \tag{4}$$

Now let us see whether wave motion can be inferred from these four equations without actually solving them. Equation (1) states that if electric field $\mathbf{E}$ is changing with time at some point, then magnetic field $\mathbf{H}$ has curl at that point; therefore $\mathbf{H}$ varies spatially in a direction normal to its orientation direction. Also, if $\mathbf{E}$ is changing with time, then $\mathbf{H}$ will in general also change with time, although not necessarily in the same way. Next, we see from Eq. (2) that a time-varying **H** generates **E**, which, having curl, varies spatially in the direction normal to its orientation. We now have once more a changing electric field, our original hypothesis, but this field is present a small distance away from the point of the original disturbance. We might guess (correctly) that the velocity with which the effect moves away from the original point is the velocity of light, but this must be checked by a more detailed examination of Maxwell's equations.

We postulate the existence of a *uniform plane wave,* in which both fields, **E** and **H**, lie in the *transverse plane*—that is, the plane whose normal is the direction of propagation. Furthermore, and by definition, both fields are of constant magnitude in the transverse plane. For this reason, such a wave is sometimes called a *transverse electromagnetic* (TEM) wave. The required spatial variation of both fields in the direction normal to their orientations will therefore occur only in the direction of travel—or normal to the transverse plane. Assume, for example, that $\mathbf{E} = E_x \mathbf{a}_x$, or that the electric field is *polarized* in the x direction. If we further assume that wave travel is in the z direction, we allow spatial variation of **E** only with z. Using Eq. (2), we note that with these restrictions, the curl of **E** reduces to a single term:

$$\nabla \times \mathbf{E} = \frac{\partial E_x}{\partial z}\mathbf{a}_y = -\mu_0 \frac{\partial \mathbf{H}}{\partial t} = -\mu_0 \frac{\partial H_y}{\partial t}\mathbf{a}_y \tag{5}$$

The direction of the curl of **E** in (5) determines the direction of **H**, which we observe to be along the y direction. Therefore, in a uniform plane wave, the directions of **E** and **H** and the direction of travel are mutually orthogonal. Using the y-directed magnetic field, and the fact that it varies only in z, simplifies Eq. (1) to read

$$\nabla \times \mathbf{H} = -\frac{\partial H_y}{\partial z}\mathbf{a}_x = \epsilon_0 \frac{\partial \mathbf{E}}{\partial t} = \epsilon_0 \frac{\partial E_x}{\partial t}\mathbf{a}_x \tag{6}$$

Equations (5) and (6) can be more succinctly written:

$$\frac{\partial E_x}{\partial z} = -\mu_0 \frac{\partial H_y}{\partial t} \tag{7}$$

$$\frac{\partial H_y}{\partial z} = -\epsilon_0 \frac{\partial E_x}{\partial t} \tag{8}$$

These equations compare directly with the telegraphist's equations for the lossless transmission line [Eqs. (20) and (21) in Chapter 11]. Further manipulations of (7) and (8) proceed in the same manner as was done with the telegraphist's equations. Specifically, we differentiate (7) with respect to z, obtaining:

$$\frac{\partial^2 E_x}{\partial z^2} = -\mu_0 \frac{\partial^2 H_y}{\partial t \partial z} \tag{9}$$

Then, (8) is differentiated with respect to t:

$$\frac{\partial^2 H_y}{\partial z \partial t} = -\epsilon_0 \frac{\partial^2 E_x}{\partial t^2} \tag{10}$$

Substituting (10) into (9) results in

$$\frac{\partial^2 E_x}{\partial z^2} = \mu_0 \epsilon_0 \frac{\partial^2 E_x}{\partial t^2} \tag{11}$$

This equation, in direct analogy to Eq. (13) in Chapter 11, we identify as the wave equation for our x-polarized TEM electric field in free space. From Eq. (11), we further identify the propagation velocity:

$$\nu = \frac{1}{\sqrt{\mu_0 \epsilon_0}} = 3 \times 10^8 \text{ m/s} = c \tag{12}$$

where c denotes the velocity of light in free space. A similar procedure, involving differentiating (7) with t and (8) with z, yields the wave equation for the magnetic field; it is identical in form to (11):

$$\frac{\partial^2 H_y}{\partial z^2} = \mu_0 \epsilon_0 \frac{\partial^2 H_y}{\partial t^2} \tag{13}$$

Animations

As was discussed in Chapter 11, the solution to equations of the form of (11) and (13) will be forward- and backward-propagating waves having the general form [in this case for Eq. (11)]:

$$E_x(z,t) = f_1(t - z/\nu) + f_2(t + z/\nu) \tag{14}$$

where again f_1 amd f_2 can be any function whose argument is of the form $t \pm z/\nu$.

From here, we immediately specialize to sinusoidal functions of a specified frequency and write the solution to (11) in the form of forward- and backward-propagating cosines. Since the waves are sinusoidal, we denote their velocity as the *phase velocity*, ν_p. The waves are written as

$$\begin{aligned} E_x(z,t) &= \mathcal{E}_x(z,t) + \mathcal{E}'_x(z,t) \\ &= |E_{x0}| \cos\left[\omega(t - z/\nu_p) + \phi_1\right] + |E'_{x0}| \cos\left[\omega(t + z/\nu_p) + \phi_2\right] \\ &= \underbrace{|E_{x0}| \cos\left[\omega t - k_0 z + \phi_1\right]}_{\text{forward } z \text{ travel}} + \underbrace{|E'_{x0}| \cos\left[\omega t + k_0 z + \phi_2\right]}_{\text{backward } z \text{ travel}} \end{aligned} \tag{15}$$

In writing the second line of (15), we have used the fact that the waves are traveling in free space, in which case the phase velocity, $\nu_p = c$. Additionally, the *wavenumber* in free space in defined as

$$k_0 \equiv \frac{\omega}{c} \text{ rad/m} \tag{16}$$

In a manner consistant with our transmission line studies, we refer to the solutions expressed in (15) as the *real instantaneous* forms of the electric field. They are the mathematical representations of what one would experimentally measure. The terms ωt and $k_0 z$, appearing in (15), have units of angle and are usually expressed in radians. We know that ω is the radian time frequency, measuring phase shift *per unit time;* it has units of *rad/s*. In a similar way, we see that k_0 will be interpreted as a *spatial* frequency, which in the present case measures the phase shift *per unit distance* along the z direction in rad/m. We note that k_0 is the phase constant for lossless propagation of uniform plane waves in free space. The *wavelength* in free space is the distance over which the spatial phase shifts by 2π radians, assuming fixed time, or

$$k_0 z = k_0 \lambda = 2\pi \quad \rightarrow \quad \boxed{\lambda = \frac{2\pi}{k_0} \quad \text{(free space)}} \tag{17}$$

The manner in which the waves propagate is the same as we encountered in transmission lines. Specifically, suppose we consider some point (such as a wave crest) on the forward-propagating cosine function of Eq. (15). For a crest to occur, the argument of the cosine must be an integer multiple of 2π. Considering the mth crest of the wave, the condition becomes

$$k_0 z = 2m\pi$$

So let us now consider the point on the cosine that we have chosen, and see what happens as time is allowed to increase. Our requirement is that the entire cosine argument be the same multiple of 2π for all time, in order to keep track of the chosen point. Our condition becomes

$$\omega t - k_0 z = \omega(t - z/c) = 2m\pi \tag{18}$$

As time increases, the position z must also increase in order to satisfy (18). The wave crest (and the entire wave) moves in the positive z direction at phase velocity c (in free space). Using similar reasoning, the wave in Eq. (15) having cosine argument $(\omega t + k_0 z)$ describes a wave that moves in the negative z direction, since as time increases, z must now decrease to keep the argument constant. For simplicity, we will restrict our attention in this chapter to only the positive z traveling wave.

As was done for transmission line waves, we express the real instantaneous fields of Eq. (15) in terms of their phasor forms. Using the forward-propagating field in (15), we write

$$\boxed{\mathcal{E}_x(z,t) = \frac{1}{2}\underbrace{|E_{x0}|e^{j\phi_1}}_{E_{x0}} e^{-jk_0 z} e^{j\omega t} + c.c. = \frac{1}{2}E_{xs}e^{j\omega t} + c.c. = \text{Re}[E_{xs}e^{j\omega t}]} \tag{19}$$

where $c.c.$ denotes the complex conjugate, and where we identify the *phasor electric field* as $E_{xs} = E_{x0}e^{-jk_0 z}$. As indicated in (19), E_{x0} is the *complex* amplitude (which includes the phase, ϕ_1).

EXAMPLE 12.1

Let us express $\mathcal{E}_y(z,t) = 100\cos(10^8 t - 0.5z + 30°)$ V/m as a phasor.

Solution. We first go to exponential notation,

$$\mathcal{E}_y(z,t) = \text{Re}\left[100e^{j(10^8 t - 0.5z + 30°)}\right]$$

and then drop Re and suppress $e^{j10^8 t}$, obtaining the phasor

$$E_{ys}(z) = 100e^{-j0.5z + j30°}$$

Note that a mixed nomenclature is used for the angle in this case; that is, $0.5z$ is in radians, while 30° is in degrees. Given a scalar component or a vector expressed as a phasor, we may easily recover the time-domain expression.

EXAMPLE 12.2

Given the complex amplitude of the electric field of a uniform plane wave, $\mathbf{E}_0 = 100\mathbf{a}_x + 20\angle 30^\circ \mathbf{a}_y$ V/m, construct the phasor and real instantaneous fields if the wave is known to propagate in the forward z direction in free space and has frequency of 10 MHz.

Solution. We begin by constructing the general phasor expression:

$$\mathbf{E}_s(z) = \left[100\mathbf{a}_x + 20e^{j30^\circ}\mathbf{a}_y\right] e^{-jk_0 z}$$

where $k_0 = \omega/c = 2\pi \times 10^7/3 \times 10^8 = 0.21$ rad/m. The real instantaneous form is then found through the rule expressed in Eq. (19):

$$\begin{aligned}\mathcal{E}(z,t) &= \text{Re}\left[100e^{-j0.21z}e^{j2\pi\times 10^7 t}\mathbf{a}_x + 20e^{j30^\circ}e^{-j0.21z}e^{j2\pi\times 10^7 t}\mathbf{a}_y\right]\\ &= \text{Re}\left[100e^{j(2\pi\times 10^7 t - 0.21z)}\mathbf{a}_x + 20e^{j(2\pi\times 10^7 t - 0.21z + 30^\circ)}\mathbf{a}_y\right]\\ &= 100\cos(2\pi\times 10^7 t - 0.21z)\mathbf{a}_x + 20\cos(2\pi\times 10^7 t - 0.21z + 30^\circ)\mathbf{a}_y\end{aligned}$$

It is evident that taking the partial derivative of any field quantity with respect to time is equivalent to multiplying the corresponding phasor by $j\omega$. As an example, we can express Eq. (8) (using sinusoidal fields) as

$$\frac{\partial \mathcal{H}_y}{\partial z} = -\epsilon_0 \frac{\partial \mathcal{E}_x}{\partial t} \tag{20}$$

where, in a manner consistent with (19):

$$\mathcal{E}_x(z,t) = \frac{1}{2}E_{xs}(z)\,e^{j\omega t} + c.c. \qquad \text{and} \qquad \mathcal{H}_y(z,t) = \frac{1}{2}H_{ys}(z)\,e^{j\omega t} + c.c. \tag{21}$$

On substituting the fields in (21) into (20), the latter equation simplifies to

$$\frac{dH_{ys}(z)}{dz} = -j\omega\epsilon_0 E_{xs}(z) \tag{22}$$

In obtaining this equation, we note first that the complex conjugate terms in (21) produce their own separate equation, redundant with (22); second, the $e^{j\omega t}$ factors, common to both sides, have divided out; third, the partial derivative with z becomes the total derivative, since the phasor, H_{ys}, depends only on z.

We next apply this result to Maxwell's equations, to obtain them in phasor form. Substituting the field as expressed in (21) into Eqs. (1) through (4) results in

$$\nabla \times \mathbf{H}_s = j\omega\epsilon_0 \mathbf{E}_s \tag{23}$$

$$\nabla \times \mathbf{E}_s = -j\omega\mu_0 \mathbf{H}_s \tag{24}$$

$$\nabla \cdot \mathbf{E}_s = 0 \tag{25}$$

$$\nabla \cdot \mathbf{H}_s = 0 \tag{26}$$

It should be noted that (25) and (26) are no longer independent relationships, for they can be obtained by taking the divergence of (23) and (24), respectively.

Equations (23) through (26) may be used to obtain the sinusoidal steady-state vector form of the wave equation in free space. We begin by taking the curl of both sides of (24):

$$\nabla \times \nabla \times \mathbf{E}_s = -j\omega\mu_0 \nabla \times \mathbf{H}_s = \nabla(\nabla \cdot \mathbf{E}_s) - \nabla^2 \mathbf{E}_s \tag{27}$$

where the last equality is an identity, which defines the *vector Laplacian* of $\mathbf{E}_s$:

$$\nabla^2 \mathbf{E}_s = \nabla(\nabla \cdot \mathbf{E}_s) - \nabla \times \nabla \times \mathbf{E}_s$$

From (25), we note that $\nabla \cdot \mathbf{E}_s = 0$. Using this, and substituting (23) in (27), we obtain

$$\nabla^2 \mathbf{E}_s = -k_0^2 \mathbf{E}_s \tag{28}$$

where again, $k_0 = \omega/c = \omega\sqrt{\mu_0\epsilon_0}$. Equation (28) is known as the vector Helmholtz equation in free space.[1] It is fairly formidable when expanded, even in rectangular coordinates, for three scalar phasor equations result (one for each vector component), and each equation has four terms. The x component of (28) becomes, still using the del-operator notation,

$$\nabla^2 E_{xs} = -k_0^2 E_{xs} \tag{29}$$

and the expansion of the operator leads to the second-order partial differential equation

$$\frac{\partial^2 E_{xs}}{\partial x^2} + \frac{\partial^2 E_{xs}}{\partial y^2} + \frac{\partial^2 E_{xs}}{\partial z^2} = -k_0^2 E_{xs}$$

Again, assuming a uniform plane wave in which E_{xs} does not vary with x or y, the two corresponding derivatives are zero, and we obtain

$$\frac{d^2 E_{xs}}{dz^2} = -k_0^2 E_{xs} \tag{30}$$

the solution of which we already know:

$$E_{xs}(z) = E_{x0} e^{-jk_0 z} + E'_{x0} e^{jk_0 z} \tag{31}$$

Let us now return to Maxwell's equations, (23) through (26), and determine the form of the **H** field. Given $\mathbf{E}_s$, $\mathbf{H}_s$ is most easily obtained from (24):

$$\nabla \times \mathbf{E}_s = -j\omega\mu_0 \mathbf{H}_s \tag{24}$$

which is greatly simplified for a single E_{xs} component varying only with z,

$$\frac{dE_{xs}}{dz} = -j\omega\mu_0 H_{ys}$$

[1] Hermann Ludwig Ferdinand von Helmholtz (1821–1894) was a professor at Berlin working in the fields of physiology, electrodynamics, and optics. Hertz was one of his students.

Using (31) for E_{xs}, we have

$$H_{ys} = -\frac{1}{j\omega\mu_0}\left[(-jk_0)E_{x0}e^{-jk_0z} + (jk_0)E'_{x0}e^{jk_0z}\right]$$

$$= E_{x0}\sqrt{\frac{\epsilon_0}{\mu_0}}e^{-jk_0z} - E'_{x0}\sqrt{\frac{\epsilon_0}{\mu_0}}e^{jk_0z} = H_{y0}e^{-jk_0z} + H'_{y0}e^{jk_0z} \tag{32}$$

In real instantaneous form, this becomes

$$H_y(z,t) = E_{x0}\sqrt{\frac{\epsilon_0}{\mu_0}}\cos(\omega t - k_0z) - E'_{x0}\sqrt{\frac{\epsilon_0}{\mu_0}}\cos(\omega t + k_0z) \tag{33}$$

where E_{x0} and E'_{x0} are assumed real.

In general, we find from (32) that the electric and magnetic field amplitudes of the forward-propagating wave in free space are related through

$$E_{x0} = \sqrt{\frac{\mu_0}{\epsilon_0}}H_{y0} = \eta_0 H_{y0} \tag{34a}$$

We also find the backward-propagating wave amplitudes are related through

$$E'_{x0} = -\sqrt{\frac{\mu_0}{\epsilon_0}}H'_{y0} = -\eta_0 H'_{y0} \tag{34b}$$

where the *intrinsic impedance* of free space is defined as

$$\eta_0 = \sqrt{\frac{\mu_0}{\epsilon_0}} = 377 \doteq 120\pi\ \Omega \tag{35}$$

The dimension of η_0 in ohms is immediately evident from its definition as the ratio of E (in units of V/m) to H (in units of A/m). It is in direct analogy to the characteristic impedance, Z_0, of a transmission line, where we defined the latter as the ratio of voltage to current in a traveling wave. We note that the difference between (34*a*) and (34*b*) is a minus sign. This is consistent with the transmission line analogy that led to Eqs. (25*a*) and (25*b*) in Chapter 11. Those equations accounted for the definitions of positive and negative current associated with forward and backward voltage waves. In a similar way, Eq. (34*a*) specifies that in a forward-z propagating uniform plane wave whose electric field vector lies in the positive x direction at a given point in time and space, the magnetic field vector lies in the *positive* y direction at the same space and time coordinates. In the case of a backward-z propagating wave having a positive x-directed electric field, the magnetic field vector lies in the *negative* y direction. The physical significance of this has to do with the definition of power flow in the wave, as specified through the Poynting vector, $\mathbf{S} = \mathbf{E} \times \mathbf{H}$ (in watts/m^2). The cross product of $\mathbf{E}$ with $\mathbf{H}$ must give the correct wave propagation direction, and so the need for the minus sign in (34*b*) is apparent. Issues relating to power transmission will be addressed in Section 12.3.

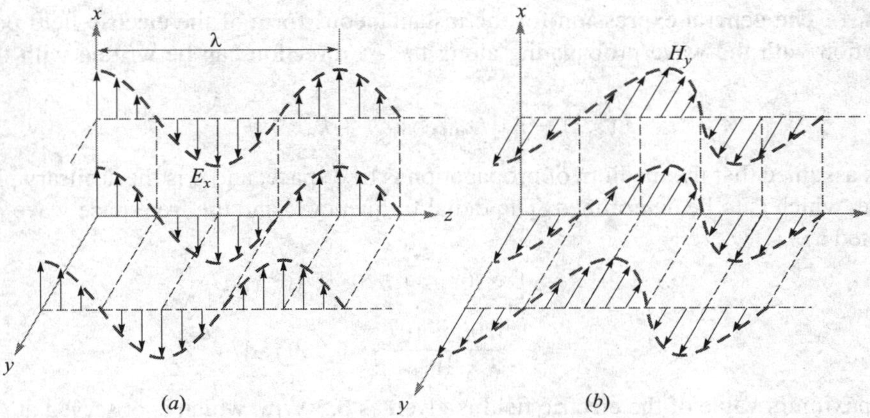

Fig. 12.1 (*a*) Arrows represent the instantaneous values of $E_{x0}\cos[\omega(t - z/c)]$ at $t = 0$ along the z axis, along an arbitrary line in the $x = 0$ plane parallel to the z axis, and along an arbitrary line in the $y = 0$ plane parallel to the z axis. (*b*) Corresponding values of H_y are indicated. Note that E_x and H_y are in phase at any point in time.

Some feeling for the way in which the fields vary in space may be obtained from Figures 12.1*a* and 12.1*b*. The electric field intensity in Figure 12.1*a* is shown at $t = 0$, and the instantaneous value of the field is depicted along three lines, the z axis and arbitrary lines parallel to the z axis in the $x = 0$ and $y = 0$ planes. Since the field is uniform in planes perpendicular to the z axis, the variation along all three of the lines is the same. One complete cycle of the variation occurs in a wavelength, λ. The values of H_y at the same time and positions are shown in Figure 12.1*b*.

A uniform plane wave cannot exist physically, for it extends to infinity in two dimensions at least and represents an infinite amount of energy. The distant field of a transmitting antenna, however, is essentially a uniform plane wave in some limited region; for example, a radar signal impinging on a distant target is closely a uniform plane wave.

Animations

Although we have considered only a wave varying sinusoidally in time and space, a suitable combination of solutions to the wave equation may be made to achieve a wave of any desired form, but which satisfies (14). The summation of an infinite number of harmonics through the use of a Fourier series can produce a periodic wave of square or triangular shape in both space and time. Nonperiodic waves may be obtained from our basic solution by Fourier integral methods. These topics are among those considered in the more advanced books on electromagnetic theory.

EXAMPLE 12.3

The maximum amplitude of the electric field of a 250 MHz uniform plane wave propagating in free space along the negative z-direction is 0.5 V/m. The electric field is polarized in the y-direction, and its maximum value is observed at $t = 0$ and $z = -12$ cm. Obtain expressions for the instantaneous forms of the electric and magnetic fields. Also specify the orientation of the magnetic field vector.

Solution. The general expression for the instantaneous form of the electric field polarized in the y-direction with the wave propagating along the $-z$ direction can be written with the help of Eq. (15) as

$$E(z,t) = \mathbf{a}_y |E_{y0}| \cos(\omega t + k_0 z + \phi) \tag{i}$$

where, it is assumed that the medium of propagation is free space, and ϕ is the arbitrary phase angle of the wave, which is to be determined. The radial frequency ω and the free space wave number k_0 are computed as

$$\omega = 2\pi \times 250 \times 10^6 \equiv 1.5708 \times 10^9 \text{ rad/s} \tag{ii}$$

$$k_0 \equiv \frac{\omega}{c} = \frac{1.5708 \times 10^9}{3 \times 10^8} = 5.236 \text{ rad/s} \tag{iii}$$

Now, the maximum value of the electric field is given as 0.5 V/m, which is observed at $t = 0$ and $z = -12$ cm. It can easily be seen from (i) that the maximum amplitude of the field would be obtained when the argument of the cosine function becomes $2n\pi$ with $n = 0, 1, 2, \ldots$. Now, without any loss of generality, the value of n can be taken as zero so that the condition for the maximum electric field can be written using (i) as

$$(\omega t_0 + k_0 z_0 + \phi) = 0 \tag{iv}$$

where t_0 and z_0 are the respective values of t and z corresponding to the maximum electric field. The values of t_0 and z_0 are given as 0 s, and -12 cm, respectively, which can be substituted in (iv) to obtain the value of phase constant, i.e.,

$$(0 - 5.236 \times 12 \times 10^{-2} + \phi) = 0 \quad \Rightarrow \phi = \frac{\pi}{5} \tag{v}$$

The values obtained using (ii), (iii) and (v) can finally be substituted in (i) to obtain the instantaneous expression for the electric field in the present situation, i.e.,

$$E(z,t) = \mathbf{a}_y 0.5 \cos\left(1.5708 \times 10^9 t + 5.236 z + \frac{\pi}{5}\right) \text{ V/m} \tag{vi}$$

Now, as the electric field is polarized along the y-direction with the wave propagating in the negative z-direction, the magnetic field should be oriented along the x-direction. This can be readily proven using the concept of Poynting vector such that ($\mathbf{E} \times \mathbf{H} \equiv E_y \mathbf{a}_y \times H_x \mathbf{a}_x$) in the present situation would provide the power flow along the $-\mathbf{a}_z$ direction as given in the problem. It is also to be noted here that as the wave is linearly polarized with the electric and magnetic field vectors always pointing towards the y- and the x-directions respectively, these two vectors would be in time phase. It basically means that the magnetic field would have the same cosine phase expression as given in (*vi*) for the electric field. The amplitude of the magnetic field can be computed with the help of (33) to (34), which in the present situation can be written as

$$\frac{|E_{y0}|}{|H_{x0}|} = 377\ \Omega \Rightarrow |H_{x0}| = \frac{0.5}{377} = 1.3 \text{ mA/m} \tag{vii}$$

The instantaneous expression for the magnetic field in the present situation can finally be written as

$$\mathbf{H}(z,t) = \mathbf{a}_x 1.3 \cos\left(1.5708 \times 10^9 t + 5.236\, z + \frac{\pi}{5}\right) \text{ mA/m} \tag{viii}$$

EXAMPLE 12.4

The time-domain expression for the magnetic field of a uniform plane wave traveling in free space is given by the following expression:

$$\mathbf{H}(z,t) = \mathbf{a}_y 2.5\cos(1.257\times 10^9 t - k_0 z)\ \text{mA/m} \tag{i}$$

Find

(*a*) the direction of wave propagation;
(*b*) the operating frequency;
(*c*) the phase constant;
(*d*) the time-domain expression for the electric field $\mathbf{E}(z,t)$ starting from the Maxwell's equations;
(*e*) the polarization of the electric field vector; and
(*f*) the phasor forms of both the electric field and the magnetic field.

Solution. For finding various parameters of the wave, the expression given in (i) can be compared with the standard expression of the magnetic field for a wave traveling in the positive z direction as given below:

$$\mathbf{H}(z,t) \equiv \mathbf{a}_y H_y(z,t) = \mathbf{a}_y |H_{yo}| \cos(wt - k_0 z + \phi) \tag{ii}$$

It can be observed by comparing (i) with the standard expression (ii) that the phase term ϕ in the present situation is zero. It basically means that the maximum amplitude of the field is attained at $t = 0$ and $z = 0$. The other parameters are obtained as follows:

(*a*) The wave is propagating along the $+z$ direction.
(*b*) The operating frequency may be computed as

$$\omega \equiv 2\times\pi\times f = 1.257\times 10^9 \Rightarrow f = \frac{1.257\times 10^9}{2\pi} = 200\ \text{MHz} \tag{iii}$$

(*c*) The phase constant is computed as

$$k_0 \equiv \frac{\omega}{c} = \frac{1.257\times 10^9}{3\times 10^8} = 4.19\,\text{rad/m} \tag{iv}$$

(*d*) The time-domain expression for the electric field can be determined from the first principle using the following Maxwell's equation:

$$\nabla\times\mathbf{H} \equiv \epsilon_0 \frac{\partial \mathbf{E}}{\partial t}$$

$$\Rightarrow \begin{bmatrix} \mathbf{a}_x & \mathbf{a}_y & \mathbf{a}_z \\ \dfrac{\partial}{\partial x} & \dfrac{\partial}{\partial y} & \dfrac{\partial}{\partial z} \\ H_x & H_y & H_z \end{bmatrix} \equiv \epsilon_0 \frac{\partial \mathbf{E}}{\partial t} \tag{v}$$

Now, it is given that the magnetic field is having only the y-component, hence $H_x = 0$; $H_z = 0$. As (i) represents a uniform plane wave traveling along the z-direction, hence there would not be any spatial variation of fields along the x and y directions. It basically means the factors $\frac{\partial}{\partial_x} = \frac{\partial}{\partial_y} = 0$, which would reduce (v) to the following form:

$$\begin{bmatrix} \mathbf{a}_x & \mathbf{a}_y & \mathbf{a}_z \\ 0 & 0 & \frac{\partial}{\partial z} \\ 0 & H_y & 0 \end{bmatrix} = \epsilon_0 \frac{\partial \mathbf{E}}{\partial t}$$

The above equation can be expanded to obtain

$$\mathbf{a}_x \left[-\frac{\partial \mathbf{H}_y}{\partial z} \right] \equiv \epsilon_0 \frac{\partial \mathbf{E}}{\partial t} \tag{vi}$$

Hence, it can be seen from (vi) that the electric field is having only the x-component as non-zero. For finding the value of the electric field, the factor on LHS of (vi) can first be calculated using (i) as follows

$$\begin{aligned} -\frac{\partial \mathbf{H}_y}{\partial z} &\equiv -\frac{\partial}{\partial z}[2.5 \cos(1.257 \times 10^9 t - k_0 z)] \\ &\equiv -2.5 \times 10^{-3}\, k_0 \sin\left(1.257 \times 10^9 t - k_0 z\right) \end{aligned} \tag{vii}$$

From (vi) and (vii), the expression for the electric field can be written as

$$\mathbf{E} \equiv \mathbf{a}_x \frac{-1}{\epsilon_0} \int_t [2.5 \times 10^{-3} k_0 \sin(1.257 \times 10^9 t - k_0 z)]dt$$

The above equation can be integrated to obtain

$$\mathbf{E} \equiv \mathbf{a}_x \frac{2.5 \times 10^{-3} k_0}{\epsilon_0 \times 1.257 \times 10^9} \cos(1.257 \times 10^9 t - k_0 z)$$

where, the constant of integration is obtained as zero using the condition that the maximum amplitude of the field is attained at $t = 0$ and $z = 0$. Meanwhile, the above equation can be further simplified by substituting the value of k_0 from (iv), which yields the final expression for the electric field as

$$\mathbf{E}(z,t) \equiv \mathbf{a}_x E_x(z,t) = \mathbf{a}_x 0.9425 \cos(1.257 \times 10^9 t - 4.19z) \text{ V/m} \tag{viii}$$

(e) It can be seen from (viii) that the electric field is polarized along the $+x$ direction. This can also be verified using the concept of Poynting vector, where it can be shown that the vector $(\mathbf{E} \times \mathbf{H} \equiv E_x \mathbf{a_x} \times H_y \mathbf{a_y})$ leads to the wave propagation and the power flow in the $+z$ direction as given in the problem.

(f) The phasor form of the electric field can be obtained by rewriting the field component corresponding to (viii) using the exponential notation as follows:

$$\begin{aligned} E_x(z,t) &\equiv \text{Re}\left[0.9425 e^{j(1.257\times 10^9 t - 4.19z)}\right] \text{ V/m} \\ &\equiv \text{Re}\left[(0.9425 e^{-j4.19z}) e^{j1.257\times 10^9 t}\right] \text{ V/m} \end{aligned} \tag{ix}$$

The form of the electric field given by (ix) can now be compared with the standard phasor notation given by (19), which yields the following phasor form of the electric field:

$$\mathbf{E}_s(z) \equiv \mathbf{a}_x\, 0.9425\, e^{-j4.19z} \quad \text{V/m} \tag{x}$$

Similarly, the phasor form of the magnetic field corresponding to (2a) can be written as

$$\mathbf{H}_s(z) \equiv \mathbf{a}_y 2.5\, e^{-j4.19z} \quad \text{mA/m} \tag{xi}$$

D12.1 The electric field amplitude of a uniform plane wave propagating in the $\mathbf{a}_z$ direction is 250 V/m. If $\mathbf{E} = E_x\mathbf{a}_x$ and $\omega = 1.00$ M rad/s, find: (*a*) the frequency; (*b*) the wavelength; (*c*) the period; (*d*) the amplitude of **H**.

Ans. 159 kHz; 1.88 km; 6.28 μs; 0.663 A/m

D12.2 Let $\mathbf{H}_s = (2\angle{-40^\circ}\mathbf{a}_x - 3\angle 20^\circ\mathbf{a}_y)e^{-j0.07z}$ A/m for a uniform plane wave traveling in free space. Find: (*a*) ω; (*b*) H_x at $P(1, 2, 3)$ at $t = 31$ ns; (*c*) $|\mathbf{H}|$ at $t = 0$ at the origin.

Ans. 21.0 M rad/s; 1.934 A/m; 3.22 A/m

12.2 WAVE PROPAGATION IN DIELECTRICS

We now extend our analytical treatment of the uniform plane wave to propagation in a dielectric of permittivity ϵ and permeability μ. The medium is assumed to be homogeneous (having constant μ and ϵ with position) and isotropic (in which μ and ϵ are invariant with field orientation). The Helmholtz equation is

$$\nabla^2\mathbf{E}_s = -k^2\mathbf{E}_s \tag{36}$$

where the wavenumber is a function of the material properties, as described by μ and ϵ:

$$k = \omega\sqrt{\mu\epsilon} = k_0\sqrt{\mu_r\epsilon_r} \tag{37}$$

For E_{xs} we have

$$\frac{d^2E_{xs}}{dz^2} = -k^2E_{xs} \tag{38}$$

An important feature of wave propagation in a dielectric is that k can be complex-valued, and as such it is referred to as the complex *propagation constant*. A general solution of (38), in fact, allows the possibility of a complex k, and it is customary to write it in terms of its real and imaginary parts in the following way:

$$jk = \alpha + j\beta \tag{39}$$

A solution to (38) will be:

$$E_{xs} = E_{x0}e^{-jkz} = E_{x0}e^{-\alpha z}e^{-j\beta z} \tag{40}$$

Multiplying (40) by $e^{j\omega t}$ and taking the real part yields a form of the field that can be more easily visualized:

$$E_x = E_{x0}e^{-\alpha z}\cos(\omega t - \beta z) \tag{41}$$

We recognize this as a uniform plane wave that propagates in the forward z direction with phase constant β, but which (for positive α) loses amplitude with increasing z according to the factor $e^{-\alpha z}$. Thus the general effect of a complex-valued k is to yield a traveling wave that changes its amplitude with distance. If α is positive, it is called the *attenuation coefficient*. If α is negative, the wave grows in amplitude with distance, and α is called the *gain coefficient*. The latter effect would occur, for example, in laser amplifiers. In the present and future discussions in this book, we will consider only passive media, in which one or more loss mechanisms are present, thus producing a positive α.

The attenuation coefficient is measured in nepers per meter (Np/m) so that the exponent of e can be measured in the dimensionless units of nepers. Thus, if $\alpha = 0.01$ Np/m, the crest amplitude of the wave at $z = 50$ m will be $e^{-0.5}/e^{-0} = 0.607$ of its value at $z = 0$. In traveling a distance $1/\alpha$ in the $+z$ direction, the amplitude of the wave is reduced by the familiar factor of e^{-1}, or 0.368.

The ways in which physical processes in a material can affect the wave electric field are described through a *complex permittivity* of the form

$$\epsilon = \epsilon' - j\epsilon'' = \epsilon_0(\epsilon_r' - j\epsilon_r'') \tag{42}$$

Two important mechanisms that give rise to a complex permittivity (and thus result in wave losses) are bound electron or ion oscillations and dipole relaxation, both of which are discussed in Appendix D. An additional mechanism is the conduction of free electrons or holes, which we will explore at length in this chapter.

Losses arising from the response of the medium to the magnetic field can occur as well, and these are modeled through a *complex permeability*, $\mu = \mu' - j\mu'' = \mu_0(\mu_r' - j\mu_r'')$. Examples of such media include *ferrimagnetic* materials, or *ferrites*. The magnetic response is usually very weak compared to the dielectric response in most materials of interest for wave propagation; in such materials $\mu \approx \mu_0$. Consequently, our discussion of loss mechanisms will be confined to those described through the complex permittivity, and we will assume that μ is entirely real in our treatment.

We can substitute (42) into (37), which results in

$$k = \omega\sqrt{\mu(\epsilon' - j\epsilon'')} = \omega\sqrt{\mu\epsilon'}\sqrt{1 - j\frac{\epsilon''}{\epsilon'}} \tag{43}$$

Note the presence of the second radical factor in (43), which becomes unity (and real) as ϵ'' vanishes. With nonzero ϵ'', k is complex, and so losses occur which are quantified through the attenuation coefficient, α, in (39). The phase constant, β (and consequently the wavelength and phase velocity), will also be affected by ϵ''. α and β are found by taking the real and imaginary parts of jk from (43). We obtain:

$$\alpha = \mathrm{Re}\{jk\} = \omega\sqrt{\frac{\mu\epsilon'}{2}}\left(\sqrt{1+\left(\frac{\epsilon''}{\epsilon'}\right)^2}-1\right)^{1/2} \tag{44}$$

$$\beta = \mathrm{Im}\{jk\} = \omega\sqrt{\frac{\mu\epsilon'}{2}}\left(\sqrt{1+\left(\frac{\epsilon''}{\epsilon'}\right)^2}+1\right)^{1/2} \tag{45}$$

We see that a nonzero α (and hence loss) results if the imaginary part of the permittivity, ϵ'', is present. We also observe in (44) and (45) the presence of the ratio ϵ''/ϵ', which is called the *loss tangent*. The meaning of the term will be demonstrated when we investigate the specific case of conductive media. The practical importance of the ratio lies in its magnitude compared to unity, which enables simplifications to be made in (44) and (45).

Whether or not losses occur, we see from (41) that the wave phase velocity is given by

$$\nu_p = \frac{\omega}{\beta} \tag{46}$$

The wavelength is the distance required to effect a phase change of 2π radians

$$\beta\lambda = 2\pi$$

which leads to the fundamental definition of wavelength,

$$\lambda = \frac{2\pi}{\beta} \tag{47}$$

Since we have a uniform plane wave, the magnetic field is found through

$$H_{ys} = \frac{E_{x0}}{\eta}e^{-\alpha z}e^{-j\beta z}$$

where the intrinsic impedance is now a complex quantity,

$$\eta = \sqrt{\frac{\mu}{\epsilon' - j\epsilon''}} = \sqrt{\frac{\mu}{\epsilon'}}\frac{1}{\sqrt{1-j(\epsilon''/\epsilon')}} \tag{48}$$

The electric and magnetic fields are no longer in phase.

A special case is that of a lossless medium, or *perfect dielectric,* in which $\epsilon'' = 0$, and so $\epsilon = \epsilon'$. From (44), this leads to $\alpha = 0$, and from (45),

$$\beta = \omega\sqrt{\mu\epsilon'} \qquad \text{(lossless medium)} \tag{49}$$

With $\alpha = 0$, the real field assumes the form

$$E_x = E_{x0}\cos(\omega t - \beta z) \tag{50}$$

We may interpret this as a wave traveling in the $+z$ direction at a phase velocity ν_p, where

$$\nu_p = \frac{\omega}{\beta} = \frac{1}{\sqrt{\mu\epsilon'}} = \frac{c}{\sqrt{\mu_r\epsilon'_r}}$$

The wavelength is

$$\lambda = \frac{2\pi}{\beta} = \frac{2\pi}{\omega\sqrt{\mu\epsilon'}} = \frac{1}{f\sqrt{\mu\epsilon'}} = \frac{c}{f\sqrt{\mu_r\epsilon'_r}} = \frac{\lambda_0}{\sqrt{\mu_r\epsilon'_r}} \quad \text{(lossless medium)} \tag{51}$$

where λ_0 is the free space wavelength. Note that $\mu_r\epsilon'_r > 1$, and therefore the wavelength is shorter and the velocity is lower in all real media than they are in free space.

Associated with E_x is the magnetic field intensity

$$H_y = \frac{E_{x0}}{\eta}\cos(\omega t - \beta z)$$

where the intrinsic impedance is

$$\eta = \sqrt{\frac{\mu}{\epsilon}} \tag{52}$$

The two fields are once again perpendicular to each other, perpendicular to the direction of propagation, and in phase with each other everywhere. Note that when **E** is crossed into **H**, the resultant vector is in the direction of propagation. We shall see the reason for this when we discuss the Poynting vector.

EXAMPLE 12.5

Let us apply these results to a 1 MHz plane wave propagating in fresh water. At this frequency, losses in water are negligible, which means that we can assume that $\epsilon'' \doteq 0$. In water, $\mu_r = 1$ and at 1 MHz, $\epsilon'_r = 81$.

Solution. We begin by calculating the phase constant. Using (45) with $\epsilon'' = 0$, we have

$$\beta = \omega\sqrt{\mu\epsilon'} = \omega\sqrt{\mu_0\epsilon_0}\sqrt{\epsilon'_r} = \frac{\omega\sqrt{\epsilon'_r}}{c} = \frac{2\pi \times 10^6\sqrt{81}}{3.0 \times 10^8} = 0.19\,\text{rad/m}$$

Using this result, we can determine the wavelength and phase velocity:

$$\lambda = \frac{2\pi}{\beta} = \frac{2\pi}{.19} = 33 \text{ m}$$

$$\nu_p = \frac{\omega}{\beta} = \frac{2\pi \times 10^6}{.19} = 3.3 \times 10^7 \text{ m/s}$$

The wavelength in air would have been 300 m. Continuing our calculations, we find the intrinsic impedance using (48) with $\epsilon'' = 0$:

$$\eta = \sqrt{\frac{\mu}{\epsilon'}} = \frac{\eta_0}{\sqrt{\epsilon'_r}} = \frac{377}{9} = 42\ \Omega$$

If we let the electric field intensity have a maximum amplitude of 0.1 V/m, then

$$E_x = 0.1\cos(2\pi 10^6 t - .19z)\ \text{V/m}$$

$$H_y = \frac{E_x}{\eta} = (2.4\times 10^{-3})\cos(2\pi 10^6 t - .19z)\ \text{A/m}$$

EXAMPLE 12.6

The time-domain expression for the electric field of a plane wave traveling in a non-magnetic medium is given by the following expression:

$$\mathbf{E}(z,t) \equiv 20\cos(5.65\times 10^9 t - 30z)\mathbf{a}_x + 15\cos\left(5.65\times 10^9 t - 30z - \frac{\pi}{4}\right)\mathbf{a}_y\ \text{V/m} \quad \text{(i)}$$

Find

(*a*) the phasor form of the electric field;
(*b*) the operating frequency;
(*c*) the dielectric constant;
(*d*) the intrinsic impedance; and
(*e*) the phasor and time domain expressions of the magnetic field.

Solution.

(*a*) The phasor form of the electric field can be obtained by rewriting (i) using the following exponential notation:

$$\mathbf{E} \equiv \text{Re}\left[\mathbf{a}_x(20e^{-j30z})e^{j(5.65\times 10^9 t)} + \mathbf{a}_y\left(15e^{-j\left(30z+\frac{\pi}{4}\right)}\right)e^{j5.65\times 10^9 t}\right]\ \text{V/m} \quad \text{(ii)}$$

The above expression can be compared with the standard phasor notation given by (19), which yields the following phasor form of the electric field:

$$\mathbf{E}_s(z) \equiv \left[20\mathbf{a}_x + 15e^{-j\frac{\pi}{4}}\mathbf{a}_y\right]e^{-j30z}\ \text{V/m} \quad \text{(iii)}$$

(*b*) The operating frequency can be computed as

$$\omega \equiv 2\times\pi\times f = 5.65\times 10^9 \Rightarrow f = \frac{5.65\times 10^9}{2\pi} = 900\ \text{MHz} \quad \text{(iv)}$$

(*c*) For determining the dielectric constant, first the propagation constant or the wave number should be computed using either (i) or (iii). In the present situation, it can be observed that the media is lossless, and so the attenuation constant α is zero. The wave number, therefore, equals the phase constant, which can be expressed using (49) as

$$k \equiv \beta = \omega\sqrt{\mu_0\epsilon_0\epsilon'_r} = 30\ \text{rad/m}$$

$$\epsilon'_r = \left(\frac{30}{5.65 \times 10^9}\right)^2 \left[\frac{1}{4\pi \times 10^{-7} \times 8.856 \times 10^{-12}}\right] = 2.53 \qquad \text{(v)}$$

(*d*) The intrinsic impedance for the lossless medium may be computed using (52)

$$\eta = \sqrt{\frac{\mu_0}{\epsilon}} \equiv = \sqrt{\frac{\mu_0}{\epsilon_0 \epsilon'_r}} \equiv \frac{120\pi}{\sqrt{\epsilon'_r}} = 237\ \Omega \qquad \text{(vi)}$$

(*e*) The phasor expression for the magnetic field can be determined using the following Maxwell's equation assuming the medium to be non-magnetic as given in the problem:

$$\nabla \times \mathbf{E}_s \equiv -j\omega\mu_0 \mathbf{H}_s$$

$$\Rightarrow \begin{bmatrix} \mathbf{a}_x & \mathbf{a}_y & \mathbf{a}_z \\ \dfrac{\partial}{\partial x} & \dfrac{\partial}{\partial y} & \dfrac{\partial}{\partial z} \\ E_{xs} & E_{ys} & E_{zs} \end{bmatrix} \equiv -j\omega\mu_0 [\mathbf{H}_{xs}\mathbf{a}_x + \mathbf{H}_{ys}\mathbf{a}_y + \mathbf{H}_{zs}\mathbf{a}_z]$$

Now, it can be observed from (iii) that the electric field has only x and y components, which makes $\mathbf{E}_z = 0$. It is also to be noted from this equation that there is no spatial variation of the field along the x and y directions, which means that the factors $\frac{\partial}{\partial x} = \frac{\partial}{\partial y} = 0$ in (vii). The above equation would then reduce to the following form:

$$\begin{bmatrix} \mathbf{a}_x & \mathbf{a}_y & \mathbf{a}_z \\ 0 & 0 & \frac{\partial}{\partial z} \\ E_{xs} & E_{ys} & 0 \end{bmatrix} \equiv -j\omega\mu_0 \mathbf{H}_s \qquad \text{(vii)}$$

The above equation can be expanded to obtain

$$\left[-\frac{(\partial E_{ys})}{\partial z}\right] \mathbf{a}_x + \left[\frac{\partial E_{xs}}{\partial z}\right] \mathbf{a}_y \equiv -j\omega\mu_0 \mathbf{H}_s \qquad \text{(viii)}$$

It can be seen from (viii) that the magnetic field has x and y components as non-zero, which can be obtained as follows:

$$\mathbf{H}_s \equiv -\frac{1}{j\omega\mu_0}\left\{-\frac{\partial}{\partial z}\left[15e^{-j\frac{\pi}{4}}e^{-j30z}\right]\mathbf{a}_x + \frac{\partial}{\partial z}[20e^{-j30z}]\mathbf{a}_y\right\}$$

$$\equiv \left[-\frac{j30}{j\omega\mu_0} \times 15e^{-j\frac{\pi}{4}}\mathbf{a}_x + \frac{j30}{j\omega\mu_0} \times 20\mathbf{a}_y\right] e^{-j30z}$$

Hence, the final expression for the magnetic field in phasor form is given as follows:

$$\mathbf{H}_s \equiv \left[-0.0634e^{-j\frac{\pi}{4}}\mathbf{a}_x + 0.0845\mathbf{a}_y\right] e^{-j30z}\ \text{A/m} \qquad \text{(ix)}$$

The magnetic field in time domain can be obtained from the above phasor expression as

$$\mathbf{H}(z,t) = -0.0634\cos\left(5.65 \times 10^9 t - 30z - \frac{\pi}{4}\right)\mathbf{a}_x$$

$$+0.0845\cos(5.65 \times 10^9 t - 30z)\mathbf{a}_y\ \text{A/m} \qquad \text{(x)}$$

D12.3 A 9.375 GHz uniform plane wave is propagating in polyethylene (see Appendix C). If the amplitude of the electric field intensity is 500 V/m and the material is assumed to be lossless, find: (*a*) the phase constant; (*b*) the wavelength in the polyethylene; (*c*) the velocity of propagation; (*d*) the intrinsic impedance; (*e*) the amplitude of the magnetic field intensity.

Ans. 295 rad/m; 2.13 cm; 1.99×10^8 m/s; 251 Ω; 1.99 A/m

EXAMPLE 12.7

We again consider plane wave propagation in water, but at the much higher microwave frequency of 2.5 GHz. At frequencies in this range and higher, dipole relaxation and resonance phenomena in the water molecules become important.[2] Real and imaginary parts of the permittivity are present, and both vary with frequency. At frequencies below that of visible light, the two mechanisms together produce a value of ϵ'' that increases with increasing frequency, reaching a maximum in the vicinity of 10^{13} Hz. ϵ' decreases with increasing frequency, reaching a minimum also in the vicinity of 10^{13} Hz. Reference 3 provides specific details. At 2.5 GHz, dipole relaxation effects dominate. The permittivity values are $\epsilon'_r = 78$ and $\epsilon''_r = 7$. From (44), we have

$$\alpha = \frac{(2\pi \times 2.5 \times 10^9)\sqrt{78}}{(3.0 \times 10^8)\sqrt{2}} \left(\sqrt{1 + \left(\frac{7}{78}\right)^2} - 1 \right)^{1/2} = 21 \text{ Np/m}$$

This first calculation demonstrates the operating principle of the *microwave oven*. Almost all foods contain water, and so they can be cooked when incident microwave radiation is absorbed and converted into heat. Note that the field will attenuate to a value of e^{-1} times its initial value at a distance of $1/\alpha = 4.8$ cm. This distance is called the *penetration depth* of the material, and of course it is frequency-dependent. The 4.8 cm depth is reasonable for cooking food, since it would lead to a temperature rise that is fairly uniform throughout the depth of the material. At much higher frequencies, where ϵ'' is larger, the penetration depth decreases, and too much power is absorbed at the surface; at lower frequencies, the penetration depth increases, and not enough overall absorption occurs. Commercial microwave ovens operate at frequencies in the vicinity of 2.5 GHz.

Using (45), in a calculation very similar to that for α, we find $\beta = 464$ rad/m. The wavelength is $\lambda = 2\pi/\beta = 1.4$ cm, whereas in free space this would have been $\lambda_0 = c/f = 12$ cm.

Using (48), the intrinsic impedance is found to be

$$\eta = \frac{377}{\sqrt{78}} \frac{1}{\sqrt{1 - j(7/78)}} = 43 + j1.9 = 43\angle 2.6^\circ \ \Omega$$

and E_x leads H_y in time by 2.6° at every point.

---

[2] These mechanisms and how they produce a complex permittivity are described in Appendix D. Additionally, the reader is referred to pp. 73–84 in Reference 1 and pp. 678–82 in Reference 2 for general treatments of relaxation and resonance effects on wave propagation. Discussions and data that are specific to water are presented in Reference 3, pp. 314–16.

We next consider the case of conductive materials, in which currents are formed by the motion of free electrons or holes under the influence of an electric field. The governing relation is $\mathbf{J} = \sigma\mathbf{E}$, where σ is the material conductivity. With finite conductivity, the wave loses power through resistive heating of the material. We look for an interpretation of the complex permittivity as it relates to the conductivity. Consider the Maxwell curl equation (23) which, using (42), becomes:

$$\nabla \times \mathbf{H}_s = j\omega(\epsilon' - j\epsilon'')\mathbf{E}_s = \omega\epsilon''\mathbf{E}_s + j\omega\epsilon'\mathbf{E}_s \tag{53}$$

This equation can be expressed in a more familiar way, in which conduction current is included:

$$\nabla \times \mathbf{H}_s = \mathbf{J}_s + j\omega\epsilon\mathbf{E}_s \tag{54}$$

We next use $\mathbf{J}_s = \sigma\mathbf{E}_s$, and interpret ϵ in (54) as ϵ'. The latter equation becomes:

$$\nabla \times \mathbf{H}_s = (\sigma + j\omega\epsilon')\mathbf{E}_s = \mathbf{J}_{\sigma s} + \mathbf{J}_{ds} \tag{55}$$

which we have expressed in terms of conduction current density, $\mathbf{J}_{\sigma s} = \sigma\mathbf{E}_s$, and displacement current density, $\mathbf{J}_{ds} = j\omega\epsilon'\mathbf{E}_s$. Comparing Eqs. (53) and (55), we find that in a conductive medium:

$$\epsilon'' = \frac{\sigma}{\omega} \tag{56}$$

Let us now turn our attention to the case of a dielectric material in which the loss is very small. The criterion by which we should judge whether or not the loss is small is the magnitude of the loss tangent, ϵ''/ϵ'. This parameter will have a direct influence on the attenuation coefficient, α, as seen from Eq. (44). In the case of conducting media, to which (56) applies, the loss tangent becomes $\sigma/\omega\epsilon'$. By inspecting (55), we see that the ratio of conduction current density to displacement current density magnitudes is

$$\frac{J_{\sigma s}}{J_{ds}} = \frac{\epsilon''}{j\epsilon'} = \frac{\sigma}{j\omega\epsilon'} \tag{57}$$

That is, these two vectors point in the same direction in space, but they are 90° out of phase in time. Displacement current density leads conduction current density by 90°, just as the current through a capacitor leads the current through a resistor in parallel with it by 90° in an ordinary electric circuit. This phase relationship is shown in Figure 12.2. The angle θ (not to be confused with the polar angle in spherical coordinates) may therefore be identified as the angle by which the displacement current density leads the total current density, and

$$\tan\theta = \frac{\epsilon''}{\epsilon'} = \frac{\sigma}{\omega\epsilon'} \tag{58}$$

The reasoning behind the term *loss tangent* is thus evident. Problem 12.16 at the end of the chapter indicates that the Q of a capacitor (its quality factor, not its charge) which incorporates a lossy dielectric is the reciprocal of the loss tangent.

If the loss tangent is small, then we may obtain useful approximations for the attenuation and phase constants, and the intrinsic impedance. The criterion for a small loss tangent is $\epsilon''/\epsilon' \ll 1$, which we say identifies the medium as a *good dielectric*. Considering a conductive material, for which $\epsilon'' = \sigma/\omega$, (43) becomes

$$jk = j\omega\sqrt{\mu\epsilon'}\sqrt{1 - j\frac{\sigma}{\omega\epsilon'}} \tag{59}$$

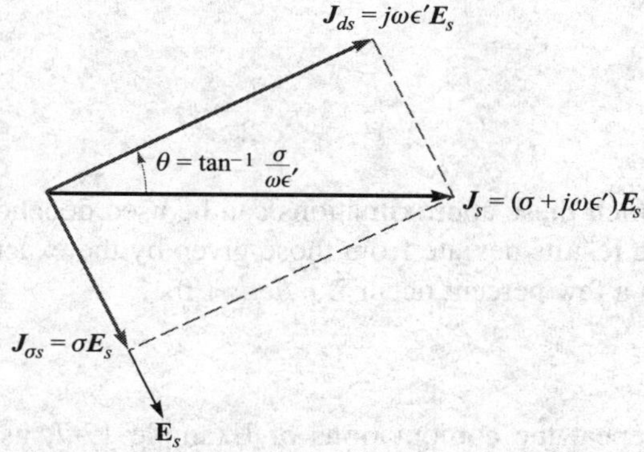

Fig. 12.2 The time-phase relationship between $\mathbf{J}_{ds}$, $\mathbf{J}_{\sigma s}$, $\mathbf{J}_s$, and $\mathbf{E}_s$. The tangent of θ is equal to $\sigma/\omega\epsilon'$, and $90° - \theta$ is the common power-factor angle, or the angle by which $\mathbf{J}_s$ leads $\mathbf{E}_s$.

We may expand the second radical using the binomial theorem

$$(1+x)^n = 1 + nx + \frac{n(n-1)}{2!}x^2 + \frac{n(n-1)(n-2)}{3!}x^3 + \cdots$$

where $|x| \ll 1$. We identify x as $-j\sigma/\omega\epsilon'$ and n as $1/2$, and thus

$$jk = j\omega\sqrt{\mu\epsilon'}\left[1 - j\frac{\sigma}{2\omega\epsilon'} + \frac{1}{8}\left(\frac{\sigma}{\omega\epsilon'}\right)^2 + \cdots\right] = \alpha + j\beta$$

Now, for a good dielectric,

$$\alpha = \text{Re}(jk) \doteq j\omega\sqrt{\mu\epsilon'}\left(-j\frac{\sigma}{2\omega\epsilon'}\right) = \frac{\sigma}{2}\sqrt{\frac{\mu}{\epsilon'}} \tag{60a}$$

and

$$\beta = \text{Im}(jk) \doteq \omega\sqrt{\mu\epsilon'}\left[1 + \frac{1}{8}\left(\frac{\sigma}{\omega\epsilon'}\right)^2\right] \tag{60b}$$

Eqs. (60a) and (60b) can be compared directly with the transmission line α and β under low-loss conditions, as expressed in Eqs. (54a) and (55b) in Chapter 11. In this comparison, we associate σ with G, μ with L, and ϵ with C. Note that in plane wave propagation in media with no boundaries, there can be no quantity that is analogous to the transmission line conductor resistance parameter, R. In many cases, the second term in (60b) is small enough, so that

$$\beta \doteq \omega\sqrt{\mu\epsilon'} \tag{61}$$

Applying the binomial expansion to (48), we obtain, for a good dielectric

$$\eta \doteq \sqrt{\frac{\mu}{\epsilon'}}\left[1 - \frac{3}{8}\left(\frac{\sigma}{\omega\epsilon'}\right)^2 + j\frac{\sigma}{2\omega\epsilon'}\right] \tag{62a}$$

or

$$\eta \doteq \sqrt{\frac{\mu}{\epsilon'}}\left(1 + j\frac{\sigma}{2\omega\epsilon'}\right) \tag{62b}$$

The conditions under which these approximations can be used depend on the desired accuracy, measured by how much the results deviate from those given by the exact formulas, (44) and (45). Deviations of no more than a few percent occur if $\sigma/\omega\epsilon' < 0.1$.

EXAMPLE 12.8

As a comparison, we repeat the computations of Example 12.7, using the approximation formulas (60*a*), (61), and (62*b*).

Solution. First, the loss tangent in this case is $\epsilon''/\epsilon' = 7/78 = 0.09$. Using (60), with $\epsilon'' = \sigma/\omega$, we have

$$\alpha \doteq \frac{\omega\epsilon''}{2}\sqrt{\frac{\mu}{\epsilon'}} = \frac{1}{2}(7 \times 8.85 \times 10^{12})(2\pi \times 2.5 \times 10^9)\frac{377}{\sqrt{78}} = 21 \text{ cm}^{-1}$$

We then have, using (61*b*),

$$\beta \doteq (2\pi \times 2.5 \times 10^9)\sqrt{78}/(3 \times 10^8) = 464 \text{ rad/m}$$

Finally, with (62*b*),

$$\eta \doteq \frac{377}{\sqrt{78}}\left(1 + j\frac{7}{2 \times 78}\right) = 43 + j1.9$$

These results are identical (within the accuracy limitations as determined by the given numbers) to those of Example 12.7. Small deviations will be found, as the reader can verify by repeating the calculations of both examples and expressing the results to four or five significant figures. As we know, this latter practice would not be meaningful because the given parameters were not specified with such accuracy. Such is often the case, since measured values are not always known with high precision. Depending on how precise these values are, one can sometimes use a more relaxed judgment on when the approximation formulas can be used by allowing loss tangent values that can be larger than 0.1 (but still less than 1).

EXAMPLE 12.9

Consider a 2.5 GHz, y-polarized uniform plane wave propagating along the positive x direction in a non-magnetic medium having a dielectric constant of 2.56, and the loss tangent as 0.002. The maximum amplitude of the electric field is 40 V/m, which is recorded at $x = 0$ and $t = 0$.

(a) Specify whether the propagating medium can be considered to be good dielectric.
(b) Find the ratio of the magnitudes of the displacement current density to the conduction current density.
(c) Find the attenuation constant, the phase constant, and the intrinsic impedance.
(d) Find the wavelength and the phase velocity.
(e) Find the phasor and time domain expressions for the electric field.
(f) Find the distance over which the amplitude of the propagating wave will reduce to 90% of its initial value.

Solution.

(a) A medium is considered a good dielectric when its loss tangent is $\epsilon''/\epsilon' \ll 1$. In the present situation, the loss tangent is given to be 0.002, which satisfies this criterion. Hence, the propagating medium in the present situation can be considered a good dielectric.

(b) The ratio of the magnitudes of the displacement current density to the conduction current density can be written using (57) and (56) as

$$\frac{|J_{ds}|}{|J_{\sigma s}|} = \frac{\omega\epsilon'}{\sigma} \equiv \frac{\epsilon'}{\epsilon''} = \frac{1}{0.002} = 500 \qquad \text{(i)}$$

Hence, it can be observed that the displacement current is quite large as compared to the conduction current at 2.5 GHz so that the conduction current can be neglected for all practical purposes. However, this situation might not be true at all the frequencies. This is especially true for conductive materials, where the imaginary part of the permittivity can be expressed as $\epsilon'' \equiv \sigma/\omega$ so that at lower frequencies, the imaginary part becomes quite significant and the conduction current can no longer be neglected.

(c) As the loss tangent is quite small making the medium a good dielectric, the approximate formulas can be used for calculating various parameters. The attenuation constant can be computed using (60a):

$$\alpha = \frac{\sigma}{2}\sqrt{\frac{\mu_0}{\epsilon'}} \equiv \frac{\omega\epsilon''}{2}\sqrt{\frac{\mu_0}{\epsilon'}} \equiv \frac{\omega}{2}\left(\frac{\epsilon''}{\epsilon'}\right)\sqrt{\mu_0\epsilon'}$$

$$= \frac{2\pi \times 2.5 \times 10^9}{2} \times 0.002 \times \sqrt{4\pi \times 10^{-7} \times 8.854 \times 10^{-12} \times 2.56} = 0.084\,\text{Np/m} \qquad \text{(ii)}$$

Similarly, the phase constant can be determined using (61):

$$\beta = \omega\sqrt{\mu_0\epsilon'} \equiv 2\pi \times 2.5 \times 10^9 \sqrt{4\pi \times 10^{-7} \times 8.854 \times 10^{-12} \times 2.56} = 83.8\ \text{rad/m} \qquad \text{(iii)}$$

Finally, the intrinsic impedance can be obtained using (62b) resulting into following expression:

$$\eta = \sqrt{\frac{\mu_0}{\epsilon'}}\left(1 + j\frac{\sigma}{2\omega\,\epsilon'}\right) = \frac{120\pi}{\sqrt{2.56}}\left(1 + j\frac{\epsilon''}{2\epsilon'}\right) = 235.6 + j0.23 \qquad \text{(iv)}$$

It can be seen from (ii), (iii) and (iv) that for good dielectrics, the attenuation constant is very small, and the intrinsic impedance is almost real having a very small value of the imaginary part.

(*d*) The wavelength will be given by

$$\lambda = \frac{2\pi}{\beta} = \frac{2\pi}{83.8} = 7.5 \text{ cm} \tag{v}$$

The phase velocity is obtained as

$$\nu_p = \frac{\omega}{\beta} = \frac{2\pi \times 2.5 \times 10^9}{83.8} = 1.87 \times 10^8 \text{ m/s} \tag{vi}$$

(*e*) The general time-domain expression for the electric field of a y-polarized uniform plane wave traveling along the $+x$ direction can be written with the help of (41) as

$$\mathbf{E}(x,t) = E_{y0}e^{-\alpha x}\cos(\omega t - \beta x)\mathbf{a}_y$$

Now, it is given that the maximum amplitude of the electric field is 40 V/m, which is recorded at $x = 0$ and $t = 0$ thus making $E_{y0} = 40$. The value of other parameters computed earlier can be substituted in the above equation to obtain the required expression for the electric field as follows

$$\mathbf{E}(x,t) = 40e^{-0.084x}\cos(5\pi \times 10^9 t - 83.8x)\mathbf{a}_y \tag{vii}$$

The phasor form of the electric field can be written as

$$\mathbf{E}_s(x) = 40e^{-0.084x}e^{-j83.8x}\mathbf{a}_y \tag{viii}$$

(*f*) The electric field at the initial position ($x = 0$) is 40 V/m. We have to find the distance x_1 at which its magnitude gets reduced to 90% of its initial value.

$$|\mathbf{E}_s(x=0)| = 40 \text{ V/m}$$
$$|\mathbf{E}_s(x=x_1)| = 40 \times e^{-0.084x_1} \equiv 0.9 \times 40 \text{ V/m}$$
$$\Rightarrow x_1 = \frac{1}{0.084}\ln\left(\frac{1}{0.9}\right) = 1.25 \text{ m}$$

Hence, the wave would travel a distance of 1.25 m before its amplitude gets reduced to 90% of its initial value.

D12.4 Given a nonmagnetic material having $\epsilon'_r = 3.2$ and $\sigma = 1.5 \times 10^{-4}$ S/m, find numerical values at 3 MHz for the: (*a*) loss tangent; (*b*) attenuation constant; (*c*) phase constant; (*d*) intrinsic impedance.

Ans. 0.28; 0.016 Np/m; 0.11 rad/m; $207\angle 7.8^\circ\ \Omega$

D12.5 Consider a material for which $\mu_r = 1$, $\epsilon'_r = 2.5$, and the loss tangent is 0.12. If these three values are constant with frequency in the range $0.5 \text{ MHz} \le f \le 100 \text{ MHz}$, calculate: (*a*) σ at 1 and 75 MHz; (*b*) λ at 1 and 75 MHz; (*c*) ν_p at 1 and 75 MHz.

Ans. 1.67×10^{-5} and 1.25×10^{-3} S/m; 190 and 2.53 m; 1.90×10^8 m/s twice

12.3 POYNTING'S THEOREM AND WAVE POWER

Illustrations

In order to find the power flow associated with an electromagnetic wave, it is necessary to develop a power theorem for the electromagnetic field known as the Poynting theorem. It was originally postulated in 1884 by an English physicist, John H. Poynting.

The development begins with one of Maxwell's curl equations, in which we assume that the medium may be conductive:

$$\nabla \times \mathbf{H} = \mathbf{J} + \frac{\partial \mathbf{D}}{\partial t} \tag{63}$$

Next, we take the scalar product of both sides of (63) with $\mathbf{E}$,

$$\mathbf{E} \cdot \nabla \times \mathbf{H} = \mathbf{E} \cdot \mathbf{J} + \mathbf{E} \cdot \frac{\partial \mathbf{D}}{\partial t} \tag{64}$$

We then introduce the following vector identity, which may be proved by expansion in rectangular coordinates:

$$\nabla \cdot (\mathbf{E} \times \mathbf{H}) = -\mathbf{E} \cdot \nabla \times \mathbf{H} + \mathbf{H} \cdot \nabla \times \mathbf{E} \tag{65}$$

Using (65) in the left side of (64) results in

$$\mathbf{H} \cdot \nabla \times \mathbf{E} - \nabla \cdot (\mathbf{E} \times \mathbf{H}) = \mathbf{J} \cdot \mathbf{E} + \mathbf{E} \cdot \frac{\partial \mathbf{D}}{\partial t} \tag{66}$$

where the curl of the electric field is given by the other Maxwell curl equation:

$$\nabla \times \mathbf{E} = -\frac{\partial \mathbf{B}}{\partial t}$$

Therefore

$$-\mathbf{H} \cdot \frac{\partial \mathbf{B}}{\partial t} - \nabla \cdot (\mathbf{E} \times \mathbf{H}) = \mathbf{J} \cdot \mathbf{E} + \mathbf{E} \cdot \frac{\partial \mathbf{D}}{\partial t}$$

or

$$-\nabla \cdot (\mathbf{E} \times \mathbf{H}) = \mathbf{J} \cdot \mathbf{E} + \epsilon \mathbf{E} \cdot \frac{\partial \mathbf{E}}{\partial t} + \mu \mathbf{H} \cdot \frac{\partial \mathbf{H}}{\partial t} \tag{67}$$

The two time derivatives in (67) can be rearranged as follows:

$$\epsilon \mathbf{E} \cdot \frac{\partial \mathbf{E}}{\partial t} = \frac{\partial}{\partial t}\left(\frac{1}{2}\mathbf{D} \cdot \mathbf{E}\right) \tag{68a}$$

and

$$\mu \mathbf{H} \cdot \frac{\partial \mathbf{H}}{\partial t} = \frac{\partial}{\partial t}\left(\frac{1}{2}\mathbf{B} \cdot \mathbf{H}\right) \tag{68b}$$

With these, Eq. (67) becomes

$$-\nabla \cdot (\mathbf{E} \times \mathbf{H}) = \mathbf{J} \cdot \mathbf{E} + \frac{\partial}{\partial t}\left(\frac{1}{2}\mathbf{D} \cdot \mathbf{E}\right) + \frac{\partial}{\partial t}\left(\frac{1}{2}\mathbf{B} \cdot \mathbf{H}\right) \tag{69}$$

Finally, we integrate (69) throughout a volume:

$$-\int_{\text{vol}} \nabla\cdot(\mathbf{E}\times\mathbf{H})\,dv = \int_{\text{vol}} \mathbf{J}\cdot\mathbf{E}\,dv + \int_{\text{vol}} \frac{\partial}{\partial t}\left(\frac{1}{2}\mathbf{D}\cdot\mathbf{E}\right)dv + \int_{\text{vol}} \frac{\partial}{\partial t}\left(\frac{1}{2}\mathbf{B}\cdot\mathbf{H}\right)dv$$

The divergence theorem is then applied to the left-hand side, thus converting the volume integral there into an integral over the surface that encloses the volume. On the right-hand side, the operations of spatial integration and time differentiation are interchanged. The final result is

$$-\oint_{\text{area}} (\mathbf{E}\times\mathbf{H})\cdot d\mathbf{S} = \int_{\text{vol}} \mathbf{J}\cdot\mathbf{E}\,d\nu + \frac{d}{dt}\int_{\text{vol}} \frac{1}{2}\mathbf{D}\cdot\mathbf{E}\,d\nu + \frac{d}{dt}\int_{\text{vol}} \frac{1}{2}\mathbf{B}\cdot\mathbf{H}\,d\nu \tag{70}$$

Equation (70) is known as Poynting's theorem. On the right-hand side, the first integral is the total (but instantaneous) ohmic power dissipated within the volume. The second integral is the total energy stored in the electric field, and the third integral is the stored energy in the magnetic field.[3] Since time derivatives are taken of the second and third integrals, those results give the time rates of increase of energy stored within the volume, or the instantaneous power going to increase the stored energy. The sum of the expressions on the right must therefore be the total power flowing *into* this volume, and so the total power flowing *out* of the volume is

$$\oint_{\text{area}} (\mathbf{E}\times\mathbf{H})\cdot d\mathbf{S} \quad \text{W} \tag{71}$$

where the integral is over the closed surface surrounding the volume. The cross product $\mathbf{E}\times\mathbf{H}$ is known as the Poynting vector, **S**,

$$\mathbf{S} = \mathbf{E}\times\mathbf{H} \quad \text{W/m}^2 \tag{72}$$

which is interpreted as an instantaneous power density, measured in watts per square meter (W/m^2). The direction of the vector **S** indicates the direction of the instantaneous power flow at a point, and many of us think of the Poynting vector as a "pointing" vector. This homonym, while accidental, is correct.[4]

Since **S** is given by the cross product of **E** and **H**, the direction of power flow at any point is normal to both the **E** and **H** vectors. This certainly agrees with our experience with the uniform plane wave, for propagation in the $+z$ direction was associated with an E_x and H_y component,

$$E_x\mathbf{a}_x \times H_y\mathbf{a}_y = S_z\mathbf{a}_z$$

In a perfect dielectric, the **E** and **H** field amplitudes are given by

$$E_x = E_{x0}\cos(\omega t - \beta z)$$

$$H_y = \frac{E_{x0}}{\eta}\cos(\omega t - \beta z)$$

[3] This is the expression for magnetic field energy that we have been anticipating since Chapter 9.

[4] Note that the vector symbol **S** is used for the Poynting vector, and is not to be confused with the differential area vector, $d\mathbf{S}$. The latter, as we know, is the product of the outward normal to the surface and the differential area.

where η is real. The power density amplitude is therefore

$$S_z = \frac{E_{x0}^2}{\eta} \cos^2(\omega t - \beta z) \tag{73}$$

In the case of a lossy dielectric, E_x and H_y are not in time phase. We have

$$E_x = E_{x0} e^{-\alpha z} \cos(\omega t - \beta z)$$

If we let

$$\eta = |\eta| \angle \theta_\eta$$

then we may write the magnetic field intensity as

$$H_y = \frac{E_{x0}}{|\eta|} e^{-\alpha z} \cos(\omega t - \beta z - \theta_\eta)$$

Thus,

$$S_z = E_x H_y = \frac{E_{x0}^2}{|\eta|} e^{-2\alpha z} \cos(\omega t - \beta z) \cos(\omega t - \beta z - \theta_\eta) \tag{74}$$

Since we are dealing with a sinusoidal signal, the time-average power density, $\langle S_z \rangle$, is the quantity that will ultimately be measured. To find this, we integrate (74) over one cycle and divide by the period $T = 1/f$. Additionally, the identity $\cos A \cos B \equiv 1/2 \cos(A+B) + 1/2 \cos(A-B)$ is applied to the integrand, and we obtain:

$$\langle S_z \rangle = \frac{1}{T} \int_0^T \frac{1}{2} \frac{E_{x0}^2}{|\eta|} e^{-2\alpha z} [\cos(2\omega t - 2\beta z - 2\theta_\eta) + \cos \theta_\eta] \, dt \tag{75}$$

The second-harmonic component of the integrand in (75) integrates to zero, leaving only the contribution from the dc component. The result is

$$\langle S_z \rangle = \frac{1}{2} \frac{E_{x0}^2}{|\eta|} e^{-2\alpha z} \cos \theta_\eta \tag{76}$$

Note that the power density attenuates as $e^{-2\alpha z}$, whereas E_x and H_y fall off as $e^{-\alpha z}$.

We may finally observe that the preceding expression can be obtained very easily by using the phasor forms of the electric and magnetic fields. In vector form, this is

$$\langle \mathbf{S} \rangle = \frac{1}{2} \text{Re}(\mathbf{E}_s \times \mathbf{H}_s^*) \quad \text{W/m}^2 \tag{77}$$

In the present case

$$\mathbf{E}_s = E_{x0} e^{-j\beta z} \mathbf{a}_x$$

and

$$\mathbf{H}_s^* = \frac{E_{x0}}{\eta^*} e^{+j\beta z} \mathbf{a}_y = \frac{E_{x0}}{|\eta|} e^{j\theta} e^{+j\beta z} \mathbf{a}_y$$

where E_{x0} has been assumed real. Eq. (77) applies to any sinusoidal electromagnetic wave and gives both the magnitude and direction of the time-average power density.

EXAMPLE 12.10

The time-domain expression for the magnetic field of a plane wave traveling in a non-magnetic medium is given by the following expression:

$$\mathbf{H}(x,\, t) \;=\; 0.2\,\cos\left(6\pi \times 10^{8}\, t \,-\, 10.2\, x\right)\, \mathbf{a}_z \;\text{A/m} \tag{i}$$

Find

(*a*) the direction of wave propagation;
(*b*) the dielectric constant, and the intrinsic impedance;
(*c*) the time-domain expression of the electric field;
(*d*) the time-average power density;
(*e*) the time-average power density making use of phasor forms of the electric and magnetic fields; and
(*f*) the net time average power crossing a rectangular plate of area (30 cm × 20 cm) oriented along the yz plane.

Solution.

(*a*) The expression of the field given in (i) can be compared with the following standard field equation for a y-polarized plane wave traveling along the z-direction,

$$\mathbf{H}(z,\, t) \;=\; H_{y0}\, e^{-\alpha z} \cos\left(\omega\, t \,-\, \beta\, z\right)\, \mathbf{a}_y \;\text{A/m} \tag{ii}$$

After comparison, it can be easily stated that the wave propagation is in the direction x, and the magnetic field is oriented along the direction z in the given problem as per (i). It can also be easily verified that the electric field is polarized along the y direction so that the Poynting vector ($\mathbf{E} \times \mathbf{H} \equiv E_y \mathbf{a}_y \times H_z \mathbf{a}_z$) in the present situation would provide wave propagation or the power flow along the $\mathbf{a}_x$ direction as given in the problem.

(*b*) After comparing (i) with (ii), we obtain

$$\beta = \; 10.2 \;\text{rad/m}\,; \quad \omega = \; 6\pi \times 10^{8} \;\text{rad/s}$$

As the media under consideration is non-magnetic and lossless, hence $\mu_r \;=\; 1, \epsilon'' = 0$. The dielectric constant can be calculated using the following expression:

$$\beta = \omega\sqrt{\mu\epsilon} \equiv \omega\sqrt{\mu_0\epsilon_0\epsilon_r'} \equiv \frac{\omega}{c}\sqrt{\epsilon_r'} = \frac{6\pi \times 10^{8}}{3 \times 10^{8}}\sqrt{\epsilon_r'} \;=\; 10.2 \;\;\Rightarrow\; \epsilon_r' = 2.63 \tag{iii}$$

where, $c \equiv 1/\sqrt{\mu_0\epsilon_0} = 3 \times 10^{8}\,\text{m/s}$ is the free space propagation velocity. The intrinsic impedance may be calculated as

$$\eta = \sqrt{\frac{\mu}{\epsilon}} \equiv \frac{\eta_0}{\sqrt{\epsilon_r'}} \equiv \frac{120\pi}{\sqrt{2.63}} = 232.5\,\Omega \tag{iv}$$

(*c*) The time-domain expression of the electric field can be computed in terms of the magnetic field given by (i), and the basic definition of the intrinsic impedance, which yields the following expression in the present situation

$$\mathbf{E} \equiv E_y\,\mathbf{a}_y \equiv \eta H_z\,\mathbf{a}_y = 46.5\cos\left(6\pi\times10^8\,t - 10.2\,x\right)\,\mathbf{a}_y\ \text{V/m} \tag{v}$$

(*d*) For computing the time-average power density, the concept of Poynting vector would be used yielding the following expression

$$E_y\,\mathbf{a}_y \times H_z\,\mathbf{a}_z \equiv S_x\,\mathbf{a}_x \tag{vi}$$

where S_x is the power density, and the direction of Poynting vector indicates the power flow along the x-direction. The time-average power density can be determined in terms of the non-zero field components as follows

$$\langle S_x\rangle = \frac{1}{T}\int_0^T S_x\,dt \equiv \frac{1}{T}\int_0^T E_y\,H_z\,dt \equiv \frac{1}{T}\int_0^T \eta\,[H_z]^2\,dt$$

$$\equiv \frac{232.5\times0.04}{T}\int_0^T \cos^2\left(6\pi\times10^8\,t - 10.2\,x\right)\,dt$$

$$\equiv \frac{9.3}{2T}\int_0^T \left[1+\cos 2\left(6\pi\times10^8\,t - 10.2\,x\right)\right]\,dt = 4.65\ \text{W/m}^2 \tag{vii}$$

(*e*) The time-average power density can be computed using the phasor forms of the field components as shown below. Let us first write down the phasor forms of the electric and magnetic field components as follows

$$\mathbf{E}_s \equiv 46.5\,e^{-j10.2x}\,\mathbf{a}_y\ \text{V/m};\ \ \mathbf{H}_s \equiv 0.2\,e^{-j10.2x}\,\mathbf{a}_z\ \text{A/m}$$

The time-average power density can then be written in the vector form using the concept of Poynting vector (77) yielding the following expression:

$$\langle\mathbf{S}\rangle = \frac{1}{2}\Re e\,(\mathbf{E}_s\times\mathbf{H}_s^*) \equiv \frac{1}{2}\,[46.5\,\mathbf{a}_y\times0.2\,\mathbf{a}_z] = 4.65\,\mathbf{a}_x\ \text{W/m}^2 \tag{viii}$$

It can be observed that the time-average power density can be computed in a simpler way using the phasor form of the field components.

(*f*) For finding the net time average power crossing a rectangular plate of area (30 cm × 20 cm) oriented along the *yz* plane, we have to note that the unit vector normal to the plane in this case would be given by

$$\mathbf{a}_n \equiv \mathbf{a}_x$$

The net time average power crossing the given rectangular plate would then be given by

$$P_{\text{avg}} \equiv \int_s \langle\mathbf{S}\rangle\cdot\mathbf{ds} \equiv 4.65\,\mathbf{a}_x\cdot\mathbf{a}_n\int_s ds \equiv 4.65\,\mathbf{a}_x\cdot\,\mathbf{a}_x 600\times10^{-4}\ \text{W/m}^2$$
$$= 27.9\,\text{mW} \tag{ix}$$

It may be noted that in the above equation, the area is given as constant (30 cm × 20 cm), and the actual time-average power density is also independent of any spatial coordinate. Hence, the integral involving the spatial coordinate can simply be replaced by the actual area given in the problem.

D12.6 At frequencies of 1, 100, and 3000 MHz, the dielectric constant of ice made from pure water has values of 4.15, 3.45, and 3.20, respectively, while the loss tangent is 0.12, 0.035, and 0.0009, also respectively. If a uniform plane wave with an amplitude of 100 V/m at $z = 0$ is propagating through such ice, find the time-average power density at $z = 0$ and $z = 10$ m for each frequency.

Ans. 27.1 and 25.7 W/m^2; 24.7 and 6.31 W/m^2; 23.7 and 8.63 W/m^2.

12.4 PROPAGATION IN GOOD CONDUCTORS: SKIN EFFECT

Interactives

As an additional study of propagation with loss, we shall investigate the behavior of a *good conductor* when a uniform plane wave is established in it. Such a material satisfies the general high-loss criterion, in which the loss tangent, $\epsilon''/\epsilon' \gg 1$. Applying this to a good conductor leads to the more specific criterion, $\sigma/(\omega\epsilon') \gg 1$. As before, we have an interest in losses that occur on wave transmission *into* a good conductor, and we will find new approximations for the phase constant, attenuation coefficient, and intrinsic impedance. New to us, however, is a modification of the basic problem, appropriate for good conductors. This concerns waves associated with electromagnetic fields existing in an external dielectric that adjoins the conductor surface; in this case, the waves propagate *along* the surface. That portion of the overall field that exists within the conductor will suffer dissipative loss arising from the conduction currents it generates. The overall field therefore attenuates with increasing distance of travel along the surface. This is the mechanism for the resistive transmission line loss that we studied in Chapter 11, and which is embodied in the line resistance parameter, R.

As implied, a good conductor has a high conductivity and large conduction currents. The energy represented by the wave traveling through the material therefore decreases as the wave propagates because ohmic losses are continuously present. When we discussed the loss tangent, we saw that the ratio of conduction current density to the displacement current density in a conducting material is given by $\sigma/\omega\epsilon'$. Choosing a poor metallic conductor and a very high frequency as a conservative example, this ratio[5] for nichrome ($\sigma \doteq 10^6$) at 100 MHz is about 2×10^8. We therefore have a situation where $\sigma/\omega\epsilon' \gg 1$, and we should be able to make several very good approximations to find α, β, and η for a good conductor.

The general expression for the propagation constant is, from (59),

$$jk = j\omega\sqrt{\mu\epsilon'}\sqrt{1 - j\frac{\sigma}{\omega\epsilon'}}$$

which we immediately simplify to obtain

$$jk = j\omega\sqrt{\mu\epsilon'}\sqrt{-j\frac{\sigma}{\omega\epsilon'}}$$

[5] It is customary to take $\epsilon' = \epsilon_0$ for metallic conductors.

or

$$jk = j\sqrt{-j\omega\mu\sigma}$$

But

$$-j = 1\ \angle{-90^\circ}$$

and

$$\sqrt{1\ \angle{-90^\circ}} = 1\ \angle{-45^\circ} = \frac{1}{\sqrt{2}}(1-j)$$

Therefore

$$jk = j(1-j)\sqrt{\frac{\omega\mu\sigma}{2}} = (1+j)\sqrt{\pi f\mu\sigma} = \alpha + j\beta \tag{78}$$

Hence

$$\boxed{\alpha = \beta = \sqrt{\pi f\mu\sigma}} \tag{79}$$

Regardless of the parameters μ and σ of the conductor or of the frequency of the applied field, α and β are equal. If we again assume only an E_x component traveling in the $+z$ direction, then

$$E_x = E_{x0}e^{-z\sqrt{\pi f\mu\sigma}}\cos\left(\omega t - z\sqrt{\pi f\mu\sigma}\right) \tag{80}$$

We may tie this field in the conductor to an external field at the conductor surface. We let the region $z > 0$ be the good conductor and the region $z < 0$ be a perfect dielectric. At the boundary surface $z = 0$, (80) becomes

$$E_x = E_{x0}\cos\omega t \qquad (z = 0)$$

This we shall consider as the source field that establishes the fields within the conductor. Since displacement current is negligible,

$$\mathbf{J} = \sigma\mathbf{E}$$

Thus, the conduction current density at any point within the conductor is directly related to **E**:

$$J_x = \sigma E_x = \sigma E_{x0}e^{-z\sqrt{\pi f\mu\sigma}}\cos\left(\omega t - z\sqrt{\pi f\mu\sigma}\right) \tag{81}$$

Equations (80) and (81) contain a wealth of information. Considering first the negative exponential term, we find an exponential decrease in the conduction current density and electric field intensity with penetration into the conductor (away from the source). The exponential factor is unity at $z = 0$ and decreases to $e^{-1} = 0.368$ when

$$z = \frac{1}{\sqrt{\pi f\mu\sigma}}$$

This distance is denoted by δ and is termed the *depth of penetration*, or the *skin depth*,

$$\boxed{\delta = \frac{1}{\sqrt{\pi f\mu\sigma}} = \frac{1}{\alpha} = \frac{1}{\beta}} \tag{82}$$

It is an important parameter in describing conductor behavior in electromagnetic fields. To get some idea of the magnitude of the skin depth, let us consider copper, $\sigma = 5.8 \times 10^7$ S/m, at several different frequencies. We have

$$\delta_{\text{Cu}} = \frac{0.066}{\sqrt{f}}$$

At a power frequency of 60 Hz, $\delta_{\text{Cu}} = 8.53$ mm. Remembering that the power density carries an exponential term $e^{-2\alpha z}$, we see that the power density is multiplied by a factor of $0.368^2 = 0.135$ for every 8.53 mm of distance into the copper.

At a microwave frequency of 10,000 MHz, δ is 6.61×10^{-4} mm. Stated more generally, all fields in a good conductor such as copper are essentially zero at distances greater than a few skin depths from the surface. Any current density or electric field intensity established at the surface of a good conductor decays rapidly as we progress into the conductor. Electromagnetic energy is not transmitted in the interior of a conductor; it travels in the region surrounding the conductor, while the conductor merely guides the waves. We shall consider guided propagation in more detail in Chapter 14.

Suppose we have a copper bus bar in the substation of an electric utility company which we wish to have carry large currents, and we therefore select dimensions of 2 by 4 inches. Then much of the copper is wasted, for the fields are greatly reduced in one skin depth, about 8.5 mm.[6] A hollow conductor with a wall thickness of about 12 mm would be a much better design. Although we are applying the results of an analysis for an infinite planar conductor to one of finite dimensions, the fields are attenuated in the finite-size conductor in a similar (but not identical) fashion.

The extremely short skin depth at microwave frequencies shows that only the surface coating of the guiding conductor is important. A piece of glass with an evaporated silver surface 3 μm thick is an excellent conductor at these frequencies.

Next, let us determine expressions for the velocity and wavelength within a good conductor. From (82), we already have

$$\alpha = \beta = \frac{1}{\delta} = \sqrt{\pi f \mu \sigma}$$

Then, since

$$\beta = \frac{2\pi}{\lambda}$$

we find the wavelength to be

$$\lambda = 2\pi\delta \tag{83}$$

Also, recalling that

$$\nu_p = \frac{\omega}{\beta}$$

we have

$$\boxed{\nu_p = \omega\delta} \tag{84}$$

[6] This utility company operates at 60 Hz.

For copper at 60 Hz, $\lambda = 5.36$ cm and $\nu_p = 3.22$ m/s, or about 7.2 mi/h! A lot of us can run faster than that. In free space, of course, a 60 Hz wave has a wavelength of 3100 mi and travels at the velocity of light.

EXAMPLE 12.11

Let us again consider wave propagation in water, but this time we will consider seawater. The primary difference between seawater and fresh water is of course the salt content. Sodium chloride dissociates in water to form Na^+ and Cl^- ions, which, being charged, will move when forced by an electric field. Seawater is thus conductive, and so it will attenuate electromagnetic waves by this mechanism. At frequencies in the vicinity of 10^7 Hz and below, the bound charge effects in water discussed earlier are negligible, and losses in seawater arise principally from the salt-associated conductivity. We consider an incident wave of frequency 1 MHz. We wish to find the skin depth, wavelength, and phase velocity. In seawater, $\sigma = 4$ S/m, and $\epsilon'_r = 81$.

Solution. We first evaluate the loss tangent, using the given data:

$$\frac{\sigma}{\omega\epsilon'} = \frac{4}{(2\pi \times 10^6)(81)(8.85 \times 10^{-12})} = 8.9 \times 10^2 \gg 1$$

Seawater is therefore a good conductor at 1 MHz (and at frequencies lower than this). The skin depth is

$$\delta = \frac{1}{\sqrt{\pi f \mu \sigma}} = \frac{1}{\sqrt{(\pi \times 10^6)(4\pi \times 10^{-7})(4)}} = 0.25 \text{ m} = 25 \text{ cm}$$

Now

$$\lambda = 2\pi\delta = 1.6 \text{ m}$$

and

$$\nu_p = \omega\delta = (2\pi \times 10^6)(0.25) = 1.6 \times 10^6 \text{ m/sec}$$

In free space, these values would have been $\lambda = 300$ m and of course $\nu = c$.

With a 25 cm skin depth, it is obvious that radio frequency communication in seawater is quite impractical. Notice however that δ varies as $1/\sqrt{f}$, so that things will improve at lower frequencies. For example, if we use a frequency of 10 Hz (in the ELF, or extremely low frequency range), the skin depth is increased over that at 1 MHz by a factor of $\sqrt{10^6/10}$, so that

$$\delta(10 \text{ Hz}) \doteq 80 \text{ m}$$

The corresponding wavelength is $\lambda = 2\pi\delta \doteq 500$ m. Frequencies in the ELF range were used for many years in submarine communications. Signals were transmitted from gigantic ground-based antennas (required because the free-space wavelength associated with 10 Hz is 3×10^7 m). The signals were then received by submarines, from which a suspended wire antenna of length shorter than 500 m is sufficient. The drawback is that signal data rates at ELF are slow enough that a single word can take several minutes to transmit. Typically, ELF signals would be used to tell the submarine to initiate emergency procedures, or to come near the surface in order to receive a more detailed message via satellite.

We next turn our attention to finding the magnetic field, H_y; associated with E_x. To do so, we need an expression for the intrinsic impedance of a good conductor. We begin with Eq. (48), Section 12.2, with $\epsilon'' = \sigma/\omega$,

$$\eta = \sqrt{\frac{j\omega\mu}{\sigma + j\omega\epsilon'}}$$

Since $\sigma \gg \omega\epsilon'$, we have

$$\eta = \sqrt{\frac{j\omega\mu}{\sigma}}$$

which may be written as

$$\eta = \frac{\sqrt{2}\angle 45^\circ}{\sigma\delta} = \frac{(1+j)}{\sigma\delta} \tag{85}$$

Thus, if we write (80) in terms of the skin depth,

$$E_x = E_{x0}e^{-z/\delta}\cos\left(\omega t - \frac{z}{\delta}\right) \tag{86}$$

then

$$H_y = \frac{\sigma\delta E_{x0}}{\sqrt{2}}e^{-z/\delta}\cos\left(\omega t - \frac{z}{\delta} - \frac{\pi}{4}\right) \tag{87}$$

and we see that the maximum amplitude of the magnetic field intensity occurs one-eighth of a cycle later than the maximum amplitude of the electric field intensity at every point.

From (86) and (87) we may obtain the time-average Poynting vector by applying (77),

$$\langle S_z \rangle = \frac{1}{2}\frac{\sigma\delta E_{x0}^2}{\sqrt{2}}e^{-2z/\delta}\cos\left(\frac{\pi}{4}\right)$$

or

$$\langle S_z \rangle = \frac{1}{4}\sigma\delta E_{x0}^2 e^{-2z/\delta}$$

We again note that in a distance of one skin depth the power density is only $e^{-2} = 0.135$ of its value at the surface.

The total average power loss in a width $0 < y < b$ and length $0 < x < L$ in the direction of the current, as shown in Figure 12.3, is obtained by finding the power crossing the conductor surface within this area,

$$P_L = \int_{\text{area}} \langle S_z \rangle da = \int_0^b \int_0^L \frac{1}{4}\sigma\delta E_{x0}^2 e^{-2z/\delta}\bigg|_{z=0} dx\,dy = \frac{1}{4}\sigma\delta b L E_{x0}^2$$

In terms of the current density J_{x0} at the surface,

$$J_{x0} = \sigma E_{x0}$$

we have

$$P_L = \frac{1}{4\sigma}\delta b L J_{x0}^2 \qquad (88)$$

Now let us see what power loss would result if the *total* current in a width b were distributed *uniformly* in one skin depth. To find the total current, we integrate the current density over the infinite depth of the conductor,

$$I = \int_0^\infty \int_0^b J_x \, dy \, dz$$

where

$$J_x = J_{x0} e^{-z/\delta} \cos\left(\omega t - \frac{z}{\delta}\right)$$

or in complex exponential notation to simplify the integration,

$$J_{xs} = J_{x0} e^{-z/\delta} e^{-jz/\delta}$$
$$= J_{x0} e^{-(1+j)z/\delta}$$

Therefore,

$$I_s = \int_0^\infty \int_0^b J_{x0} e^{-(1+j)z/\delta} dy \, dz$$
$$= J_{x0} b e^{-(1+j)z/\delta} \frac{-\delta}{1+j}\bigg|_0^\infty$$
$$= \frac{J_{x0} b \delta}{1+j}$$

and

$$I = \frac{J_{x0} b \delta}{\sqrt{2}} \cos\left(\omega t - \frac{\pi}{4}\right)$$

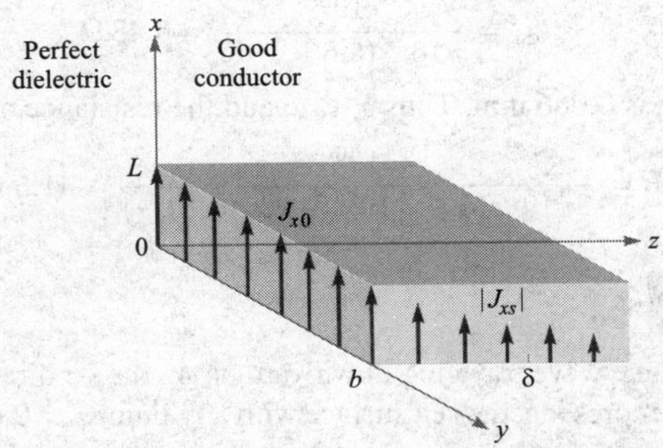

Fig. 12.3 The current density $J_x = J_{x0} e^{-z/\delta} e^{-jz/\delta}$ decreases in magnitude as the wave propagates into the conductor. The average power loss in the region $0 < x < L$, $0 < y < b$, $z > 0$, is $\delta b L J_{x0}^2 / 4\sigma$ watts.

If this current is distributed with a uniform density J' throughout the cross section $0 < y < b$, $0 < z < \delta$, then

$$J' = \frac{J_{x0}}{\sqrt{2}} \cos\left(\omega t - \frac{\pi}{4}\right)$$

The ohmic power loss per unit volume is $\mathbf{J} \cdot \mathbf{E}$, and thus the total instantaneous power dissipated in the volume under consideration is

$$P_{Li}(t) = \frac{1}{\sigma}(J')^2 bL\delta = \frac{J_{x0}^2}{2\sigma} bL\delta \cos^2\left(\omega t - \frac{\pi}{4}\right)$$

The time-average power loss is easily obtained, since the average value of the cosine-squared factor is one-half,

$$P_L = \frac{1}{4\sigma} J_{x0}^2 bL\delta \tag{89}$$

Comparing (88) and (89), we see that they are identical. Thus the average power loss in a conductor with skin effect present may be calculated by assuming that the total current is distributed uniformly in one skin depth. In terms of resistance, we may say that the resistance of a width b and length L of an infinitely thick slab with skin effect is the same as the resistance of a rectangular slab of width b, length L, and thickness δ without skin effect, or with uniform current distribution.

We may apply this to a conductor of circular cross section with little error, provided that the radius a is much greater than the skin depth. The resistance at a high frequency where there is a well-developed skin effect is therefore found by considering a slab of width equal to the circumference $2\pi a$ and thickness δ. Hence

$$R = \frac{L}{\sigma S} = \frac{L}{2\pi a \sigma \delta} \tag{90}$$

A round copper wire of 1 mm radius and 1 km length has a resistance at direct current of

$$R_{\text{dc}} = \frac{10^3}{\pi 10^{-6}(5.8 \times 10^7)} = 5.48\ \Omega$$

At 1 MHz, the skin depth is 0.066 mm. Thus $\delta \ll a$, and the resistance at 1 MHz is found by (90),

$$R = \frac{10^3}{2\pi 10^{-3}(5.8 \times 10^7)(0.066 \times 10^{-3})} = 41.5\ \Omega$$

EXAMPLE 12.12

Consider a 200 kHz plane wave traveling downward along the $+z$ direction into the sea water. The x-y plane at $z = 0$ represents the sea surface with any point $z > 0$ representing the surface into the sea water. The constitutive parameters of the sea water are given as $\epsilon_r = 81$, $\mu_r = 1$, and $\sigma = 4\,\text{S/m}$. The electric field at z = 0^+ (just below the surface inside water) is given by

$$\mathbf{E}(0^+, t) = 0.2 \cos\left(4\pi \times 10^5 t + 75^\circ\right) \mathbf{a}_x \ \text{V/m} \quad \text{(i)}$$

(*a*) For this wave, determine whether the sea water can be considered a good dielectric or a good conductor, and why?

(*b*) Calculate the propagation constant, the intrinsic impedance and the phase velocity for this medium.

(*c*) Calculate the depth at which the amplitude of **E** gets reduced to half its value at the surface.

(*d*) Obtain the time-domain expressions for the electric and magnetic fields associated with this wave in sea water ($z > 0$).

Solution.

(*a*) For determining the condition whether the medium under test can be considered a good dielectric or a good conductor at the designated frequency, we should first determine the effective loss tangent using the parameters of sea water:

$$\frac{\sigma}{\omega\,\epsilon'} \equiv \frac{\sigma}{\omega\,\epsilon_0\,\epsilon_r} = \frac{4}{2 \times \pi \times 200 \times 10^3 \times 8.854 \times 10^{-12} \times 81} = 4.4 \times 10^3 \gg 1 \quad \text{(ii)}$$

From the above equation, it is obvious that sea water can be considered a good conductor at 200 kHz. Hence, all the formulas applicable for good conductors can be applied to determine other parameters.

(*b*) For calculating other parameters, let us first calculate the skin depth applied to good conductors using (82):

$$\delta = \sqrt{\frac{1}{\pi f \mu \sigma}} \equiv \frac{1}{\sqrt{\pi \times 200 \times 10^3 \times 4\pi \times 10^{-7} \times 4}} = 0.56\,\text{m} \quad \text{(iii)}$$

Now, the attenuation constant and the phase constant can be determined using the following condition:

$$\alpha \equiv \frac{1}{\delta} = 1.78\,\text{Np/m}\,; \ \beta \equiv \frac{1}{\delta} = 1.78\,\text{rad/m} \quad \text{(iv)}$$

The propagation constant can be expressed as

$$\gamma \equiv \alpha + j\beta = (1.78 + j\,1.78)\ \text{m}^{-1} \quad \text{(v)}$$

The intrinsic impedance inside the sea water can be determined using (85) as

$$\eta \equiv \frac{(1+j)}{\sigma\,\delta} = \frac{(1+j)}{4 \times 0.56} = (0.45 + j\,0.45) \equiv 0.63\,e^{j\pi/4}\,\Omega \quad \text{(vi)}$$

Finally, the phase velocity inside the medium can be determined using (84);

$$\nu_p \equiv \omega\,\delta = 2 \times \pi \times 200 \times 10^3 \times 0.56 = 7.04 \times 10^5\ \text{m/s} \quad \text{(vii)}$$

(*c*) The amplitude of the electric field at the surface is 0.2 V/m. The depth z_1 at which this amplitude of **E** gets reduced to half its value at the surface can be obtained using the following expression:

$$0.1 = 0.2\,e^{-1.78 z_1} \Rightarrow z_1 = \frac{1}{1.78}\ \ln 2 = 39\,\text{cm}$$

(*d*) For finding the time-domain expression of the electric field, let us first write down the phasor form of (i):

$$\mathbf{E}_s(0^+) \;=\; 0.2\,\angle 75^\circ\,\mathbf{a}_x \;\equiv E_{x0}\,\mathbf{a}_x \equiv\; |E_{x0}|\,\angle 75^\circ\,\mathbf{a}_x\ \text{V/m} \qquad \text{(viii)}$$

From the above equation, it is obvious that the value of the electric field at $z = 0$ is a complex quantity. Now, the general phasor expression of the electric field for a plane wave polarized along the x-direction and propagating along the $+z$ direction can be written using (40) as

$$\mathbf{E}_s(z) = \;E_{x0}\,e^{-\alpha z}\,e^{-j\beta z}\,\mathbf{a}_x\ \text{V/m}$$

where the values of E_{x0} is given using (viii), and the values of attenuation constant and the phase constant are computed earlier. These values can finally be substituted in the above equation to obtain the expression for electric field in phasor form:

$$\mathbf{E}_s(z) \;=\; 0.2\,e^{-1.78z}\,e^{-j1.78z}\,e^{j75^\circ}\,\mathbf{a}_x\ \text{V/m} \qquad \text{(ix)}$$

The phasor form of the magnetic field in the present case can be written using the above equation as

$$\mathbf{H}_s(z) = \;\frac{0.2}{|\eta|}\;e^{-1.78z}\,e^{-j\theta_\eta}e^{j75^\circ}\,e^{-j1.78z}\,\mathbf{a}_y\ \text{A/m}$$

The value of the intrinsic impedance can be substituted from (vi) into the above equation to obtain the final expression for the magnetic field in the phasor form:

$$\mathbf{H}_s(z) = \;0.32\,e^{-1.78z}\,e^{-j1.78z+j30^\circ}\,\mathbf{a}_y\ \text{A/m} \qquad \text{(x)}$$

Finally, the time-domain expressions for the electric and magnetic fields associated with this wave in sea water ($z > 0$) can be written using (ix) and (x) as

$$\mathbf{E}(z,\,t) \;=\; 0.2\,e^{-1.78z}\,\cos\left(4\pi\times 10^5\,t\;-1.78z + 75^\circ\right)\mathbf{a}_x\ \text{V/m} \qquad \text{(xi)}$$

$$\mathbf{H}(z,\,t) = 0.32\,e^{-1.78z}\,\cos\left(4\pi\times 10^5\,t\;-1.78z + 30^\circ\right)\mathbf{a}_y\ \text{A/m} \qquad \text{(xii)}$$

EXAMPLE 12.13

A plane wave is incident normally on a good conductor, which can be considered flat and non-magnetic for all practical purposes. The velocity of the wave inside the conductor at a certain frequency is 3.142×10^5 m/s, while the skin depth for this conductor at the same frequency is given as 0.04 mm.

Determine

(*a*) the wavelength inside the medium and frequency of the wave;

(*b*) the conductivity of the material; and

(*c*) the distance, the wave travels inside the conductor at 1.5 GHz before its field intensity gets reduced to 20 dB below its initial value.

Solution.

(*a*) The skin depth can be related with the wavelength inside the conductor using the following expression:

$$\delta = \frac{1}{\beta} = \frac{\lambda}{2\pi} \quad \Rightarrow \lambda = 2\pi \times \delta = 2\pi \times 0.04 \times 10^{-3} = 0.2513\,\text{mm} \tag{i}$$

The frequency of the wave may then be computed as follows:

$$f = \frac{\nu_p}{\lambda} = \frac{3.142 \times 10^5}{0.2513 \times 10^{-3}} = 1.25\,\text{GHz} \tag{ii}$$

(*b*) The conductivity of the material may be computed in terms of the skin depth and frequency using the following expression:

$$\delta = \frac{1}{\sqrt{\pi f \mu \sigma}}$$

$$\Rightarrow \sigma = \frac{1}{\delta^2 \pi f \mu} = \frac{1}{(0.04 \times 10^{-3})^2 \times \pi \times 1.25 \times 10^9 \times 4\pi \times 10^{-7}} = 1.27 \times 10^5\,\text{S/m} \tag{iii}$$

(*c*) For determining the wave attenuation at 1.5 GHz, let us first find the attenuation constant at this frequency as follows:

$$\delta = \frac{1}{\sqrt{\pi f \mu \sigma}} \equiv \frac{1}{\alpha}$$

$$\Rightarrow \alpha = \sqrt{\pi f \mu \sigma} = \sqrt{\pi \times 1.5 \times 10^9 \times 4\pi \times 10^{-7} \times 1.27 \times 10^5} = 2.74 \times 10^4\,\text{Np/m} \tag{iv}$$

Now, let us assume that the plane wave is traveling along the direction z inside the conductor. The magnitude of the electric field inside the conductor at any arbitrary position z can then be written as

$$|\mathbf{E}(z)| = |\mathbf{E}_0|\, e^{-\alpha z} \tag{v}$$

where, $|\mathbf{E}_0|$ is the magnitude of initial value of the electric field defined at its surface. Now, we have to compute the distance $z = d$ where the field gets attenuated to 20 dB of its value at the surface. This can be mathematically expressed as follows

$$20 \log |\mathbf{E}(z = d)| - 20 \log |\mathbf{E}_0| = -20 \;\Rightarrow 20 \log \frac{|\mathbf{E}(z = d)|}{|\mathbf{E}_0|} = -20 \tag{vi}$$

The above equation can be further simplified to obtain

$$\frac{|\mathbf{E}(z = d)|}{|\mathbf{E}_0|} = 10^{-1} = 0.1 \tag{vii}$$

The above equation can then be substituted in (v) to obtain the following expression:

$$\frac{|\mathbf{E}(z = d)|}{|\mathbf{E}_0|} = 0.1 \equiv e^{-\alpha d} \quad \Rightarrow d = \frac{1}{\alpha} \ln(10)$$

Hence, the distance at which the field intensity gets reduced to 20 dB below its initial value at 1.5 GHz is given by

$$d = \frac{1}{2.74 \times 10^4} \ln(10) = 0.084 \, \text{mm} \tag{vi}$$

D12.7 A steel pipe is constructed of a material for which $\mu_r = 180$ and $\sigma = 4 \times 10^6$ S/m. The two radii are 5 and 7 mm, and the length is 75 m. If the total current $I(t)$ carried by the pipe is $8 \cos \omega t$ A, where $\omega = 1200\pi$ rad/s, find: (*a*) the skin depth; (*b*) the effective resistance; (*c*) the dc resistance; (*d*) the time-average power loss.

Ans. 0.766 mm; 0.557 Ω; 0.249 Ω; 17.82 W

12.5 WAVE POLARIZATION

Animations

In the previous sections, we have treated uniform plane waves in which the electric and magnetic field vectors were assumed to lie in fixed directions. Specifically, with the wave propagating along the z axis, **E** was taken to lie along x, which then required **H** to lie along y. This orthogonal relationship between **E**, **H**, and **S** is always true for a uniform plane wave. The directions of **E** and **H** within the plane perpendicular to $\mathbf{a}_z$ may change, however, as functions of time and position, depending on how the wave was generated or on what type of medium it is propagating through. Thus a complete description of an electromagnetic wave would not only include parameters such as its wavelength, phase velocity, and power, but also a statement of the instantaneous orientation of its field vectors. We define the *wave polarization* as the electric field vector orientation as a function of time, at a fixed point in space. A more complete characterization of a wave's polarization would in fact include specifying the field orientation at *all* points because some waves demonstrate spatial variations in their polarization. Specifying only the electric field direction is sufficient, since magnetic field is readily found from **E** using Maxwell's equations.

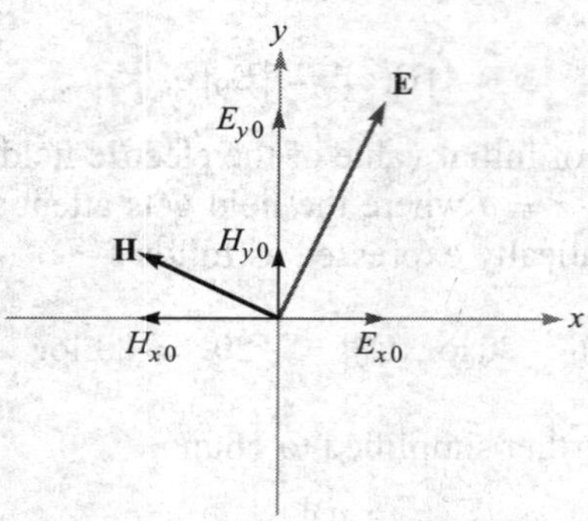

Fig. 12.4 Electric and magnetic field configuration for a general linearly polarized plane wave propagating in the forward z direction (out of the page). Field components correspond to those in Eqs. (91) and (92).

In the waves we have previously studied, **E** was in a fixed straight orientation for all times and positions. Such a wave is said to be *linearly polarized*. We have taken **E** to lie along the x axis, but the field could be oriented in any fixed direction in the xy plane and be linearly polarized. For positive z propagation, the wave would in general have its electric field phasor expressed as

$$\mathbf{E}_s = (E_{x0}\mathbf{a}_x + E_{y0}\mathbf{a}_y)e^{-\alpha z}e^{-j\beta z} \tag{91}$$

where E_{x0} and E_{y0} are constant amplitudes along x and y. The magnetic field is readily found by determining its x and y components directly from those of $\mathbf{E}_s$. Specifically, $\mathbf{H}_s$ for the wave of Eq. (91) is

$$\mathbf{H}_s = [H_{x0}\mathbf{a}_x + H_{y0}\mathbf{a}_y]\,e^{-\alpha z}e^{-j\beta z} = \left[-\frac{E_{y0}}{\eta}\mathbf{a}_x + \frac{E_{x0}}{\eta}\mathbf{a}_y\right]e^{-\alpha z}e^{-j\beta z} \tag{92}$$

The two fields are sketched in Figure 12.4. The figure demonstrates the reason for the minus sign in the term involving E_{y0} in Eq. (92). The direction of power flow, given by $\mathbf{E}\times\mathbf{H}$, is in the positive z direction in this case. A component of $\mathbf{E}$ in the positive y direction would require a component of $\mathbf{H}$ in the negative x direction—thus the minus sign. Using (91) and (92), the power density in the wave is found using (77):

$$\begin{aligned}\langle \mathbf{S}_z\rangle &= \frac{1}{2}\text{Re}\{\mathbf{E}_s\times\mathbf{H}_s^*\} = \frac{1}{2}\text{Re}\{E_{x0}H_{y0}^*(\mathbf{a}_x\times\mathbf{a}_y) + E_{y0}H_{x0}^*(\mathbf{a}_y\times\mathbf{a}_x)\}e^{-2\alpha z}\\ &= \frac{1}{2}\text{Re}\left\{\frac{E_{x0}E_{x0}^*}{\eta^*} + \frac{E_{y0}E_{y0}^*}{\eta^*}\right\}e^{-2\alpha z}\mathbf{a}_z\\ &= \frac{1}{2}\text{Re}\left\{\frac{1}{\eta^*}\right\}(|E_{x0}|^2 + |E_{y0}|^2)e^{-2\alpha z}\mathbf{a}_z \text{ W/m}^2\end{aligned}$$

This result demonstrates the idea that our linearly polarized plane wave can be considered as two distinct plane waves having x and y polarizations, whose electric fields are combining *in phase* to produce the total $\mathbf{E}$. The same is true for the magnetic field components. This is a critical point in understanding wave polarization, in that *any polarization state can be described in terms of mutually perpendicular components of the electric field and their relative phasing*.

EXAMPLE 12.14

The electric field of a plane wave travelling in a non-magnetic medium is given by the following expression:

$$\mathbf{E} = 0.5\cos\left(0.5\,\pi\times10^9\,t - 7.6\,z\right)\mathbf{a}_x + 0.866\cos\left(0.5\,\pi\times10^9\,t - 7.6\,z\right)\mathbf{a}_y \text{V/m} \tag{i}$$

Find

(*a*) the polarization of the wave (linear, elliptical or circular);
(*b*) the direction of wave propagation;
(*c*) the dielectric constant, and the intrinsic impedance;
(*d*) the phasor expression of the magnetic field; and
(*e*) the time-average power density.

Solution.

(a) It can be observed from (i) that the orientation of the electric field remains fixed with passage of time. In other words, although the electric field has two components, but these two components are always in the time phase. Hence, the wave is *linearly polarized*. To find the actual direction of the electric field vector, let us write (i) in phasor form as follows

$$\mathbf{E}_s = [0.5\,\mathbf{a}_x + 0.866\mathbf{a}_y]\,e^{-j7.6z}\,\mathrm{V/m} \tag{ii}$$

Hence, the electric field vector lies in the *xy* plane, and it is oriented at an angle $\phi \equiv \tan^{-1}\left(\dfrac{0.866}{0.5}\right) = 60°$ to the x-axis.

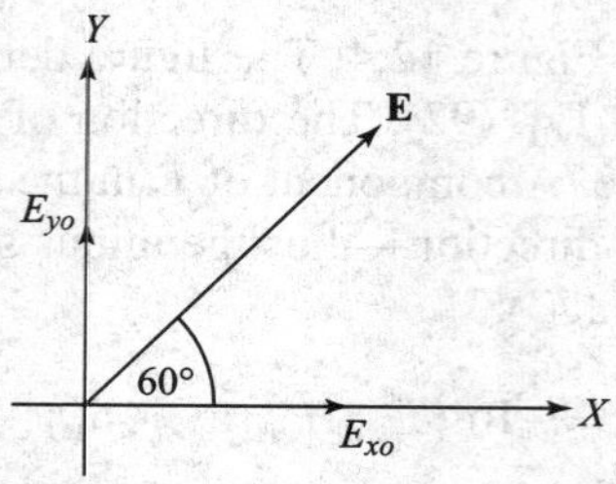

Fig. 12.5 Orientation of electric field.

(b) The wave is propagating in the positive z-direction as easily observed from (i) and (ii).

(c) From (i), the phase constant and the radial frequency are obtained as

$$\beta = 7.6\,\mathrm{rad/m}\,; \quad \omega = 0.5\pi \times 10^9\,\mathrm{rad/s}$$

The media under consideration is given to be non-magnetic and lossless, hence $\mu_r = 1, \alpha = 0, \epsilon'' = 0$. The dielectric constant can be calculated using the following expression:

$$\beta \equiv \frac{\omega}{c}\sqrt{\epsilon_r'} = \frac{0.5\,\pi \times 10^9}{3 \times 10^8}\sqrt{\epsilon_r'} = 7.6 \quad \Rightarrow \epsilon_r' = 2.11 \tag{iii}$$

where, $c \equiv 3 \times 10^8\,\mathrm{m/s}$ is the free-space propagation velocity. The intrinsic impedance may be calculated as

$$\eta = \sqrt{\frac{\mu}{\epsilon}} \equiv \frac{\eta_0}{\sqrt{\epsilon_r'}} \equiv \frac{120\pi}{\sqrt{2.11}} = 259.5\,\Omega \tag{iv}$$

(d) The phasor expression of the magnetic field corresponding to the electric field given by (ii) may be computed as follows

$$\begin{aligned}\mathbf{H}_s &= \frac{1}{\eta}\,[0.5\,\mathbf{a}_z \times \mathbf{a}_x + 0.866\,\mathbf{a}_z \times \mathbf{a}_y]\,e^{-j7.6z}\,\mathrm{A/m} \\ &\equiv [1.9\,\mathbf{a}_y - 3.3\,\mathbf{a}_x]\,e^{-j7.6z}\,\mathrm{mA/m} \end{aligned} \tag{v}$$

where, the value of intrinsic impedance is substituted from (iv) and the wave is considered to be propagating along the positive $\mathbf{a}_z$ direction.

(e) The time-average power density associated with the wave can be computed using the phasor forms of the electric and magnetic field vectors as follows:

$$\begin{aligned}\langle \mathbf{S} \rangle &= \frac{1}{2}\,\mathrm{Re}\,(\mathbf{E}_s \times \mathbf{H}_s^*) \\ &\equiv \frac{1}{2}\,\mathrm{Re}\left\{[0.5\,\mathbf{a}_x + 0.866\,\mathbf{a}_y]\,e^{-j7.6z} \times 10^{-3}\,[1.9\,\mathbf{a}_y - 3.3\,\mathbf{a}_x]\,e^{+j7.6z}\right\} \\ &\equiv 1.9\mathbf{a}_z\ \mathrm{mW/m^2}\end{aligned} \qquad \text{(vi)}$$

It is to be observed that although both the electric and magnetic fields have two different components, but the net power flow due to all the non-zero electric and magnetic field components is along the positive z-direction as described earlier.

We next consider the effect of a phase difference, ϕ, between E_{x0} and E_{y0}, where $\phi < \pi/2$. For simplicity, we will consider propagation in a lossless medium. The total field in phasor form is

$$\mathbf{E}_s = (E_{x0}\mathbf{a}_x + E_{y0}e^{j\phi}\mathbf{a}_y)e^{-j\beta z} \qquad (93)$$

Again, to aid in visualization, we convert this wave to real instantaneous form by multiplying by $e^{j\omega t}$ and taking the real part:

$$\mathbf{E}(z,t) = E_{x0}\cos(\omega t - \beta z)\mathbf{a}_x + E_{y0}\cos(\omega t - \beta z + \phi)\mathbf{a}_y \qquad (94)$$

where we have assumed that E_{x0} and E_{y0} are real. Suppose we set $t = 0$, in which case (94) becomes [using $\cos(-x) = \cos(x)$]

$$\mathbf{E}(z,0) = E_{x0}\cos(\beta z)\mathbf{a}_x + E_{y0}\cos(\beta z - \phi)\mathbf{a}_y \qquad (95)$$

The component magnitudes of $\mathbf{E}(z,0)$ are plotted as functions of z in Figure 12.6. Since time is fixed at zero, the wave is frozen in position. An observer can move along the z axis, measuring the component magnitudes and thus the orientation of the total electric field at each point. Let's consider a crest of E_x, indicated as point a in Figure 12.6. If ϕ were zero, E_y would have a crest at the same location. Since ϕ is not zero (and positive), the crest of E_y that would otherwise occur at point a is now displaced to point b farther down z. The two points are separated by distance ϕ/β. E_y thus *lags behind* E_x when we consider the *spatial* dimension.

Now suppose the observer stops at some location on the z axis, and time is allowed to move forward. Both fields now move in the positive z direction, as (94) indicates. But point b reaches the observer first, followed by point a. So we see that E_y *leads* E_x when we consider the *time* dimension. In either case (fixed t and varying z, or vice versa) the observer notes that the net field rotates about the z axis while its magnitude changes. Considering a starting point in z and t, at which the field has a given orientation and magnitude, the wave will return to the same orientation and magnitude at a distance of one wavelength in z (for fixed t) or at a time $t = 2\pi/\omega$ later (at a fixed z).

For illustration purposes, if we take the length of the field vector as a measure of its magnitude, we find that at a fixed position, the tip of the vector traces out the shape of an ellipse over time $t = 2\pi/\omega$. The wave is said to be *elliptically polarized*. Elliptical polarization is in fact the most general polarization state of a wave, since it encompasses any magnitude and phase difference between E_x and E_y. Linear polarization is a special case of elliptical polarization in which the phase difference is zero.

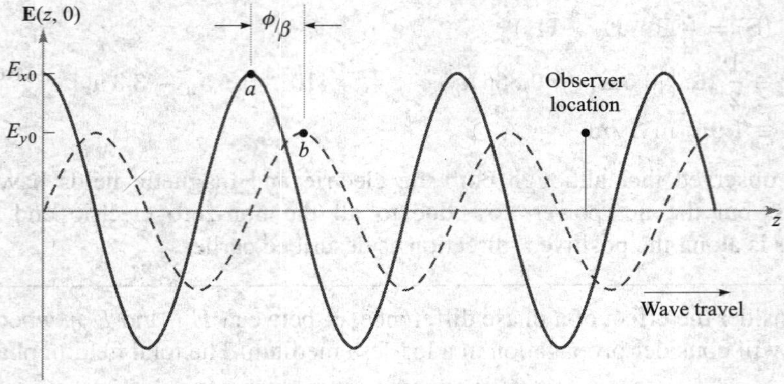

Fig. 12.6 Plots of the electric field component magnitudes in Eq. (95) as functions of z. Note that the y component lags behind the x component in z. As time increases from zero, both waves travel to the right, as per Eq. (94). Thus to an observer at a fixed location, the y component leads in time.

Another special case of elliptical polarization occurs when $E_{x0} = E_{y0} = E_0$ and when $\phi = \pm\pi/2$. The wave in this case exhibits *circular polarization*. To see this, we incorporate these restrictions into Eq. (94) to obtain

$$\mathbf{E}(z,t) = E_0[\cos(\omega t - \beta z)\mathbf{a}_x + \cos(\omega t - \beta z \pm \pi/2)\mathbf{a}_y]$$

$$= E_0[\cos(\omega t - \beta z)\mathbf{a}_x \mp \sin(\omega t - \beta z)\mathbf{a}_y] \tag{96}$$

If we consider a fixed position along z (such as $z = 0$) and allow time to vary, (96), with $\phi = +\pi/2$, becomes

$$\mathbf{E}(0,t) = E_0[\cos(\omega t)\mathbf{a}_x - \sin(\omega t)\mathbf{a}_y] \tag{97}$$

If we choose $-\pi/2$ in (96), we obtain

$$\mathbf{E}(0,t) = E_0[\cos(\omega t)\mathbf{a}_x + \sin(\omega t)\mathbf{a}_y] \tag{98}$$

The field vector of Eq. (98) rotates in the counterclockwise direction in the xy plane, while maintaining constant amplitude E_0, and so the tip of the vector traces out a circle. Figure 12.7 shows this behavior.

Choosing $+\pi/2$ leads to (97), whose field vector rotates in the clockwise direction. The *handedness* of the circular polarization is associated with the rotation and propagation directions in the following manner: The wave exhibits *left circular polarization* (l.c.p.) if, when orienting the left hand with the thumb in the direction of propagation, the fingers curl in the rotation direction of the field with time. The wave exhibits *right circular polarization* (r.c.p.) if, with the right-hand thumb

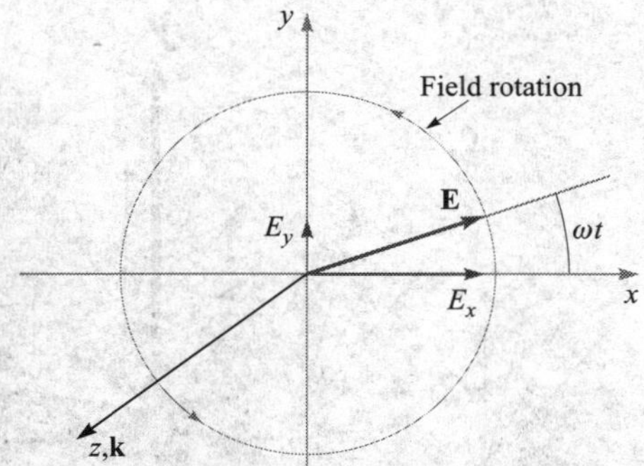

Fig. 12.7 Electric field in the xy plane of a right circularly polarized plane wave, as described by Eq. (98). As the wave propagates in the forward z direction, the field vector rotates counterclockwise in the xy plane.

in the propagation direction, the fingers curl in the field rotation direction.[7] Thus, with forward z propagation, (97) describes a left circularly polarized wave, and (98) describes a right circularly polarized wave. The same convention is applied to elliptical polarization, in which the descriptions *left elliptical polarization* and *right elliptical polarization* are used.

Using (96), the instantaneous angle of the field from the x direction can be found for any position along z through

$$\theta(z,t) = \tan^{-1}\left(\frac{E_y}{E_x}\right) = \tan^{-1}\left(\frac{\mp\sin(\omega t - \beta z)}{\cos(\omega t - \beta z)}\right) = \mp(\omega t - \beta z) \tag{99}$$

where again the minus sign (yielding l.c.p. for positive z travel) applies for the choice of $\phi = +\pi/2$ in (96); the plus sign (yielding r.c.p. for positive z travel) is used if $\phi = -\pi/2$. If we choose $z = 0$, the angle becomes simply ωt, which reaches 2π (one complete rotation) at time $t = 2\pi/\omega$. If we chose $t = 0$ and allow z to vary, we form a "corkscrew"-like field pattern. One way to visualize this is to consider a spiral staircase–shaped pattern, in which the field lines (stairsteps) are perpendicular to the z (or staircase) axis. The relationship between this spatial field pattern and the resulting time behavior at fixed z as the wave propagates is shown in an artist's conception in Figure 12.8.

The handedness of the polarization is changed by reversing the pitch of the corkscrew pattern. The spiral staircase model is only a visualization aid. It must be remembered that the wave is still a uniform plane wave whose fields at any position along z are infinite in extent over the transverse plane.

There are many uses of circularly polarized waves. Perhaps the most obvious advantage is that reception of a wave having circular polarization does not depend on the antenna orientation in the plane normal to the propagation direction. Dipole antennas, for example, are required to be oriented along the electric field direction of the signal they receive. If circularly polarized signals

[7] This convention is reversed by some workers (most notably in optics), who emphasize the importance of the *spatial* field configuration. Note that r.c.p. by our definition is formed by propagating a spatial field that is in the shape of a *left-handed* screw, and for that reason it is sometimes called left circular polarization (see Figure 12.8). Left circular polarization as we define it results from propagating a spatial field in the shape of a right-handed screw, and it is called right circular polarization by the spatial enthusiasts. Caution is obviously necessary in interpreting what is meant when polarization handedness is stated in an unfamiliar text.

Fig. 12.8 Representation of a right circularly polarized wave. The electric field vector (in white) will rotate toward the y axis as the entire wave moves through the xy plane in the direction of k. This counterclockwise rotation (when looking toward the wave source) satisfies the temporal right-handed rotation convention as described in the text. The wave, however, appears as a left-handed screw, and for this reason it is called left circular polarization in the other convention.

are transmitted, the receiver orientation requirements are relaxed considerably. In optics, circularly polarized light can be passed through a polarizer of any orientation, thus yielding linearly polarized light in any direction (although one loses half the original power this way). Other uses involve treating linearly polarized light as a superposition of circularly polarized waves, to be described next.

Circularly polarized light can be generated using an *anisotropic* medium—a material whose permittivity is a function of electric field direction. Many crystals have this property. A crystal orientation can be found such that along one direction (say, the x axis), the permittivity is lowest, while along the orthogonal direction (y axis), the permittivity is highest. The strategy is to input a linearly polarized wave with its field vector at 45 degrees to the x and y axes of the crystal. It will thus have equal-amplitude x and y components in the crystal, and these will now propagate in the z direction at different speeds. A phase difference (or *retardation*) accumulates between the components as they propagate, which can reach $\pi/2$ if the crystal is long enough. The wave at the output thus becomes circularly polarized. Such a crystal, cut to the right length and used in this manner, is called a *quarter-wave plate,* since it introduces a relative phase shift of $\pi/2$ between E_x and E_y, which is equivalent to $\lambda/4$.

It is useful to express circularly polarized waves in phasor form. To do this, we note that (96) can be expressed as

$$\mathbf{E}(z,t) = \text{Re}\Big\{ E_0 e^{j\omega t} e^{-j\beta z} \Big[\mathbf{a}_x + e^{\pm j\pi/2} \mathbf{a}_y \Big] \Big\}$$

Using the fact that $e^{\pm j\pi/2} = \pm j$, we identify the phasor form as:

$$\mathbf{E}_s = E_0(\mathbf{a}_x \pm j\mathbf{a}_y)e^{-j\beta z} \tag{100}$$

where the plus sign is used for left circular polarization and the minus sign for right circular polarization. If the wave propagates in the negative z direction, we have

$$\mathbf{E}_s = E_0(\mathbf{a}_x \pm j\mathbf{a}_y)e^{+j\beta z} \tag{101}$$

where in this case the positive sign applies to right circular polarization and the minus sign to left circular polarization. The student is encouraged to verify this.

EXAMPLE 12.15

The electric field of a 100 MHz plane wave propagating in a non-magnetic medium is given by the following expression:

$$\mathbf{E}_s = 2\left[\mathbf{a}_y - j\,\mathbf{a}_z\right] e^{-j\frac{5\pi}{3}x}\,\mathrm{V/m} \tag{i}$$

(*a*) Compute the dielectric constant and intrinsic impedance of the medium.
(*b*) Determine the wavelength of the wave.
(*c*) Describe the polarization of the wave (linear/circular/elliptical) using appropriate diagram.
(*d*) Find the corresponding magnetic field in phasor form.

Solution.

(*a*) It can be observed from (i) that the phase constant is given by

$$\beta = \frac{5\pi}{3} \equiv \frac{\omega}{c}\sqrt{\epsilon_r'} = \frac{2\,\pi \times 10^8}{3 \times 10^8}\sqrt{\epsilon_r'} = \frac{5\pi}{3} \quad \Rightarrow \epsilon_r' = 6.25 \tag{ii}$$

The intrinsic impedance may be calculated as

$$\eta = \sqrt{\frac{\mu}{\epsilon}} \equiv \frac{\eta_0}{\sqrt{\epsilon_r'}} \equiv \frac{120\pi}{\sqrt{6.25}} = 150.8\,\Omega \tag{iii}$$

(*b*) The wavelength would be given by

$$\lambda = \frac{2\pi}{\beta} \equiv \frac{2\pi}{5\pi/3} = 1.2\,\mathrm{m} \tag{iv}$$

(*c*) For finding the polarization of the wave, let us first convert the phasor form of the field given by (i) into instantaneous form as follows:

$$\mathbf{E}(x,t) = \mathrm{Re}\left[\mathbf{E}_s e^{j\omega t}\right] \equiv \mathrm{Re}\left[2\,(\mathbf{a}_y - j\,\mathbf{a}_z)\,e^{-j\frac{5\pi}{3}x}e^{j\omega t}\right]$$

$$\equiv \text{Re}\left[2\left(\mathbf{a}_y + e^{-j\frac{\pi}{2}}\,\mathbf{a}_z\right)e^{-j\frac{5\pi}{3}x}e^{j\omega t}\right] \text{V/m}$$

$$\equiv 2\left[\cos\left(2\pi\times 10^8 t - \frac{5\pi}{3}x\right)\mathbf{a}_y + \cos\left(2\pi\times 10^8 t - \frac{5\pi}{3}x - \frac{\pi}{2}\right)\mathbf{a}_z\right] \text{V/m}$$

$$\Rightarrow \mathbf{E}(x,t) \equiv 2\left[\cos\left(2\pi\times 10^8 t - \frac{5\pi}{3}x\right)\mathbf{a}_y + \sin\left(2\pi\times 10^8 t - \frac{5\pi}{3}x\right)\mathbf{a}_z\right] \text{V/m} \quad \text{(v)}$$

It is also to be observed from (i) that the wave is propagating along the positive x-direction. Now, for finding the orientation of the electric field vector as a function of time, let us try to put $x = 0$ in the above equation, which yields the following expression:

$$\Rightarrow \mathbf{E}(0,t) \equiv 2\left[\cos(\omega t)\,\mathbf{a}_y + \sin(\omega t)\,\mathbf{a}_z\right] \text{V/m} \quad \text{(vi)}$$

where, $\omega \equiv 2\pi \times 10^8$ rad/s is substituted in the above equation for convenience.

For plotting (vi), we first compute the electric field components at various instants as follows:

S. No.	ωt	$\mathbf{E}(0,t)$
1.	0	$2\,\mathbf{a}_y$
2.	$\frac{\pi}{2}$	$2\,\mathbf{a}_z$
3.	π	$-2\,\mathbf{a}_y$
4.	$\frac{3\pi}{2}$	$-2\,\mathbf{a}_z$
5.	2π	$2\,\mathbf{a}_y$

The electric field components at different time intervals can now be plotted in the *yz* plane as shown below, where it can be observed that the tip of the field vector traces a *circle* of 2 V/m radius over time $t \equiv 2\pi/\omega$, thus making the polarization circular.

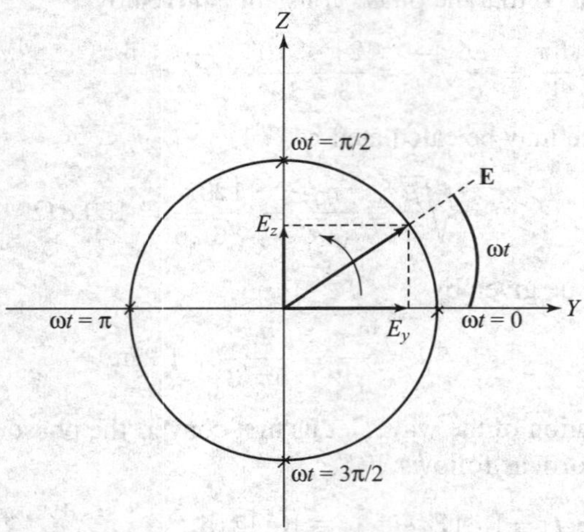

Fig. 12.9 Right circular polarization.

From Figure 12.9, it can be seen that for the propagation in the positive x-direction, the wave exhibits a *right circular polarization*. You should note that the positive x-direction is denoted by the thumb of the right hand flowing out of the paper with fingers representing the rotation direction of the electric field with time. It should also be noted that the electric field maintains a constant amplitude of 2 V/m while rotating in the counterclockwise direction in the *yz* plane.

(*d*) For determining the magnetic field in phasor form, let us try to compute the two components of the magnetic field separately. For this purpose, let us rewrite the electric field in the following form:

$$\mathbf{E}_s = \left[2e^{-j\frac{5\pi}{3}x}\mathbf{a}_y - j\,2e^{-j\frac{5\pi}{3}x}\mathbf{a}_z\right] \equiv [E_{sy}\mathbf{a}_y - jE_{sz}\mathbf{a}_z] \text{ V/m}$$

Now, the wave is propagating along the positive x-direction. Hence, the y component of the electric field will be associated with the z-component of the magnetic field, and the z-component of the electric field will be associated with the y-component of the magnetic field. The direction of the corresponding magnetic field vector will be decided by the Poynting vector using the condition that the power flow is always along the positive x-direction. This leads to the following sets of equations:

$$E_{sy}\,\mathbf{a}_y \times H_{sz}\,\mathbf{a}_z \equiv S_x\,\mathbf{a}_x$$

$$E_{sz}\,\mathbf{a}_z \times -H_{sy}\,\mathbf{a}_y \equiv S_x\,\mathbf{a}_x$$

The magnitudes of the individual components of the magnetic field can be computed as follows:

$$H_{sz} \equiv \frac{E_{sy}}{\eta} = \frac{2e^{-j\frac{5\pi}{3}x}}{150.8} = 1.33\,e^{-j\frac{5\pi}{3}x}\text{mA/m}\,;\; H_{sy} \equiv \frac{E_{sz}}{\eta} = 1.33\,e^{-j\frac{5\pi}{3}x}\text{mA/m}$$

Hence, the total magnetic field in the present situation can be written using the following expression:

$$\begin{aligned}\mathbf{H}_s &= [H_{sz}\,\mathbf{a}_z + j\,H_{sy}\,\mathbf{a}_y] \text{ A/m}\\ &\equiv 1.33\,[\mathbf{a}_z + j\,\mathbf{a}_y]\;e^{-j\frac{5\pi}{3}x}\text{ mA/m}\end{aligned}$$

EXAMPLE 12.16

Let us consider the result of superimposing left and right circularly polarized fields of the same amplitude, frequency, and propagation direction, but where a phase shift of δ radians exists between the two.

Solution. Taking the waves to propagate in the $+z$ direction, and introducing a relative phase, δ, the total phasor field is found, using (100):

$$\mathbf{E}_{sT} = \mathbf{E}_{sR} + \mathbf{E}_{sL} = E_0[\mathbf{a}_x - j\mathbf{a}_y]e^{-j\beta z} + E_0[\mathbf{a}_x + j\mathbf{a}_y]e^{-j\beta z}e^{j\delta}$$

Grouping components together, this becomes

$$\mathbf{E}_{sT} = E_0[(1 + e^{j\delta})\mathbf{a}_x - j(1 - e^{j\delta})\mathbf{a}_y]e^{-j\beta z}$$

Factoring out an overall phase term, $e^{j\delta/2}$, we obtain

$$\mathbf{E}_{sT} = E_0 e^{j\delta/2}\left[(e^{-j\delta/2} + e^{j\delta/2})\mathbf{a}_x - j(e^{-j\delta/2} - e^{j\delta/2})\mathbf{a}_y\right]e^{-j\beta z}$$

From Euler's identity, we find that $e^{j\delta/2} + e^{-j\delta/2} = 2\cos\delta/2$, and $e^{j\delta/2} - e^{-j\delta/2} = 2j\sin\delta/2$. Using these relations, we obtain

$$\mathbf{E}_{sT} = 2E_0[\cos(\delta/2)\mathbf{a}_x + \sin(\delta/2)\mathbf{a}_y]e^{-j(\beta z - \delta/2)} \tag{102}$$

We recognize (102) as the electric field of a *linearly polarized* wave, whose field vector is oriented at angle $\delta/2$ from the x axis.

EXAMPLE 12.17

The magnetic field of a plane wave propagating in free space is given by the following expression:

$$\mathbf{H}_s = 0.05\,[\mathbf{a}_y - j\,\mathbf{a}_x]\,e^{-j6\pi z}\ \text{A/m} \tag{i}$$

(a) Find the frequency and direction of wave propagation.

(b) Find the corresponding electric field in phasor form.

(c) Determine the polarization of the wave (linear/circular/elliptical), and the sense of polarization (right handed/left handed).

(d) Compute the time-average power density associated with the wave.

Solution.

(a) It can be easily established that the wave is propagating along the *positive* z-direction by comparing (i) with the standard phasor form of the equation for the field. The frequency of the wave may be computed as

$$\beta = 6\pi \equiv \frac{2\pi f}{c} \Rightarrow f = \frac{6\pi \times 3 \times 10^8}{2\pi} = 900\ \text{MHz} \tag{ii}$$

where c is the velocity of wave in free space as given in the problem.

(b) Let us try to compute the electric field here from the first principle using the Maxwell's equation applied to free space as follows

$$\nabla \times \mathbf{H}_s \equiv j\omega\epsilon_0\mathbf{E}_s$$

$$\Rightarrow \begin{bmatrix} \mathbf{a}_x & \mathbf{a}_y & \mathbf{a}_z \\ \dfrac{\partial}{\partial x} & \dfrac{\partial}{\partial y} & \dfrac{\partial}{\partial z} \\ H_{xs} & H_{ys} & H_{zs} \end{bmatrix} \equiv j\omega\epsilon_0[E_x\mathbf{a}_x + E_y\mathbf{a}_y + E_z\mathbf{a}_z]$$

In the next step, we can substitute the value of different components of the magnetic field as given in the problem, and we can also note that there is no variation of the field in the x or y directions, which basically means that $\frac{\partial}{\partial x} = \frac{\partial}{\partial y} = 0$ in the above equation. These substitutions lead to the following expression:

$$\Rightarrow \begin{bmatrix} \mathbf{a}_x & \mathbf{a}_y & \mathbf{a}_z \\ 0 & 0 & \frac{\partial}{\partial z} \\ -j0.05e^{-j6\pi z} & 0.05e^{-j6\pi z} & 0 \end{bmatrix} \equiv j\omega\epsilon_0[E_x\mathbf{a}_x + E_y\mathbf{a}_y + E_z\mathbf{a}_z]$$

The above equation can then be expanded to obtain

$$-0.05\left\{\frac{\partial}{\partial z}[e^{-j6\pi z}]\mathbf{a}_x + j\frac{\partial}{\partial z}[e^{-j6\pi z}]\mathbf{a}_y\right\} = j\omega\epsilon_0[E_x\mathbf{a}_x + E_y\mathbf{a}_y + E_z\mathbf{a}_z]$$
$$\Rightarrow 0.3\pi[j\mathbf{a}_x - \mathbf{a}_y]e^{-j6\pi z} = j\omega\epsilon_0[E_x\mathbf{a}_x + E_y\mathbf{a}_y + E_z\mathbf{a}_z]$$

The individual terms on both sides of the above equation can finally be compared to obtain the following electric field components:

$$E_x = \frac{j}{j\omega\epsilon_0}0.3\pi e^{-j6\pi z} \equiv \frac{0.3\pi e^{-j6\pi z}}{2\pi \times 900 \times 10^6 \times 8.854 \times 10^{-12}} = 18.8e^{-j6\pi z};$$

$$E_y = -\frac{1}{j}\frac{0.3\pi}{\omega\epsilon_0}e^{-j6\pi z} = j18.8\,e^{-j6\pi z} \quad E_z = 0;$$

Finally, the individual electric field components can be combined together to obtain the phasor form of the electric field as follows:

$$\mathbf{E}_s = 18.8\,[\mathbf{a}_x + j\,\mathbf{a}_y]\,e^{-j6\pi z}\ \text{V/m} \tag{iii}$$

(*c*) For finding the polarization of the wave, let us first convert the phasor form of the field given by (iii) into instantaneous form as follows:

$$\mathbf{E}(z,t) = \text{Re}\left[\mathbf{E}_s e^{j\omega t}\right] \equiv \mathbf{Re}\left[18.8\,(\mathbf{a}_x + j\,\mathbf{a}_y)\,e^{-j6\pi z}e^{j\omega t}\right]$$

$$\equiv \text{Re}\left[18.8\left(\mathbf{a}_x + e^{j\frac{\pi}{2}}\,\mathbf{a}_y\right)\,e^{-j6\pi z}e^{j\omega t}\right]\text{V/m}$$

$$\equiv 18.8\left\{\cos(\omega t - 6\pi z)\ \mathbf{a}_x + \cos\left[(\omega t - 6\pi z) + \frac{\pi}{2}\right]\mathbf{a}_y\right\}\text{V/m}$$

$$\Rightarrow \mathbf{E}(z,t) \equiv 18.8\,[\cos(\omega t - 6\pi z)\,\mathbf{a}_x - \sin(\omega t - 6\pi z)\,\mathbf{a}_y]\ \text{V/m} \tag{iv}$$

Now, for finding the orientation of the electric field vector as a function of time at a fixed position, let us put $z = 0$ in the above equation, which yields the following simplified expression:

$$\Rightarrow \mathbf{E}(0,t) \equiv 18.8\,[\cos(\omega\,t)\ \mathbf{a}_x - \sin(\omega\,t)\,\mathbf{a}_y]\,\text{V/m} \tag{v}$$

where, $\omega \equiv 1.8\,\pi \times 10^9$ rad/s is given in the problem.
Equation (v) can be plotted by computing the electric field components at various instant of time as follows:

S. No.	ωt	$\mathbf{E}(0,t)$
1.	0	$18.8\,\mathbf{a}_x$
2.	$\frac{\pi}{2}$	$-18.8\,\mathbf{a}_y$
3.	π	$-18.8\,\mathbf{a}_x$
4.	$\frac{3\pi}{2}$	$18.8\,\mathbf{a}_y$
5.	2π	$18.8\,\mathbf{a}_x$

The electric field components at different time intervals can now be plotted in the *xy* plane as shown below, where it can be observed that the tip of the field vector traces a *circle* over time $t \equiv 2\pi/\omega$ thus making the polarization circular. The amplitude of the electric field remains constant at 18.8 V/m as observed from the above table.

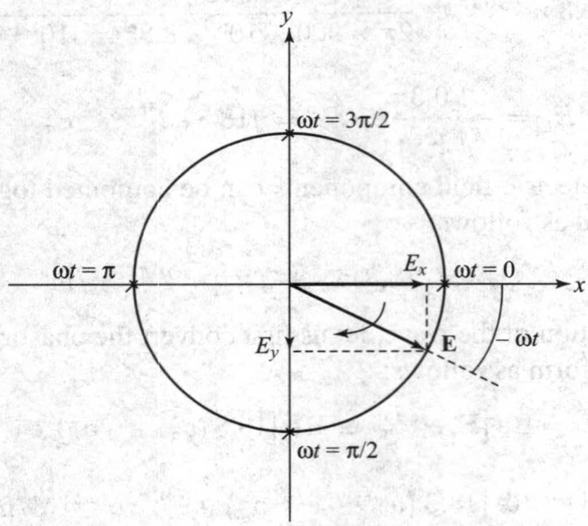

Fig. 12.10 Left circular polarization.

From Figure 12.10, it can be observed that for the propagation along the positive z direction (out of the paper), the wave exhibits a *left circular polarization*. You can also verify this by noting that the left-hand thumb flows out of the paper representing the direction of wave propagation when its fingers are oriented along the rotation direction of the electric field with time.

(d) The time-average power density associated with the wave can be computed using the phasor forms of the electric and magnetic field vectors as follows:

$$\begin{aligned}\langle \mathbf{S} \rangle &= \frac{1}{2}\Re e\,(\mathbf{E}_s \times \mathbf{H}_s^*) \\ &\equiv \frac{1}{2}\ \mathrm{Re}\{18.8\,[\mathbf{a}_x + j\,\mathbf{a}_y]\,e^{-j6\pi z} \times\ 0.05\,[\mathbf{a}_y + j\,\mathbf{a}_x]\,e^{j6\pi z}\} \qquad \text{(vi)} \\ &\equiv \frac{0.94}{2}\ \mathrm{Re}\{[\mathbf{a}_x + j\,\mathbf{a}_y] \times [\mathbf{a}_y + j\,\mathbf{a}_x]\} \\ &\equiv 0.94\,\mathbf{a}_z\ \mathrm{W/m^2}\end{aligned}$$

It can be observed from the above equation that the net power flow is along the positive z direction as described earlier.

Example 12.16 shows that any linearly polarized wave can be expressed as the sum of two circularly polarized waves of opposite handedness, where the linear polarization direction is determined by the relative phase difference between the two waves. Such a representation is convenient (and necessary) when considering, for example, the propagation of linearly polarized light through media which contain organic molecules. These often exhibit spiral structures having left- or right-handed pitch, and they will thus interact differently with left- or right-hand circular polarization. As a result, the left circular component can propagate at a different speed than the right circular component, and so the two waves will accumulate a phase difference as they propagate. As a result, the direction of the linearly polarized field vector at the output of the material will differ from the direction that it had at the input. The extent of this rotation can be used as a measurement tool to aid in material studies.

Polarization issues will become extremely important when we consider wave reflection in Chapter 13.

References

1. Balanis, C. A. *Advanced Engineering Electromagnetics*. New York: John Wiley & Sons, 1989.
2. International Telephone and Telegraph Co., Inc. *Reference Data for Radio Engineers.* 7th ed. Indianapolis, Ind.: Howard W. Sams & Co., 1985. This handbook has some excellent data on the properties of dielectric and insulating materials.
3. Jackson, J. D. *Classical Electrodynamics*. 3d ed. New York: John Wiley & Sons, 1999.
4. Ramo, S., J. R. Whinnery, and T. Van Duzer. *Fields and Waves in Communication Electronics*. 3d ed. New York: John Wiley & Sons, 1994.

CHAPTER 12 Problems

Quizzes

12.1 Show that $E_{xs} = Ae^{j(k_0 z+\phi)}$ is a solution of the vector Helmholtz equation, Eq. (30), for $k_0 = \omega\sqrt{\mu_0\epsilon_0}$ and any ϕ and A.

12.2 A 100 MHz uniform plane wave propagates in a lossless medium for which $\epsilon_r = 5$ and $\mu_r = 1$. Find: (*a*) ν_p; (*b*) β; (*c*) λ; (*d*) $\mathbf{E}_s$; (*e*) $\mathbf{H}_s$; (*f*) $\langle \mathbf{S} \rangle$.

12.3 An $\mathbf{H}$ field in free space is given as $\mathcal{H}(x,t) = 10\cos(10^8 t - \beta x)\mathbf{a}_y$ A/m. Find: (*a*) β; (*b*) λ; (*c*) $\mathcal{E}(x,t)$ at $P(0.1, 0.2, 0.3)$ at $t = 1$ ns.

12.4 Given $\mathcal{E}(z,t) = E_0 e^{-\alpha z}\sin(\omega t - \beta z)\mathbf{a}_x$, and $\eta = |\eta| e^{j\phi}$, find: (*a*) $\mathbf{E}_s$; (*b*) $\mathbf{H}_s$; (*c*) $\langle \mathbf{S} \rangle$.

12.5 A 150 MHz uniform plane wave in free space is described by $\mathbf{H}_s = (4 + j10)(2\mathbf{a}_x + j\mathbf{a}_y)e^{-j\beta z}$ A/m. (*a*) Find numerical values for ω, λ, and β. (*b*) Find $\mathcal{H}(z,t)$ at $t = 1.5$ ns, $z = 20$ cm. (*c*) What is $|E|_{\max}$?

12.6 A linearly polarized plane wave in free space has electric field given by $\mathcal{E}(z,t) = (25\mathbf{a}_x - 30\mathbf{a}_z)\cos(\omega t - 50y)$ V/m. Find: (*a*) ω; (*b*) $\mathbf{E}_s$; (*c*) $\mathbf{H}_s$; (*d*) $\langle \mathbf{S} \rangle$.

12.7 The phasor magnetic field intensity for a 400 MHz uniform plane wave propagating in a certain lossless material is $(2\mathbf{a}_y - j5\mathbf{a}_z)e^{-j25x}$ A/m. Knowing that the maximum amplitude of **E** is 1500 V/m, find β, η, λ, ν_p, ϵ_r, μ_r, and $\mathcal{H}(x, y, z, t)$.

12.8 Let the fields $\mathcal{E}(z, t) = 1800\cos(10^7\pi t - \beta z)\mathbf{a}_x$ V/m and $\mathcal{H}(z, t) = 3.8\cos(10^7\pi t - \beta z)\mathbf{a}_y$ A/m represent a uniform plane wave propagating at a velocity of 1.4×10^8 m/s in a perfect dielectric. Find: (*a*) β; (*b*) λ; (*c*) η; (*d*) μ_r; (*e*) ϵ_r.

12.9 A certain lossless material has $\mu_r = 4$ and $\epsilon_r = 9$. A 10 MHz uniform plane wave is propagating in the $\mathbf{a}_y$ direction with $E_{x0} = 400$ V/m and $E_{y0} = E_{z0} = 0$ at $P(0.6, 0.6, 0.6)$ at $t = 60$ ns. (*a*) Find β, λ, ν_p, and η. (*b*) Find $E(t)$. (*c*) Find $H(t)$.

12.10 In a medium characterized by intrinsic impedance $\eta = |\eta|e^{j\phi}$, a linearly polarized plane wave propagates, with magnetic field given as $\mathbf{H}_s = (H_{0y}\mathbf{a}_y + H_{0z}\mathbf{a}_z)e^{-\alpha x}e^{-j\beta x}$. Find: (*a*) $\mathbf{E}_s$; (*b*) $\mathcal{E}(x, t)$; (*c*) $\mathcal{H}(x, t)$; (*d*) $\langle \mathbf{S} \rangle$.

12.11 A 2 GHz uniform plane wave has an amplitude $E_{y0} = 1.4$ kV/m at $(0, 0, 0, t = 0)$ and is propagating in the $\mathbf{a}_z$ direction in a medium where $\epsilon'' = 1.6 \times 10^{-11}$ F/m, $\epsilon' = 3.0 \times 10^{-11}$ F/m, and $\mu = 2.5$ μH/m. Find: (*a*) E_y at $P(0, 0, 1.8\text{ cm})$ at 0.2 ns; (*b*) H_x at P at 0.2 ns.

12.12 The plane wave $\mathbf{E}_s = 300e^{-jkx}\mathbf{a}_y$ V/m is propagating in a material for which $\mu = 2.25$ μH/m, $\epsilon' = 9$ pF/m, and $\epsilon'' = 7.8$ pF/m. If $\omega = 64$ M rad/s, find: (*a*) α; (*b*) β; (*c*) ν_p; (*d*) λ; (*e*) η; (*f*) $\mathbf{H}_s$; (*g*) $\mathcal{E}(3, 2, 4, 10\text{ ns})$.

12.13 Let $jk = 0.2 + j1.5$ m^{-1} and $\eta = 450 + j60\,\Omega$ for a uniform plane propagating in the $\mathbf{a}_z$ direction. If $\omega = 300$ M rad/s, find μ, ϵ', and ϵ'' for the medium.

12.14 A certain nonmagnetic material has the material constants $\epsilon'_r = 2$ and $\epsilon''/\epsilon' = 4 \times 10^{-4}$ at $\omega = 1.5$ Grad/s. Find the distance a uniform plane wave can propagate through the material before: (*a*) it is attenuated by 1 Np; (*b*) the power level is reduced by one-half; (*c*) the phase shifts 360°.

12.15 A 10 GHz radar signal may be represented as a uniform plane wave in a sufficiently small region. Calculate the wavelength in centimeters and the attenuation in nepers per meter if the wave is propagating in a nonmagnetic material for which: (*a*) $\epsilon'_r = 1$ and $\epsilon''_r = 0$; (*b*) $\epsilon'_r = 1.04$ and $\epsilon''_r = 9.00 \times 10^{-4}$; (*c*) $\epsilon'_r = 2.5$ and $\epsilon''_r = 7.2$.

12.16 The power factor of a capacitor is defined as the cosine of the impedance phase angle, and its Q is ωCR, where R is the parallel resistance. Assume an idealized parallel-plate capacitor having a dielectric characterized by σ, ϵ', and μ_r. Find both the power factor and Q in terms of the loss tangent.

12.17 Let $\eta = 250 + j30\,\Omega$ and $jk = 0.2 + j2\text{m}^{-1}$ for a uniform plane wave propagating in the $\mathbf{a}_z$ direction in a dielectric having some finite conductivity. If $|E_s| = 400$ V/m at $z = 0$, find: (*a*) $\langle \mathbf{S} \rangle$ at $z = 0$ and $z = 60$ cm; (*b*) the average ohmic power dissipation in watts per cubic meter at $z = 60$ cm.

12.18 Given: a 100 MHz uniform plane wave in a medium known to be a good dielectric. The phasor electric field is $\mathcal{E}_s = 4e^{-0.5z}e^{-j20z}\mathbf{a}_x$ V/m. Determine: (*a*) ϵ'; (*b*) ϵ''; (*c*) η; (*d*) $\mathbf{H}_s$; (*e*) $\langle \mathbf{S} \rangle$; (*f*) the power in watts that is incident on a rectangular surface measuring 20 m × 30 m at $z = 1$ km.

12.19 Perfectly conducting cylinders with radii of 8 mm and 20 mm are coaxial. The region between the cylinders is filled with a perfect dielectric for which $\epsilon = 10^{-9}/4\pi$ F/m and $\mu_r = 1$. If $\mathcal{E}$ in this region is $(500/\rho)\cos(\omega t - 4z)\mathbf{a}_\rho$ V/m, find: (*a*) ω, with the help of Maxwell's equations in cylindrical coordinates; (*b*) $\mathcal{H}(\rho, z, t)$; (*c*) $\langle \mathbf{S}(\rho, z, t)\rangle$; (*d*) the average power passing through every cross section $8 < \rho < 20$ mm, $0 < \phi < 2\pi$.

12.20 If $\mathbf{E}_s = 60\frac{\sin\theta}{r}e^{-j2r}\mathbf{a}_\theta$ V/m and $\mathbf{H}_s = \frac{\sin\theta}{4\pi r}e^{-j2r}\mathbf{a}_\phi$ A/m in free space, find the average power passing outward through the surface $r = 10^6$, $0 < \theta < \pi/3$, $0 < \phi < 2\pi$.

12.21 The cylindrical shell, 1 cm$< \rho <$ 1.2 cm, is composed of a conducting material for which $\sigma = 10^6$ S/m. The external and internal regions are nonconducting. Let $H_\phi = 2000$ A/m at $\rho = 1.2$ cm. (*a*) Find **H** everywhere. (*b*) Find **E** everywhere. (*c*) Find $\langle \mathbf{S}\rangle$ everywhere.

12.22 The inner and outer dimensions of a coaxial copper transmission line are 2 and 7 mm, respectively. Both conductors have thicknesses much greater than δ. The dielectric is lossless and the operating frequency is 400 MHz. Calculate the resistance per meter length of the: (*a*) inner conductor; (*b*) outer conductor; (*c*) transmission line.

12.23 A hollow tubular conductor is constructed from a type of brass having a conductivity of 1.2×10^7 S/m. The inner and outer radii are 9 and 10 mm, respectively. Calculate the resistance per meter length at a frequency of: (*a*) dc; (*b*) 20 MHz; (*c*) 2 GHz.

12.24 (*a*) Most microwave ovens operate at 2.45 GHz. Assume that $\sigma = 1.2 \times 10^6$ S/m and $\mu_r =$ 500 for the stainless steel interior, and find the depth of penetration. (*b*) Let $E_s = 50\angle 0°$ V/m at the surface of the conductor, and plot a curve of the amplitude of E_s versus the angle of E_s as the field propagates into the stainless steel.

12.25 A good conductor is planar in form, and it carries a uniform plane wave that has a wavelength of 0.3 mm and a velocity of 3×10^5 m/s. Assuming the conductor is nonmagnetic, determine the frequency and the conductivity.

12.26 The dimensions of a certain coaxial transmission line are $a = 0.8$ mm and $b = 4$ mm. The outer conductor thickness is 0.6 mm, and all conductors have $\sigma = 1.6 \times 10^7$ S/m. (*a*) Find R, the resistance per unit length at an operating frequency of 2.4 GHz. (*b*) Use information from Sections 6.4 and 9.10 to find C and L, the capacitance and inductance per unit length, respectively. The coax is air-filled. (*c*) Find α and β if $\alpha + j\beta = \sqrt{j\omega C(R + j\omega L)}$.

12.27 The planar surface $z = 0$ is a brass-Teflon interface. Use data available in Appendix C to evaluate the following ratios for a uniform plane wave having $\omega = 4 \times 10^{10}$ rad/s: (*a*) $\alpha_{\text{Tef}}/\alpha_{\text{brass}}$; (*b*) $\lambda_{\text{Tef}}/\lambda_{\text{brass}}$; (*c*) $v_{\text{Tef}}/v_{\text{brass}}$.

12.28 A uniform plane wave in free space has electric field vector given by $\mathbf{E}_s = 10e^{-j\beta x}\mathbf{a}_z + 15e^{-j\beta x}\mathbf{a}_y$ V/m. (*a*) Describe the wave polarization; (*b*) Find $\mathbf{H}_s$; (*c*) determine the average power density in the wave in W/m^2.

12.29 Consider a left circularly polarized wave in free space that propagates in the forward z direction. The electric field is given by the appropriate form of Eq. (100). (*a*) Determine the magnetic field phasor, $\mathbf{H}_s$; (*b*) determine an expression for the average power density in the wave in W/m^2 by direct application of Eq. (77).

12.30 The electric field of a uniform plane wave in free space is given by $\mathbf{E}_s = 100(\mathbf{a}_z + j\mathbf{a}_x)e^{-j50y}$. Determine: (*a*) f; (*b*) $\mathbf{H}_s$; (*c*) $\langle \mathbf{S}\rangle$. (*d*) Describe the polarization of the wave.

12.31 A linearly polarized uniform plane wave, propagating in the forward z direction, is input to a lossless *anisotropic* material, in which the dielectric constant encountered by waves polarized along $y(\epsilon_{ry})$ differs from that seen by waves polarized along $x(\epsilon_{rx})$. Suppose $\epsilon_{rx} = 2.15$, $\epsilon_{ry} = 2.10$, and the wave electric field at input is polarized at 45° to the positive x and y axes. (*a*) Determine, in terms of the free space wavelength, λ, the shortest length of the material such that the wave as it emerges from the output and is circularly polarized. (*b*) Will the output wave be right or left circularly polarized?

12.32 Suppose that the length of the medium of Problem 12.31 is made to be *twice* that determined in the problem. Describe the polarization of the output wave in this case.

12.33 Given a wave for which $\mathbf{E}_s = 15e^{-j\beta z}\mathbf{a}_x + 18e^{-j\beta z}e^{j\phi}\mathbf{a}_y$ V/m in a medium characterized by complex intrinsic impedance, η: (*a*) Find $\mathbf{H}_s$; (*b*) determine the average power density in W/m^2.

12.34 Given a general elliptically polarized wave as per Eq. (93):

$$\mathbf{E}_s = [E_{x0}\mathbf{a}_x + E_{y0}e^{j\phi}\mathbf{a}_y]e^{-j\beta z}$$

(*a*) Show, using methods similar to those of Example 12.16, that a linearly polarized wave results when superimposing the given field and a phase-shifted field of the form:

$$\mathbf{E}_s = [E_{x0}\mathbf{a}_x + E_{y0}e^{-j\phi}\mathbf{a}_y]e^{-j\beta z}e^{j\delta}$$

where δ is a constant. (*b*) Find δ in terms of ϕ such that the resultant wave is linearly polarized along x.

13

Plane Wave Reflection and Dispersion

In Chapter 12, we learned how to mathematically represent uniform plane waves as functions of frequency, medium properties, and electric field orientation. We also learned how to calculate the wave velocity, attenuation, and power. In this chapter we consider wave reflection and transmission at planar boundaries between different media. Our study will allow any orientation between the wave and boundary and will also include the important cases of multiple boundaries. We will also study the practical case of waves that carry power over a finite band of frequencies, as would occur, for example, in a modulated carrier. We will consider such waves in *dispersive* media, in which some parameter that affects propagation (permittivity for example) varies with frequency. The effect of a dispersive medium on a signal is of great importance because the signal envelope will change its shape as it propagates. As a result, detection and faithful representation of the original signal at the receiving end become problematic. Consequently, dispersion and attenuation must both be evaluated when establishing maximum allowable transmission distances.

13.1 REFLECTION OF UNIFORM PLANE WAVES AT NORMAL INCIDENCE

We first consider the phenomenon of reflection which occurs when a uniform plane wave is incident on the boundary between regions composed of two different materials. The treatment is specialized to the case of *normal incidence*—in which the wave propagation direction is perpendicular to the boundary. In later sections, we remove this restriction. Expressions will be found for the wave that is reflected from the interface and for that which is transmitted from one region into the other. These results are directly related to impedance-matching problems in ordinary transmission lines, as we have already encountered in Chapter 11. They are also applicable to waveguides, which we will study in Chapter 14.

We again assume that we have only a single vector component of the electric field intensity. Referring to Figure 13.1, we define region 1 (ϵ_1, μ_1) as the half-space for which $z < 0$; region 2 (ϵ_2, μ_2) is the half-space for which $z > 0$. Initially we establish a wave in region 1, traveling in the $+z$ direction, and linearly polarized along x.

$$\mathcal{E}_{x1}^+(z,t) = E_{x10}^+ e^{-\alpha_1 z} \cos(\omega t - \beta_1 z)$$

In phasor form, this is

$$E_{xs1}^+(z) = E_{x10}^+ e^{-jkz} \tag{1}$$

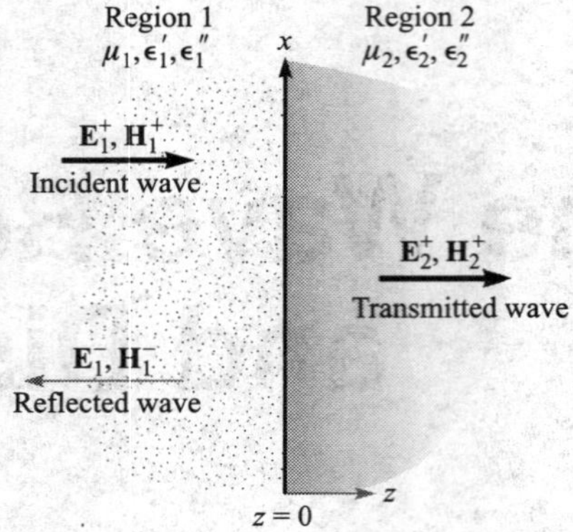

Fig. 13.1 A plane wave incident on a boundary establishes reflected and transmitted waves having the indicated propagation directions. All fields are parallel to the boundary, with electric fields along x and magnetic fields along y.

where we take E_{x10}^+ as real. The subscript 1 identifies the region, and the superscript + indicates a positively traveling wave. Associated with $E_{xs1}^+(z)$ is a magnetic field in the y direction,

$$H_{ys1}^+(z) = \frac{1}{\eta_1} E_{x10}^+ e^{-jk_1 z} \tag{2}$$

where k_1 and η_1 are complex unless ϵ_1'' (or σ_1) is zero. This uniform plane wave in region 1 that is traveling toward the boundary surface at $z = 0$ is called the *incident* wave. Since the direction of propagation of the incident wave is perpendicular to the boundary plane, we describe it as normal incidence.

We now recognize that energy may be transmitted across the boundary surface at $z = 0$ into region 2 by providing a wave moving in the $+z$ direction in that medium. The phasor electric and magnetic fields for this wave are

$$E_{xs2}^+(z) = E_{x20}^+ e^{-jk_2 z} \tag{3}$$

$$H_{ys2}^+(z) = \frac{1}{\eta_2} E_{x20}^+ e^{-jk_2 z} \tag{4}$$

This wave which moves away from the boundary surface into region 2 is called the *transmitted* wave. Note the use of the different propagation constant k_2 and intrinsic impedance η_2.

Now we must satisfy the boundary conditions at $z = 0$ with these assumed fields. With **E** polarized along x, the field is tangent to the interface, and therefore the **E** fields in regions 1 and 2 must be equal at $z = 0$. Setting $z = 0$ in (1) and (3) would require that $E_{x10}^+ = E_{x20}^+$. **H**, being y-directed, is also a tangential field, and must be continuous across the boundary (no current sheets are present in real media). When we let $z = 0$ in (2) and (4), we find that we must have $E_{x10}^+/\eta_1 = E_{x20}^+/\eta_2$. Since $E_{x10}^+ = E_{x20}^+$, then $\eta_1 = \eta_2$. But this is a very special condition that does not fit the facts in general, and we are therefore unable to satisfy the boundary conditions with only an incident and a transmitted wave. We require a wave traveling away from the boundary in region 1, as shown in Figure 13.1; this is the *reflected* wave,

$$E^-_{xs1}(z) = E^-_{x10}\, e^{jk_1 z} \tag{5}$$

$$H^-_{xs1}(z) = -\frac{E^-_{x10}}{\eta_1}\, e^{jk_1 z} \tag{6}$$

where E^-_{x10} may be a complex quantity. Since this field is traveling in the $-z$ direction, $E^-_{xs1} = -\eta_1 H^-_{ys1}$, for the Poynting vector shows that $\mathbf{E}^-_1 \times \mathbf{H}^-_1$ must be in the $-\mathbf{a}_z$ direction.

The boundary conditions are now easily satisfied, and in the process the amplitudes of the transmitted and reflected waves may be found in terms of E^+_{x10}. The total electric field intensity is continuous at $z = 0$,

$$E_{xs1} = E_{xs2} \qquad (z = 0)$$

or

$$E^+_{xs1} + E^-_{xs1} = E^+_{xs2} \qquad (z = 0)$$

Therefore

$$E^+_{x10} + E^-_{x10} = E^+_{x20} \tag{7}$$

Furthermore,

$$H_{ys1} = H_{ys2} \qquad (z = 0)$$

or

$$H^+_{ys1} + H^-_{ys1} = H^+_{ys2} \qquad (z = 0)$$

and therefore

$$\frac{E^+_{x10}}{\eta_1} - \frac{E^-_{x10}}{\eta_1} = \frac{E^+_{x20}}{\eta_2} \tag{8}$$

Solving (8) for E^+_{x20} and substituting into (7), we find

$$E^+_{x10} + E^-_{x10} = \frac{\eta_2}{\eta_1} E^+_{x10} - \frac{\eta_2}{\eta_1} E^-_{x10}$$

or

$$E^-_{x10} = E^+_{x10} \frac{\eta_2 - \eta_1}{\eta_2 + \eta_1}$$

The ratio of the amplitudes of the reflected and incident electric fields defines the *reflection coefficient,* designated by Γ,

$$\Gamma = \frac{E^-_{x10}}{E^+_{x10}} = \frac{\eta_2 - \eta_1}{\eta_2 + \eta_1} = |\Gamma| e^{j\phi} \tag{9}$$

It is evident that as η_1 or η_2 may be complex, Γ will also be complex, and so we include a reflective phase shift, ϕ. The interpretation of Eq. (9) is identical to that used with transmission lines [Eq. (73), Chapter 11].

The relative amplitude of the transmitted electric field intensity is found by combining (9) and (7) to yield the *transmission coefficient*, τ,

$$\tau = \frac{E_{x20}^+}{E_{x10}^+} = \frac{2\eta_2}{\eta_1 + \eta_2} = 1 + \Gamma = |\tau| e^{j\phi_t} \tag{10}$$

whose form and interpretation are consistent with the usage in transmission lines [Eq. (75), Chapter 11].

Let us see how these results may be applied to several special cases. We first let region 1 be a perfect dielectric and region 2 be a perfect conductor. Then we apply Eq. (48), Chapter 12, with $\epsilon_2'' = \sigma_2/\omega$, obtaining

$$\eta_2 = \sqrt{\frac{j\omega\mu_2}{\sigma_2 + j\omega\epsilon_2'}} = 0$$

in which zero is obtained since $\sigma_2 \to \infty$. Therefore, from (10),

$$E_{x20}^+ = 0$$

No time-varying fields can exist in the perfect conductor. An alternate way of looking at this is to note that the skin depth is zero.

Since $\eta_2 = 0$, Eq. (9) shows that

$$\Gamma = -1$$

and

$$E_{x10}^+ = -E_{x10}^-$$

The incident and reflected fields are of equal amplitude, and so all the incident energy is reflected by the perfect conductor. The fact that the two fields are of opposite sign indicates that at the boundary (or at the moment of reflection), the reflected field is shifted in phase by 180° relative to the incident field. The total $\mathbf{E}$ field in region 1 is

$$\begin{aligned} E_{xs1} &= E_{xs1}^+ + E_{xs1}^- \\ &= E_{x10}^+ e^{-j\beta_1 z} - E_{x10}^+ e^{j\beta_1 z} \end{aligned}$$

where we have let $jk_1 = 0 + j\beta_1$ in the perfect dielectric. These terms may be combined and simplified,

$$\begin{aligned} E_{xs1} &= (e^{-j\beta_1 z} - e^{j\beta_1 z}) E_{x10}^+ \\ &= -j2 \sin(\beta_1 z) E_{x10}^+ \end{aligned} \tag{11}$$

Multiplying (11) by $e^{j\omega t}$ and taking the real part, we obtain the real instantaneous form:

Animations

$$\mathcal{E}_{x1}(z,t) = 2E_{x10}^+ \sin(\beta_1 z)\sin(\omega t) \tag{12}$$

We recognize this total field in region 1 as a standing wave, obtained by combining two waves of equal amplitude traveling in opposite directions. We first encountered standing waves in transmission lines, but in the form of counter-propagating voltage waves (see Example 11.1).

Again, we compare the form of (12) to that of the incident wave,

$$\mathcal{E}_{x1}(z,t) = E_{x10}^{+}\cos(\omega t - \beta_1 z) \tag{13}$$

Here we see the term $\omega t - \beta_1 z$ or $\omega(t - z/\nu_{p1})$, which characterizes a wave traveling in the $+z$ direction at a velocity $\nu_{p1} = \omega/\beta_1$. In (12), however, the factors involving time and distance are separate trigonometric terms. Whenever $\omega t = m\pi$, $\mathcal{E}_{x1}$ is zero at all positions. On the other hand, spatial nulls in the standing wave pattern occur for all times wherever $\beta_1 z = m\pi$, which in turn occurs when $m = (0, \pm 1, \pm 2, \ldots)$. In such cases,

$$\frac{2\pi}{\lambda_1}z = m\pi$$

and the null locations occur at

$$z = m\frac{\lambda_1}{2}$$

Thus $E_{x1} = 0$ at the boundary $z = 0$ and at every half-wavelength from the boundary in region 1, $z < 0$, as illustrated in Figure 13.2.

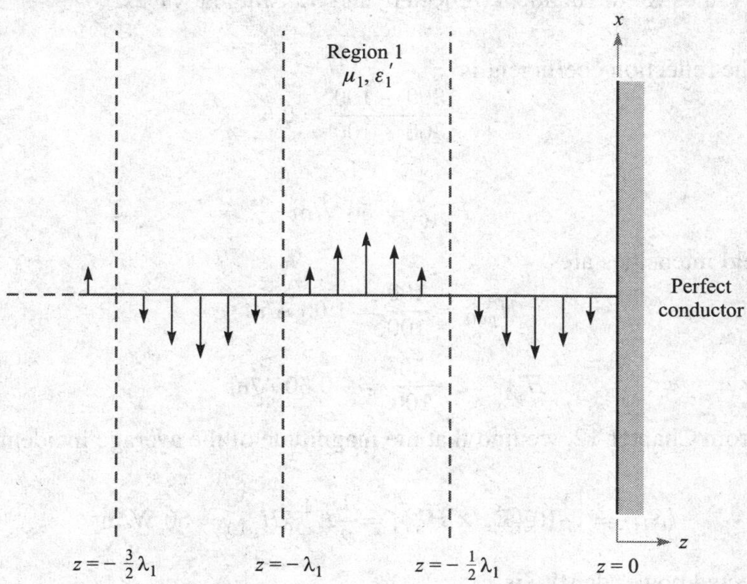

Fig. 13.2 The instantaneous values of the total field E_{x1} are shown at $t = \pi/2$. $E_{x1} = 0$ for all time at multiples of one half-wavelength from the conducting surface.

Since $E_{xs1}^{+} = \eta_1 H_{ys1}^{+}$ and $E_{xs1}^{-} = -\eta_1 H_{ys1}^{-}$, the magnetic field is

$$H_{ys1} = \frac{E_{x10}^{+}}{\eta_1}(e^{-j\beta_1 z} + e^{j\beta_1 z})$$

or

$$H_{y1}(z,t) = 2\,\frac{E_{x10}^+}{\eta_1}\cos(\beta_1 z)\cos(\omega t) \tag{14}$$

This is also a standing wave, but it shows a maximum amplitude at the positions where $E_{x1} = 0$. It is also 90° out of time phase with E_{x1} everywhere. As a result, the average power as determined through the Poynting vector [Eq. (77), Chapter 12] is zero in the forward and backward directions.

Let us now consider perfect dielectrics in both regions 1 and 2; η_1 and η_2 are both real positive quantities and $\alpha_1 = \alpha_2 = 0$. Equation (9) enables us to calculate the reflection coefficient and find E_{x1}^- in terms of the incident field E_{x1}^+. Knowing E_{x1}^+ and E_{x1}^-, we then find H_{y1}^+ and H_{y1}^-. In region 2, E_{x2}^+ is found from (10), and this then determines H_{y2}^+.

EXAMPLE 13.1

As a numerical example we select

$$\eta_1 = 100\ \Omega$$
$$\eta_2 = 300\ \Omega$$
$$E_{x10}^+ = 100\ \text{V/m}$$

and calculate values for the incident, reflected, and transmitted waves.

Solution. The reflection coefficient is

$$\Gamma = \frac{300 - 100}{300 + 100} = 0.5$$

and thus

$$E_{x10}^- = 50\ \text{V/m}$$

The magnetic field intensities are

$$H_{y10}^+ = \frac{100}{100} = 1.00\ \text{A/m}$$

$$H_{y10}^- = -\frac{50}{100} = -0.50\ \text{A/m}$$

Using Eq. (77) from Chapter 12, we find that the magnitude of the average incident power density is

$$\langle S_{1i}\rangle = \left|\frac{1}{2}\text{Re}\{\mathbf{E}_s \times \mathbf{H}_s^*\}\right| = \frac{1}{2}E_{x10}^+ H_{y10}^+ = 50\ \text{W/m}^2$$

The average reflected power density is

$$\langle S_{1r}\rangle = -\frac{1}{2}E_{x10}^- H_{y10}^- = 12.5\ \text{W/m}^2$$

In region 2, using (10),

$$E_{x20}^+ = \tau E_{x10}^+ = 150\ \text{V/m}$$

and

$$H_{y20}^+ = \frac{150}{300} = 0.500\ \text{A/m}$$

Therefore, the average power density that is transmitted through the boundary into region 2 is

$$\langle S_2 \rangle = \frac{1}{2} E_{x20}^{+} H_{y20}^{+} = 37.5 \text{ W/m}^2$$

We may check and confirm the power conservation requirement:

$$\langle S_{1i} \rangle = \langle S_{1r} \rangle + \langle S_2 \rangle$$

A general rule on the transfer of power through reflection and transmission can be formulated. We consider the same field vector and interface orientations as before, but allow for the case of complex impedances. For the incident power density, we have

$$\langle S_{1i} \rangle = \frac{1}{2} \text{Re}\{E_{xs1}^{+} H_{ys1}^{+*}\} = \frac{1}{2} \text{Re}\left\{ E_{x10}^{+} \frac{1}{\eta_1^*} E_{x10}^{+*} \right\} = \frac{1}{2} \text{Re}\left\{ \frac{1}{\eta_1^*} \right\} \left| E_{x10}^{+} \right|^2$$

The reflected power density is then

$$\langle S_{1r} \rangle = -\frac{1}{2} \text{Re}\{E_{xs1}^{-} H_{ys1}^{-*}\} = \frac{1}{2} \text{Re}\left\{ \Gamma E_{x10}^{+} \frac{1}{\eta_1^*} \Gamma^* E_{x10}^{+*} \right\} = \frac{1}{2} \text{Re}\left\{ \frac{1}{\eta_1^*} \right\} \left| E_{x10}^{+} \right|^2 |\Gamma|^2$$

We thus find the general relation between the reflected and incident power:

$$\langle S_{1r} \rangle = |\Gamma|^2 \langle S_{1i} \rangle \tag{15}$$

In a similar way, we find the transmitted power density:

$$\langle S_2 \rangle = \frac{1}{2} \text{Re}\{E_{xs2}^{+} H_{ys2}^{+*}\} = \frac{1}{2} \text{Re}\left\{ \tau E_{x10}^{+} \frac{1}{\eta_2^*} \tau^* E_{x10}^{+*} \right\} = \frac{1}{2} \text{Re}\left\{ \frac{1}{\eta_2^*} \right\} \left| E_{x10}^{+} \right|^2 |\tau|^2$$

and so we see that the incident and transmitted power densities are related through

$$\langle S_2 \rangle = \frac{\text{Re}\{1/\eta_2^*\}}{\text{Re}\{1/\eta_1^*\}} |\tau|^2 \langle S_{1i} \rangle = \left| \frac{\eta_1}{\eta_2} \right|^2 \left(\frac{\eta_2 + \eta_2^*}{\eta_1 + \eta_1^*} \right) |\tau|^2 \langle S_{1i} \rangle \tag{16}$$

Equation (16) is a relatively complicated way to calculate the transmitted power, unless the impedances are real. It is easier to take advantage of energy conservation by noting that whatever power is not reflected must be transmitted. Eq. (15) can be used to find

$$\langle S_2 \rangle = (1 - |\Gamma|^2) \langle S_{1i} \rangle \tag{17}$$

As would be expected (and which must be true), Eq. (17) can also be derived from Eq. (16).

EXAMPLE 13.2

A uniform plane wave propagating in free space in the positive z direction at the frequency of 1.5 GHz with the electric field amplitude of $E_{x10}^{+} = 5$ V/m is incident normally on the air-dielectric interface. The dielectric medium characterized by $\epsilon_r = 8.0$, $\mu_r = 2.0$, $\epsilon_r'' = \mu_r'' = 0$ can be considered semi-infinite. Obtain the time-domain expressions for the incident, reflected, and transmitted electric and magnetic fields.

Solution. The incident wave is propagating in the positive z direction, and is polarized along the x-direction as shown in Figure 3.13.

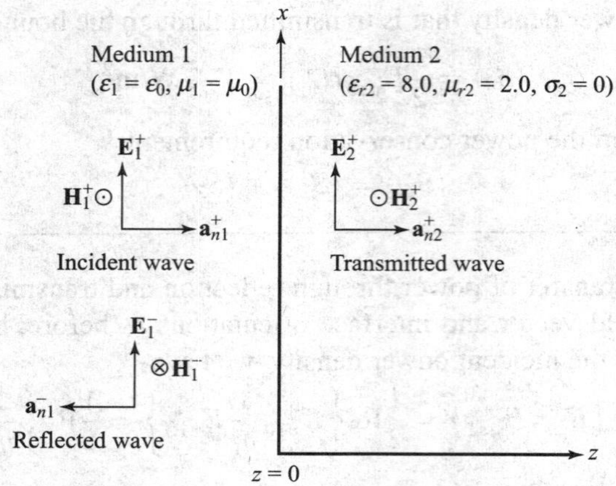

Fig. 13.3 Plane interface between the free-space/magnetic media.

The general expression for the incident wave in this situation may be written as

$$\mathbf{E}_1^+(z,t) \equiv \mathbf{a}_x E_{x10}^+ \cos(\omega t - \beta_1 z) = \mathbf{a}_x 5\cos(3\pi \times 10^9 t - 31.4z)\ \text{V/m}$$

where, $\beta_1 \equiv k_0 = \omega/c$ is the free-space wave number and $\omega = 2\pi f$ is the radial frequency. Now, in order to find out other field components, let us first determine the impedances in medium 1 and 2. The medium 1 is free space, and the medium 2 is a lossless nonmagnetic medium. The impedances in medium 1 and medium 2 may be calculated as follows:

$$\eta_1 \equiv \eta_0 \equiv \sqrt{\frac{\mu_0}{\epsilon_0}} = 120\ \pi\Omega$$

$$\eta_2 \equiv \sqrt{\frac{\mu_2}{\epsilon_2}} = \eta_0\sqrt{\frac{\mu_{r2}}{\epsilon_{r2}}} = 120\ \pi\sqrt{\frac{2.0}{8.0}} = 60\ \pi\Omega$$

The reflection coefficient may be calculated as

$$\Gamma \equiv \frac{\eta_2 - \eta_1}{\eta_2 + \eta_1} = \frac{60\ \pi - 120\ \pi}{60\ \pi + 120\ \pi} = -0.33$$

Similarly, the transmission coefficient is calculated as

$$\tau \equiv \frac{2\eta_2}{\eta_2 + \eta_1} = \frac{2 \times 60\pi}{60\pi + 120\ \pi} = 0.67$$

Now, as the incident wave is propagating in the positive z-direction with the electric field being oriented along the x-direction, the magnetic field would be oriented along the positive y-direction with its value being given by

$$\mathbf{H}_1^+(z,t) \equiv \mathbf{a}_y \frac{E_{x10}^+}{\eta_0}\cos(\omega t - \beta_1 z) = \mathbf{a}_y 13.3\cos(3\pi \times 10^9 t - 31.4z)\ \text{mA/m}$$

The reflected wave will propagate in the negative z-direction in medium 1, and the reflected electric and magnetic fields can be determined using the following relationships:

$$\mathbf{E}_1^-(z,t) \equiv \mathbf{a}_x E_{x10}^- \cos(\omega t + \beta_1 z) = -\mathbf{a}_x 1.65\cos(3\pi \times 10^9 t + 31.4z)\ \text{V/m}$$

$$\mathbf{H}_1^-(z,t) \equiv -\mathbf{a}_y \frac{E_{x10}^-}{\eta_0}\cos(\omega t + \beta_1 z) = \mathbf{a}_y 4.4\cos(3\pi \times 10^9 t + 31.4z)\ \text{mA/m}$$

where $E_{x10}^- \equiv \Gamma E_{x10}^+$ is the amplitude of the reflected electric field. It should be noted that the amplitude of the reflected magnetic field $H_{y10}^- \equiv -E_{x10}^+/\eta_0$ is taken as negative so that the net power flow for the reflected wave represented by the Poynting vector is along the $-z$ direction.

The transmitted wave will propagate in the positive z direction in medium 2, and the expressions for the electric and magnetic fields corresponding to the transmitted wave are determined as follows

$$\mathbf{E}_2^+(z,t) \equiv \mathbf{a}_x E_{x20}^+ \cos(\omega t - \beta_2 z) = \mathbf{a}_x 3.35\cos(3\pi \times 10^9 t - 125.7z)\ \text{V/m}$$

$$\mathbf{H}_2^+(z,t) \equiv \mathbf{a}_y H_{y20}^+ \cos(\omega t + \beta_2 z) = \mathbf{a}_y 17.8\cos(3\pi \times 10^9 t - 125.7z)\ \text{mA/m}$$

where $E_{x20}^+ \equiv \tau E_{x10}^+$ is the amplitude of the transmitted electric field, and the amplitude of the transmitted magnetic field is given by $H_{y20}^+ \equiv E_{x20}^+/\eta_2$. The signs for the transmitted electric and magnetic fields are chosen in such a way that the net power flow for the transmitted wave is along the $+z$ direction.

EXAMPLE 13.3

A uniform plane wave propagating in free space is characterized by the following expression:

$$\mathbf{E}_s^+(z) = 1.5\, e^{-j1.571z}\mathbf{a}_y\ \text{V/m}$$

It is incident normally upon a semi-infinite lossless nonmagnetic medium having $\epsilon_r = 6.25$. Determine

(a) the reflection and transmission coefficients;

(b) the amplitudes of the reflected and transmitted electric and magnetic fields; and

(c) the magnitudes of the time-average power density for the incident, the reflected, and the transmitted waves.

Solution.

(a) It can be seen from the given expression that the incident electric field is polarized along the y-direction, while the wave is traveling in direction z as shown in Figure 13.4.

The medium 1 is free space, and the medium 2 is a lossless nonmagnetic medium. The impedances in medium 1 and medium 2 can be calculated as follows:

$$\eta_1 \equiv \eta_0 \equiv \sqrt{\frac{\mu_0}{\epsilon_0}} = 120\,\pi\,\Omega \quad \eta_2 \equiv \sqrt{\frac{\mu_2}{\epsilon_2}} = \frac{\eta_0}{\sqrt{\epsilon_{r2}}} = \frac{120\,\pi}{\sqrt{6.25}} = 48\,\pi\,\Omega$$

The reflection coefficient may be calculated as

$$\Gamma \equiv \frac{\eta_2 - \eta_1}{\eta_2 + \eta_1} = \frac{48\pi - 120\pi}{48\pi + 120\pi} = -0.43$$

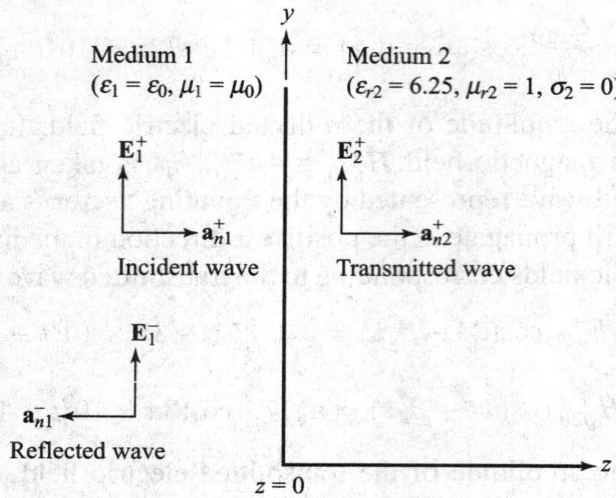

Fig. 13.4 Interface between two media.

Similarly, the transmission coefficient is calculated as

$$\tau \equiv \frac{2\eta_2}{\eta_2 + \eta_1} = \frac{2 \times 48\pi}{48\pi + 120\pi} = 0.57$$

(*b*) For calculating the amplitudes of the reflected and transmitted fields, let us first compare the incident electric field with the standard expression as follows:

$$E_{ys1}^+(z) \equiv E_{y10}^+ e^{-jk_1 \mathbf{a}_{n1}^+ \cdot \mathbf{r}} = 1.5e^{-j1.571z}\ \mathrm{V/m}$$

where $\mathbf{a}_{n1}^+ \equiv \mathbf{a}_z$ represents unit vector along the direction of incident wave (positive z direction), $\mathbf{r} \equiv z\,\mathbf{a}_z$ represents the radius vector in the present situation, subscript 1 is used for the medium 1 which is free space in the present situation, and $E_{y10}^+ = 1.5\ \mathrm{V/m}$ represents the amplitude of the incident electric field. The amplitude of the reflected electric field is then computed as

$$E_{y10}^- \equiv \Gamma E_{y10}^+ = -0.43 \times 1.5 = -0.645\ \mathrm{V/m}$$

Similarly, the amplitudes of the incident and the reflected magnetic fields can be computed as

$$H_{x10}^+ \equiv -\frac{E_{y10}^+}{\eta_1} = -\frac{1.5}{120\pi} = -4\ \mathrm{mA/m}$$

$$H_{x10}^- \equiv \frac{E_{y10}^-}{\eta_1} = \frac{-0.645}{120\pi} = -1.7\ \mathrm{mA/m}$$

The amplitudes of the transmitted electric and magnetic fields in region 2 may be calculated as follows

$$E_{y20}^+ \equiv \tau E_{y10}^+ = 0.57 \times 1.5 = 0.855\ \mathrm{V/m}$$

$$H_{x20}^+ \equiv -\frac{E_{y20}^+}{\eta_2} = -\frac{0.855}{48\pi} = -5.7\ \mathrm{mA/m}$$

It is to be noted here that the sign of the amplitudes of the magnetic field is chosen in such a way that the Poynting vector is directed towards the power flow for the corresponding wave. For example, the reflected power in the present situation flows along the negative z ($\mathbf{a}_{n1}^- \equiv -\mathbf{a}_z$) direction, which is achieved by the cross product of the y-component of the electric field (E_{y10}^-) with the x-component of the magnetic field (H_{x10}^+) provided that both the field components have the same sign. On the contrary, the transmitted waves travel towards the positive z ($a_{n2}^+ \equiv a_z$) direction, hence the sign of the magnetic field (H_{x20}^+) should be taken as opposite to that of the electric field (E_{y20}^+).

(c) The magnitude of the time-average power density for the incident wave may be determined using the following expression:

$$\langle S_{1i}\rangle = \left|\frac{1}{2}\mathrm{Re}\ (\mathbf{E}_{s1}^+ \times \mathbf{H}_{s1}^{+*})\right| \equiv \left|\frac{1}{2}E_{y10}^+ H_{x10}^+\right| = \frac{1}{2} \times 1.5 \times 4 \times 10^{-3} = 3\ \mathrm{mW/m^2}$$

Similarly, the magnitude of the time-average power density for the reflected wave is determined as

$$\langle S_{1r}\rangle = \left|\frac{1}{2}\mathrm{Re}\ (\mathbf{E}_{s1}^- \times \mathbf{H}_{s1}^{-*})\right| \equiv \left|\frac{1}{2}E_{y10}^- H_{x10}^-\right| = \frac{1}{2} \times 0.645 \times 1.7 \times 10^{-3} = 0.55\ \mathrm{mW/m^2}$$

Finally, the magnitude of the time-average power density for the transmitted wave can be determined using the following expression:

$$\langle S_{2t}\rangle = \left|\frac{1}{2}\mathrm{Re}\ (\mathbf{E}_{s2}^+ \times \mathbf{H}_{s2}^{+*})\right| \equiv \left|\frac{1}{2}E_{y20}^+ H_{x20}^+\right| = \frac{1}{2} \times 0.855 \times 5.7 \times 10^{-3} = 2.45\ \mathrm{mW/m^2}$$

It can be observed from the above expressions that

$$\langle S_{1i}\rangle = \langle S_{1r}\rangle + \langle S_{2t}\rangle$$

which satisfies the energy conservation principle stating that the total incident power is equal to sum of the reflected power plus the transmitted power.

D13.1 A 1 MHz uniform plane wave is normally incident onto a freshwater lake ($\epsilon_r' = 78$, $\epsilon_r'' = 0$, $\mu_r = 1$). Determine the fraction of the incident power that is (a) reflected and (b) transmitted. (c) Determine the amplitude of the electric field that is transmitted into the lake.

Ans. 0.63; 0.37; 0.20 V/m.

13.2 STANDING WAVE RATIO

In cases where $|\Gamma| < 1$, some energy is transmitted into the second region and some is reflected. Region 1 therefore supports a field that is composed of both a traveling wave and a standing wave. We encountered this situation previously in transmission lines, in which partial reflection occurs at the load. Measurements of the voltage standing wave ratio and the locations of voltage minima or maxima enabled the determination of an unknown load impedance or established the extent to which the load impedance was matched to that of the line (Section 11.10). Similar measurements can be performed on the field amplitudes in plane wave reflection.

Using the same fields investigated in the previous section, we combine the incident and reflected electric field intensities. Medium 1 is assumed to be a perfect dielectric ($\alpha_1 = 0$), but region 2 may be any material. The total electric field phasor in region 1 will be

$$E_{x1T} = E_{x1}^+ + E_{x1}^- = E_{x10}^+ e^{-j\beta_1 z} + \Gamma E_{x10}^+ e^{j\beta_1 z} \tag{18}$$

where the reflection coefficient is as expressed in (9):

$$\Gamma = \frac{\eta_2 - \eta_1}{\eta_2 + \eta_1} = |\Gamma| e^{j\phi}$$

We allow for the possibility of a complex reflection coefficient by including its phase, ϕ. This is necessary because although η_1 is real and positive for a lossless medium, η_2 will in general be complex. Additionally, if region 2 is a perfect conductor, η_2 is zero, and so ϕ is equal to π; if η_2 is real and less than η_1, ϕ is also equal to π; and if η_2 is real and greater than η_1, ϕ is zero.

Incorporating the phase of Γ into (18), the total field in region 1 becomes

$$E_{x1T} = \left(e^{-j\beta_1 z} + |\Gamma| e^{j(\beta_1 z + \phi)}\right) E_{x10}^+ \tag{19}$$

The maximum and minimum field amplitudes in (19) are z-dependent, and are subject to measurement. Their ratio, as found for voltage amplitudes in transmission lines (Section 11.10), is the *standing wave ratio,* denoted by s. We have a maximum when each term in the larger parentheses in (19) has the same phase angle; so, for E_{x10}^+ positive and real,

$$|E_{x1T}|_{\max} = (1 + |\Gamma|) E_{x10}^+ \tag{20}$$

and this occurs where

$$-\beta_1 z = \beta_1 z + \phi + 2m\pi \qquad (m = 0, \pm 1, \pm 2, \ldots) \tag{21}$$

Therefore

$$z_{\max} = -\frac{1}{2\beta_1}(\phi + 2m\pi) \tag{22}$$

Note that an electric field maximum is located at the boundary plane ($z = 0$) if $\phi = 0$; moreover, $\phi = 0$ when Γ is real and positive. This occurs for real η_1 and η_2 when $\eta_2 > \eta_1$. Thus there is a field maximum at the boundary surface when the intrinsic impedance of region 2 is greater than that of region 1 and both impedances are real. With $\phi = 0$, maxima also occur at $z_{\max} = -m\pi/\beta_1 = -m\lambda_1/2$.

For the perfect conductor $\phi = \pi$, and these maxima are found at $z_{\max} = -\pi/(2\beta_1), -3\pi/(2\beta_1)$, or $z_{\max} = -\lambda_1/4, -3\lambda_1/4$, and so forth.

The minima must occur where the phase angles of the two terms in the larger parentheses in (19) differ by 180°, thus

$$|E_{x1T}|_{\min} = (1 - |\Gamma|) E_{x10}^+ \tag{23}$$

and this occurs where

$$-\beta_1 z = \beta_1 z + \phi + \pi + 2m\pi \qquad (m = 0, \pm 1, \pm 2, \ldots) \tag{24}$$

or

$$z_{\min} = -\frac{1}{2\beta_1}(\phi + (2m + 1)\pi) \tag{25}$$

The minima are separated by multiples of one half-wavelength (as are the maxima), and for the perfect conductor the first minimum occurs when $-\beta_1 z = 0$, or at the conducting surface. In general, an electric field minimum is found at $z = 0$ whenever $\phi = \pi$; this occurs if $\eta_2 < \eta_1$ and both are real. The results are mathematically identical to those found for the transmission line study in Section 11.10. Figure 11.6 in that chapter provides a visualization.

Further insights can be obtained by working with Eq. (19) and rewriting it in real instantaneous form. The steps are identical to those taken in Chapter 11, Eqs. (81) through (84). We find the total field in region 1 to be

$$\mathcal{E}_{x1T}(z,t) = \underbrace{(1-|\Gamma|)E_{x10}^{+}\cos(\omega t - \beta_1 z)}_{\text{traveling wave}} + \underbrace{2|\Gamma|E_{x10}^{+}\cos(\beta_1 z + \phi/2)\cos(\omega t + \phi/2)}_{\text{standing wave}} \qquad (26)$$

The field expressed in Eq. (26) is the sum of a traveling wave of amplitude $(1-|\Gamma|)E_{x10}^{+}$ and a standing wave having amplitude $2|\Gamma|E_{x10}^{+}$. The portion of the incident wave that reflects and back-propagates in region 1 interferes with an equivalent portion of the incident wave to form a standing wave. The rest of the incident wave (that does not interfere) is the traveling wave part of (26). The maximum amplitude observed in region 1 is found where the amplitudes of the two terms in (26) add directly to give $(1+|\Gamma|)E_{x10}^{+}$. The minimum amplitude is found where the standing wave achieves a null, leaving only the traveling wave amplitude of $(1-|\Gamma|)E_{x10}^{+}$. The fact that the two terms in (26) combine in this way with the proper phasing can be confirmed by substituting z_{max} and z_{min}, as given by (22) and (25).

EXAMPLE 13.4

To illustrate some of these results, let us consider a 100 V/m, 3 GHz wave that is propagating in a material having $\epsilon'_{r1} = 4$, $\mu_{r1} = 1$, and $\epsilon''_r = 0$. The wave is normally incident on another perfect dielectric in region 2, $z > 0$, where $\epsilon'_{r2} = 9$ and $\mu_{r2} = 1$ (Figure 13.5). We seek the locations of the maxima and minima of **E**.

Solution. We calculate $\omega = 6\pi \times 10^9$ rad/s, $\beta_1 = \omega\sqrt{\mu_1\epsilon_1} = 40\pi$ rad/m, and $\beta_2 = \omega\sqrt{\mu_2\epsilon_2} = 60\pi$ rad/m. Although the wavelength would be 10 cm in air, we find here that $\lambda_1 = 2\pi/\beta_1 = 5$ cm, $\lambda_2 = 2\pi/\beta_2 = 3.33$ cm, $\eta_1 = 60\pi\,\Omega$, $\eta_2 = 40\pi\,\Omega$, and $\Gamma = (\eta_2 - \eta_1)/(\eta_2 + \eta_1) = -0.2$. Since Γ is real and negative ($\eta_2 < \eta_1$), there will be a minimum of the electric field at the boundary, and it will be repeated at half-wavelength (2.5 cm) intervals in dielectric 1. From (23), we see that $|E_{x1T}|_{\text{min}} = 80$ V/m.

Maxima of **E** are found at distances of 1.25, 3.75, 6.25, ... cm from $z = 0$. These maxima all have amplitudes of 120 V/m, as predicted by (20).

There are no maxima or minima in region 2 because there is no reflected wave there.

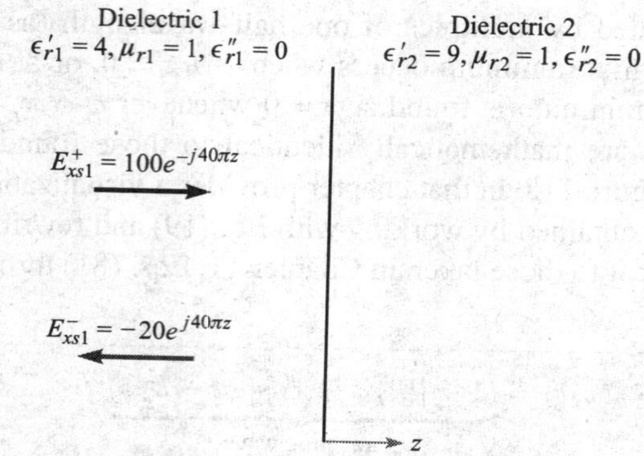

Fig. 13.5 An incident wave, $E^+_{xs1} = 100e^{-j40\pi z}$ V/m, is reflected with a reflection coefficient $\Gamma = -0.2$. Dielectric 2 is infinitely thick.

The ratio of the maximum to minimum amplitudes is the standing wave ratio:

$$s = \frac{|E_{x1T}|_{\max}}{|E_{x1T}|_{\min}} = \frac{1 + |\Gamma|}{1 - |\Gamma|} \tag{27}$$

Since $|\Gamma| < 1$, s is always positive and greater than or equal to unity. For the preceding example,

$$s = \frac{1 + |-0.2|}{1 - |-0.2|} = \frac{1.2}{0.8} = 1.5$$

If $|\Gamma| = 1$, the reflected and incident amplitudes are equal, all the incident energy is reflected, and s is infinite. Planes separated by multiples of $\lambda_1/2$ can be found on which E_{x1} is zero at all times. Midway between these planes, E_{x1} has a maximum amplitude twice that of the incident wave.

If $\eta_2 = \eta_1$, then $\Gamma = 0$, no energy is reflected, and $s = 1$; the maximum and minimum amplitudes are equal.

If one-half the incident power is reflected, $|\Gamma|^2 = 0.5$, $|\Gamma| = 0.707$, and $s = 5.83$.

EXAMPLE 13.5

Consider a 500 MHz, x-polarized uniform plane wave of amplitude 7.5 V/m propagating along the positive z direction in a nonmagnetic lossless dielectric medium as shown in Figure 13.6. The wave is normally incident on a perfect conductor at the plane $z = 0$. The first maximum of the total electric field is observed at a distance of 5 cm from the dielectric-conductor boundary in region 1.

Find:

(*a*) the position of the first two minima of the total electric field in region 1;
(*b*) the position of the first minima and the first maxima of the total magnetic field in region 1;
(*c*) the relative permittivity, the propagation constant, and the intrinsic impedance of the medium represented by region 1; and
(*d*) the time-domain expressions of the total electric and magnetic fields in region 1.

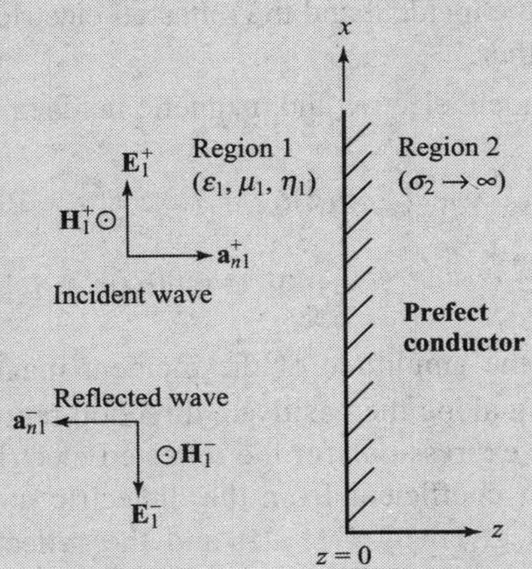

Fig. 13.6 Plane wave incident normally on a perfect conductor.

Solution.

(*a*) As the plane wave is incident on a perfect conductor at the plane $z = 0$, hence the minima or null of the total electric field in region 1 will be observed at locations defined by $z \equiv -m\lambda_1/2$; $(m = 0, 1, 2, \ldots)$, where λ_1 is the wavelength in medium 1, and negative sign is taken to consider the fact that the region 1 is defined by half-space for which $z< 0$. Similarly, the maxima of the total electric field in region 1 will be observed at locations defined by $z \equiv -(2m+1)\lambda_1/4$; $(m = 0, 1, 2, \ldots)$.

Now, it is given that the first maximum of the total electric field is observed at a distance of 5 cm from the dielectric-conductor boundary in region 1. Hence, the wavelength in the region 1 may be computed as follows:

$$-5 \equiv -\lambda_1/4 \Rightarrow \lambda_1 = 20 \text{ cm}$$

The first minimum of the electric field in region 1 will be observed at $z = 0$.

The second minimum of the electric field in region 1 will be observed at a distance of $z = \lambda_1/2 = 10$ cm from the interface.

(*b*) The total magnetic field will have maximum value at those positions where the electric field is minimum and vice versa. Hence, the first minimum of the total magnetic field will be observed at a distance of 5 cm from the dielectric-conductor boundary in region 1. Similarly, the position of the first maximum of the total magnetic field will be at the plane $z = 0$.

(*c*) The relative permittivity, the propagation constant and the intrinsic impedance of the region 1 representing a nonmagnetic lossless dielectric medium may be determined as follows:

$$\beta_1 \equiv \frac{2\pi}{\lambda_1} = \frac{2\pi}{20} = 0.1\pi \text{ rad/cm}$$

$$\beta_1 \equiv \frac{2\pi f}{c}\sqrt{\epsilon_{r1}} \Rightarrow \epsilon_{r1} = \left[\frac{c \times \beta_1}{2\pi f}\right]^2 = \left[\frac{3 \times 10^{10} \times 0.1\pi}{2\pi \times 500 \times 10^6}\right]^2 = 9.0$$

$$\eta_1 \equiv \sqrt{\frac{\mu_1}{\epsilon_1}} = \frac{\eta_0}{\sqrt{\epsilon_{r1}}} = \frac{120\pi}{\sqrt{9.0}} = 40\pi\ \Omega$$

(*d*) The time-domain expressions of the total electric and magnetic fields in region 1 can be obtained by determining separately the incident and the reflected electric and magnetic fields in region 1, and then adding them together.

The expressions for the incident electric and magnetic fields in region 1 for the present situation may be written as follows:

$$\mathbf{E}_1^+(z,t) \equiv E_{x10}^+ \cos(\omega t - \beta_1 z)\mathbf{a}_x = 7.5\cos(\pi \times 10^9 t - 0.1\pi z)\mathbf{a}_x \text{ V/m}$$

$$\mathbf{H}_1^+(z,t) \equiv \frac{E_{y10}^+}{\eta_1}\cos(\omega t - \beta_1 z)\mathbf{a}_y = 0.06\cos(\pi \times 10^9 t - 0.1\pi z)\mathbf{a}_y \text{ A/m}$$

where $H_{y10}^+ \equiv E_{x10}^+/\eta_1$ is the amplitude of the incident magnetic field in region 1, and the incident wave is propagating along the positive z direction represented by $\mathbf{a}_{n1}^+ \equiv \mathbf{a}_z$.

Now, in order to obtain the expressions for the reflected electric and magnetic fields, it should be noted that the reflection coefficient from the dielectric conductor interface assuming the conductor to be perfect is given by $\Gamma = -1$, and the reflected wave will propagate along the $\mathbf{a}_{n1}^- \equiv -\mathbf{a}_z$ direction in region 1. The reflected electric and magnetic fields can then be determined using the following relationships:

$$\mathbf{E}_1^-(z,t) \equiv E_{x10}^- \cos(\omega t + \beta_1 z)\mathbf{a}_x = -7.5\cos(\pi \times 10^9 t + 0.1\pi z)\mathbf{a}_x \text{ V/m}$$

$$\mathbf{H}_1^-(z,t) \equiv -\frac{E_{x10}^-}{\eta_1}\cos(\omega t + \beta_1 z)\mathbf{a}_y = 0.06\cos(\pi \times 10^9 t + 0.1\pi z)\mathbf{a}_y \text{ A/m}$$

where $E_{x10}^- \equiv \Gamma E_{x10}^+ = -7.5$ V/m is the amplitude of the reflected electric field, and $H_{y10}^- \equiv -E_{x10}^+/\eta_1 = 0.06$ A/m is the amplitude of the reflected magnetic field. It may be noted from above equations that the net power flow for the reflected wave represented by the Poynting vector $(E_{x10}^-\mathbf{a}_x \times \mathbf{a}_y H_{y10}^-)$ is along the $\mathbf{a}_{n1}^- \equiv -\mathbf{a}_z$ direction in region 1.

Once the expressions for the incident and the reflected electric fields are obtained, the total electric field in the region 1 can simply be obtained by adding these components as follows:

$$\begin{aligned}\mathbf{E}_1(z,t) &\equiv \mathbf{E}_1^+(z,t) + \mathbf{E}_1^-(z,t)\\ &= 7.5[\cos(\pi \times 10^9 t - 0.1\pi z) - \cos(\pi \times 10^9 t - 0.1\pi z)]\mathbf{a}_x \text{ V/m}\\ &= 15.0\sin(\pi \times 10^9 t)\sin(0.1\pi z)\mathbf{a}_x \text{ V/m}\end{aligned}$$

where we have used the following trigonometric relationship:

$$[\cos(A-B) - \cos(A+B)] = 2\sin(A)\sin(B)$$

Similarly, the total magnetic field in the region 1 can be obtained by adding the incident and the reflected magnetic fields components as follows:

$$\begin{aligned}\mathbf{H}_1(z,t) &\equiv \mathbf{H}_1^+(z,t) + \mathbf{H}_1^-(z,t)\\ &= 0.06\,[\cos(\pi \times 10^9 t - 0.1\pi z) + \cos(\pi \times 10^9 t + 0.1\pi z)]\mathbf{a}_y \text{ A/m}\\ &= 0.12\cos(\pi \times 10^9 t)\cos(0.1\pi z)]\mathbf{a}_x \text{ A/m}\end{aligned}$$

where the trigonometric relationship $[\cos(A-B) + \cos(A+B)] = 2\cos A \cos B$ is used.

We can verify from the above expressions that at $z = 0$, the electric field is having null, while the magnetic field is having maximum value at any instant. Similarly, at a distance of $z = 5$ cm $\equiv \lambda_1/4$ from the interface in region 1, the magnetic field is having null while the electric field is having maximum value at any instant. It can also be verified that the time-domain forms of the total electric and magnetic field in region 1 derived here are similar to the general expression given by equations (13) and (14) in the text.

D13.2 What value of s results when $\Gamma = \pm 1/2$?

Ans. 3

Since the standing wave ratio is a ratio of amplitudes, the relative amplitudes as measured by a probe permit its use to determine s experimentally.

EXAMPLE 13.6

A uniform plane wave in air partially reflects from the surface of a material whose properties are unknown. Measurements of the electric field in the region in front of the interface yield a 1.5 m spacing between maxima, with the first maximum occurring 0.75 m from the interface. A standing wave ratio of 5 is measured. Determine the intrinsic impedance, η_u, of the unknown material.

Solution. The 1.5 m spacing between maxima is $\lambda/2$, which implies that a wavelength is 3.0 m, or $f = 100$ MHz. The first maximum at 0.75 m is thus at a distance of $\lambda/4$ from the interface, which means that a field minimum occurs at the boundary. Thus Γ will be real and negative. We use (27) to write

$$|\Gamma| = \frac{s-1}{s+1} = \frac{5-1}{5+1} = \frac{2}{3}$$

So

$$\Gamma = -\frac{2}{3} = \frac{\eta_u - \eta_0}{\eta_u + \eta_0}$$

which we solve for η_u to obtain

$$\eta_u = \frac{1}{5}\eta_0 = \frac{377}{5} = 75.4\,\Omega$$

EXAMPLE 13.7

A uniform plane wave propagating along the positive x direction in a nonmagnetic lossless medium 1 is incident normally on a semi-infinite medium 2 characterized by $\epsilon'_{r2} = 5$, $\sigma_2 = 0.5\ S/m$, $\mu'_{r2} = 1.2$, $\mu''_{r2} = 0$ as shown in Figure 13.7. The incident electric field is given by the following expression:

$$\mathbf{E}_1^+(x,t) = 5\cos(5\pi \times 10^9 t - 83.8x)\mathbf{a}_y \text{ V/m}$$

Find

(a) the reflection and transmission coefficients;

(b) the phasor expressions for the total electric and magnetic fields in medium 1;

(c) the time-average power density in medium 2; and

(d) the total impedance in medium 1 at a distance of 2 cm from the boundary.

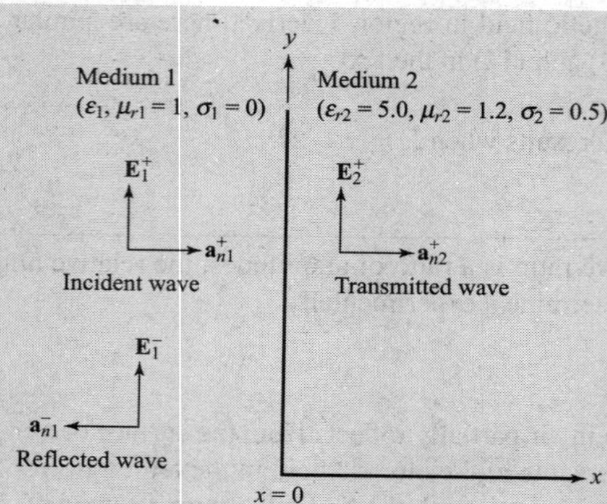

Fig. 13.7 Plane wave incident normally on a semi-infinite magnetic medium.

Solution.

(a) The incident wave is traveling in the positive x direction in medium 1, and the electric field is polarized along the y direction as can be seen from the given expression. As the medium 1 is given to be nonmagnetic lossless type, its properties can be determined as follows:

$$\beta_1 \equiv \frac{\omega}{c}\sqrt{\epsilon_{r1}} = 83.8 \text{ rad/m} \Rightarrow \epsilon_{r1} = \left(\frac{\beta_1 c}{\omega}\right)^2 \equiv \left(\frac{83.8 \times 3 \times 10^8}{5\pi \times 10^9}\right)^2 = 2.56$$

The impedance of medium 1 may then be calculated as follows:

$$\eta_1 \equiv \frac{\eta_0}{\sqrt{\epsilon_{r1}}} = \frac{120\pi}{\sqrt{2.56}} = 235.6\ \Omega$$

The regions 2 is a lossy medium, hence the general expression given by (48) of Chapter 12 may be used to determine its impedance as follows:

$$\eta_2 \equiv \sqrt{\frac{\mu_2}{\epsilon_2}} = \eta_0\sqrt{\frac{\mu'_{r2}}{\epsilon'_{r2} - j\epsilon''_{r2}}} = \eta_0\sqrt{\frac{\mu'_{r2}}{\epsilon'_{r2} - j\sigma_2/\omega\epsilon_0}}$$

$$= 120\pi\sqrt{\frac{1.2}{5 - j0.5/(5\pi \times 10^9 \times 8.854 \times 10^{-12})}} = (158.4 + j51)\Omega$$

Once the impedances for both the regions are calculated, the reflection and the transmission coefficients may be determined as follows:

$$\Gamma \equiv \frac{\eta_2 - \eta_1}{\eta_2 + \eta_1} = \frac{158.4 + j51 - 235.6}{158.4 + j51 + 235.6} = -0.176 + j0.152 \equiv 0.2329e^{j2.429}$$

$$\tau \equiv \frac{2\eta_2}{\eta_2 + \eta_1} = \frac{2 \times (158.4 + j51)}{158.4 + j51 + 235.6} = 0.824 + j0.152 \equiv 0.838e^{j0.183}$$

(b) For determining the phasor expressions of the total electric and magnetic fields in region 1, the incident and the reflected electric and magnetic fields may be separately determined. The incident electric and magnetic fields are given as

$$\mathbf{E}_{s1}^{+}(x) \equiv E_{ys1}^{+}(x)\mathbf{a}_y \equiv E_{y10}^{+}e^{-jk_1x}\mathbf{a}_y = 5.0e^{-j83.8x}\mathbf{a}_y \text{ V/m}$$

$$\mathbf{H}_{s1}^{+}(x) \equiv H_{zs1}^{+}(x)\mathbf{a}_z \equiv \frac{E_{y10}^{+}}{\eta_1}e^{-jk_1x}\mathbf{a}_z = 0.0212e^{-j83.8x}\mathbf{a}_z \text{ A/m}$$

Similarly, the reflected electric and magnetic fields may be computed by noting that the reflected wave will propagate in the negative x direction as follows:

$$\mathbf{E}_{s1}^{-}(x) \equiv E_{y10}^{-}e^{jk_1x}\mathbf{a}_y \equiv \Gamma E_{y10}^{+}e^{jk_1x}\mathbf{a}_y = 1.16e^{j2.429}e^{j83.8x}\mathbf{a}_y \text{ V/m}$$

$$\mathbf{H}_{s1}^{-}(x) \equiv H_{zs1}^{-}(x)\mathbf{a}_z \equiv -\frac{E_{y10}^{-}}{\eta_1}e^{jk_1x}\mathbf{a}_z = -4.9e^{j2.429}e^{j83.8x}\mathbf{a}_z \text{ mA/m}$$

where $E_{y10}^{-} \equiv \Gamma E_{y10}^{+}$ is the amplitude of the reflected electric field, and the direction of the reflected magnetic field is selected in such a way that the resultant Poynting vector $(\mathbf{E}_{s1}^{-} \times \mathbf{H}_{s1}^{-})$ is along the $\mathbf{a}_{n1}^{-} \equiv -\mathbf{a}_x$ direction.

The total electric and magnetic fields in medium 1 are simply given by summation of the incident and the reflected fields as follows:

$$\mathbf{E}_{s1}(x) \equiv \mathbf{E}_{s1}^{+}(x) + \mathbf{E}_{s1}^{-}(x) \equiv 5.0^{-j83.8x}[1 + \Gamma e^{j167.7x}]\mathbf{a}_y$$

$$\mathbf{H}_{s1}(x) \equiv \mathbf{H}_{s1}^{+}(x) + \mathbf{H}_{s1}^{-}(x) \equiv 0.0212e^{-j83.8x}[1 - \Gamma e^{j167.7x}]\mathbf{a}_z$$

where $\Gamma \equiv 0.2329e^{j2.429}$ is the reflection coefficient as determined earlier.

(c) For calculating the time-average power density in medium 2, let us first compute the time-average power density of the incident wave in medium 1, which may be obtained using the following expression:

$$\langle \mathbf{S}_{1i} \rangle = \frac{1}{2}\text{Re}\ (\mathbf{E}_{s1}^{+} \times \mathbf{H}_{s1}^{+*}) \equiv \frac{1}{2} \times 5 \times 0.0212\ [\mathbf{a}_y \times \mathbf{a}_z] = 53\ \mathbf{a}_x \text{ mW/m}^2$$

Similarly, the magnitude of the time-average power density for the reflected wave in medium 1 may be determined as follows:

$$\langle \mathbf{S}_{1r} \rangle = \frac{1}{2}\text{Re}\ (\mathbf{E}_{s1}^{-} \times \mathbf{H}_{s1}^{-*}) \equiv -\frac{1}{2} \times 1.16 \times 4.9 \times 10^{-3}\ [\mathbf{a}_y \times \mathbf{a}_z] = -2.8\mathbf{a}_x \text{ mW/m}^2$$

The negative sign in the above expression testifies that the reflected power is propagating in the $-\mathbf{a}_x$ direction as described earlier. It is to be noted that the magnitude of the reflected power density can also be obtained using the following expression:

$$\langle S_{1r} \rangle \equiv |\Gamma|^2 \langle S_{1i} \rangle = (0.2329)^2 \times 0.053 = 2.8 \text{ mW/m}^2$$

which provides the same value as mentioned earlier. Now, the principle of power conservation can be used to calculate the time-average power density in medium 2 as follows:

$$\langle S_2 \rangle \equiv \left[1 - |\Gamma|^2\right] \langle S_{1i} \rangle = \left[1 - (0.2329)^2\right] \times 0.053 = 50.2 \text{ mW/m}^2$$

It should be noted that since the medium 2 is semi-infinite, hence there is no reflected wave in this medium and the average power density computed above corresponds to the transmitted wave only. You should note that the transmitted power density is medium 2 may also be determined using the expression given by Eq. (16) as follows:

$$\langle S_2 \rangle \equiv \left|\frac{\eta_1}{\eta_2}\right|^2 \left(\frac{\eta_2 + \eta_2^*}{\eta_1 + \eta_1^*}\right) |\tau|^2 \langle S_{1i} \rangle$$

$$= \left|\frac{235.6}{158.4 + j51}\right|^2 \times \frac{158.4 + 158.4}{235.6 + 235.6} \times 0.838^2 \times 53 = 50.2 \text{ mW/m}^2$$

where the computed values of various quantities are substituted. It can clearly be noted that although the same value of the transmitted power can be obtained using both the expressions, the expression based on the power conservation is simpler.

(*d*) The total impedance in medium 1 at a distance of 2 cm from the boundary can be determined by finding the values of the total electric and magnetic fields at this position as given below:

$$E_{sy1}(x) \equiv 5.0^{-j83.8x}[1 + \Gamma e^{j167.7x}]$$

$$H_{sz1}(x) \equiv 0.0212e^{-j83.8x}[1 - \Gamma e^{j167.7x}]$$

$$\eta\,|_{x=-2\text{cm}} = \frac{E_{sy1}\,|_{(x=-2\text{ cm})}}{H_{sz1}\,|_{(x=-2\text{ cm})}} = \frac{5.0e^{j83.8\times0.02}[1 + 0.2329e^{j2.429}e^{-j167.7\times0.02}]}{0.0212e^{j83.8\times0.02}[1 - 0.2329e^{j2.429}e^{-j167.7\times0.02}]}$$

$$= (288.2 - j113.5)\Omega$$

13.3 WAVE REFLECTION FROM MULTIPLE INTERFACES

So far we have treated the reflection of waves at the single boundary that occurs between semi-infinite media. In this section, we consider wave reflection from materials that are finite in extent, such that we must consider the effect of the front and back surfaces. Such a two-interface problem would occur, for example, for light incident on a flat piece of glass. Additional interfaces are present if the glass is coated with one or more layers of dielectric material for the purpose (as we will see) of reducing reflections. Such problems in which more than one interface is involved are frequently encountered; single-interface problems are in fact more the exception than the rule.

Consider the general situation shown in Figure 13.8, in which a uniform plane wave propagating in the forward z direction is normally incident from the left onto the interface between regions 1 and 2; these have intrinsic impedances η_1 and η_2. A third region of impedance η_3 lies beyond region 2, and so a second interface exists between regions 2 and 3. We let the second interface location occur at $z = 0$, and so all positions to the left will be described by values of z that are negative. The width of the second region is l, so the first interface will occur at position $z = -l$.

When the incident wave reaches the first interface, events occur as follows: A portion of the wave reflects, while the remainder is transmitted, to propagate toward the second interface. There, a portion is transmitted into region 3, while the rest reflects and returns to the first interface; there it is again partially reflected. This reflected wave then combines with additional transmitted energy from region 1, and the process repeats. We thus have a complicated sequence of multiple reflections that occur within region 2, with partial transmission at each bounce. To analyze the situation in this way would involve keeping track of a very large number of reflections; this would be necessary when studying the *transient* phase of the process, where the incident wave first encounters the interfaces.

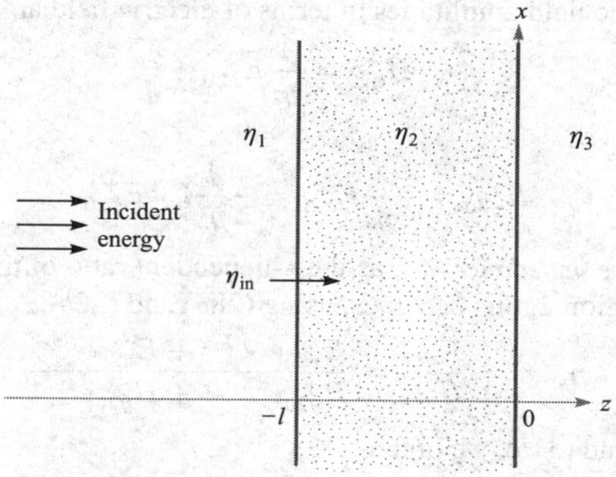

Fig. 13.8 Basic two-interface problem, in which the impedances of regions 2 and 3, along with the finite thickness of region 2, are accounted for in the input impedance at the front surface, η_{in}.

If the incident wave is left on for all time, however, a *steady-state* situation is eventually reached, in which: (1) an overall fraction of the incident wave is reflected from the two-interface configuration and back-propagates in region 1 with a definite amplitude and phase; (2) an overall fraction of the incident wave is transmitted through the two interfaces and forward-propagates in the third region; (3) a net backward wave exists in region 2, consisting of all reflected waves from the second interface; and (4) a net forward wave exists in region 2, which is the superposition of the transmitted wave through the first interface and all waves in region 2 that have reflected from the first interface and are now forward-propagating. The effect of combining many co-propagating waves in this way is to establish a single wave which has a definite amplitude and phase, determined through the sums of the amplitudes and phases of all the component waves. In steady state, we thus have a total of five waves to consider. These are the incident and net reflected waves in region 1, the net transmitted wave in region 3, and the two counterpropagating waves in region 2.

The situation is analyzed in the same manner as that used in the analysis of finite-length transmission lines (Section 11.11). Let us assume that all regions are composed of lossless media, and consider the two waves in region 2. If we take these as x-polarized, their electric fields add to yield

$$E_{xs2} = E_{x20}^{+}\, e^{-j\beta_2 z} + E_{x20}^{-}\, e^{j\beta_2 z} \tag{28a}$$

where $\beta_2 = \omega\sqrt{\epsilon_{r2}}/c$, and where the amplitudes, E_{x20}^{+} and E_{x20}^{-}, are complex. The y-polarized magnetic field is similarly written, using complex amplitudes:

$$H_{ys2} = H_{y20}^{+}\, e^{-j\beta_2 z} + H_{y20}^{-}\, e^{j\beta_2 z} \tag{28b}$$

We now note that the forward and backward electric field amplitudes in region 2 are related through the reflection coefficient at the second interface, Γ_{23}, where

$$\Gamma_{23} = \frac{\eta_3 - \eta_2}{\eta_3 + \eta_2} \tag{29}$$

We thus have

$$E_{x20}^{-} = \Gamma_{23} E_{x20}^{+} \tag{30}$$

We then write the magnetic field amplitudes in terms of electric field amplitudes through

$$H_{y20}^{+} = \frac{1}{\eta_2} E_{x20}^{+} \tag{31a}$$

and

$$H_{y20}^{-} = -\frac{1}{\eta_2} E_{x20}^{-} = -\frac{1}{\eta_2} \Gamma_{23} E_{x20}^{+} \tag{31b}$$

We now define the *wave impedance,* η_w, as the z-dependent ratio of the total electric field to the total magnetic field. In region 2, this becomes, using (28a) and (28b),

$$\eta_w(z) = \frac{E_{xs2}}{H_{ys2}} = \frac{E_{x20}^{+} e^{-j\beta_2 z} + E_{x20}^{-} e^{j\beta_2 z}}{H_{y20}^{+} e^{-j\beta_2 z} + H_{y20}^{-} e^{j\beta_2 z}}$$

Then, using (30), (31a), and (31b), we obtain

$$\eta_w(z) = \eta_2 \left[\frac{e^{-j\beta_2 z} + \Gamma_{23} e^{j\beta_2 z}}{e^{-j\beta_2 z} - \Gamma_{23} e^{j\beta_2 z}} \right]$$

Now, using (29) and Euler's identity, we have

$$\eta_w(z) = \eta_2 \times \frac{(\eta_3 + \eta_2)(\cos\beta_2 z - j\sin\beta_2 z) + (\eta_3 - \eta_2)(\cos\beta_2 z + j\sin\beta_2 z)}{(\eta_3 + \eta_2)(\cos\beta_2 z - j\sin\beta_2 z) - (\eta_3 - \eta_2)(\cos\beta_2 z + j\sin\beta_2 z)}$$

This is easily simplified to yield

$$\eta_w(z) = \eta_2 \frac{\eta_3 \cos\beta_2 z - j\eta_2 \sin\beta_2 z}{\eta_2 \cos\beta_2 z - j\eta_3 \sin\beta_2 z} \tag{32}$$

We now use the wave impedance in region 2 to solve our reflection problem. Of interest to us is the net reflected wave amplitude at the first interface. Since tangential **E** and **H** are continuous across the boundary, we have

$$E_{xs1}^{+} + E_{xs1}^{-} = E_{xs2} \qquad (z = -l) \tag{33a}$$

and

$$H_{ys1}^{+} + H_{ys1}^{-} = H_{ys2} \qquad (z = -l) \tag{33b}$$

Then, in analogy to (7) and (8), we may write

$$E_{x10}^{+} + E_{x10}^{-} = E_{xs2}(z = -l) \tag{34a}$$

and

$$\frac{E_{x10}^{+}}{\eta_1} - \frac{E_{x10}^{-}}{\eta_1} = \frac{E_{xs2}(z = -l)}{\eta_w(-l)} \tag{34b}$$

where E_{x10}^{+} and E_{x10}^{-} are the amplitudes of the incident and reflected fields. We call $\eta_w(-l)$ the *input impedance,* η_{in}, to the two-interface combination. We now solve (34a) and (34b) together, eliminating E_{xs2}, to obtain

$$\frac{E_{x10}^{-}}{E_{x10}^{+}} = \Gamma = \frac{\eta_{\text{in}} - \eta_1}{\eta_{\text{in}} + \eta_1} \tag{35}$$

To find the input impedance, we evaluate (32) at $z = -l$, resulting in

$$\eta_{in} = \eta_2 \frac{\eta_3 \cos \beta_2 l + j\eta_2 \sin \beta_2 l}{\eta_2 \cos \beta_2 l + j\eta_3 \sin \beta_2 l} \tag{36}$$

Equations (35) and (36) are general results that enable us to calculate the net reflected wave amplitude and phase from two parallel interfaces between lossless media.[1] Note the dependence on the interface spacing, l, and on the wavelength as measured in region 2, characterized by β_2. Of immediate importance to us is the fraction of the incident power that reflects from the dual interface and back-propagates in region 1. As we found earlier, this fraction will be $|\Gamma|^2$. Also of interest is the transmitted power, which propagates away from the second interface in region 3. It is simply the remaining power fraction, which is $1 - |\Gamma|^2$. The power in region 2 stays constant in steady state; power leaves that region to form the reflected and transmitted waves, but is immediately replenished by the incident wave. We have already encountered an analogous situation involving cascaded transmission lines, which culminated in Eq. (101) in Chapter 11.

An important result of situations involving two interfaces is that it is possible to achieve total transmission in certain cases. From (35), we see that total transmission occurs when $\Gamma = 0$, or when $\eta_{in} = \eta_1$. In this case, as in transmission lines, we say that the input impedance is *matched* to that of the incident medium. There are a few methods of accomplishing this.

As a start, suppose that $\eta_3 = \eta_1$, and region 2 is of such thickness that $\beta_2 l = m\pi$, where m is an integer. Now $\beta_2 = 2\pi/\lambda_2$, where λ_2 is the wavelength *as measured in region 2*. Therefore

$$\frac{2\pi}{\lambda_2} l = m\pi$$

or

$$l = m\frac{\lambda_2}{2} \tag{37}$$

With $\beta_2 l = m\pi$, the second region thickness is an integer multiple of half-wavelengths as measured in that medium. Equation (36) now reduces to $\eta_{in} = \eta_3$. Thus the general effect of a multiple half-wave thickness is to render the second region immaterial to the results on reflection and transmission. Equivalently, we have a single-interface problem involving η_1 and η_3. Now, with $\eta_3 = \eta_1$, we have a matched input impedance, and there is no net reflected wave. This method of choosing the region 2 thickness is known as *half-wave matching*. Its applications include, for example, antenna housings on airplanes known as *radomes,* which form a part of the fuselage. The antenna, inside the aircraft, can transmit and receive through this layer, which can be shaped to enable good aerodynamic characteristics. Note that the half-wave matching condition no longer applies as we deviate from the wavelength that satisfies it. When this is done, the device reflectivity increases (with increased wavelength deviation), so it ultimately acts as a bandpass filter.

[1] For convenience, (34*a*) and (34*b*) have been written for a specific time at which the incident wave amplitude, E^+_{x10}, occurs at $z = -l$. This establishes a zero-phase reference at the front interface for the incident wave, and so it is from this reference that the reflected wave phase is determined. Equivalently, we have repositioned the $z = 0$ point at the front interface. Eq. (36) allows this because it is only a function of the interface spacing, l.

Often, it is convenient to express the dielectric constant of the medium through the *refractive index* (or just index), n, defined as

$$n = \sqrt{\epsilon_r} \tag{38}$$

Characterizing materials by their refractive indices is primarily done at optical frequencies (on the order of 10^{14} Hz), whereas at much lower frequencies, a dielectric constant is traditionally specified. Since ϵ_r is complex in lossy media, the index will also be complex. Rather than complicate the situation in this way, we will restrict our use of the refractive index to cases involving lossless media, having $\epsilon_r'' = 0$, and $\mu_r = 1$. Under lossless conditions, we may write the plane wave phase constant and the material intrinsic impedance in terms of the index through

$$\beta = k = \omega\sqrt{\mu_0\epsilon_0}\sqrt{\epsilon_r} = \frac{n\omega}{c} \tag{39}$$

and

$$\eta = \frac{1}{\sqrt{\epsilon_r}}\sqrt{\frac{\mu_0}{\epsilon_0}} = \frac{\eta_0}{n} \tag{40}$$

Finally, the phase velocity and wavelength in a material of index n are

$$\nu_p = \frac{c}{n} \tag{41}$$

and

$$\lambda = \frac{\nu_p}{f} = \frac{\lambda_0}{n} \tag{42}$$

where λ_0 is the wavelength in free space. It is obviously important not to confuse the index n with the similar-appearing Greek η (intrinsic impedance), which has an entirely different meaning.

Another application, typically seen in optics, is the *Fabry-Perot interferometer*. This, in its simplest form, consists of a single block of glass or other transparent material of index n, whose thickness, l, is set to transmit wavelengths which satisfy the condition $\lambda = \lambda_0/n = 2l/m$. Often we want to transmit only one wavelength, not several, as (37) would allow. We would therefore like to assure that adjacent wavelengths that are passed through the device are separated as far as possible, so that only one will lie within the input power spectrum. In terms of wavelength as measured in the material, this separation is in general given by

$$\lambda_{m-1} - \lambda_m = \Delta\lambda_f = \frac{2l}{m-1} - \frac{2l}{m} = \frac{2l}{m(m-1)} \doteq \frac{2l}{m^2}$$

Note that m is the number of half-wavelengths in region 2, or $m = 2l/\lambda = 2nl/\lambda_0$, where λ_0 is the desired free-space wavelength for transmission. Thus

$$\Delta\lambda_f \doteq \frac{\lambda_2^2}{2l} \tag{43a}$$

In terms of wavelength measured in free space, this becomes

$$\Delta\lambda_{f0} = n\Delta\lambda_f \doteq \frac{\lambda_0^2}{2nl} \tag{43b}$$

$\Delta\lambda_{f0}$ is known as the *free spectral range* of the Fabry-Perot interferometer in terms of free-space wavelength separation. The interferometer can be used as a narrow-band filter (transmitting a desired wavelength and a narrow spectrum around this wavelength) if the spectrum to be filtered is narrower than the free spectral range.

EXAMPLE 13.8

Suppose we wish to filter an optical spectrum of full width $\Delta\lambda_{s0} = 50$ nm (measured in free space), whose center wavelength, λ_0, is in the red part of the visible spectrum at 600 nm, where one nm (nanometer) is 10^{-9} m. A Fabry-Perot filter is to be used, consisting of a lossless glass plate in air, having refractive index $n = 1.45$. We need to find the required range of glass thicknesses such that multiple wavelength orders will not be transmitted.

Solution. We require that the free spectral range be greater than the optical spectral width, or $\Delta\lambda_{f0} > \Delta\lambda_s$. Using (43*b*)

$$l < \frac{\lambda_0^2}{2n\Delta\lambda_{s0}}$$

So

$$l < \frac{600^2}{2(1.45)(50)} = 2.5 \times 10^3\text{nm} = 2.5\ \mu\text{m}$$

where 1μm (micrometer) $= 10^{-6}$ m. Fabricating a glass plate of this thickness or less is somewhat ridiculous to contemplate. Instead, what is often used is an airspace of thickness on this order, between two thick plates whose surfaces on the sides opposite the airspace are antireflection coated. This is in fact a more versatile configuration because the wavelength to be transmitted (and the free spectral range) can be adjusted by varying the plate separation.

Next we remove the restriction $\eta_1 = \eta_3$ and look for a way to produce zero reflection. Returning to Eq. (36), suppose we set $\beta_2 l = (2m-1)\pi/2$, or an odd multiple of $\pi/2$. This means that

$$\frac{2\pi}{\lambda_2}l = (2m-1)\frac{\pi}{2} \qquad (m = 1, 2, 3, \ldots)$$

or

$$\boxed{l = (2m-1)\frac{\lambda_2}{4}} \tag{44}$$

The thickness is an odd multiple of a quarter-wavelength as measured in region 2. Under this condition (36) reduces to

$$\eta_{\text{in}} = \frac{\eta_2^2}{\eta_3} \tag{45}$$

Typically, we choose the second region impedance to allow matching between given impedances η_1 and η_3. To achieve total transmission, we require that $\eta_{\text{in}} = \eta_1$, so that the required second region impedance becomes

$$\eta_2 = \sqrt{\eta_1 \eta_3} \tag{46}$$

With the conditions given by (44) and (46) satisfied, we have performed *quarter-wave matching*. The design of antireflective coatings for optical devices is based on this principle.

EXAMPLE 13.9

A violet light with wavelength of 0.4 μm is incident normally on an air-dielectric interface at $z = 0$. The dielectric medium is considered semi-infinite, and its properties are given by $\epsilon'_{r2} = 3.5$, $\epsilon''_{r2} = 0$, $\mu'_{r2} = 1.0$, $\mu''_{r2} = 0$.

Determine

(a) the location of the first electric field maximum and the first electric field minimum in medium '1' away from the interface;

(b) the standing wave ratio; and

(c) the fraction of incident power transmitted into the dielectric medium.

Solution.

(a) Let us first calculate the impedances for both the mediums. As the first medium is air, hence its impedance will simply be equal to that of free space, which would be given by

$$\eta_1 \equiv \eta_0 = \sqrt{\frac{\mu_0}{\epsilon_0}} = 377\Omega$$

Similarly, the impedance for medium 2 may be computed as follows:

$$\eta_2 \equiv \sqrt{\frac{\mu_2}{\epsilon_0}} = \frac{\eta_0}{\sqrt{\epsilon_{r2}}} = \frac{377}{\sqrt{3.5}} = 201.5\Omega$$

Next, the reflection coefficient may be determined as

$$\Gamma \equiv |\Gamma| e^{j\varphi} \equiv \frac{\eta_2 - \eta_1}{\eta_2 - \eta_1} = \frac{201.5 - 377}{201.5 + 377} = -0.303e \equiv 0.303e^{j\pi}$$

Now, since the phase of the reflection coefficient ϕ is equal to π in the present situation, the first electric field minimum would occur directly at the interface $z = 0$. The location of the first electric field maximum would be at a distance of $\lambda_1/4$ from the interface, which may be computed as follows:

$$z_1, \text{max} \equiv -\frac{\lambda_1}{4} = \frac{0.4}{4} = -0.1\ \mu\text{m}$$

(b) The standing wave ratio would be given by

$$s \equiv \frac{1 + |\Gamma|}{1 - |\Gamma|} = \frac{1 + 0.303}{1 - 0.303} = 1.87$$

(*c*) The fraction of incident power transmitted into the dielectric medium may be calculated as follows

$$P_t \equiv [1 - |\Gamma|^2]P_{in} = [1 - |0.303|^2]P_{in} = 0.9082P_{in}$$

Hence, 90.8% of the total incident power will be transmitted into the dielectric medium.

EXAMPLE 13.10

We wish to coat a glass surface with an appropriate dielectric layer to provide total transmission from air to the glass at a free-space wavelength of 570 nm. The glass has refractive index $n_3 = 1.45$. Determine the required index for the coating and its minimum thickness.

Solution. The known impedances are $\eta_1 = 377\,\Omega$ and $\eta_3 = 377/1.45 = 260\,\Omega$. Using (46) we have

$$\eta_2 = \sqrt{(377)(260)} = 313\,\Omega$$

The index of region 2 will then be

$$n_2 = \left(\frac{377}{313}\right) = 1.20$$

The wavelength in region 2 will be

$$\lambda_2 = \frac{570}{1.20} = 475\,\text{nm}$$

The minimum thickness of the dielectric layer is then

$$l = \frac{\lambda_2}{4} = 119\,\text{nm} = 0.119\,\mu\text{m}$$

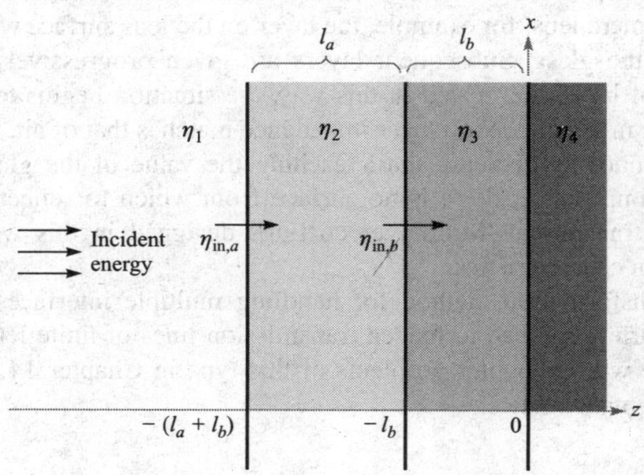

Fig. 13.9 A three-interface problem in which input impedance $\eta_{\text{in},a}$ is transformed back to the front interface to form input impedance $\eta_{\text{in},b}$.

The procedure in this section for evaluating wave reflection has involved calculating an effective impedance at the first interface, η_{in}, which is expressed in terms of the impedances that lie beyond the front surface. This process of *impedance transformation* is more apparent when we consider problems involving more than two interfaces.

For example, consider the three-interface situation shown in Figure 13.9, where a wave is incident from the left in region 1. We wish to determine the fraction of the incident power that is reflected and back-propagates in region 1 and the fraction of the incident power that is transmitted into region 4. To do this, we need to find the input impedance at the front surface (the interface between regions 1 and 2). We start by transforming the impedance of region 4 to form the input impedance at the boundary between regions 2 and 3. This is shown as $\eta_{\text{in},b}$ in Figure 13.9. Using (36), we have

$$\eta_{\text{in},b} = \eta_3 \frac{\eta_4 \cos \beta_3 l_b + j\eta_3 \sin \beta_3 l_b}{\eta_3 \cos \beta_3 l_b + j\eta_4 \sin \beta_3 l_b} \tag{47}$$

We have now effectively reduced the situation to a two-interface problem in which $\eta_{\text{in},b}$ is the impedance of all that lies beyond the second interface. The input impedance at the front interface, $\eta_{\text{in},a}$, is now found by transforming $\eta_{\text{in},b}$ as follows:

$$\eta_{\text{in},a} = \eta_2 \frac{\eta_{\text{in},b} \cos \beta_2 l_a + j\eta_2 \sin \beta_2 l_a}{\eta_2 \cos \beta_2 l_a + j\eta_{\text{in},b} \sin \beta_2 l_a} \tag{48}$$

The reflected power fraction is now $|\Gamma|^2$, where

$$\Gamma = \frac{\eta_{\text{in},a} - \eta_1}{\eta_{\text{in},a} + \eta_1}$$

The fraction of the power transmitted into region 4 is, as before, $1 - |\Gamma|^2$. The method of impedance transformation can be applied in this manner to any number of interfaces. The process, although tedious, is easily handled by a computer.

The motivation for using multiple layers to reduce reflection is that the resulting structure is less sensitive to deviations from the design wavelength if the impedances (or refractive indices) are arranged to progressively increase or decrease from layer to layer. For multiple layers to antireflection coat a camera lens, for example, the layer on the lens surface would be of impedance very close to that of the glass. Subsequent layers are given progressively higher impedances. With a large number of layers fabricated in this way, the situation begins to approach (but never reaches) the ideal case, in which the top layer impedance matches that of air, while the impedances of deeper layers continuously decrease until reaching the value of the glass surface. With this continuously varying impedance, there is no surface from which to reflect, and so light of any wavelength is totally transmitted. Multilayer coatings designed in this way produce excellent broadband transmission characteristics.

The impedance transformation method for handling multiple interfaces applies not only to plane waves at boundaries, but also to loaded transmission lines of finite length, and to cascaded transmission lines. We will encounter problems of this type in Chapter 14, which we will solve using exactly the same mathematics.

D13.1 A uniform plane wave in air is normally incident on a dielectric slab of thickness $\lambda_2/4$, and intrinsic impedance $\eta_2 = 260\ \Omega$. Determine the magnitude and phase of the reflection coefficient.

Ans. 0.356; 180°

EXAMPLE 13.11

An anti-reflection dielectric coating is to be applied to a glass surface whose electrical properties are given by $\epsilon'_{r3} = 5.1$, $\epsilon''_{r3} = 0$, $\mu'_{r3} = 1.0$, $\mu''_{r3} = 0$. The dielectric coating should result into zero reflection of the yellow beam light with a free-space wavelength of 0.6 μm, which is incident normally on the interface as shown in Figure 13.10. The glass medium can be considered semi-infinite for all practical purposes.

(*a*) Determine the dielectric constant and the minimum thickness required for the coating.

(*b*) If now a red beam with a free-space wavelength of 0.75 μm is incident on the interface, what would be resultant reflection coefficient at the air-dielectric interface?

(*c*) How can the dielectric coating designed in part (a) be changed so that it produces zero reflection for the red beam with a free-space wavelength of 0.75 μm?

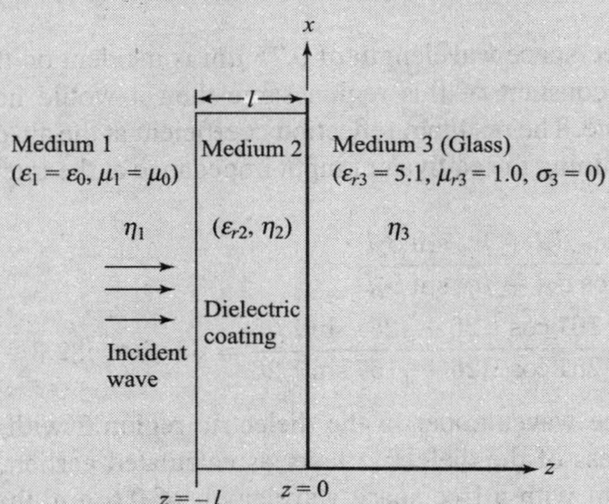

Fig. 13.10 The anti-reflection dielectric coating applied on the glass.

Solution.

(*a*) For obtaining zero reflection at the air-dielectric interface ($z = -l$) shown in Figure 13.10, the impedance of the dielectric layer (region 2) should be geometric mean of the impedances of the free space (region 1) and the glass (region 3), i.e.,

$$\eta_2 \equiv \sqrt{\eta_1 \eta_3}$$

Now, the first medium is air, whose impedance will be simply equal to that of free space, i.e.,

$$\eta_1 \equiv \eta_0 = \sqrt{\frac{\mu_0}{\epsilon_0}} = 377\ \Omega$$

The impedance of the region 3 represented by glass may be computed as follows:

$$\eta_3 \equiv \sqrt{\frac{\mu_3}{\epsilon_3}} = \frac{\eta_0}{\sqrt{\epsilon'_{r3}}} = \frac{377}{\sqrt{5.1}} = 167\ \Omega$$

Hence, the impedance of the dielectric layer will be given by

$$\eta_2 \equiv \sqrt{377 \times 167} = 251\ \Omega$$

As both the regions 1 and 3 are lossless and nonmagnetic, hence the dielectric material required for the coating can also be chosen to be of similar type. The dielectric constant of the required coating in this case may be calculated as follows:

$$\eta_2 \equiv \sqrt{\frac{\mu_2}{\epsilon_2}} = \frac{\eta_0}{\sqrt{\epsilon_{r2}}} = 251\ \Omega \quad \Rightarrow \epsilon_{r2} = \left(\frac{377}{251}\right)^2 = 2.25$$

The minimum thickness required for the coating will be given by

$$t = \frac{\lambda_2}{4} \equiv \frac{\lambda_2}{4\sqrt{\epsilon_{r2}}} = \frac{0.6}{4\sqrt{2.25}} = 0.10\ \mu\text{m}$$

Where λ_2 is the wavelength in the dielectric medium, and λ_0 is the free-space wavelength of the yellow beam.

(*b*) If a red beam with a free-space wavelength of 0.75 μm is incident on the air-dielectric interface keeping the dielectric constant of this region same then it would no longer result into zero reflection at this interface. The resultant reflection coefficient at the air-dielectric interface can be obtained by first determining the equivalent input impedance at the interface using the following relationship:

$$\begin{aligned}\eta_{\text{in}} &\equiv \eta_2 \frac{\eta_3 \cos \beta_2 l + j\eta_2 \sin \beta_2 l}{\eta_2 \cos \beta_2 l + j\eta_3 \sin \beta_2 l} \\ &= 251 \times \frac{167 \cos 1.26 + j251 \sin 1.26}{251 \cos 1.26 + j167 \sin 1.26} = 337.5 + j82.3 \equiv 347.4\, e^{j0.239}\end{aligned}$$

where $\beta_2 = 2\pi/\lambda_2$ is the wavenumber in the dielectric region 2 with $\lambda_2 \equiv 0.75/\sqrt{2.25}$, and l = 0.1 μm is the thickness of the dielectric layer as calculated earlier. It may be easily proved that for the yellow beam with a free-space wavelength of 0.6 μm, the input impedance using the above formula will simply reduce to the free-space impedance having a value of 377 Ω. Once the input impedance is calculated using the above formula, the reflection coefficient at the air-dielectric interface may be calculated as follows:

$$\Gamma|_{z=-l} \equiv \frac{\eta_{\text{in}} - \eta_0}{\eta_{\text{in}} + \eta_0} = \frac{337.5 - 377 + j82.3}{337.5 + 377 + j82.3} \equiv 0.127 e^{j1.904}$$

(*c*) The zero reflection for the red beam would be obtained when the thickness of the dielectric coating becomes a multiple of the quarter wavelength corresponding to the red beam. If the dielectric constant of the material is assumed to be same as in part (a), and minimum thickness is considered, then the parameters for the coating would be as follows:

$$\epsilon_{r2} = 2.25 \qquad t_{\min}\big|_{(red\,beam)} = \frac{\lambda_2}{4} \equiv \frac{\lambda_0}{4\sqrt{\epsilon_{r2}}} = \frac{0.75}{4\sqrt{2.25}} = 0.125\ \mu\text{m}$$

13.4 PLANE WAVE PROPAGATION IN GENERAL DIRECTIONS

In this section we will learn how to mathematically describe uniform plane waves that propagate in any direction. Our motivation for doing this is our need to address the problem of incident waves on boundaries that are not perpendicular to the propagation direction. Such problems of *oblique incidence* generally occur, with normal incidence being a special case. Addressing such problems requires (as always) that we establish an appropriate coordinate system. With the boundary positioned in the x, y plane, for example, the incident wave will propagate in a direction that could involve all three coordinate axes, whereas with normal incidence, we were only concerned with propagation along z. We need a mathematical formalism that will allow for the general direction case.

Let us consider a wave that propagates in a lossless medium, with propagation constant $\beta = k = \omega\sqrt{\mu\epsilon}$. For simplicity, we consider a two-dimensional case, where the wave travels in a direction between the x and z axes. The first step is to consider the propagation constant as a *vector,* **k**, indicated in Figure 13.11. The direction of **k** is the propagation direction, which is the same as the direction of the Poynting vector in our case.[2] The magnitude of **k** is the phase shift per unit distance

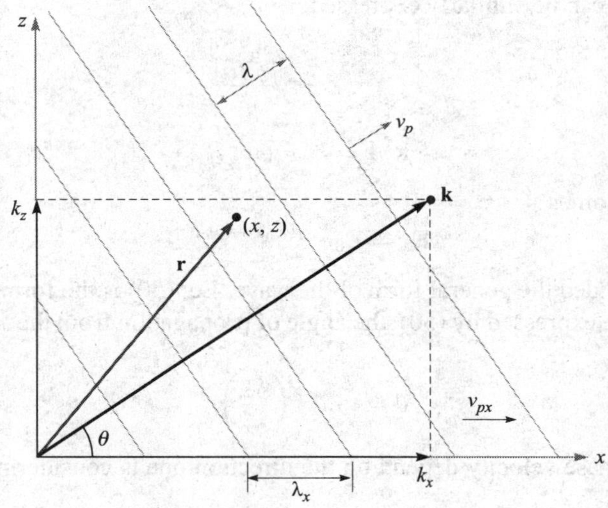

Fig. 13.11 Representation of a uniform plane wave with wavevector **k** at angle θ to the x axis. The phase at point (x, z) is given by $\mathbf{k} \cdot \mathbf{r}$. Planes of constant phase (shown as lines perpendicular to **k**) are spaced by wavelength λ but have wider spacing when measured along the x or z axes.

[2] We assume here that the wave is in an isotropic medium, where the permittivity and permeability do not change with field orientation. In anisotropic media (where ϵ and/or μ depend on field orientation), the directions of the Poynting vector and **k** may differ.

along that direction. Part of the process of characterizing a wave involves specifying its phase at any spatial location. For the waves we have considered that propagate along the z axis, this was accomplished by the factor $e^{\pm jkz}$ in the phasor form. To specify the phase in our two-dimensional problem, we make use of the vector nature of **k** and consider the phase at a general location (x, z) described through the position vector **r**. The phase at that location, referenced to the origin, is given by the projection of **k** along **r** times the magnitude of **r**, or just $\mathbf{k} \cdot \mathbf{r}$. If the electric field is of magnitude E_0, we can thus write down the phasor form of the wave in Figure 13.11 as

$$\mathbf{E}_s = \mathbf{E}_0 e^{-j\mathbf{k}\cdot\mathbf{r}} \tag{49}$$

The minus sign in the exponent indicates that the phase along **r** moves in time in the direction of increasing **r**. Again, the wave power flow in an isotropic medium occurs in the direction along which the phase shift per unit distance is maximum—or along **k**. The vector **r** serves as a means to measure phase at any point using **k**. This construction is easily extended to three dimensions by allowing **k** and **r** to each have three components.

In our two-dimensional case of Figure 13.11, we can express **k** in terms of its x and z components:

$$\mathbf{k} = k_x\mathbf{a}_x + k_z\mathbf{a}_z$$

The position vector, **r**, can be similarly expressed:

$$\mathbf{r} = x\mathbf{a}_x + z\mathbf{a}_z$$

so that

$$\mathbf{k}\cdot\mathbf{r} = k_x x + k_z z$$

Equation (49) now becomes

$$\mathbf{E}_s = \mathbf{E}_0 e^{-j(k_x x + k_z z)} \tag{50}$$

Whereas Eq. (49) provided the general form of the wave, Eq. (50) is the form that is specific to the situation. Given a wave expressed by (50), the angle of propagation from the x axis is readily found through

$$\theta = \tan^{-1}\left(\frac{k_z}{k_x}\right)$$

The wavelength and phase velocity depend on the direction one is considering. In the direction of **k**, these will be

$$\lambda = \frac{2\pi}{k} = \frac{2\pi}{(k_x^2 + k_z^2)^{1/2}}$$

and

$$\nu_p = \frac{\omega}{k} = \frac{\omega}{(k_x^2 + k_z^2)^{1/2}}$$

If, for example, we consider the x direction, these quantities will be

$$\lambda_x = \frac{2\pi}{k_x}$$

and

$$\nu_{px} = \frac{\omega}{k_x}$$

Note that both λ_x and ν_{px} are greater than their counterparts along the direction of **k**. This result, at first surprising, can be understood through the geometry of Figure 13.11. The diagram shows a series of phase fronts (planes of constant phase) which intersect **k** at right angles. The phase shift between adjacent fronts is set at 2π in the figure; this corresponds to a spatial separation along the **k** direction of one wavelength, as shown. The phase fronts intersect the x axis, and we see that *along* x the front separation is greater than it was along **k**. λ_x is the spacing between fronts along x and is indicated on the figure. The phase velocity along x is the velocity of the intersection points between the phase fronts and the x axis. Again, from the geometry, we see that this velocity must be faster than the velocity along **k** and will, of course, exceed the speed of light in the medium. This does not constitute a violation of special relativity, however, since the energy in the wave flows in the direction of **k** and not along x or z. The wave frequency is $f = \omega/2\pi$ and is invariant with direction. Note, for example, that in the directions we have considered,

$$f = \frac{\nu_p}{\lambda} = \frac{\nu_{px}}{\lambda_x} = \frac{\omega}{2\pi}$$

EXAMPLE 13.12

Consider a 50 MHz uniform plane wave having electric field amplitude 10 V/m. The medium is lossless, having $\epsilon_r = \epsilon_r' = 9.0$ and $\mu_r = 1.0$. The wave propagates in the x, y plane at a 30° angle to the x axis and is linearly polarized along z. Write down the phasor expression for the electric field.

Solution. The propagation constant magnitude is

$$k = \omega\sqrt{\mu\epsilon} = \frac{\omega\sqrt{\epsilon_r}}{c} = \frac{2\pi \times 50 \times 10^6(3)}{3 \times 10^8} = 3.2\ \text{m}^{-1}$$

The vector **k** is now

$$\mathbf{k} = 3.2(\cos 30\mathbf{a}_x + \sin 30\mathbf{a}_y) = 2.8\mathbf{a}_x + 1.6\mathbf{a}_y\ \text{m}^{-1}$$

Then

$$\mathbf{r} = x\,\mathbf{a}_x + y\,\mathbf{a}_y$$

With the electric field directed along z, the phasor form becomes

$$\mathbf{E}_s = E_0 e^{-j\mathbf{k}\cdot\mathbf{r}}\,\mathbf{a}_z = 10e^{-j(2.8x+1.6y)}\,\mathbf{a}_z$$

D13.4 For Example 13.12, calculate λ_x, λ_y, ν_{px}, and ν_{py}.

Ans. 2.2 m; 3.9 m; 1.1×10^8 m/s; 2.0×10^8 m/s.

13.5 PLANE WAVE REFLECTION AT OBLIQUE INCIDENCE ANGLES

We now consider the problem of wave reflection from plane interfaces, in which the incident wave propagates at some angle to the surface. Our objectives are (1) to determine the relation between incident, reflected, and transmitted angles, and (2) to derive reflection and transmission coefficients that are functions of the incident angle and wave polarization. We will also show that cases exist in which total reflection or total transmission may occur at the interface between two dielectrics if the angle of incidence and the polarization are appropriately chosen.

The situation is illustrated in Figure 13.12, in which the incident wave direction and position-dependent phase are characterized by wavevector $\mathbf{k}_1^+$. The angle of incidence is the angle between $\mathbf{k}_1^+$ and a line that is normal to the surface (the x axis in this case). The incidence angle is shown as θ_1. The reflected wave, characterized by wavevector $\mathbf{k}_1^-$, will propagate away from the interface at angle θ_1'. Finally, the transmitted wave, characterized by $\mathbf{k}_2$, will propagate into the second region at angle θ_2 as shown. One would suspect (from previous experience) that the incident and reflected angles are equal ($\theta_1 = \theta_1'$), which is correct. We need to show this, however, to be complete.

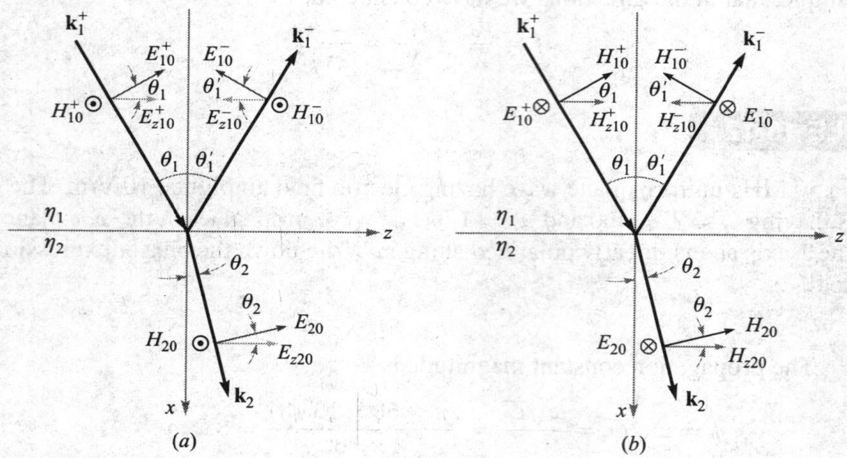

Fig. 13.12 Geometries for plane wave incidence at angle θ_1 onto an interface between dielectrics having intrinsic impedances η_1 and η_2. The two polarization cases are shown: (*a*) p-polarization (or TM), with $\mathbf{E}$ in the plane of incidence; (*b*) s-polarization (or TE), with $\mathbf{E}$ perpendicular to the plane of incidence.

The two media are lossless dielectrics, characterized by intrinsic impedances η_1 and η_2. We will assume, as before, that the materials are nonmagnetic, and thus have permeability μ_0. Consequently, the materials are adequately described by specifying their dielectric constants, ϵ_{r1} and ϵ_{r2}, or their refractive indices, $n_1 = \sqrt{\epsilon_{r1}}$ and $n_2 = \sqrt{\epsilon_{r2}}$.

In Figure 13.12, two cases are shown which differ by the choice of electric field orientation. In Figure 13.12a, the $\mathbf{E}$ field is polarized in the plane of the page, with $\mathbf{H}$ therefore perpendicular to the page and pointing outward. In this illustration, the plane of the page is also the *plane of incidence,* which is more precisely defined as the plane spanned by the incident $\mathbf{k}$ vector and the normal to the surface. With $\mathbf{E}$ lying in the plane of incidence, the wave is said to have *parallel polarization* or to be *p-polarized* ($\mathbf{E}$ is parallel to the incidence plane). Note that while $\mathbf{H}$ is perpendicular to the incidence plane, it lies parallel (or transverse) to the interface. Consequently, another name for this type of polarization is *transverse magnetic,* or TM polarization.

Figure 13.12*b* shows the situation in which the field directions have been rotated by 90°. Now **H** lies in the plane of incidence, whereas **E** is perpendicular to the plane. Since **E** is used to define polarization, the configuration is called *perpendicular polarization,* or is said to be *s-polarized.*[3] **E** is also parallel to the interface, and so the case is also called *transverse electric,* or TE polarization. We will find that the reflection and transmission coefficients will differ for the two polarization types, but that reflection and transmission angles will not depend on polarization. We only need to consider s- and p-polarizations because any other field direction can be constructed as some combination of s and p waves.

Our desired knowledge of reflection and transmission coefficients, as well as how the angles relate, can be found through the field boundary conditions at the interface. Specifically, we require that the transverse components of **E** and **H** be continuous across the interface. These were the conditions we used to find Γ and τ for normal incidence ($\theta_1 = 0$), which is in fact a special case of our current problem. We will consider the case of p-polarization (Figure 13.12*a*) first. To begin, we write down the incident, reflected, and transmitted fields in phasor form, using the notation developed in Section 13.4:

$$\mathbf{E}_{s1}^{+} = \mathbf{E}_{10}^{+} e^{-j\mathbf{k}_1^{+}\cdot\mathbf{r}} \tag{51}$$

$$\mathbf{E}_{s1}^{-} = \mathbf{E}_{10}^{-} e^{-j\mathbf{k}_1^{-}\cdot\mathbf{r}} \tag{52}$$

$$\mathbf{E}_{s2} = \mathbf{E}_{20} e^{-j\mathbf{k}_2\cdot\mathbf{r}} \tag{53}$$

where

$$\mathbf{k}_1^{+} = k_1(\cos\theta_1\,\mathbf{a}_x + \sin\theta_1\,\mathbf{a}_z) \tag{54}$$

$$\mathbf{k}_1^{-} = k_1(-\cos\theta_1'\,\mathbf{a}_x + \sin\theta_1'\,\mathbf{a}_z) \tag{55}$$

$$\mathbf{k}_2 = k_2(\cos\theta_2\,\mathbf{a}_x + \sin\theta_2\,\mathbf{a}_z) \tag{56}$$

and where

$$\mathbf{r} = x\,\mathbf{a}_x + z\,\mathbf{a}_z \tag{57}$$

The wavevector magnitudes are $k_1 = \omega\sqrt{\epsilon_{r1}}/c = n_1\omega/c$ and $k_2 = \omega\sqrt{\epsilon_{r2}}/c = n_2\omega/c$.

Now, to evaluate the boundary condition that requires continuous tangential electric field, we need to find the components of the electric fields (z components) that are parallel to the interface. Projecting all **E** fields in the z direction, and using (51) through (57), we find

$$E_{zs1}^{+} = E_{z10}^{+} e^{-j\mathbf{k}_1^{+}\cdot\mathbf{r}} = E_{10}^{+}\cos\theta_1 e^{-jk_1(x\cos\theta_1 + z\sin\theta_1)} \tag{58}$$

$$E_{zs1}^{-} = E_{z10}^{-} e^{-j\mathbf{k}_1^{-}\cdot\mathbf{r}} = E_{10}^{-}\cos\theta_1' e^{jk_1(x\cos\theta_1' - z\sin\theta_1')} \tag{59}$$

$$E_{zs2} = E_{z20} e^{-j\mathbf{k}_2\cdot\mathbf{r}} = E_{20}\cos\theta_2 e^{-jk_2(x\cos\theta_2 + z\sin\theta_2)} \tag{60}$$

The boundary condition for continuous tangential electric field now reads:

---

[3] The *s* designation is an abbreviation for the German *senkrecht,* meaning *perpendicular*. The *p* in *p-polarized* is an abbreviation for the German word for parallel, which is *parallel.*

$$E^+_{zs1} + E^-_{zs1} = E_{zs2} \quad (\text{at } x = 0)$$

We now substitute Eqs. (58) through (60) into (61) and evaluate the result at $x = 0$ to obtain

$$E^+_{10} \cos\theta_1 \, e^{-jk_1 z \sin\theta_1} + E^-_{10} \cos\theta'_1 \, e^{-jk_1 z \sin\theta'_1} = E_{20} \cos\theta_2 \, e^{-jk_2 z \sin\theta_2} \tag{61}$$

Note that E^+_{10}, E^-_{10}, and E_{20} are all constants (independent of z). Further, we require that (61) hold for all values of z (everywhere on the interface). For this to occur, it must follow that all the phase terms appearing in (61) are equal. Specifically,

$$k_1 z \sin\theta_1 = k_1 z \sin\theta'_1 = k_2 z \sin\theta_2$$

From this, we see immediately that $\theta'_1 = \theta_1$, or the angle of reflection is equal to the angle of incidence. We also find that

$$k_1 \sin\theta_1 = k_2 \sin\theta_2 \tag{62}$$

Equation (62) is known as *Snell's law of refraction.* Since in general, $k = n\omega/c$, we can rewrite (62) in terms of the refractive indices:

$$n_1 \sin\theta_1 = n_2 \sin\theta_2 \tag{63}$$

Equation (63) is the form of Snell's law that is most readily used for our present case of nonmagnetic dielectrics. Equation (62) is a more general form which would apply, for example, to cases involving materials with different permeabilities as well as different permittivities. In general, we would have $k_1 = (\omega/c)\sqrt{\mu_{r1}\epsilon_{r1}}$ and $k_2 = (\omega/c)\sqrt{\mu_{r2}\epsilon_{r2}}$.

Having found the relations between angles, we next turn to our second objective, which is to determine the relations between the amplitudes, E^+_{10}, E^-_{10}, and E_{20}. To accomplish this, we need to consider the other boundary condition, requiring tangential continuity of $\mathbf{H}$ at $x = 0$. The magnetic field vectors for the p-polarized wave are all negative y-directed. At the boundary, the field amplitudes are related through

$$H^+_{10} + H^-_{10} = H_{20} \tag{64}$$

Then, when we use the fact that $\theta'_1 = \theta_1$ and invoke Snell's law, (61) becomes

$$E^+_{10} \cos\theta_1 + E^-_{10} \cos\theta_1 = E_{20} \cos\theta_2 \tag{65}$$

Using the medium intrinsic impedances, we know, for example, that $E^+_{10}/H^+_{10} = \eta_1$ and $E^+_{20}/H^+_{20} = \eta_2$. Equation (64) can be written as follows:

$$\frac{E^+_{10}\cos\theta_1}{\eta_{1p}} - \frac{E^-_{10}\cos\theta_1}{\eta_{1p}} = \frac{E^+_{20}\cos\theta_2}{\eta_{2p}} \tag{66}$$

Note the minus sign in front of the second term in (66), which results from the fact that $E^-_{10}\cos\theta_1$ is negative (from Figure 13.12*a*), whereas H^-_{10} is positive (again from the figure). When we write Eq. (66), *effective impedances,* valid for p-polarization, are defined through

$$\eta_{1p} = \eta_1 \cos\theta_1 \tag{67}$$

and

$$\eta_{2p} = \eta_2 \cos\theta_2 \tag{68}$$

Using this representation, Eqs. (65) and (66) are now in a form that enables them to be solved together for the ratios E_{10}^-/E_{10}^+ and E_{20}/E_{10}^+. Performing analogous procedures to those used in solving (7) and (8), we find the reflection and transmission coefficients:

$$\Gamma_p = \frac{E_{10}^-}{E_{10}^+} = \frac{\eta_{2p} - \eta_{1p}}{\eta_{2p} + \eta_{1p}} \tag{69}$$

$$\tau_p = \frac{E_{20}}{E_{10}^+} = \frac{2\eta_{2p}}{\eta_{2p} + \eta_{1p}} \left(\frac{\cos\theta_1}{\cos\theta_2} \right) \tag{70}$$

A similar procedure can be carried out for s-polarization, referring to Figure 13.12*b*. The details are left as an exercise; the results are

$$\Gamma_s = \frac{E_{y10}^-}{E_{y10}^+} = \frac{\eta_{2s} - \eta_{1s}}{\eta_{2s} + \eta_{1s}} \tag{71}$$

Illustrations

$$\tau_s = \frac{E_{y20}}{E_{y10}^+} = \frac{2\eta_{2s}}{\eta_{2s} + \eta_{1s}} \tag{72}$$

where the effective impedances for s-polarization are

$$\eta_{1s} = \eta_1 \sec\theta_1 \tag{73}$$

and

$$\eta_{2s} = \eta_2 \sec\theta_2 \tag{74}$$

Equations (67) through (74) are what we need to calculate wave reflection and transmission for either polarization, and at any incident angle.

EXAMPLE 13.13

A uniform plane wave is incident from air onto glass at an angle from the normal of 30°. Determine the fraction of the incident power that is reflected and transmitted for (*a*) p-polarization and (*b*) s-polarization. Glass has refractive index $n_2 = 1.45$.

Solution. First, we apply Snell's law to find the transmission angle. Using $n_1 = 1$ for air, we use (63) to find

$$\theta_2 = \sin^{-1}\left(\frac{\sin 30}{1.45} \right) = 20.2^\circ$$

Now, for p-polarization:

$$\eta_{1p} = \eta_1 \cos 30 = (377)(.866) = 326\,\Omega$$

$$\eta_{2p} = \eta_2 \cos 20.2 = \frac{377}{1.45}(.938) = 244\,\Omega$$

Then, using (69), we find

$$\Gamma_p = \frac{244 - 326}{244 + 326} = -0.144$$

The fraction of the incident power that is reflected is

$$\frac{P_r}{P_{inc}} = |\Gamma_p|^2 = .021$$

The transmitted fraction is then

$$\frac{P_t}{P_{inc}} = 1 - |\Gamma_p|^2 = .979$$

For s-polarization, we have

$$\eta_{1s} = \eta_1 \sec 30 = 377/.866 = 435\,\Omega$$

$$\eta_{2s} = \eta_2 \sec 20.2 = \frac{377}{1.45(.938)} = 277\,\Omega$$

Then, using (71):

$$\Gamma_s = \frac{277 - 435}{277 + 435} = -.222$$

The reflected power fraction is thus

$$|\Gamma_s|^2 = .049$$

The fraction of the incident power that is transmitted is

$$1 - |\Gamma_s|^2 = .951$$

EXAMPLE 13.14

A uniform plane wave propagating in free space represented by region 1 has the following expression for the incident electric field:

$$\mathbf{E}_{s1}^{+} = 15e^{-j(0.3x+0.4z)\pi}\mathbf{a_y}\ \mathrm{V/m}$$

The wave is incident from free space on a perfect conductor at $z = 0$.

Determine

(*a*) the wave number and frequency of the wave;

(*b*) the incident angle;

(*c*) the phasor expressions of the electric and magnetic fields for the reflected wave;

(*d*) the total electric and magnetic fields in region 1; and

(*e*) the time-average power density for the incident wave and the total wave.

Solution.

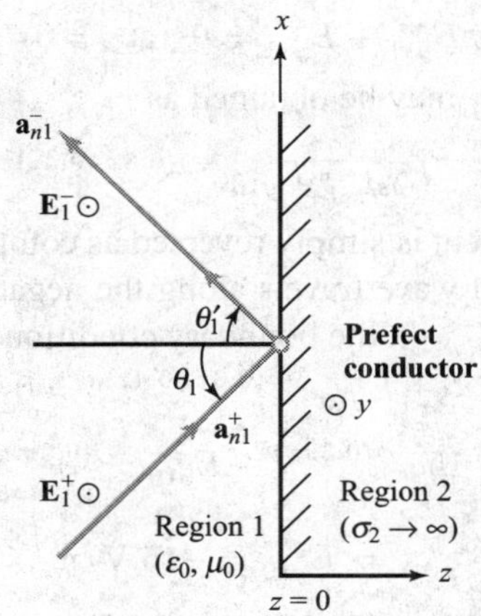

Fig. 13.13 The plane wave incident at angle θ_1 onto an air-conductor interface (s-polarization)

(*a*) The incident electric field can be written as

$$E_{ys1}^+ = 15e^{-j(0.3x+0.4z)\pi} \equiv 15e^{-j\mathbf{k}_1^+\cdot\mathbf{r}} \equiv 15e^{-j(0.3\pi\,\mathbf{a}_x+0.4\pi\,\mathbf{a}_z)\cdot(x\mathbf{a}_x+z\mathbf{a}_z)}$$

It is obvious from the above expression that the electric field is totally perpendicular to the plane of incidence, and hence it represents a case of perpendicular polarization as shown in Figure 13.13. The wave number can be computed as

$$k_1 \equiv |\mathbf{k}_1^+| = \sqrt{k_x^2 + k_z^2} = \pi\sqrt{0.3^2 + 0.4^2} = 0.5\pi\,\text{rad/m}$$

The frequency may be calculated as

$$f = \frac{c}{\lambda} = \frac{c}{(2\pi/k_1^+)} = \frac{3\times10^8\times0.5\pi}{2\pi} = 75\,\text{MHz}$$

where c is the velocity of light in free space.

(*b*) For determining the incident angle θ_1, let us first rewrite the wave number as follows:

$$\mathbf{k}_1^+ = 0.3\pi\,\mathbf{a}_x + 0.4\pi\,\mathbf{a}_z \equiv k_1\mathbf{a}_{n1}^+ \equiv 0.5\pi(0.6\mathbf{a}_x + 0.8\mathbf{a}_z)\ \text{rad/m}$$

where $\mathbf{a}_{n1}^+ \equiv (0.6\mathbf{a}_x + 0.8\mathbf{a}_z)$ may be considered as the unit vector along the direction of $\mathbf{k}_1^+$. Now, as the incident angle is defined as the angle between $\mathbf{k}_1^+$ and a line normal to the surface as shown in Figure 13.13, it may be computed as follows:

$$\cos\theta_1 = \mathbf{a}_z\cdot\mathbf{a}_{n1}^+ \equiv \mathbf{a}_z\cdot(0.6\mathbf{a}_x + 0.8\mathbf{a}_z) = 0.8$$

$$\Rightarrow \theta_1 = \cos^{-1}0.8 = 36.9°$$

(*c*) The electric field for the reflected wave may be obtained by applying the boundary condition for the tangential component at the surface $z = 0$ and noting that the transmission component of the electric field would be zero because of the presence of perfect conductor. This results into the following expression:

$$E_{ys1}^{+} + E_{ys1}^{-} = 0 \;\; (\text{at } z = 0)$$

The general expression for E_{ys1}^{-} may be obtained as

$$E_{ys1}^{-} = E_{y10}^{-} e^{-j(0.3x-0.4z)\pi}$$

where the sign of the z component is simply reversed as compared to that of the incidence wave due to the fact that the reflected wave travels along the negative z axis as shown in figure. The substitution of both E_{ys1}^{+} and E_{ys1}^{-} in the boundary condition yields:

$$15e^{-j(0.3x-0.4z)\pi} + E_{y10}^{-} e^{-j(0.3x-0.4z)\pi} = 0 \quad (\text{at } z = 0)$$

$$\Rightarrow 15e^{-j(0.3\pi x)} + E_{y10}^{-} e^{-j(0.3\pi x)} = 0$$

$$\Rightarrow E_{y10}^{-} = -15 \text{ V/m}$$

Hence, the phasor expression of the reflected electric field would be given by

$$\mathbf{E}_{s1}^{-} = -15e^{-j(0.3x-0.4z)\pi}\mathbf{a}_y \text{ V/m}$$

The above expression of the reflected electric field may also be written as

$$E_{ys1}^{-} = -15e^{-j(0.3x-0.4z)\pi} = -15e^{-j\mathbf{k}_1^{-}\cdot\mathbf{r}} = -15e^{-j(0.3\pi\mathbf{a}_x - 0.4\pi\mathbf{a}_z)\cdot(x\mathbf{a}_x + z\mathbf{a}_z)}$$

The wave number for the reflected wave may then be written as follows

$$\mathbf{k}_1^{-} = 0.3\pi\mathbf{a}_x - 0.4\pi\mathbf{a}_z \equiv k_1\, \mathbf{a}_{n1}^{-} \equiv 0.5\pi(0.6\mathbf{a}_x - 0.8\mathbf{a}_z) \text{ rad/m}$$

where $\mathbf{a}_{n1}^{-}$ is the unit vector along the direction of $\mathbf{k}_1^{-}$. The magnetic field for the reflected wave may be obtained by noting that the Poynting vector ($\mathbf{E}_{s1}^{-} \times \mathbf{H}_{s1}^{-}$) for the reflected wave is oriented along the $\mathbf{a}_{n1}^{-}$ direction

$$\mathbf{H}_{s1}^{-} = \frac{1}{\eta_1}\mathbf{a}_{n1}^{-} \times \mathbf{E}_{s1}^{-}$$

The above expression may be verified for the *normal* incidence case where the reflected wave is propagating along the negative z direction, which would result into the H_{s1}^{-} component oriented along the negative y direction if the E_{s1}^{-} component is assumed to be oriented along the positive x direction. Hence, the phasor expression of the reflected magnetic field may be computed using the above expression as follows:

$$\begin{aligned}
\mathbf{H}_{s1}^{-} &= \frac{1}{\eta_0}(0.6\mathbf{a}_x - 0.8\mathbf{a}_z) \times (-15e^{-j(0.3x-0.4z)\pi})\mathbf{a}_y \text{ A/m} \\
&= -\frac{15}{377}e^{-j(0.3x-0.4z)\pi}(0.6\mathbf{a}_x - 0.8\mathbf{a}_z) \times \mathbf{a}_y \\
&= -0.04e^{-j(0.3x-0.4z)\pi}(0.6\mathbf{a}_z + 0.8\mathbf{a}_x) \text{ A/m}
\end{aligned}$$

(*d*) The total electric field in region 1 may be calculated as

$$\begin{aligned}\mathbf{E}_{s1}(z) &= \mathbf{E}_{s1}^{-} + \mathbf{E}_{s1}^{-} = 15[e^{-j(0.3x+0.4z)\pi} - e^{-j(0.3x-0.4z)\pi}]\mathbf{a}_y \\ &= 15[e^{-j0.3\pi x}[e^{-j0.4\pi z} - e^{j0.4\pi z}]\mathbf{a}_y \\ &= 15[e^{-j0.3\pi x}[-2j\sin(0.4\pi z)]\mathbf{a}_y \\ &= -30je^{-j0.3\pi x}\sin(0.4\pi z)\mathbf{a}_y \text{ V/m}\end{aligned}$$

It is interesting to observe here that while there is a standing wave along the direction z, there is also a component of the traveling wave along the direction x. The standing wave occurs along the direction z in region 1 mainly because of the presence of conducting boundary at $z = 0$. The magnetic field for the incident wave may be calculated with the help of the following equation as explained earlier in part (c):

$$\begin{aligned}\mathbf{H}_{s1}^{+} &= \frac{1}{\eta_1}\mathbf{a}_{n1}^{+} \times \mathbf{E}_{s1}^{+} \\ &= \frac{1}{\eta_0}(0.6\mathbf{a}_x + 0.8\mathbf{a}_z) \times (15e^{-j(0.3x+0.4z)\pi})\mathbf{a_y} \text{ A/m} \\ &= 0.04e^{-j(0.3x+0.4z)\pi}(0.6\mathbf{a}_z - 0.8\mathbf{a}_x) \text{ A/m}\end{aligned}$$

where $\mathbf{a}_{n1}^{+}$ is the unit vector along the direction of $\mathbf{k}_1^{+}$ as computed earlier.
The total magnetic field in region 1 may then be calculated as follows:

$$\begin{aligned}\mathbf{H}_{s1}(z) &\equiv \mathbf{H}_{s1}^{+} + \mathbf{H}_{s1}^{-} \\ &= 0.04e^{-j(0.3x+0.4z)\pi}(0.6\mathbf{a}_z - 0.8\mathbf{a}_x) - 0.04e^{-j(0.3x-0.4z)\pi}(0.6\mathbf{a}_z + 0.8\mathbf{a}_x) \\ &= 0.024e^{-j0.3\pi x}[e^{-j0.4\pi z} - e^{j0.4\pi z}]\mathbf{a}_z - 0.032e^{-j0.3\pi x}[e^{-j0.4\pi z} + e^{j0.4\pi z}]\mathbf{a}_x\end{aligned}$$

$$\Rightarrow \quad \mathbf{H}_{s1}(z) = -[0.064\cos(0.4\pi z)\mathbf{a}_x + 0.048j\sin(0.4\pi z)\mathbf{a}_z]e^{-j0.3\pi x}$$

(*e*) The time-average power density for the incident wave is given by

$$\begin{aligned}\langle \mathbf{S}_1^{+} \rangle &= \frac{1}{2}\text{Re}\,(\mathbf{E}_{s1}^{+} \times \mathbf{H}_{s1}^{+*}) \\ &\equiv \frac{1}{2}[15e^{-j(0.3x+0.4z)\pi}\mathbf{a}_y \times (0.6\mathbf{a}_z - 0.8\mathbf{a}_x)0.04e^{+j(0.3x+0.4z)\pi}] \\ &\equiv (0.18\mathbf{a}_x + 0.24\mathbf{a}_z) \text{ W/m}^2\end{aligned}$$

It can be observed from the above equation that the incident power flows along both the positive x and positive z directions, which can also be verified from Figure 13.13.

In order to compute the time-average power density for the *total* wave in region 1, firstly the conjugate of the magnetic field in region 1 may be determined as follows:

$$\mathbf{H}_{s1}^{*}(z) = 0.048je^{0.3\pi x}\sin(0.4\pi z)\mathbf{a}_z - 0.064e^{j0.3\pi x}\cos(0.4\pi z)\mathbf{a}_x$$

The time-average power density for the *total* wave in region 1 can now be computed as follows:

$$\langle \mathbf{S}_1 \rangle = \frac{1}{2}\text{Re}\,(\mathbf{E}_{s1} \times \mathbf{H}_{s1}^*)$$

$$= \frac{1}{2}\text{Re}\{-30je^{-j0.3\pi x}\sin(0.4\pi z)\,\mathbf{a}_y \times [0.048je^{+j0.3\pi x}\sin(0.4\pi z)\,\mathbf{a}_z$$

$$-0.064e^{j0.3\pi x}\cos(0.4\pi z)\,\mathbf{a}_x]\}$$

$$= \frac{1}{2}\text{Re}\{1.44\sin^2(0.4\pi z)\,\mathbf{a}_x - j1.92\sin^2(0.4\pi z)\,\mathbf{a}_z\}$$

$$\Rightarrow \langle \mathbf{S}_1 \rangle = 0.72\sin^2(0.4\pi z)\,\mathbf{a}_x\ \text{W/m}^2$$

The above equation indicates there is some power flow along the x direction. However, there is no net power flow along the z direction.

EXAMPLE 13.15

A uniform plane wave in air with the following expression for the incident electric field

$$\mathbf{E}_{s1}^+ = (4.1\mathbf{a}_x - 2.9\mathbf{a}_z)e^{-j(6.0x+8.6z)}\ \text{V/m}$$

is incident on the air-dielectric interface. The dielectric medium occupied by the region $z > 0$ is characterized by nonmagnetic lossless dielectric with $\mu_{r2} = 1.0$, $\epsilon_{r2} = 2.25$, $\sigma_2 = 0$.

Find

(*a*) the type of polarization (perpendicular or parallel);
(*b*) the incident angle;
(*c*) the amplitude of the incident electric field;
(*d*) the frequency of the wave;
(*e*) the transmitted angle;
(*f*) the expressions for the reflected electric and magnetic fields;
(*g*) the expression for the transmitted magnetic field.

Solution.

(*a*) It can be seen from the given expression that the incident electric field has both the x and the z components, and it is lying in the plane of incidence. Hence, the wave can be said to have parallel polarization or is p-polarized.

(*b*) The expression of the given incident electric field can be compared with the standard expression of the electric field given by the following expression:

$$\mathbf{E}_{s1}^+ = \mathbf{E}_{10}^+ e^{-jk_1\mathbf{a}_{n1}^+\cdot\mathbf{r}} \equiv \mathbf{E}_{10}^+(\cos\theta_1\mathbf{a}_x - \sin\theta_1\mathbf{a}_z)e^{-jk_1(x\sin\theta_1 + z\cos\theta_1)}\ \text{V/m}$$

where k_1 is the wave number inside the free space (medium 1), $\mathbf{a}_{n1}^+ \equiv (\sin\theta_1\mathbf{a}_x + \cos\theta_1\mathbf{a}_z)$ is the unit vector along the direction of propagation of the incident wave as shown in Figure 13.14, $\mathrm{r} \equiv (x\mathbf{a}_x + z\mathbf{a}_z)$ is the radius vector, E_{10}^+ is the amplitude of the incident electric field, and θ_1

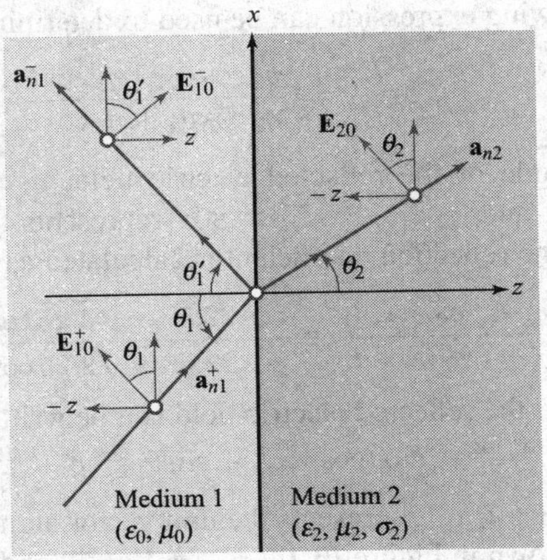

Fig. 13.14 The plane wave incident at angle θ_1 onto an air-dielectric interface (*p*-polarization).

represents the incident angle. A comparison of the above equation with the given equation yields following value of the incident angle:

$$E_{10}^+ \cos\theta_1 = 4.1; \quad E_{10}^+ \sin\theta_1 = 2.9$$

$$\Rightarrow \tan\theta_1 = \frac{2.9}{4.1}; \quad \Rightarrow \theta_1 = \tan^{-1}\left(\frac{2.9}{4.1}\right) = 35^\circ$$

(*c*) The amplitude of the incident electric field can be computed as

$$E_{10}^+ \cos 35^\circ = 4.1 \qquad \Rightarrow E_{10}^+ = \frac{4.1}{\cos 35^\circ} = 5.0 \text{ V/m}$$

(*d*) First of all, the wave number inside the free space (medium 1) can be computed by comparing the incident electric field with the standard expression as follows:

$$k_1 \sin 35^\circ = 6.0 \Rightarrow k_1 = \frac{6.0}{\sin 35^\circ} = 10.46 \text{ rad/m} \equiv \frac{2\pi \text{f}}{\text{c}}$$

Hence, the frequency of the wave can be determined as

$$f \equiv \frac{ck_1}{2\pi} = \frac{3 \times 10^8 \times 10.46}{2\pi} = 500 \text{ MHz}$$

(*e*) The transmitted angle θ_2 can be obtained by applying the Snell's law as follows:

$$n_1 \sin\theta_1 = n_2 \sin\theta_2$$

$$\Rightarrow \sin\theta_2 = \frac{n_1}{n_2}\sin\theta_1 = \frac{1}{\sqrt{2.25}}\sin 35^\circ$$

$$\Rightarrow \theta_2 = 22.5^\circ$$

(*f*) In order to derive the expression for the reflected electric field, first of all the reflection coefficient should be determined. As the present configuration here represents a parallel polarized wave, the following expression can be used to determine the reflection coefficient:

$$\Gamma_p \equiv \frac{E_{10}^-}{E_{10}^+} = \frac{\eta_2 \cos\theta_2 - \eta_1 \cos\theta_1}{\eta_2 \cos\theta_2 + \eta_1 \cos\theta_1}$$

where E_{10}^- is the amplitude of the reflected electric field, $\eta_1 = \eta_0 \equiv 120\,\pi$ is the intrinsic impedance of free space, and $\eta_2 = \eta_0/\sqrt{\epsilon_{r2}} \equiv 80\pi$ represents the intrinsic impedance of the dielectric region. Hence the reflection coefficient is calculated as

$$\Gamma_p = \frac{\cos\theta_2 - \sqrt{\epsilon_{r2}}\cos\theta_1}{\cos\theta_2 + \sqrt{\epsilon_{r2}}\cos\theta_1} \equiv \frac{\cos 22.5^\circ - \sqrt{2.25}\cos 35^\circ}{\cos 22.5^\circ + \sqrt{2.25}\cos 35^\circ} = -0.142$$

The general expression for the reflected electric field can be written as

$$\mathbf{E}_{s1}^- = \mathbf{E}_{10}^- e^{-jk_1\mathbf{a}_{n1}^-\cdot\mathbf{r}} \equiv E_{10}^-(\cos\theta_1'\mathbf{a}_x + \sin\theta_1'\mathbf{a}_z)e^{-jk_1(x\sin\theta_1' - z\cos\theta_1')}\ \text{V/m}$$

where $\mathbf{a}_{n1}^- \equiv (\sin\theta_1'\mathbf{a}_x - \cos\theta_1'\mathbf{a}_z)$ represents the unit vector along the direction of propagation of the reflected wave as shown in Figure 13.14, $E_{10}^- = \Gamma_p E_{10}^+$, and $\theta_1' \equiv \theta_1$ is the reflected angle. The values of various quantities can now be substituted in the above equation to obtain the expression for the reflected electric field in the present situation:

$$\begin{aligned}\mathbf{E}_{s1}^- &= -0.142 \times 5(\cos 35^\circ\mathbf{a}_x + \sin 35^\circ\mathbf{a}_z)e^{-j10.46(x\sin 35^\circ - z\cos 35^\circ)}\text{V/m}\\ &= -(0.5816\mathbf{a}_x + 0.4072\mathbf{a}_z)e^{j(-6.0x+8.6z)}\text{V/m}\end{aligned}$$

The expression for the reflected magnetic field can be obtained using the concept of the impedance and the Poynting vector as follows:

$$\begin{aligned}\mathbf{H}_{s1}^- &\equiv \frac{1}{\eta_1}[\mathbf{a}_{n1}^- \times \mathbf{E}_{s1}^-]\\ &\equiv \frac{1}{120\pi}\{(\sin 35^\circ\mathbf{a}_x - \cos 35^\circ\mathbf{a}_z) \times [-(0.5816\mathbf{a}_x + 0.4072\mathbf{a}_z)e^{j(-6.0x+8.6z)}]\}\\ &\equiv \frac{-1}{120\pi}[(-0.2336\mathbf{a}_y - 0.4764\mathbf{a}_y)e^{j(-6.0x+8.6z)}]\\ &= \frac{0.71}{120\pi}\mathbf{a}_y\, e^{j(-6.0x+8.6z)}\end{aligned}$$

From the above expression, it can be ascertained that the magnetic field is perpendicular to the plane of incidence for the parallel polarization.

(*g*) For deriving expressions of the transmitted fields, the transmission coefficient should first be determined as follows:

$$\tau_p \equiv \frac{E_{20}}{E_{10}^+} = \frac{2\eta_2\cos\theta_1}{\eta_2\cos\theta_2 + \eta_1\cos\theta_1} \equiv \frac{2\cos\theta_1}{\cos\theta_2 + (\eta_1/\eta_2)\cos\theta_1} = \frac{2\cos\theta_1}{\cos\theta_2 + \sqrt{\epsilon_{r2}}\cos\theta_1}$$

where E_{20} represents the amplitude of the transmitted electric field. The values of various quantities can now be substituted in the above equation to obtain the transmission coefficient:

$$\tau_p \equiv \frac{2\cos 35^\circ}{\cos 22.5^\circ + \sqrt{2.25}\cos 35^\circ} = 0.761$$

The general expression for the transmitted electric field in the present situation can be written as

$$\mathbf{E}_{s2} = \mathbf{E}_{20}e^{-jk_1\mathbf{a}_{n2}\cdot\mathbf{r}} \equiv E_{20}(\cos\theta_2\mathbf{a}_x - \sin\theta_2\mathbf{a}_z)e^{-jk_2(x\sin\theta_2 + z\cos\theta_2)}\ \text{V/m}$$

where,

$$k_2 \equiv \frac{2\pi f\sqrt{\epsilon_{r2}}}{c} = \frac{2\pi \times 500 \times 10^6 \times \sqrt{2.25}}{3 \times 10^8} = 15.7 \text{ rad/m}$$

is the wave number inside the medium 2, and $\mathbf{a}_{n2} \equiv (\sin\theta_2\,\mathbf{a}_x + \cos\theta_2\,\mathbf{a}_z)$ represents the unit vector along the direction of propagation of the transmitted wave as shown in Figure 13.14. The values of various parameters can now be substituted in the above equation to obtain the actual expression for the transmitted electric field as follows:

$$\mathbf{E}_{s2} \equiv 0.761 \times 5(\cos 22.5°\mathbf{a}_x - \sin 22.5°\mathbf{a}_z)e^{-j15.7(x\sin 22.5° + z\cos 22.5°)} \text{ V/m}$$
$$\equiv (3.515a_x - 1.456a_z)e^{-j(6x+14.5z)} \text{ V/m}$$

In Example 13.13, reflection coefficient values for the two polarizations were found to be negative. The meaning of a negative reflection coefficient is that the component of the reflected electric field that is parallel to the interface will be directed opposite the incident field component when both are evaluated at the boundary.

This effect is also observed when the second medium is a perfect conductor. In this case, we know that the electric field inside the conductor must be zero. Consequently, $\eta_2 = E_{20}/H_{20} = 0$, and the reflection coefficients will be $\Gamma_p = \Gamma_s = -1$. Total reflection occurs, regardless of the incident angle or polarization.

13.6 TOTAL REFLECTION AND TOTAL TRANSMISSION OF OBLIQUELY INCIDENT WAVES

Now that we have methods available to us for solving problems involving oblique incidence reflection and transmission, we can explore the special cases of *total reflection* and *total transmission*. We look for special combinations of media, incidence angles, and polarizations that produce these properties. To begin, we identify the necessary condition for total reflection. We want total *power* reflection, so that $|\Gamma|^2 = \Gamma\Gamma^* = 1$, where Γ is either Γ_p or Γ_s. The fact that this condition involves the possibility of a complex Γ allows some flexibility. For the incident medium, we note that η_{1p} and η_{1s} will always be real and positive. On the other hand, when we consider the second medium, η_{2p} and η_{2s} involve factors of $\cos\theta_2$ or $1/\cos\theta_2$, where

$$\cos\theta_2 = \left[1 - \sin^2\theta_2\right]^{1/2} = \left[1 - \left(\frac{n_1}{n_2}\right)^2 \sin^2\theta_1\right]^{1/2} \tag{75}$$

where Snell's law has been used. We observe that $\cos\theta_2$, and hence η_{2p} and η_{2s}, become imaginary whenever $\sin\theta_1 > n_2/n_1$. Let us consider parallel polarization, for example. Under conditions of imaginary η_{2p}, (69) becomes

$$\Gamma_p = \frac{j|\eta_{2p}| - \eta_{1p}}{j|\eta_{2p}| + \eta_{1p}} = -\frac{\eta_{1p} - j|\eta_{2p}|}{\eta_{1p} + j|\eta_{2p}|} = -\frac{Z}{Z^*}$$

where $Z = \eta_{1p} - j|\eta_{2p}|$. *We can therefore see that* $\Gamma_p\Gamma_p^* = 1$, *meaning total power reflection, whenever* η_{2p} *is imaginary*. The same will be true whenever η_{2p} is zero, which will occur when $\sin\theta_1 = n_2/n_1$.

We thus have our condition for total reflection, which is

$$\sin\theta_1 \geq \frac{n_2}{n_1} \tag{76}$$

From this condition arises the *critical angle* of total reflection, θ_c, defined through

$$\sin\theta_c = \frac{n_2}{n_1} \tag{77}$$

The total reflection condition can thus be more succinctly written as

$$\theta_1 \geq \theta_c \qquad \text{(for total reflection)} \tag{78}$$

Note that for (76) and (77) to make sense, it must be true that $n_2 < n_1$, or the wave must be incident from a medium of higher refractive index than that of the medium beyond the boundary. For this reason, the total reflection condition is sometimes called total *internal* reflection; it is often seen (and applied) in optical devices such as beam-steering prisms, where light within the glass structure totally reflects from glass-air interfaces.

EXAMPLE 13.16

A prism is to be used to turn a beam of light by 90°, as shown in Figure 13.15. Light enters and exits the prism through two antireflective (AR-coated) surfaces. Total reflection is to occur at the back surface, where the incident angle is 45° to the normal. Determine the minimum required refractive index of the prism material if the surrounding region is air.

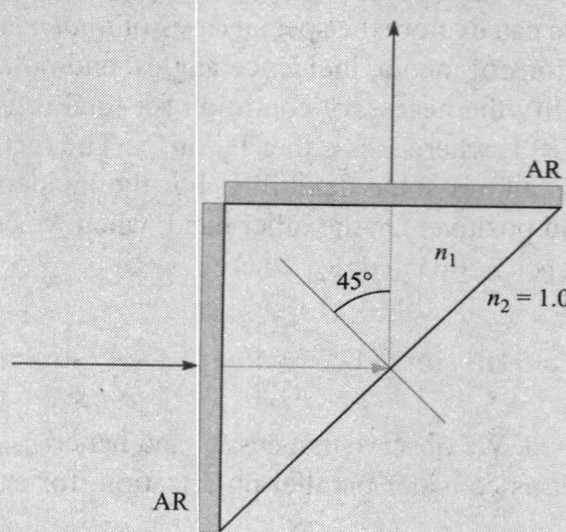

Fig. 13.15 Beam-steering prism for Example 13.16.

Solution. Considering the back surface, the medium beyond the interface is air, with $n_2 = 1.00$. Since $\theta_1 = 45°$, (76) is used to obtain

$$n_1 \geq \frac{n_2}{\sin 45} = \sqrt{2} = 1.41$$

Since fused silica glass has refractive index $n_g = 1.45$, it is a suitable material for this application and is in fact widely used.

EXAMPLE 13.17

A uniform plane wave propagating in a nonmagnetic lossless dielectric medium is incident on free space at the critical angle of 30°. Find the transmission angle, the refractive index and the relative permittivity of the dielectric medium.

Solution. The wave is incident here from a medium of high refractive index than that of the medium beyond the boundary, which is free space. Let us assume that the incident angle and the transmitted angle are given by θ_1 and θ_2, respectively. As the angle of incidence is given here as the critical angle, Snell's law can be applied to obtain following equation:

$$n_1 \sin\theta_1 = n_2 \sin\theta_2$$

$$\Rightarrow n_1 \sin\theta_c = n_2 \sin\theta_t$$

where θ_t is the transmitted angle corresponding to the incidence angle θ_c. Now, as per Eq. (77), the critical angle is given by following relationship:

$$\sin\theta_c = \frac{n_2}{n_1}; \quad n_2 < n_1$$

The above equation can only be satisfied if $\sin\theta_t = 1$. Hence, the transmission angle would be given by $\theta_t \equiv \sin^{-1}(1) = 90°$. It basically means that when the wave is incident at the critical angle from a higher refractive index to a lower refractive index medium, it would just be confined to the interface as shown in Figure 13.16.

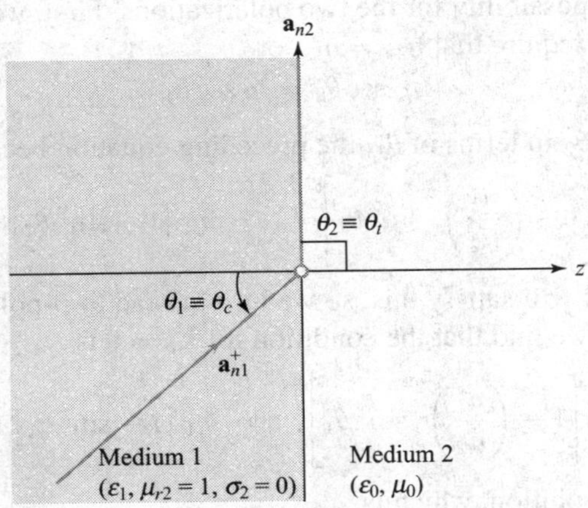

Fig. 13.16 A plane wave incident at a critical angle from a denser to a lighter medium.

However, if the incident angle is increased beyond the critical angle, i.e., $\theta_1 > \theta_c$, then the wave will be totally reflected in the first medium.

Now, as the second medium is free space, hence the refractive index of the second medium would be $n_2 = 1$. The refractive index of the first medium can then be calculated as:

$$n_1 = \frac{1}{\sin\theta_c} \equiv \frac{1}{\sin 30^\circ} = 2$$

Finally, the relative permittivity can be determined as

$$\epsilon_{r1} = n_1^2 = 4.0$$

Another important application of total reflection is in *optical waveguides*. These, in their simplest form, are constructed of three layers of glass, in which the middle layer has a slightly higher refractive index than the outer two. Figure 13.17 shows the basic structure. Light, propagating from left to right, is confined to the middle layer by total reflection at the two interfaces, as shown. Optical fiber waveguides are constructed on this principle, in which a cylindrical glass core region of small radius is surrounded coaxially by a lower-index cladding glass material of larger radius. Basic waveguiding principles as applied to metallic and dielectric structures will be presented in Chapter 14.

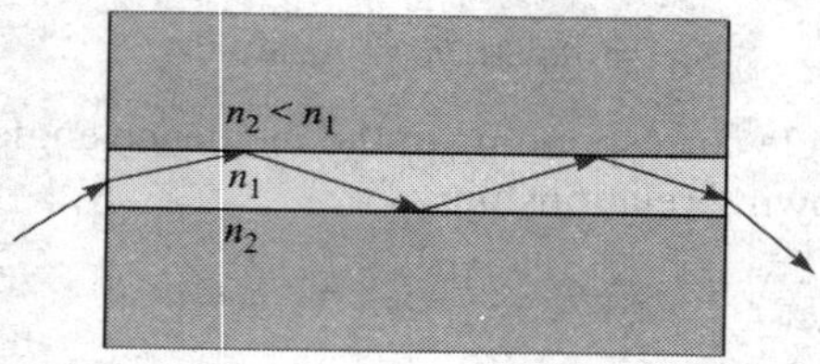

Fig. 13.17 A dielectric slab waveguide (symmetric case), showing light confinement to the center material by total reflection.

We next consider the possibility of *total transmission*. In this case the requirement is simply that $\Gamma = 0$. We investigate this possibility for the two polarizations. First, we consider s-polarization. If $\Gamma_s = 0$, then from (71) we require that $\eta_{2s} = \eta_{1s}$, or

$$\eta_2 \sec\theta_2 = \eta_1 \sec\theta_1$$

Using Snell's law to write θ_2 in terms of θ_1, the preceding equation becomes

$$\eta_2 \left[1 - \left(\frac{n_1}{n_2}\right)^2 \sin^2\theta_1\right]^{-1/2} = \eta_1 \left[1 - \sin^2\theta_1\right]^{-1/2}$$

There is no value of θ_1 that will satisfy this, so we turn instead to p-polarization. Using (67), (68), and (69), with Snell's law, we find that the condition for $\Gamma_p = 0$ is

$$\eta_2 \left[1 - \left(\frac{n_1}{n_2}\right)^2 \sin^2\theta_1\right]^{1/2} = \eta_1 \left[1 - \sin^2\theta_1\right]^{1/2}$$

This equation does have a solution, which is

$$\sin\theta_1 = \sin\theta_B = \frac{n_2}{\sqrt{n_1^2 + n_2^2}} \tag{79}$$

where we have used $\eta_1 = \eta_0/n_1$ and $\eta_2 = \eta_0/n_2$. We call this special angle θ_B, where total transmission occurs, the *Brewster angle* or *polarization angle*. The latter name comes from the fact that if light having both s- and p-polarization components is incident at $\theta_1 = \theta_B$, the p component will be totally transmitted, leaving the partially reflected light entirely s-polarized. At angles that are slightly off the Brewster angle, the reflected light is still predominantly s-polarized. Most reflected light that we see originates from horizontal surfaces (such as the surface of the ocean), and so the light has mostly horizontal polarization. Polaroid sunglasses take advantage of this fact to reduce glare, for they are made to block the transmission of horizontally polarized light while passing light that is vertically polarized.

EXAMPLE 13.18

Light is incident from air to glass at Brewster's angle. Determine the incident and transmitted angles.

Solution. Since glass has refractive index $n_2 = 1.45$, the incident angle will be

$$\theta_1 = \theta_B = \sin^{-1}\left(\frac{n_2}{\sqrt{n_1^2 + n_2^2}}\right) = \sin^{-1}\left(\frac{1.45}{\sqrt{1.45^2 + 1}}\right) = 55.4^\circ$$

The transmitted angle is found from Snell's law, through

$$\theta_2 = \sin^{-1}\left(\frac{n_1}{n_2}\sin\theta_B\right) = \sin^{-1}\left(\frac{n_1}{\sqrt{n_1^2 + n_2^2}}\right) = 34.6^\circ$$

Note from this exercise that $\sin\theta_2 = \cos\theta_B$, which means that the sum of the incident and refracted angles at the Brewster condition is always 90°.

EXAMPLE 13.19

Show that for the plane interface joining two nonmagnetic lossless dielectric media, the Brewster angle does not exist in case of perpendicular polarization. Describe whether the situation would change if both the media are considered to be magnetic.

Solution. The Brewster angle is defined as the incident angle at which there is no reflection or total transmission. For the perpendicular polarization, this condition can be obtained by putting

$$\Gamma_s \equiv \frac{\eta_2 \cos\theta_{\mathrm{Bs}} - \eta_1 \cos\theta_2}{\eta_2 \cos\theta_{\mathrm{Bs}} + \eta_1 \cos\theta_2} = 0$$

where $\theta_{\mathrm{Bs}} \equiv \theta_1$ is the incident angle corresponding to no reflection for the perpendicular polarization or s-polarization. The above equation can be simplified to obtain the following condition:

$$\eta_2 \cos\theta_{Bs} - \eta_1 \cos\theta_2 = 0$$

The transmitted angle θ_2 can be expressed in terms of the incident angle θ_{Bs} using the the generalized form of Snell's law given by (62), i.e.,

$$k_1 \sin\theta_{Bs} \equiv k_2 \sin\theta_2 \Rightarrow \cos\theta_2 = \sqrt{1 - \left(\frac{k_1}{k_2}\right)^2 \sin^2(\theta_{Bs})}$$

The above expression can now be combined with the previous equation to obtain

$$\cos\theta_{Bs} \equiv \frac{\eta_1}{\eta_2}\cos\theta_2 \equiv \frac{\eta_1}{\eta_2}\sqrt{1 - \left(\frac{k_1}{k_2}\right)^2 \sin^2(\theta_{Bs})}$$

$$\Rightarrow \eta_2^2[1 - \sin^2(\theta_{Bs})] = \eta_1^2\left[1 - \left(\frac{k_1}{k_2}\right)^2 \sin^2(\theta_{Bs})\right]$$

$$\Rightarrow \sin^2(\theta_{Bs}) = \frac{(\eta_2^2 - \eta_1^2)}{\left[\eta_2^2 - \eta_1^2\left(\frac{k_1}{k_2}\right)^2\right]}$$

The above equation represents the general expression for the Brewster angle in case of perpendicular polarization.

Now, when the media on both sides of the interface are considered to be nonmagnetic and lossless then the above equation can be simplified as follows:

$$\sin^2(\theta_{Bs}) = \frac{\left(\frac{\eta_0^2}{\epsilon_{r2}} - \frac{\eta_0^2}{\epsilon_{r1}}\right)}{\left[\frac{\eta_0^2}{\epsilon_{r2}} - \frac{\eta_0^2}{\epsilon_{r1}} \times \frac{\epsilon_{r1}}{\epsilon_{r2}}\right]} \equiv \frac{\left(\frac{1}{\epsilon_{r2}} - \frac{1}{\epsilon_{r1}}\right)}{\left[\frac{1}{\epsilon_{r2}} - \frac{1}{\epsilon_{r2}}\right]}$$

The above equation is clearly not defined. Hence, it can be concluded that the Brewster angle does not exist in case of perpendicular polarization if the media on both sides of the interface are considered nonmagnetic dielectrics.

Let us now suppose that both the media are magnetic, but still lossless. Under this situation, the equation for the Brewster angle can be written as

$$\sin^2(\theta_{Bs}) = \frac{\left(\eta_0^2\frac{\mu_{r2}}{\epsilon_{r2}} - \eta_0^2\frac{\mu_{r1}}{\epsilon_{r1}}\right)}{\left[\eta_0^2\frac{\mu_{r2}}{\epsilon_{r2}} - \eta_0^2\frac{\mu_{r1}}{\epsilon_{r1}} \times \frac{\mu_{r1}\,\epsilon_{r1}}{\mu_{r2}\,\epsilon_{r2}}\right]}$$

$$\equiv \frac{\left(\frac{\mu_{r2}}{\epsilon_{r2}} - \frac{\mu_{r1}}{\epsilon_{r1}}\right)}{\left[\frac{\mu_{r2}}{\epsilon_{r2}} - \frac{\mu_{r1}^2}{\mu_{r2}\epsilon_{r2}}\right]} \equiv \frac{\left(1 - \frac{\mu_{r1}\,\epsilon_{r2}}{\mu_{r2}\,\epsilon_{r1}}\right)}{\left[1 - \frac{\mu_{r1}^2}{\mu_{r2}^2}\right]}$$

As a special case, if we have a situation such that $\epsilon_{r1} = \epsilon_{r2}$ but $\mu_{r1} \neq \mu_{r2}$, then the above equation reduces to the following simplified expression:

$$\sin^2(\theta_{Bs}) \equiv \frac{1}{\left[1 + \frac{\mu_{r1}}{\mu_{r2}}\right]}$$

$$\Rightarrow \theta_{\rm Bs} \equiv \sin^{-1}\sqrt{\frac{1}{\left(1+\frac{\mu_{r1}}{\mu_{r2}}\right)}}$$

The above equation is, however, of limited interest as practically it is usually not possible to have a situation where two media have same permittivity but different permeability.

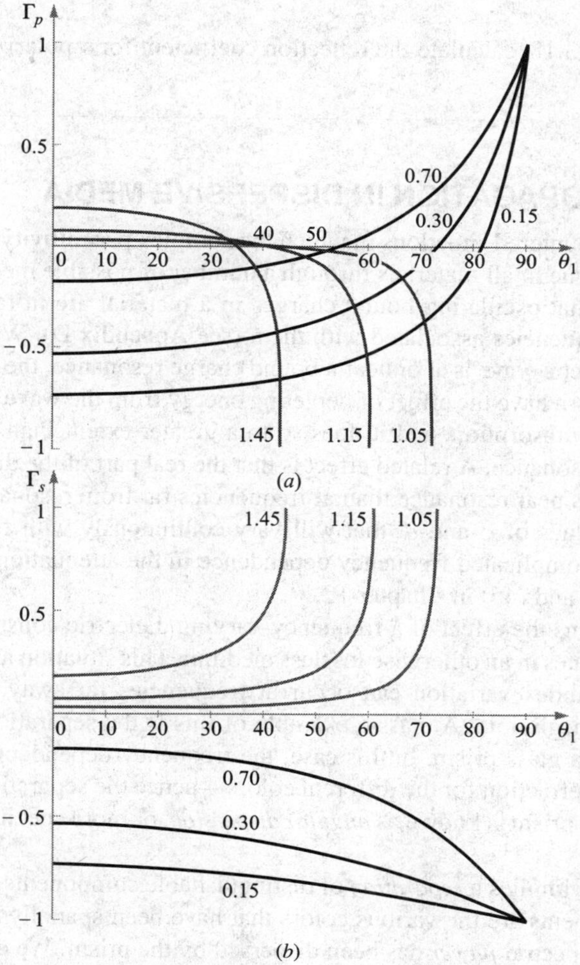

Fig. 13.18 (*a*) Plots of Γ_p [Eq. (69)] as functions of the incident angle, θ_1, as shown in Figure 13.12*a*. Curves are shown for selected values of the refractive index ratio, n_1/n_2. Both media are lossless and have $\mu_r = 1$. Thus $\eta_1 = \eta_0/n_1$ and $\eta_2 = \eta_0/n_2$. (*b*) Plots of Γ_s [Eq. (71)] as functions of the incident angle, θ_1, as shown in Figure 13.7*b*. As in Figure 13.18*a*, the media are lossless, and curves are shown for selected n_1/n_2.

Many of the results we have seen in this section are summarized in Figure 13.18, in which Γ_p and Γ_s, from (69) and (71), are plotted as functions of the incident angle, θ_1. Curves are shown for selected values of the refractive index ratio, n_1/n_2. For all plots in which $n_1/n_2 > 1$, Γ_s and Γ_p achieve values of ± 1 at the critical angle. At larger angles, the reflection coefficients become imaginary (and are not shown) but nevertheless retain magnitudes of unity. The occurrence of the Brewster angle is evident in the curves for Γ_p (Figure 13.18*a*) because all curves cross the θ_1 axis. This behavior is not seen in the Γ_s functions because Γ_s is positive for all values of θ_1 when $n_1/n_2 > 1$.

D13.5 In Example 13.18, calculate the reflection coefficient for s-polarized light.

Ans. -0.355

13.7 WAVE PROPAGATION IN DISPERSIVE MEDIA

In Chapter 12, we encountered situations in which the complex permittivity of the medium depends on frequency. This is true in all materials through a number of possible mechanisms. One of these, mentioned earlier, is that oscillating bound charges in a material are in fact harmonic oscillators that have resonant frequencies associated with them (see Appendix D). When the frequency of an incoming electromagnetic wave is at or near a bound charge resonance, the wave will induce strong oscillations; these in turn have the effect of depleting energy from the wave in its original form. The wave thus experiences absorption, and it does so to a greater extent than it would at a frequency that is detuned from resonance. A related effect is that the real part of the dielectric constant will be different at frequencies near resonance than at frequencies far from resonance. In short, resonance effects give rise to values of ϵ' and ϵ'' that will vary continuously with frequency. These in turn will produce a fairly complicated frequency dependence in the attenuation and phase constants as expressed in Eqs. (44) and (45) in Chapter 12.

This section concerns the effect of a frequency-varying dielectric constant (or refractive index) on a wave as it propagates in an otherwise lossless medium. This situation arises quite often because significant refractive index variation can occur at frequencies far away from resonance, where absorptive losses are negligible. A classic example of this is the separation of white light into its component colors by a glass prism. In this case, the frequency-dependent refractive index results in different angles of refraction for the different colors—hence the separation. The color separation effect produced by the prism is known as *angular dispersion,* or more specifically, *chromatic* angular dispersion.

The term *dispersion* implies a *separation* of distinguishable components of a wave. In the case of the prism, the components are the various colors that have been spatially separated. An important point here is that the spectral *power* has been dispersed by the prism. We can illustrate this idea by considering what it would take to measure the difference in refracted angles between, for example, blue and red light. One would need to use a power detector with a very narrow aperture, as shown in Figure 13.19. The detector would be positioned at the locations of the blue and red light from the prism, with the narrow aperture allowing essentially one color at a time (or light over a very narrow spectral range) to pass through to the detector. The detector would then measure the power in what we could call a "spectral packet," or a very narrow slice of the total power spectrum. The smaller the aperture, the narrower the spectral width of the packet, and the greater the precision in

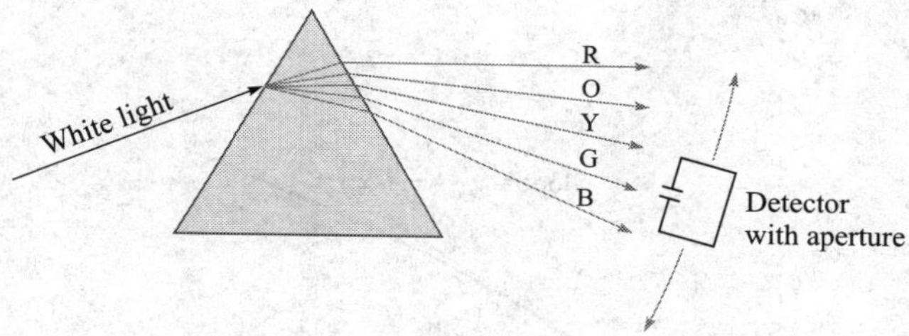

Fig. 13.19 The angular dispersion of a prism can be measured using a movable device which measures both wavelength and power. The device senses light through a small aperture, thus improving wavelength resolution.

the measurement.[4] It is important for us to think of wave power as subdivided into spectral packets in this way because it will figure prominently in our interpretation of the main topic of this section, which is wave dispersion *in time*.

We now consider a lossless nonmagnetic medium in which the refractive index varies with frequency. The phase constant of a uniform plane wave in this medium will assume the form

$$\beta(\omega) = k = \omega\sqrt{\mu_0\epsilon(\omega)} = n(\omega)\frac{\omega}{c} \tag{80}$$

If we take $n(\omega)$ to be a monotonically increasing function of frequency (as is usually the case), a plot of ω versus β would look something like the curve shown in Figure 13.20. Such a plot is known as an *ω-β diagram* for the medium. Much can be learned about how waves propagate in the material by considering the shape of the ω-β curve.

Suppose we have two waves at two frequencies, ω_a and ω_b, which are co-propagating in the material and whose amplitudes are equal. The two frequencies are labeled on the curve in Figure 13.20, along with the frequency midway between the two, ω_0. The corresponding phase constants, β_a, β_b, and β_0, are also labeled. The electric fields of the two waves are linearly polarized in the same direction (along x, for example), while both waves propagate in the forward z direction. The waves will thus interfere with each other, producing a resultant wave whose field function can be found simply by adding the **E** fields of the two waves. This addition is done using the complex fields:

$$E_{c,\text{net}}(z,t) = E_0[e^{-j\beta_a z}e^{j\omega_a t} + e^{-j\beta_b z}e^{j\omega_b t}]$$

Note that we must use the full complex forms (with frequency dependence retained) as opposed to the phasor forms, since the waves are at different frequencies. Next, we factor out the term $e^{-j\beta_0 z}e^{j\omega_0 t}$:

$$\begin{aligned} E_{c,\text{net}}(z,t) &= E_0 e^{-j\beta_0 z}e^{j\omega_0 t}[e^{j\Delta\beta z}e^{-j\Delta\omega t} + e^{-j\Delta\beta z}e^{j\Delta\omega t}] \\ &= 2E_0 e^{-j\beta_0 z}e^{j\omega_0 t}\cos(\Delta\omega t - \Delta\beta z) \end{aligned} \tag{81}$$

where

$$\Delta\omega = \omega_0 - \omega_a = \omega_b - \omega_0$$

[4] To perform this experiment, one would need to measure the wavelength as well. To do this, the detector would likely be located at the output of a spectrometer or monochrometer whose input slit performs the function of the bandwidth-limiting aperture.

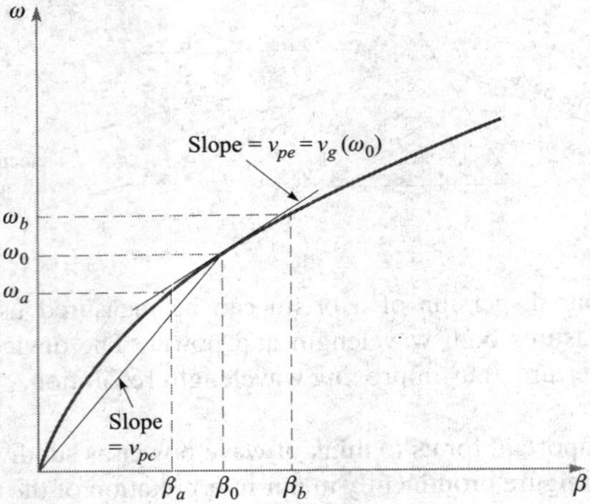

Fig. 13.20 ω-β diagram for a material in which the refractive index increases with frequency. The slope of a line tangent to the curve at ω_0 is the group velocity at that frequency. The slope of a line joining the origin to the point on the curve at ω_0 is the phase velocity at ω_0.

and

$$\Delta\beta = \beta_0 - \beta_a = \beta_b - \beta_0$$

The preceding expression for $\Delta\beta$ is approximately true as long as $\Delta\omega$ is small. This can be seen from Figure 13.20 by observing how the shape of the curve affects $\Delta\beta$, given uniform frequency spacings.

The real instantaneous form of (81) is found through

$$\mathcal{E}_{\text{net}}(z,t) = \text{Re}\{E_{c,\text{net}}\} = 2E_0 \cos(\Delta\omega t - \Delta\beta z)\cos(\omega_0 t - \beta_0 z) \tag{82}$$

If $\Delta\omega$ is fairly small compared to ω_0, we recognize (82) as a carrier wave at frequency ω_0 that is sinusoidally modulated at frequency $\Delta\omega$. The two original waves are thus "beating" together to form a slow modulation, as one would hear when the same note is played by two slightly out-of-tune musical instruments. The resultant wave is shown in Figure 13.21.

Of interest to us are the phase velocities of the carrier wave and the modulation envelope. From (82), we can immediately write these down as:

$$\nu_{pc} = \frac{\omega_0}{\beta_0} \qquad \text{(carrier velocity)} \tag{83}$$

$$\nu_{pe} = \frac{\Delta\omega}{\Delta\beta} \qquad \text{(envelope velocity)} \tag{84}$$

Referring to the ω-β diagram, Figure 13.20, we recognize the carrier phase velocity as the slope of the straight line that joins the origin to the point on the curve whose coordinates are ω_0 and β_0. We recognize the envelope velocity as a quantity that approximates the slope of the ω-β curve at the location of an operation point specified by (ω_0, β_0). The envelope velocity in this case is thus

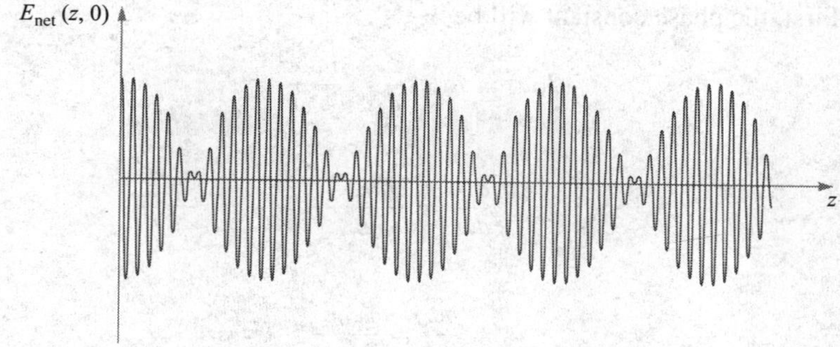

Fig. 13.21 Plot of the total electric field strength as a function of z (with $t = 0$) of two co-propagating waves having different frequencies, ω_a and ω_b, as per Eq. (81). The rapid oscillations are associated with the carrier frequency, $\omega_0 = (\omega_a + \omega_b)/2$. The slower modulation is associated with the envelope or "beat" frequency, $\Delta\omega = (\omega_b - \omega_a)/2$.

somewhat less than the carrier velocity. As $\Delta\omega$ becomes vanishingly small, the envelope velocity is exactly the slope of the curve at ω_0. We can therefore state the following for our example:

$$\lim_{\Delta\omega \to 0} \frac{\Delta\omega}{\Delta\beta} = \left.\frac{d\omega}{d\beta}\right|_{\omega_0} = \nu_g(\omega_0) \tag{85}$$

The quantity $d\omega/d\beta$ is called the *group velocity* function for the material, $\nu_g(\omega)$. When evaluated at a specified frequency ω_0, it represents the velocity of a group of frequencies within a spectral packet of vanishingly small width, centered at frequency ω_0. In stating this, we have extended our two-frequency example to include waves that have a continuous frequency spectrum. Each frequency component (or packet) is associated with a group velocity at which the energy in that packet propagates. Since the slope of the ω-β curve changes with frequency, group velocity will obviously be a function of frequency. The *group velocity dispersion* of the medium is, to the first order, the rate at which the slope of the ω-β curve changes with frequency. It is this behavior that is of critical practical importance to the propagation of modulated waves within dispersive media and to understanding the extent to which the modulation envelope may degrade with propagation distance.

EXAMPLE 13.20

Consider a medium in which the refractive index varies linearly with frequency over a certain range:

$$n(\omega) = n_0 \frac{\omega}{\omega_0}$$

Determine the group velocity and the phase velocity of a wave at frequency ω_0.

Solution. First, the phase constant will be

$$\beta(\omega) = n(\omega)\frac{\omega}{c} = \frac{n_0\omega^2}{\omega_0 c}$$

Now

$$\frac{d\beta}{d\omega} = \frac{2n_0\omega}{\omega_0 c}$$

so that

$$\nu_g = \frac{d\omega}{d\beta} = \frac{\omega_0 c}{2n_0\omega}$$

The group velocity at ω_0 is

$$\nu_g(\omega_0) = \frac{c}{2n_0}$$

The phase velocity at ω_0 will be

$$\nu_p(\omega_0) = \frac{\omega}{\beta(\omega_0)} = \frac{c}{n_0}$$

13.8 PULSE BROADENING IN DISPERSIVE MEDIA

To see how a dispersive medium affects a modulated wave, let us consider the propagation of an electromagnetic pulse. Pulses are used in digital signals, where the presence or absence of a pulse in a given time slot corresponds to a digital "one" or "zero." The effect of the dispersive medium on a pulse is to broaden it in time. To see how this happens, we consider the pulse *spectrum,* which is found through the Fourier transform of the pulse in time domain. In particular, suppose the pulse shape in time is Gaussian, and has electric field given at position $z = 0$ by

$$E(0,t) = E_0 e^{-\frac{1}{2}(t/T)^2} e^{j\omega_0 t} \qquad (86)$$

where E_0 is a constant, ω_0 is the carrier frequency, and T is the characteristic half-width of the pulse envelope; this is the time at which the pulse *intensity,* or magnitude of the Poynting vector, falls to $1/e$ of its maximum value (note that intensity is proportional to the square of the electric field). The frequency spectrum of the pulse is the Fourier transform of (86), which is

$$E(0,\omega) = \frac{E_0 T}{\sqrt{2\pi}} e^{-\frac{1}{2}T^2(\omega-\omega_0)^2} \qquad (87)$$

Note from (87) that the frequency displacement from ω_0 at which the spectral *intensity* (proportional to $|E(0,\omega)|^2$) falls to $1/e$ of its maximum is $\Delta\omega = \omega - \omega_0 = 1/T$.

Figure 13.22a shows the Gaussian intensity spectrum of the pulse, centered at ω_0, where the frequencies corresponding to the $1/e$ spectral intensity positions, ω_a and ω_b, are indicated. Figure 13.22b shows the same three frequencies marked on the ω-β curve for the medium. Three lines are drawn that are tangent to the curve at the three frequency locations. The slopes of the lines

Animations

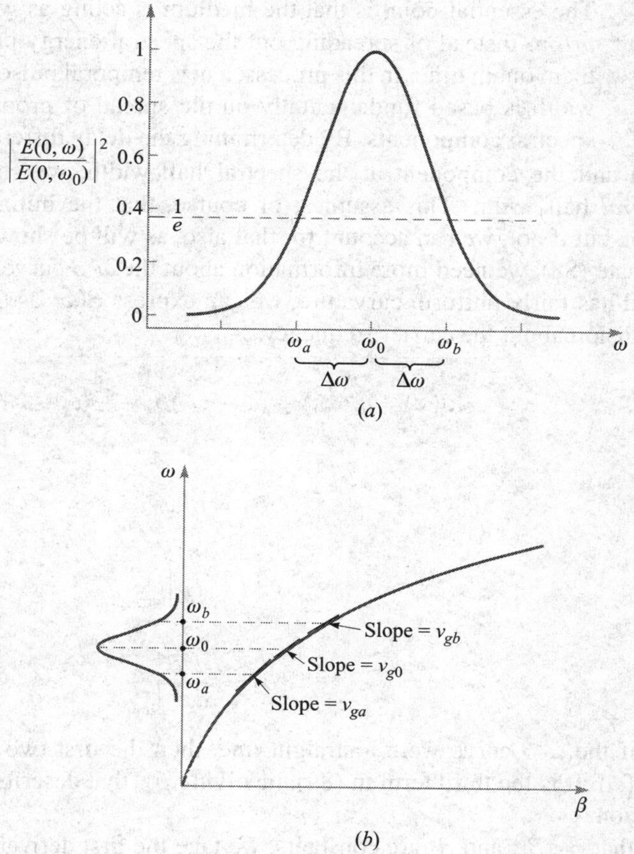

Animations

Fig. 13.22 (a) Normalized power spectrum of a Gaussian pulse, as determined from Eq. (86). The spectrum is centered at carrier frequency ω_0 and has $1/e$ half-width, $\Delta\omega$. Frequencies ω_a and ω_b correspond to the $1/e$ positions on the spectrum. (b) The spectrum of Figure 13.22a as shown on the ω-β diagram for the medium. The three frequencies specified in Figure 13.22a are associated with three different slopes on the curve, resulting in different group delays for the spectral components.

indicate the group velocities at ω_a, ω_b, and ω_0, indicated as ν_{ga}, ν_{gb}, and ν_{g0}. We can think of the pulse spreading in time as resulting from the differences in propagation times of the spectral energy packets that make up the pulse spectrum. Since the pulse spectral energy is highest at the center frequency, ω_0, we can use this as a reference point about which further spreading of the energy will occur. For example, let us consider the difference in arrival times (group delays) between the frequency components, ω_0 and ω_b, after propagating through a distance z of the medium:

$$\Delta\tau = z\left(\frac{1}{\nu_{gb}} - \frac{1}{\nu_{g0}}\right) = z\left(\left.\frac{d\beta}{d\omega}\right|_{\omega_b} - \left.\frac{d\beta}{d\omega}\right|_{\omega_0}\right) \tag{88}$$

Animations

The essential point is that the medium is acting as what could be called a *temporal prism*. Instead of spreading out the spectral energy packets spatially, it is spreading them out in time. In this process, a new temporal pulse envelope is constructed whose width is based fundamentally on the spread of propagation delays of the different spectral components. By determining the delay difference between the peak spectral component and the component at the spectral half-width, we construct an expression for the new *temporal* half-width. This assumes, of course, that the initial pulse width is negligible in comparison, but if not, we can account for that also, as will be shown later on.

To evaluate (88), we need more information about the ω-β curve. If we assume that the curve is smooth and has fairly uniform curvature, we can express $\beta(\omega)$ as the first three terms of a Taylor series expansion about the carrier frequency, ω_0:

$$\beta(\omega) \doteq \beta(\omega_0) + (\omega - \omega_0)\beta_1 + \frac{1}{2}(\omega - \omega_0)^2\beta_2 \tag{89}$$

where

$$\beta_0 = \beta(\omega_0)$$

$$\beta_1 = \left.\frac{d\beta}{d\omega}\right|_{\omega_0} \tag{90}$$

and

$$\beta_2 = \left.\frac{d^2\beta}{d\omega^2}\right|_{\omega_0} \tag{91}$$

Note that if the ω-β curve were a straight line, then the first two terms in (89) would precisely describe $\beta(\omega)$. It is the third term in (89), involving β_2, that describes the curvature and ultimately the dispersion.

Noting that β_0, β_1, and β_2 are constants, we take the first derivative of (89) with respect to ω to find

$$\frac{d\beta}{d\omega} = \beta_1 + (\omega - \omega_0)\beta_2 \tag{92}$$

We now substitute (92) into (88) to obtain

$$\Delta\tau = [\beta_1 + (\omega_b - \omega_0)\beta_2]\,z - [\beta_1 + (\omega_0 - \omega_0)\beta_2]\,z = \Delta\omega\beta_2 z = \frac{\beta_2 z}{T} \tag{93}$$

where $\Delta\omega = (\omega_b - \omega_0) = 1/T$. β_2, as defined in Eq. (91), is the *dispersion parameter*. Its units are in general time^2/distance that is, pulse spread in time per unit spectral bandwidth, per unit distance. In optical fibers, for example, the units most commonly used are picoseconds^2/kilometer (psec^2/km). β_2 can be determined when we know how β varies with frequency, or it can be measured.

If the initial pulse width is very short compared to $\Delta\tau$, then the broadened pulse width at location z will be simply $\Delta\tau$. If the initial pulse width is comparable to $\Delta\tau$, then the pulse width at z can be found through the convolution of the initial Gaussian pulse envelope of width T with a Gaussian envelope whose width is $\Delta\tau$. Thus in general, the pulse width at location z will be

$$T' = \sqrt{T^2 + (\Delta\tau)^2} \tag{94}$$

EXAMPLE 13.21

An optical fiber link is known to have dispersion $\beta_2 = 20\ \text{ps}^2/\text{km}$. A Gaussian light pulse at the input of the fiber is of initial width $T = 10$ ps. Determine the width of the pulse at the fiber output if the fiber is 15 km long.

Solution. The pulse spread will be

$$\Delta\tau = \frac{\beta_2 z}{T} = \frac{(20)(15)}{10} = 30\ \text{ps}$$

So the output pulse width is

$$T' = \sqrt{(10)^2 + (30)^2} = 32\ \text{ps}$$

Animations

An interesting by-product of pulse broadening through chromatic dispersion is that the broadened pulse is *chirped*. This means that the instantaneous frequency of the pulse varies monotonically (either increases or decreases) with time over the pulse envelope. This again is just a manifestation of the broadening mechanism, in which the spectral components at different frequencies are spread out in time as they propagate at different group velocities. We can quantify the effect by calculating the group delay, τ_g, as a function of frequency, using (92). We obtain:

$$\tau_g = \frac{z}{\nu_g} = z\frac{d\beta}{d\omega} = (\beta_1 + (\omega - \omega_0)\beta_2)\, z \tag{95}$$

This equation tells us that the group delay will be a linear function of frequency and that higher frequencies will arrive at later times if β_2 is positive. We refer to the chirp as positive if the lower frequencies lead the higher frequencies in time [requiring a positive β_2 in (95)]; chirp is negative if the higher frequencies lead in time (negative β_2). Figure 13.23 shows the broadening effect and illustrates the chirping phenomenon.

D13.6 For the fiber link of Example 13.21, a 20 ps pulse is input instead of the 10 ps pulse in the example. Determine the output pulsewidth.

Ans. 25 ps

As a final point, we note that the pulse bandwidth, $\Delta\omega$, was found to be $1/T$. This is true as long as the Fourier transform of the pulse *envelope* is taken, as was done with (86) to obtain (87). In that case, E_0 was taken to be a constant, and so the only time variation arose from the carrier wave and the Gaussian envelope. Such a pulse, whose frequency spectrum is obtained only from the pulse envelope, is known as *transform-limited*. In general, however, additional frequency bandwidth may be present since E_0 may vary with time for one reason or another (such as phase noise that could be present on the carrier). In these cases, pulse broadening is found from the more general expression

$$\Delta\tau = \Delta\omega\beta_2 z \tag{96}$$

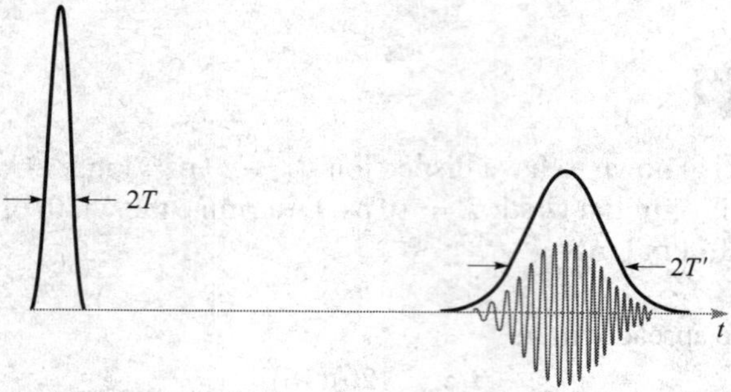

Fig. 13.23 Gaussian pulse intensities as functions of time (smooth curves) before and after propagation through a dispersive medium, as exemplified by the ω-β diagram of Figure 13.22*b*. The electric field oscillations are shown under the second trace to demonstrate the chirping effect as the pulse broadens. Note the reduced amplitude of the broadened pulse, which occurs because the pulse energy (the area under the intensity envelope) is constant.

where $\Delta\omega$ is the net spectral bandwidth arising from all sources. Clearly, transform-limited pulses are preferred in order to minimize broadening because these will have the smallest spectral width for a given pulse width.

References

1. DuBroff, R. E., S. V. Marshall, and G. G. Skitek. *Electromagnetic Concepts and Applications*. 4th ed. Englewood Cliffs, N. J.: Prentice-Hall, 1996. Chapter 9 of this text develops the concepts presented here, with additional examples and applications.
2. Iskander, M. F. *Electromagnetic Fields and Waves*. Englewood Cliffs, N. J.: Prentice-Hall, 1992. The multiple interface treatment in Chapter 5 of this text is particularly good.
3. Harrington, R. F. *Time-Harmonic Electromagnetic Fields*. New York: McGraw-Hill, 1961. This advanced text provides a good overview of general wave reflection concepts in Chapter 2.
4. Marcuse, D. *Light Transmission Optics*. New York: Van Nostrand Reinhold, 1982. This intermediate-level text provides detailed coverage of optical waveguides and pulse propagation in dispersive media.

CHAPTER 13 Problems

Quizzes

13.1 A uniform plane wave in air, $E^+_{x1} = E^+_{x10}\cos(10^{10}t - \beta z)$ V/m, is normally incident on a copper surface at $z = 0$. What percentage of the incident power density is transmitted into the copper?

13.2 The plane $z = 0$ defines the boundary between two dielectrics. For $z < 0$, $\epsilon_{r1} = 5$, $\epsilon''_{r1} = 0$, and $\mu_1 = \mu_0$. For $z > 0$, $\epsilon'_{r2} = 3$, $\epsilon''_{r2} = 0$, and $\mu_2 = \mu_0$. Let $E^+_{x1} = 200\cos(\omega t - 15z)$ V/m and find (*a*) ω; (*b*) $\langle \mathbf{S}^+_1 \rangle$; (*c*) $\langle \mathbf{S}^-_1 \rangle$; (*d*) $\langle \mathbf{S}^+_2 \rangle$.

13.3 A uniform plane wave in region 1 is normally incident on the planar boundary separating regions 1 and 2. If $\epsilon_1'' = \epsilon_2'' = 0$, while $\epsilon_{r1}' = \mu_{r1}^3$ and $\epsilon_{r2}' = \mu_{r2}^3$, find the ratio $\epsilon_{r2}'/\epsilon_{r1}'$ if 20% of the energy in the incident wave is reflected at the boundary. There are two possible answers.

13.4 A 10 MHz uniform plane wave having an initial average power density of 5 W/m^2 is normally incident from free space onto the surface of a lossy material in which $\epsilon_2''/\epsilon_2' = 0.05$, $\epsilon_{r2}' = 5$, and $\mu_2 = \mu_0$. Calculate the distance into the lossy medium at which the transmitted wave power density is down by 10 dB from the initial 5 W/m^2.

13.5 The region $z < 0$ is characterized by $\epsilon_r' = \mu_r = 1$ and $\epsilon_r'' = 0$. The total **E** field here is given as the sum of two uniform plane waves, $\mathbf{E}_s = 150\, e^{-j10z}\mathbf{a}_x + (50\angle 20^\circ)\, e^{j10z}\mathbf{a}_x$ V/m. (*a*) What is the operating frequency? (*b*) Specify the intrinsic impedance of the region $z > 0$ that would provide the appropriate reflected wave. (*c*) At what value of z, $-10\,\text{cm} < z < 0$, is the total electric field intensity a maximum amplitude?

13.6 Region 1, $z < 0$, and region 2, $z > 0$, are described by the following parameters: $\epsilon_1' = 100$ pF/m, $\mu_1 = 25\ \mu$H/m, $\epsilon_1'' = 0$, $\epsilon_2' = 200$ pF/m, $\mu_2 = 50\ \mu$H/m, and $\epsilon_2''/\epsilon_2' = 0.5$. If $\mathbf{E}_1^+ = 5e^{-\alpha_1 z}\cos(4\times 10^9 t - \beta_1 z)\mathbf{a}_x$ V/m, find: (*a*) α_1; (*b*) β_1; (*c*) $\langle \mathbf{S}_1^+ \rangle$; (*d*) $\langle \mathbf{S}_1^- \rangle$; (*e*) $\langle \mathbf{S}_2^+ \rangle$.

13.7 The semi-infinite regions $z < 0$ and $z > 1$ m are free space. For $0 < z < 1$ m, $\epsilon_r' = 4$, $\mu_r = 1$, and $\epsilon_r'' = 0$. A uniform plane wave with $\omega = 4\times 10^8$ rad/s is traveling in the $\mathbf{a}_z$ direction toward the interface at $z = 0$. (*a*) Find the standing wave ratio in each of the three regions. (*b*) Find the location of the maximum $|\mathbf{E}|$ for $z < 0$ that is nearest to $z = 0$.

13.8 A wave starts at point a, propagates 100 m through a lossy dielectric for which $\alpha = 0.5$ Np/m, reflects at normal incidence at a boundary at which $\Gamma = 0.3 + j0.4$, and then returns to point a. Calculate the ratio of the final power to the incident power after this round trip.

13.9 Region 1, $z < 0$, and region 2, $z > 0$, are both perfect dielectrics ($\mu = \mu_0$, $\epsilon'' = 0$). A uniform plane wave traveling in the $\mathbf{a}_z$ direction has a radian frequency of 3×10^{10} rad/s. Its wavelengths in the two regions are $\lambda_1 = 5$ cm and $\lambda_2 = 3$ cm. What percentage of the energy incident on the boundary is: (*a*) reflected; (*b*) transmitted? (*c*) What is the standing wave ratio in region 1?

13.10 In Figure 13.1, let region 2 be free space, while $\mu_{r1} = 1$, $\epsilon_{r1}'' = 0$, and ϵ_{r1}' is unknown. Find ϵ_{r1}' if: (*a*) the amplitude of $\mathbf{E}_1^-$ is one-half that of $\mathbf{E}_1^+$; (*b*) $\langle \mathbf{S}_1^- \rangle$ is one-half of $\langle \mathbf{S}_1^+ \rangle$; (*c*) $|\mathbf{E}_1|_{\text{min}}$ is one-half of $|\mathbf{E}_1|_{\text{max}}$.

13.11 A 150 MHz uniform plane wave is normally incident from air onto a material whose intrinsic impedance is unknown. Measurements yield a standing wave ratio of 3 and the appearance of an electric field minimum at 0.3 wavelengths in front of the interface. Determine the impedance of the unknown material.

13.12 A 50 MHz uniform plane wave is normally incident from air onto the surface of a calm ocean. For seawater, $\sigma = 4$ S/m, and $\epsilon_r' = 78$. (*a*) Determine the fractions of the incident power that are reflected and transmitted. (*b*) Qualitatively, how will these answers change (if at all) as the frequency is increased?

13.13 A right-circularly polarized plane wave is normally incident from air onto a semi-infinite slab of plexiglas ($\epsilon_r' = 3.45$, $\epsilon_r'' = 0$). Calculate the fractions of the incident power that are reflected and transmitted. Also, describe the polarizations of the reflected and transmitted waves.

13.14 A left-circularly polarized plane wave is normally incident onto the surface of a perfect conductor. (*a*) Construct the superposition of the incident and reflected waves in phasor form.

(*b*) Determine the real instantaneous form of the result of part (*a*). (*c*) Describe the wave that is formed.

13.15 Consider these regions in which $\epsilon'' = 0$: region 1, $z < 0$, $\mu_1 = 4\ \mu$H/m, and $\epsilon'_1 = 10$ pF/m; region 2, $0 < z < 6$ cm, $\mu_2 = 2\ \mu$H/m, $\epsilon'_2 = 25$ pF/m; region 3, $z > 6$ cm, $\mu_3 = \mu_1$, and $\epsilon'_3 = \epsilon'_1$. (*a*) What is the lowest frequency at which a uniform plane wave incident from region 1 onto the boundary at $z = 0$ will have no reflection? (*b*) If $f = 50$ MHz, what will the standing wave ratio be in region 1?

13.16 A uniform plane wave in air is normally incident onto a lossless dielectric plate of thickness $\lambda/8$ and of intrinsic impedance $\eta = 260\ \Omega$. Determine the standing wave ratio in front of the plate. Also find the fraction of the incident power that is transmitted to the other side of the plate.

13.17 Repeat Problem 13.16 for the cases in which the frequency is (*a*) doubled and (*b*) quadrupled. Assume that the slab impedance is independent of frequency.

13.18 A uniform plane wave is normally incident onto a slab of glass ($n = 1.45$) whose back surface is in contact with a perfect conductor. Determine the reflective phase shift at the front surface of the glass if the glass thickness is: (*a*) $\lambda/2$; (*b*) $\lambda/4$; (*c*) $\lambda/8$.

13.19 You are given four slabs of lossless dielectric, all with the same intrinsic impedance, η, known to be different from that of free space. The thickness of each slab is $\lambda/4$, where λ is the wavelength as measured in the slab material. The slabs are to be positioned parallel to one another, and the combination lies in the path of a uniform plane wave, normally incident. The slabs are to be arranged such that the airspaces between them are either zero, one-quarter wavelength, or one-half wavelength in thickness. Specify an arrangement of slabs and airspaces such that (*a*) the wave is totally transmitted through the stack, and (*b*) the stack presents the highest reflectivity to the incident wave. Several answers may exist.

13.20 The 50 MHz plane wave of Problem 13.12 is incident onto the ocean surface at an angle to the normal of 60°. Determine the fractions of the incident power that are reflected and transmitted for (*a*) s-polarization, and (*b*) p-polarization.

13.21 A right-circularly polarized plane wave in air is incident at Brewster's angle onto a semi-infinite slab of plexiglas ($\epsilon'_r = 3.45$, $\epsilon''_r = 0$). (*a*) Determine the fractions of the incident power that are reflected and transmitted. (*b*) Describe the polarizations of the reflected and transmitted waves.

13.22 A dielectric waveguide is shown in Figure 13.24 with refractive indices as labeled. Incident light enters the guide at angle ϕ from the front surface normal as shown. Once inside, the light totally reflects at the upper $n_1 - n_2$ interface, where $n_1 > n_2$. All subsequent reflections from the upper and lower boundaries will be total as well, and so the light is confined to the guide. Express, in terms of n_1 and n_2, the maximum value of ϕ such that total confinement will occur, with $n_0 = 1$. The quantity $\sin\phi$ is known as the *numerical aperture* of the guide.

13.23 Suppose that ϕ in Figure 13.24 is Brewster's angle, and that θ_1 is the critical angle. Find n_0 in terms of n_1 and n_2.

13.24 A *Brewster prism* is designed to pass p-polarized light without any reflective loss. The prism of Figure 13.25 is made of glass ($n = 1.45$) and is in air. Considering the light path shown, determine the apex angle α.

13.25 In the Brewster prism of Figure 13.25, determine for s-polarized light the fraction of the incident power that is transmitted through the prism.

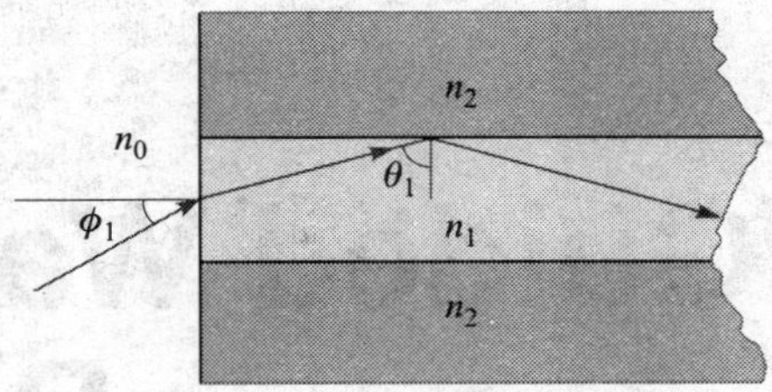

Fig. 13.24 See Problems 13.22 and 13.23.

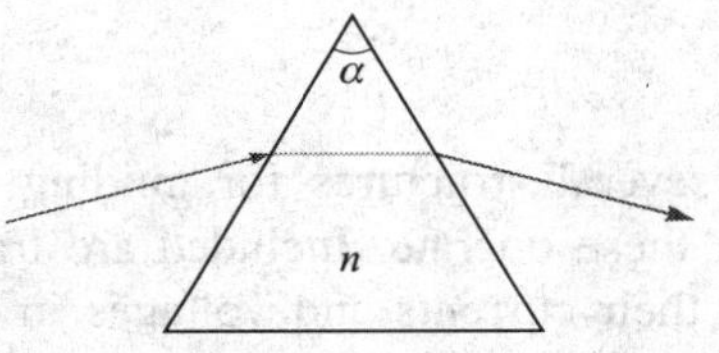

Fig. 13.25 See Problems 13.24 and 13.25.

13.26 Show how a single block of glass can be used to turn a p-polarized beam of light through 180°, with the light suffering (in principle) zero reflective loss. The light is incident from air, and the returning beam (also in air) may be displaced sideways from the incident beam. Specify all pertinent angles and use $n = 1.45$ for glass. More than one design is possible here.

13.27 Using Eq. (79) in Chapter 12 as a starting point, determine the ratio of the group and phase velocities of an electromagnetic wave in a good conductor. Assume conductivity does not vary with frequency.

13.28 Over a small wavelength range, the refractive index of a certain material varies approximately linearly with wavelength as $n(\lambda) \doteq n_a + n_b(\lambda - \lambda_a)$, where n_a, n_b and λ_a are constants, and where λ is the free-space wavelength. (*a*) Show that $d/d\omega = -(2\pi c/\omega^2)d/d\lambda$. (*b*) Using $\beta(\lambda) = 2\pi n/\lambda$, determine the wavelength-dependent (or independent) group delay over a unit distance. (*c*) Determine β_2 from your result of part (*b*). (*d*) Discuss the implications of these results, if any, on pulse broadening.

13.29 A $T = 5$ ps transform-limited pulse propagates in a dispersive medium for which $\beta_2 = 10$ ps^2/km. Over what distance will the pulse spread to twice its initial width?

13.30 A $T = 20$ ps transform-limited pulse propagates through 10 km of a dispersive medium for which $\beta_2 = 12$ ps^2/km. The pulse then propagates through a second 10 km medium for which $\beta_2 = -12$ ps^2/km. Describe the pulse at the output of the second medium and give a physical explanation for what happened.

14

Guided Waves and Radiation

In this chapter, we investigate several structures for guiding electromagnetic waves, and we explore the principles by which these operate. Included are transmission lines, which we first explored from the viewpoint of their currents and voltages in Chapter 11, and which we now revisit from a fields point of view using the plane-wave approach. The generalized wave equation approach to analyze these transmission-line structures will be discussed in the next chapter. Our main focus in this chapter is to introduce several structures, which are commonly being used to guide electromagnetic waves for various applications. Broadly defined, a waveguide is a structure through which electromagnetic waves can be transmitted from point to point and within which the fields are confined to a certain extent. A transmission line fits this description, but it is a special case that employs two conductors, and it propagates a purely TEM field configuration. Waveguides, in general, depart from these restrictions and may employ any number of conductors and dielectrics—or as we will see, dielectrics alone and no conductors.

Next, we explore the basic concepts of electromagnetic wave radiation using antennas. An antenna is any device that radiates electromagnetic energy into space, where the energy originates from a source that feeds the antenna through a transmission line or waveguide. The antenna thus serves as an interface between the confining line and space when used as a transmitter—or between space and the line when used as a receiver.

The chapter begins with a presentation of several transmission line structures, with emphasis on obtaining expressions for the primary constants, L, C, G, and R, for high- and low-frequency operating regimes. Next, we begin our study of waveguides by first taking a broad view of waveguide devices to obtain a physical understanding of how they work and the conditions under which they are used. We then explore the simple parallel-plate structure and distinguish between its operation as a transmission line and as a waveguide. In this device, the concept of waveguide modes is developed, as are the conditions under which these will occur. We will study here the electric and magnetic field configurations of the guided modes using simple plane wave models. We will then study more complicated structures, including dielectric slab guides, and optical fibers. The detailed analysis of the guided modes inside the parallel plate structure and the single conductor waveguide structure using the wave equation approach will be carried out in the next chapter.

Our study of antennas will include the derivation of the radiated fields from an elemental dipole, beginning with the retarded vector potentials that we studied in Chapter 10. We will address issues that include the efficiency of power radiation from an antenna and the parameters that govern this.

14.1 TRANSMISSION LINE FIELDS AND PRIMARY CONSTANTS

We begin by establishing the equivalence between transmission line operations when considering voltage and current, from the point of view of the fields within the line. Consider for example the parallel-plate line shown in Figure 14.1. In the line, we assume that the plate spacing, d, is much less than the line width, b (into the page), so electric and magnetic fields can be assumed to be uniform within any transverse plane. Lossless propagation is also assumed. Figure 14.1 shows the side view, which includes the propagation axis z. The fields, along with the voltage and current, are shown at an instant in time.

The voltage and current in phasor form are:

$$V_s(z) = V_0 e^{-j\beta z} \tag{1a}$$

$$I_s(z) = \frac{V_0}{Z_0} e^{-j\beta z} \tag{1b}$$

where $Z_0 = \sqrt{L/C}$. The electric field in a given transverse plane at location z is just the parallel-plate capacitor field:

$$E_{sx}(z) = \frac{V_s}{d} = \frac{V_0}{d} e^{-j\beta z} \tag{2a}$$

The magnetic field is equal to the surface current density, assumed uniform, on either plate [Eq. (95), Chapter 8]:

$$H_{sy}(z) = K_{sz} = \frac{I_s}{b} = \frac{V_0}{bZ_0} e^{-j\beta z} \tag{2b}$$

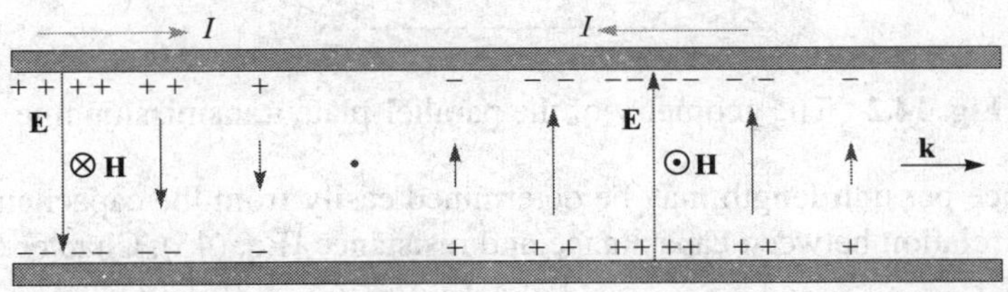

Fig. 14.1 A transmission-line wave represented by voltage and current distributions along the length is associated with transverse electric and magnetic fields, forming a TEM wave.

The two fields, both uniform, orthogonal, and lying in the transverse plane, are identical in form to those of a uniform plane wave. As such, they are transverse electromagnetic (TEM) fields, also known simply as transmission-line fields. They differ from the fields of the uniform plane wave only in that they exist within the interior of the line, and nowhere else.

The power flow down the line is found through the time-average Poynting vector, integrated over the line cross section. Using ($2a$) and ($2b$), we find:

$$P_z = \int_0^b \int_0^d \frac{1}{2}\text{Re}\{E_{xs}H_{ys}^*\}dx dy = \frac{1}{2}\frac{V_0}{d}\frac{V_0^*}{bZ_0^*}(bd) = \frac{|V_0|^2}{2Z_0^*} = \frac{1}{2}\text{Re}\{V_s I_s^*\} \tag{3}$$

The power transmitted by the line is one of the most important quantities that we wish to know from a practical standpoint. Eq. (3) shows that this can be obtained consistently through the line fields, or through the voltage and current. As would be expected, this consistency is maintained when losses are included. The fields picture is in fact advantageous, and is generally preferred, since it is easy to incorporate dielectric loss mechanisms (other than conductivity) in addition to the dispersive properties of the dielectric. The transmission-line fields are also needed to produce the primary constants, as we now demonstrate for the parallel-plate line and other selected line geometries.

We assume the line is filled with dielectric having permittivity ϵ', conductivity σ, and permeability μ, usually μ_0 (Figure 14.2). The upper and lower plate thickness is t, which, along with the plate width b and plate conductivity σ_c, is used to evaluate the resistance per unit length parameter R under low-frequency conditions. We will, however, consider high-frequency operation, in which the skin effect gives an effective plate thickness or skin depth δ that is much less than t.

First, the capacitance and conductance per unit length are simply those of the parallel-plate structure, assuming static fields. Using Eq. (27) from Chapter 6, we find

$$C = \frac{\epsilon' b}{d} \tag{4}$$

The value of permittivity used should be appropriate for the range of operating frequencies considered.

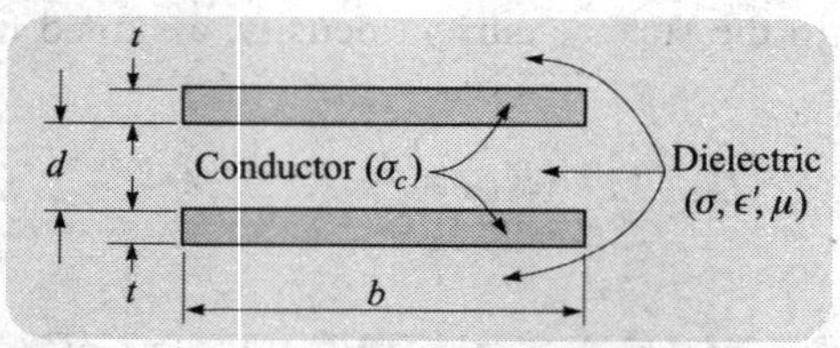

Fig. 14.2 The geometry of the parallel-plate transmission line.

The conductance per unit length may be determined easily from the capacitance expression by use of the simple relation between capacitance and resistance [Eq. (45), Chapter 6]:

$$G = \frac{\sigma}{\epsilon'} C = \frac{\sigma b}{d} \tag{5}$$

The evaluation of L and R involves the assumption of a well-developed skin effect such that $\delta < t$. Consequently, the inductance is primarily external because the magnetic flux within either conductor is negligible compared to that between conductors. Therefore,

$$L \doteq L_{\text{ext}} = \frac{\mu d}{b} \tag{6}$$

Note that $L_{\text{ext}} C = \mu\epsilon' = 1/\nu_p^2$, and we are therefore able to evaluate the external inductance for any transmission line for which we know the capacitance and insulator characteristics.

The last of the four parameters that we need is the resistance R per unit length. If the frequency is very high and the skin depth δ is very small, then we obtain an appropriate expression for R by distributing the total current uniformly throughout a depth δ. The skin effect resistance (through both conductors in series over a unit length) is

$$R = \frac{2}{\sigma_c \delta b} \tag{7}$$

Finally, it is convenient to include the common expression for the characteristic impedance of the line here with the parameter formulas:

$$Z_0 = \sqrt{\frac{L_{\text{ext}}}{C}} = \sqrt{\frac{\mu}{\epsilon'}\frac{d}{b}} \tag{8}$$

If necessary, a more accurate value may be obtained from Eq. (47), Chapter 11. Note that when substituting (8) into (2*b*), and using (2*a*), we obtain the expected relation for a TEM wave, $E_{xs} = \eta H_{ys}$, where $\eta = \sqrt{\mu/\epsilon'}$.

D14.1 Parameters for the planar transmission line shown in Figure 14.2 are $b = 6$ mm, $d = 0.25$ mm, $t = 25$ mm, $\sigma_c = 55 \times 10^7$S/m, $\epsilon' = 25$ pF/m, $\mu = \mu_0$, and $\sigma/\omega\epsilon' = 0.03$. If the operating frequency is 750 MHz, calculate: (*a*) α; (*b*) β; (*c*) Z_0.

Ans. 0.47 Np/m; 26 rad/m; $9.3 < 0.7^\circ\ \Omega$

14.1.1 Coaxial (High Frequencies)

We next consider a coaxial cable in which the dielectric has an inner radius a and outer radius b (Figure 14.3). The capacitance per unit length, obtained as Eq. (29) of Section 6.4, is

$$C = \frac{2\pi\epsilon'}{\ln(b/a)} \tag{9}$$

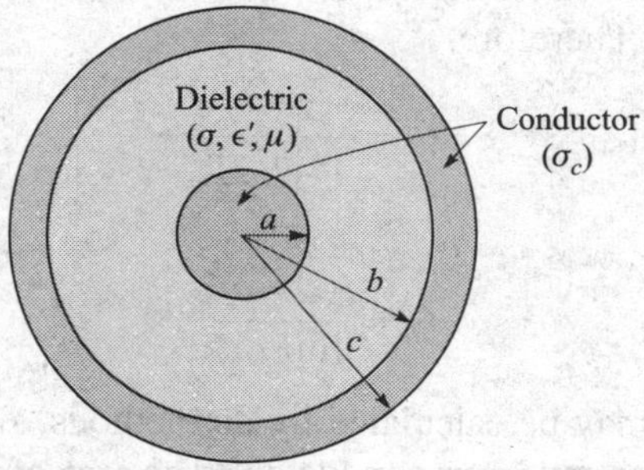

Fig. 14.3 Coaxial transmission-line geometry.

Again, using Eq. (45) from Chapter 6, the conductance is

$$G = \frac{2\pi\sigma}{\ln(b/a)} \tag{10}$$

where σ is the conductivity of the dielectric between the conductors at the operating frequency.

The inductance per unit length was computed for the coaxial cable as Eq. (50) in Section 9.10,

$$L_{\text{ext}} = \frac{\mu}{2\pi}\ln(b/a) \tag{11}$$

Again, this is an external inductance, for the small skin depth precludes any appreciable magnetic flux within the conductors.

For a circular conductor of radius a and conductivity σ_c, we let Eq. (90) of Section 12.4 apply to a unit length, obtaining

$$R_{\text{inner}} = \frac{1}{2\pi a\delta\sigma_c}$$

There is also a resistance for the outer conductor, which has an inner radius b. We assume the same conductivity σ_c and the same value of skin depth δ, leading to

$$R_{\text{outer}} = \frac{1}{2\pi b\delta\sigma_c}$$

Since the line current flows through these two resistances in series, the total resistance is the sum:

$$R = \frac{1}{2\pi\delta\sigma_c}\left(\frac{1}{a} + \frac{1}{b}\right) \tag{12}$$

Finally, the characteristic impedance, assuming low losses, is

$$Z_0 = \sqrt{\frac{L_{\text{ext}}}{C}} = \frac{1}{2\pi}\sqrt{\frac{\mu}{\epsilon'}}\ln\frac{b}{a} \tag{13}$$

14.1.2 Coaxial (Low Frequencies)

We now obtain the coaxial line parameter values at very low frequencies where there is no appreciable skin effect and the current is assumed to be distributed uniformly throughout the conductor cross sections.

We first note that the current distribution in the conductor does not affect either the capacitance or conductance per unit length. Therefore,

$$C = \frac{2\pi\epsilon'}{\ln(b/a)} \tag{14}$$

and

$$G = \frac{2\pi\sigma}{\ln(b/a)} \tag{15}$$

The resistance per unit length may be calculated by dc methods, $R = l/(\sigma_c S)$, where l= 1 m and σ_c is the conductivity of the outer and inner conductors. The area of the center conductor is πa^2 and that of the outer is $\pi\,(c^2 - b^2)$. Adding the two resistance values, we have

$$R = \frac{1}{\sigma_c\pi}\left(\frac{1}{a^2} + \frac{1}{c^2 - b^2}\right) \tag{16}$$

Only one of the four parameter values remains to be found, the inductance per unit length. The external inductance that we calculated at high frequencies is the greatest part of the total inductance. To it, however, must be added smaller terms representing the internal inductances of the inner and outer conductors.

At very low frequencies where the current distribution is uniform, the internal inductance of the center conductor is the subject of Problem 43 in Chapter 9; the relationship is also given as Eq. (62) in Section 9.10:

$$L_{a,\text{int}} = \frac{\mu}{8\pi} \tag{17}$$

The determination of the internal inductance of the outer shell is a more difficult problem, and most of the work was requested in Problem 36 in Chapter 9. There, we found that the energy stored per unit length in an outer cylindrical shell of inner radius b and outer radius c with uniform current distribution is

$$W_H = \frac{\mu I^2}{16\pi(c^2 - b^2)}\left(b^2 - 3c^2 + \frac{4c^2}{c^2 - b^2}\ln\frac{c}{b}\right)$$

Thus the internal inductance of the outer conductor at very low frequencies is

$$L_{bc,\text{int}} = \frac{\mu}{8\pi(c^2 - b^2)}\left(b^2 - 3c^2 + \frac{4c^2}{c^2 - b^2}\ln\frac{c}{b}\right) \tag{18}$$

At low frequencies the total inductance is obtained by adding (11), (17), and (18):

$$L = \frac{\mu}{2\pi}\left[\ln\frac{b}{a} + \frac{1}{4} + \frac{1}{4(c^2 - b^2)}\left(b^2 - 3c^2 + \frac{4c^2}{c^2 - b^2}\ln\frac{c}{b}\right)\right] \tag{19}$$

14.1.3 Coaxial (Intermediate Frequencies)

There still remains the frequency interval where the skin depth is neither very much larger than nor very much smaller than the radius. In this case, the current distribution is governed by Bessel functions, and both the resistance and internal inductance are complicated expressions. Values are tabulated in the handbooks, and it is necessary to use them for very small conductor sizes at high frequencies and for larger conductor sizes used in power transmission at low frequencies.[1]

D14.2 The dimensions of a coaxial transmission line are a = 4 mm, b = 175 mm, and c = 20 mm. The conductivity of the inner and outer conductors is 2×10^7 S/m, and the dielectric properties are μ_r = $1\epsilon'_r$ = 3, and $\sigma_{/}\omega\epsilon'$ = 0.025. Assume that the loss tangent is constant with frequency. Determine:(a) $LCRG$ and Z_0 at 150 MHz; (b) L and R at 60 Hz.

Ans. 0.30 μH/m, 113 pF/m, 0.27 Ω/m, 2.7 mS/m, 51 Ω; 0.36 μH/m, 1.16 mΩ/m

14.1.4 Two-Wire (High Frequencies)

For the two-wire transmission line of Figure 14.4 with conductors of radius a and conductivity σ_c with center-to-center separation d in a medium of permeability μ, permittivity ϵ', and conductivity σ_c, the capacitance was found in Chapter 6 [Eq. (40), Section 6.5] to be

$$C = \frac{\pi\epsilon'}{\cosh^{-1}(d/2a)} \tag{20}$$

or

$$C \doteq \frac{\pi\epsilon'}{\ln(d/a)} \quad (a \ll d)$$

[1] Bessel functions are discussed within the context of optical fiber in Section 14.7. The current distribution, internal inductance, and internal resistance of round wires is discussed (with numerical examples) in Weeks, pp. 35–44. See the References at the end of this chapter.

The external inductance may be found from $L_{\text{ext}}C = \mu\epsilon'$. It is

$$L_{\text{ext}} = \frac{\mu}{\pi}\cosh^{-1}(d/2a) \tag{21}$$

or

$$L_{\text{ext}} \doteq \frac{\mu}{\pi}\ln(d/a) \quad (a \ll d)$$

The conductance per unit length may be written immediately from an inspection of the capacitance expression,

$$G = \frac{\pi\sigma}{\cosh^{-1}(d/2a)} \tag{22}$$

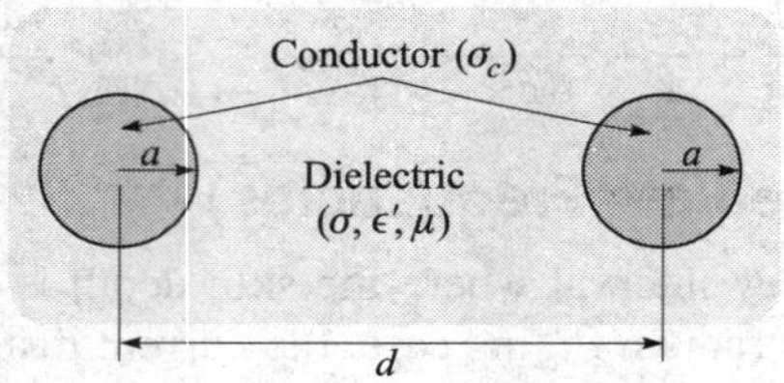

Fig. 14.4 The geometry of the two-wire transmission line.

The resistance per unit length is twice that of the center conductor of the coax,

$$R = \frac{1}{\pi a \delta \sigma_c} \tag{23}$$

Finally, using the capacitance and the external inductance expressions, we obtain a value for the characteristic impedance,

$$Z_0 = \sqrt{\frac{L_{\text{ext}}}{C}} = \frac{1}{\pi}\sqrt{\frac{\mu}{\epsilon}}\cosh^{-1}(d/2a) \tag{24}$$

14.1.5 Two-Wire (Low Frequencies)

At low frequencies where a uniform current distribution may be assumed, we again must modify the L and R expressions, but not those for C and G. The latter two are again expressed by (20) and (22):

$$C = \frac{\pi\epsilon'}{\cosh^{-1}(d/2a)}$$

$$G = \frac{\pi\sigma}{\cosh^{-1}(d/2a)}$$

The inductance per unit length must be increased by twice the internal inductance of a straight round wire,

$$L = \frac{\mu}{\pi}\left[\frac{1}{4} + \cosh^{-1}(d/2a)\right] \tag{25}$$

The resistance becomes twice the dc resistance of a wire of radius a, conductivity σ_c, and unit length:

$$R = \frac{2}{\pi a^2 \sigma_c} \tag{26}$$

D14.3 The conductors of a two-wire transmission line each have a radius of 0.8 mm and a conductivity of 3×10^7 S/m. They are separated by a center-to-center distance of 0.8 cm in a medium for which $\epsilon'_r = 2.5 \mu_r = 1$, and $\sigma = 4 \times 10^{-9}$ S/m. If the line operates at 60 Hz, find: (*a*) δ; (*b*) C; (*c*) G; (*d*) L; (*e*) R.

Ans. 1.2 cm; 30 pF/m; 5.5 nS/m; 1.02 μH/m; 0.033 Ω/m

14.1.6 Microstrip Line (Low Frequencies)

Microstrip line is one example of a class of configurations involving planar conductors of finite widths on or within dielectric substrates; they are usually employed as device interconnects for microelectronic circuitry. The microstrip configuration, shown in Figure 14.5, consists of a dielectric (assumed lossless) of thickness d and of permittivity $\epsilon' = \epsilon_r \epsilon_0$, sandwiched between a conducting ground plane and a narrow conducting strip of width w. The region above the top strip is air (assumed here) or a dielectric of lower permittivity.

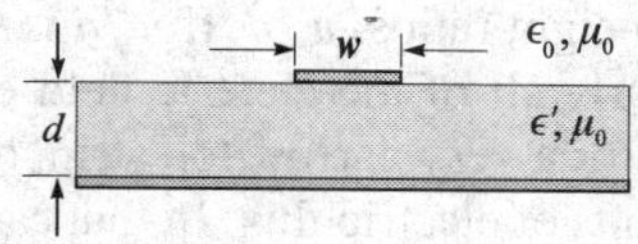

Fig. 14.5 Microstrip line geometry.

The structure approaches the case of the parallel-plate line if $w >> d$. In a microstrip, such an assumption is generally not valid, and so significant charge densities exist on both surfaces of the upper conductor. The resulting electric field, originating at the top conductor and terminating on the bottom conductor, will exist within both substrate *and* air regions. The same is true for the magnetic field, which circulates around the top conductor. This electromagnetic field configuration cannot propagate as a purely TEM wave because wave velocities within the two media will differ. Instead, waves having z components of **E** and **H** occur, with the z-component magnitudes established so that the air and dielectric fields do achieve equal phase velocities (the reasoning behind this will be explained in Section 14.6). Analyzing the structure while allowing for the special fields is complicated, but it is usually permissible to approach the problem under the assumption of negligible z components. This is the *quasi* TEM approximation, in which the static fields (obtainable through numerical solution of Laplace's equation, for example) are used to evaluate the primary constants. Accurate results are obtained at low frequencies (below 1 or 2 GHz). At higher frequencies, results obtained through the static fields can still be used but in conjunction with appropriate modifying functions. We will consider the simple case of low-frequency operation and assume lossless propagation.[2]

[2] The high-frequency case is treated in detail in Edwards (Reference 2).

To begin, it is useful to consider the microstrip line characteristics when the dielectric is *not* present. Assuming that both conductors have very small thicknesses, the internal inductance will be negligible, and so the phase velocity within the air-filled line, ν_{p0}, will be

$$\nu_{p0} = \frac{1}{\sqrt{L_{\text{ext}} C_0}} = \frac{1}{\sqrt{\mu_0 \epsilon_0}} = c \tag{27a}$$

where C_0 is the capacitance of the air-filled line (obtained from the electric field for that case), and c is the velocity of light. With the dielectric in place, the capacitance changes, *but the inductance does not,* provided the dielectric permeability is μ_0. Using (27*a*), the phase velocity now becomes

$$\nu_p = \frac{1}{\sqrt{L_{\text{ext}} C}} = c\sqrt{\frac{C_0}{C}} = \frac{c}{\sqrt{\epsilon_{r,\text{eff}}}} \tag{27b}$$

where the *effective dielectric constant* for the microstrip line is

$$\epsilon_{r,\text{eff}} = \frac{C}{C_0} = \left(\frac{c}{\nu_p}\right)^2 \tag{28}$$

It is implied from (28) that the microstrip capacitance C would result if both the air and substrate regions were filled homogeneously with material having dielectric constant $\epsilon_{r,\text{eff}}$. The effective dielectric constant is a convenient parameter to use because it provides a way of unifying the effects of the dielectric and the conductor geometry. To see this, consider the two extreme cases involving large and small width-to-height ratios, w/d. If w/d is very large, then the line resembles the parallel-plate line, in which nearly all of the electric field exists within the dielectric. In this case $\epsilon_{r,\text{eff}} \doteq \epsilon_r$. On the other hand, for a very narrow top strip, or small w/d, the dielectric and air regions contain roughly equal amounts of electric flux. In that case, the effective dielectric constant approaches its minimum, given by the average of the two dielectric constants. We therefore obtain the range of allowed values of $\epsilon_{r,\text{eff}}$:

$$\frac{1}{2}(\epsilon_r + 1) < \epsilon_{r,\text{eff}} < \epsilon_r \tag{29}$$

The physical interpretation of $\epsilon_{r,\text{eff}}$ is that it is a *weighted average* of the dielectric constants of the substrate and air regions, with the weighting determined by the extent to which the electric field fills either region. We may thus write the effective dielectric constant in terms of a *field filling factor,* q, for the substrate:

$$\epsilon_{r,\text{eff}} = 1 + q(\epsilon_r - 1) \tag{30}$$

where $0.5 < q < 1$. With large w/d, $q \to 1$; with small w/d, $q \to 0.5$.

Now, the characteristic impedances of the air-filled line and the line with dielectric substrate are respectively, $Z_0^{\text{air}} = \sqrt{L_{\text{ext}}/C_0}$ and $Z_0 = \sqrt{L_{\text{ext}}/C}$ using (28), we find

$$Z_0 = \frac{Z_0^{\text{air}}}{\sqrt{\epsilon_{r,\text{eff}}}} \tag{31}$$

A procedure for obtaining the characteristic impedance would be to first evaluate the air-filled impedance for a given w/d. Then, knowing the effective dielectric constant, determine the actual impedance using (31). Another problem would be to determine the required ratio w/d for a given substrate material in order to achieve a desired characteristic impedance.

Detailed analyses have led to numerous approximation formulas for the evaluation of $\epsilon_{r,\text{eff}}$, Z_0^{air}, and Z_0 within different regimes (again, see Reference 2 and the references therein). For example, with dimensions restricted such that $1.3 < w/d < 3.3$, applicable formulas include:

$$Z_0^{\text{air}} \doteq 60 \ln \left[4\left(\frac{d}{w}\right) + \sqrt{16\left(\frac{d}{w}\right)^2 + 2} \right] \quad \frac{w}{d} < 3.3 \tag{32}$$

and

$$\epsilon_{r,\text{eff}} \doteq \frac{\epsilon_r + 1}{2} + \frac{\epsilon_r - 1}{2}\left(1 + 10\frac{d}{w}\right)^{-0.555} \quad \frac{w}{d} > 1.3 \tag{33}$$

Or, if a line is to be fabricated having a desired value of Z_0, the effective dielectric constant (from which the required w/d can be obtained) is found through:

$$\epsilon_{r,\text{eff}} \doteq \epsilon_r [0.96 + \epsilon_r(0.109 - 0.004\epsilon_r)\,(\log_{10}(10 + Z_0) - 1]^{-1} \quad \frac{w}{d} > 1.3 \tag{34}$$

D14.4 A microstrip line is fabricated on a lithium niobate substrate ($\epsilon_r = 4.8$) of 1 mm thickness. If the top conductor is 2 mm wide, find (*a*) $\epsilon_{r,\text{eff}}$; (*b*) Z_0; (*c*) ν_p.

Ans. 3.6; 47 Ω; 16×10^8 m/s

14.2 BASIC WAVEGUIDE OPERATION

Waveguides assume many different forms that depend on the purpose of the guide and on the frequency of the waves to be transmitted. The simplest form (in terms of analysis) is the parallel-plate guide shown in Figure 14.6. Other forms are the hollow-pipe guides, including the rectangular waveguide of Figure 14.7, and the cylindrical guide, shown in Figure 14.8. Dielectric waveguides, used primarily at optical frequencies, include the slab waveguide of Figure 14.9 and the optical fiber, shown in Figure 14.10. Each of these structures possesses certain advantages over the others, depending on the application and the frequency of the waves to be transmitted. All guides, however, exhibit the same basic operating principles, which we will explore in this section.

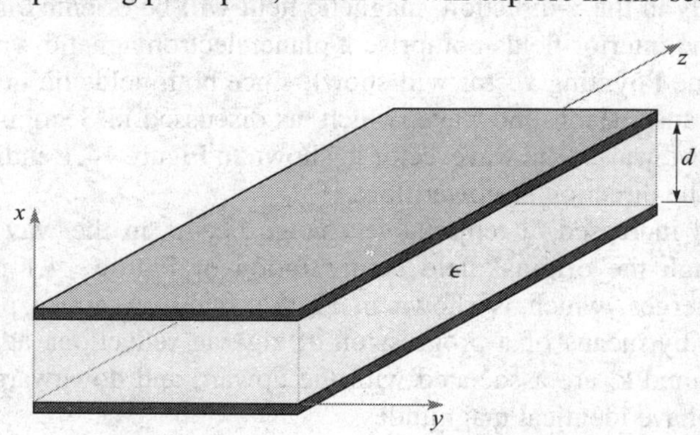

Fig. 14.6 Parallel-plate waveguide, with metal plates at $x = 0, d$. Between the plates is a dielectric of permittivity ϵ.

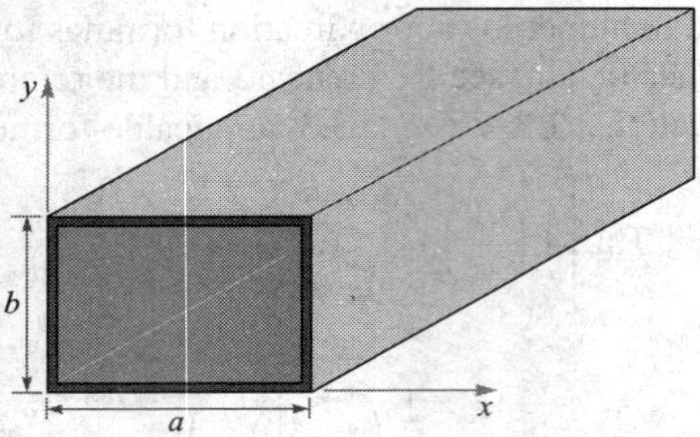

Fig. 14.7 Rectangular waveguide.

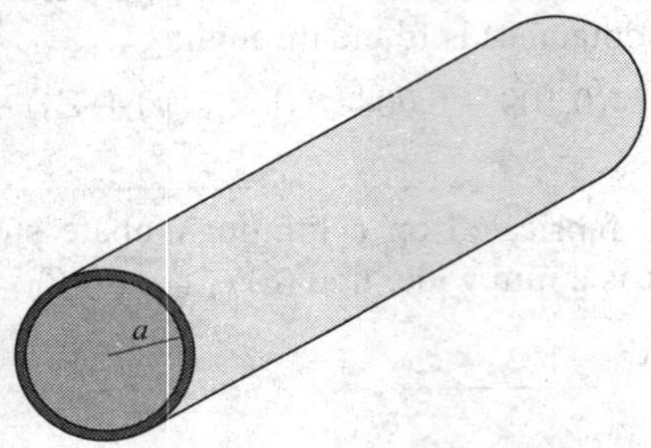

Fig. 14.8 Cylindrical waveguide.

To develop an understanding of waveguide behavior, we consider the parallel-plate waveguide of Figure 14.6. At first, we recognize this as one of the transmission-line structures that we investigated in Section 14.1. So the first question that arises is: how does a waveguide differ from a transmission line to begin with? The difference lies in the form of the electric and magnetic fields within the line. To see this, consider again Figure 14.1, which shows the fields when the line operates as a transmission line. As we saw earlier, a sinusoidal voltage wave, with voltage applied between conductors, leads to an electric field that is directed vertically between the conductors as shown. Since current flows only in the z-direction, magnetic field will be oriented in and out of the page (in the y-direction). The interior fields comprise a plane electromagnetic wave which propagates in the z-direction (as the Poynting vector will show), since both fields lie in the transverse plane. We refer to this as a transmission-line wave, which, as discussed in Section 14.1, is a transverse electromagnetic, or TEM, wave. The wavevector **k**, shown in Figure 14.1, indicates the direction of wave travel as well as the direction of power flow.

As the frequency is increased, a remarkable change occurs in the way the fields progagate down the line. Although the original field configuration of Figure 14.1 may still be present, another possibility emerges, which is shown in Figure 14.11. Again, a plane wave is guided in the z direction, but by means of a progression of zig-zag reflections at the upper and lower plates. Wavevectors $\mathbf{k}_u$ and $\mathbf{k}_d$ are associated with the upward and downward-propagating waves, respectively, and these have identical magnitudes,

$$|\mathbf{k}_u| = |\mathbf{k}_d| = k = \omega\sqrt{\mu\epsilon}$$

For such a wave to propagate, all upward-propagating waves must be *in phase* (as must be true of all downward-propagating waves). This condition can only be satisfied at certain discrete angles

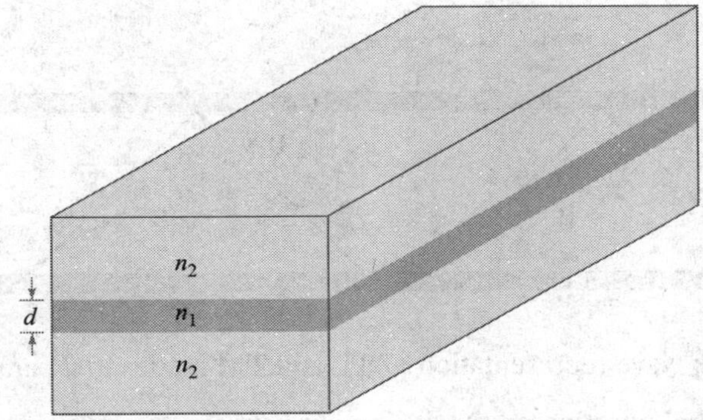

Fig. 14.9 Symmetric dielectric slab waveguide, with slab region (refractive index n_1) surrounded by two dielectrics of index $n_2 < n_1$.

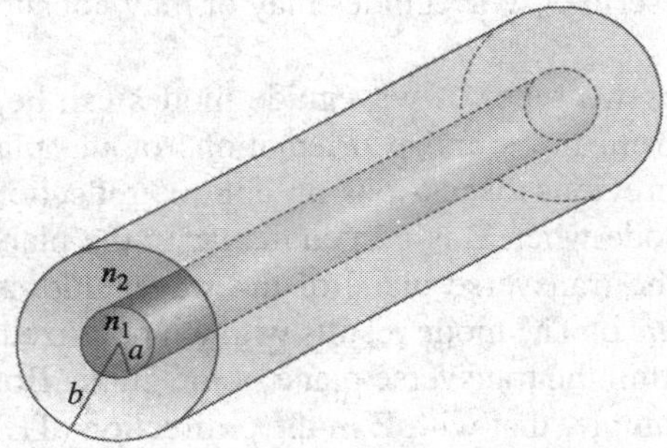

Fig. 14.10 Optical fiber waveguide, with the core dielectric ($\rho < a$) of refractive index n_1. The cladding dielectric ($a < \rho < b$) is of index $n_2 < n_1$.

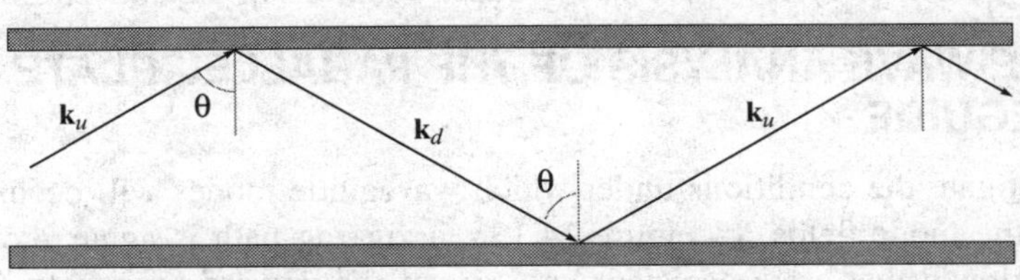

Fig. 14.11 In a parallel-plate waveguide, plane waves can propagate by oblique reflection from the conducting walls. This produces a waveguide mode that is not TEM.

of incidence, shown as θ in Figure 14.11. An allowed value of θ, along with the resulting field configuration, comprise a *waveguide mode* of the structure. Associated with each guided mode is a *cutoff frequency*. If the operating frequency is below the cutoff frequency, the mode will not propagate. If above cutoff, the mode propagates. The TEM mode, however, has no cutoff; it will be supported at any frequency. At a given frequency, the guide may support several modes, the quantity of which depends on the plate separation and on the dielectric constant of the interior medium, as will be shown. The number of modes increases as the frequency is raised.

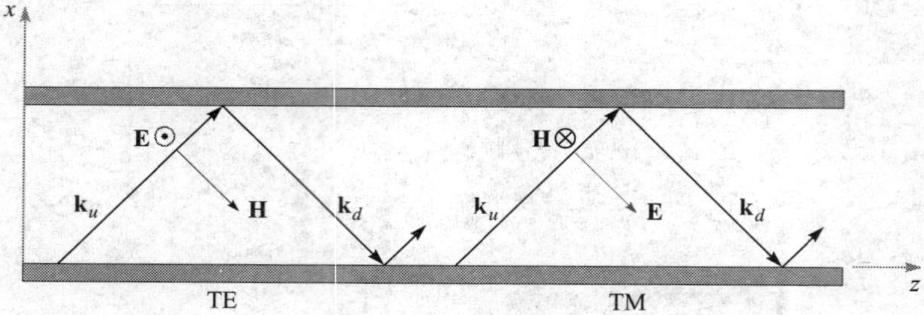

Fig. 14.12 Plane wave representation of TE and TM modes in a parallel-plate guide.

So to answer our initial question on the distinction between transmission lines and waveguides, we can state the following: Transmission lines consist of two or more conductors and as a rule will support TEM waves (or something which could approximate such a wave). A waveguide may consist of one or more conductors, or no conductors at all, and will support waveguide modes of forms similar to those just described. Waveguides may or may not support TEM waves, depending on the design.

In the parallel-plate guide, two types of waveguide modes can be supported. These are shown in Figure 14.12 as arising from the s and p orientations of the plane wave polarizations. In a manner consistent with our previous discussions on oblique reflection (Section 13.5), we identify a *transverse electric* or TE mode when **E** is perpendicular to the plane of incidence (s-polarized); this positions **E** parallel to the transverse plane of the waveguide, as well as to the boundaries. Similarly, a *transverse magnetic* or TM mode results with a p polarized wave; the entire **H** field is in the y direction and is thus within the transverse plane of the guide. Both possibilities are illustrated in Figure 14.12. Note, for example, that with **E** in the y direction (TE mode), **H** will have x- and z components. Likewise, a TM mode will have x- and z components of **E**.[3] In any event, the reader can verify from the geometry of Figure 14.12 that it is not possible to achieve a purely TEM mode for values of θ other than 90°. Other wave polarizations are possible that lie between the TE and TM cases, but these can always be expressed as super positions of TE and TM modes.

14.3 PLANE WAVE ANALYSIS OF THE PARALLEL-PLATE WAVEGUIDE

Let us now investigate the conditions under which waveguide modes will occur, using our plane wave model for the mode fields. In Figure 14.13*a*, a zig-zag path is again shown, but this time phase fronts are drawn that are associated with two of the upward-propagating waves. The first wave has reflected twice (at the top and bottom surfaces) to form the second wave (the downward-propagating phase fronts are not shown). Note that the phase fronts of the second wave do not coincide with those of the first wave, and so the two waves are out of phase. In Figure 14.13*b*, the wave angle has been adjusted so that the two waves are now in phase. Having satisfied this condition for the two waves, we will find that *all* upward-propagating waves will have coincident phase fronts. The same condition will automatically occur for all downward-propagating waves. This is the requirement to establish a guided mode.

[3] Other types of modes can exist in other structures (not the parallel-plate guide) in which *both* **E** and **H** have $z$components. These are known as *hybrid* modes, and they typically occur in guides with cylindrical cross sections, such as the optical fiber.

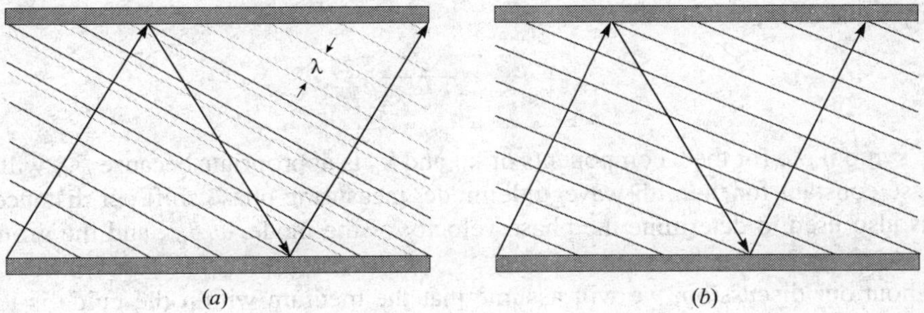

Fig. 14.13 (a) Plane wave propagation in a parallel-plate guide in which the wave angle is such that the upward-propagating waves are not in phase. (b) The wave angle has been adjusted so that the upward waves are in phase, resulting in a guided mode.

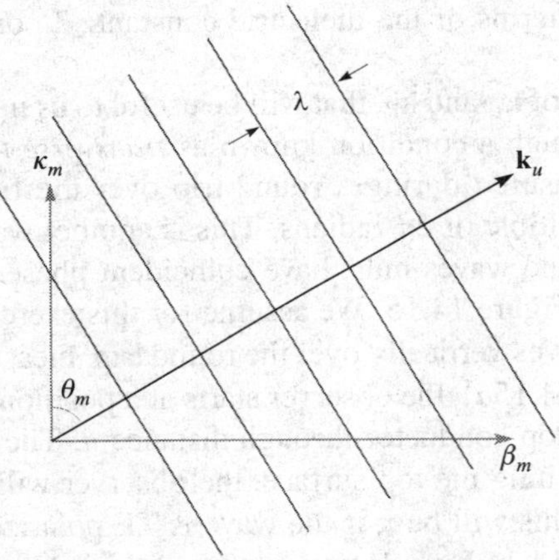

Fig. 14.14 The components of the upward wavevector are κ_m and β_m, the transverse and axial phase constants. To form the downward wavevector k_d, the direction of κ_m is reversed.

In Figure 14.14 we show the wavevector, $\mathbf{k}_u$, and its components, along with a series of phase fronts. A drawing of this kind for $\mathbf{k}_d$ would be the same, except the x component, κ_m, would be reversed. In Section 13.4, we measured the phase shift per unit distance along the x and z directions by the components, k_x and k_z, which varied continuously as the direction of $\mathbf{k}$ changed. In our discussion of waveguides, we introduce a different notation, where κ_m and β_m are used for k_x and k_z. The subscript m is an integer indicating the *mode number*. This provides a subtle hint that β_m and κ_m will assume only certain discrete values that correspond to certain allowed directions of $\mathbf{k}_u$ and $\mathbf{k}_d$, such that our coincident phase front requirement is satisfied.[4] From the geometry we see

[4] Subscripts (m) are not shown on $\mathbf{k}_u$ and $\mathbf{k}_d$, but are understood. Changing m does not affect the magnitudes of these vectors, only their directions.

that for any value of m,

$$\beta_m = \sqrt{k^2 - \kappa_m^2} \tag{35}$$

Use of the symbol β_m for the z components of $\mathbf{k}_u$ and $\mathbf{k}_d$ is appropriate because β_m will ultimately be the phase constant for the mth waveguide mode, measuring phase shift per distance down the guide; it is also used to determine the phase velocity of the mode, ω/β_m, and the group velocity, $d\omega/d\beta_m$.

Throughout our discussion, we will assume that the medium within the guide is lossless and nonmagnetic, so that

$$k = \omega\sqrt{\mu_0\epsilon'} = \frac{\omega\sqrt{\epsilon'_r}}{c} = \frac{\omega n}{c} \tag{36}$$

which we express either in terms of the dielectric constant, ϵ'_r, or the refractive index, n, of the medium.

It is κ_m, the x component of $\mathbf{k}_u$ and $\mathbf{k}_d$, that will be useful to us in quantifying our requirement on coincident phase fronts through a condition known as *transverse resonance*. This condition states that the net phase shift measured during a round trip over the full transverse dimension of the guide must be an integer multiple of 2π radians. This is another way of stating that all upward (or downward-) propagating plane waves must have coincident phases. The various segments of this round trip are illustrated in Figure 14.15. We assume for this exercise that the waves are frozen in time and that an observer moves vertically over the round trip, measuring phase shift along the way. In the first segment (Figure 14.15a), the observer starts at a position just above the lower conductor and moves vertically to the top conductor through distance d. The measured phase shift over this distance is $\kappa_m d$ rad. On reaching the top surface, the observer will note a possible phase shift on reflection (Figure 14.15b). This will be π if the wave is TE polarized and will be zero if the wave is TM polarized (see Figure 14.16 for a demonstration of this). Next, the observer moves along the reflected wave phases down to the lower conductor and again measures a phase shift of $\kappa_m d$ (Figure 14.15c). Finally, after including the phase shift on reflection at the bottom conductor, the observer is back at the original starting point and is noting the phase of the next upward-propagating wave.

The total phase shift over the round trip is required to be an integer multiple of 2π:

$$\kappa_m d + \phi + \kappa_m d + \phi = 2m\pi \tag{37}$$

where ϕ is the phase shift on reflection at each boundary. Note that with $\phi = \pi$ (TE waves) or 0 (TM waves) the net reflective phase shift over a round trip is 2π or 0, regardless of the angle of incidence. Thus the reflective phase shift has no bearing on the current problem, and we may simplify (37) to read

$$\kappa_m = \frac{m\pi}{d} \tag{38}$$

which is valid for *both* TE and TM modes. Note from Figure 14.14 that $\kappa_m = k\cos\theta_m$. Thus the wave angles for the allowed modes are readily found from (38) with (36):

$$\theta_m = \cos^{-1}\left(\frac{m\pi}{kd}\right) = \cos^{-1}\left(\frac{m\pi c}{\omega n d}\right) = \cos^{-1}\left(\frac{m\lambda}{2nd}\right) \tag{39}$$

where λ is the wavelength in free space.

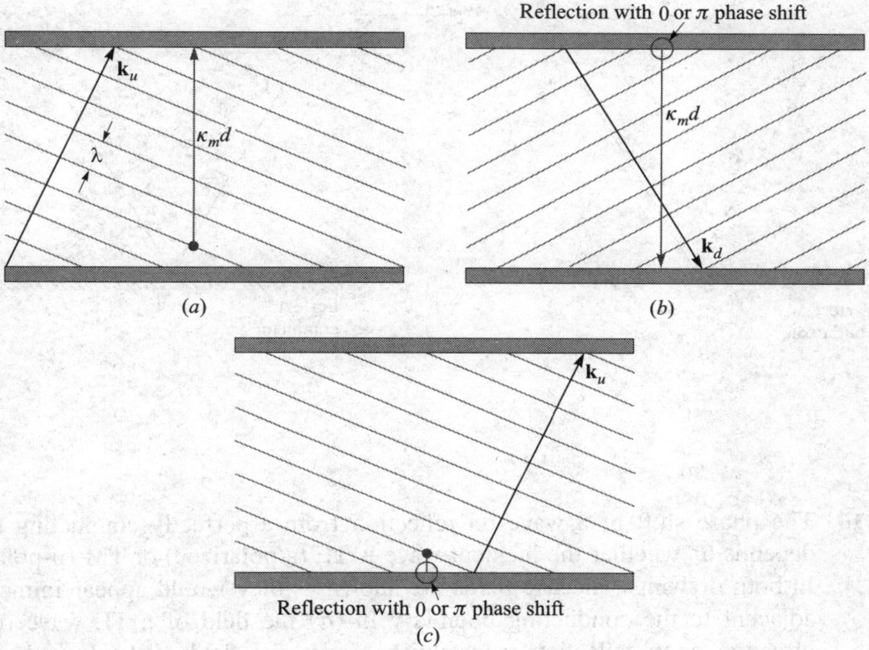

Fig. 14.15 The net phase shift over a round trip in the parallel-plate guide is found by first measuring the transverse phase shift between plates of the initial upward wave (a); next, the transverse phase shift in the reflected (downward) wave is measured, while accounting for the reflective phase shift at the top plate (b); finally the phase shift on reflection at the bottom plate is added, thus returning to the starting position, but with a new upward wave (c). Transverse resonance occurs if the phase at the final point is the same as that at the starting point (the two upward waves are in phase).

We can next solve for the phase constant for each mode, using (35) with (38):

$$\beta_m = \sqrt{k^2 - \kappa_m^2} = k\sqrt{1 - \left(\frac{m\pi}{kd}\right)^2} = k\sqrt{1 - \left(\frac{m\pi c}{\omega n d}\right)^2} \tag{40}$$

We define the radian *cutoff frequency* for mode m as

$$\omega_{\text{cm}} = \frac{m\pi c}{nd} \tag{41}$$

so that (40) becomes

$$\beta_m = \frac{n\omega}{c}\sqrt{1-\left(\frac{\omega_{cm}}{\omega}\right)^2} \tag{42}$$

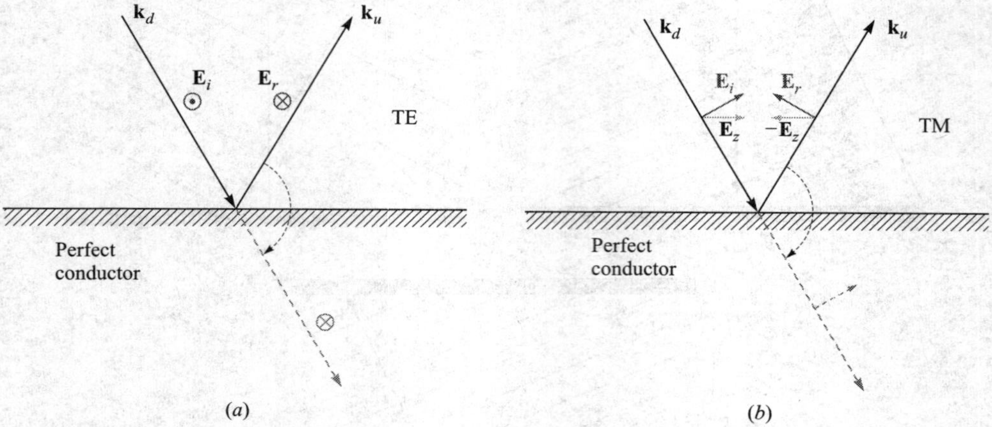

Fig. 14.16 The phase shift of a wave on reflection from a perfectly conducting surface depends on whether the incident wave is TE (s-polarized) or TM (p-polarized). In both drawings, electric fields are shown as they would appear immediately adjacent to the conducting boundary In (a) the field of a TE wave reverses direction upon reflection to establish a zero net field at the boundary. This constitutes a π phase shift, as is evident by considering a fictitious transmitted wave (dashed line) formed by a simple rotation of the reflected wave into alignment with the incident wave. In (b) an incident TM wave experiences a reversal of the *z component* of its electric field. The resultant field of the reflected wave, however has not been phase-shifted; rotating the reflected wave into alignment with the incident wave (dashed line) shows this.

The significance of the cutoff frequency is readily seen from (42): If the operating frequency ω is greater than the cutoff frequency for mode m, then that mode will have phase constant β_m that is real-valued, and so the mode will propagate. For $\omega < \omega_{cm}$, β_m is imaginary, and the mode does not propagate.

Associated with the cutoff frequency is the *cutoff wavelength,* λ_{cm}, defined as the free-space wavelength at which cutoff for mode $m$occurs. This will be

$$\lambda_{cm} = \frac{2\pi c}{\omega_{cm}} = \frac{2nd}{m} \tag{43}$$

Note, for example, that in an air-filled guide ($n = 1$) the wavelength at which the lowest-order mode first starts to propagate is $\lambda_{c1} = 2d$, or the plate separation is one-half wavelength. Mode m will propagate whenever $\omega > \omega_{cm}$, or equivalently whenever $\lambda < \lambda_{cm}$. Use of the cutoff wavelength enables us to construct a second useful form of Eq. (42):

$$\beta_m = \frac{2\pi n}{\lambda}\sqrt{1-\left(\frac{\lambda}{\lambda_{cm}}\right)^2} \tag{44}$$

EXAMPLE 14.1

A parallel-plate waveguide has plate separation $d = 1$ cm and is filled with teflon having dielectric constant $\epsilon'_r = 2.1$. Determine the maximum operating frequency such that only the TEM mode will propagate. Also find the range of frequencies over which the TE_1 and $TM_1 (m = 1)$ modes will propagate, and no higher-order modes.

Solution. Using (41), the cutoff frequency for the first waveguide mode ($m = 1$) will be

$$f_{c1} = \frac{\omega_{c1}}{2\pi} = \frac{2.99 \times 10^{10}}{2\sqrt{2.1}} = 1.03 \times 10^{10}\,\text{Hz} = 10.3\,\text{GHz}$$

To propagate only TEM waves, we must have $f < 10.3$ GHz. To allow TE_1 and TM_1 (along with TEM) only, the frequency range must be $\omega_{c1} < \omega < \omega_{c2}$, where $\omega_{c2} = 2\omega_{c1}$, from (41). Thus, the frequencies at which we will have the $m = 1$ modes and TEM will be 10.3 GHz $< f <$ 20.6 GHz.

EXAMPLE 14.2

In the parallel-plate guide of Example 14.1, the operating wavelength is $\lambda = 2$ mm. How many waveguide modes will propagate?

Solution. For mode m to propagate, the requirement is $\lambda < \lambda_{cm}$. For the given waveguide and wavelength, the inequality becomes, using (43),

$$2\text{ mm} < \frac{2\sqrt{2.1}\,(10\text{ mm})}{m}$$

from which

$$m < \frac{2\sqrt{2.1}(10\text{ mm})}{2\text{ mm}} = 14.5$$

Thus the guide will support modes at the given wavelength up to order $m = 14$. Since there will be a TE and a TM mode for each value of m, this gives, not including the TEM mode, a total of 28 guided modes that are above cutoff.

The field configuration for a given mode can be found through the superposition of the fields of all the reflected waves. We can do this for the TE waves, for example, by writing the electric field phasor in the guide in terms of incident and reflected fields through

$$E_{ys} = E_0 e^{-j\mathbf{k}_u\cdot\mathbf{r}} - E_0 e^{-j\mathbf{k}_d\cdot\mathbf{r}} \tag{45}$$

where the wavevectors, $\mathbf{k}_u$ and $\mathbf{k}_d$, are indicated in Figure 14.12. The minus sign in front of the second term arises from the π phase shift on reflection. From the geometry depicted in Figure 14.14, we write

$$\mathbf{k}_u = \kappa_m \mathbf{a}_x + \beta_m \mathbf{a}_z \tag{46}$$

and

$$\mathbf{k}_d = -\kappa_m \mathbf{a}_x + \beta_m \mathbf{a}_z \tag{47}$$

Then, using

$$\mathbf{r} = x\mathbf{a}_x + z\mathbf{a}_z$$

Equation (45) becomes

$$E_{ys} = E_0(e^{-j\kappa_m x} - e^{jk_m x})e^{-j\beta_m z} = 2jE_0 \sin(\kappa_m x)e^{-j\beta_m z} = E_0' \sin(\kappa_m x)e^{-j\beta_m z} \tag{48}$$

where the plane wave amplitude, E_0, and the overall phase are absorbed into E_0'. In real instantaneous form, (48) becomes

$$E_y(z,t) = \text{Re}(E_{ys}e^{j\omega t}) = E_0' \sin(\kappa_m x)\cos(\omega t - \beta_m z) \qquad \text{(TE mode above cutoff)} \tag{49}$$

We interpret this as a wave that propagates in the positive z direction (down the guide) while having a field profile that varies with x.[5] The TE mode field is the *interference pattern* resulting from the superposition of the upward and downward plane waves. Note that if $\omega < \omega_{cm}$, then (42) yields an imaginary value for β_m, which we may write as $-j|\beta_m| = -j\alpha_m$. Equations (48) and (49) then become

$$E_{ys} = E_0' \sin(\kappa_m x)e^{-\alpha_m z} \tag{50}$$

$$E(z,t) = E_0' \sin(\kappa_m x)e^{-\alpha_m z}\cos(\omega t) \quad \text{(TE mode below cutoff)} \tag{51}$$

This mode does not propagate, but simply oscillates at frequency ω, while exhibiting a field pattern that decreases in strength with increasing z. The attenuation coefficient, α_m, is found from (42) with $\omega < \omega_{cm}$:

$$\alpha_m = \frac{n\omega_{cm}}{c}\sqrt{1 - \left(\frac{\omega}{\omega_{cm}}\right)^2} = \frac{2\pi n}{\lambda_{cm}}\sqrt{1 - \left(\frac{\lambda_{cm}}{\lambda}\right)^2} \tag{52}$$

We note from (39) and (41) that the plane wave angle is related to the cutoff frequency and cutoff wavelength through

$$\cos\theta_m = \frac{\omega_{cm}}{\omega} = \frac{\lambda}{\lambda_{cm}} \tag{53}$$

So we see that at cutoff ($\omega = \omega_{cm}$), $\theta_m = 0$, and the plane waves are just reflecting back and forth over the cross section; they are making no forward progress down the guide. As ω is increased beyond cutoff (or λ is decreased), the wave angle increases, approaching 90° as ω approaches infinity (or as λ approaches zero). From Figure 14.14, we have

$$\beta_m = k\sin\theta_m = \frac{n\omega}{c}\sin\theta_m \tag{54}$$

[5] We can also interpret this field as that of a standing wave in x while it is a traveling wave in z.

and so the phase velocity of mode m will be

$$\nu_{pm} = \frac{\omega}{\beta_m} = \frac{c}{n \sin\theta_m} \tag{55}$$

The velocity minimizes at c/n for all modes, approaching this value at frequencies far above cutoff; ν_{pm} approaches infinity as the frequency is reduced to approach the cutoff frequency. Again, phase velocity is the speed of the phases in the z direction, and the fact that this velocity may exceed the speed of light in the medium is not a violation of relativistic principles, as discussed in Section 13.7.

The energy will propagate at the group velocity, $\nu_g = d\omega/d\beta$. Using (42), we have

$$\nu_{gm}^{-1} = \frac{d\beta_m}{d\omega} = \frac{d}{d\omega}\left[\frac{n\omega}{c}\sqrt{1 - \left(\frac{\omega_{cm}}{\omega}\right)^2}\right] \tag{56}$$

The derivative is straightforward. Carrying it out and taking the reciprocal of the result yields:

$$\nu_{gm} = \frac{c}{n}\sqrt{1 - \left(\frac{\omega_{cm}}{\omega}\right)^2} = \frac{c}{n}\sin\theta_m \tag{57}$$

Group velocity is thus identified as the projection of the velocity associated with $\mathbf{k}_u$ or $\mathbf{k}_d$ into the z direction. This will be less than or equal to the velocity of light in the medium, c/n, as expected.

EXAMPLE 14.3

In the guide of Example 14.1, the operating frequency is 25 GHz. Consequently, modes for which $m = 1$ and $m = 2$ will be above cutoff. Determine the *group delay difference* between these two modes over a distance of 1 cm. This is the difference in propagation times between the two modes when energy in each propagates over the 1 cm distance.

Solution. The group delay difference is expressed as

$$\Delta t = \left(\frac{1}{\nu_{g2}} - \frac{1}{\nu_{g1}}\right) \text{ (s/cm)}$$

From (57), along with the results of Example 14.1, we have

$$\nu_{g1} = \frac{c}{\sqrt{2.1}}\sqrt{1 - \left(\frac{10.3}{25}\right)^2} = 0.63c$$

$$\nu_{g2} = \frac{c}{\sqrt{2.1}}\sqrt{1 - \left(\frac{20.6}{25}\right)^2} = 0.39c$$

Then

$$\Delta t = \frac{1}{c}\left[\frac{1}{.39} - \frac{1}{.63}\right] = 3.3 \times 10^{-11} \text{ s/cm} = 33 \text{ ps/cm}$$

This computation gives a rough measure of the *modal dispersion* in the guide, applying to the case of having only two modes propagating. A pulse, for example, whose center frequency is 25 GHz would have its energy divided between the two modes. The pulse would broaden by approximately 33 ps/cm of propagation distance as the energy in the modes separates. If, however, we include the TEM mode (as we really must), then the broadening will be even greater. The group velocity for TEM will be $c/\sqrt{2.1}$. The group delay difference of interest will then be between the TEM mode and the m= 2 mode (TE or TM). We would therefore have

$$\Delta t_{\text{net}} = \frac{1}{c}\left[\frac{1}{.39} - 1\right] = 52 \text{ ps/cm}$$

D14.5 Determine the wave angles θ_m for the first four modes (m = 1, 2, 3, 4) in a parallel-plate guide with d = 2 cm, ϵ'_r = 1, and f = 30 GHz.

Ans. 76°; 60°; 41°; 0°

D14.6 A parallel-plate guide has plate spacing d = 5 mm and is filled with glass (n = 1.45). What is the maximum frequency at which the guide will operate in the TEM mode only?

Ans. 20.7 GHz

D14.7 A parallel-plate guide having d= 1 cm is filled with air. Find the cutoff wavelength for the m= 2 mode (TE or TM).

Ans. 1 cm

In Section 14.1, it was observed that at higher frequencies the loss in any transmission line would increase due to increase in the series resistance per unit length, R. Hence, there is an upper limit on the frequency of operation of typical two-wire transmission lines. Above this frequency, the single conductor hollow metal waveguides are usually employed as described in the next chapter. However, at very high frequencies, it becomes difficult to use even these hollow metal guides. The reason is that the transverse dimensions of these metal waveguides become quite small at very high frequencies, and hence the skin effect losses again become problematic due to decrease in the skin depth and the metal surface area. In addition, the guides become too difficult to fabricate, with machining tolerances becoming more stringent at such high frequencies. So, at very high frequencies, we look for another type of structure where the skin-effect losses can be minimized.

14.4 PLANAR DIELECTRIC WAVEGUIDES

When skin effect losses become excessive, a good way to remove them is to remove the metal in the structure entirely and use interfaces between dielectrics for the confining surfaces. We thus obtain a *dielectric waveguide*; a basic form, the *symmetric slab waveguide*, is shown in Figure 14.17. The structure is so named because of its vertical symmetry about the z axis. The guide is assumed to have width in y much greater than the slab thickness d, so the problem becomes two-dimensional, with fields presumed to vary with x and z while being independent of y. The slab guide works in very much the same way as the parallel-plate waveguide, except wave reflections occur at the interfaces between dielectrics, having different refractive indices, n_1 for the slab and n_2 for the surrounding regions above and below. In the dielectric guide, total reflection is needed, so the incident angle must exceed the critical angle. Consequently, as discussed in Section 13.6, the slab index, n_1, must be greater than that of the surrounding materials, n_2. Dielectric guides differ from conducting guides in that power is not completely confined to the slab but resides partially above and below.

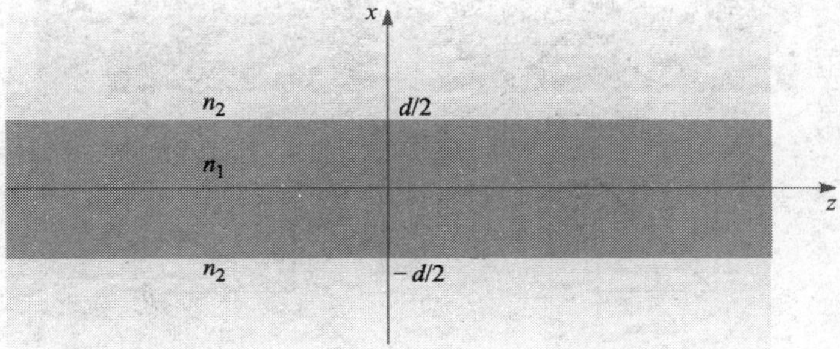

Fig. 14.17 Symmetric dielectric slab waveguide structure, in which waves propagate along z. The guide is assumed to be infinite in the y-direction, thus making the problem two-dimensional.

Dielectric guides are used primarily at optical frequencies (on the order of 10^{14} Hz). Again, guide transverse dimensions must be kept on the order of a wavelength to achieve operation in a single mode. A number of fabrication methods can be used to accomplish this. For example, a glass plate can be doped with materials that will raise the refractive index. The doping process allows materials to be introduced only within a thin layer adjacent to the surface that is a few micrometers thick.

To understand how the guide operates, consider Figure 14.18, which shows a wave propagating through the slab by multiple reflections, but where *partial transmission* into the upper and lower regions occurs at each bounce. Wavevectors are shown in the middle and upper regions, along with their components in the x and z directions. As we found in Chapter 13, the z components (β) of all wavevectors are equal, as must be true if the field boundary conditions at the interfaces are to be satisfied for all positions and times. Partial transmission at the boundaries is, of course, an undesirable situation, since power in the slab will eventually leak away. We thus have a *leaky wave* propagating in the structure, whereas we need to have a guided mode. Note that in either case, we still have the two possibilities of wave polarization, and the resulting mode designation—either TE or TM.

Total power reflection at the boundaries for TE or TM waves implies, respectively, that $|\Gamma_s|^2$ or $|\Gamma_p|^2$ is unity, where the reflection coefficients are given in Eqs. (71) and (69) in Chapter 13:

$$\Gamma_s = \frac{\eta_{2s} - \eta_{1s}}{\eta_{2s} + \eta_{1s}} \tag{58}$$

and

$$\Gamma_p = \frac{\eta_{2p} - \eta_{1p}}{\eta_{2p} + \eta_{1p}} \tag{59}$$

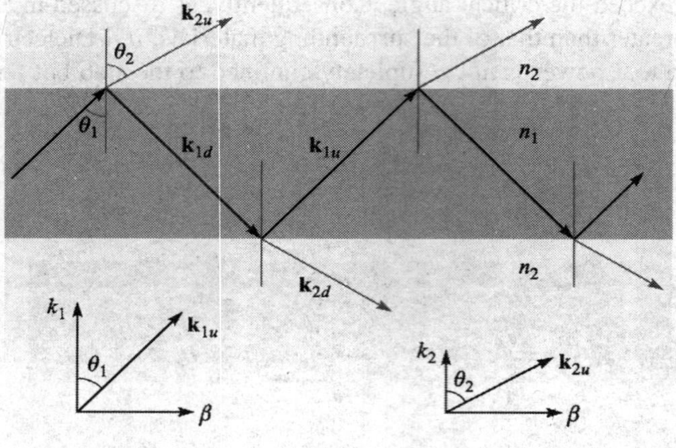

Fig. 14.18 Plane wave geometry of a leaky wave in a symmetric slab waveguide. For a guided mode, total reflection occurs in the interior, and the x components of $\mathbf{k}_{2u}$ and $\mathbf{k}_{2d}$ are imaginary.

As discussed in Section 13.6, we require that the effective impedances, η_{2s} or η_{2p}, be purely imaginary, zero, or infinite if (58) or (59) is to have unity magnitudes. Knowing that

$$\eta_{2s} = \frac{\eta_2}{\cos\theta_2} \tag{60}$$

and

$$\eta_{2p} = \eta_2 \cos\theta_2 \tag{61}$$

the requirement is that $\cos\theta_2$ be zero or imaginary, where, from Eq. (75), Section 13.6,

$$\cos\theta_2 = [1 - \sin^2\theta_2]^{1/2} = \left[1 - \left(\frac{n_1}{n_2}\right)^2 \sin^2\theta_1\right]^{1/2} \tag{62}$$

As a result, we require that

$$\theta_1 \geq \theta_c \tag{63}$$

where the critical angle is defined through

$$\sin\theta_c = \frac{n_2}{n_1} \tag{64}$$

Now, from the geometry of Figure 14.18, we can construct the field distribution of a TE wave in the guide using plane wave superposition. In the slab region $(-d/2 < x < d/2)$, we have

$$E_{y1s} = E_0 e^{-j\mathbf{k}_{1u}\cdot\mathbf{r}} \pm E_0 e^{-j\mathbf{k}_{1d}\cdot\mathbf{r}} \quad \left(-\frac{d}{2} < x < \frac{d}{2}\right) \tag{65}$$

where

$$\mathbf{k}_{1u} = \kappa_1 \mathbf{a}_x + \beta \mathbf{a}_z \tag{66}$$

and

$$\mathbf{k}_{1d} = -\kappa_1 \mathbf{a}_x + \beta \mathbf{a}_z \tag{67}$$

The second term in (65) may either add to or subtract from the first term, since either case would result in a symmetric intensity distribution in the x direction. We expect this because the guide is symmetric. Now, using $\mathbf{r} = x\mathbf{a}_x + z\mathbf{a}_z$, (65) becomes

$$E_{y1s} = E_0[e^{j\kappa_1 x} + e^{-j\kappa_1 x}]e^{-j\beta z} = 2E_0 \cos(\kappa_1 x)e^{-j\beta z} \tag{68}$$

for the choice of the plus sign in (65), and

$$E_{y1s} = E_0[e^{j\kappa_1 x} - e^{-j\kappa_1 x}]e^{-j\beta z} = 2jE_0 \sin(\kappa_1 x)e^{-j\beta z} \tag{69}$$

if the minus sign is chosen. Since $\kappa_1 = n_1 k_0 \cos\theta_1$, we see that larger values of κ_1 imply smaller values of θ_1 at a given frequency. In addition, larger κ_1 values result in a greater number of spatial oscillations of the electric field over the transverse dimension, as (68) and (69) show. We found similar behavior in the parallel-plate guide. In the slab waveguide, as with the parallel-plate guide, we associate higher-order modes with increasing values of κ_1.[6]

In the regions above and below the slab, waves propagate according to wave-vectors $\mathbf{k}_{2u}$ and $\mathbf{k}_{2d}$ as shown in Figure 14.18. Above the slab, for example ($x > d/2$), the TE electric field will be of the form

$$E_{y2s} = E_{02}e^{-j\mathbf{k}_2\cdot\mathbf{r}} = E_{02}e - j^{\kappa_2 x}e^{-j\beta z} \tag{70}$$

However, $\kappa_2 = n_2 k_0 \cos\theta_2$, where $\cos\theta_2$, given in (62), is imaginary. We may therefore write

$$\kappa_2 = -j\gamma_2 \tag{71}$$

where γ_2 is real and is given by (using 62)

$$\gamma_2 = j\kappa_2 = jn_2k_0\cos\theta_2 = jn_2k_0(-j)\left[\left(\frac{n_1}{n_2}\right)^2 \sin^2\theta_1 - 1\right]^{1/2} \tag{72}$$

Equation (70) now becomes

$$E_{\gamma 2s} = E_{02}e^{-\gamma_2(x-d/2)}e^{-j\beta z} \quad \left(x > \frac{d}{2}\right) \tag{73}$$

where the x variable in (70) has been replaced by $x-(d/2)$ to position the field magnitude, E_{02}, at the boundary. Using similar reasoning, the field in the region below the lower interface, where x is

[6] It would be appropriate to add the mode number subscript, m, to κ_1, κ_2, β, and θ_1, since, as was true with the metal guides, we will obtain discrete values of these quantities. To keep notation simple, the m subscript is suppressed, and we will assume it to be understood. Again, the subscripts 1 and 2 in this section indicate respectively the slab and surrounding *regions,* and have nothing to do with mode number.

negative, and where $\mathbf{k}_{2d}$ is involved, will be

$$E_{\gamma 2s} = E_{02} e^{\gamma_2 (x+d/2)} e^{-j\beta z} \quad \left(x < -\frac{d}{2} \right) \tag{74}$$

The fields expressed in (73) and (74) are those of *surface* waves. Note that they propagate in the $z$direction only, according to $e^{-j\beta z}$, but simply reduce in amplitude with increasing $|x|$, according to the $e^{-\gamma_2 (x-d/2)}$ term in (73) and the $e^{\gamma_2 (x+d/2)}$ term in (75). These waves represent a certain fraction of the total power in the mode, and so we see an important fundamental difference between dielectric waveguides and metal waveguides: in the dielectric guide, the fields (and guided power) exist over a cross section that extends beyond the confining boundaries, and in principle they exist over an infinite cross section. In practical situations, the exponential decay of the fields above and below the boundaries is typically sufficient to render the fields negligible within a few slab thicknesses from each boundary.

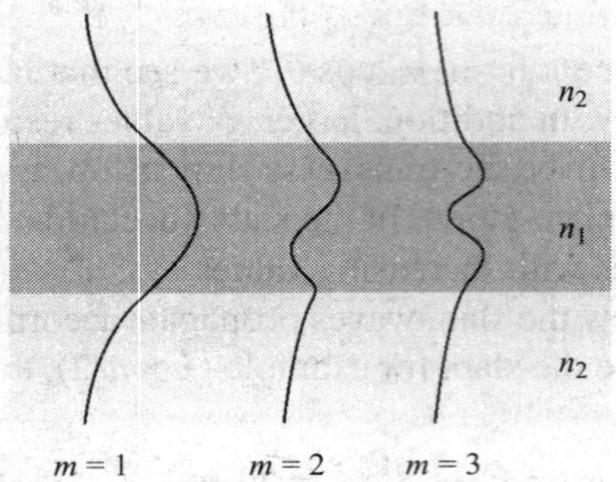

Fig. 14.19 Electric field amplitude distributions over the transverse plane for the first three TE modes in a symmetric slab waveguide.

The total electric field distribution is composed of the field in all three regions and is sketched in Figure 14.19 for the first few modes. Within the slab, the field is oscillatory and is of a similar form to that of the parallel-plate waveguide. The difference is that the fields in the slab guide do not reach zero at the boundaries but connect to the evanescent fields above and below the slab. The restriction is that the TE fields on either side of a boundary (being tangent to the interface) must match at the boundary. Specifically,

$$E_{y1s}|_{x=\pm d/2} = E_{y2s}|_{x=\pm d/2} \tag{75}$$

Applying this condition to (68), (69), (73), and (74) results in the final expressions for the TE electric field in the symmetric slab waveguide, for the cases of even and odd symmetry:

$$E_{se}(\text{even TE}) = \begin{cases} E_{0e} \cos(\kappa_1 x) e^{-j\beta z} & \left(-\frac{d}{2} < x < \frac{d}{2}\right) \\ E_{0e} \cos\left(\kappa_1 \frac{d}{2}\right) e^{-\gamma_2 (x-d/2)} e^{-j\beta z} & \left(x > \frac{d}{2}\right) \\ E_{0e} \cos\left(\kappa_1 \frac{d}{2}\right) e^{\gamma_2 (x+d/2)} e^{-j\beta z} & \left(x < -\frac{d}{2}\right) \end{cases} \tag{76}$$

$$E_{so}(\text{odd TE}) = \begin{cases} E_{0o}\sin(\kappa_1 x)e^{-j\beta z} & \left(-\frac{d}{2} < x < \frac{d}{2}\right) \\ E_{0o}\sin\left(\kappa_1 \frac{d}{2}\right)e^{-\gamma_2(x-d/2)}e^{-j\beta z} & \left(x > \frac{d}{2}\right) \\ -E_{0e}\sin\left(\kappa_1 \frac{d}{2}\right)e^{\gamma_2(x+d/2)}e^{-j\beta z} & \left(x < -\frac{d}{2}\right) \end{cases} \tag{77}$$

Solution of the wave equation yields (as it must) results identical to these. The reader is referred to References 2 for the details. The magnetic field for the TE modes will consist of x- and z components, as was true for the parallel-plate guide. Finally, the TM mode fields will be nearly the same in form as those of TE modes, but with a simple rotation in polarization of the plane wave components by 90°. Thus, in TM modes, H_y will result, and it will have the same form as E_y for TE, as presented in (76) and (77).

Apart from the differences in the field structures, the dielectric slab waveguide operates in a manner that is qualitatively similar to the parallel-plate guide. Again, a finite number of discrete modes will be allowed at a given frequency, and this number increases as frequency increases. Higher-order modes are characterized by successively smaller values of θ_1.

An important difference in the slab guide occurs at cutoff for any mode. We know that $\theta = 0$ at cutoff in the metal guides. In the dielectric guide at cutoff, the wave angle, θ_1, is equal to the *critical angle*, θ_c. Then, as the frequency of a given mode is raised, its θ_1 value increases beyond θ_c in order to maintain transverse resonance, while maintaining the same number of field oscillations in the transverse plane.

As wave angle increases, however, the character of the evanescent fields changes significantly. This can be understood by considering the wave angle dependence on evanescent decay coefficient, γ_2, as given by (72). Note in that equation that as θ_1 increases (as frequency goes up), γ_2 also increases, leading to a more rapid falloff of the fields with increasing distance above and below the slab. The mode therefore becomes more tightly confined to the slab as frequency is raised. Also, at a given frequency, lower-order modes, having smaller wave angles, will have lower values of γ_2 as (72) indicates. Consequently, when considering several modes propagating together at a single frequency, the higher-order modes will carry a greater percentage of their power in the upper and lower regions surrounding the slab than will modes of lower order.

One can determine the conditions under which modes will propagate by using the transverse resonance condition, as we did with the parallel-plate guide. We perform the transverse round trip analysis in the slab region in the same manner that was done in Section 14.3, and obtain an equation similar to (37):

$$\kappa_1 d + \phi_{TE} + \kappa_1 d + \phi_{TE} = 2m\pi \tag{78}$$

for TE waves and

$$\kappa_1 d + \phi_{TM} + \kappa_1 d + \phi_{TM} = 2m\pi \tag{79}$$

for the TM case. Equations (78) and (79) are called the *eigenvalue equations* for the symmetric dielectric slab waveguide. The phase shifts on reflection, ϕ_{TE} and ϕ_{TM}, are the phases of the reflection coefficients, Γ_s and Γ_p, given in (58) and (59). These are readily found, but they turn out to be functions of θ_1. As we know, κ_1 also depends on θ_1, but in a different way than ϕ_{TE} and ϕ_{TM}. Consequently, (78) and (79) are *transcendental* in θ_1, and they cannot be solved in closed form. Instead, numerical or graphical methods must be used (see References 3 or 4). Emerging from this solution process, however, is a fairly simple cutoff condition for any TE or TM mode:

$$k_0 d\sqrt{n_1^2 - n_2^2} \geq (m-1)\pi \qquad (m = 1, 2, 3, \ldots) \tag{80}$$

For mode m to propagate, (80) must hold. The physical interpretation of the mode number m is again the number of half-cycles of the electric field (for TE modes) or magnetic field (for TM modes) that occur over the transverse dimension. The lowest-order mode ($m = 1$) is seen to have no cutoff—it will propagate from zero frequency on up. We will thus achieve single-mode operation (actually a single pair of TE and TM modes) if we can assure that the $m = 2$ modes are below cutoff. Using (80), our single-mode condition will thus be

$$k_0 d\sqrt{n_1^2 - n_2^2} < \pi \tag{81}$$

Using $k_0 = 2\pi/\lambda$, the wavelength range over which single-mode operation occurs is

$$\lambda > 2d\sqrt{n_1^2 - n_2^2} \tag{82}$$

EXAMPLE 14.4

A symmetric dielectric slab waveguide is to guide light at wavelength $\lambda = 130\ \mu$m. The slab thickness is to be $d = 500\ \mu$m, and the refractive index of the surrounding material is $n_2 =$ 1450. Determine the maximum allowable refractive index of the slab material that will allow single TE and TM mode operation.

Solution. Equation (82) can be rewritten in the form,

$$n_1 < \sqrt{\left(\frac{\lambda}{2d}\right)^2 + n_2^2}$$

So

$$n_1 < \sqrt{\left(\frac{1.30}{2(5.00)}\right)^2 + (1.450)^2} = 1.456$$

Clearly, fabrication tolerances are very exacting when constructing dielectric guides for single-mode operation!

D14.8 A 0.5 mm thick slab of glass ($n_1 = 1.45$) is surrounded by air ($n_2 = 1$). The slab guides infrared light at wavelength $\lambda = 10\ \mu$m. How many TE and TM modes will propagate?

Ans. 2102

14.5 OPTICAL FIBER

Optical fiber works on the same principle as the dielectric slab waveguide, except of course for the round cross section. A *step index* fiber is shown in Figure 14.10, in which a high index *core* of radius $a$is surrounded by a lower-index *cladding* of radius b. Light is confined to the core through the mechanism of total reflection, but again some fraction of the power resides in the cladding as well. As we found in the slab guide, the cladding power again moves in toward the core as frequency is raised. Additionally, as is true in the slab waveguide, the fiber supports a mode that has no cutoff.

Analysis of the optical fiber is complicated. This is mainly because of the round cross section, along with the fact that it is generally a three-dimensional problem; the slab waveguide had only two dimensions to be concerned about. It is possible to analyze the fiber using rays within the core that reflect from the cladding boundary as light progresses down the fiber. We did this with the slab guide and obtained results fairly quickly. The method is difficult in fiber, however, because ray paths are complicated. There are two types of rays in the core: (1) those that pass through the fiber axis (z axis), known as *meridional rays*, and (2) those that avoid the axis but progress in a spiral-like path as they propagate down the guide. These are known as *skew rays*; their analysis, although possible, is tedious. Fiber modes are developed that can be associated with the individual ray types, or with combinations thereof, but it is easier to obtain these by solving the wave equation directly. Our purpose in this section is to provide a first exposure to the optical fiber problem (and to avoid an excessively long treatment). To accomplish this, we will solve the simplest case in the quickest way.

The simplest fiber configuration is that of a step index, but with the core and cladding indices of values that are very close, that is $n_1 \doteq n_2$. This is the *weak-quidance* condition, whose simplifying effect on the analysis is significant. We already saw how core and cladding indices in the slab waveguide need to be very close in value in order to achieve single-mode or few-mode operation. Fiber manufacturers have taken this result to heart, such that the weak-guidance condition is in fact satisfied by most commercial fibers today. Typical dimensions of a single-mode fiber are between 5 and 10 μm for the core diameter, with the cladding diameter usually 125 μm. Refractive index differences between core and cladding are typically a small fraction of a percent.

The main result of the weak-guidance condition is that a set of modes appears in which each mode is *linearly polarized*. This means that light having x-polarization, for example, will enter the fiber and establish itself in a mode or in a set of modes that preserve the x-polarization. Magnetic field is essentially orthogonal to **E**, and so it would in that case lie in the y direction. The $z$components of both fields, although present, are too weak to be of significance; the nearly equal core and cladding indices lead to ray paths that are essentially parallel to the guide axis—deviating only slightly. In fact, we may write for a given mode, $E_x \doteq \eta H_y$, when η is approximated as the intrinsic impedance of the cladding. Therefore, in the weak-guidance approximation, the fiber mode fields are treated as plane waves (nonuniform, of course). The designation for these modes is LP_{lm}, meaning linearly polarized, with integer order parameters ℓ and m. The latter express the numbers of variations over the two dimensions in the circular transverse plane. Specifically, ℓ, the *azimuthal mode number,* is one-half the number of power density maxima (or minima) that occur at a given radius as ϕ varies from 0 to $2\pi \cdot m$, the *radial mode number*, expresses the number of maxima that occur along a radial line (at constant ϕ) that extends from zero to infinity.

Although we may assume a linearly polarized field in a rectangular coordinate system, we are obliged to work in cylindrical coordinates for obvious reasons. In a manner that reminds us of Chapter 7, it is possible to write the x-polarized phasor electric field within a weakly guiding cylindrical fiber as a product of three functions, each of which varies with one of the coordinate variables, ρ, ϕ, and z:

$$E_{xs}(\rho, \phi, z) = \sum_i R_i(\rho)\Phi_i(\phi)\exp(-j\beta_i z) \tag{83}$$

Each term in the summation is an individual mode of the fiber. Note that the z function is just the propagation term, $e^{-j\beta z}$, since we are assuming an infinitely long lossless fiber.

The wave equation in this case is given by $\nabla^2 \epsilon_{sx} + k^2 E_{sx} = 0$, which for cylindrical coordinates can be written in the following format:

$$\frac{1}{\rho}\frac{\partial}{\partial \rho}\left(\rho \frac{\partial^2 E_{xs}}{\partial \rho}\right) + \frac{1}{\rho^2}\frac{\partial^2 E_{xs}}{\partial \phi^2} + (k^2 - \beta^2)E_{xs} = 0 \tag{84}$$

where we recognize that the $\partial^2/\partial z^2$ operation, when applied to (83), leads to a factor of $-\beta^2$. We now substitute a single term of (83) into (84) [since each term in (83) should alone satisfy the wave equation]. Dropping the subscript i, expanding the radial derivative, and rearranging terms, we obtain:

$$\underbrace{\frac{\rho^2}{R}\frac{d^2R}{d\rho^2} + \frac{\rho}{R}\frac{dR}{d\rho} + \rho^2(k^2 - \beta^2)}_{\ell^2} = \underbrace{-\frac{1}{\Phi}\frac{d^2\Phi}{d\phi^2}}_{\ell^2} \tag{85}$$

We note that the left-hand side of (85) varies only with ρ, whereas the right-hand side varies only with ϕ. Since the two variables are independent, it must follow that each side of the equation must be equal to a constant. Calling this constant ℓ^2, as shown, we may write separate equations for each side; the variables are now separated:

$$\frac{d^2\Phi}{d\phi^2} + \ell^2\Phi = 0 \tag{86a}$$

$$\frac{d^2R}{d\rho^2} + \frac{1}{\rho}\frac{dR}{d\rho} + \left[k^2 - \beta^2 - \frac{\ell^2}{\rho^2}\right]R = 0 \tag{86b}$$

The solution of (86*a*) is of the form of the sine or cosine of ϕ:

$$\Phi(\phi) = \begin{cases} \cos\ (\ell\phi + \alpha) \\ \sin\ (\ell\phi + \alpha) \end{cases} \tag{87}$$

where α is a constant. The form of (87) dictates that ℓ must be an integer, since the same mode field must occur in the transverse plane as ϕ is changed by 2π radians. Since the fiber is round, the orientation of the x and y axes in the transverse plane is immaterial, so we may choose the cosine function and set $\alpha = 0$. We will thus use $\Phi(\phi) = \cos(\ell\phi)$.

The solution of (86*b*) to obtain the radial function is more complicated. Eq. (86*b*) is a form of Bessel's equation, whose solutions are Bessel functions of various forms. The key parameter is the function $\beta_t \equiv (k^2 - \beta^2)^{1/2}$, the square of which appears in (86b). Note that β_t will differ in the two regions: Within the core $(\rho < a)\beta_t = \beta_{t1} = (n_1^2 k_0^2 - \beta^2)^{1/2}$; within the cladding $(\rho > a)$, we have $\beta_t = \beta_{t2} = \left(n_2^2 R_0^2 - \beta^2\right)^{1/2}$. Depending on the relative magnitudes of k and β, β_t may be real or imaginary. These possibilities lead to two solution forms of (86b):

$$R(\rho) = \begin{cases} AJ_\ell(\beta_t \rho) & \beta_t \text{ real} \\ BK_\ell(|\beta_t|\rho) & \beta_t \text{ imaginary} \end{cases} \tag{88}$$

where A and B are constants. $J_\ell(\beta_t\rho)$ is the ordinary Bessel function of the first kind, of order ℓ and of argument $\beta_t\rho K_l(|\beta_t|\rho)$ is the modified Bessel function of the second kind, of order l, and having argument $|\beta_t|\rho$. The first two orders of each of these functions are illustrated in Figures 14.20a and b. In our study, it is necessary to know the precise zero crossings of the J_0 and J_1 functions. Those shown in Figure 14.20a are as follows: For J_0, the zeros are 2.405, 5.520, 8.654, 11.792, and 14.931. For J_1, the zeros are 0, 3.832, 7.016, 10.173, and 13.324. Other Bessel function types would contribute to the solutions in Eq. (88), but these exhibit nonphysical behavior with radius and are not included.

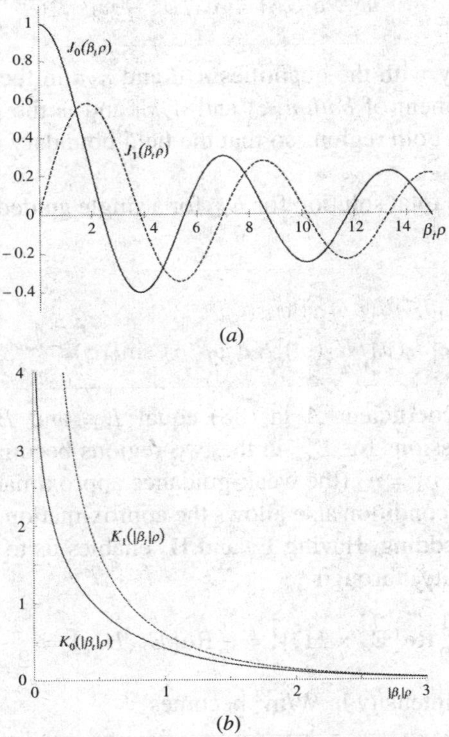

Fig. 14.20 Ordinary Bessel functions of the first kind, of orders 0 and 1, and of argument $\beta_t\rho$, where β_t is real. (b) Modified Bessel functions of the second kind, of orders 0 and 1, and of argument $|\beta_t|\rho$, where β_t is imaginary.

We next need to determine which of the two solutions is appropriate for each region. Within the core ($\rho < a$) we expect to get an oscillatory solution for the field—much in the same manner as we found in the slab waveguide. Therefore, we assign the ordinary Bessel function solutions to that region by requiring that $\beta_{t1} = (n_1^2 k_0^2 - \beta^2)^{1/2}$ is real. In the cladding ($\rho > a$), we expect surface waves that decrease in amplitude with increasing radius away from the core/cladding boundary. The Bessel Kfunctions provide this behavior and will apply if β_{t2} is imaginary. Requiring this, we may therefore write $|\beta_{t2}| = (\beta^2 - n_2^2 k_0^2)^{1/2}$. The diminishing field amplitude with increasing radius within the cladding allows us to neglect the effect of the outer cladding boundary (at $\rho = b$), as fields there are presumed too weak for this boundary to have any effect on the mode field.

Since β_{t1} and β_{t2} are in units of m^{-1}, it is convenient to normalize these quantities (while making them dimensionless) by multiplying both by the core radius, a. Our new normalized parameters become

$$u \equiv a\beta_{t1} = a\sqrt{n_1^2 k_0^2 - \beta^2} \tag{89a}$$

$$w \equiv a|\beta_{r2}| = a\sqrt{\beta^2 - n_2^2 k_0^2} \tag{89b}$$

u and w are in direct analogy with the quantities $\kappa_1 d$ and $\kappa_2 d$ in the slab waveguide. As in those parameters, β is the $z$component of *both* $n_1 k_0$ and $n_2 k_0$ and is the phase constant of the guided mode. β must be the same in both regions so that the field boundary conditions will be satisfied at $\rho = a$ for all z and t.

We may now construct the total solution for E_{xs} for a single guided mode, using (83) along with (87), (88), (89a), and (89b):

$$E_{xs} = \begin{cases} E_0 J_\ell(\beta_t(u\rho/a)\cos(\ell\phi)e^{-j\beta z} & \rho \leq a \\ E_0[J_\ell(u)/K_\ell(w)]K_\ell(w\rho/a)\sin(\ell\phi)e^{-j\beta z} & \rho \geq a \end{cases} \tag{90}$$

Note that we have let the coefficient A in (88) equal E_0, and $B = E_0[J_\ell(u)/K_\ell(w)]$. These choices assure that the expressions for E_{xs} in the two regions become equal at $\rho = a$, a condition approximately true as long as $n_1 \doteq n_2$ (the weak-guidance approximation).

Again, the weak-guidance condition also allows the approximation $H \doteq E/\eta$, with η taken as the intrinsic impedance of the cladding. Having $\mathbf{E}_s$ and $\mathbf{H}_s$ enables us to find the LP$_{lm}$ mode average power density (or light intensity) through

$$|\langle \mathbf{S} \rangle| \left|\frac{1}{2}\text{Re}\{\mathbf{E}_s \times \mathbf{H}_s^*\}\right| = \frac{1}{2}\text{Re}\{E_{xs}H_{ys}^*\} = \frac{1}{2\eta}|E_{xs}|^2 \tag{91}$$

Using (90) in (91), the mode intensity in W/m^2 becomes

$$I_{\ell m} = I_0 J_\ell^2\left(\frac{u\rho}{a}\right)\cos^2(\ell\phi) \qquad \rho \leq a \tag{92a}$$

$$I_{\ell m} = I_0 \left(\frac{J_\ell(u)}{K_\ell(w)} \right)^2 K_\ell^2 \left(\frac{w\rho}{a} \right) \cos^2(\ell\phi) \qquad \rho \geq a \tag{92b}$$

where I_0 is the peak intensity value. The role of the azimuthal mode number ℓ, as evident in (92a) and (92b), is to determine the number of intensity variations around the circle, $0 < \phi < 2\pi$; it also determines the order of the Bessel functions that are used. The influence of the radial mode number, m, is not immediately apparent in (92a) and (92b). Briefly stated, $m$determines the range of allowed values of u that occur in the Bessel function, $J(u\rho/a)$. The greater the value of m, the greater the allowed values of u; with larger u, the Bessel function goes through more oscillations over the range $0 <\rho< a$, and so more radial intensity variations occur with larger m. In the slab waveguide, the mode number (also m) determines the allowed ranges of κ_1. As we saw in Section 14.4, increasing κ_1 at a given frequency means that the slab ray propagates closer to the normal (smaller θ_1), and so more spatial oscillations of the field occur in the transverse direction (larger m).

The final step in the analysis is to obtain an equation from which values of mode parameters (u, w and β, for example) can be determined for a given operating frequency and fiber construction. In the slab waveguide, two equations, (78) and (79), were found using transverse resonance arguments, and these were associated with TE and TM waves in the slab. In our fiber, we do not apply transverse resonance directly, but rather *implicitly,* by requiring that all fields satisfy the boundary conditions at the core/cladding interface, $\rho = a$.[7] We have already applied conditions on the transverse fields to obtain Eq. (90). The remaining condition is continuity of the $z$components of **E** and **H**. In the weak-guidance approximation, we have neglected all $z$components, but we will consider them now for this last exercise. Using Faraday's law in point form, continuity of H_{zs} at $\rho = a$ is the same as the continuity of the z component of $\nabla\times \mathbf{E}_s$, provided that $\mu = \mu_0$ (or is the same value) in both regions. Specifically

$$(\nabla \times \mathbf{E}_{s1})_z|_{\rho=a} = (\nabla \times \mathbf{E}_{s2})_z|_{\rho=a} \tag{93}$$

The procedure begins by expressing the electric field in (90) in terms of ρ and ϕ components and then applying (93). This is a lengthy procedure and is left as an exercise (or may be found in Reference 4). The result is the eigenvalue equation for LP modes in the weakly guiding step index fiber:

$$\frac{J_{\ell-1}(u)}{J_\ell(u)} = -\frac{w}{u}\frac{K_{\ell-1}(w)}{K_\ell(w)} \tag{94}$$

This equation, like (78) and (79), is transcendental, and it must be solved for u and w numerically or graphically. This exercise in all of its aspects is beyond the scope of our treatment. Instead, we will obtain from (94) the conditions for cutoff for a given mode and some properties of the most important mode—that which has no cutoff, and which is therefore the mode that is present in single-mode fiber.

The solution of (94) is facilitated by noting that u and w can be combined to give a new parameter that is independent of β and depends only on the fiber construction and on the operating frequency.

[7] Recall that the equations for reflection coefficient (58) and (59), from which the phase shift on reflection used in transverse resonance is determined, originally came from the application of the field boundary conditions.

This new parameter, called the *normalized frequency*, or V number, is found using (89a) and (89b):

$$V \equiv \sqrt{u^2 + w^2} = ak_0\sqrt{n_1^2 - n_2^2} \tag{95}$$

We note that an increase in V is accomplished through an increase in core radius, frequency, or index difference.

The cutoff condition for a given mode can now be found from (94) in conjunction with (95). To do this, we note that cutoff in a dielectric guide means that total reflection at the core/cladding boundary just ceases, and power just begins to propagate radially, away from the core. The effect on the electric field of Eq. (90) is to produce a cladding field that no longer diminishes with increasing radius. This occurs in the modified Bessel function, $K(w\rho/a)$, when $w = 0$. This is our general cutoff condition, which we now apply to (94), whose right-hand side becomes zero when $w = 0$. This leads to cutoff values of u and V (u_c and V_c), and, by (95), $u_c = V_c$. Eq. (94) at cutoff now becomes

$$J_{\ell-1}(V_c) = 0 \tag{96}$$

Finding the cutoff condition for a given mode is now a matter of finding the appropriate zero of the relevant ordinary Bessel function, as determined by (96). This gives the value of V at cutoff for that mode.

For example, the lowest-order mode is the simplest in structure; therefore it has no variations in ϕ and one variation (one maximum) in ρ. The designation for this mode is therefore LP_{01}, and with $\ell = 0$, (96) gives the cutoff condition as $J_{-1}(V_c) = 0$.Since $J_{-1} = J_1$ (true only for the J_1 Bessel function), we take the first zero of J_1, which is $V_c(01) = 0$. The LP_{01} mode therefore has no cutoff and will propagate at the exclusion of all other modes provided V for the fiber is greater than zero but less than V_c for the next-higher-order mode. By inspecting Figure 14.20a, we see that the next Bessel function zero is 2.405 (for the J_0 function). Therefore, $\ell - 1 = 0$ in (95), and so $\ell = 1$ for the next-higher-order mode. Also, we use the lowest value of ml($m = 1$), and the mode is therefore identified as LP_{11}. Its cutoff V is V_c (11) = 2405. If $m = 2$ were to be chosen instead, we would obtain the cutoff V number for the LP_{12} mode. We use the next zero of the J_0 function, which is 5.520, or V_c(12) = 5520. In this way, the radial mode number, m, numbers the zeros of the Bessel function of order $\ell - 1$, taken in order of increasing value.

When we follow the reasoning just described, the condition for single-mode operation in a step index fiber is found to be

$$V < V_c(11) = 2.405 \tag{97}$$

Then, using (95) along with $k_0 = 2\pi/\lambda$, we find

$$\lambda > \lambda_c = \frac{2\pi a}{2.405}\sqrt{n_1^2 - n_2^2} \tag{98}$$

as the requirement on free-space wavelength to achieve single-mode operation in a step index fiber. The similarity to the single-mode condition in the slab waveguide [Eq. (82)] is apparent. The *cutoff wavelength,* λ_c, is that for the LP_{11} mode. Its value is quoted as a specification of most commercial single-mode fiber.

EXAMPLE 14.5

The cutoff wavelength of a step index fiber is quoted as $\lambda_c = 120\ \mu$m. If the fiber is operated at wavelength $\lambda = 155\ \mu$m, what is V?

Solution. Using (96) and (98), we find

$$V = 2.405\frac{\lambda_c}{\lambda} = 2.405\left(\frac{1.20}{1.55}\right) = 1.86$$

The intensity profiles of the first two modes can be found using (92a) and (92b), having determined u and w values for each mode from (94). For LP_{01} we find

$$I_{01} = \begin{cases} I_0 J_0^2(\mu_{01}\rho/a) & \rho \leq a \\ I_0\left(\frac{J_0(u_{01})}{K_0(w_{01})}\right)^2 K_0^2(w_{01}\rho/a) & \rho \geq a \end{cases} \tag{99}$$

and for LP_{11} we find

$$I_{11} = \begin{cases} I_0 J_1^2(\mu_{11}\rho/a)\cos^2\phi & \rho \leq a \\ I_0\left(\frac{J_1(u_{11})}{K_1(w_{11})}\right)^2 K_1^2(w_{11}\rho/a)\cos^2\phi & \rho \geq a \end{cases} \tag{100}$$

The two intensities for a single V value are plotted as functions of radius at $\phi = 0$ in Figure 14.21. We again note the lower confinement of the higher-order mode to the core, as was true in the slab waveguide.

As V increases (accomplished by increasing the frequency, for example), existing modes become more tightly confined to the core, while new modes of higher order may begin to propagate. The behavior of the lowest-order mode with changing V is depicted in Figure 14.22, where we again note that the mode becomes more tightly confined as V increases. In determining the intensities, Eq. (94) must in general be solved numerically to obtain u and w. Various analytic approximations to the exact numerical solution exist, the best of which is the Rudolf-Neumann formula for the LP_{01} mode, valid over the range $1.3 < V < 3.5$:

$$w_{01} = \doteq 1.1428\,V - 0.9960 \tag{101}$$

Having w_{01}, u_{01} can be found from (95), knowing V.

Another important simplification for the LP_{01} mode is the approximation of its intensity profile by a Gaussian function. An inspection of any of the intensity plots of Figure 14.22 shows a resemblence to a Gaussian, which would be expressed as

$$I_{01} \approx I_0 e^{-2\rho^2/\rho_0^2} \tag{102}$$

where ρ_0, termed the *mode field radius*, is defined as the radius from the fiber axis at which the mode intensity falls to $1/e^2$ times its on-axis value. This radius depends on frequency, and most generally on V. A similar approximation can be made for the fundamental symmetric slab guide

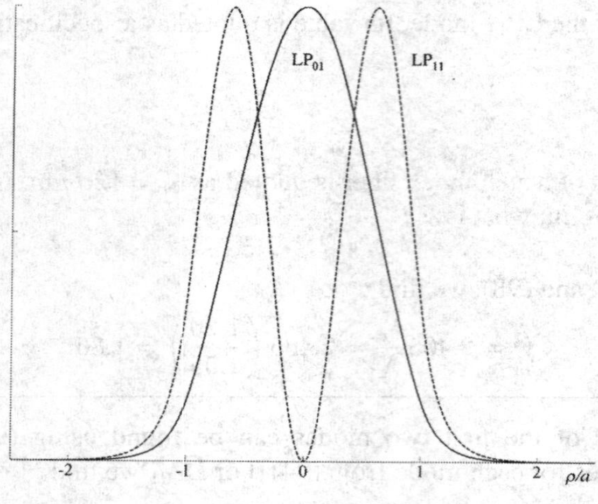

Fig. 14.21 Intensity plots from Eqs. (99) and (100) of the first two LP modes in a weakly guiding step index fiber as functions of normalized radius, ρ/a. Both functions were evaluated at the same operating frequency; the relatively weak confinement of the LP_{11} mode compared to that of LP_{01} is evident.

mode intensity. In step index fiber, the best fit between the Gaussian approximation and the actual mode intensity as given in (99) is given by the Marcuse formula:

$$\frac{\rho_0}{a} \approx 0.65 + \frac{1.619}{V^{3/2}} + \frac{2.879}{V^6} \tag{103}$$

The mode field radius (at a quoted wavelength) is another important specification (along with the cutoff wavelength) of commercial single-mode fiber. It is important to know for several reasons: First, in splicing or connecting two single-mode fibers together, the lowest connection loss will be attained if both fibers have the same mode field radius, and if the fiber axes are precisely aligned. Different radii or displaced axes result in increased loss, but this can be calculated and compared with measurement. Alignment tolerance (allowable deviation from precise axis alignment) is relaxed somewhat if the fibers have larger mode field radii. Second, a smaller mode field radius means that the fiber is less likely to suffer loss as a result of bending. A loosely confined mode tends to radiate away more as the fiber is bent. Finally, mode field radius is directly related to the mode phase constant, β, since if u and w are known (found from ρ_0), β can be found from (89a) or (89b). Therefore, knowledge of how β changes with frequency (leading to the quantification of dispersion) can be found by measuring the change in mode field radius with frequency. Again, References 4 and 5 (and references therein) provide more detail.

D14.9 For the fiber of Example 14.5, the core radius is given as $a = 50\ \mu$m. Find the mode field radius at wavelengths (a) 1.55 μm; (b) 1.30 μm.

Ans. 6.78 μm; 5.82 μm

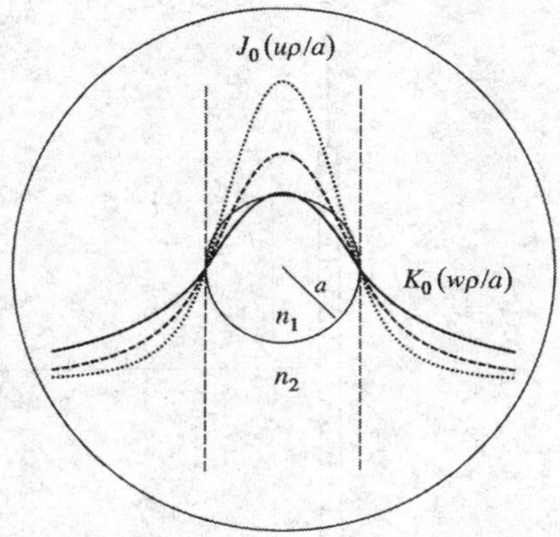

Fig. 14.22 Intensity plots for the LP_{01} mode in a weakly guiding step index fiber. Traces are shown for $V = 1.0$ (solid), $V = 1.2$ (dashed), and $V = 1.5$ (dotted), corresponding to increases in frequency in those proportions. Dashed vertical lines indicate the core/cladding boundary, at which for all three cases, the J_0 radial dependence in the core connects to the K_0 radial dependence in the cladding, as demonstrated in Eq. (99). The migration of mode power toward the fiber axis as frequency increases is evident.

14.6 BASIC ANTENNA PRINCIPLES

In this final section we explore some concepts on radiation of electromagnetic energy from a simple dipole antenna. A complete discussion of antennas and their applications would require several chapters or entire books. Our purpose is to produce a fundamental understanding of how electromagnetic fields radiate from *current distributions*. So for the first time, we shall have the specific field which results from a specific time-varying *source*. In the discussion of waves and fields in bulk media and in waveguides, only the wave motion in the medium was investigated, and the sources of the fields were not considered. The current distribution in a conductor was a similar problem, although we did at least relate the current to an assumed electric field intensity at the conductor surface. This might be considered as a source, but it is not a very practical one, for it is infinite in extent.

We now assume a current filament (of infinitesimally small cross section) as the source, positioned within an infinite lossless medium. The filament is taken as a differential length, but we shall be able to extend the results easily to a filament which is short compared to a wavelength, specifically less than about one-quarter of a wavelength overall. The differential filament is shown at the origin and is oriented along the z axis in Figure 14.23. The positive sense of the current is taken in the $\mathbf{a}_z$ direction. We assume a uniform current $I_0 \cos \omega t$ in this short length d and do not concern ourselves at present with the apparent discontinuity at each end.

We shall not attempt at this time to discover the "source of the source," but we shall merely assume that the current distribution cannot be changed by any field which it produces.

The first step is the application of the retarded vector magnetic potential expression, as presented in Section 10.5,

$$\mathbf{A} = \int \frac{\mu [I] d\mathbf{L}}{4\pi R}$$

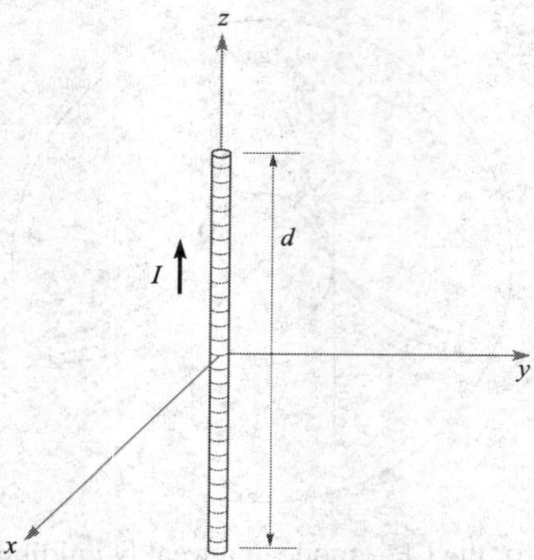

Fig. 14.23 A differential current filament of length d carries a current $I = I_0 \cos \omega t$.

where $[I]$ is a function of the retarded time $t - R/\nu$. When a single frequency is used to drive the antenna, ν is the phase velocity at that frequency. Since no integration is required for the very short filament assumed, we have

$$\mathbf{A} = \frac{\mu[I]d}{4\pi R}\mathbf{a}_z$$

Only the z component of $\mathbf{A}$ is present, for the current is only in the $\mathbf{a}_z$ direction. At any point P distant R from the origin, the vector potential is retarded by R/ν and

$$I = I_0 \cos \omega t$$

becomes

$$[I] = I_0 \cos\left[\omega\left(t - \frac{R}{\nu}\right)\right]$$

$$[I_s] = I_0\, e^{-j\omega R/\nu}$$

Thus

$$A_{zs} = \frac{\mu I_0 d}{4\pi R} e^{-j\omega R/\nu}$$

Using a mixed coordinate system for the moment, let us replace R by the small r of the spherical coordinate system and then determine which spherical components are represented by A_{zs}. Figure 14.24 helps to determine that

$$A_{rs} = A_{zs} \cos\theta$$

$$A_{\theta s} = -A_{zs} \sin\theta$$

and therefore

$$A_{rs} = \frac{\mu I_0 d}{4\pi r} \cos\theta e^{-j\omega r/\nu}$$

$$A_{\theta s} = -\frac{\mu I_0 d}{4\pi r} \sin\theta e^{-j\omega r/\nu}$$

From these two components of the vector magnetic potential at P we may find $\mathbf{B}_s$ or $\mathbf{H}_s$ from the definition of $\mathbf{A}_s$,

$$\mathbf{B}_s = \mu\mathbf{H}_s = \nabla\times\mathbf{A}_s$$

by merely taking the indicated partial derivatives. Thus

$$H_{\phi s} = \frac{1}{\mu r}\frac{\partial}{\partial r}(rA_{\theta s}) - \frac{1}{\mu r}\frac{\partial A_{rs}}{\partial\theta}$$

$$H_{rs} = H_{\theta s} = 0$$

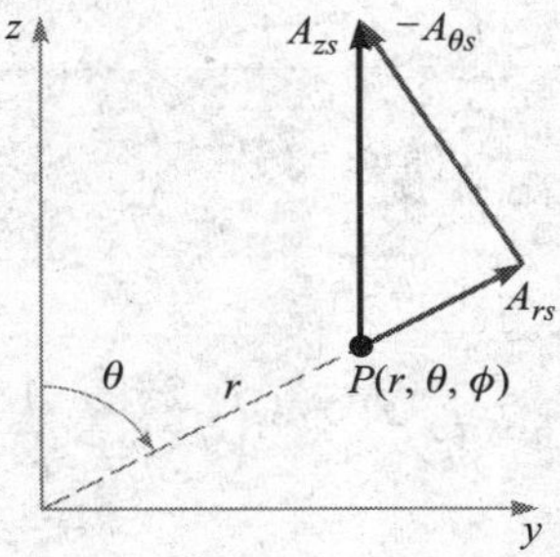

Fig. 14.24 The resolution of A_{zs} at $P(r,\theta,\phi)$ into the two spherical components A_{rs} and $A_{\theta s}$. The sketch is arbitrarily drawn in the $\phi = 90°$ plane.

and

$$H_{\phi s} = \frac{I_0 d}{4\pi}\sin\theta e^{-j\omega r/\nu}\left(j\frac{\omega}{\nu r} + \frac{1}{r^2}\right)$$

The components of the electric field which must be associated with this magnetic field are found from the point form of Ampére's circuital law as it applies to a region in which conduction and convection current are absent,

$$\nabla\times\mathbf{H} = \frac{\partial\mathbf{D}}{\partial t}$$

or in complex notation,

$$\nabla\times\mathbf{H}_s = j\omega\epsilon\mathbf{E}_s$$

Expansion of the curl in spherical coordinates leads to

$$E_{rs} = \frac{1}{j\omega\epsilon}\frac{1}{r\sin\theta}\frac{\partial}{\partial\theta}(H_{\phi s}\sin\theta)$$

$$E_{\theta s} = \frac{1}{j\omega\epsilon}\left(-\frac{1}{r}\right)\frac{\partial}{\partial\theta}(rH_{\phi s})$$

or

$$E_{rs} = \frac{I_0 d}{2\pi} \cos\theta e^{-j\omega r/\nu} \left(\frac{1}{\epsilon \nu r^2} + \frac{1}{j\omega\epsilon r^3} \right)$$

$$E_{\theta s} = \frac{I_0 d}{4\pi} \sin\theta e^{-j\omega r/\nu} \left(\frac{j\omega}{\epsilon \nu^2 r} + \frac{1}{\epsilon \nu r^2} + \frac{1}{j\omega\epsilon r^3} \right)$$

In order to simplify the interpretation of the terms enclosed in parentheses, we make the substitutions $\omega = 2\pi f$, $f\lambda = \nu$, $\nu = 1\sqrt{\mu\epsilon}$, and $\eta = \sqrt{\mu/\epsilon}$, producing

$$H_{\phi s} = \frac{I_0 d}{4\pi} \sin\theta e^{-j2\pi r/\lambda} \left(j\frac{2\pi}{\lambda r} + \frac{1}{r^2} \right) \tag{104}$$

$$E_{rs} = \frac{I_0 d\eta}{2\pi} \cos\theta e^{-j2\pi r/\lambda} \left(\frac{1}{r^2} + \frac{\lambda}{j2\pi r^3} \right) \tag{105}$$

$$E_{\theta s} = \frac{I_0 d\eta}{4\pi} \sin\theta e^{-j2\pi r/\lambda} \left(j\frac{2\pi}{\lambda r} + \frac{1}{r^2} + \frac{\lambda}{j2\pi r^3} \right) \tag{106}$$

These three equations are indicative of the reason that so many problems involving antennas are solved by experimental rather than theoretical methods. They have resulted from three general steps: an integration (atypically trivial) and two differentiations. These steps are sufficient to cause the simple current element and its simple current expression to "blow up" into the complicated field described by (104) to (106). In spite of this complexity, several interesting observations are possible.

We might notice first the $e^{-j2\pi r/\lambda}$ factor appearing with each component. This indicates propagation outward from the origin in the positive r direction with a phase factor $\beta = 2\pi/\lambda$; thus the wavelength is λ and the velocity $v = 1\sqrt{\mu\epsilon}$. We use the term *wavelength* now in a somewhat broader sense than the original definition, which identified the wavelength of a uniform plane wave with the distance between two points, measured in the direction of propagation, at which the wave has identical instantaneous values. Here there are additional complications caused by the terms enclosed in parentheses, which are complex functions of r. These variations must now be neglected in determining the wavelength. This is equivalent to a determination of the wavelength at a large distance from the origin, and we may demonstrate this by sketching the H_ϕ component as a function of r under the following conditions:

$$I_0 d = 4\pi \quad \theta = 90^\circ \quad t = 0 \quad f = 300 \text{ MHz}$$

$$\nu = 3 \times 10^8 \text{ m/s (free space)} \quad \lambda = 1 \text{ m}$$

Therefore

$$H_{\phi s} = \left(j\frac{2\pi}{r} + \frac{1}{r^2} \right) e^{-j2\pi r}$$

and the real part may be determined at $t = 0$,

$$H_\phi = \sqrt{\left(\frac{2\pi}{r}\right)^2 + \frac{1}{r^4}}\cos[(\tan^{-1} 2\pi r) - 2\pi r]$$

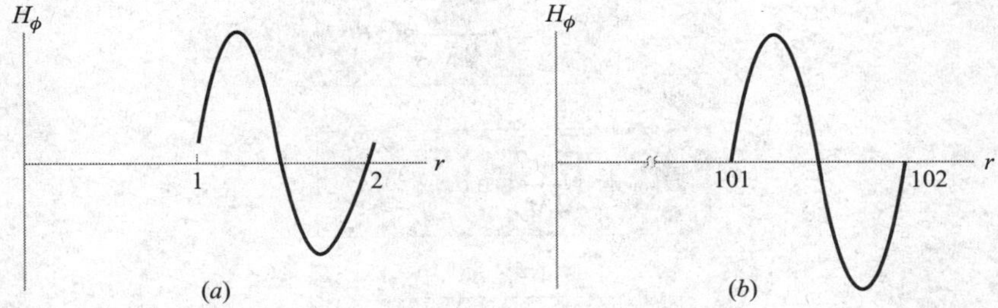

Fig. 14.25 The instantaneous amplitude of H_ϕ for the special case of a current element having $I_0 d = 4\pi$ and $\lambda = 1$ is plotted at $\theta = 90°$ and $t = 0$ (a) in the region $1 \le r \le 2$ near the antenna, and (b) in the region $101 \le r \le 102$ distant from the antenna. The left curve is noticeably nonsinusoidal, since if it were a sinusoid, its end points should reach the r axis precisely at 1 and 2.

Knowing that cos $(a - b)$ = cos a cos b + sin a sin b and that cos $(\tan_x^{-1}) = 1/\sqrt{1+x^2}$, we may simplify this result to

$$H_\phi = \frac{1}{r^2}(\cos 2\pi r + 2\pi r \sin 2\pi r)$$

Values obtained from this last equation are plotted against r in the range $1 \le r \le 2$ in Figure 14.25a; the curve is noticeably nonsinusoidal. At $r = 1$, $H_\phi = 1$, while at $r = 2$, one wavelength greater, $H_\phi = 0.25$. Moreover, the curve crosses the axis (with positive slope) at $r = 1 - 0.0258$ and $r = 2 - 0.0127$, again a distance not equal to a wavelength. If a similar sketch is made in the range $101 \le r \le 102$, shown in Figure 14.25b on a different amplitude scale, an essentially sinusoidal wave is obtained, and the instantaneous values of H_ϕ at $r = 101$ and $r = 102$ are 0.000 099 8 and 0.000 099 6. The maximum amplitudes of the positive and negative portions of the waveform differ by less than 1 percent, and we may say that for all practical purposes the wave in this region is a uniform plane wave having a sinusoidal variation with distance (and time, of course) and a well-defined wavelength. This wave evidently carries energy away from the differential antenna, and we shall calculate this power shortly.

Continuing the investigation of (104) to (106), let us now take a more careful look at the expressions containing terms varying as $1/r^3$, $1/r^2$, and $1/r$. At points very close to the current element the $1/r^3$ term must be dominant. In the numerical example we have used, the relative values of the terms in $1/r^3$, $1/r^2$, and $1/r$ in the $E_{\theta s}$ expression are about 250, 16, and 1, respectively, when r is 1 cm. The variation of an electric field as $1/r^3$ should remind us of the *electrostatic* field of the dipole (Chapter 4). This term represents energy stored in a reactive (capacitive) field, and it does not contribute to the radiated power. The inverse-square term in the $H_{\phi s}$ expression is similarly important only in the region very near to the current element and corresponds to the *induction* field of the dc element given by the Biot-Savart law. At distances corresponding to 10 or more wavelengths from the oscillating current element, all terms except the inverse-distance $(1/r)$

term may be neglected, and the *distant* or *radiation* fields become

$$E_{rs} = 0$$

Illustrations

$$E_{\theta s} = j\frac{I_0 d\eta}{2\lambda r}\sin\theta e^{-j2\pi r/\lambda} \tag{107}$$

$$H_{\phi s} = j\frac{I_0 d}{2\lambda r}\sin\theta e^{-j2\pi r/\lambda} \tag{108}$$

or

$$E_{\theta s} = \eta H_{\phi s}$$

The relationship between $E_{\theta s}$ and $H_{\phi s}$ is thus seen to be that between the electric and magnetic fields of the uniform plane wave, thus substantiating the conclusion we reached when investigating the wavelength.

The variation of both radiation fields with the polar angle θ is the same; the fields are maximum in the equatorial plane (x, y plane) of the current element and vanish off the ends of the element. The variation with angle may be shown by plotting a *vertical pattern* (assuming a vertical orientation of the current element) in which the relative magnitude of $E_{\theta s}$ is plotted against θ for a constant r. The pattern is usually shown on polar coordinates, as in Figure 14.26. A *horizontal pattern* may also be plotted for more complicated antenna systems, and it shows the variation of field intensity with ϕ. The horizontal pattern of the current element is a circle centered at the origin since the field is not a function of the azimuth angle.

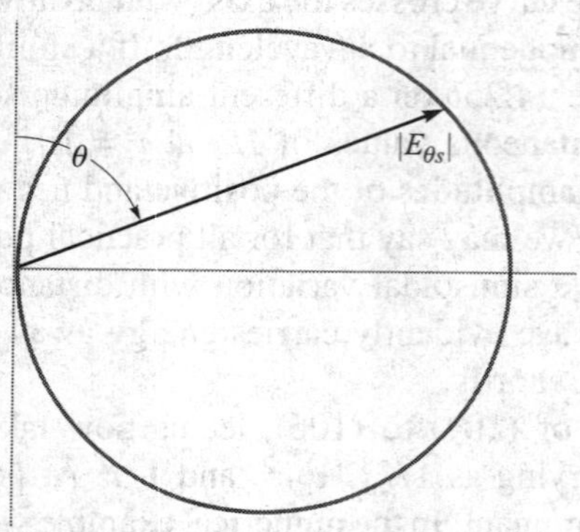

Fig. 14.26 The polar plot of the vertical pattern of a vertical current element. The crest amplitude of $E_{\theta s}$ is plotted as a function of the polar angle θ at a constant distance r. The locus is a circle.

In order to obtain a quantitative expression for the power radiated, we need to apply the Poynting vector $\mathbf{S} = \mathbf{E} \times \mathbf{H}$ developed in Section 12.3. The instantaneous expressions for the radiation components of the electric and magnetic field intensities are

$$\mathcal{E}_\theta = \eta H_\phi$$

$$\mathcal{H}_\phi = -\frac{I_0 d}{2\lambda r}\sin\theta \sin\left(\omega t - \frac{2\pi r}{\lambda}\right)$$

and thus

$$\mathcal{S}_r = \mathcal{E}_\theta \mathcal{H}_\phi = \left(\frac{I_0 d}{2\lambda r}\right)^2 \eta \sin^2\theta \sin^2\left(\omega t - \frac{2\pi r}{\lambda}\right)$$

The total (in space) instantaneous (in time) power crossing the surface of a sphere of radius r_0 is then

$$\mathcal{P} = \int_{\phi=0}^{2\pi}\int_{\theta=0}^{\pi} \mathcal{S}_r r_0^2 \sin\theta d\theta d\phi$$
$$= \left(\frac{I_0 d}{\lambda}\right)^2 \eta \frac{2\pi}{3}\sin^2\left(\omega t - \frac{2\pi r_0}{\lambda}\right)$$

and the time-average power is given by one-half the maximum amplitude,

$$P_{\text{av}} = \left(\frac{I_0 d}{\lambda}\right)^2 \eta \frac{\pi}{3} = 40\pi^2 \left(\frac{I_0 d}{\lambda}\right)^2$$

where $\eta = 120\pi\Omega$ in free space.

This is the same power as that which would be dissipated in a resistance R_{rad} by the current I_0 in the absence of any radiation, where

$$P_{\text{av}} = \frac{1}{2} I_0^2 R_{\text{rad}}$$

$$R_{\text{rad}} = \frac{2P_{\text{av}}}{I_0^2} = 80\pi^2 \left(\frac{d}{\lambda}\right)^2 \tag{109}$$

If we assume the differential length is 0.01λ, then R_{rad} is about 0.08 Ω. This small resistance is probably comparable to the *ohmic* resistance of a practical antenna, and thus the efficiency of the antenna might be unsatisfactorily low. Effective matching to the source also becomes very difficult to achieve, for the input reactance of a short antenna is much greater in magnitude than the input resistance R_{rad}. This is the basis for the statement that an effective antenna should be an appreciable fraction of a wavelength long.

The actual current distribution on a thin linear antenna is very nearly sinusoidal, even for antennas that may be several wavelengths long. Note that if the conductors of an open-circuited two-wire transmission line are folded back 90°, the standing-wave distribution on the line is sinusoidal. The current is zero at each end and maximizes one-quarter wavelength from each end, and the current continues to vary in this manner toward the center. The current at the center, therefore, will be very small for an antenna whose length is an integral number of wavelengths, but it will be equal to the maximum found at any point on the antenna if the antenna length is $\lambda/2$, $3\lambda/2$, $5\lambda/2$, and so forth.

On a short antenna, then, we see only the first portion of the sine wave; the amplitude of the current is zero at each end and increases approximately in a linear manner to a maximum value of I_0 at the center. This is suggested by Figure 14.27. Note that this antenna has identical currents in the two halves and it may be fed conveniently by a two-wire line, where the currents in the two conductors are equal in amplitude but opposite in direction. The gap at the feed point is small and has negligible effects. A symmetrical antenna of this type is called a *dipole*. The linear current variation with distance is a reasonable assumption for antennas having an overall length less than about one-quarter wavelength.

Interactives

It is possible to extend the analysis of the differential current element to the short dipole if we assume that the length is short enough that retardation effects may be neglected. That is, we consider that signals arriving at any field point P from the two ends of the antenna are in phase. The average current along the antenna is $I_0/2$, where I_0 is the input current at the center terminals. Thus, the electric and magnetic field intensities will be one-half the values given in (107) and (108), and there are no changes in the vertical and horizontal patterns. The power will be one-quarter of its previous value, and thus the radiation resistance will also be one-quarter of the value given by (109).

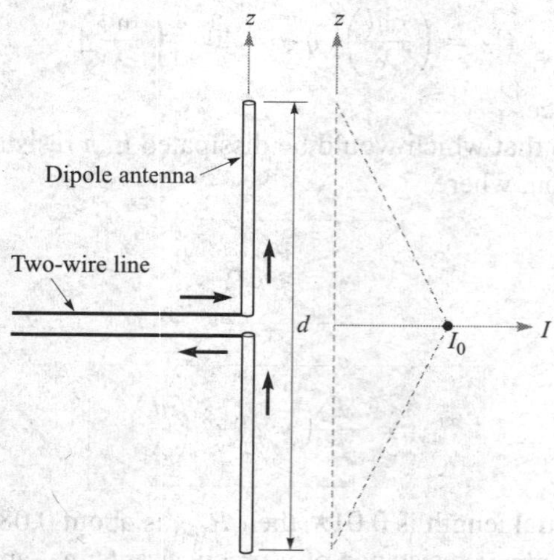

Fig. 14.27 A short antenna ($d < \lambda/4$) has a linear current distribution and may be driven by a two-wire line.

If we try to improve our results by assuming a sinusoidal variation of current amplitude with distance along the antenna, and if the effects of retardation are included in the analysis, then the integrations required to find **A** and the power radiated become enormously more difficult. Since we want to hurry along toward the last page, let us merely note that for the world's most popular antenna, the half-wave dipole ($d = \lambda/2$), the following results are eventually obtained:

$$E_{\theta s} = \frac{I_0 \eta}{2\pi r} \frac{\cos\left(\frac{\pi}{2}\cos\theta\right)}{\sin\theta} \tag{110}$$

$$H_{\phi s} = \frac{E_{\theta s}}{\eta} \tag{111}$$

$$R_{\text{rad}} = 30\left[\frac{(2\pi)^2}{2\cdot 2!} - \frac{(2\pi)^4}{4\cdot 4!} + \frac{(2\pi)^6}{6\cdot 6!} - \frac{(2\pi)^8}{8\cdot 8!} + \cdots\right] = 73.1\ \Omega \tag{112}$$

Let us compare this accurate value with results obtained by more approximate means. Suppose we first attempt to find the radiation resistance by assuming a uniform current distribution and neglecting the effects of retardation. The result is obtained from (109) with $d/\lambda = 1/2$; $R_{\text{rad}} = 20\pi^2 = 197.4\ \Omega$. This is much greater than 73.1 Ω, but we have also assumed a much greater current on the antenna than is actually present.

The result may be improved by considering a linear current distribution while still ignoring retardation. The average current is half the maximum value, the power is one-quarter, and the radiation resistance drops to $5\pi^2$ or 49.3 Ω. Now the result is too small, primarily because the average value of a triangular wave is less than the average value of a sine wave.

Finally, if we assume a sinusoidal current distribution, we have an average value of $2/\pi$ times the maximum, and the radiation resistance comes to $(2/\pi)^2(20\pi^2)$, or 80 Ω. This is reasonably close to the true value, and the discrepancy lies in neglecting retardation. In a linear antenna, the effect of retardation is always one of cancellation, and therefore its consideration must always lead to smaller values of radiation resistance. This decrease is of relatively small magnitude here (from 80 to 73.1 Ω) because the current elements tending to cancel each other are those at the ends of the dipole, and these are of small amplitude; moreover, the cancellation is greatest in a direction along the antenna axis where all radiation fields are zero for a linear antenna.

Familiar antennas that fall into the dipole classification are the elements used in the common TV and FM receiving antennas.

As a final example of a practical antenna, let us collect a few facts about the *monopole* antenna. This is one-half of a dipole plus a perfectly conducting plane, as shown in Figure 14.28*a*. The image principle discussed in Section 5.5 provides the image shown in Figure 14.28*b* and assures us that the fields above the plane are the same for the monopole and the dipole. Hence, the expressions of (107) and (108) are equally valid for the monopole. The Poynting vector is therefore also the same above the plane, but the integration to find the total power radiated is extended through but one-half the volume. Thus the power radiated and the radiation resistance for the monopole are half the corresponding values for the dipole. As an example, a monopole with an assumed uniform current distribution has $R_{\text{rad}} = 40\pi^2(d/\lambda)^2$; a triangular current leads to $R_{\text{rad}} = 10\pi^2(d/\lambda)^2$; and the sinusoidal current distribution of a $\lambda/4$ monopole leads to $R_{\text{rad}} = 36.5\ \Omega$.

Monopole antennas may be driven by a coaxial cable below the plane, having its center conductor connected to the antenna through a small hole, and having its outer conductor connected to the plane. If the region below the plane is inaccessible or inconvenient, the coax may be laid on top of the plane and its outer conductor connected to it. Examples of this type of antenna include AM broadcasting towers and CB antennas.

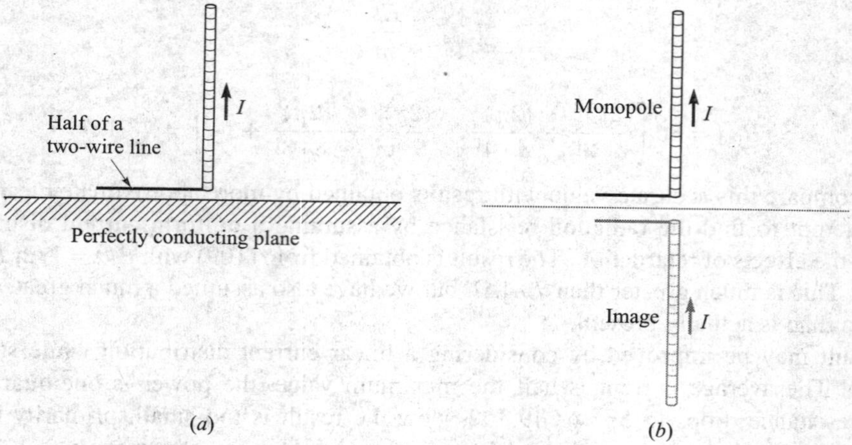

Fig. 14.28 (*a*) An ideal monopole is always associated with a perfectly conducting plane. (*b*) The monopole plus its image form a dipole.

D14.10 Calculate values for the curve shown in Figure 14.29*a* at r = 1, 1.2, 1.4, 1.6, 1.8, and 2.

Ans. 1.00; 5.19; 2.23; −2.62; −3.22; 0.25

D14.11 A short antenna with a uniform current distribution in air has $Id = 3 \times 10^{-4}$ A·m and $\lambda = 10$ cm. Find $|E_{\theta s}|$ at $\theta = 90°$, $\phi = 0°$, and r =: (*a*) 2 cm; (*b*) 20 cm; (*c*) 200 cm.

Ans. 25 V/m; 2.8 V/m; 0.28 V/m

D14.12 The monopole antenna of Figure 14.28*a* has a length $d/2 = 0.080$ m and may be assumed to carry a triangular current distribution for which the feed current I_0 is 16.0 A at a frequency of 375 MHz in free space. At point $P(r = 400 \text{ m}, \theta = 60°, \phi = 45°)$ find: (*a*) $H_{\phi s}$; (*b*) $E_{\theta s}$; (*c*) the amplitude of $\mathcal{P}_r$.

Ans. j 17 mA/m; j0.65 V/m; 1.1 mW/m^2

References

1. Weeks, W. L. *Transmission and Distribution of Electrical Energy*. New York: Harper and Row, 1981. Line parameters for various configurations of power transmission and distribution systems are discussed in Chapter 2, along with typical parameter values.
2. Edwards, T. C. *Foundations for Microstrip Circuit Design*. Chichester, N.Y.: Wiley-Interscience, 1981. Chapters 3 and 4 provide an excellent treatment of microstrip lines, with many design formulas.

3. Marcuse, D. *Theory of Dielectric Optical Waveguides*. 2d ed. New York: Academic Press, 1990. This book provides a very general and complete discussion of dielectric slab waveguides, plus other types.

4. Buck, J. A. *Fundamentals of Optical Fibers*. 2d ed. New York: Wiley-Interscience, 2004. Symmetric slab dielectric guides and weakly guiding fibers are emphasized in this book by one of the coauthors.

5. Weeks, W. L. *Antenna Engineering*. New York: McGraw-Hill, 1968. This excellent text probably contains more information about antennas than you want to know.

6. Jordan, E. C., and K. G. Balmain. *Electromagnetic Waves and Radiating Systems*. 2d ed. Englewood Cliffs, N.J.: Prentice-Hall, 1968. This classic text provides an excellent treatment of waveguides and antennas.

CHAPTER 14 Problems

Quizzes

14.1 The dimensions of the outer conductor of a coaxial cable are b and c, $c > b$. Assume $\sigma = \sigma_c$ and let $\mu = \mu_0$. Find the magnetic energy stored per unit length in the region $b < r < c$ for a uniformly distributed total current I flowing in the opposite directions in the inner and outer conductors.

14.2 The conductors of a coaxial transmission line are copper ($\sigma_c = 58 \times 10^7$ S/m), and the dielectric is polyethylene ($\epsilon'_r = 2.26$, $\sigma/\omega\epsilon' = 0.0002$). If the inner radius of the outer conductor is 4 mm, find the radius of the inner conductor so that: (*a*) $Z_0 = 50\ \Omega$; (*b*) $C = 100$ pF/m; (*c*) $L = 0.2\mu$H/m. A lossless line can be assumed.

14.3 Two aluminum-clad steel conductors are used to construct a two-wire transmission line. Let $\sigma_{\text{Al}} = 38 \times 10^7$ S/m, $\sigma_{\text{St}} = 5 \times 10^6$ S/m, and $\mu_{\text{St}} = 100\ \mu$H/m. The radius of the steel wire is 0.5 in., and the aluminum coating is 0.05 in. thick. The dielectric is air, and the center-to-center wire separation is 4 in. Find C, L, G, and R for the line at 10 MHz.

14.4 Each conductor of a two-wire transmission line has a radius of 0.5 mm; their center-to-center separation is 0.8 cm. Let $f = 150$ MHz, and assume σ and σ_c are zero. Find the dielectric constant of the insulating medium if: (*a*) $Z_0 = 300\Omega$; (*b*) $C = 20$ pF/m; (*c*) $\nu_p = 26 \times 10^8$ m/s.

14.5 Pertinent dimensions for the transmission line shown in Figure 14.2 are $b = 3$ mm and $d = 0.2$ mm. The conductors and the dielectric are nonmagnetic. (*a*) If the characteristic impedance of the line is 15 Ω, find ϵ'_r. Assume a low-loss dielectric. (*b*) Assume copper conductors and operation at 2×10^8 rad/s. If $RC = GL$, determine the loss tangent of the dielectric.

14.6 A transmission line constructed from perfect conductors and an air dielectric is to have a maximum dimension of 8 mm for its cross section. The line is to be used at high frequencies. Specify the dimensions if it is: (*a*) a two-wire line with $Z_0 = 300\ \Omega$; (*b*) a planar line with $Z_0 = 15\ \Omega$; (*c*)a 72 Ω coax having a zero-thickness outer conductor.

14.7 A microstrip line is to be constructed using a lossless dielectric for which $\epsilon'_r = 7.0$. If the line is to have a 50 Ω characteristic impedance, determine: (*a*) ϵ_{reff}; (*b*) w/d.

14.8 Two microstrip lines are fabricated end-to-end on a 2 mm thick wafer of lithium niobate ($\epsilon'_r = 4.8$). Line 1 is of 4 mm width; line 2 (unfortunately) has been fabricated with a 5 mm width. Determine the power loss in dB for waves transmitted through the junction.

14.9 A parallel-plate waveguide is known to have a cutoff wavelength for the $m = 1$ TE and TM modes of $\lambda_{c1} = 41$ mm. The guide is operated at wavelength $\lambda = 1.0$ mm. How many modes propagate?

14.10 A parallel-plate guide is to be constructed for operation in the TEM mode only over the frequency range $0 < f < 3$ GHz. The dielectric between plates is to be teflon ($\epsilon'_r = 2.1$). Determine the maximum allowable plate separation, d.

14.11 A parallel-plate guide is partially filled with two lossless dielectrics (Figure 14.29) where $\epsilon'_{r1} = 40$, $\epsilon'_{r2} = 2.1$, and $d = 1$ cm. At a certain frequency, it is found that the TM_1 mode propagates through the guide without suffering any reflective loss at the dielectric interface. (*a*) Find this frequency. (*b*) Is the guide operating at a single TM mode at the frequency found in part (*a*)? *Hint*: Remember Brewster's angle?

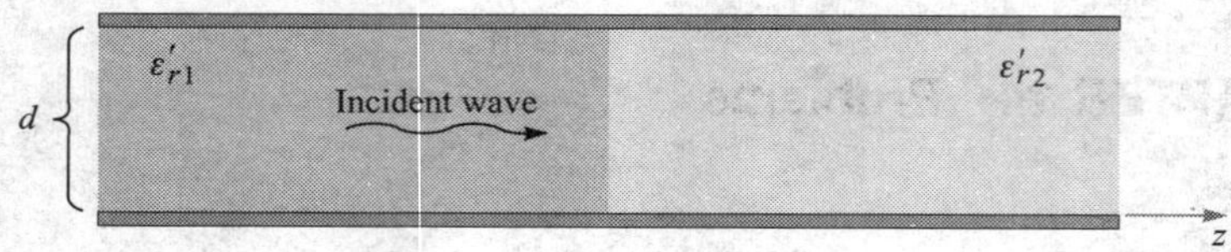

Fig. 14.29 See Problems 14.11 and 14.12.

14.12 In the guide of Figure 14.29, it is found that $m = 1$ modes propagating from left to right totally reflect at the interface, so that no power is transmitted into the region of dielectric constant ϵ'_{r2}. (*a*) Determine the range of frequencies over which this will occur. (*b*) Does your part (*a*) answer in any way relate to the cutoff frequency for m= 1 modes in either region? *Hint*: Remember the critical angle?

14.13 A symmetric dielectric slab waveguide has a slab thickness $d = 10\ \mu$m, with $n_1 = 1.48$ and $n_2 = 1.45$. If the operating wavelength is $\lambda = 1.3\ \mu$m, what modes will propagate?

14.14 A symmetric slab waveguide is known to support only a single pair of TE and TM modes at wavelength $\lambda = 1.55\ \mu$m. If the slab thickness is 5 μm, what is the maximum value of n_1 if $n_2 = 3.30$?

14.15 In a symmetric slab waveguide, $n_1 = 1.45$, $n_2 = 1.50$, and $d = 10\ \mu$m. (*a*) What is the phase velocity of the $m = 1$ TE or TM mode at cutoff? (*b*) How will your part (*a*) result change for higher-order modes (if at all)?

14.16 An *asymmetric* slab waveguide is shown in Figure 14.30. In this case, the regions above and below the slab have unequal refractive indices, where $n_1 > n_3 > n_2$. (*a*) Write, in terms of the appropriate indices, an expression for the minimum possible wave angle, θ_1, that a guided mode may have. (*b*) Write an expression for the maximum phase velocity a guided mode may have in this structure, using given or known parameters.

14.17 A step index optical fiber is known to be single mode at wavelengths $\lambda > 12\ \mu$m. Another fiber is to be fabricated from the same materials, but it is to be single mode at wavelengths $\lambda > 0.63\ \mu$m. By what percentage must the core radius of the new fiber differ from the old one, and should it be larger or smaller?

14.18 Is the mode field radius greater than or less than the fiber core radius in single-mode step index fiber?

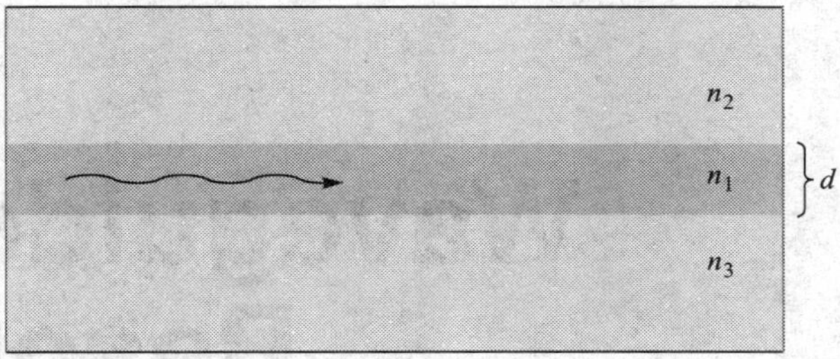

Fig. 14.30 See Problem 14.16.

14.19 The mode field radius of a step index fiber is measured as 4.5 μm at free-space wavelength λ = 130 μm. If the cutoff wavelength is specified as λ_c = 1.20 μm, find the expected mode field radius at λ = 155 μm.

14.20 A short dipole carrying current $I_0 \cos \omega t$ in the $\mathbf{a}_z$ direction is located at the origin in free space. (*a*) If β = 1 rad/m, r = 2 m, θ = 45°, ϕ = 0, and t = 0, give a unit vector in rectangular components that shows the instantaneous direction of **E**. (*b*) What fraction of the total average power is radiated in the belt $80° < \theta < 100°$?

14.21 Prepare a curve, r versus θ in polar coordinates, showing the locus in the ϕ = 0 plane where: (*a*) the radiation field $|E_{\theta s}|$ is one-half of its value at $r = 10_4$ m, $\theta = \pi/2$; (*b*) the average radiated power density (S_r) is one-half of its value at $r = 10_4$ m, $\theta = \pi/2$.

14.22 Two short antennas at the origin in free space carry identical currents of 5 cos $\omega t A$, one in the $\mathbf{a}_z$ direction, one in the $\mathbf{a}_y$ direction. Let $\lambda = 2\pi$ m and d = 0.1 m. Find $\mathbf{E}_s$ at the distant point: (*a*) (x= 0, y= 1000, z= 0); (*b*) (0, 0, 1000); (*c*) (1000, 0, 0). (*d*) Find **E** at (1000, 0, 0) at t= 0. (e) Find $|\mathbf{E}|$ at (1000, 0, 0) at t= 0.

14.23 A short current element has $d = 0.03\lambda$. Calculate the radiation resistance for each of the following current distributions: (*a*) uniform, I_0; (*b*) linear, $I(z) = I_0(0.5d - |z|)/0.5d$; (*c*) step, I_0 for $0 < |z| < 0.25d$ and $0.5I_0$ for $0.25\,d < |z| < 0.5d$.

14.24 A dipole antenna in free space has a linear current distribution. If the length d is 0.02λ, what value of I_0 is required to: (*a*) provide a radiation-field amplitude of 100 mV/m at a distance of 1 mi. at $\theta = 90°$; (*b*) radiate a total power of 1 W?

14.25 A monopole antenna in free space, extending vertically over a perfectly conducting plane, has a linear current distribution. If the length of the antenna is 0.01λ, what value of I_0 is required to: (*a*) provide a radiation-field amplitude of 100 mV/m at a distance of 1 mi. at $\theta = 90°$; (*b*) radiate a total power of 1 W?

14.26 The radiation field of a certain short vertical current element is $E_{\theta s} = (20/r)\sin\theta e^{-j10\pi r}$ V/m if it is located at the origin in free space. (*a*) Find $E_{\theta s}$ at $P(r = 100, \theta = 90°, \phi = 30°)$. (*b*) Find $E_{\theta s}$ at $P(100, 90°, 30°)$ if the vertical element is located at $A(0.1, 90°, 90°)$. (*c*) Find $E_{\theta s}$ at $P(100, 90°, 30°)$ if identical vertical elements are located at $A(0.1, 90°90°)$ and $B(0.1, 90°, 270°)$.

15

Waveguides and Resonators

The waveguide is a quite general type of structure used to guide and confine the electromagnetic waves for transmission from one point to another. If we go by this definition, then even the transmission lines discussed in earlier chapters can be classified under the category of waveguides. Hence, the question arises that in what way the waveguides are different from transmission lines, and why they should be studied separately. The answer is very simple. Traditionally, the transmission lines represent special types of structures employing two conductors, which allow the propagation of pure TEM waves along these lines under certain conditions as discussed in earlier chapters. The waveguides on the other hand can use either two conductors or a single conductor for the wave transmission and confinement, and can support the TE and TM mode propagation apart from the TEM mode propagation. It may be mentioned here that even the two-conductor lines can support higher order TE and TM modes if some parameters such as the frequency of operation and the spacing between the conductors are not within the specified limits. Hence, the term *waveguide* for the analysis can be applied under two most general scenarios. The first scenario is that the guiding structure consists of two conductors, but the analysis should be carried out for higher order TE and TM field configurations. The second typical situation is that of a guiding structure consisting of a single conductor, such as a hollow metallic pipe of a uniform cross section supporting any TE or TM mode. In this chapter, we limit ourselves to the study of metallic waveguides having rectangular cross section due to their practical importance. However, the overall formulation provided in the text can be generalized for any other type of guiding structure having a uniform cross section.

The parallel plate waveguide was already discussed in the previous chapter using the simple plane-wave model, where various parameters such as the mode number, the cutoff frequency, etc., were briefly described. In this chapter, we use the wave equation approach to analyze the parallel plate waveguide, and accordingly field expressions for various modes supported by the parallel plate waveguide are derived. In the next step, the rectangular waveguide is analyzed using the generalized field theory approach, and the field expressions for various TE and TM modes starting from Maxwell's equations are obtained. The field configurations for some basic modes are plotted, and a number of parameters describing fields of waveguide structures are defined. Lastly, we present a brief discussion about resonators in the microwave frequency range, where the rectangular cavity resonator is described. The major practical applications of waveguides and cavity resonators are briefly provided at the end.

15.1 PARALLEL-PLATE WAVEGUIDE

We start our discussion in this section with the configuration of the parallel plate waveguide shown in Figure 15.1. This structure can also be termed as the parallel-plate transmission line, which was briefly discussed in Chapter 14. However, in this chapter, it is described as the parallel plate waveguide to stress the fact that we are mainly interested here in deriving the expressions for the higher order TE and TM field configurations supported by this structure rather than the basic TEM mode usually considered for the analysis of transmission lines.

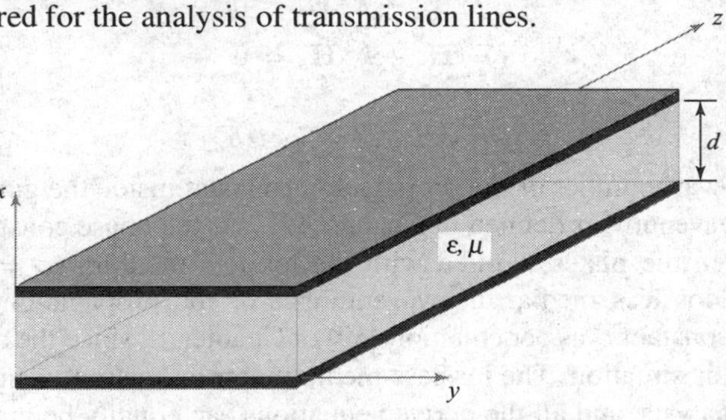

Fig. 15.1 The parallel-plate waveguide geometry.

The geometry shown in Figure 15.1 is made of two perfectly conducting plates separated by a distance d, and the medium between two plates is filled with a lossless dielectric having permittivity ϵ, and permeability μ. The width of each plate along the y-direction is assumed to be very large as compared to the plate separation d. This assumption leads to the fact that there is no field variation along the y-direction thereby ignoring the fringing fields at the edges of the plates. The wave is assumed to be propagating along the z direction guided by the lower and upper plates.

The parallel plate waveguide shown in Figure 15.1 was already discussed in the last chapter using the simple plane wave model. In this chapter, our focus is to use the wave equation approach to analyze this geometry, which would facilitate the determination of various possible electric and magnetic field configurations, also known as *modes*, for this structure. The wave equation approach requires the solution of Maxwell's equations introduced in Chapter 10, subject to the boundary conditions discussed in chapters 5 and 9 applied to both conducting plates of Figure 15.1. The analysis here is carried out by assuming the phasor forms of the field quantities, which is actually valid for the time harmonic electromagnetic fields as discussed in earlier chapters.

The Maxwell's equations in the phasor form for the region between two conducting plates can be written as

$$\nabla \times \mathbf{H}_s = j\omega\epsilon\,\mathbf{E}_s \tag{1a}$$

$$\nabla \times \mathbf{E}_s = -j\omega\mu\mathbf{H}_s \tag{1b}$$

$$\nabla \cdot \mathbf{E}_s = 0 \tag{1c}$$

$$\nabla \cdot \mathbf{B}_s = 0 \tag{1d}$$

where it is assumed that the medium between the two plates is source free ($\rho_v = \mathbf{J} = 0$), and simple (homogeneous, isotropic and linear). It is to be noted here that the assumption of source free medium is valid as long as the region of interest is far away from all the sources, which are responsible for generating these fields. This kind of situation occurs quite often in solving the real electromagnetic field problems as one is more interested to know about the propagation effects of these fields rather than knowing the details about how these field are generated. Meanwhile, equations (1*a*) to (1*d*) can be combined using the procedure explained in Chapter 12 to obtain the following vector wave equations for the electric and magnetic fields

$$\nabla^2\mathbf{E}_s + k^2\,\mathbf{E}_s = 0 \tag{2a}$$

$$\nabla^2\mathbf{H}_s + k^2\,\mathbf{H}_s = 0 \tag{2b}$$

where,

$$k \equiv \beta = \omega\sqrt{\mu\epsilon} = k_0\,\sqrt{\mu_r\,\epsilon_r} \tag{3}$$

with k being the total wavenumber or the propagation constant inside the guided medium, $k_0 = \frac{\omega}{c}$ being the free space wavenumber defined in Chapter 12, β is the phase constant, and it is assumed that the region between the plates is filled with the lossless medium ($\alpha = 0$). It is to be noted from (3) that for the lossless media, the wavenumber or the propagation constant k becomes identical to the phase constant β as per equation (39) of Chapter 12 since the attenuation constant α becomes zero under this situation. The lossless medium approximation is, however, assumed here only to simplify the analysis, and all the derived equations can equally be made valid for the lossy medium by making the propagation constant k to be complex. Meanwhile, it can be observed that equations (2*a*) and (2*b*) can be decomposed into the individual components depending upon the type of coordinate system used for the particular geometry. The parallel plates shown in Figure 15.1 are considered to be rectangular in shape, and hence the Cartesian coordinate system can be quite conveniently employed here to find the individual electric and magnetic field components using (1) and (2) by applying appropriate boundary conditions. Meanwhile, if the waves guided by the parallel plate geometry are considered to be uniform plane waves traveling along the +z direction, then the variation along the direction of propagation can be described as

$$e^{-j\beta_z z} \tag{4}$$

where β_z is the propagation constant inside the guided medium along the direction of propagation (z). It is to be noted that if the uniform plane wave is traveling in the unbounded lossless medium, then the propagation constant along the direction of propagation β_z would be same as the total propagation constant $\beta \equiv k$ of the medium. However, for the propagation inside the guided medium, this is not true as would be evident in the later part of this chapter. The advantage of writing the z-dependence of waves into the explicit form given by (4) is two fold. The first advantage is that the field dependence along the direction of propagation (z-direction) can be separated from the field dependence along the transverse direction (either x or y), i.e.,

$$\mathbf{E}_s\,(x, y, z) = \mathbf{E}_s^t\,(x, y)\;e^{-j\beta_z z} \tag{5a}$$

$$\mathbf{H}_s\,(x, y, z) = \mathbf{H}_s^t\,(x, y)\;e^{-j\beta_z z} \tag{5b}$$

where $\mathbf{E}_s^t$ ($\mathbf{H}_s^t$) now represents the phasor form of the electric (magnetic) field vector, which depends only on the *transverse* or the cross-sectional coordinates (x, y). The second advantage of separating the axial and the transverse dependence is that the partial derivatives with respect to the axial dimension z of all the field components can be simplified by multiplying the corresponding field component with the propagation constant along the direction of propagation β_z, i.e.,

$$\frac{\partial}{\partial z}\left[E_{\xi s}(x,y,z)\right] \equiv -j\,\beta_z E_{\xi s}(x,y,z) \quad \frac{\partial^2}{\partial z^2}\left[E_{\xi s}(x,y,z)\right] \equiv -\beta_z^2 E_{\xi s}(x,y,z) \tag{6a}$$

$$\frac{\partial}{\partial z}\left[H_{\xi s}(x,y,z)\right] \equiv -j\,\beta_z H_{\xi s}(x,y,z) \quad \frac{\partial^2}{\partial z^2}\left[H_{\xi s}(x,y,z)\right] \equiv -\beta_z^2 H_{\xi s}(x,y,z) \tag{6b}$$

where $E_{\xi s}$ ($H_{\xi s}$) represents the individual component of the electric (magnetic) field with $\xi = x, y$ or z for the cartesian coordinate system. It is to be noted that in the most general case, each of the six field components ($E_{xs}, E_{ys}, E_{zs}, H_{xs}, H_{ys}, H_{zs}$) can have dependence on all the three coordinates x, y and z. Now, once the z-dependence of fields and their partial derivatives with respect to z are defined using (5) and (6), we can separate out the transverse (x, y) and the longitudinal (z) components of the Laplacian operator ∇^2 in order to obtain the modified form of (2) for the propagation inside the lossless guided medium, i.e.,

$$\frac{\partial^2 \mathbf{E}_s}{\partial x^2} + \frac{\partial^2 \mathbf{E}_s}{\partial y^2} + \left(k^2 - \beta_z^2\right)\mathbf{E}_s = 0 \tag{7a}$$

$$\frac{\partial^2 \mathbf{H}_s}{\partial x^2} + \frac{\partial^2 \mathbf{H}_s}{\partial y^2} + \left(k^2 - \beta_z^2\right)\mathbf{H}_s = 0 \tag{7b}$$

where, we have made use of (6). The scalar form of (7) can be written for each individual component of the electric and magnetic fields by representing the corresponding field vector in terms of its components. Now, after separating the axial and the transverse dependence of the electric and magnetic fields, the two Maxwell's curl equations described in Chapter 12 can be expanded as follows

$$\nabla \times \mathbf{E}_s = -j\omega\mu\mathbf{H}_s \qquad \nabla \times \mathbf{H}_s = j\omega\epsilon\mathbf{E}_s$$

$$\begin{vmatrix} \mathbf{a}_x & \mathbf{a}_y & \mathbf{a}_z \\ \frac{\partial}{\partial x} & \frac{\partial}{\partial y} & -j\beta_z \\ E_{xs} & E_{ys} & E_{zs} \end{vmatrix} = -j\omega\mu(H_{xs}\mathbf{a}_x + H_{ys}\mathbf{a}_y + H_{zs}\mathbf{a}_z) \tag{8}$$

$$\begin{vmatrix} \mathbf{a}_x & \mathbf{a}_y & \mathbf{a}_z \\ \frac{\partial}{\partial x} & \frac{\partial}{\partial y} & -j\beta_z \\ H_{xs} & H_{ys} & H_{zs} \end{vmatrix} = j\omega\epsilon(E_{xs}\mathbf{a}_x + E_{ys}\mathbf{a}_y + E_{zs}\mathbf{a}_z) \tag{9}$$

where the respective curl equations have been expanded in the rectangular coordinate system using the approach given in Appendix A.1, and the partial differentiation with respect to z, i.e., $\frac{\partial}{\partial z}$ has been replaced with $-j\beta_z$ in conformity with (6). Now, the individual components on two sides of both the curl equations can be compared to obtain the following sets of equations:

$$\frac{\partial E_{zs}^t}{\partial y} + j\beta_z E_{ys}^t = -j\omega\mu H_{xs}^t \tag{10a}$$

$$\frac{\partial E_{zs}^t}{\partial x} + j\beta_z E_{xs}^t = j\omega\mu H_{ys}^t \tag{10b}$$

$$\frac{\partial E_{ys}^t}{\partial x} - \frac{\partial E_{xs}^t}{\partial y} = -j\omega\mu H_{zs}^t \tag{10c}$$

$$\frac{\partial H_{zs}^t}{\partial y} + j\beta_z H_{ys}^t = j\omega\epsilon E_{xs}^t \tag{11a}$$

$$\frac{\partial H_{zs}^t}{\partial x} + j\beta_z H_{xs}^t = -j\omega\epsilon E_{ys}^t \tag{11b}$$

$$\frac{\partial H_{ys}^t}{\partial x} - \frac{\partial H_{xs}^t}{\partial y} = j\omega\epsilon E_{zs}^t \tag{11c}$$

It is to be noted that in equations (10*a*) to (11*c*), all the field components have dependence now only on the x and y coordinates. This is because of the fact that the z-dependence expressed by the common factor $e^{-j\beta_z z}$ can be cancelled on both sides from all the above sets of equations. Now, before moving further, the whole philosophy of writing (7) to (11) in the given form should be clearly understood. The conventional way of solving the wave equation for any arbitrary order mode is to first obtain the solution for the corresponding longitudinal field component of the electric (E_z) or the magnetic field (H_z) using (7). It basically means that in the present situation, first of all the E_z and H_z components are to be determined using the scalar form of (7) subject to the boundary condition which is dependent upon the type of structure to be analyzed. The next step involves the computation of the transverse field components of both the electric and magnetic fields, which basically requires expressing these transverse field components in terms of the corresponding longitudinal components. This is achieved by combining various equations of (10) and (11) in order to write the expression for each component in the following form:

$$E^t_{xs} = -\frac{j}{\beta_z}\left[j\omega\mu H^t_{ys} - \frac{\partial E^t_{zs}}{\partial x}\right] \quad (12a) \qquad H^t_{ys} = -\frac{j}{\beta_z}\left[j\omega\epsilon E^t_{xs} - \frac{\partial H^t zs}{\partial y}\right] \quad (13a)$$

$$E^t_{ys} = \frac{j}{\beta_z}\left[j\omega\mu H^t_{xs} + \frac{\partial E^t_{zs}}{\partial y}\right] \quad (12b) \qquad H^t_{xs} = \frac{j}{\beta_z}\left[j\omega\epsilon E^t_{ys} + \frac{\partial H^t_{zs}}{\partial x}\right] \quad (13b)$$

In the next step, the corresponding equations of (12) and (13) containing the desired field components can be combined again in order to obtain the expression for the particular component. For example, the x-component of the electric field can be expressed in terms of the longitudinal field components of the electric and magnetic fields by combining (12*a*) and (13*a*) as follows:

$$\begin{aligned} E^t_{xs} &= -\frac{j\omega\mu}{\beta_z^2}\left[j\omega\epsilon E^t_{xs} - \frac{\partial H^t_{zs}}{\partial y}\right] + \frac{j}{\beta_z}\frac{\partial E^t_{zs}}{\partial x} \\ &\equiv \frac{\omega^2\mu\epsilon}{\beta_z^2}E^t_{xs} + \frac{j}{\beta_z^2}\left[\omega\mu\frac{\partial H^t_{zs}}{\partial y} + \beta_z\frac{\partial E^t_{zs}}{\partial x}\right] \\ \Rightarrow E^t_{xs} &= \frac{-j}{\omega^2\mu\epsilon - \beta_z^2}\left[\beta_z\frac{\partial E^t_{zs}}{\partial x} + \omega\mu\frac{\partial H^t_{zs}}{\partial y}\right] = \frac{-j}{k^2 - \beta_z^2}\left[\beta_z\frac{\partial E^t_{zs}}{\partial x} + \omega\mu\frac{\partial H^t_{zs}}{\partial y}\right] \end{aligned} \quad (14)$$

where we have made use of (3) to define k for the lossless medium filled between the two plates. A similar kind of procedure as explained above can be employed to express all the other transverse field components in terms of the longitudinal components, which ultimately yields following sets of expressions

$$E^t_{xs} = \frac{-j}{k^2 - \beta_z^2}\left[\beta_z\frac{\partial E^t_{zs}}{\partial x} + \omega\mu\frac{\partial H^t_{zs}}{\partial y}\right] \quad (15a) \qquad H^t_{xs} = \frac{j}{k^2 - \beta_z^2}\left[\omega\epsilon\frac{\partial E^t_{zs}}{\partial y} - \beta_z\frac{\partial H^t_{zs}}{\partial x}\right] \quad (16a)$$

$$E^t_{ys} = \frac{-j}{k^2 - \beta_z^2}\left[\beta_z\frac{\partial E^t_{zs}}{\partial y} - \omega\mu\frac{\partial H^t_{zs}}{\partial x}\right] \quad (15b) \qquad H^t_{ys} = \frac{-j}{k^2 - \beta_z^2}\left[\omega\epsilon\frac{\partial E^t_{zs}}{\partial x} + \beta_z\frac{\partial H^t_{zs}}{\partial y}\right] \quad (16b)$$

It can be observed from above sets of equations that once the longitudinal (z) components of the electric and magnetic field for any given geometry or the structure is obtained using (7) subject to the appropriate boundary conditions, all the other field components for the corresponding geometry can be computed with the help of (15*a*) to (16*b*). In a most general case, we can have both the electric and the magnetic field components along the direction of propagation (z), and in this case all the transverse field components can be obtained by the superposition of fields corresponding to E_{zs} and H_{zs} components as seen from (15*a*) to (16*b*). However, in most of the situations, we usually have either the electric field or the magnetic field component along the direction of propagation with a special situation arising when both the longitudinal field components along the direction of propagation are absent as discussed in the last section. This kind of nomenclature of field components for waves propagating along the z-direction can, therefore, be summarized as follows:

- **TE (Transverse Electric) mode:** The waves characterizing the TE mode have no component of the electric field along the direction of propagation, i.e., $E_z = 0$. These types of waves are defined by the fact that the electric field is fully transverse to the direction of propagation.
- **TM (Transverse Magnetic) mode:** The waves characterizing the TM mode have no component of the magnetic field along the direction of propagation, i.e., $H_z = 0$. These types of waves are defined by the fact that the magnetic field is fully transverse to the direction of propagation.
- **TEM (Transverse Electromagnetic) mode:** The waves characterizing the TEM mode have neither the electric nor the magnetic field component along the direction of propagation, i.e., $E_z = H_z = 0$. These waves are defined by the fact that both the electric and the magnetic fields are fully transverse to the direction of propagation.

As discussed in Section 14.2, the TEM mode, also known as the transmission line mode, is usually supported by two conductor transmission lines. However, the TE and TM waveguide modes can be supported by two conductor lines as well as by a specially designed single conductor line, which will be discussed in detail in next few sections.

Now, let us carry out the detailed analysis of various modes supported by the two conductor parallel plate waveguide shown in Figure 15.1.

15.1.1 Transverse Electric Mode for the Parallel-Plate Geometry

We start the discussion with TE modes characterized by $E_z = 0$ and $H_z \neq 0$. It is obvious from the sets of equations given by (15) that first of all the H_z component is to be determined by solving the second order differential equation (7*b*) subject to the boundary condition for the parallel-plate waveguide shown in Figure 15.1. As mentioned earlier, the width of each plate in Figure 15.1 along the y-direction can be assumed to be very large so that the fringing fields at the edges of the plates are ignored which makes $\frac{\partial^2}{\partial y^2} \equiv 0$ in (7*b*). Hence, the differential equation for the H_z component can be written in the following form by reducing the vector wave equation (7*b*) to the scalar wave equation

$$\frac{\partial^2 H_{zs}(x,z)}{\partial x^2} + \left(k^2 - \beta_z^2\right) H_{zs}(x,z) = 0 \tag{17a}$$

$$\Rightarrow \frac{d^2 H_{zs}^t(x)}{dx^2} + \left(k^2 - \beta_z^2\right) H_{zs}^t(x) = 0 \tag{17b}$$

$$\Rightarrow \frac{d^2 H_{zs}^t(x)}{dx^2} + k_x^2 H_{zs}^t(x) = 0 \tag{17c}$$

where $H_{zs}(x,z) \equiv H_{zs}^t(x)\, e^{-j\beta_z z}$ so that the common factor $e^{-j\beta_z z}$ can be taken out from (17*a*) leading to (17*b*), and

$$k_x^2 = k^2 - \beta_z^2 \tag{18}$$

is defined here for convenience. The exact value of k_x, which may be considered as the wavenumber along the x-direction, depends upon the particular geometry and is determined by applying appropriate boundary conditions as will be shown later. It is to be noted here that the partial derivative is replaced with the full derivative in (17*b*) as $H^t_{zs}(x)$ is now exclusive function of x only. Now, it can be observed that (17*c*) is a second-order simple harmonic differential equation whose general solution can be written in the following form:

$$H^t_{zs}(x) \equiv A_1 \cos(k_x x) + B_1 \sin(k_x x) \tag{19}$$

where A_1 and B_1 are arbitrary constants whose values are to be determined by applying the boundary condition(s) corresponding to the particular geometry. It can be easily verified that (19) is the solution of (17*c*) as the form of $H^t_{zs}(x)$ given by (19) satisfies the differential equation (17*c*). It should be noted here that in the most general case, the solution of (17*c*) may also involve exponential functions of the form $e^{\pm jk_x x}$. However, the exponential functions are normally used to represent traveling waves, and hence they are not considered in (19) since it is explicitly mentioned that the wave in the present situation is propagating along the $+z$ direction. For computing constants A_1 and B_1 of (19), the boundary conditions on various field components corresponding to the parallel plate geometry of Figure 15.1 are to be applied. If the top and bottom plates are considered to be made of perfect electric conductor then the boundary condition would require that the tangential electric fields on both the plates are zero as discussed earlier for the static case in Chapter 5. The tangential electric fields on the two plates are represented by E_y and E_z, which ultimately provides the following boundary conditions:

$$E^t_{ys}(x = 0) \equiv E^t_{ys}(x = d) = 0 \tag{20}$$

as the E_{zs} component is already zero for the TE case discussed here. Now, in order to apply the boundary condition given by (20), first of all the general expressions of all the field components applicable for the parallel plate geometry shown in Figure 15.1 should be determined. These field expressions are obtained by substituting $E_z = 0$ in (15) and (16), which ultimately provides the following non-zero transverse components of the electric and magnetic fields in terms of the longitudinal component of the magnetic field component $H^t_{zs}(x)$

$$E^t_{xs} = \frac{-j\omega\mu}{k^2 - \beta_z^2}\frac{\partial H^t_{zs}}{\partial y} \equiv 0 \quad (21a) \qquad H^t_{xs} = \frac{-j\beta_z}{k^2 - \beta_z^2}\frac{\partial H^t_{zs}}{\partial x} \quad (22a)$$

$$E^t_{ys} = \frac{j\omega\mu}{k^2 - \beta_z^2}\frac{\partial H^t_{zs}}{\partial x} \quad (21b) \qquad H^t_{ys} = \frac{-j\beta_z}{k^2 - \beta_z^2}\frac{\partial H^t_{zs}}{\partial y} \equiv 0 \quad (22b)$$

where, we have applied the condition that the fields are not varying along the y-direction due to fact that the width of each plate in Figure 15.1 along the y-direction is assumed to be very large, i.e., $\frac{\partial}{\partial y} \equiv 0$. It is obvious from above sets of equations that the only non-zero field components are E^t_{ys} and H^t_{xs} in the present situation. Now, in order to apply the boundary condition as defined by (20), let us substitute the value of $H^t_{zs}(x)$ given by (19) in (21*b*), i.e.,

$$\begin{aligned} E^t_{ys}(x = 0) &\equiv \frac{j\omega\mu}{k^2 - \beta_z^2} k_x \left[B_1 \cos(k_x x) - A_1 \sin(k_x x)\right]_{x=0} = 0 \\ &\Rightarrow \left[\frac{j\omega\mu}{k^2 - \beta_z^2}\right] k_x\, B_1 = 0 \\ &\Rightarrow B_1 = 0 \end{aligned} \tag{23}$$

The above formulation follows from the fact that the constant k_x in a general case cannot be made equal to zero as it would amount to no variation of fields in the x-direction as per (19). The value of B_1 computed above can then be substituted in (19), and accordingly the boundary condition defined by (20) at $x = d$ can be applied for E^t_{ys} resulting into the following equation:

$$
\begin{aligned}
E^t_{ys}(x=d) &\equiv \frac{j\omega\mu}{\beta_z^2-k^2}k_x\,[A_1\sin(k_x x)]_{x=d}=0\\
&\Rightarrow\left[\frac{j\omega\mu}{\beta_z^2-k^2}\right]k_x\,A_1\,\sin(k_x d)=0\\
&\Rightarrow\sin(k_x d)=0\equiv\sin(m\pi)\qquad\text{where } m=1,2,\ldots
\end{aligned}
\tag{24}
$$

where, m is an integer as defined above, and it has again been made use of the fact that the constant A_1 in a general case can not be made equal to zero since it would lead to all the field components being equal to zero as per (19) and (21). The integer $m=0$ is not considered in (24) as it would make the component $H^t_{zs}(x)$ a constant as per (19) resulting into all the transverse field components being equal to zero, which would not allow any real power flow along the direction of propagation (z direction). It is to be observed here that k_x can only assume some discrete values $m=1,2,3\ldots$ also known as *eigen* values. The field configuration corresponding to each m is assumed to represent a particular mode, which basically means that in a general guided media we can have infinite number of modes. Meanwhile, using (24) and (18), the propagation constant along the direction of wave travel (z-direction) can be defined as follows:

$$k_x=\frac{m\pi}{d}\tag{25a}$$

$$\beta_z=\sqrt{k^2-\left(\frac{m\pi}{d}\right)^2}\equiv\sqrt{\omega^2\mu\epsilon-\left(\frac{m\pi}{d}\right)^2}\tag{25b}$$

Now, before moving further, let us try to clarify few points regarding the constants k_x and β_z defined above. It is obvious from (25) and (3) that if $k_x=0$, then the z-directed propagation constant β_z becomes equal to the total propagation constant $\beta\equiv k$ for the lossless medium, and the wave propagation along the z-direction in this particular case can be defined as $e^{-j\beta z}$. This situation now becomes identical to that of the wave propagation in an unbounded lossless media as discussed in Chapter 12. The situation here can be understood from the fact that when $k_x=0$, then there is no variation of any field component in either x or y direction leading to spatial variation of field only in the direction of travel, which defines a plane wave or TEM wave as described in Chapter 12. Hence, it can be postulated that the propagation defined by the $e^{-j\beta_z z}$ term in (4) represents a more general situation, which reduces to the special case of a TEM wave or plane wave if there is no spatial variations of the field components along the direction perpendicular to the direction of travel. Now, once the propagation constant and wavenumbers for the parallel plate waveguide geometry are defined, the non-zero field components for the general TE to z mode corresponding to this geometry can be derived to have the following expressions:

$$H_{zs}(x,z)=A_m\cos\left(\frac{m\pi}{d}x\right)e^{-j\beta_z z}\tag{26a}$$

$$E_{ys}(x,z)=-\frac{j\omega\mu}{k_x}A_m\sin\left(\frac{m\pi}{d}x\right)e^{-j\beta_z z}\tag{26b}$$

$$H_{xs}(x,z)=\frac{j\beta_z}{k_x}A_m\sin\left(\frac{m\pi}{d}x\right)e^{-j\beta_z z}\tag{26c}$$

where, the value of k_x is defined using (25), the subscript m is added to the factor A to emphasize the fact that the amplitude of each mode corresponding to a particular number m might be different, and the dependence of field components along the z direction is added using (4). It should be noted that k_x and β_z in (26) for all the field components can assume only discrete values as per (25) with each value of m specifying a particular field configuration or TE_m *mode*.

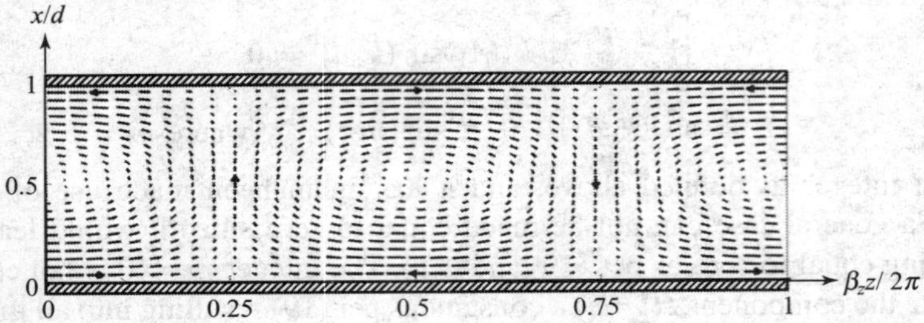

Fig. 15.2 Magnetic field lines for the TE_1 mode in the parallel-plate waveguide.

The magnetic field plot for the first possible TE mode in the parallel plate waveguide is shown in Figure 15.2. The field lines in this plot are obtained using (26*a*) and (26*c*) by converting the phasor quantities into their corresponding instantaneous forms, and then plotting the magnetic field in the xz plane at $t = 0$. As the magnetic field for the TE case has both x and z components, the plot shown in Figure 15.2 has component along both x and z directions. It can, however, be seen that near $x = 0$ and $x = d$, the field lines become almost horizontal as the x component of the magnetic field becomes close to zero as per (26*c*). Similarly, near $\beta_z z = \pi/2, 3\pi/2$, the field lines become close to vertical since the z component of the magnetic field becomes close to zero as per (26*a*). It should be noted that the conversion of (26) into instantaneous form would result into $\cos(\beta_z z)$ dependence for the H_z, and $\sin(\beta_z z)$ dependence for the H_x component. The magnetic field plot for the second TE mode in the parallel plate waveguide is shown in Figure 15.3. The

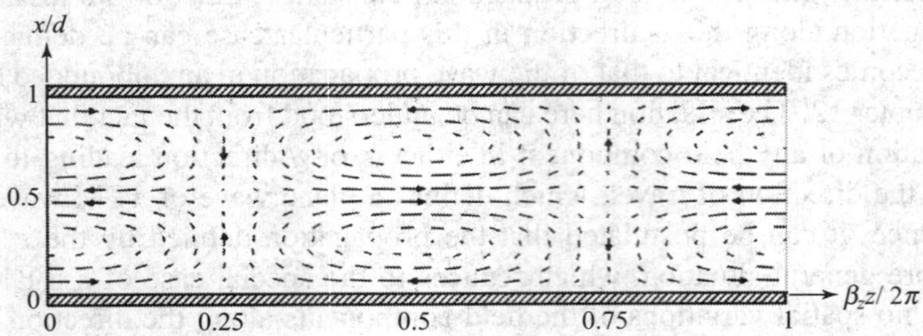

Fig. 15.3 Magnetic field lines for the TE_2 mode in the parallel plate waveguide.

field lines for the TE_2 mode shown in Figure 15.3 can be explained on similar lines as that of the TE_1 mode discussed above with the only difference that there are now 2 half-wave variations of the field along the x-direction as compared to one half-wave variation for the TE_1 case. For both TE_1 and TE_2 modes, the field lines would repeat after every 2π radians in the $\beta_z z$ direction. It should be noted that the electric field lines for the TE case has only y component as per (26*b*), and hence they are not shown here as they would be perpendicular to xz plane.

For the guided media, one important parameter is the cutoff frequency $f_c \equiv \frac{\omega_c}{2\pi}$ which is defined as the frequency at which there is no real power flow along the direction of propagation. The cutoff condition is obtained by making $\beta_z = 0$ in (25*b*), i.e.,

$$\beta_z^2 = 0 \equiv \omega_c^2 \mu\epsilon - \left(\frac{m\pi}{d}\right)^2 \tag{27}$$

The above equation after some simplification yields the following expression for the cutoff frequency f_c of the TE_m mode in the parallel plate waveguide:

$$f_c = \frac{m}{2\,d\,\sqrt{\mu\epsilon}} \qquad m = 1, 2, 3, \ldots \tag{28}$$

The mode having the lowest cutoff frequency is called the dominant mode. As the value of m in the above equation cannot be taken as zero, hence the lowest order TE mode possible in the parallel plate waveguide is the TE_1 mode. Once the cutoff frequency is defined, the propagation constant for the parallel plate geometry supporting a TE mode can be reformulated with the help of (25) and (28) depending upon the situation whether the operating frequency f is greater than or smaller than the cutoff frequency f_c. Let us first consider the situation having $f > f_c$, which yields the following equation:

$$\beta_z = \beta\sqrt{1-\left(\frac{k_c}{k}\right)^2} \qquad k > k_c \tag{29a}$$

$$\Rightarrow \beta_z = \beta\sqrt{1-\left(\frac{f_c}{f}\right)^2} \qquad f > f_c \tag{29b}$$

where $k_c = \omega_c\sqrt{\mu\epsilon} \equiv \frac{m\pi}{d}$ is the cutoff wavenumber as per (27), and we have made use of (28). It can be observed from (29*b*) that if the actual frequency f is greater than the cutoff frequency f_c then the propagation constant β_z in this particular situation would be a purely real quantity thus allowing the wave propagation along direction z as per (4).

Now, let us consider the situation where the operating frequency f becomes less than the cutoff frequency f_c, i.e., $\left(\frac{f_c}{f} > 1\right)$ yielding the following equation:

$$\beta_z = -j\beta\sqrt{\left(\frac{f_c}{f}\right)^2 - 1} \equiv -j\,\alpha_z \qquad f < f_c \tag{30}$$

where $\alpha_z \equiv j\beta_z$ represents a positive real quantity, and the negative sign is chosen for β_z in (30) to stress the fact that for this particular situation the wave propagation represented by $e^{-j\beta_z z} \equiv e^{-\alpha_z z}$ should represent a decaying wave in the passive guided media. The waves characterized by $f < f_c$ as per (30) are usually called the *evanescent* waves as they represent the non-propagating waves decaying very fast in the guided media. Hence, it can be concluded that for the real power transfer in the guided media, the operating frequency should be greater than the cutoff frequency. In this chapter, the majority of the discussion is related to the propagating waves assuming the working frequency f to be greater than the cutoff frequency f_c unless stated otherwise.

Apart from the propagation constant and the cutoff frequency, the other important parameter quite useful to define the propagation characteristics is the wave impedance defined in Chapter 12. The wave impedance for the propagating wave traveling along the positive z-direction in the present situation can be defined as

$$Z_w^{\text{TE}}(z) \equiv -\frac{E_{ys}}{H_{xs}} = \frac{\omega\mu}{\beta_z} = \frac{\eta}{\sqrt{1-\left(\frac{f_c}{f}\right)^2}} \qquad f > f_c \tag{31}$$

where $\eta = \frac{\omega\mu}{\beta} \equiv \sqrt{\frac{\mu}{\epsilon}}$ is the intrinsic impedance of the medium filling the region between the two parallel plates. It can be observed from the above expression that the wave impedance for the propagating wave ($f > f_c$) is a purely real quantity if the parallel-plate waveguide is filled with the lossless media, and its value is greater than the intrinsic impedance of the medium. It can be easily proved that for non-propagating waves ($f < f_c$), the wave impedance becomes a purely imaginary quantity indicating that there is no real power transfer. The negative sign is added in (31) to stress the fact that the power transfer is along the $+z$-direction, which can be explained with the help of Poynting vector concept introduced in Chapter 12.

Another important parameter to characterize the propagating waves is the phase velocity introduced in Chapter 12, which is defined as the velocity with which the equiphase plane moves along the direction of propagation. In the present situation, the expression for the phase velocity can be written as

$$v_p \equiv \frac{\omega}{\beta_z} = \frac{v}{\sqrt{1-\left(\frac{f_c}{f}\right)^2}} \qquad f > f_c \tag{32}$$

where, $v \equiv \frac{\omega}{\beta} = \frac{1}{\sqrt{\mu\epsilon}}$ is the phase velocity inside the unbounded media supporting a plane wave. It can be seen from the above equation that for the bounded media supporting higher order modes, the phase velocity becomes greater than the velocity of light. However, it does not violate the Einstein's theory of relativity since the phase velocity is not the velocity with which the actual energy propagates. The electromagnetic energy propagates in the direction of propagation with a velocity known as the group velocity introduced in Chapter 13,

$$v_g = \frac{d\omega}{d\beta_z} \equiv \frac{1}{d\beta_z/d\omega} = v\sqrt{1-\left(\frac{f_c}{f}\right)^2} \qquad f > f_c \tag{33}$$

It is obvious from the above equation that the group velocity for the propagating waves supporting higher order modes becomes less than the velocity of light in the media. It is readily seen from the above two equations that $v_p\, v_g = v^2$ indicating that the product of the phase velocity and the group velocity for the higher order mode is equal to square of the velocity of the electromagnetic wave inside the unbounded media assuming a plane wave propagation.

15.1.2 Transverse Magnetic Mode for the Parallel-plate Geometry

The analysis for the TM modes supported by the parallel plate waveguide geometry can be carried out on similar lines as that of TE modes discussed in the last section. The TM modes are characterized by $H_z = 0$ and $E_z \neq 0$ for the waves propagating along the z direction. The E_z component should first be determined here by solving the following second-order differential equation derived from (7*b*) subject to appropriate boundary condition for the parallel plate waveguide

$$\frac{d^2\, E^t_{zs}(x)}{dx^2} + k_x^2\, E^t_{zs}(x) = 0 \tag{34}$$

where $E_{zs}(x, z) \equiv E^t_{zs}(x)\, e^{-j\beta_z z}$, the factor k_x is defined using (18), and the width of each plate along the y-direction is again assumed to be very large making $\frac{\partial^2}{\partial y^2} \equiv 0$. The general solution of the simple harmonic differential equation (34) for waves not traveling along the x-direction can be written in the following form:

$$E^t_{zs}(x) \equiv A_2 \cos(k_x x) + B_2 \sin(k_x x) \tag{35}$$

where A_2 and B_2 are arbitrary constants whose values are to be determined by applying the boundary condition(s), which requires that the tangential electric fields on both the plates should be zero. The non-zero tangential electric fields on the two plates in the present situation are represented by E_y and E_z providing following sets of boundary conditions:

$$E^t_{zs}(x=0) \equiv E^t_{zs}(x=d) = 0 \tag{36a}$$

$$E^t_{ys}(x=0) \equiv E^t_{ys}(x=d) = 0 \tag{36b}$$

Now, let us first apply the boundary condition defined by (36*a*) for the $E^t_{zs}(x)$ at $x=0$ given by (35) yielding the following equation:

$$E^t_{zs}(x=0) \equiv A_2 = 0 \tag{37}$$

The value of A_2 computed above is then substituted in (35), and the boundary condition defined by (36*a*) at $x=d$ is applied for the E^t_{zs} resulting into the following equation:

$$\begin{aligned} &E^t_{zs}(x=d) \equiv B_2 \sin(k_x d) = 0 \\ &\Rightarrow \sin(k_x d) \equiv \sin(m\pi) \qquad \text{where } m = 0, 1, 2, \ldots \end{aligned} \tag{38}$$

where, m is an integer as defined above, and the above solution is based on the fact that the constant B_2 cannot be zero since it would make all the field components to be zero. The integer $m=0$ is included here in contrary to the TE case as it does not make all the transverse field components to be zero, which would be evident when all the field components are derived. It is to be noted that k_x in this case also takes only discrete values known as *eigen* values, which is related with β_z and the total propagation constant as per (25). The transverse field components for this case are obtained by substituting $H_z = 0$ in (15) and (16), which yields the following sets of equations:

$$E^t_{xs} = \frac{-j\beta_z}{k^2-\beta_z^2}\frac{\partial E^t_{zs}}{\partial x} \tag{39a}$$

$$H^t_{xs} = \frac{j\omega\epsilon}{k^2-\beta_z^2}\frac{\partial E^t_{zs}}{\partial y} \equiv 0 \tag{40a}$$

$$E^t_{ys} = \frac{j\beta_z}{k^2-\beta_z^2}\frac{\partial E^t_{zs}}{\partial y} \equiv 0 \tag{39b}$$

$$H^t_{ys} = \frac{-j\omega\epsilon}{k^2-\beta_z^2}\frac{\partial E^t_{zs}}{\partial x} \tag{40b}$$

where, again the condition $\frac{\partial}{\partial y} \equiv 0$ is applied by noting that the fields are not varying along the y-direction. It can be observed from above sets of equations that the non-zero transverse field components for the TM case are E_x and H_y. The above sets of equations can now be combined with (35), (38) and (25) to obtain the following non-zero field components for the general TM to z-mode satisfying the parallel-plate waveguide geometry:

$$E_{zs}(x,z) = B_m \sin\left(\frac{m\pi}{d}x\right) e^{-j\beta_z z} \tag{41a}$$

$$E_{xs}(x,z) = -\frac{j\beta_z}{k_x} B_m \cos\left(\frac{m\pi}{d}x\right) e^{-j\beta_z z} \tag{41b}$$

$$H_{ys}(x,z) = -\frac{j\omega\epsilon}{k_x} B_m \cos\left(\frac{m\pi}{d}x\right) e^{-j\beta_z z} \tag{41c}$$

where, the subscript m is added to the constant B to emphasize the fact that the amplitude of each field solution represented by a particular mode m might be different, and we have made use of (18) to arrive at the above expressions.

For plotting the field lines in the parallel plate waveguide corresponding to TM case, it should be noted that the magnetic field lines for the TM mode has only y component as per (41*c*). As the H_y component is fully perpendicular to xz plane, only the electric field is plotted here for the

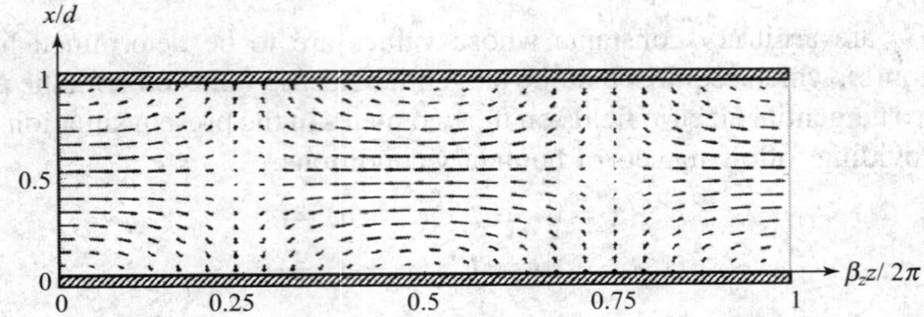

Fig. 15.4 Electric field lines for the TM_1 mode in the parallel plate waveguide.

TM mode. The electric field lines for the TM_1 mode are shown in Figure 15.4, which are obtained using (41*a*) and (41*b*) by converting the phasor quantities into their corresponding instantaneous forms and choosing $t = 0$. It can be observed from Figure 15.4 that the electric field for the TM_1 mode has component along both x and z directions. It can, however, be seen that at $x = d/2$, the field lines become fully horizontal as the x-component of the electric field becomes zero as per (41*b*). Similarly, near $\beta_z z = \pi/2,\ 3\pi/2$, the field lines become close to vertical since the z-component of the instantaneous electric field having cos $(\beta_z z)$ dependence becomes close to zero as per (41*a*). The electric field plot for the TM_2 mode in the parallel plate waveguide is shown in Figure 15.5. The field lines for the TM_2 mode shown in Figure 15.5 have similar dependence along the z-direction as that of TM_1 mode shown in Figure 15.4. However, the electric field now has 2 half wave variation along the x direction with the field lines becoming fully horizontal at $x = d/4$ and $x = 3d/4$. The direction of field lines at $x = d/4$ and $x = 3d/4$, however, becomes opposite to each other which is due to sin $\left(\frac{2\pi}{d}x\right)$ dependence as per (41*a*). The field lines for all the TM modes would repeat after every 2π radian in the $\beta_z z$ direction similar to the TE case.

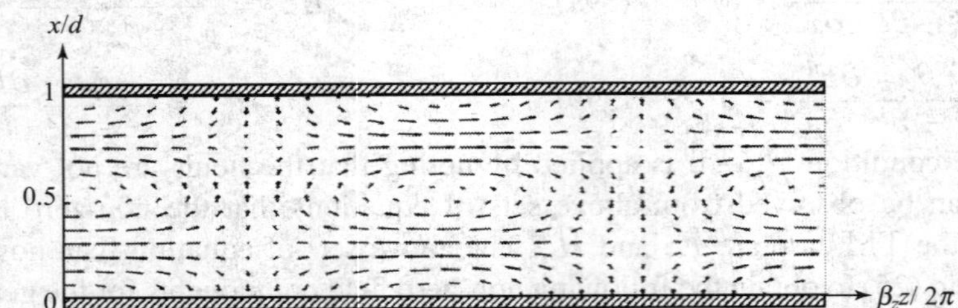

Fig. 15.5 Electric field lines for the TM_2 mode in the parallel plate waveguide.

The expressions for the cutoff frequency f_c, the z-directed propagation constant β_z, the phase and group velocities of the parallel-plate waveguide corresponding to the TM case are same as that of the TE case described in the last section. However, as the value of m for the TM case can be taken as zero, the lowest order TM mode possible in the parallel-plate waveguide is the TM_0 mode. The TM_0 mode is the *dominant* mode of the parallel-plate waveguide, which actually represents the TEM mode as shown in the next section.

The wave impedance for the parallel plate waveguide supporting a TM propagating wave along the $+z$-direction can be defined as

$$Z_w^{TM}(z) \equiv \frac{E_{xs}}{H_{ys}} = \frac{\beta_z}{\omega\epsilon} = \eta\sqrt{1 - \left(\frac{f_c}{f}\right)^2} \tag{42}$$

where $\eta \equiv \sqrt{\frac{\mu}{\epsilon}}$ is the intrinsic impedance of the medium filling the region between the two parallel plates. It can be observed from the above equation that the wave impedance for the TM propagating wave $(f > f_c)$ is a real quantity with its value being less than the intrinsic impedance of the medium.

15.1.3 TEM Waves for the Parallel-plate Geometry

It was discussed in Chapter 12 that the transverse electromagnetic (TEM) waves are characterized by the fact that all the field components are perpendicular to the direction of propagation. It was also mentioned in Chapter 14 that the transmission lines having two or more conductors can support TEM waves. Now, as the parallel-plate waveguide shown in Figure 15.1 consists of two conductors, hence it should also support TEM waves apart from TE and TM waves discussed in previous sections. The governing equations for the TEM waves supported by the parallel-plate waveguide can be derived from the TM case discussed in the last section by substituting $m = 0$, which is possible as mentioned earlier. The procedure in this section is to first revisit some of the equations derived in previous sections for the special case of $m = 0$ before deriving all the field equations for the TEM waves.

From (38) and (25), the propagation constant along the z direction for $m = 0$ representing a TEM wave in the parallel-plate waveguide can be written as

$$\beta_z \equiv \beta = \omega\sqrt{\mu\epsilon} \tag{43}$$

The above equation is based on the fact that the wave number along the x-direction $k_x \equiv \frac{m\pi}{d}$ becomes zero for the TEM wave indicating no spatial variation of fields along the transverse direction thus making the propagation constant along the direction of propagation (β_z) to be same as the wave number of the media. The electric field along the direction of propagation E_{zs} also becomes zero for this case as per (41*a*). Now, since both E_z and H_z become zero for the wave travelling along the z-direction and the propagation constant for the lossless media filled between the plates becomes equal to the wave number, the present situation represents a TEM to z-wave. The only non-zero field components for the TEM to z mode satisfying the parallel-plate waveguide geometry are E_x and H_y as per (41). These field components can, however, be not directly computed using (41) as they become undefined because of zero in the denominator $(k_x = 0)$. It looks rather more appropriate to use the basic equations (10) and (11) to compute these non-zero field components resulting into the following expressions:

$$\beta E_{xs}^t = \omega\mu H_{ys}^t \tag{44a}$$

$$\beta H_{ys}^t = \omega\epsilon E_{xs}^t \tag{44b}$$

$$\frac{\partial H_{ys}^t}{\partial x} = 0 \tag{44c}$$

where, the conditions $E_z \equiv H_z = 0$ and $\frac{\partial}{\partial y} \equiv 0$ are applied, and we have made use of (43). From (44*a*) and (44*b*), the wave impedance along the z direction can be defined as follows:

$$Z_w^{\text{TEM}}(z) \equiv \frac{E_{xs}}{H_{ys}} = \frac{\beta}{\omega\epsilon} \equiv \frac{\omega\mu}{\beta} = \eta \equiv \sqrt{\frac{\mu}{\epsilon}} \tag{45}$$

where, η is the intrinsic impedance as mentioned earlier. Hence for the TEM wave, the wave impedance is same as the intrinsic impedance of the medium. Meanwhile, it can be noted from (44) that the field components in the present situation become independent of x direction. As the fields are already assumed to be independent of y direction, the non-zero transverse field components E_x and H_y of the parallel-plate geometry supporting a TEM mode propagation can hence be expressed as exclusive functions of z only resulting into the following expressions:

$$E_{xs}(z) = E_0\, e^{-j\beta z} \tag{46a}$$

$$H_{ys}(x, z) = \frac{E_0}{\eta}\, e^{-j\beta z} \tag{46b}$$

where, E_0 is an arbitrary constant, and we have made use of (45). It can be clearly observed that (46*a*) and (46*b*) are now similar in form as those of (2a) and (2b) introduced in Chapter 14 for TEM waves although (46*a*) and (46*b*) have been derived here from the first principle using the electromagnetic wave equation approach. It is obvious from (46) that for the TEM mode in the parallel-plate waveguide, the electric field is oriented along the x-direction while the magnetic field has only y component which is fully perpendicular to the xz-plane. The electric field for the TEM mode obtained using (46*a*) would have sinusoidal variation along the z-direction, which is plotted in Figure 15.6. It can be observed from Figure 15.6 that the electric field for the TEM mode is fully vertical directed between the lower and upper plate just like the parallel-plate capacitor.

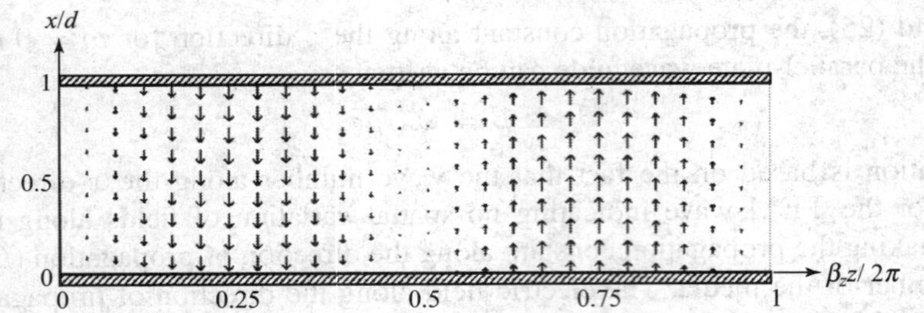

Fig. 15.6 Electric field lines for the TEM mode in the parallel-plate waveguide.

As the dominant mode in the parallel-plate waveguide shown is Figure 15.1 is the TEM wave, the dimensions of the waveguide and the frequency of operation can be selected in such a way that the guide is mostly operated in the dominant mode. The cutoff frequency f_c for the TEM wave obviously becomes zero as per (28), thus making the phase velocity and the group velocity to be equal.

EXAMPLE 15.1

Determine the maximum possible plate separation d in case of a parallel-plate waveguide so that only the dominant mode propagates inside the guide over the frequency range starting from dc to 8 GHz. The spacing between the plates is filled with Polyethylene having a dielectric constant of 2.25.

Solution. The dominant mode inside the parallel-plate waveguide is the TEM mode. If we want to ensure that only the dominant TEM mode propagates up to 8 GHz, then the cutoff frequency of all the higher order modes should be more than 8 GHz. However, if we can show that the cutoff frequency of the first waveguide mode is more than 8 GHz, then it automatically ensures that all the other higher order modes would not propagate inside the guide since their cutoff frequencies would be much higher.

Now, in case of a parallel-plate waveguide, the first waveguide mode is either TE_1 or TM_1 with the cutoff frequency given by the following expression:

$$f_c^{TE_1} \equiv f_c^{TM_1} = \frac{1}{2d\sqrt{\mu\epsilon}} \equiv \frac{c}{2d\sqrt{\epsilon_r}} > 8 \text{ GHz}$$

$$\Rightarrow d < \frac{3 \times 10^8}{2 \times \sqrt{2.25} \times 8 \times 10^9} < 1.25 \text{ cm}$$

where, $c \equiv 1/\sqrt{\mu_0\epsilon_0}$ is the velocity of the electromagnetic wave in free space. It means that in order to ensure the TEM mode propagation over the frequency range starting from dc to 8 GHz, the separation between the plates should be less than 1.25 cm.

EXAMPLE 15.2

A parallel-plate waveguide with the plate separation $d = 2$ cm is operated at 15 GHz. The medium between the plates is filled with Teflon having $\epsilon_r' \equiv 2.1, \epsilon_r'' \equiv 0$.
Determine

(a) the total number of modes propagating inside the guide at the operating frequency;

(b) the maximum frequency of operation which allows only the TEM mode propagation inside the guide; and

(c) the time average power per unit width in the parallel-plate waveguide corresponding to the TEM mode if the amplitude of the electric field between the plates is 10 V/m.

Solution.

(a) For determining the number of modes propagating inside the guide at 15 GHz, let us first determine the cutoff frequency of few higher order waveguide modes as follows:

$$f_c^{TE_1} \equiv f_c^{TM_1} = \frac{1}{2d\sqrt{\mu\epsilon}} \equiv \frac{c}{2d\sqrt{\epsilon_r'}} = \frac{3 \times 10^8}{2 \times 2 \times 10^{-2} \times \sqrt{2.1}} = 5.17 \text{ GHz}$$

$$f_c^{TE_2} \equiv f_c^{TM_2} = \frac{2}{2d\sqrt{\mu\epsilon}} \equiv \frac{c}{d\sqrt{\epsilon_r'}} = \frac{3 \times 10^8}{2 \times 10^{-2} \times \sqrt{2.1}} = 10.3 \text{ GHz}$$

$$f_c^{TE_3} \equiv f_c^{TM_3} = \frac{3}{2d\sqrt{\mu\epsilon}} \equiv \frac{3c}{2d\sqrt{\epsilon_r'}} = \frac{3 \times 3 \times 10^8}{2 \times 2 \times 10^{-2} \times \sqrt{2.1}} = 15.5 \text{ GHz}$$

where, $c \equiv 1/\sqrt{\mu_0\epsilon_0}$ is the velocity of the electromagnetic wave in free space. From the above equations, it is obvious that the cutoff frequency of the third waveguide mode ($m = 3$) is higher than the operating frequency (15 GHz). It is also obvious that the cutoff frequency of all the other higher order modes will be more than the cutoff frequency of the third mode. Now, the propagation of any m^{th} waveguide mode inside the guide requires that its cutoff frequency should be less than the operating frequency, i.e. $f > f_c^m$.
It can be concluded from the above discussion that at the operating frequency of 15 GHz, only the first two TE and TM waveguide modes will propagate apart from the TEM modes. Hence, the total number of modes propagating at 15 GHz in the specified parallel-plate line is 5 (TEM, TE_1, TM_1, TE_2, TM_2).

(*b*) If we want to ensure that only the TEM mode propagates inside the parallel-plate line, then the maximum operating frequency should be less than the cutoff frequency of the first waveguide mode. Hence, in the present situation, the maximum frequency of operation which allows only the TEM mode propagation inside the parallel-plate waveguide can be taken as 5 GHz (< 5.17 GHz).

(*c*) For calculating the time average power in the parallel-plate waveguide corresponding to the TEM mode, let us first determine the average power density. The phasor expressions of the electric and magnetic fields corresponding to the TEM mode inside the parallel-plate line are given by equations (46*a*) and (46*b*), which can be used to calculate the time average power density as follows:

$$E_{xs}(z) = E_0 e^{-j\beta z}, \; H_{xy}(z) = \frac{E_0}{\eta} e^{-j\beta z}$$

$$\langle \mathbf{S}^{\text{TEM}} \rangle = \frac{1}{2}\text{Re}\left[E_{sx}\,\mathbf{a}_x \times \mathbf{a}_y\, H^*_{sy}\right]$$

$$\equiv \tfrac{1}{2}\text{Re}\left[E_0\, e^{-j\beta z} \times \tfrac{E_0^*}{\eta}\, e^{\beta z}\right] \mathbf{a}_z = \frac{1}{2}\frac{|E_0|^2}{\eta}\mathbf{a}_z \text{ W/m}^2$$

where E_0 represents the amplitude of the electric field, $\eta \equiv \sqrt{\mu/\epsilon}$ is the intrinsic impedance of the medium occupying the region between the plates, and the wave propagation is assumed along the +z direction. You should note that if the parallel-plate line is filled with the lossless medium, then the intrinsic impedance η will be a purely real quantity. Meanwhile, the power density along the z-direction corresponding to the TEM mode can be determined from the above expression as

$$\langle S_z^{\text{TEM}} \rangle = \frac{1}{2}\frac{|E_0|^2}{(\eta_0/\sqrt{\epsilon_r'})} \equiv \frac{1}{2}\frac{(10)^2 \times \sqrt{2.1}}{377} = 0.192 \text{ W/m}^2$$

where η_0 is the free-space intrinsic impedance. The average power flow will then be given by

$$P_{\text{avg}} \equiv \int_s \langle \mathbf{S}^{\text{TEM}} \rangle \cdot \mathbf{ds} \equiv 0.192\mathbf{a}_z \cdot \mathbf{a}_z \int_0^w \int_0^2 dx dy = 3.84 \times w \text{ (mW)}$$

where w represents the width of the parallel-plate line. Hence, the time average power per unit width in the parallel-plate waveguide corresponding to the TEM mode is given by

$$P_{\text{avg},w} = 3.84 \text{ mW/m}$$

EXAMPLE 15.3

A parallel-plate waveguide made of two infinitely large perfect conductors allows propagation of waves along the z-direction. The conductors are separated by $d = 5$ cm, and the waveguide is operated at 3 GHz. The medium between the plates is filled with the lossless dielectric Polystyrene having $\epsilon_r' \equiv 2.25$.

Determine

(*a*) the cutoff frequency of the TEM, TE_1, TM_1, TE_2, and TM_2 modes;

(*b*) the propagation constant along the direction of propagation for all the above modes;

(*c*) the wave impedance for all the propagating modes;

(d) the phase velocity of all the propagating modes;
(e) the group velocity of all the propagating modes; and
(f) the wavelength inside the guided media for all the propagating modes.

Solution.

(a) The cutoff frequency of the TEM and other higher order waveguide modes may be determined as follows:

$$f_c^{\text{TEM}} \equiv f_c^{\text{TM}_0} = \frac{0}{2d\sqrt{\mu\epsilon}} = 0$$

$$f_c^{\text{TE}_1} \equiv f_c^{\text{TM}_1} = \frac{1}{2d\sqrt{\mu\epsilon}} \equiv \frac{c}{2d\sqrt{\epsilon_r'}} = \frac{3\times10^8}{2\times5\times10^{-2}\times\sqrt{2.25}} = 2.0\text{ GHz}$$

$$f_c^{\text{TE}_2} \equiv f_c^{\text{TM}_2} = \frac{2}{2d\sqrt{\mu\epsilon}} \equiv \frac{c}{d\sqrt{\epsilon_r'}} = \frac{3\times10^8}{5\times10^{-2}\times\sqrt{2.25}} = 4.0\text{ GHz}$$

where, $c \equiv 1/\sqrt{\mu_0\epsilon_0}$ is velocity of the electromagnetic wave in free space. It may be noted from above calculations that at the operating frequency of 3 GHz, the TE_1 and TM_1 waveguide modes along with the TEM mode will propagate along the parallel-plate line in the present situation.

(b) As the medium between the plates is filled with the lossless dielectric, the propagation constant will be same as the phase constant. Hence, the propagation constant along the direction of propagation for all the above modes may be computed as follows:

$$\beta_z^{\text{TEM}} \equiv \beta = \omega\sqrt{\mu_0\epsilon_0\epsilon_r} \equiv \left(\frac{2\pi f}{c}\right)\sqrt{\epsilon_r'} = \frac{2\times\pi\times3\times10^9}{3\times10^8}\sqrt{2.25} = 94.25\text{ m}^{-1}$$

$$\beta_z^{\text{TE}_1} \equiv \beta_z^{\text{TM}_1} = \beta\sqrt{1-\left(\frac{f_c^{TE_1}}{f}\right)^2} \equiv 94.25\sqrt{1-\left(\frac{2}{3}\right)^2} = 70.25\text{ m}^{-1}$$

$$\beta_z^{\text{TE}_2} \equiv \beta_z^{\text{TM}_2} = \beta\sqrt{1-\left(\frac{f_c^{TE_2}}{f}\right)^2} \equiv 94.25\sqrt{1-\left(\frac{4}{3}\right)^2} = -j83.12\text{ m}^{-1}$$

It is to be noted here that the TE_2 and TM_2 modes are non-propagating modes in the parallel-plate waveguide at 3 GHz, and hence the propagation constants for these two modes become fully imaginary quantities. The signs are chosen to be negative for these two modes so that they represent decaying waves as explained earlier in the text.

(c) The propagating modes in the parallel-plate waveguide are the TEM, TE_1 and TM_1 modes at 3 GHz as described earlier. The wave impedance for all these propagating modes may be calculated as follows:

$$Z_w^{\text{TEM}} = \frac{\beta_z^{\text{TEM}}}{\omega\epsilon} \equiv \frac{\beta}{\omega\epsilon_0\epsilon_r} = \frac{\eta_0}{\sqrt{\epsilon_r'}} = \frac{377}{\sqrt{2.25}} = 251.3\ \Omega$$

$$Z_w^{\text{TE}_1} = \frac{\omega\mu}{\beta_z^{\text{TE}_1}} \equiv \frac{\eta_0}{\sqrt{\epsilon_r'}}\frac{1}{\sqrt{1-\left(\frac{f_c^{\text{TE}_1}}{f}\right)^2}} = \frac{251.3}{\sqrt{1-\left(\frac{2}{3}\right)^2}} = 337.2\ \Omega$$

$$Z_w^{\mathrm{TM}_1} = \frac{\beta_z^{\mathrm{TM}_1}}{\omega\epsilon} \equiv \frac{\eta_0}{\sqrt{\epsilon_r'}}\sqrt{1-\left(\frac{f_c^{\mathrm{TM}_1}}{f}\right)^2} = 251.3 \times \sqrt{1-\left(\frac{2}{3}\right)^2} = 187.3\,\Omega$$

where $\eta_0 \equiv \sqrt{\mu_0/\epsilon_0}$ is the free space intrinsic impedance.

(d) The phase velocity of all the propagating modes may be calculated using the following expressions:

$$v_{ph}^{\mathrm{TEM}} \equiv \frac{v}{\sqrt{1-\left(\frac{f_c^{\mathrm{TEM}}}{f}\right)^2}} = \frac{c}{\sqrt{\epsilon_r'}} = \frac{3\times10^8}{\sqrt{2.25}} = 2\times10^8\mathrm{ms}^{-1}$$

$$v_{ph}^{\mathrm{TE}_1} \equiv \frac{c}{\sqrt{\epsilon_r'}\sqrt{1-\left(\frac{f_c^{\mathrm{TE}_1}}{f}\right)^2}} = \frac{3\times10^8}{\sqrt{2.25}\sqrt{1-\left(\frac{2}{3}\right)^2}} = 2.68\times10^8\ \mathrm{ms}^{-1}$$

$$v_{ph}^{\mathrm{TM}_1} \equiv v_{ph}^{\mathrm{TE}_1} = 2.68\times10^8\ \mathrm{ms}^{-1}$$

(e) The group velocity all the propagating modes may be calculated as

$$v_g^{\mathrm{TEM}} \equiv v\sqrt{1-\left(\frac{f_c^{\mathrm{TEM}}}{f}\right)^2} = \frac{c}{\sqrt{\epsilon_r'}} = \frac{3\times10^8}{\sqrt{2.25}} = 2\times10^8\ \mathrm{ms}^{-1}$$

$$v_g^{\mathrm{TE}_1} \equiv v\sqrt{1-\left(\frac{f_c^{\mathrm{TE}_1}}{f}\right)^2} = \frac{3\times10^8}{\sqrt{2.25}}\sqrt{1-\left(\frac{2}{3}\right)^2} = 1.49\times10^8\ \mathrm{ms}^{-1}$$

$$v_g^{\mathrm{TM}_1} \equiv v_g^{\mathrm{TE}_1} = 1.49\times10^8\ \mathrm{ms}^{-1}$$

It can be observed here that for the TEM mode, the group velocity is same as the phase velocity which is equal to the velocity of the wave inside the unguided media. However, for all the other higher order waveguide propagating modes, the phase velocity is more than the velocity of the wave inside the unguided media. On the contrary, the group velocity is less than the velocity of the wave inside the unguided media for all the other higher order propagating modes

(f) The wavelength inside the guided media for all the propagating modes may be computed using the following expressions:

$$\lambda_g^{\mathrm{TEM}} = \frac{2\pi}{\beta_z^{\mathrm{TEM}}} \equiv \frac{2\pi}{\beta} \equiv \lambda = \frac{2\pi}{94.25} = 6.67\ \mathrm{cm}$$

$$\lambda_g^{\mathrm{TE}_1} = \frac{2\pi}{\beta_z^{\mathrm{TE}_1}} \equiv \frac{\lambda}{\sqrt{1-\left(\frac{f_c^{\mathrm{TE}_1}}{f}\right)^2}} = \frac{6.67}{\sqrt{1-\left(\frac{2}{3}\right)^2}} = 8.95\ \mathrm{cm}$$

$$\lambda_g^{\mathrm{TM}_1} \equiv \lambda_g^{\mathrm{TE}_1} = 8.95\ \mathrm{cm}$$

where $\lambda = 6.67$ cm is the wavelength in the unbounded dielectric media. It can be observed that for any waveguide mode, the wavelength inside the guided media is more than the free-space wavelength. However, for the TEM wave propagation, the wavelength inside the guided media is same as the free-space wavelength.

D15.1 Determine the group velocity of the m = 1 (TE or TM) mode in an air-filled parallel-plate guide with d = 0.5 cm at f = (a) 30 GHz, (b) 60 GHz, and (c) 100 GHz.

Ans. 0; 2.6 $\times 10^8$ m/s; 2.9 $\times 10^8$ m/s

D15.2 A TE mode in a parallel-plate guide is observed to have three maxima in its electric field pattern between x = 0 and $x = d$. What is the value of m?

Ans. 3

15.2 RECTANGULAR WAVEGUIDES

In the last section, we discussed about the parallel-plate waveguide in detail, and accordingly various expressions representing the wave propagation behaviour along this guide were derived. It was also mentioned that the parallel plate waveguide could support the TEM mode apart from the TE and TM modes because of its typical two conductor geometry. However, there are certain types of structures having single hollow conductors, which can not support TEM waves. These structures are widely being used in the microwave frequency range, and the rectangular waveguide having a uniform rectangular cross-section throughout its length represents such type of structure as briefly mentioned in Chapter 14. The geometry of a typical rectangular waveguide structure is shown in Figure 15.7, where the electromagnetic wave propagation is still considered to be along the z-direction. It may be mentioned here that the single hollow conductors having a circular cross section, known as the *circular waveguides*, are also available. However, the rectangular waveguides are more commonly being used because of certain advantages such as the widest possible bandwidth for a single mode of operation, and hence we limit our discussion to rectangular waveguides in this book.

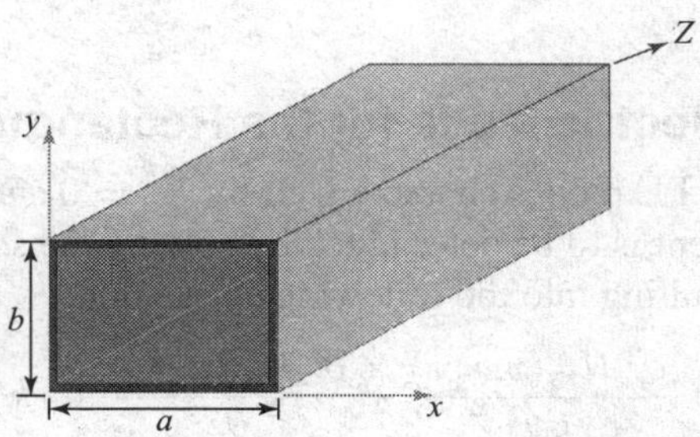

Fig. 15.7 The rectangular waveguide geometry.

For the parallel-plate waveguide geometry discussed in the last section, the width of each conducting plate was considered to be very large thereby limiting the spatial variation of fields only along one of the transverse directions. The parallel-plate waveguide shown in Figure 15.1 may be transformed into the rectangular waveguide by making the width of each plate to be finite, and placing two conductors oriented vertically at two different horizontal positions. The resulting rectangular waveguide geometry is shown in Figure 15.7, where it can be observed that the cross section of the structure now represents a rectangular geometry having width a along the x-direction and height b along the y-direction. The enclosure of the waveguide geometry along both the x and y-directions has the consequence that the electromagnetic fields have now spatial variation in both x and y-directions apart from the longitudinal z-direction. The walls of the guide are assumed to be made of perfect electric conductors, and the inner region of the guide is filled with a lossless dielectric having permittivity ϵ, and permeability μ. Our basic aim here is to derive the electric and magnetic field expressions for the arbitrary TE and TM modes since the basic TEM mode propagation for this kind of geometry is ruled out.

The TE and TM mode expressions for the rectangular waveguide geometry can be obtained by solving the Maxwell's equation subject to the boundary conditions imposed on all the four conducting walls shown in Figure 15.7. The overall procedure is similar to the methodology adopted in the last section to derive various mode configurations for the parallel plate structure except that the new sets of boundary conditions have to be imposed, and the spatial variation of fields in all three directions should be taken into account.

For simplifying the analysis, the inner source-free region of the rectangular waveguide is assumed to be filled with a simple (homogeneous, isotropic and linear), lossless medium. The total wave number inside the waveguide is denoted by symbol k as per (3) assuming the attenuation constant α to be zero. The propagation of waves along the z-direction is described by $e^{-j\beta_z z}$ factor as per (4) with β_z being the propagation constant inside the waveguide along the direction of propagation (z). The solution of TE and TM waves supported by the rectangular waveguide is obtained in two basic steps as discussed in the last section. The first step involves the determination of the E_z or the H_z component using the scalar form of (7) subject to the boundary condition appropriate for the rectangular waveguide geometry. In the second step, the transverse components of both the electric and magnetic fields are computed in terms of the corresponding longitudinal field components using the generalized expressions (15) and (16), where again the boundary conditions corresponding to the rectangular waveguide geometry are to be applied in order to arrive at the specific solution. In the following paragraphs, the analysis of TE and TM waves describing the propagation behavior of various modes supported by the rectangular waveguide geometry is given in detail.

15.2.1 Transverse Electric Mode for the Rectangular Waveguide

We start the discussion with TE modes characterized by $E_z = 0$ and $H_z \neq 0$. In this case, first of all the $H_{zs}(x, y, z)$ component is to be determined by solving the scalar form of the second-order differential equation (2b) resulting into the following expression:

$$\frac{\partial^2 H_{zs}(x,y,x)}{\partial x^2} + \frac{\partial^2 H_{zs}(x,y,z)}{\partial y^2} + \frac{\partial^2 H_{zs}(x,y,z)}{\partial z^2} + k^2 H_{zs}(x,y,z) = 0 \tag{47}$$

where, we have made use of (3) assuming the waveguide to be filled with the lossless media. Now, the solution of (47) is not straightforward as the longitudinal field component H_{zs} is function of all three variables x, y and z. One of the methods to solve such types of differential equations is the method known as the separation of variables, which may be considered as the modified form of approach introduced in Chapter 12 to solve the Laplace's equation involving potential function of two variables. In order to use this method, let us assume that the field component H_{zs} can be expressed as the product of three quantities as follows:

$$H_{zs}(x,y,z) \equiv X(x)\,Y(y)\,Z(z) \tag{48}$$

where, X, Y and Z are *exclusive* functions of x, y, and z, respectively. The above equation can next be substituted into (47) to obtain

$$Y\,Z\,\frac{d^2\,X(x)}{dx^2} + X\,Z\,\frac{d^2\,Y(y)}{dy^2} + X\,Y\,\frac{d^2\,Z(z)}{dz^2} + k^2\,X(x)Y(y)Z(z) = 0 \tag{49}$$

where, the partial derivatives are replaced with full derivatives as $X(x)$, $Y(y)$ and $Z(z)$ depend only on x, y and z, respectively. The above equation can then be simplified as follows:

$$\frac{1}{X(x)}\frac{d^2\,X(x)}{dx^2} + \frac{1}{Y(y)}\frac{d^2\,Y(y)}{dy^2} + \frac{1}{Z(z)}\frac{d^2\,Z(z)}{dz^2} + k^2 = 0 \tag{50a}$$

$$\frac{1}{X(x)}\frac{d^2\,X(x)}{dx^2} + \frac{1}{Y(y)}\frac{d^2\,Y(y)}{dy^2} + \frac{1}{Z(z)}\frac{d^2\,Z(z)}{dz^2} = -\,k^2 \tag{50b}$$

Now, the right-hand side of (50*b*) represents the total wave number of the media defined by (3) having no spatial dependence. Moreover, each term on left-hand side of (50*b*) involves only single variable x, y or z. The above equation can be satisfied for all values of x, y and z only under the condition when each individual term on left-hand side of (50*b*) becomes a constant. Using this concept, (50*b*) can now be separated into three equations yielding

$$\frac{1}{X(x)}\frac{d^2\,X(x)}{dx^2} = -\,k_x^2 \tag{51a}$$

$$\frac{1}{Y(y)}\frac{d^2\,Y(y)}{dy^2} = -\,k_y^2 \tag{51b}$$

$$\frac{1}{Z(z)}\frac{d^2\,Z(z)}{dz^2} = -\,\beta_z^2 \tag{51c}$$

where, the individual constants k_x and k_y may be considered as the wave numbers along the x- and y-directions, respectively, while the β_z represents the propagation constant along the direction of propagation as mentioned earlier. The three constants may be related with the total propagation constant using the above equation as follows:

$$k_x^2 + k_y^2 + \beta_z^2 = k^2 \equiv \omega^2\,\mu\epsilon \tag{52}$$

Meanwhile, the three equations of (51) may be rewritten as

$$\frac{d^2\,X(x)}{dx^2} + k_x^2 X(x) = 0 \tag{53a}$$

$$\frac{d^2\,Y(y)}{dy^2} + k_y^2 Y(y) = 0 \tag{53b}$$

$$\frac{d^2\,Z(z)}{dz^2} + \beta_z^2 Z(z) = 0 \tag{53c}$$

It can now clearly be observed that (53*a*) to (53*c*) represent second-order simple harmonic differential equations whose general solution can usually be written in the form of sinusoidal or exponential functions. As the wave is assumed to be travelling along the z-direction, the exponential function of the form $e^{\pm j\beta_z z}$ can be used to represent the solution of (53*c*). For the solution of (53*a*) or (53*b*), the general sin or cos functions of the form similar to (19) can be used to stress the fact that there would be standing waves along x or y directions as there is no real wave propagation along these two directions. Using this concept, the general solution of (48) can be written as follows:

$$H_{zs}(x, y, z) \equiv [A_1 \cos(k_x x) + B_1 \sin(k_x x)] [A_2 \cos(k_y y) + B_2 \sin(k_y y)] e^{-j\beta_z z} \tag{54}$$

where, the wave is considered to be travelling along the $+z$-direction, and A_1, B_1, A_2, B_2, are arbitrary constants whose values are to be determined by applying the appropriate boundary conditions satisfying the rectangular waveguide geometry. It can readily be verified that (54) is the solution of the differential equation (47). In the next step, the transverse and longitudinal spatial variation can be separated by rewriting (54) into the following form:

$$H^t_{zs}(x, y) \equiv [A_1 \cos(k_x x) + B_1 \sin(k_x x)] [A_2 \cos(k_y y) + B_2 \sin(k_y y)] \tag{55}$$

where, $H_{zs}(x, y, z) \equiv H^t_{zs}(x, y)\, e^{-j\beta_z z}$, and (55) now looks similar in form as (19) with the difference that the field variation in the present situation is along both x- and y-directions.

Now, in order to determine various constants of (55), the boundary conditions on field components corresponding to the rectangular waveguide geometry of Figure 15.7 are to be applied. If all the four walls of the rectangular waveguide are considered to be made of perfect electric conductor, the boundary condition would require that the tangential electric fields on all the metallic walls should be zero as mentioned earlier. The tangential electric fields on the top and bottom walls in Figure 15.7 are E_x and E_z components while the tangential fields on side walls are E_y and E_z components leading to the following boundary conditions:

$$E^t_{xs}(y = 0) \equiv E^t_{xs}(y = b) = 0 \tag{56a}$$

$$E^t_{ys}(x = 0) \equiv E^t_{ys}(x = a) = 0 \tag{56b}$$

where, we have made use of the fact that the E_{zs} component is zero for the TE case. Now, in order to apply the boundary condition given by (56), the expressions of various field components for the rectangular waveguide geometry should first be determined. These transverse field expressions are still given by (21) and (22) with the difference that the fields in the present situation are varying along both x- and y-directions. Now, let us first apply the boundary conditions at $x = 0$ by determining the E^t_{ys} components using (21):

$$\begin{aligned} E^t_{ys}(x, y) &\equiv \frac{j\omega\mu}{k^2 - \beta_z^2} \frac{\partial H^t_{zs}}{\partial x} \\ &= \frac{j\omega\mu}{k^2 - \beta_z^2} k_x \, [A_2 \cos(k_y y) + B_2 \sin(k_y y)] \, [-A_1 \sin(k_x x) + B_1 \cos(k_x x)]_{x=0} = 0 \\ &\Rightarrow \frac{j\omega\mu}{k^2 - \beta_z^2} k_x \, [A_2 \cos(k_y y) + B_2 \sin(k_y y)] \, B_1 = 0 \\ &\Rightarrow B_1 = 0 \end{aligned} \tag{57}$$

where, we have made use of the fact that the terms related to the spatial variation in the y-direction in the above equation can not be made equal to zero as our aim here is to determine the condition for the field variation in the x-direction. The factor k_x cannot be made zero as it would amount to no variation of field along the x-direction. The value of B_1 computed above is next substituted in (55) in order to apply the boundary condition defined by (56*b*) at $x = a$ resulting into the following equation:

$$E_{ys}^t(x,y) = \frac{j\omega\mu}{k^2-\beta_z^2}k_x\,[A_2\cos(k_y y)+B_2\sin(k_y y)]\,[-A_1\sin(k_x x)]_{x=a} = 0$$

$$\Rightarrow \sin(k_x a) = 0 \equiv \sin(m\pi) \tag{58}$$

$$\Rightarrow k_x = \frac{m\pi}{a} \qquad \text{where } m = 0,1,2,\ldots$$

After applying boundary conditions at $x = 0$ and $x = a$, the boundary conditions at $y = 0$ and $y = b$ are to be applied. For this purpose, let us find the expression of E_{xs}^t using (21*b*) and (55) as follows:

$$E_{xs}^t(x,y) \equiv -\frac{j\omega\mu}{k^2-\beta_z^2}\frac{\partial H_{zs}^t}{\partial y}$$

$$= -\frac{j\omega\mu}{k^2-\beta_z^2}k_y\,[A_1\cos(k_x x)]\,[-A_2\sin(k_y y)+B_2\cos(k_y y)]_{y=0} = 0 \tag{59}$$

$$\Rightarrow -\frac{j\omega\mu}{\beta^2-\beta_z^2}k_y\,[A_1\cos(k_x x)]\,B_2 = 0$$

$$\Rightarrow B_2 = 0$$

The above formulation follows from the fact that the factor k_y and other terms related to the spatial variation in the x-direction in a general case cannot be made equal to zero as the boundary condition is applied here for the y-direction. The value of B_2 computed above can next be substituted in (55), and then the boundary condition defined by (56*a)* at $y = b$ can be applied for E_{xs}^t resulting into the following equation:

$$E_{xs}^t(y=b) \equiv -\frac{j\omega\mu}{k^2-\beta_z^2}k_y\,[A_1\cos(k_x x)]\,[-A_2\sin(k_y y)]_{y=b} = 0$$

$$\Rightarrow \sin(k_y b) = 0 \equiv \sin(n\pi) \tag{60}$$

$$\Rightarrow k_y = \frac{n\pi}{b} \qquad \text{where } n = 0,1,2,\ldots$$

where, n is an integer as defined above, and it has again been made use of the fact that the constant A_2 in a general case cannot be made equal to zero. It is to be noted that both k_x and k_y can assume only discrete values with $m = 0,1,2,3\ldots$ and $n = 0,1,2,3\ldots$ known as *eigen* values. The field configuration corresponding to each m, n represents a particular mode, which means that the rectangular waveguide can also support infinite number of modes just like the parallel-plate geometry. The propagation constant along the direction of wave travel (β_z) in the rectangular waveguide can be written in the following form with the help of (52), (58) and (60):

$$\beta_z = \sqrt{k^2-(k_x^2+k_y^2)} \equiv \sqrt{\omega^2\mu\epsilon - \left[\left(\frac{m\,\pi}{a}\right)^2+\left(\frac{n\,\pi}{b}\right)^2\right]} \tag{61}$$

It can be seen from (61) and (52) that if $k_x = k_y = 0$, then the z-directed propagation constant β_z becomes equal to the total propagation constant β defining the wave propagation along the z direction as $e^{-j\beta z}$. The cutoff condition for the rectangular waveguide geometry may be obtained by making $\beta_z = 0$ in (61), i.e.,

$$0 = k_c^2 - (k_x^2+k_y^2) \equiv \omega_c^2\,\mu\epsilon - \left[\left(\frac{m\,\pi}{a}\right)^2+\left(\frac{n\,\pi}{b}\right)^2\right] \tag{62}$$

which provides the cutoff wave number as follows:

$$k_c^2 \equiv k_x^2 + k_y^2 = k^2 - \beta_z^2 = \left(\frac{m\,\pi}{a}\right)^2 + \left(\frac{n\,\pi}{b}\right)^2 \tag{63}$$

The cutoff frequency $f_c \equiv \omega_c/2\pi$ of the TE_{mn} mode in the rectangular waveguide from the above expression may be computed as

$$f_c = \frac{1}{2\sqrt{\mu\epsilon}}\sqrt{\left(\frac{m}{a}\right)^2 + \left(\frac{n}{b}\right)^2} \qquad m = 0, 1, 2, 3, \ldots; n = 0, 1, 2, 3, \ldots; \tag{64}$$

It can be seen from the above equation that although m and n can attain the values starting from 0, both the integers cannot have the value of 0 *simultaneously*. In other words the condition $m = n = 0$, which would make the cutoff frequency f_c to be zero is totally ruled out. The reason for avoiding this condition would be evident in the later part of the discussion when the field expressions are derived. If the geometry of the rectangular waveguide shown in Figure 15.7 is chosen in such as way that $a > b$, then the mode having the lowest cutoff frequency becomes TE_{10} mode which is the *dominant* mode for the rectangular waveguide geometry. The propagation constant in the direction of propagation (β_z) for the rectangular waveguide can be determined using (29) and (30) depending upon whether the operating frequency f is more or less than the cutoff frequency f_c given by (64). As discussed for the parallel-plate geometry, the actual propagation of wave occurs only when the operating frequency f becomes greater than the cutoff frequency f_c making the propagation constant β_z to be a purely real quantity.

Now, once the propagation constants for the rectangular waveguide geometry are defined, the non-zero field components for the general TE to z mode corresponding to this geometry can be derived using (21), (22) and (54) to have the following expressions:

$$H_{zs}(x, y, z) = A_{mn} \cos\left(\frac{m\,\pi}{a}x\right) \cos\left(\frac{n\,\pi}{b}y\right) e^{-j\beta_z z} \tag{65a}$$

$$E_{xs}(x, y, z) = A_{mn} \frac{j\omega\mu}{k_c^2}\frac{n\pi}{b} \cos\left(\frac{m\,\pi}{a}x\right) \sin\left(\frac{n\,\pi}{b}y\right) e^{-j\beta_z z} \tag{65b}$$

$$E_{ys}(x, y, z) = -A_{mn} \frac{j\omega\mu}{k_c^2}\frac{m\pi}{a} \sin\left(\frac{m\,\pi}{a}x\right) \cos\left(\frac{n\,\pi}{b}y\right) e^{-j\beta_z z} \tag{65c}$$

$$H_{xs}(x, y, z) = A_{mn} \frac{j\beta_z}{k_c^2}\frac{m\pi}{a} \sin\left(\frac{m\,\pi}{a}x\right) \cos\left(\frac{n\,\pi}{b}y\right) e^{-j\beta_z z} \tag{65d}$$

$$H_{ys}(x, y, z) = A_{mn} \frac{j\beta_z}{k_c^2}\frac{n\pi}{b} \cos\left(\frac{m\,\pi}{a}x\right) \sin\left(\frac{n\,\pi}{b}y\right) e^{-j\beta_z z} \tag{65e}$$

where, $k_c^2 = k^2 - \beta_z^2$ as per (61) and (63), and A_{mn} represents the strength of the particular m, n field configuration or mode. It can be noted from above sets of equations that $m = n \neq 0$ although both m and n can attain values starting from 0. The reason is that the condition $m = n = 0$ would result into all the transverse field components being equal to zero as per (65*b*) to (65*e*), which would ultimately not allow any real power flow along the direction of propagation (z-direction). It should also be noted that k_c and β_z in above sets of equations can assume only discrete values with each value of m and n specifying a particular field configuration or TE_{mn} *mode* as mentioned earlier. The mode numbers m and n represent the half wave field variations along x- and y-directions, respectively.

The wave impedance for the TE_{mn} propagating waves traveling along the + z-direction for the rectangular waveguide in the present situation can be defined as

$$Z_w^{\text{TE}_{mn}} \equiv \frac{E_{xs}}{H_{ys}} \equiv -\frac{E_{ys}}{H_{xs}} = \frac{\omega\mu}{\beta_z} = \frac{\eta}{\sqrt{1-\left(\frac{f_c}{f}\right)^2}} \qquad f > f_c \tag{66}$$

where η is the intrinsic impedance of the medium filling the inner region of the waveguide, and f_c is given by (64). It is again to be noted that the wave impedance for the propagating wave ($f > f_c$) is a real quantity indicating the actual power flow along z-direction. The expressions for the phase velocity and the group velocity are still given by (32) and (33) with the cutoff frequency f_c given by (64).

As mentioned earlier, the dominant mode for the rectangular waveguide is TE_{10} mode for the condition $a > b$. The rectangular waveguide is usually operated in the TE_{10} mode for all practical purposes, and the waveguide cross-sectional dimensions are chosen in such a way that the single-mode propagates over maximum possible bandwidth.

As the TE_{10} mode is most commonly used for the rectangular waveguide geometry, the non-zero field configurations corresponding to this mode are given below:

$$H_{zs}(x,z) = A_{10}\cos\left(\frac{\pi}{a}x\right)e^{-j\beta_z z} \tag{67a}$$

$$E_{ys}(x,z) = -A_{10}\frac{j\omega\mu\, a}{\pi}\sin\left(\frac{\pi}{a}x\right)e^{-j\beta_z z} \tag{67b}$$

$$H_{xs}(x,z) = A_{10}\frac{j\beta_z\, a}{\pi}\sin\left(\frac{\pi}{a}x\right)e^{-j\beta_z z} \tag{67c}$$

where, $k_c = \frac{\pi}{a}$ for TE_{10} mode is substituted in the above equation. The cutoff frequency for the TE_{10} mode is calculated as

$$f_c^{\text{TE}_{10}} = \frac{1}{2\,a\sqrt{\mu\epsilon}} \equiv \frac{v}{2\,a} \equiv \frac{v}{\lambda_c} \tag{68}$$

where, v is the velocity in the unbounded media, and $\lambda_c = 2a$ is the cutoff wave number for the TE_{10} mode. It is obvious from above equation that the operating frequency ($f > f_c$) of the rectangular waveguide will depend upon the broad dimension a of the waveguide for the TE_{10} dominant mode of operation. In other words, the dimension a of the waveguide is chosen in such way the cutoff frequency for the TE_{10} mode is less than the lowest frequency at which the waveguide is to be operated. As the waveguide is usually operated in such a way that only a single mode propagates over the whole band, the highest frequency of operation will depend upon the cutoff frequency of the next higher order mode.

The field lines for the dominant TE_{10} mode in the rectangular waveguide are shown in Figure 15.8, which are obtained using (67*a*) to (67*c*) by converting the phasor quantities into their corresponding instantaneous forms and then considering $t = 0$. The electric field is plotted in the xy-plane while the magnetic field is plotted in the xz-plane as shown in Figures 15.8(*a*) and 15.8(*b*), respectively. The strength and the direction of the field with respect to position in these plots are represented by arrows, with the arrow size being directly proportional to the field strength. For example, the electric field is fully oriented along the y-direction in Figure 15.8(*a*) having maximum value at the center ($x = a/2$), and going to zero along the boundary at $x = 0$ and $x = a$ satisfying the boundary condition that the tangential component should be zero at these two perfectly conducting walls. Now, you might recall that the electric field lines between

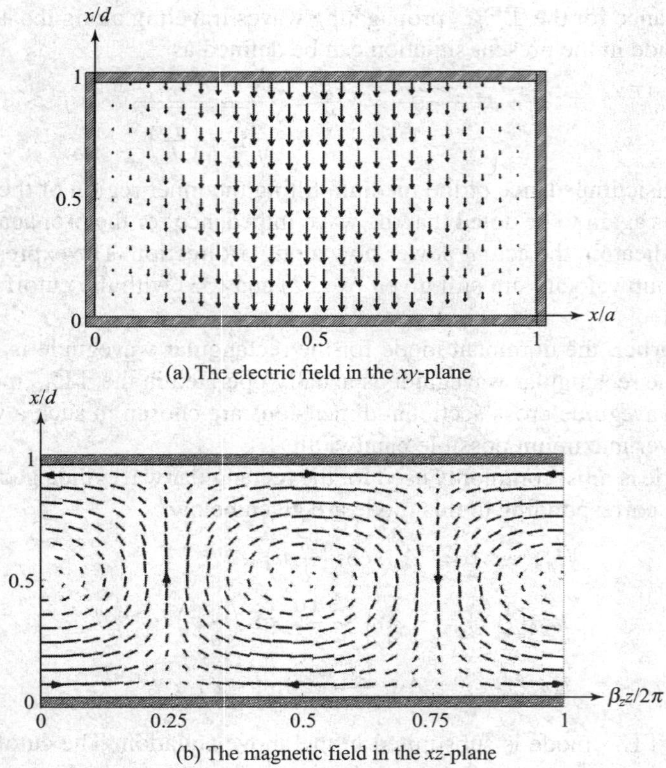

(a) The electric field in the *xy*-plane

(b) The magnetic field in the *xz*-plane

Fig. 15.8 The field lines for the TE_{10} mode in the rectangular waveguide.

two parallel-plate conductors always start from one of the conductors and terminate at the other conductor as also seen from Figure 15.6. In that sense, we can say that the electric field lines shown in Figure 15.8(a) in the xy-plane are complete, and hence no component of the electric field would exist along the direction of propagation (z-direction) which makes this plot corresponding to the TE mode. It is also easy to observe from Figure 15.8(a) that only y-component of the electric field exists with one half wavelength variation of the field along the x-axis and no variation along the y-direction. Similarly, a close analysis of Figure 15.8(b) shows that both x- and z-components of the magnetic field components are present, and that these field lines form a closed loop in the xz-plane. Now, you know that the magnetic field lines always exist in the closed loop which basically means that the magnetic lines are complete for this particular mode in the xz-plane. If you try to plot the magnetic field lines for this mode in the xy-plane, then you would see completely horizontal lines which would signify that an additional component of the magnetic field should exist along the z-direction in order to complete the closed loop. It is mainly due to this reason that the magnetic field lines are plotted in the xz-plane for this particular mode.

EXAMPLE 15.4

An air-filled rectangular waveguide has dimensions a = 2 cm and b = 1 cm. Determine the range of frequencies over which the guide will operate single mode (TE_{10}).

Solution. Since the guide is air-filled, $n = 1$, and (82) gives, for $m = 1$: $f_c(10) = \omega_c/2\pi = c/2a = 3 \times 10^{10}/2(2) = 7.5$ GHz

The next higher-order mode will be either TE_{20} or TE_{01}, which, from (82) and (88), will have the same cutoff frequency since $a = 2b$. This frequency will be twice that found for TE_{10}, or 15 GHz. Thus, the operating frequency range over which the guide will be single mode is 7.5 GHz $< f <$ 15 GHz.

EXAMPLE 15.5

The cross-sectional dimensions of a rectangular waveguide is expressed as $a \times b$ with a being the broader dimension of the guide ($a > b$). The waveguide is filled with air, and is to be operated in the dominant mode. It is also required that the bandwidth, defining the frequency range over which only the dominant mode propagates, should be maximized. Explain the procedure to choose the dimensions a and b of the guide satisfying the above conditions.

Solution. The dominant mode of the rectangular waveguide ($a > b$) is the TE_{10} mode, whose cutoff frequency assuming the guide to be filled with air is given by

$$f_c^{TE_{10}} = \frac{c}{2a} \equiv \frac{c}{\lambda_c^{TE_{10}}} \tag{1i}$$

where, c is the velocity of the electromagnetic wave in free space, and $\lambda_c^{TE_{10}} \equiv 2a$ is the cutoff wavelength for the dominant mode. Now, if the waveguide is to be operated in the dominant mode, then the operating frequency should be greater than the cutoff frequency given by (1i), i.e.,

$$f \geq f_c^{TE_{10}} \Rightarrow \frac{c}{\lambda} \geq \frac{c}{\lambda_c^{TE_{10}}}$$

$$\Rightarrow \lambda \leq \lambda_c^{TE_{10}} \tag{1ii}$$

where, $\lambda \equiv c/f$ is the free-space wavelength corresponding to the operating frequency. It basically means that for the dominant-mode propagation, the broader dimension of the guide a should be chosen as per the specified cutoff frequency or the cutoff wavelength given by (1i), and (1ii), respectively. As a typical example, let us assume that you want to design a waveguide such that the cutoff frequency is 6.55 GHz, which would then give the broader dimension of the guide as

$$f_c^{TE_{10}} = 6.55\text{ GHz} \Rightarrow \lambda_c^{TE_{10}} \equiv 2a = \frac{3 \times 10^{10}}{6.55 \times 10^9} = 4.58\text{ cm}$$

$$\Rightarrow a = \frac{4.58}{2} = 2.29\text{ cm} \tag{1iii}$$

Now, once the dimension a is fixed, the criterion for choosing the dimension b is to be evolved. The basic condition here is to maximize the bandwidth or the frequency range over which *only* the dominant mode will propagate in the guide. This condition will be met when the maximum operating frequency of the guide is less than the cutoff frequency of the next possible mode in the waveguide. The next sets of possible modes in the rectangular waveguide are TE_{01}, TE_{20}, TE_{11}/TM_{11}, where TE_{11}/TM_{11} are degenerate modes having same cutoff frequency. The cutoff frequencies of these modes can be determined as follows:

$$f_c^{TE_{20}} = \frac{c}{a} \equiv 2 f_c^{TE_{10}} = 13.1\text{ GHz} \tag{1iv}$$

It is obvious from the above equation that the cutoff frequency of the TE_{20} mode is twice the cut-off frequency of the dominant mode, and it does not depend upon the dimension b. Hence, if it can be ensured that the TE_{20} mode is the next possible mode in the waveguide after the dominant mode, then the maximum bandwidth of one octave $\left(\frac{f_c^{TE20}}{f_c^{TE10}} = 2\right)$ for the dominant mode propagation can be obtained in the guide. However, in order to ensure this condition, it is necessary that the cut-off frequencies of all the other higher order modes should be greater that of the TE_{20} mode as shown below:

$$\left[f_c^{\text{TE}_{01}} \equiv \frac{c}{2b}\right] \geq \left[f_c^{\text{TE}_{20}} \equiv \frac{c}{a}\right]$$

$$\Rightarrow \quad \frac{a}{b} \geq 2; \text{ or, } \; a \geq 2b \tag{1v}$$

Similarly, the condition satisfying the TE_{11}/TM_{11} mode can be evolved as follows:

$$\left[f_c^{\text{TE}_{11}} \equiv f_c^{\text{TM}_{11}} = \frac{c}{2}\sqrt{\frac{1}{a^2}+\frac{1}{b^2}}\right] \geq \left[f_c^{\text{TE}_{20}} \equiv \frac{c}{a}\right]$$

$$\Rightarrow \quad \sqrt{\frac{1}{a^2}+\frac{1}{b^2}} \geq \frac{2}{a}; \text{ or, } \sqrt{1+\frac{a^2}{b^2}} \geq 2 \text{ or, } \left(\frac{a^2}{b^2}\right) \geq 3 \quad \text{or,} a \geq 1.73b \tag{1vi}$$

Now, from $(1v)$ and $(1vi)$, it can be seen that if the criterion $a \geq 2b$ is satisfied, then the cutoff frequency of all other higher order modes in the rectangular waveguide will be greater than that of the TE_{20} mode, and the bandwidth of the dominant-mode propagation can be maximized.

In summary, we can say that the broader dimension a of the rectangular waveguide $(a > b)$ is determined by the cutoff frequency of the dominant TE_{10} mode. For maximizing the bandwidth of the dominant mode propagation in the guide, the dimension b of the guide should be chosen in such a way that $b \leq \frac{a}{2}$. The maximum bandwidth obtained in this case is typical one octave $\left(\frac{f_c^{\text{TE}_{20}}}{f_c^{\text{TE}_{10}}} = 2\right)$, where $f_c^{\text{TE}_{20}}$ is the cut-off frequency of the next possible mode in the guide.

It should be noted that the actual frequency range of operation of the waveguide ranges between 25% above the cutoff frequency of the dominant mode, and 90–95 % of the cutoff frequency of the next possible mode. For example, in the present situation, the frequency range of operation of the waveguide will be given by

$$1.25 f_c^{\text{TE}_{10}} \leq f \leq 0.95 f_c^{\text{TE}_{20}} \text{ GHz}$$

$$\Rightarrow 1.25 \times 6.55 \text{ GHz} \leq f \leq 0.95 \times 13.1 \text{ GHz}$$

$$\Rightarrow 8.2\text{GHz} \leq f \leq 12.4 \text{ GHz}$$

You should note that the X-band rectangular waveguide (2.286 cm × 1.016 cm) shown in Fig. 15.14 is actually operated in the above frequency range, and its cross-sectional dimensions satisfy the conditions given in this problem.

Now, as the dominant TE_{10} mode is commonly being used for most of the practical applications, it is useful to briefly discuss about the coupling mechanism specially suited to this particular mode. The coupling mechanism can be briefly described as the process to excite the particular mode inside the waveguide from the external source in such a way that maximum energy is transferred from the source to the waveguide. As most of the low power microwave test sources have the coaxial output, the coupling for the dominant-mode excitation in the rectangular waveguide usually involves the effective transfer of power from a coaxial line into the waveguide supporting the TE_{10}

mode by means of electric field coupling. This is typically achieved by selecting a thin linear probe, usually extension of the inner conductor of the coaxial line, and placing it at the position of maximum electric field. The orientation of the probe is made parallel to the electric field line of the waveguide, which would be along the y-direction for the dominant mode excitation. Now, as obvious from Figure 15.8(a), the electric field is maximum at $x = a/2$ in the xy-plane and hence the vertical probe should be placed at this position. For determining the probe position along the z-direction, the convention is to put a short plate at $z = 0$ and then place the probe at $\lambda_g/4$ away from the short as shown in Figure 15.9(a) where λ_g is the guide wavelength. You may recall from discussions in earlier chapters that the electric field will be zero at the short position, and hence the maximum of the electric field is observed at a distance of $\lambda_g/4$ away from the short. The coupling mechanism can be understood in a clear way with the help of Figure 15.9(b), which shows a typical coaxial-waveguide adapter for the Ka-band operation having cross-sectional dimensions of 7.11×3.55 mm. The coaxial probe in this figure is inserted into a piece of rectangular waveguide at $x = a/2$, and at $z = \lambda_g/4$ from the short end as explained above. The design of the coaxial-waveguide adapter basically requires the efficient transmission of the microwave power from the coaxial line supporting the TEM mode into the rectangular waveguide supporting the TE_{10} mode.

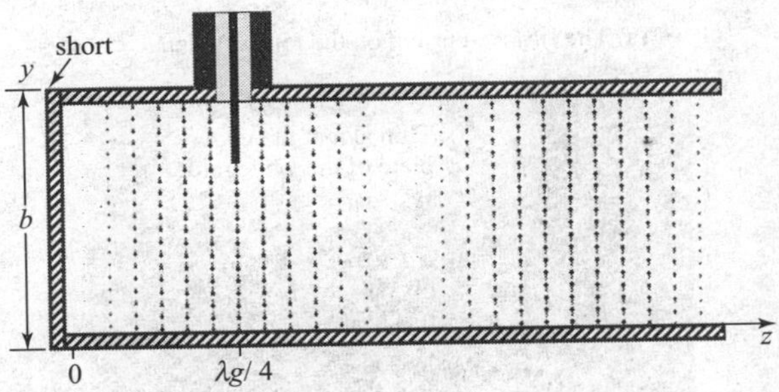

(a) The electric field in the yz-plane after placing a short at $z = 0$

(b) The coaxial-waveguide adapter for the Ka band (26.5–40 GHz)

Fig. 15.9 The coupling mechanism to excite the TE_{10} mode in the rectangular waveguide.

It is also interesting to plot the linear surface current density $\mathbf{J_s} \equiv -\mathbf{a}_y \times \mathbf{H}$ for the TE_{10} mode on the top wall of the rectangular waveguide as shown in Figure 15.10(a), which is obtaine using instantaneous forms of (67a) and (67c) at $t = 0$. It can be observed from Figure 15.10(a) th the surface current on top wall (xz-plane) of the rectangular waveguide becomes almost horizo at $x = a/2$ for the dominant TE_{10} mode. This is very useful for many practical applications s a very thin horizontal slot can be cut at the centre of the top wall of the rectangular waveguide supporting the dominant-mode propagation without causing much radiation into free space. The reason is that a horizontal slot cut at the centre of the top wall of the waveguide is parallel to the surface current at that place as shown in Figure 15.10(*b*), and hence the current distribution on this surface will not be significantly disturbed provided that the slot is quite thin.

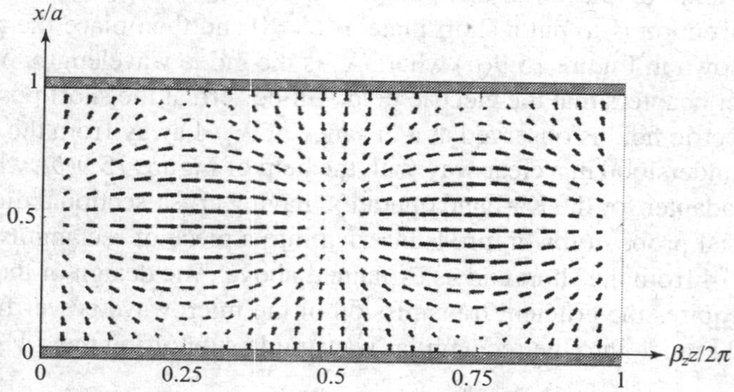

(a) The surface current on the top (*xz*) plane

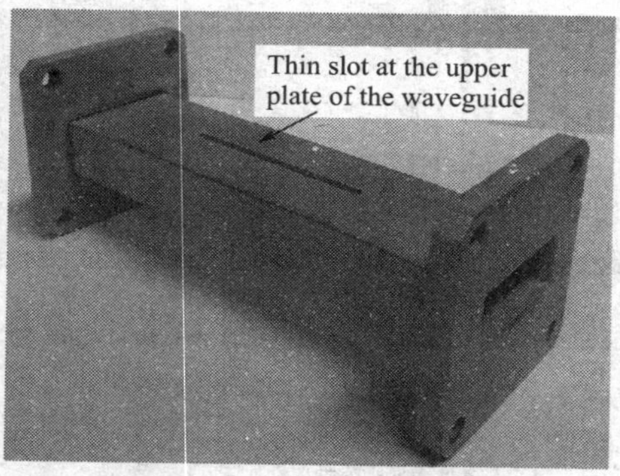

(b) The X-band rectangular waveguide with a slot at its upper wall

Fig. 15.10 The surface current plot for the TE_{10} mode in the rectangular waveguide.

EXAMPLE 15.6

Design an air-filled rectangular waveguide whose cross-sectional dimensions are given as $(a \times b)$ with a being the broader dimension of the guide $(a > b)$. The waveguide is to be operated in the dominant mode with the operating frequency starting from 26.5 GHz. It is lesired to have the widest possible bandwidth for the dominant-mode propagation, and the est operating frequency should be 25% above the cutoff frequency of the dominant mode. nine the highest operating frequency of the designed waveguide, if it is required that st operating frequency should be 95% of the cut-off frequency of the next possible guide. Please give all dimensions in mm.

Solution. The dominant mode of the air-filled rectangular waveguide ($a > b$) is the TE_{10} mode, whose cutoff frequency is given by

$$f_c^{TE_{10}} = \frac{c}{2a} \equiv \frac{3 \times 10^{11}}{2a} \tag{2i}$$

where a is in mm. Now, it is given that the lowest operating frequency (26.5 GHz) should be 25% above the cut-off frequency of the dominant mode, which yields

$$f_{\text{lowest}} = 26.5 \text{ GHz} \equiv 1.25 \times f_c^{TE_{10}} \equiv \frac{1.25 \times 3 \times 10^{11}}{2a}$$

$$\Rightarrow a \equiv \frac{1.25 \times 3 \times 10^{11}}{2 \times 26.5 \times 10^9} \text{ or, } a = 7.075 \text{ mm} \tag{2ii}$$

Once, the value of the broader dimension a of the guide is determined, the next step involves finding the dimension b of the guide. It is required that the bandwidth over which the dominant mode propagates should be maximum. We have seen in the previous example that to obtain the maximum possible bandwidth of the dominant mode, the next possible order mode should be the TE_{20} mode. It also requires that the cutoff frequency of all other higher order modes should be greater than that of the TE_{20} mode. This can only be achieved when the criterion $a \geq 2b$ is satisfied as discussed earlier, which provides the dimension b as

$$\frac{a}{b} \geq 2; \text{ or, } b \leq \frac{a}{2}$$

$$\Rightarrow b = \frac{7.075}{2} = 3.5375 \text{ mm} \tag{2iii}$$

It should be noted that theoretically b can also attain values less than $a/2$ satisfying the above criterion. However, any decrease in b would decrease the power-handling capability of the waveguide. Hence, it is preferable to choose the maximum value of b which satisfies the bandwidth criterion.

The dimensions of the designed waveguide are, therefore, 7.075 mm × 3.537 mm, which completes the design.

Now, it is required to determine the highest operating frequency of the designed waveguide with the condition that it should be 95% of the cut-off frequency of the next possible (TE_{20}) mode of the guide. Hence, the highest frequency can be obtained as

$$f_{\text{highest}} = 0.95 \times f_c^{TE_{20}} \equiv 0.95 \times \frac{c}{a} \equiv \frac{0.95 \times 3 \times 10^{11}}{7.075} = 40.0 \text{ GHz} \tag{2iv}$$

Hence, the designed waveguide will work in the frequency range of 26.5 GHz–40 GHz satisfying all the criteria specified above. You should note that the dimensions of the designed waveguide here actually correspond to the Ka-band waveguide shown in Fig. 15.9(b). This waveguide is also designated as the WR-28 as described in Fig. 15.5, which comes from the fact that the broader dimension a of the waveguide in this case is 0.28 inch. Similarly, the X-band waveguide shown in Fig. 15.14 is designated as the WR-90 since the broader dimension a of the waveguide in that case becomes 0.90 inch.

Now, after observing the field plots for the dominant mode, the field patterns for some other modes can also be investigated. The field patterns for the TE_{20} mode in the rectangular waveguide are plotted in Figure 15.11, which are obtained using instantaneous forms of (65a) to (65e)

$t = 0$ by substituting $m = 2$ and $n = 0$. The plots for the TE_{20} mode can be explained on similar lines as those of TE_{10} mode except for the fact that now there are two half-wavelength variations of the electric field along the x-axis with no variation along the y-direction as shown in Figure 15.11(a). Similarly, the magnetic field lines still exist in the closed loop in the xz-plane as shown in Figure 15.11(a). However, there are now two loops along the x-direction corresponding to $x = 2$ as obvious from Figure 15.11(b), which is in contrast to only one loop seen in Figure 15.8(b) corresponding to the TE_{10} mode.

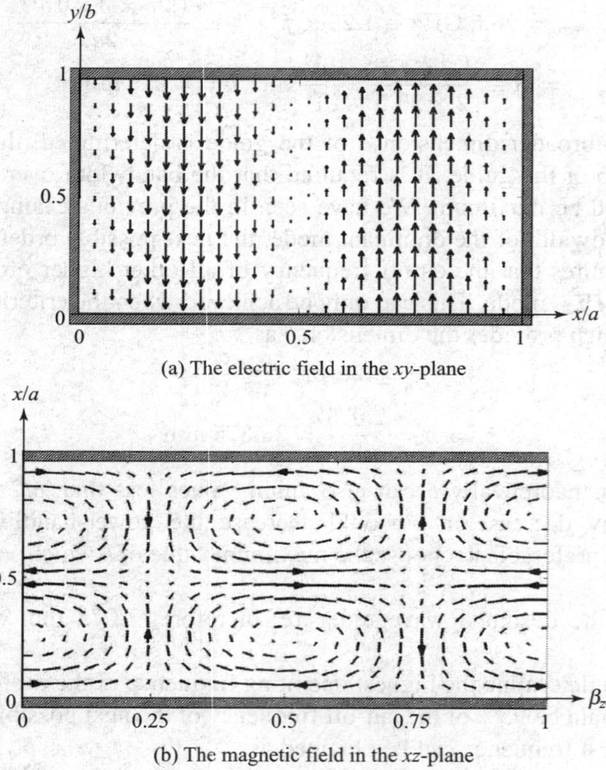

Fig. 15.11 The field lines for the TE_{20} mode in the rectangular waveguide.

Lastly, the field lines are plotted here for the TE_{11} mode as shown in Figure 15.12, which are obtained using instantaneous forms of (65a) to (65e) at $t = 0$ by substituting $m = 1$ and $n = 1$. It can be clearly observed from Figure 15.12(a) that now both E_x and E_y components are present, and hence the electric field lines are not vertical any more in the xy-plane as was the case for TE_{10} and TE_{20} modes.

However, the electric field lines in the xy-plane shown in Figure 15.12(a) may still be looked ...n as originating from one conductor and terminating at the other conductor. This basically ... that no component of the electric field exists along the z-direction, and hence this plot may ...dered to be correspond to the TE mode having one half wavelength variation of the field ... x and y-directions. It can be interpreted from Figure 15.12(a) that near $x = 0$ and ... the electric field lines becomes almost horizontal making E_y component to be close

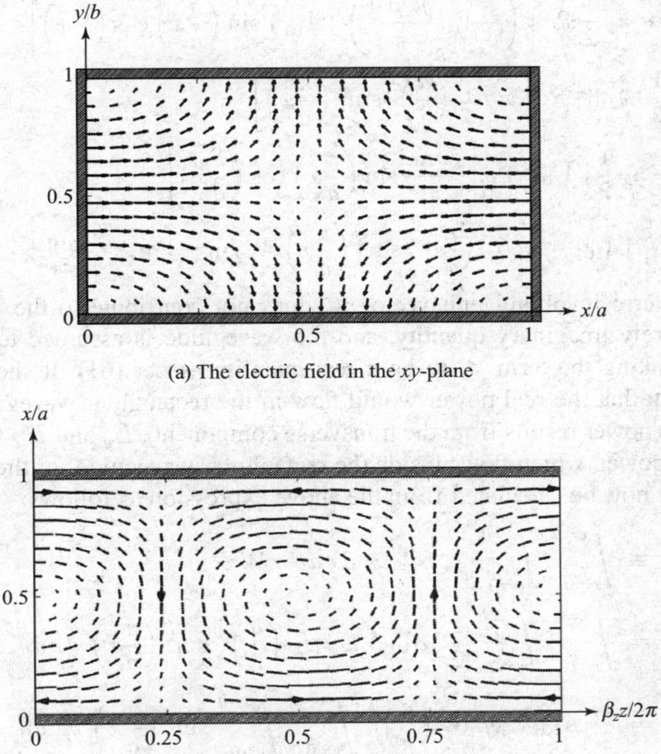

(a) The electric field in the *xy*-plane

(b) The magnetic field in the *xz*-plane

Fig. 15.12 The field lines for the TE_{11} mode in the rectangular waveguide.

to zero. Similarly, near $y = 0$ and $y = b$ region, the electric field lines becomes almost vertical making E_x component to be zero. Similarly, both the E_x and E_y components become zero at the centre of the plot $(x = a/2, y = b/2)$. The magnetic field plot for the TE_{11} mode in the xz-plane is shown in Figure 15.12(*b*), which is same as Figure 15.8(*b*) since the variation of magnetic field in the xz-plane would be same for both TE_{10} and TE_{11} modes. It can be easily verified that the plots shown in Figure 15.12(*b*) satisfy (65) for $m = n = 1$. All the mode plots shown above in the xz-plane have been obtained at $y = b/2$ unless stated otherwise.

After plotting the field patterns for various TE modes, it is useful to calculate the average power density in the rectangular waveguide for the TE_{10} mode using the concept introduced in Chapter 12 as follows:

$$
\begin{aligned}
<\mathbf{S}^{10}> &= \frac{1}{2}\Re\mathrm{e}\,[\mathbf{E}_s \times \mathbf{H}_s^*]\ \mathrm{W/m^2} \\
&\equiv \frac{1}{2}\,\Re\mathrm{e}\,[E_{ys}\,\mathbf{a}_y \times \mathbf{a}_x\,H_{xs}^* + E_{ys}\,\mathbf{a}_y \times \mathbf{a}_z\,H_{zs}^*] \\
&= -\mathbf{a}_z\,\frac{1}{2}\,\Re\mathrm{e}\left[\left(-A_{10}\,\frac{j\omega\mu\,a}{\pi}\right)\left(A_{10}^*\,\frac{-j\beta_z\,a}{\pi}\right)\sin^2\left(\frac{\pi}{a}x\right)\right]
\end{aligned}
$$

$$
\begin{aligned}
&+ \mathbf{a}_x \frac{1}{2} \Re e \left[\left(-A_{10} \frac{j\omega\mu\, a}{\pi} \right) (A_{10}^*) \sin\left(\frac{\pi}{a}x\right) \cos\left(\frac{\pi}{a}x\right) \right] \\
&= \mathbf{a}_z \frac{1}{2} |A_{10}|^2 \Re e \left[\omega\mu\beta_z \left(\frac{a}{\pi}\right)^2 \sin^2\left(\frac{\pi}{a}x\right) \right] \\
&\quad - \mathbf{a}_x \frac{1}{2} |A_{10}|^2 \Re e \left[\frac{j\omega\mu a}{\pi} \sin\left(\frac{\pi}{a}x\right) \cos\left(\frac{\pi}{a}x\right) \right] \\
&= \mathbf{a}_z \frac{1}{2} |A_{10}|^2 \,\omega\mu\,\beta_z \left(\frac{a}{\pi}\right)^2 \sin^2\left(\frac{\pi}{a}x\right) \text{ W/m}^2 = \mathbf{a}_z < S_z^{10} > \text{ W/m}^2
\end{aligned}
\tag{69}
$$

where, the second term involving unit vector $\mathbf{a}_x$ does not contribute to the real power transfer as it becomes a purely imaginary quantity, and the waveguide is assumed to be filled with the lossless material making the term β_z to be a real quantity as per (61). It should be noted from the above expression that the real power would flow in the rectangular waveguide only along the z-direction, and this power results from the transverse components E_y and H_x for the TE_{10} mode. The actual average power, which flows inside the rectangular waveguide for the TE_{10} mode along the z-direction, may now be calculated from the above expression as follows:

$$
\begin{aligned}
P_z^{10} &\equiv \int_{x=0}^{x=a} \int_{y=0}^{y=b} < S_z^{10} > dx\, dy \quad W \\
&= \int_{x=0}^{x=a} \int_{y=0}^{y=b} \frac{1}{2} |A_{10}|^2 \,\omega\mu\,\beta_z \left(\frac{a}{\pi}\right)^2 \sin^2\left(\frac{\pi}{a}x\right) dx\, dy \quad W \\
&= \frac{1}{2} |A_{10}|^2 \,\omega\mu\,\beta_z \left(\frac{a}{\pi}\right)^2 \int_{x=0}^{x=a} \int_{y=0}^{y=b} \sin^2\left(\frac{\pi}{a}x\right) dx\, dy \quad W \\
&= \frac{1}{2} |A_{10}|^2 \,\omega\mu\,\beta_z \left(\frac{a}{\pi}\right)^2 \frac{ab}{2} \quad W = \frac{\omega\mu\,\beta_z\, a^3\, b}{4\,\pi^2} |A_{10}|^2 \quad W
\end{aligned}
\tag{70}
$$

EXAMPLE 15.7

The z-component of magnetic field for the dominant mode propagating in an air-filled rectangular waveguide in the z-direction at 10 GHz is given by the following expression:

$$H_{zs}(xz) = 10.0 \cos(43.74\pi x) e^{-j\beta_z z} \text{ A/m} \tag{3i}$$

Find

(a) the cutoff wavenumber;
(b) the broader dimension of the guide;
(c) the cutoff frequency;
(d) the propagation constant along the direction of wave travel;
(e) the phase velocity;
(f) the group velocity;
(g) the guide wavelength;

(*h*) the wave impedance; and

(*i*) the time average power flowing through the guide.

Solution. As the dominant mode in the rectangular waveguide is the TE_{10} mode, the magnetic field expression given in (3*i*) corresponds to H_{zs} component of the TE_{10} mode. The various parameters can, therefore, be determined by comparing (3*i*) with the standard expression of the H_{zs} component corresponding to the TE_{10} mode, i.e.,

$$H_{zs}(xz) \equiv A_{10}\cos\left(\frac{\pi}{a}x\right)e^{-j\beta_z z}\ \text{A/m}$$

(*a*) The cutoff wavenumber is given by

$$k_c^{TE_{10}} \equiv k_x \equiv \frac{\pi}{a} = 43.74\pi\ \text{rad/m}$$

(*b*) The broader dimension of the guide may be calculated as

$$a = \frac{\pi}{k_c^{TE_{10}}} \equiv \frac{1}{43.74} = 2.2286\ \text{cm}$$

(*c*) The cutoff frequency may determined as

$$k_c^{TE_{10}} \equiv 2\pi f_c^{TE_{10}}\sqrt{\mu_0\epsilon_0}$$

$$\Rightarrow f_c^{TE_{10}} \equiv \frac{k_c^{TE_{10}} \times c}{2\pi} = \frac{43.74\pi \times 3\times 10^8}{2\pi} = 6.56\ \text{GHz}$$

where $c \equiv 1/\sqrt{\mu_0\epsilon_0}$ is the velocity of wave in free space.

(*d*) The propagation constant along the direction of wave travel for the dominant mode in the air-filed waveguide at 10 GHz can be computed as

$$\beta_z = \sqrt{\omega^2\mu_0\epsilon_0 - \left(k_c^{TE_{10}}\right)^2} \equiv \sqrt{\left(\frac{2\pi f}{c}\right)^2 - (43.74\pi)^2}$$

$$\equiv \sqrt{\left(\frac{2\times\pi\times 10\times 10^9}{3\times 10^8}\right)^2 - (43.74\pi)^2} = 158.06\ \text{m}^{-1}$$

It is to be noted that β_z actually represents the phase constant along the direction of propagation. However, the waveguide in the present situation is filled with air (lossless media), and hence the phase constant becomes identical to the propagation constant since the attenuation constant is zero.

(*e*) The phase velocity is given by

$$v_p = \frac{\omega}{\beta_z} \equiv \frac{c}{\sqrt{1-\left(\frac{f_c^{TE_{10}}}{f}\right)^2}} = \frac{3\times 10^8}{\sqrt{1-\left(\frac{6.56}{10}\right)^2}} = 3.975\times 10^8\ \text{ms}^{-1}$$

(*f*) The group velocity may be calculated as

$$v_g = c\sqrt{1-\left(\frac{f_c^{TE_{10}}}{f}\right)^2} = 3\times 10^8 \times \sqrt{1-\left(\frac{6.56}{10}\right)^2} = 2.264\times 10^8\ \text{ms}^{-1}$$

(*g*) The guide wavelength is actually the wavelength inside the guided media, which is given by the following expression:

$$\lambda_g = \frac{2\pi}{\beta_z} \equiv \frac{\lambda}{\sqrt{1-\left(\frac{f_c^{\mathrm{TE}_{10}}}{f}\right)^2}} = \frac{(3\times10^8/10\times10^9)}{\sqrt{1-\left(\frac{6.56}{10}\right)^2}} = 0.0397 \text{ m}$$

where $\lambda = \frac{c}{f} = 0.03$ m is the wavelength in the unbounded dielectric media. It can be observed that the guide wavelength is more than the free space wavelength.

(*h*) The wave impedance for the TE mode corresponding to the wave propagation along the z-direction is given by the following expression:

$$Z_w^{\mathrm{TE}_{10}}(z) = \frac{\omega\mu}{\beta_z} \equiv \frac{\eta}{\sqrt{1-\left(\frac{f_c^{\mathrm{TE}_{10}}}{f}\right)^2}} = \frac{377}{\sqrt{1-\left(\frac{6.56}{10}\right)^2}} = 499.5\ \Omega$$

where $\eta \equiv \sqrt{\mu_0/\epsilon_0}$ is the free-space intrinsic impedance.

(*i*) The time average power flow along the z-direction inside the rectangular waveguide for the TE_{10} mode is given by the following expression as per equation (70)

$$P_z^{\mathrm{TE}_{10}} \equiv \frac{\omega\mu_0\beta_z a^3 b}{4\pi^2}|A_{10}|^2 \text{ W}$$

$$= \frac{2\pi\times10\times10^9\times4\pi\times10^{-7}\times158.06\times(0.0229)^3\times0.0114}{4\pi^2}|10|^2 = 4.328 \text{ W}$$

where $b \equiv a/2$ is assumed to achieve the maximum bandwidth for the dominant mode.

It can be seen from the above equation that the power flow in the waveguide depends upon the cross-sectional dimensions a and b apart from some other parameters. For the dominant TE_{10} mode operation, the broader dimension a of the waveguide is decided by the cutoff frequency as mentioned earlier. Hence, the maximum power capacity of the waveguide can be increased by increasing the dimension b. However, increasing the dimension b beyond a certain value decreases the cutoff frequency of the next higher order mode, thereby limiting the frequency range over which the waveguide can be operated in a single mode. Typically, the dimensions of the rectangular waveguide are chosen in such a way that $a = 2b$, which is a compromise between the maximum power handling capacity and the bandwidth.

Another useful parameter to characterize the transmission of electromagnetic waves inside the rectangular waveguide is the attenuation constant. The rectangular waveguide discussed till now was considered to be made of perfect electric conductor ($\sigma \to \infty$), and it was assumed to be filled with a lossless dielectric ($\epsilon'' \equiv 0$). Although these approximations simplify the analysis, they might not be valid under all situations giving rise to attenuation in rectangular waveguides for propagating waves. The attenuation in waveguides can occur from two effects, and accordingly it may be classified as the conductor loss and the dielectric loss defining the total attenuation constant as follows:

$$\alpha = \alpha_c + \alpha_d \tag{71}$$

where, α_c and α_d are the attenuation constants due to the conductor loss, and the dielectric loss, respectively. The conductor loss occurs when the walls of the waveguide are not made of perfect electric conductors giving rise to finite value of electrical conductivity. The attenuation constant due to imperfect conductor walls depends upon the field configuration giving rise to different values for various modes. As a matter of fact, if the conductors are assumed to have finite conductivity, then the boundary conditions used in the above analysis would also be different thereby changing the field configurations derived earlier. However, the attenuation constant due to finite conductive of the waveguide walls is usually derived by assuming that the transverse field almost remains constant, and the imperfect conductor only causes some power loss along the direction of propagation. It may be, however, mentioned here that for most of the applications at RF and microwave frequencies the conductors can be assumed to be perfect. Hence, we assume here the value of α_c to be zero, and try to compute the attenuation due to dielectric loss only.

The attenuation due to lossy dielectrics placed inside the rectangular waveguide is of much significance at microwave frequencies, and hence is considered in this section. When the waveguide is filled with the lossy dielectric, (3) and (52) relating the z-direction propagation constant with wave numbers along various directions should be slightly modified. Under this condition, the concept of complex propagation constant introduced in Chapter 12 can be used as follows:

$$j\,k_z = \alpha_d + j\,\beta_z \tag{72a}$$

$$\Rightarrow k_z = \beta_z - j\,\alpha_d \tag{72b}$$

where, α_d now represents the attenuation constant due to lossy dielectric present inside the waveguide. It is to be noted that the above definition of complex propagation constant would modify the z-dependence of all the field components as $e^{-jk_z z} \equiv e^{-j\beta_z z}\, e^{-\alpha_d z}$ showing the attenuation along the direction of propagation. Now, let us assume that the test material filled inside the waveguide is a nonmagnetic lossy dielectric $(\mu' = \mu_0,\, \mu'' = 0)$, which is actually true for most of the microwave applications. Under this situation, the effect of dielectric losses can be considered by making the permittivity to be a complex quantity ($\epsilon^* = \epsilon' - j\epsilon''$) as follows:

$$\begin{aligned} k_z &= \sqrt{k^2 - k_c^2} \equiv \sqrt{\omega^2\mu\,(\epsilon' - j\epsilon'') - k_c^2} \\ &= \sqrt{\omega^2\mu\,\epsilon' - k_c^2}\left[1 - j\,\frac{\omega^2\mu\,\epsilon''}{\omega^2\mu\,\epsilon' - k_c^2}\right]^{1/2} \\ &\cong \sqrt{\omega^2\mu\,\epsilon' - k_c^2}\left[1 - \frac{j}{2}\,\frac{\omega^2\mu\,\epsilon''}{\omega^2\mu\,\epsilon' - k_c^2}\right] \\ &\cong \sqrt{\omega^2\mu\,\epsilon' - k_c^2} - \frac{j}{2}\,\frac{\omega^2\mu\,\epsilon''}{\sqrt{\omega^2\mu\,\epsilon' - k_c^2}} \end{aligned} \tag{73}$$

where, we have made use of (61) and (63), and β_z is replaced with k_z to stress the fact that the propagation constant now becomes a complex quantity. The last row of the above equation follows by expanding the function under bracket using the binomial theorem, and retaining only the first two terms of the expansion. This approximation is satisfied for good dielectrics so that the loss tangent $\frac{\epsilon''}{\epsilon'}$ becomes very small as mentioned in Chapter 12, and the condition $\omega^2\mu\,\epsilon'' << (\omega^2\mu\,\epsilon' - k_c^2)$ is satisfied. Now, after comparing (73) with (72), it can be established that β_z is still given by same expression as (61) with ϵ being replaced by ϵ'. The attenuation constant α_d can be simplified using (73) and (72):

$$\begin{aligned}\alpha_d &= \frac{1}{2}\frac{\omega^2\mu\epsilon''}{\sqrt{\omega^2\mu\epsilon' - k_c^2}} \\ &\equiv \frac{1}{2}\frac{\omega^2\mu\epsilon''}{\sqrt{\omega^2\mu\epsilon' - \omega_c^2\mu\epsilon'}} \\ &\equiv \frac{1}{2}\frac{\omega^2\mu\epsilon''}{\omega\sqrt{\mu\epsilon'}\sqrt{1-\left(\frac{\omega_c}{\omega}\right)^2}} \\ &\equiv \frac{1}{2}\sqrt{\frac{\mu}{\epsilon'}}\frac{\omega\epsilon''}{\sqrt{1-\left(\frac{f_c}{f}\right)^2}}\end{aligned} \tag{74}$$

The final expression for the attenuation constant due to dielectric losses inside the rectangular waveguide can be written in the following form:

$$\alpha \equiv \alpha_d = \frac{\sigma_d}{2}\sqrt{\frac{\mu}{\epsilon'}}\frac{1}{\sqrt{1-\left(\frac{f_c}{f}\right)^2}} \tag{75}$$

where, $\sigma_d \equiv \omega\epsilon''$ is the equivalent conductivity for the dielectric medium filled inside the waveguide, and the waveguide walls are assumed to be made of perfect electric conductor making $\alpha_c = 0$. The above equation becomes identical to the attenuation constant for the unbounded good dielectric media if $f_c = 0$ as evident from (60a) of Chapter 12.

Lastly, it is sometimes more convenient to rewrite the field equations for the TE_{10} mode given by (67) into the following form:

$$E_{ys}(x,z) = E_0 \sin\left(\frac{\pi}{a}x\right) e^{-j\beta_z z} \tag{76a}$$

$$H_{xs}(x,z) = -\frac{E_0}{Z_w^{TE_{mn}}} \sin\left(\frac{\pi}{a}x\right) e^{-j\beta_z z} \tag{76b}$$

$$H_{zs}(x,z) = E_0\frac{j\pi}{\omega\mu a} \cos\left(\frac{\pi}{a}x\right) e^{-j\beta_z z} \tag{76c}$$

where, $E_0 \equiv -A_{10}\frac{j\omega\mu a}{\pi}$, and the value of $Z_w^{\text{TE}_{mn}}$ is substituted using (66).

EXAMPLE 15.8

A 20 cm long rectangular waveguide section filled with a Teflon sample ($\epsilon_r' = 2.1$, $\tan\delta = 2.0 \times 10^{-4}$) is operating at 2.45 GHz. The walls of the guide are made of perfect electric conductor (PEC), and its inner cross-sectional dimensions are given as $a = 60$ mm and $b = 30$ mm. The waveguide is connected at one end to a magnetron source providing an output of 800 W at 2.45 GHz. Calculate the attenuation constant and the power dissipated inside the waveguide section assuming the dominant mode propagation.

Solution. As the waveguide walls are made of a perfect electric conductor, the conductor loss will be zero resulting in $\alpha_c = 0$. Hence, the power dissipation inside the guide will be due to dielectric loss only. For calculating the power dissipation, let us first determine the attenuation constant due to the dielectric loss assuming the dominant mode propagation, which is given by the following formula:

$$\alpha_d = \frac{1}{2}\sqrt{\frac{\mu}{\epsilon'}}\frac{\omega\epsilon_0\epsilon''_r}{\sqrt{1-\left(\frac{f_c}{f}\right)^2}} \equiv \frac{1}{2}\frac{\eta_0}{\sqrt{\epsilon'_r}}\frac{\omega\epsilon_0\epsilon'_r \tan\delta}{\sqrt{1-\left(\frac{f_c}{f}\right)^2}} \text{ Np/m}$$

where $\epsilon''_r \equiv \epsilon'_r \tan\delta$ is the imaginary part of the complex relative permittivity, $\eta_0 \equiv 377\ \Omega$ is the free-space intrinsic impedance, and $\epsilon_0 \equiv (10^{-9}/36\pi)$ is the free-space permittivity.

The cutoff frequency in the present situation may be determined as

$$f_c^{\mathrm{TE}_{10}} \equiv \frac{v}{2a} \equiv \frac{c}{\sqrt{\epsilon'_r}2a} = \frac{3\times 10^8}{\sqrt{2.1}\times 2\times 60\times 10^{-3}} = 1.725 \text{ GHz}$$

where $v \equiv c/\sqrt{\epsilon'_r}$ is the velocity of wave in the unbounded dielectric medium.

The attenuation constant due to the dielectric loss may then be calculated from the above equation as

$$\alpha_d = \frac{1}{2}\frac{377}{\sqrt{2.1}}\frac{2\times\pi\times 2.45\times 10^9\times 10^{-9}\times 2.1\times 2\times 10^{-4}}{36\times\pi\times\sqrt{1-\left(\frac{1.725}{2.45}\right)^2}} = 0.0105 \text{ Np/m}$$

Now, the power at input of the rectangular waveguide is given as 800 W. The power at its output after passing through the 20 cm long waveguide section filled with Teflon may be calculated as

$$P_{\mathrm{out}} \equiv P_{\mathrm{in}}\, e^{-2\alpha_d z} = 800\times e^{-2\times 0.0105\times 20\times 10^{-2}} = 796.65 \text{ W}$$

where the wave propagation is assumed along the z-direction.

Hence, the power dissipated inside the 20 cm long waveguide section filled with Teflon is given by

$$P_d \equiv P_{\mathrm{in}} - P_{\mathrm{out}} = 800 - 796.65 = 3.35 \text{ W}$$

15.2.2 Transverse Magnetic Mode for the Rectangular Waveguide

The analysis for TM modes characterized by $H_z = 0$ ($E_z \neq 0$) inside the rectangular waveguide can be carried out on similar lines as that of TE modes described in the last section. In this case, the $E_{zs}(x, y, z)$ component is to be determined by solving the scalar form of the following second-order differential equation derived from ($2a$)

$$\frac{\partial^2 E_{zs}(x,y,x)}{\partial x^2} + \frac{\partial^2 E_{zs}(x,y,z)}{\partial y^2} + \frac{\partial^2 E_{zs}(x,y,z)}{\partial z^2} + k^2 E_{zs}(x,y,z) = 0 \qquad (77)$$

where, the waveguide is considered to be filled with the lossless media. For obtaining the solution of (77), the field component E_{zs} can be expressed as the product of three quantities as

$$E_{zs}(x,y,z) \equiv X(x)\, Y(y)\, Z(z) \qquad (78)$$

where, X, Y, and Z are *exclusive* functions of x, y, and z, respectively as discussed earlier. The above equation can be substituted into (77) in order to obtain three independent equations using the method proposed in the last section, and accordingly the total propagation constant k may be defined in terms of wave numbers along the x- and y-directions using (52). Using the procedure described earlier, the general solution of (78) assuming the wave to be travelling along the $+z$-direction can be written into the following form:

$$E^t_{zs}(x,y) \equiv [A_1\cos(k_x x) + B_1\sin(k_x x)]\,[A_2\cos(k_y y) + B_2\sin(k_y y)] \qquad (79)$$

where, $E_{zs}(x, y, z) \equiv E^t_{zs}(x, y)\, e^{-j\beta_z z}$, and A_1, B_1, A_2, B_2, are arbitrary constants whose values are to be determined by applying the appropriate boundary conditions for the TM mode satisfying the rectangular waveguide geometry. The boundary conditions still require that the tangential electric fields on all the four perfect electric walls are made zero leading to the following equations:

$$E^t_{zs}(y=0) \equiv E^t_{zs}(y=b) = 0 \qquad (80a)$$

$$E^t_{zs}(x=0) \equiv E^t_{zs}(x=a) = 0 \qquad (80b)$$

It is to be noted that in the present situation, we can directly apply the boundary condition for the E_z component as it is tangential to the top and bottom walls as well as to both the side walls. It can be readily proved that applying the boundary conditions for the tangential E_x component on the top and bottom walls, and for the tangential E_y component on both the side walls leads to the same sets of equations.

Now, let us first apply the boundary conditions at $x = 0$ for the E^t_{zs} components using (79):

$$\begin{aligned} E^t_{zs}(x=0) &\equiv [A_2 \cos(k_y y) + B_2 \sin(k_y y)]\, [A_1 \cos(k_x x) + B_1 \sin(k_x x)]_{x=0} = 0 \\ &\Rightarrow [A_2 \cos(k_y y) + B_2 \sin(k_y y)]\, A_1 = 0 \\ &\Rightarrow A_1 = 0 \end{aligned} \qquad (81)$$

where, k_y and other terms related to the spatial variation in the y-direction in a general case cannot be made equal to zero as our aim here is to exclusively find the field dependence along the x-direction using the above boundary condition. The value of A_1 computed above is next substituted in (79) in order to apply the boundary condition defined by (80b) at $x = a$

$$\begin{aligned} E^t_{zs}(x=a) &= [A_2 \cos(k_y y) + B_2 \sin(k_y y)]\, [B_1 \sin(k_x x)]_{x=a} = 0 \\ \Rightarrow k_x &= \frac{m\pi}{a} \qquad \text{where } m = 1, 2, \ldots \end{aligned} \qquad (82)$$

The boundary condition at $y = 0$ yields $A_2 = 0$ in (81). Finally, the boundary condition defined by (80a) at $y = b$ can be applied as follows

$$\begin{aligned} E^t_{zs}(y=b) &\equiv \left[B_1 \sin\left(\frac{m\pi}{a}x\right)\right] [B_2 \sin(k_y y)]_{y=b} = 0 \\ &\Rightarrow \sin(k_y b) = 0 \equiv \sin(n\pi) \qquad \text{where } n = 1, 2, \ldots \\ &\Rightarrow k_y = \frac{n\pi}{b} \qquad \text{where } n = 1, 2, \ldots \end{aligned} \qquad (83)$$

where, n is an integer, and the above formulation follows from the fact that the constant B_2 in a general case cannot be made equal to zero. It is to be noted that for the TM mode in the rectangular waveguide, neither m nor n can attain the value of 0 since it would make $E_z = 0$ resulting into all the field components to be vanished. It is also obvious that both m and n assume only discrete values and the field configuration corresponding to each m, n represents a particular TM_{mn} mode in the rectangular waveguide geometry. The expressions for the propagation constant along the direction of wave travel (β_z) and the cutoff frequency (f_c) are same as derived earlier for the TE case. The lowest possible TM mode in the rectangular waveguide geometry is the TM_{11} mode. It should be noted that the two modes having the same cutoff frequencies are called degenerate modes. Thus the TE_{mn} and TM_{mn} modes are degenerate modes.

The non-zero field components for the general TM to z-mode corresponding to the rectangular waveguide geometry can finally be derived with the help of (15), (16), (79) and above boundary conditions to have the following expressions:

$$E_{zs}(x,y,z) = B_{mn}\sin\left(\frac{m\pi}{a}x\right)\sin\left(\frac{n\pi}{b}y\right)e^{-j\beta_z z} \tag{84a}$$

$$E_{xs}(x,y,z) = -B_{mn}\frac{j\beta_z}{k_c^2}\frac{m\pi}{a}\cos\left(\frac{m\pi}{a}x\right)\sin\left(\frac{n\pi}{b}y\right)e^{-j\beta_z z} \tag{84b}$$

$$E_{ys}(x,y,z) = -B_{mn}\frac{j\beta_z}{k_c^2}\frac{n\pi}{b}\sin\left(\frac{m\pi}{a}x\right)\cos\left(\frac{n\pi}{b}y\right)e^{-j\beta_z z} \tag{84c}$$

$$H_{xs}(x,y,z) = B_{mn}\frac{j\omega\epsilon}{k_c^2}\frac{n\pi}{b}\sin\left(\frac{m\pi}{a}x\right)\cos\left(\frac{n\pi}{b}y\right)e^{-j\beta_z z} \tag{84d}$$

$$H_{ys}(x,y,z) = -B_{mn}\frac{j\omega\epsilon}{k_c^2}\frac{m\pi}{a}\cos\left(\frac{m\pi}{a}x\right)\sin\left(\frac{n\pi}{b}y\right)e^{-j\beta_z z} \tag{84e}$$

where, B_{mn} represents the strength of the particular $\text{TM}_{m,n}$ mode and the values of different coefficients obtained in (82) and (83) have been utilized. It can be observed from above sets of equations that both m and n always attain non-zero values as the condition of either $m = 0$ or $n = 0$ would make all the field components vanish.

The field lines for the TM modes can be plotted using the same procedure as that of TE modes described in the last section. The field plots for the TM_{11} mode, which is the lowest possible order mode for the TM case, are shown in Figure 15.13. These plots are obtained using instantaneous forms of (84*a*) to (84*e*) at $t = 0$ by substituting $m = 1$ and $n = 1$. It is easy to observe from Figure 15.13(a) that both x- and y-components of the magnetic field are present, which ultimately form a closed loop in the xy-plane. Hence, it can be said that the magnetic field lines are complete for this particular mode in the xy-plane indicating no component of the magnetic field present along the direction of propagation. It can easily be shown that the electric field lines for any TM mode in the xy-plane would not be complete indicting the presence of an E field component along the z-direction. The electric field lines for the TM_{11} mode are plotted in the yz-plane in Figure 15.13(b), which can be verified using (84*a*) and (84*c*).

The wave impedance for the TM_{mn} propagating modes traveling along the + z-direction in the rectangular waveguide can be derived using (84) to have the following form:

$$Z_w^{\text{TM}_{mn}} \equiv \frac{E_{xs}}{H_{ys}} \equiv -\frac{E_{ys}}{H_{xs}} = \frac{\beta_z}{\omega\epsilon} = \eta\sqrt{1-\left(\frac{f_c}{f}\right)^2} \qquad f > f_c \tag{85}$$

where, η is the intrinsic impedance of the medium filling the inner region of the waveguide, and f_c is given by (64).

The expression for the attenuation constant in the TM case assuming the waveguide walls to be made of perfect conductors is still given by (75). The expressions for the phase velocity and the group velocity in case of rectangular waveguides are given by the same equations as derived earlier for the parallel-plate waveguide with f_c given by (68).

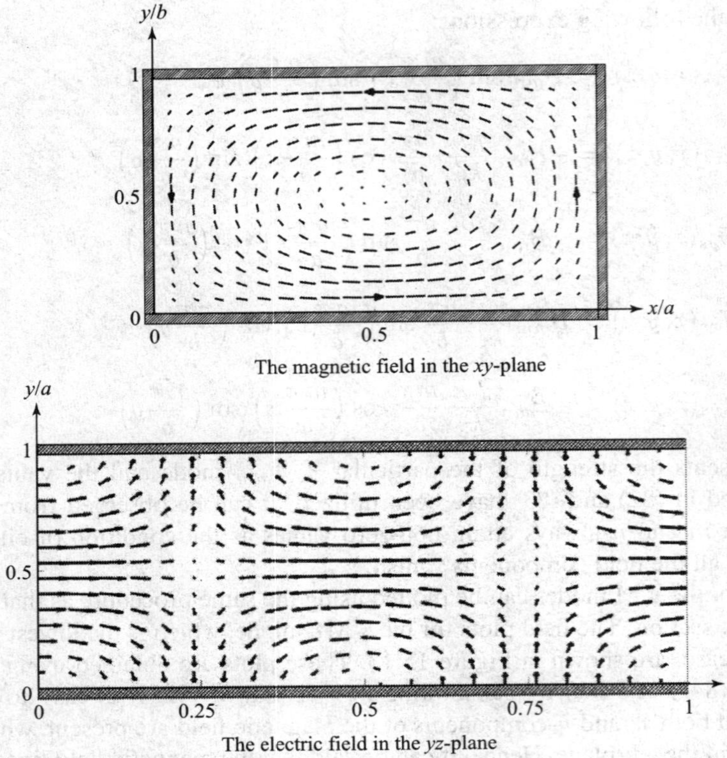

Fig. 15.13 The field lines for the TM_{11} mode in the rectangular waveguide.

EXAMPLE 15.9

The electric field in an air-filled rectangular waveguide at 3 GHz is given by the following expression:

$$E_{zs}(xyz) = 15.0 \sin(5\pi x) \sin(12.5\pi x)\, e^{-j\beta_z z} \text{ V/m} \tag{4i}$$

The cross-sectional dimensions of the waveguide are given as a = 20 cm and b = 8 cm. Find

(*a*) the mode of propagation;
(*b*) the cutoff frequency;
(*c*) the phase constant along direction of propagation;
(*d*) the guide wavelength; and
(*e*) the wave impedance along direction of propagation.

Solution.

(*a*) First, we have to determine whether the mode is of TE or TM type. From the field expression given in (4*i*), it is obvious that the wave is traveling along the z-direction. It basically means that the electric field (E_z component) is present along the direction of propagation which makes the field configuration of TM type.

In the next step, the E_z given by ($4i$) can be compared with the following standard expression of the electric field:

$$E_{zs}(xyz) = E_0 \sin\left(\frac{m\pi x}{a}\right) \sin\left(\frac{n\pi y}{b}\right) e^{-j\beta_z z} \text{ V/m} \qquad (4ii)$$

After comparing, one gets following equations:

$$\frac{m\pi}{a} = 5\pi \Rightarrow m = 5a = 5 \times 20 \times 10^{-2} = 1$$

Similarly, n can be calculated as

$$\frac{n\pi}{b} = 12.5\pi \Rightarrow n = 12.5b = 12.5 \times 8 \times 10^{-2} = 1$$

Hence, the mode of propagation in the present situation is TM_{11}.

(b) The cutoff frequency may be calculated as

$$f_c^{\text{TM}_{11}} \equiv \frac{c}{2}\sqrt{\left(\frac{m}{a}\right)^2 + \left(\frac{n}{b}\right)^2} = \frac{3 \times 10^8}{2}\sqrt{\left(\frac{100}{20}\right)^2 + \left(\frac{100}{8}\right)^2} = 2.019 \text{ GHz}$$

where $c \equiv 1/\sqrt{\mu_0\epsilon_0}$ is the velocity of wave in free-space.

(c) The phase constant along the direction of wave travel for the TM_{11} mode in the air-filed waveguide at 3 GHz may be determined as

$$\beta_z = \frac{\omega}{c}\sqrt{1 - \left(\frac{f_c^{\text{TM}_{11}}}{f}\right)^2} \equiv \frac{2 \times \pi \times 3 \times 10^9}{3 \times 10^8}\sqrt{1 - \left(\frac{2.019}{3}\right)^2}$$

$$= 46.47 \text{ rad/m}$$

(d) The guide wavelength for the TM_{11} mode propagation may be computed using the following expression:

$$\lambda_g = \frac{2\pi}{\beta_z} = \frac{2\pi}{46.47} = 13.52 \text{ cm}$$

(e) The wave impedance for the TM_{11} mode corresponding to the wave propagation along the z-direction in the air-filled waveguide is given by the following expression:

$$Z_w^{\text{TM}_{11}} = \frac{\beta_z}{\omega\epsilon_0} \equiv \eta_0\sqrt{1 - \left(\frac{f_c^{\text{TM}_{11}}}{f}\right)^2} = 377 \times \sqrt{1 - \left(\frac{2.019}{3}\right)^2} = 278.8\ \Omega$$

where $\eta_0 \equiv \sqrt{\mu_0/\epsilon_0}$ is the free-space intrinsic impedance.

D15.3 Specify the minimum width, a, and the maximum height, b, of an air-filled rectangular guide so that it will operate single mode over the frequency range 15 GHz $< f <$ 20 GHz.

Ans. 1 cm; 0.75 cm

After providing the analysis of rectangular waveguides in detail, it is important to know their application in real world. To answer this question, let us first consider the operation of simple transmission lines at frequencies high enough so that other waveguide modes apart from the fundamental TEM mode start appearing. The onset of higher order guided modes in a transmission line, known as *moding*, is in fact a problem that needs to be avoided because the signal distortion may result. A signal that is input to such a line will find its power divided in some proportions among the various modes. The signal power in each mode propagates at a group velocity unique to that mode. With the power thus distributed, distortion will occur over sufficient distances as the signal components among the modes lose synchronization with each other, owing to the different delay times (group delays) associated with the different modes. We encountered this concept in Example 14.3.

The preceding problem of modal dispersion in transmission lines is avoided by assuring that only the TEM mode propagates, and that all other higher order waveguide modes are below cutoff. This is accomplished either by using the line dimensions that are smaller than one-half the signal wavelength, or by assuring an upper limit to the operating frequency in a given line. This is also the reason why all kinds of transmission lines and associated connectors are operated below a certain frequency usually specified by the manufacturer. The actual scenario is, however, more complicated than this.

In Section 14.1, we saw that increasing the frequency increases the loss in transmission lines as a result of the skin effect. For example, in case of parallel-plate transmission lines, the losses at sufficiently higher frequencies would increase due to the increase in the series resistance per unit length, $R\left(R = \frac{2}{b\,\sigma_c\,\delta} \equiv \frac{2}{b}\sqrt{\frac{\pi f \mu}{\sigma_c}}\right)$. One can compensate this increase in R up to certain extent by increasing one or more dimensions in the line cross section, as shown for example in equations (7) and (12) of Chapter 14. However, an increase in dimension of the line cross section beyond a certain limit would give rise to the possibility of higher order mode generation along the line. Typically, the increasing loss with increasing frequency will render the transmission line useless even before the onset of moding, and the line dimensions cannot be increased to reduce losses without considering the possibility of moding. This limitation on dimensions also limits the power-handling capability of the line because the voltage at which dielectric breakdown occurs decreases with decreasing conductor separation. Consequently, the use of conventional transmission lines as frequencies are increased beyond a certain point becomes undesirable, for losses will become excessive, and the limitation on dimensions will limit the power-handling capability. Instead, we look to other guiding structures, among which the rectangular waveguide is a viable option.

The important fundamental difference between the rectangular waveguide (or any hollow pipe guide) and the transmission line is that the rectangular guide will not support a TEM mode. The fact that the guide is formed from a completely enclosed metal structure means that any electric field distribution in the transverse plane must exhibit variations in the plane. This is necessary in order to apply the boundary condition which requires that the tangential component of the electric field must be zero at the surface of perfect electric conductors. This can be understood by looking at the electric field lines of the TE_{10} mode corresponding to the rectangular waveguide shown in Figure 15.8, where it can be clearly observed that the electric field is varying along the x direction in the transverse plane in order to satisfy the boundary condition requiring the tangential component E_y to vanish at $x = 0$ and $x = a$. Now, since $\mathbf{E}$ varies in the transverse plane, the computation of $\mathbf{H}$ through $\nabla \times \mathbf{E} = -j\omega\mu\mathbf{H}$ must lead to a z-component of $\mathbf{H}$, and so we cannot have a TEM mode. We cannot find any other orientation of a completely transverse $\mathbf{E}$ in the waveguide that will allow a completely transverse $\mathbf{H}$. Since the rectangular guide will not support a TEM mode, it will not operate until the frequency exceeds the cutoff frequency of the lowest order guided mode of the structure. Thus, it must be constructed of large enough size to accomplish this for a given frequency; the required transverse dimensions will consequently be larger than those of a transmission line that is designed to support only the TEM mode. The increased size, coupled with the fact that there is more conductor surface area than in a transmission line of equal volume, means that losses will be substantially lower in the rectangular waveguide structure. Additionally, the guides will support more power at a given electric field strength than a typical transmission line, since the rectangular waveguide will have a higher cross-sectional area.

Still, the hollow pipe guides must operate in a single mode in order to avoid the signal distortion problems arising from multimode transmission. This means that the guides must be of dimension such that they operate above the cutoff frequency of the lowest order mode, but below the cutoff frequency of the next higher order mode, as discussed earlier in the text. It is mainly due to this reason that a number of standard waveguides are practically available, with each designated waveguide operating in a particular frequency band. The details about various designated waveguides including their dimensions and the operating frequency bands can be found in some of the references given at the end of this chapter. The cross-sectional dimensions of a designated rectangular waveguide are chosen in such a way that the maximum bandwidth is achieved over which the TE_{10} mode can propagate without compromising much on its power handling capability. It can be easily shown that if $a \sim 2b$, then the maximum bandwidth for the TE_{10} mode propagation in the rectangular waveguide is achieved. If the narrow dimension of the waveguide b is increased much beyond this limit, then it would further increase the power handling capability of the guide but then the bandwidth corresponding to TE_{10} mode propagation would be reduced.

From the above discussion, it is obvious that the rectangular waveguides should inherently be used for high-power and high-frequency applications. The waveguides are quite effectively being used to transmit the electromagnetic energy, and design various components in the higher microwave and millimeter wave frequency range as the dimensions of the guide become conveniently small. One of its practical usage for high-power applications is in the microwave oven, which employs a high-power magnetron source at 2.45 GHz and uses a section of rectangular waveguide to couple the microwave power from source to the main chamber of the oven. Apart from the high-power applications, the waveguides are practically also quite often being used for low power applications such as the estimation of the dielectric and magnetic properties of various test samples in the microwave frequency range. Typically, this procedure requires precisely machined sample of the test specimen to be fitted inside a rectangular waveguide as shown in Figure 15.14(b), which depicts a Teflon sample fitted inside the section of a X-band rectangular waveguide having the cross-sectional dimensions as 22.86×10.16 mm. The waveguide loaded with the test sample is then connected to two ports of a network analyzer through a coaxial-waveguide adapter and a section of X-band waveguide at each port as shown in Figure 15.14(a) in order to measure the reflection and transmission coefficients. The network analyzer is a modern day microwave measuring equipment, which has in-built low power microwave source, receiver and detector in order to facilitate the measurement of reflection and transmission coefficients of the test specimen over a wide frequency range. The access to the network analyzer is provided to the outside world through coaxial connectors, and hence a pair of coaxial-waveguide adapters are required to carry out measurement using the waveguide. Once the reflection and transmission coefficients of the test sample loaded in the rectangular waveguide are measured, the dielectric and magnetic properties

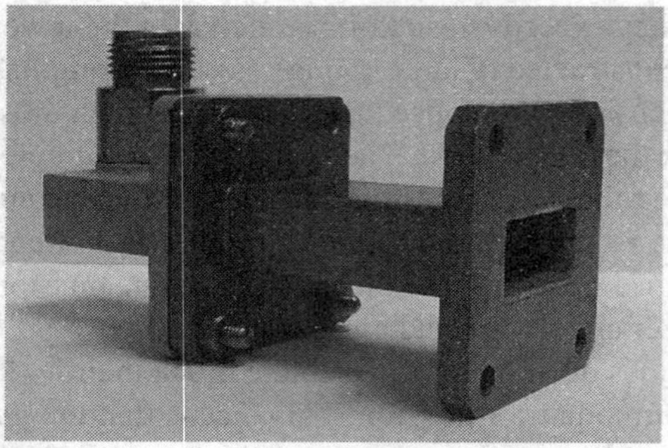

(a) The X-band coaxial-waveguide adapter attached with the waveguide section

(b) The X-band rectangular waveguide loaded with the Teflon sample

Fig. 15.14 The X-band (8.2 to 12.4 GHz) coaxial-waveguide adapter and the waveguide sample holder.

of the test specimen can be estimated using simple mathematical algorithms available in some advanced text provided at the end of this chapter. Lastly, from the application point of view, the typical rectangular waveguide can be converted into a rectangular pyramidal horn antenna by flaring it along both the cross-sectional dimensions. Figure 15.15 shows a Ka-band rectangular pyramidal horn antenna, which can be thought of as the extension of a typical Ka-band rectangular waveguide by opening it along both the wider and narrower dimensions. The pyramidal horn antennas are quite often being used for various applications in the industry because of their simpler design and lower cost.

15.3 RESONATORS

The transmission lines and waveguides considered so far in this book are mainly used to transmit electromagnetic energy between two different points in space. The waveguide media considered in this chapter was bounded along the transverse plane, but was still assumed to be unbounded along the longitudinal direction thereby facilitating the propagation of wave in this direction. The resonators are structures mainly used to confine the electromagnetic energy without any real propagation of waves. You are already aware of the resonance circuits composed of R, L, and

Fig. 15.15 The WR-28 (26.50–40 GHz) rectangular pyramidal horn antenna.

C elements at lower frequencies. You also know that the resonance phenomenon is frequency dependent with the resonant frequency defined in terms of L and C values. Now, the R, L, and C lumped circuit elements cannot be realized in the microwave frequency range (1 GHz $< f$ $<$ 30 GHz) since the wavelength becomes comparable to the dimension of the circuit elements (30 cm $< \lambda <$ 1 cm) leading to the problem of stray electromagnetic fields and radiations. In the higher frequency range, it is more convenient to use a finite length of transmission line as the circuit element where the input impedance of a terminated line can be shown to possess an inductive or capacitive behaviour in the specified frequency band. The transmission lines can be converted into resonating structures by terminating both the ends using short circuits, and providing coupling through small openings at the input and output. However, at microwave frequencies, it is more common to covert a piece of rectangular waveguide into resonator by placing short plates at both the ends as discussed below.

15.3.1 Rectangular Cavity Resonators

The most commonly used resonating structure at microwave frequencies is the rectangular cavity resonator obtained by shorting both the ends of a rectangular waveguide using metallic plates as shown in Figure 15.16. This kind of structure is also some times simply termed as the cavity resonator since the complete enclosure made by the six metallic walls on each side forms a cavity in Figure 15.16. A piece of rectangular waveguide having cross-sectional dimensions as $a \times b$ with a and b being the width and height of the guide, respectively ($a > b$) can be converted into a rectangular cavity resonator by placing two perfectly conducting plates at $z = 0$ and $z = d$ as obvious from Figure 15.16. The rectangular cavity resonator can also support infinite number of TE and TM modes just like the rectangular waveguide with the only difference that there would not be any propagation of waves inside the cavity. It basically means that in the rectangular cavity, there will also be standing waves along the z-direction apart from the x- and y-directions. The extra boundary condition imposed on the rectangular cavity shown in Figure 15.16 requires that the tangential component of the electric field should vanish at both the metallic plates introduced at $z = 0$ and $z = d$ as follows:

$$E_{xs}(z = 0) \equiv E_{xs}(z = d) = 0 \tag{86a}$$

$$E_{ys}(z = 0) \equiv E_{ys}(z = d) = 0 \tag{86b}$$

The above boundary condition is in addition to the boundary conditions imposed at $x = 0$, a and at $y = 0$, b given by (56) and (80). The field expressions for the rectangular cavity shown in Figure 15.16 can be obtained after solving the Maxwell's equations and applying all the above boundary conditions. However, rather than deriving all the field expressions starting from Maxwell's equations, it looks more convenient to directly modify the field expressions of the corresponding TE and TM modes by imposing additional boundary condition given by (86).

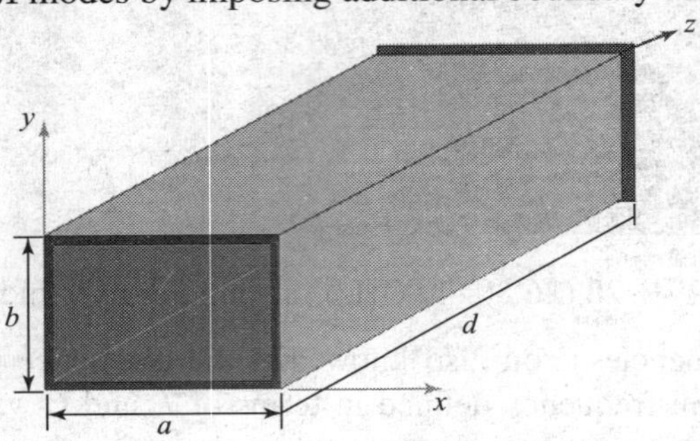

Fig. 15.16 The rectangular cavity resonator.

Assuming the rectangular waveguide cavity supporting the TE_{mnp} mode, the transverse electric field components corresponding to this geometry can be derived to have the following expression using (65):

$$E_{xs}(x, y, z) = A_{mn} \frac{j\omega\mu}{k_c^2} \frac{n\pi}{b} \cos\left(\frac{m\,\pi}{a}x\right) \sin\left(\frac{n\,\pi}{b}y\right) \left[A_3 e^{-jk_{zc}z} + B_3 e^{jk_{zc}z}\right] \tag{87a}$$

$$E_{ys}(x, y, z) = -A_{mn} \frac{j\omega\mu}{k_c^2} \frac{m\pi}{a} \sin\left(\frac{m\,\pi}{a}x\right) \cos\left(\frac{n\,\pi}{b}y\right) \left[A_3 e^{-jk_{zc}z} + B_3 e^{jk_{zc}z}\right] \tag{87b}$$

where, k_c is defined as per (61) and (63), and k_{zc} may be regarded as the z-directed wave number inside the cavity which is written in lieu of β_z to stress the fact that there is no propagation of waves inside the cavity. The introduction of conducting walls at $z = 0$ and $z = d$ results into multiple reflections from these two ends leading to presence of both the positive and negative z-directed waves inside the cavity shown in (87*a*) and (87*b*). The superposition of these + and − z-directed waves should satisfy the boundary condition given by (86), which would also facilitate the determination of constants A_3 and B_3 in (87). If we apply the boundary condition at $z = 0$ in either (87*a*) or (87*b*) then it would lead to

$$[A_3 + B_3] = 0 \quad \Rightarrow A_3 = -B_3 \tag{88}$$

The substitution of above constant in either (87*a*) or (87*b*) would result into a $\sin(k_{zc}z)$ function providing the variation in z-direction for both E_x and E_y components. In the next step, the boundary condition at $z = d$ is to be applied for both E_x and E_y components resulting into

$$\begin{aligned} &[A_3 \sin(k_{zc}z)]_{z=d} = 0 \\ &\Rightarrow \sin(k_{zc}d) = 0 \equiv \sin(p\pi) \\ &\Rightarrow k_{zc} = \frac{p\pi}{d} \qquad \text{where } p = 1, 2, \ldots \end{aligned} \tag{89}$$

where, p is an integer defining the half wavelength field variation along the z-direction, and it has been made use of the fact that the constant A_3 can not be made equal to zero in a general case. The values of constants A_3 and B_3, and the wavenumber k_{zc} obtained above may finally be substituted into (87) to obtain the non-zero electric field expression for the TE_{mnp} mode

$$E_{xs}(x,y,z) = A_{mnp}\,\frac{j\omega\mu}{k_c^2}\,\frac{n\pi}{b}\,\cos\left(\frac{m\,\pi}{a}x\right)\,\sin\left(\frac{n\,\pi}{b}y\right)\,\sin\left(\frac{p\,\pi}{d}z\right) \tag{90a}$$

$$E_{ys}(x,y,z) = -\,A_{mnp}\,\frac{j\omega\mu}{k_c^2}\,\frac{m\pi}{a}\,\sin\left(\frac{m\,\pi}{a}x\right)\,\cos\left(\frac{n\,\pi}{b}y\right)\,\sin\left(\frac{p\,\pi}{d}z\right) \tag{90b}$$

where, $A_{mnp} \equiv -2\,j A_{mn}\,A_3$ is now a new constant representing a particular TE_{mnp} mode inside the rectangular cavity, and

$$k_c^2 = \left(\frac{m\pi}{a}\right)^2 + \left(\frac{n\pi}{b}\right)^2 \tag{91}$$

is defined as per (63). An infinite number of TE_{mnp} resonant modes would be excited inside the rectangular waveguide cavity, which can be defined using (52) as follows:

$$\omega_r^2\,\mu\epsilon = \left(\frac{m\pi}{a}\right)^2 + \left(\frac{n\pi}{b}\right)^2 + \left(\frac{p\pi}{d}\right)^2 \tag{92}$$

where, $k_{zc} = \frac{p\pi}{d}$ is used. The resonant frequency of the rectangular cavity supporting the TE_{mnp} mode is then derived from the above equation:

$$f_{r,mnp} = \frac{1}{2\sqrt{\mu\epsilon}}\sqrt{\left(\frac{m}{a}\right)^2 + \left(\frac{n}{b}\right)^2 + \left(\frac{p}{d}\right)^2} \tag{93}$$

The magnetic field components for the TE_{mnp} mode inside the rectangular cavity are obtained in terms of the corresponding electric field components using one of the Maxwell's curl equation as follows:

$$\nabla \times \mathbf{E}_s = -j\omega\mu\mathbf{H}_s$$

$$\Rightarrow \begin{vmatrix} \mathbf{a}_x & \mathbf{a}_y & \mathbf{a}_z \\ \frac{\partial}{\partial x} & \frac{\partial}{\partial y} & \frac{\partial}{\partial z} \\ E_{xs} & E_{ys} & 0 \end{vmatrix} = -j\omega\mu(H_{xs}\mathbf{a}_x + H_{ys}\mathbf{a}_y + H_{zs}\mathbf{a}_z) \tag{94}$$

The individual components on both sides of the above curl equation may be compared to obtain the following sets of expressions for the magnetic field components:

$$H_{xs} = \frac{1}{j\omega\mu}\,\frac{\partial E_{ys}}{\partial z} \tag{95a}$$

$$H_{ys} = -\frac{1}{j\omega\mu}\,\frac{\partial E_{xs}}{\partial z} \tag{95b}$$

$$H_{zs} = \frac{-1}{j\omega\mu}\left[\frac{\partial E_{ys}}{\partial x} - \frac{\partial E_{xs}}{\partial y}\right] \tag{95c}$$

The above sets of equations may then be combined with (90) to obtain the magnetic field expression for the TE_{mnp} mode:

$$H_{xs}(x,y,z) = -\,\frac{A_{mnp}}{k_c^2}\left(\frac{m\pi}{a}\right)\left(\frac{p\pi}{d}\right)\sin\left(\frac{m\,\pi}{a}x\right)\,\cos\left(\frac{n\,\pi}{b}y\right)\,\cos\left(\frac{p\,\pi}{d}z\right) \tag{96a}$$

$$H_{ys}(x,y,z) = -\frac{A_{mnp}}{k_c^2}\left(\frac{n\pi}{b}\right)\left(\frac{p\pi}{d}\right)\cos\left(\frac{m\pi}{a}x\right)\sin\left(\frac{n\pi}{b}y\right)\cos\left(\frac{p\pi}{d}z\right) \tag{96b}$$

$$H_{zs}(x,y,z) = A_{mnp}\cos\left(\frac{m\pi}{a}x\right)\cos\left(\frac{n\pi}{b}y\right)\sin\left(\frac{p\pi}{d}z\right) \tag{96c}$$

where, we have made use of (91). Equations (90) and (96) provide the non-zero field expressions for the TE_{mnp} mode inside the rectangular cavity resonator shown in Figure 15.16. A similar kind of procedure can be used to obtain the field expressions of the TM_{mnp} mode inside the rectangular cavity resonator, where the field expressions given by (84) for the TM_{mn} can be appropriately modified by applying additional boundary conditions.

As the operating mode of the rectangular waveguide is usually the dominant TE_{10} mode, the rectangular waveguide cavity most commonly being used for various applications is operated in the TE_{10p} mode. The nonzero field expressions for the TE_{10p} mode inside the rectangular cavity resonator can be obtained with the help of (90) and (96) as follows:

$$E_{ys}^{10p}(x,y,z) = -A_{10p}\frac{j\omega\mu a}{\pi}\sin\left(\frac{\pi}{a}x\right)\sin\left(\frac{p\pi}{d}z\right) \tag{97a}$$

$$H_{xs}^{10p}(x,y,z) = -A_{10p}\left(\frac{pa}{d}\right)\sin\left(\frac{\pi}{a}x\right)\cos\left(\frac{p\pi}{d}z\right) \tag{97b}$$

$$H_{zs}^{10p}(x,y,z) = A_{10p}\cos\left(\frac{\pi}{a}x\right)\sin\left(\frac{p\pi}{d}z\right) \tag{97c}$$

where, $k_c \equiv \frac{\pi}{a}$ for the TE_{10p} mode has been employed. The field lines for the TE_{105} mode in the rectangular cavity are shown in Figure 15.17, which are obtained using (97*a*) to (97*c*). The electric field is plotted in the yz-plane while the magnetic field is plotted in the xz-plane as shown in Figures 15.17(*a*) and 15.17(*b*), respectively. It should be mentioned that the field plots for this mode in the xy plane would be same as shown earlier for the TE_{10} rectangular waveguide case. It can be observed from Figure 15.17(*a*) that the electric field is maximum at the centre of the cavity length ($z = d/2$) for the TE_{105} mode. As a matter of fact, for any TE_{10p} mode with p being odd, the maxima of the electric field will occur at the centre of the cavity length. You may also recall that for the TE_{10} mode inside the rectangular waveguide, the electric field is maximum at $x = a/2$. This basically means that for any TE_{10p} mode inside the rectangular cavity with p being odd, the electric field will be maximum at its centre defined by $x = a/2$, $z = d/2$. This behaviour of the electric field pattern inside the rectangular cavity is exploited to measure the dielectric constant of low loss materials by placing a small sample at the centre of the cavity parallel to its narrow wall as discussed in the later part of the text. The basic idea here is that if a very thin nonmagnetic low loss sample is placed at the position of the maximum electric field by aligning it parallel to the field, then the electric field pattern inside the cavity will be disturbed by a very small amount and hence the *perturbation* approximation can be applied.

Meanwhile, the resonant frequency for the TE_{10p} mode of a rectangular cavity resonator may be written with the help of (93) as follows:

$$f_{r,10p} = \frac{1}{2\sqrt{\mu\epsilon}}\sqrt{\left(\frac{1}{a}\right)^2 + \left(\frac{p}{d}\right)^2} \tag{98}$$

Apart from the resonant frequency, the other important parameter quite useful for any resonator or cavity is the quality factor or the Q factor. The Q factor of an ideal cavity resonator may be

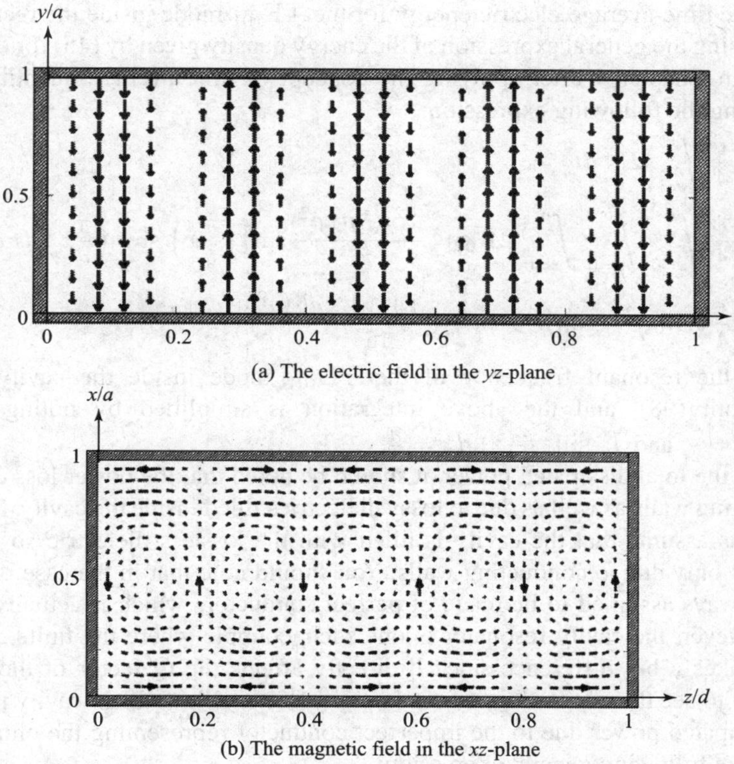

Fig. 15.17 The field lines for the TE_{105} mode in the rectangular cavity resonator.

considered to be infinite if its walls are made of perfect electric conductor and the cavity is filled with a lossless dielectric. However, under actual situation, both these conditions are not satisfied resulting into a finite value of Q, which in a general case may be defined as

$$\begin{aligned} Q &= 2\pi f \, \frac{\text{Time-average stored energy at resonance}}{\text{Time-average dissipated power}} \\ &= 2\pi f \, \frac{< W_E > + < W_M >}{< P_d >} \\ &\equiv 4\pi f \, \frac{< W_E >}{< P_d >} \end{aligned} \tag{99}$$

where, $< W_E >$ and $< W_M >$ are the time-average electric and magnetic stored energies, respectively at resonance and $< P_d >$ is the time-average dissipated power. It is to be noted that at the resonance condition, the electric and magnetic stored energies are equal, and hence the total average stored energy may be determined by computing the average energy stored either in the electric field or in the magnetic field, i.e., $< W_E > + < W_M > \equiv 2 < W_E > \equiv 2 < W_M >$. It is obvious from the above equation that the Q factor of the cavity depends upon the field configuration, which makes it difficult to derive the general expression for any arbitrary mnp mode. For simplicity, we derive here the expression of the Q factor for the TE_{101} mode. However, the overall technique presented here can also be applied for other modes after applying appropriate modifications. The time-average electric energy for the TE_{101} mode inside the rectangular cavity can be obtained using the general expression of the energy density given by (45) for the electrostatic field introduced in Chapter 4 after applying the concept of time average quantity discussed in Chapter 11 yielding the following expression:

$$\begin{aligned} < W_E > &= \frac{\epsilon}{4} \int_{vol} |E|^2 \, dv \\ &= \frac{\epsilon}{4} \int_{z=0}^{d} \int_{y=0}^{b} \int_{x=0}^{a} |A_{101}|^2 \, \frac{\omega_{r,101}^2 \mu^2 \, a^2}{\pi^2} \, \sin^2\left(\frac{\pi}{a}x\right) \, \sin^2\left(\frac{\pi}{d}z\right) dx \, dy \, dz \\ &= \frac{\epsilon}{4} \, |A_{101}|^2 \, f_{r,101}^2 \mu^2 \, b \, d \, a^3 = \frac{|A_{101}|^2}{16} \, \frac{ab\left(a^2 + d^2\right)\mu}{d} \end{aligned} \tag{100}$$

where, $f_{r,101}$ is the resonant frequency of the TE_{101} mode inside the cavity whose value is substituted from (98), and the above integration is simplified by noting the fact that $\int_0^a \sin^2\left(\frac{\pi}{a}x\right) dx = \frac{a}{2}$, and $\int_0^d \sin^2\left(\frac{\pi}{d}z\right) dz = \frac{d}{2}$.

For computing the total dissipated power, it should be noted that the power loss could be due to imperfect conducting walls as well as due to lossy dielectrics filled inside the cavity. For simplifying the analysis, let us assume that the cavity is filled with the lossless dielectric so that the power dissipation occurs only due to conducting walls. You should note that in the case of a waveguide, the walls were always assumed to be made of perfect conductors which is actually true for most applications. However, the cavity resonator is one such example where the finite conductivity of metallic walls makes a big difference since it directly affects the Q factor of the empty cavity. Hence, it is useful to see the effect of finite conductivity of metallic walls in cavity resonators. The time-average dissipated power due to the imperfect conductor representing the ohmic loss can be computed using the following general expression:

$$< P_d > = \frac{R_s}{2} \int\!\!\int_S |\mathbf{J}_s|^2 \, ds \tag{101}$$

where, $R_s \equiv \frac{1}{\sigma_c \delta}$ is the surface resistance of the conductor determined using equation (85) of Chapter 12 with δ being the skin depth, and $\mathbf{J}_s$ is the linear current density induced on the surface of the conductor defined using

$$\mathbf{J}_s = \mathbf{a}_n \times \mathbf{H} \text{ A/m} \tag{102}$$

with $\mathbf{a}_n$ being the unit vector normal to the surface. Now, there are a total of six metallic walls in the cavity, and hence the total dissipated power should be obtained by adding the power lost in each of these six walls. However, for the cavity shown is Figure 15.16, the dissipated power on the top wall ($y = b$) would be same as that on the bottom wall ($y = 0$), that on the back wall ($z = 0$) would be same as on the front wall ($z = d$). Similarly, the power dissipated on two side walls ($x = 0$) and ($x = a$) would also be same. Hence, the total dissipated power in the present situation can be obtained by computing the power loss on the back ($z = 0$), the bottom ($y = 0$), and one of the side walls ($x = 0$) followed by a multiplication with a factor of two, i.e.,

$$< P_d > = 2\left[< P_d >_{(z=0)} + < P_d >_{(y=0)} + < P_d >_{(x=0)}\right]$$

$$= R_s\left[\iint_{\text{back}} |\mathbf{J}_{s,\text{back}}|^2\, ds + \iint_{\text{bottom}} |\mathbf{J}_{s,\text{bottom}}|^2\, ds + \iint_{\text{side}} |\mathbf{J}_{s,\text{side}}|^2\, ds\right] \quad (103)$$

where, we have made use of (101). Let us now try to determine the current density on each of the metallic walls in order to evaluate the above expression. The current density on the back wall may be computed as

$$\mathbf{J}_{s,\text{back}} \equiv [\mathbf{a}_n \times \mathbf{H}]_{z=0} = \left[\mathbf{a}_z \times \mathbf{a}_x\, H_{xs}^{101} + \mathbf{a}_z \times \mathbf{a}_z\, H_{zs}^{101}\right]_{z=0}$$
$$= \mathbf{a}_y\left[H_{xs}^{101}\right]_{z=0} = -\mathbf{a}_y \frac{a\, A_{101}}{d} \sin\left(\frac{\pi}{a}x\right) \quad (104)$$

where, we have made use of (102) and (97). The above equation yields

$$\iint_{\text{back}} |\mathbf{J}_{s,\text{back}}|^2\, ds \equiv \int_{y=0}^{b}\int_{x=0}^{a} \mathbf{J}_{s,\text{back}} \cdot \mathbf{J}^*_{s,\text{back}}\, dx\, dy$$
$$= \int_{y=0}^{b}\int_{x=0}^{a} |A_{101}|^2 \left(\frac{a}{d}\right)^2 \sin^2\left(\frac{\pi}{a}x\right) dx\, dy \quad (105)$$
$$= |A_{101}|^2 \left(\frac{a^3\, b}{2\, d^2}\right)$$

Similarly, the current density on the left side wall may be computed as

$$\mathbf{J}_{s,\text{side}} \equiv [\mathbf{a}_n \times \mathbf{H}]_{x=0} = \left[\mathbf{a}_x \times \mathbf{a}_x\, H_{xs}^{101} + \mathbf{a}_x \times \mathbf{a}_z\, H_{zs}^{101}\right]_{x=0}$$
$$= -\mathbf{a}_y\left[H_{zs}^{101}\right]_{x=0} = -\mathbf{a}_y\, A_{101} \sin\left(\frac{\pi}{d}z\right) \quad (106)$$

which yields

$$\iint_{\text{side}} |\mathbf{J}_{s,\text{side}}|^2\, ds \equiv \int_{z=0}^{d}\int_{y=0}^{b} |A_{101}|^2 \sin^2\left(\frac{\pi}{d}z\right) dy\, dz$$
$$= |A_{101}|^2 \left(\frac{b\, d}{2}\right) \quad (107)$$

Lastly, the current density on the bottom wall may be computed as

$$\mathbf{J}_{s,\text{bottom}} \equiv [\mathbf{a}_n \times \mathbf{H}]_{y=0} = \left[\mathbf{a}_y \times \mathbf{a}_x\, H_{xs}^{101} + \mathbf{a}_y \times \mathbf{a}_z\, H_{zs}^{101}\right]_{y=0}$$
$$= \left[-\mathbf{a}_z\, H_{xs}^{101} + \mathbf{a}_x\, H_{zs}^{101}\right]_{y=0} \quad (108)$$
$$= \left[\mathbf{a}_z \left(\frac{a\, A_{101}}{d}\right) \sin\left(\frac{\pi}{a}x\right) \cos\left(\frac{\pi}{d}z\right) + \mathbf{a}_x\, A_{101} \cos\left(\frac{\pi}{a}x\right) \sin\left(\frac{\pi}{d}z\right)\right]$$

which provides

$$\begin{aligned}\iint_{\text{bottom}} |\mathbf{J}_{s,\text{bottom}}|^2 \, ds &\equiv \int_{z=0}^{d}\int_{x=0}^{a} |A_{101}|^2 \left(\frac{a}{d}\right)^2 \sin^2\left(\frac{\pi}{a}x\right)\cos^2\left(\frac{\pi}{d}z\right) dx\,dz \\ &+ \int_{z=0}^{d}\int_{x=0}^{a} |A_{101}|^2 \cos^2\left(\frac{\pi}{a}x\right)\sin^2\left(\frac{\pi}{d}z\right) dx\,dz \\ &= |A_{101}|^2\left(\frac{a^3}{4\,d}\right) + |A_{101}|^2\left(\frac{a\,d}{4}\right)\end{aligned} \tag{109}$$

Finally, (105), (107) and (109) may be substituted in (103) to obtain the total dissipated power in the rectangular cavity for the TE_{101} mode:

$$\begin{aligned}< P_d > &= R_s\,|A_{101}|^2\left[\left(\frac{a^3\,b}{2\,d^2}\right) + \left(\frac{b\,d}{2}\right) + \left(\frac{a^3}{4\,d}\right) + \left(\frac{a\,d}{4}\right)\right] \\ &= R_s\,|A_{101}|^2\left[\frac{a^3\,(2b+d)\, +\, d^3\,(2b+a)}{4\,d^2}\right]\end{aligned} \tag{110}$$

The expression for the Q factor corresponding to the TE_{101} mode inside the rectangular cavity can now be obtained using (99), (100), and (110):

$$\begin{aligned}[Q]_{101}^{\text{TE}} &= 4\pi f_{r,101}\,\frac{< W_E >}{< P_d >} = \frac{abd\,(a^2+d^2)\,\pi\mu\,f_{r,101}}{R_s\,[a^3\,(2b+d)\, +\, d^3\,(2b+a)]} \\ &= \frac{\pi\,b\,\eta\,(a^2+d^2)^{3/2}}{2\,R_s\,[a^3\,(2b+d)\, +\, d^3\,(2b+a)]}\end{aligned} \tag{111}$$

where, η represents the intrinsic impedance of the medium filled inside the cavity, and the value of $f_{r,101}$ is substituted from (98) by making $p = 1$. It can now be observed from (111) that the Q factor of the rectangular cavity depends upon its physical dimensions as well as the electrical properties of the metallic walls and the material filled inside the cavity. If the cavity wall is assumed to be made of perfect electric conductor, then the value of σ_c would be infinity and the value of $R_s \equiv \frac{1}{\sigma_c\,\delta}$ would be close to zero making the value of Q to be close to infinity. Practically, the cavity is manufactured using very good conductor such as Aluminum or Copper having σ_c of the order of 10^7 S/m and hence the Q of a metallic cavity can possess quite high value. It should be, however, noted that (111) usually provides the unloaded Q of the cavity as the cavity was assumed to be filled with the lossless dielectric material. Under actual situation, if the cavity is loaded with a lossy dielectric material then the Q of the cavity would be less than the corresponding unloaded value.

One of the practical applications of the rectangular cavity is to measure the dielectric constant and the loss tangent of a low loss dielectric sample. The method is usually known as the *cavity perturbation technique*, which is based on the assumption that the insertion of a small piece of low loss dielectric sample inside a cavity does not appreciably change the field configuration inside the cavity. This procedure is usually carried out by designing a rectangular cavity working in the TE_{10p} mode with p being odd, and inserting the test sample at the centre of the cavity as mentioned earlier. The overall method is based on simple fact that if the designed cavity is loaded with a small piece of dielectric sample then it would slightly change the resonant frequency as well as the Q factor of the cavity. The dielectric constant and the loss tangent of the test specimen can be obtained in terms of change in the resonant frequency and the Q factor of the cavity after insertion of the sample using simple analytical expressions available in some advanced text given at the end of this

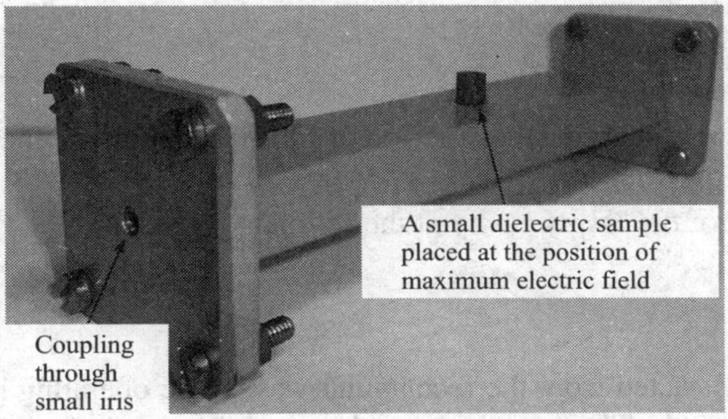

Fig. 15.18 The X-band rectangular cavity designed for dielectric measurement.

chapter. A typical X-band TE_{105} rectangular cavity resonator is shown in Figure 15.18, which has been designed to measure the dielectric properties of the specimen by inserting it at the centre of the cavity through a small opening in the top wall as shown in this figure. The small hole in the cavity coincides with the position of the maximum electric field inside the cavity, which is obvious from our earlier discussions. The advantage of placing the test specimen at the maximum electric field position is that it provides maximum sensitivity during the measurement, and hence even very small dielectric samples can be accurately characterized. The particular mode in the designed cavity is excited by providing magnetic coupling through small hole at the centre known as *iris* at both the ends of the cavity as shown in the above figure. This kind of magnetic coupling through iris is usually common under situations where the microwave energy is to be efficiently transferred between two waveguides of similar types. In the present situation, the iris coupling is desirable as the energy is to be transferred from the coaxial-waveguide adapters to the rectangular cavity both having same type of waveguide. It should be noted that the cavity method to measure the dielectric properties of materials is simpler than that of the waveguide approach discussed in the last section because of the fact that it only requires the measurement of real quantities such as the resonant frequency and the Q factor of the cavity. On the contrary, the rectangular waveguide approach usually requires both the amplitude and phase of the reflection and transmission coefficients of the test specimen filled inside the rectangular waveguide for accurate determination of the dielectric properties. The rectangular waveguide method can, however, provide measurement of dielectric properties of data over a wide band in comparison to the cavity technique which works at a single frequency.

EXAMPLE 15.10

A piece of X-band rectangular waveguide (22.86 $\times$ 10.16 mm) operating in the dominant mode is converted into a cavity by placing two shorting plates at $z = 0$ and $z = d$. The cavity is to be designed for the measurement of dielectric constant of materials by inserting a test sample at the centre of the cavity. The inner walls of the cavity are made of copper ($\sigma_c = 5.8 \times 10^7\ \text{Sm}^{-1}$), and it is desired to keep the length of the cavity to be minimum.

Find:

(a) the operating mode of the cavity;

(b) the size of the cavity satisfying the criterion that the resonant frequency of the empty cavity is 10 GHz;

(c) the quality factor of the empty cavity at the resonant frequency.

Solution.

(a) As the cavity is constructed from the rectangular waveguide operating in the dominant (TE_{10}) mode, it will operate in the TE_{10p} mode. Now, the cavity is to be designed for testing of dielectric samples by placing them at its centre, hence the length of the cavity should be chosen in such as way that the electric field is maximum at the centre ($z = d/2$). This can be achieved if the integer p is chosen to be odd, i.e., $p = 1, 3, 5$, etc. However, it is also desired to make the length of the cavity to be minimum which makes $p = 1$. Hence the operating mode of the cavity satisfying the given situations is TE_{101}.

(b) The cavity is made from the X-band waveguide, and, hence, the dimensions of the cavity along x and y-directions are $a = 22.86$ mm and $b = 10.16$ mm, respectively. For obtaining the length of the cavity, we can make use of the given condition that the resonant frequency of the empty cavity should be 10 GHz. The resonant frequency of the empty rectangular cavity working in the TE_{101} mode is given by the following expression:

$$f_r^{TE_{101}} \equiv \frac{c}{2}\sqrt{\left(\frac{1}{a}\right)^2 + \left(\frac{1}{d}\right)^2} = \frac{3\times 10^8}{2}\sqrt{\left(\frac{1}{22.86\times 10^{-3}}\right)^2 + \left(\frac{1}{d}\right)^2} = 10.0 \text{ GHz}$$

$$\Rightarrow \sqrt{1.9136\times 10^3 + \left(\frac{1}{d}\right)^2} = \frac{2\times 10^{10}}{3\times 10^8} = 66.667$$

$$\Rightarrow \left(\frac{1}{d}\right)^2 = 2.5308\times 10^3$$

$$\Rightarrow d = 19.85 \text{ mm}$$

Hence, the size of the required rectangular cavity, which will resonate in the TE_{101} mode at 10 GHz is $22.86 \times 10.16 \times 19.85$ mm.

(c) The quality factor of the empty cavity operating in the TE_{101} mode at the resonant frequency of 10 GHz is given by the following expression:

$$Q^{TE_{101}} \equiv \frac{abd\left(a^2+d^2\right)\pi\mu_0 f_r^{TE_{101}}}{R_s\left[a^3(2b+d)+d^3(2b+a)\right]}$$

The value of R_s at the resonant frequency of 10 GHz for the cavity made of copper is computed as

$$R_s = \sqrt{\frac{\pi f\mu_0}{\sigma_c}} = \sqrt{\frac{\pi\times 10^{10}\times 4\pi\times 10^{-7}}{5.8\times 10^7}} = 0.0261\ \Omega$$

Hence, the Q-factor is calculated as

$$Q_{101}^{\rm TE} \equiv \frac{0.02286 \times 0.01016 \times 0.01985 \left(0.02286^2 + 0.01985^2\right) \pi \times 4\pi \times 10^{-7} \times 10^{10}}{0.0261 \left[0.02286^3 \left(2 \times 0.01016 + 0.01985\right) + 0.01985^3 \left(2 \times 0.01016 + 0.02286\right)\right]}$$

$$= 10,337$$

EXAMPLE 15.11

The dimensions of a rectangular cavity filled with Polystyrene ($\epsilon_r' = 2.25, \epsilon_r'' = 0$) are a = 5 cm, b = 2.5 cm, and d = 10 cm. The walls of the cavity are made of Brass ($\sigma_c = 1.5 \times 10^7$ Sm^{-1}).

Find:

(a) the three lowest order modes for a finite size cavity;

(b) the time-average electric and magnetic stored energies at the resonant frequency for the lowest order mode; and

(c) the power dissipated inside the cavity for the lowest order mode.

You can assume the amplitude of the z-component of the magnetic field for the lowest order mode to be 0.2 A/m.

Solution.

(a) The general expression for the resonant frequency of the rectangular cavity is given by the following equation:

$$f_{r,mnp} \equiv \frac{c}{2\sqrt{\epsilon_r'}}\sqrt{\left(\frac{m}{a}\right)^2 + \left(\frac{n}{b}\right)^2 + \left(\frac{p}{d}\right)^2} = \frac{3 \times 10^8}{2b \times \sqrt{2.25}}\sqrt{\left(\frac{m}{2}\right)^2 + \left(\frac{n}{1}\right)^2 + \left(\frac{p}{4}\right)^2}$$

$$= \frac{3 \times 10^8}{2 \times 2.5 \times 10^{-2} \times \sqrt{2.25}}\sqrt{\left(\frac{m}{2}\right)^2 + \left(\frac{n}{1}\right)^2 + \left(\frac{p}{4}\right)^2}$$

$$= 4.0\sqrt{\left(\frac{m}{2}\right)^2 + \left(\frac{n}{1}\right)^2 + \left(\frac{p}{4}\right)^2} \text{ (GHz)}$$

where we have made use of the fact that $a = 2b$, and $d = 4b$ in the present situation. The order of various modes supported by the cavity will be decided by integers m, n and p. The lowest value of p will be 1 for a finite length of the cavity. Now, the rectangular cavity can, in general, support either TE or TM modes. You know from the rectangular waveguide concept that the lowest order TM mode is TM_{11} mode as neither m nor n can be made zero for the TM mode. Hence the lowest order TM mode supported by the rectangular cavity will be TM_{111} mode. Similarly, for the TE mode in the rectangular waveguide, both m and n can be zero but they both cannot be zero simultaneously, i.e. $m = n = 0$ is ruled out. Hence some of the lower order TE modes in the cavity would be TE_{101}, TE_{011}, TE_{201}, etc.

Now, let us find the resonant frequency of the above-specified modes in order to find the first three lower order modes using the above equation.

$$f_r^{\mathrm{TE}_{101}} = 4.0\sqrt{\left(\frac{1}{2}\right)^2 + \left(\frac{0}{1}\right)^2 + \left(\frac{1}{4}\right)^2} = 2.236 \text{ GHz}$$

$$f_r^{\mathrm{TE}_{011}} = 4.0\sqrt{\left(\frac{0}{2}\right)^2 + \left(\frac{1}{1}\right)^2 + \left(\frac{1}{4}\right)^2} = 4.123 \text{ GHz}$$

$$f_r^{\mathrm{TE}_{201}} = 4.0\sqrt{\left(\frac{2}{2}\right)^2 + \left(\frac{0}{1}\right)^2 + \left(\frac{1}{4}\right)^2} = 4.123 \text{ GHz} \equiv f_r^{\mathrm{TE}_{011}}$$

$$f_r^{\mathrm{TE}_{111}} = 4.0\sqrt{\left(\frac{1}{2}\right)^2 + \left(\frac{1}{1}\right)^2 + \left(\frac{1}{4}\right)^2} = 4.583 \text{ GHz}$$

$$f_r^{\mathrm{TM}_{111}} \equiv f_r^{TE_{111}} = 4.583 \text{ GHz}$$

Hence, the three lowest order modes of the finite size rectangular cavity in the present situation are

$$\mathrm{TE}_{101},\ \mathrm{TE}_{201}/\mathrm{TE}_{011},\ \mathrm{TE}_{111}/\mathrm{TM}_{111}$$

where $\mathrm{TE}_{201}/\mathrm{TE}_{011}$ and $\mathrm{TE}_{111}/\mathrm{TM}_{111}$ are pairs of degenerate modes having the same resonant frequency.

It can be observed from above that the TE_{101} is the lowest order mode, which may be also termed as the dominant mode. You may also note that the dimensions of the given rectangular cavity may be arranged in the following order:

$$d > a > b$$

(*b*) At the resonant frequency, the time-average electric and magnetic stored energies would be equal. Hence, let us calculate here the time average electric stored energy in the cavity for the lowest order (TE_{101}) mode using (100) as follows:

$$< W_E^{\mathrm{TE}_{101}} > \equiv \frac{\epsilon}{4}\left|A_{101}\right|^2 \left(f_r^{\mathrm{TE}_{101}}\right)^2 \mu_0^2 bda^3$$

$$\equiv \frac{8.856\times10^{-12}\times 2.25}{4}\left|0.2\right|^2 \times \left(2.236\times10^9\right)^2 \times \left(4\pi\times10^{-7}\right)^2$$

$$\times \left(2.5\times10^{-2}\right) \times \left(10\times10^{-2}\right) \times \left(5\times10^{-2}\right)^3$$

$$\equiv 4.916\times10^{-13} \text{ J}$$

where A_{101} = 0.2 A/m represents the amplitude of the z-component of the magnetic field as given in the problem. The time average magnetic stored energy in the cavity for the TE_{101} mode at the resonant frequency would be same as that of the electric stored energy, i.e.

$$< W_M^{\mathrm{TE}_{101}} > \equiv < W_E^{\mathrm{TE}_{101}} > \equiv 4.916\times10^{-13} J$$

(*c*) As the cavity is filled with the lossless dielectric, the power loss will be only due to the imperfect conductor walls of the cavity. The power dissipated inside the cavity due to the conductor loss of the walls for the lowest order (TE_{101}) mode is given by (110), i.e.

$$< P_d^{TE_{101}} > \equiv R_s \left| A_{101} \right|^2 \left[\frac{a^3 (2b+d) + d^3 (2b+a)}{4d^2} \right]$$

The value of R_s at the resonant frequency corresponding to the TE_{101} mode for the cavity walls made of Brass ($\sigma_c = 1.5 \times 10^7$ Sm^{-1}) may be computed as

$$R_s = \sqrt{\frac{\pi f \mu_0}{\sigma_c}} = \sqrt{\frac{\pi \times 2.236 \times 10^9 \times 4\pi \times 10^{-7}}{1.5 \times 10^7}} = 0.0243\ \Omega$$

Hence, the power dissipation may be calculated by substituting the value of R_s in the above-defined equations as follows:

$$< P_d^{TE_{101}} > \equiv 0.0243 \times |0.2|^2 \left[\frac{(0.05)^3 (2 \times 0.025 + 0.10) + (0.1)^3 (2 \times 0.025 + 0.05)}{4 \times (0.1)^2} \right]$$

$$= 2.886 \times 10^{-6}\ \text{W}$$

References

1. Constantine A Balanis, *Advanced Engineering Electromagnetics*, John Wiley & Sons, 1989. This advanced text provides a detailed analysis of the rectangular as well the circular waveguides and cavities. The designations of various standard rectangular waveguides including their dimensions and frequency bands are provided in Chapter 8.
2. L. F. Chen, C. K. Ong, C. P. Neo, V. V. Varadan, and V. K. Varadan, *Microwave Electronics: Measurement and Materials Characterization*, John Wiley & Sons, 2004. This book gives a good overview of the rectangular waveguide and the rectangular cavity technique for the electromagnetic characterization of materials in the microwave frequency band.

CHAPTER 15 Problems

15.1 A lossless parallel-plate waveguide is known to propagate the $m = 2$ TE and TM modes at frequencies as low as 10 GHz. If the plate separation is 1 cm, determine the dielectric constant of the medium between plates.

15.2 A $d = 1$ cm parallel-plate guide is made with glass ($n = 1.45$) between plates. If the operating frequency is 32 GHz, which modes will propagate?

15.3 For the guide of Problem 15.2, and at the 32 GHz frequency, determine the difference between the group delays of the highest-order mode (TE or TM) and the TEM mode. Assume a propagation distance of 10 cm.

15.4 The cutoff frequency of the $m = 1$ TE and TM modes in an air-filled parallel-plate guide is known to be $f_{c1} = 7.5$ GHz. The guide is used at wavelength $\lambda = 1.5$ cm. Find the group velocity of the $m = 2$ TE and TM modes.

15.5 A rectangular waveguide has dimensions $a = 6$ cm and $b = 4$ cm. (*a*) Over what range of frequencies will the guide operate single mode? (*b*) Over what frequency range will the guide support *both* TE_{10} and TE_{01} modes and no others?

15.6 Two rectangular waveguides are joined end-to-end. The guides have identical dimensions, where $a = 2b$. One guide is air-filled; the other is filled with a lossless dielectric characterized by ϵ_r'. (*a*) Determine the maximum allowable value of ϵ_r' such that single-mode operation can be simultaneously assured in *both* guides at some frequency. (*b*) Write an expression for the frequency range over which single-mode operation will occur in both guides; your answer should be in terms of ϵ_r', guide dimensions as needed, and other known constants.

15.7 An air-filled rectangular waveguide is to be constructed for single-mode operation at 15 GHz. Specify the guide dimensions, a and b, such that the design frequency is 10 percent higher than the cutoff frequency for the TE_{10} mode, while being 10 percent lower than the cutoff frequency for the next-higher-order mode.

15.8 Using the relation $\langle S \rangle = \frac{1}{2}\,\mathrm{Re}\{\mathbf{E}_s \times \mathbf{H}_s^*\}$ show that the average power density in the TE_{10} mode in a rectangular waveguide is given by

$$\langle S \rangle = \frac{\beta_{10}}{2\omega\mu} E_0^2 \sin^2(\kappa_{10} x)\mathbf{a}_z \ \mathrm{W/m^2}$$

15.9 Integrate the result of Problem 15.8 over the guide cross section, $0 < x < a, 0 < y < b$, to show that the average power in watts transmitted down the guide is given as

$$P_{\mathrm{av}} = \frac{\beta_{10} ab}{4\omega\mu} E_0^2 = \frac{ab}{4\eta} E_0^2 \sin\theta_{10} \ \mathrm{W}$$

where $\eta = \sqrt{\mu/\epsilon}$ and θ_{10} is the wave angle associated with the TE_{10} mode. Interpret.

15.10 Show that the group dispersion parameter, $d^2\beta/d\omega^2$, for a given mode in a parallel-plate or rectangular waveguide is given by

$$\frac{d^2\beta}{d\omega^2} = -\frac{n}{\omega c}\left(\frac{\omega_c}{\omega}\right)^2 \left[1 - \left(\frac{\omega_c}{\omega}\right)^2\right]^{-3/2}$$

where ω_c is the radian cutoff frequency for the mode in question [note that the first derivative form was already found, resulting in Eq. (57)].

15.11 Consider a transform-limited pulse of center frequency $f = 10$ GHz, and of full-width $2T = 1.0$ ns. The pulse propagates in a lossless single-mode rectangular guide which is air-filled and in which the 10 GHz operating frequency is 1.1 times the cutoff frequency of the TE_{10} mode. Using the result of Problem 15.4, determine the length of guide over which the pulse broadens to twice its initial width. What simple step can be taken to reduce the amount of pulse broadening in this guide, while maintaining the same initial pulse width? Additional background for this problem is found in Section 13.6.

A

Vector Analysis

A.1 GENERAL CURVILINEAR COORDINATES

Let us consider a general orthogonal coordinate system in which a point is located by the intersection of three mutually perpendicular surfaces (of unspecified form or shape),

$$u = \text{constant}$$
$$\nu = \text{constant}$$
$$w = \text{constant}$$

where u, ν, and w are the variables of the coordinate system. If each variable is increased by a differential amount and three more mutually perpendicular surfaces are drawn corresponding to these new values, a differential volume is formed which is closely a rectangular parallelepiped. Since u, ν, and w need not be measures of length, such as, for example, the angle variables of the cylindrical and spherical coordinate systems, each must be multiplied by a general function of u, ν, and w in order to obtain the differential sides of the parallelepiped. Thus we define the scale factors h_1, h_2, and h_3 each as a function of the three variables u, ν, and w and write the lengths of the sides of the differential volume as

$$dL_1 = h_1 du$$
$$dL_2 = h_2 d\nu$$
$$dL_3 = h_3 dw$$

In the three coordinate systems discussed in Chapter 1, it is apparent that the variables and scale factors are

$$\begin{array}{llll} \text{Rectangular:} & u = x & \nu = y & w = z \\ & h_1 = 1 & h_2 = 1 & h_3 = 1 \\ \text{Cylindrical:} & u = \rho & \nu = \phi & w = z \\ & h_1 = 1 & h_2 = \rho & h_3 = 1 \\ \text{Spherical:} & u = r & \nu = \theta & w = \phi \\ & h_1 = 1 & h_2 = r & h_3 = r \sin\theta \end{array} \tag{A.1}$$

The choice of u, ν, and w has been made so that $\mathbf{a}_u \times \mathbf{a}_\nu = \mathbf{a}_w$ in all cases. More involved expressions for h_1, h_2, and h_3 are to be expected in other less familiar coordinate systems.[1]

A.2 DIVERGENCE, GRADIENT, AND CURL IN GENERAL CURVILINEAR COORDINATES

If the method used to develop divergence in Sections 3.4 and 3.5 is applied to the general curvilinear coordinate system, the flux of the vector **D** passing through the surface of the parallelepiped whose unit normal is $\mathbf{a}_u$ is

$$D_{u0} dL_2 dL_3 + \frac{1}{2}\frac{\partial}{\partial u}(D_u dL_2 dL_3) du$$

or

$$D_{u0} h_2 h_3 d\nu\, dw + \frac{1}{2}\frac{\partial}{\partial u}(D_u h_2 h_3 d\nu\, dw) du$$

and for the opposite face it is

$$-D_{u0} h_2 h_3 d\nu\, dw + \frac{1}{2}\frac{\partial}{\partial u}(D_u h_2 h_3 d\nu\, dw) du$$

giving a total for these two faces of

$$\frac{\partial}{\partial u}(D_u h_2 h_3 d\nu\, dw) du$$

Since u, ν, and w are independent variables, this last expression may be written as

$$\frac{\partial}{\partial u}(h_2 h_3 D_u) du\, d\nu\, dw$$

and the other two corresponding expressions obtained by a simple permutation of the subscripts and of u, ν, and w. Thus the total flux leaving the differential volume is

$$\left[\frac{\partial}{\partial u}(h_2 h_3 D_u) + \frac{\partial}{\partial v}(h_3 h_1 D_\nu) + \frac{\partial}{\partial w}(h_1 h_2 D_w)\right] du\, d\nu\, dw$$

and the divergence of **D** is found by dividing by the differential volume

$$\nabla \cdot \mathbf{D} = \frac{1}{h_1 h_2 h_3}\left[\frac{\partial}{\partial u}(h_2 h_3 D_u) + \frac{\partial}{\partial \nu}(h_3 h_1 D_\nu) + \frac{\partial}{\partial w}(h_1 h_2 D_w)\right] \tag{A.2}$$

The components of the gradient of a scalar V may be obtained (following the methods of Section 4.6) by expressing the total differential of V,

$$dV = \frac{\partial V}{\partial u} du + \frac{\partial V}{\partial \nu} d\nu + \frac{\partial V}{\partial w} dw$$

in terms of the component differential lengths, $h_1 du$, $h_2 d\nu$, and $h_3 dw$,

$$dV = \frac{1}{h_1}\frac{\partial V}{\partial u} h_1 du + \frac{1}{h_2}\frac{\partial V}{\partial \nu} h_2 d\nu + \frac{1}{h_3}\frac{\partial V}{\partial w} h_3 dw$$

---

[1] The variables and scale factors are given for nine orthogonal coordinate systems on pp. 50–59 in J. A. Stratton, *Electromagnetic Theory*. New York: McGraw-Hill, 1941. Each system is also described briefly.

Then, since

$$d\mathbf{L} = h_1 du\mathbf{a}_u + h_2 d\nu\mathbf{a}_\nu + h_3 dw\mathbf{a}_w \qquad \text{and} \qquad dV = \nabla V \cdot d\mathbf{L}$$

we see that

$$\nabla V = \frac{1}{h_1}\frac{\partial V}{\partial u}\mathbf{a}_u + \frac{1}{h_2}\frac{\partial V}{\partial \nu}\mathbf{a}_\nu + \frac{1}{h_3}\frac{\partial V}{\partial w}\mathbf{a}_w \tag{A.3}$$

The components of the curl of a vector **H** are obtained by considering a differential path first in a $u =$ constant surface and finding the circulation of **H** about that path, as discussed for rectangular coordinates in Section 8.3. The contribution along the segment in the $\mathbf{a}_\nu$ direction is

$$H_{\nu 0}h_2 d\nu - \frac{1}{2}\frac{\partial}{\partial w}(H_\nu h_2 d\nu)dw$$

and that from the oppositely directed segment is

$$-H_{\nu 0}h_2 d\nu - \frac{1}{2}\frac{\partial}{\partial w}(H_\nu h_2 d\nu)dw$$

The sum of these two parts is

$$-\frac{\partial}{\partial w}(H_\nu h_2 d\nu)dw$$

or

$$-\frac{\partial}{\partial w}(h_2 H_\nu)d\nu\, dw$$

and the sum of the contributions from the other two sides of the path is

$$\frac{\partial}{\partial \nu}(h_3 H_w)d\nu\, dw$$

Adding these two terms and dividing the sum by the enclosed area, $h_2 h_3 d\nu\, dw$, we see that the $\mathbf{a}_u$ component of curl **H** is

$$(\nabla \times \mathbf{H})_u = \frac{1}{h_2 h_3}\left[\frac{\partial}{\partial \nu}(h_3 H_w) - \frac{\partial}{\partial w}(h_2 H_\nu)\right]$$

and the other two components may be obtained by cyclic permutation. The result is expressible as a determinant,

$$\nabla \times \mathbf{H} = \begin{vmatrix} \dfrac{\mathbf{a}_u}{h_2 h_3} & \dfrac{\mathbf{a}_\nu}{h_3 h_1} & \dfrac{\mathbf{a}_w}{h_1 h_2} \\ \dfrac{\partial}{\partial u} & \dfrac{\partial}{\partial \nu} & \dfrac{\partial}{\partial w} \\ h_1 H_u & h_2 H_\nu & h_3 H_w \end{vmatrix} \tag{A.4}$$

The Laplacian of a scalar is found by using (A.2) and (A.3):

$$\nabla^2 V = \nabla \cdot \nabla V = \frac{1}{h_1 h_2 h_3}\left[\frac{\partial}{\partial u}\left(\frac{h_2 h_3}{h_1}\frac{\partial \nu}{\partial u}\right) + \frac{\partial}{\partial \nu}\left(\frac{h_3 h_1}{h_2}\frac{\partial V}{\partial \nu}\right) + \frac{\partial}{\partial w}\left(\frac{h_1 h_2}{h_3}\frac{\partial V}{\partial w}\right)\right] \tag{A.5}$$

Equations (A.2) to (A.5) may be used to find the divergence, gradient, curl, and Laplacian in any orthogonal coordinate system for which h_1, h_2, and h_3 are known.

Expressions for $\nabla \cdot \mathbf{D}$, ∇V, $\nabla \times \mathbf{H}$, and $\nabla^2 V$ are given in rectangular, circular cylindrical, and spherical coordinate systems inside the back cover.

A.3 VECTOR IDENTITIES

The vector identities that follow may be proved by expansion in rectangular (or general curvilinear) coordinates. The first two identities involve the scalar and vector triple products, the next three are concerned with operations on sums, the following three apply to operations when the argument is multiplied by a scalar function, the next three apply to operations on scalar or vector products, and the last four concern the second-order operations.

$$(\mathbf{A} \times \mathbf{B}) \cdot \mathbf{C} \equiv (\mathbf{B} \times \mathbf{C}) \cdot \mathbf{A} \equiv (\mathbf{C} \times \mathbf{A}) \cdot \mathbf{B} \tag{A.6}$$

$$\mathbf{A} \times (\mathbf{B} \times \mathbf{C}) \equiv (\mathbf{A} \cdot \mathbf{C})\mathbf{B} - (\mathbf{A} \cdot \mathbf{B})\mathbf{C} \tag{A.7}$$

$$\nabla \cdot (\mathbf{A} + \mathbf{B}) \equiv \nabla \cdot \mathbf{A} + \nabla \cdot \mathbf{B} \tag{A.8}$$

$$\nabla (V + W) \equiv \nabla V + \nabla W \tag{A.9}$$

$$\nabla \times (\mathbf{A} + \mathbf{B}) \equiv \nabla \times \mathbf{A} + \nabla \times \mathbf{B} \tag{A.10}$$

$$\nabla \cdot (V\mathbf{A}) \equiv \mathbf{A} \cdot \nabla V + V \nabla \cdot \mathbf{A} \tag{A.11}$$

$$\nabla (VW) \equiv V \nabla W + W \nabla V \tag{A.12}$$

$$\nabla \times (V\mathbf{A}) \equiv \nabla V \times \mathbf{A} + V \nabla \times \mathbf{A} \tag{A.13}$$

$$\nabla \cdot (\mathbf{A} \times \mathbf{B}) \equiv \mathbf{B} \cdot \nabla \times \mathbf{A} - \mathbf{A} \cdot \nabla \times \mathbf{B} \tag{A.14}$$

$$\nabla (\mathbf{A} \cdot \mathbf{B}) \equiv (\mathbf{A} \cdot \nabla)\mathbf{B} + (\mathbf{B} \cdot \nabla)\mathbf{A} + \mathbf{A} \times (\nabla \times \mathbf{B}) + \mathbf{B} \times (\nabla \times \mathbf{A}) \tag{A.15}$$

$$\nabla \times (\mathbf{A} \times \mathbf{B}) \equiv \mathbf{A} \nabla \cdot \mathbf{B} - \mathbf{B} \nabla \cdot \mathbf{A} + (\mathbf{B} \cdot \nabla)\mathbf{A} - (\mathbf{A} \cdot \nabla)\mathbf{B} \tag{A.16}$$

$$\nabla \cdot \nabla V \equiv \nabla^2 V \tag{A.17}$$

$$\nabla \cdot \nabla \times \mathbf{A} \equiv 0 \tag{A.18}$$

$$\nabla \times \nabla V \equiv 0 \tag{A.19}$$

$$\nabla \times \nabla \times \mathbf{A} \equiv \nabla (\nabla \cdot \mathbf{A}) - \nabla^2 \mathbf{A} \tag{A.20}$$

B Units

We shall describe first the International System (abbreviated SI, for Système International d'Unités), which is used in this book and is now standard in electrical engineering and much of physics. It has also been officially adopted as the international system of units by many countries, including the United States.[1]

The fundamental unit of length is the meter, which was defined in the latter part of the nineteenth century as the distance between two marks on a certain platinum-iridium bar. The definition was improved in 1960 by relating the meter to the wavelength of the radiation emitted by the rare gas isotope krypton 86 under certain specified conditions. This so-called krypton meter was accurate to four parts per billion, a value leading to negligible uncertainties in constructing skyscrapers or building highways, but capable of causing an error greater than one meter in determining the distance to the moon. The meter was redefined in 1983 in terms of the velocity of light. At that time the velocity of light was specified to be an auxiliary constant with an *exact* value of 299 792 458 meters per second. As a result, the latest definition of the meter is the distance light travels in a vacuum in $1/299\,792\,458$ of a second. If greater accuracy is achieved in measuring c, that value will remain 299 792 458 m/s, but the length of the meter will change.

It is evident that our definition of the meter is expressed in terms of the second, the fundamental unit of time. The second is defined as 9 192 631 770 periods of the transition frequency between the hyperfine levels $F = 4$, $m_F = 0$, and $F = 3$, $m_F = 0$ of the ground state $^2s_{1/2}$ of the atom of cesium 133, unperturbed by external fields. This definition of the second, complex though it may be, permits time to be measured with an accuracy better than one part in 10^{13}.

The standard mass of one kilogram is defined as the mass of an international standard in the form of a platinum-iridium cylinder at the International Bureau of Weights and Measures at Sèvres, France.

The unit of temperature is the kelvin, defined by placing the triple-point temperature of water at 273.16 kelvins.

A fifth unit is the candela, defined as the luminous intensity of an omnidirectional radiator at the freezing temperature of platinum (2042 K) having an area of $1/600\,000$ square meter and under a pressure of 101 325 newtons per square meter.

[1] The International System of Units was adopted by the Eleventh General Conference on Weights and Measures in Paris in 1960, and it was officially adopted for scientific usage by the National Bureau of Standards in 1964. It is a metric system, which interestingly enough is the only system which has ever received specific sanction from Congress. This occurred first in 1966 and then again in 1975 with the Metric Conversion Act, which provides for "voluntary conversion" to the metric system. No specific time was specified, however, and we can assume that it will still be a few years before the bathroom scale reads mass in kilograms and Miss America is a 90–60–90.

The last of the fundamental units is the ampere. Before explicitly defining the ampere, we must first define the newton. It is defined in terms of the other fundamental units from Newton's third law as the force required to produce an acceleration of one meter per second per second on a one-kilogram mass. We now may define the ampere as that constant current, flowing in opposite directions in two straight parallel conductors of infinite length and negligible cross section, separated one meter in vacuum, that produces a repulsive force of 2×10^{-7} newton per meter length between the two conductors. The force between the two parallel conductors is known to be

$$F = \mu_0 \frac{I^2}{2\pi d}$$

and thus

$$2 \times 10^{-7} = \mu_0 \frac{1}{2\pi}$$

or

$$\mu_0 = 4\pi \times 10^{-7} \qquad (\text{kg} \cdot \text{m/A}^2 \cdot \text{s}^2, \text{or H/m})$$

We thus find that our definition of the ampere has been formulated in such a way as to assign an exact, simple numerical value to the permeability of free space.

Returning to the International System, the units in which the other electric and magnetic quantities are measured are given in the body of the text at the time each quantity is defined, and all of them can be related to the basic units already defined. For example, our work with the plane wave in Chapter 12 shows that the velocity with which an electromagnetic wave propagates in free space is

$$c = \frac{1}{\sqrt{\mu_0 \epsilon_0}}$$

and thus

$$\epsilon_0 = \frac{1}{\mu_0 c^2} = \frac{1}{4\pi 10^{-7} c^2} = 8.854\,187\,817 \times 10^{-12} \text{ F/m}$$

It is evident that the numerical value of ϵ_0 depends upon the defined value of the velocity of light in vacuum, 299 792 458 m/s.

The units are also given in Table B.1 for easy reference. They are listed in the same order in which they are defined in the text.

Table B.1 Names and units of the electric and magnetic quantities in the International System (in the order in which they appear in the text)

Symbol	Name	Unit	Abbreviation
v	Velocity	meter/second	m/s
F	Force	newton	N
Q	Charge	coulomb	C
r, R	Distance	meter	m
ϵ_0, ϵ	Permittivity	farad/meter	F/m

Table B.1 *(Continued)*

Symbol	Name	Unit	Abbreviation
E	Electric field intensity	volt/meter	V/m
ρ_ν	Volume charge density	coulomb/meter^3	C/m^3
ν	Volume	meter^3	m^3
ρ_L	Linear charge density	coulomb/meter	C/m
ρ_S	Surface charge density	coulomb/meter^2	C/m^2
Ψ	Electric flux	coulomb	C
D	Electric flux density	coulomb/meter^2	C/m^2
S	Area	meter^2	m^2
W	Work, energy	joule	J
L	Length	meter	m
V	Potential	volt	V
p	Dipole moment	coulomb-meter	C·m
I	Current	ampere	A
J	Current density	ampere/meter^2	A/m^2
μ_e, μ_h	Mobility	meter^2/volt-second	m^2/V·s
e	Electronic charge	coulomb	C
σ	Conductivity	siemens/meter	S/m
R	Resistance	ohm	Ω
P	Polarization	coulomb/meter^2	C/m^2
$\chi_{e,m}$	Susceptibility		
C	Capacitance	farad	F
R_s	Sheet resistance	ohm per square	Ω
H	Magnetic field intensity	ampere/meter	A/m
K	Surface current density	ampere/meter	A/m
B	Magnetic flux density	tesla (or weber/meter^2)	T (or Wb/m^2)
μ_0, μ	Permeability	henry/meter	H/m
Φ	Magnetic flux	weber	Wb
V_m	Magnetic scalar potential	ampere	A
A	Vector magnetic potential	weber/meter	Wb/m
T	Torque	newton-meter	N·m
m	Magnetic moment	ampere-meter^2	A·m^2
M	Magnetization	ampere/meter	A/m
$\mathcal{R}$	Reluctance	ampere-turn/weber	A·t/Wb

Table B.1 *(Continued)*

Symbol	Name	Unit	Abbreviation
L	Inductance	henry	H
M	Mutual inductance	henry	H
ω	Radian frequency	radian/second	rad/s
c	Velocity of light	meter/second	m/s
λ	Wavelength	meter	m
η	Intrinsic impedance	ohm	Ω
k	Wave number	meter^{-1}	m^{-1}
α	Attenuation constant	neper/meter	Np/m
β	Phase constant	radian/meter	rad/m
f	Frequency	hertz	Hz
S	Poynting vector	watt/meter^2	W/m^2
P	Power	watt	W
δ	Skin depth	meter	m
Γ	Reflection coefficient		
s	Standing wave ratio		
γ	Propagation constant	meter^{-1}	m^{-1}
G	Conductance	siemen	S
Z	Impedance	ohm	Ω
Y	Admittance	siemen	S
Q	Quality factor		

Finally, other systems of units have been used in electricity and magnetism. In the electrostatic system of units (esu), Coulomb's law is written for free space,

$$F = \frac{Q_1 Q_2}{R^2} \qquad \text{(esu)}$$

The permittivity of free space is assigned the value of unity. The gram and centimeter are the fundamental units of mass and distance, and the esu system is therefore a cgs system. Units bearing the prefix *stat-* belong to the electrostatic system of units.

In a similar manner, the electromagnetic system of units (emu) is based on Coulomb's law for magnetic poles, and the permeability of free space is unity. The prefix *ab-* identifies emu units. When electric quantities are expressed in esu units, magnetic quantities in emu units, and both appear in the same equation (such as Maxwell's curl equations), the velocity of light appears explicitly. This follows from noting that in esu $\epsilon_0 = 1$, but $\mu_0 \epsilon_0 = 1/c^2$, and therefore $\mu_0 = 1/c^2$, and in emu $\mu_0 = 1$, and hence $\epsilon_0 = 1/c^2$. Thus, in this intermixed system known as the Gaussian system of units,

$$\nabla \times \mathbf{H} = 4\pi\mathbf{J} + \frac{1}{c}\frac{\partial \mathbf{D}}{\partial t} \qquad \text{(Gaussian)}$$

Other systems include the factor 4π explicitly in Coulomb's law, and it then does not appear in Maxwell's equations. When this is done, the system is said to be rationalized. Hence the Gaussian system is an unrationalized cgs system (when rationalized it is known as the Heaviside-Lorentz system), and the International System we have used throughout this book is a rationalized mks system.

Table B.2 Conversion of International to Gaussian and other units (use $c = 2.997\,924\,58 \times 10^8$)

Quantity	1 mks unit	= Gaussian units	= Other units
d	1 m	10^2 cm	39.37 in.
F	1 N	10^5 dyne	0.2248 lb_f
W	1 J	10^7 erg	0.7376 ft-lb_f
Q	1 C	$10c$ statC	0.1 abC
ρ_ν	1 C/m^3	$10^{-5}c$ statC/cm^3	10^{-7} abC/cm^3
D	1 C/m^2	$4\pi 10^{-3}c$ (esu)	$4\pi 10^{-5}$ (emu)
E	1 V/m	$10^4/c$ statV/cm	10^6 abV/cm
V	1 V	$10^6/c$ statV	10^8 abV
I	1 A	0.1 abA	$10c$ statA
H	1 A/m	$4\pi 10^{-3}$ oersted	$0.4\pi c$ (esu)
V_m	1 A·t	0.4π gilbert	$40\pi c$ (esu)
B	1 T	10^4 gauss	$100/c$ (esu)
Φ	1 Wb	10^8 maxwell	$10^6/c$ (esu)
A	1 Wb/m	10^6 maxwell/cm	
R	1 Ω	10^9 abΩ	$10^5/c^2$ statΩ
L	1 H	10^9 abH	$10^5/c^2$ statH
C	1 F	$10^{-5}c^2$ statF	10^{-9} abF
σ	1 S/m	10^{-11} abS/cm	$10^{-7}c^2$ statS/cm
μ	1 H/m	$10^7/4\pi$ (emu)	$10^3/4\pi c^2$ (esu)
ϵ	1 F/m	$4\pi 10^{-7}c^2$ (esu)	$4\pi 10^{-11}$ (emu)

Table B.2 gives the conversion factors between the more important units of the International System (or rationalized mks system) and the Gaussian system, and several other assorted units.

Table B.3 lists the prefixes used with any of the SI units, their abbreviations, and the power of ten each represents. Those checked are widely used. Both the prefixes and their abbreviations are written without hyphens, and therefore $10^{-6}\,\text{F} = 1$ microfarad $= 1\mu\text{F}$ $= 1000$ nanofarads $= 1000$ nF, and so forth.

Table B.3 Standard prefixes used with SI units

Prefix	Abbrev.	Meaning	Prefix	Abbrev.	Meaning
atto-	a-	10^{-18}	deka-	da-	10^{1}
femto-	f-	10^{-15}	hecto-	h-	10^{2}
pico-	p-	10^{-12}	kilo-	k-	10^{3}
nano-	n-	10^{-9}	mega-	M-	10^{6}
micro-	μ-	10^{-6}	giga-	G-	10^{9}
milli-	m-	10^{-3}	tera-	T-	10^{12}
centi-	c-	10^{-2}	peta-	P-	10^{15}
deci-	d-	10^{-1}	exa-	E-	10^{18}

C Material Constants

Table C.1 lists typical values of the relative permittivity ϵ'_r or dielectric constant for common insulating and dielectric materials, along with representative values for the loss tangent. The values should only be considered representative for each material, and they apply to normal temperature and humidity conditions and to very low audio frequencies. Most of them have been taken from *Reference Data for Radio Engineers.*[1] *The Standard Handbook for Electrical Engineers*,[2] and von Hippel,[3] and these volumes may be referred to for further information on these and other materials.

Table C.2 gives the conductivity for a number of metallic conductors, for a few insulating materials, and for several other materials of general interest. The values have been taken from the references listed previously, and they apply at zero frequency and at room temperature. The listing is in the order of decreasing conductivity.

Some representative values of the relative permeability for various diamagnetic, paramagnetic, ferrimagnetic, and ferromagnetic materials are listed in Table C.3. They have been extracted from the references listed previously, and the data for the ferromagnetic materials is only valid for very low magnetic flux densities. Maximum permeabilities may be an order of magnitude higher.

Values are given in Table C.4 for the charge and rest mass of an electron, the permittivity and permeability of free space, and the velocity of light.[4]

Table C.1 ϵ'_r and ϵ''/ϵ'

Material	ϵ'_r	ϵ''/ϵ'
Air	1.0005	
Alcohol, ethyl	25	0.1
Aluminum oxide	8.8	0.000 6
Amber	2.7	0.002
Bakelite	4.74	0.022
Barium titanate	1200	0.013

[1] See References for Chapter 11.
[2] See References for Chapter 5.
[3] von Hippel, A. R. *Dielectric Materials and Applications.* Cambridge, Mass. and New York: The Technology Press of the Massachusetts Institute of Technology and John Wiley & Sons, 1954.
[4] Cohen, E. R., and B. N. Taylor. *The 1986 Adjustment of the Fundamental Physical Constants.* Elmsford, N.Y.: Pergamon Press, 1986.

Table C.1 *(Continued)*

Material	ϵ'_r	ϵ''/ϵ'
Carbon dioxide	1.001	
Ferrite (NiZn)	12.4	0.000 25
Germanium	16	
Glass	4–7	0.002
Ice	4.2	0.05
Mica	5.4	0.000 6
Neoprene	6.6	0.011
Nylon	3.5	0.02
Paper	3	0.008
Plexiglas	3.45	0.03
Polyethylene	2.26	0.000 2
Polypropylene	2.25	0.000 3
Polystyrene	2.56	0.000 05
Porcelain (dry process)	6	0.014
Pyranol	4.4	0.000 5
Pyrex glass	4	0.000 6
Quartz (fused)	3.8	0.000 75
Rubber	2.5–3	0.002
Silica or SiO_2 (fused)	3.8	0.000 75
Silicon	11.8	
Snow	3.3	0.5
Sodium chloride	5.9	0.000 1
Soil (dry)	2.8	0.05
Steatite	5.8	0.003
Styrofoam	1.03	0.000 1
Teflon	2.1	0.000 3
Titanium dioxide	100	0.001 5
Water (distilled)	80	0.04
Water (sea)		4
Water (dehydrated)	1	0
Wood (dry)	1.5–4	0.01

Table C.2 σ

Material	σ, S/m	Material	σ, S/m
Silver	6.17×10^7	Nichrome	0.1×10^7
Copper	5.80×10	Graphite	7×10^4
Gold	4.10×10^7	Silicon	2300
Aluminum	3.82×10	Ferrite (typical)	100
Tungsten	1.82×10^7	Water (sea)	5
Zinc	1.67×10	Limestone	10^{-2}
Brass	1.5×10^7	Clay	5×10^{-3}
Nickel	1.45×10	Water (fresh)	10^{-3}
Iron	1.03×10^7	Water (distilled)	10^{-4}
Phosphor bronz	$1 \times 10'$	Soil (sandy)	10^{-5}
Solder	0.7×10^7	Granite	10^{-6}
Carbon steel	0.6×10	Marble	10^{-8}
German silver	0.3×10^7	Bakelite	10^{-9}
Manganin	0.227×10	Porcelain (dry process	10^{-10}
Constantan	0.226×10^7	Diamond	2×10^{-13}
Germanium	0.22×10	Polystyrene	10^{-16}
Stainless steel	0.11×10^7	Quartz	10^{-17}

Table C.3 μ_r

Material	μ_r	Material	μ_r
Bismuth	0.999 998 6	Powdered iron	100
Paraffin	0.999 999 42	Machine steel	300
Wood	0.999 999 5	Ferrite (typical)	1000
Silver	0.999 999 81	Permalloy 45	2500
Aluminum	1.000 000 65	Transformer iron	3000
Beryllium	1.000 000 79	Silicon iron	3500
Nickel chloride	1.000 04	Iron (pure)	4000
Manganese sulfate	1.000 1	Mumetal	20 000
Nickel	50	Sendust	30 000
Cast iron	60	Supermalloy	100 000
Cobalt	60		

Table C.4 Physical constants

Quantity	Value
Electron charge	$e = (1.602\,177\,33 \pm 0.000\,000\,46) \times 10^{-19}$ C
Electron mass	$m = (9.109\,389\,7 \pm 0.000\,005\,4) \times 10^{-31}$ kg
Permittivity of free space	$\epsilon_0 = 8.854\,187\,817 \times 10^{-12}$ F/m
Permeability of free space	$\mu_0 = 4\pi 10^{-7}$ H/m
Velocity of light	$c = 2.997\,924\,58 \times 10^{8}$ m/s

D

Origins of the Complex Permittivity

As we learned in Chapter 5, a dielectric can be modeled as an arrangement of atoms and molecules in free space, which can be polarized by an electric field. The field forces positive and negative bound charges to separate against their Coulomb attractive forces, thus producing an array of microscopic dipoles. The molecules can be arranged in an ordered and predictable manner (such as in a crystal) or may exhibit random positioning and orientation, as would occur in an amorphous material or a liquid. The molecules may or may not exhibit permanent dipole moments (existing before the field is applied), and if they do, they will usually have random orientations throughout the material volume. As discussed in Section 6.1, the displacement of charges in a regular manner, as induced by an electric field, gives rise to a macroscopic polarization, **P**, defined as the dipole moment per unit volume:

$$\mathbf{P} = \lim_{\Delta\nu \to 0} \frac{1}{\Delta\nu} \sum_{i=1}^{N\Delta\nu} \mathbf{p_i} \tag{D.1}$$

where N is the number of dipoles per unit volume and $\mathbf{p_i}$ is the dipole moment of the *i*th atom or molecule, found through

$$\mathbf{p_i} = Q_i \mathbf{d}_i \tag{D.2}$$

Q_i is the positive one of the two bound charges composing dipole i, and $\mathbf{d}_i$ is the distance between charges, expressed as a vector from the negative to the positive charge. Again, borrowing from Section 6.1, the electric field and the polarization are related through

$$\mathbf{P} = \epsilon_0 \chi_e \mathbf{E} \tag{D.3}$$

where the electric susceptibility, χ_e, forms the more interesting part of the dielectric constant:

$$\epsilon_r = 1 + \chi_e \tag{D.4}$$

Therefore, to understand the nature of ϵ_r, we need to understand χ_e, which in turn means that we need to explore the behavior of the polarization, **P**.

Here, we consider the added complications of how the dipoles respond to a time-harmonic field that propagates as a wave through the material. The result of applying such a forcing function is that *oscillating* dipole moments are set up, and *these in turn establish a polarization wave that propagates through the material.* The effect is to produce a polarization function, $\mathbf{P}(z, t)$, having the same functional form as the driving field, $\mathbf{E}(z, t)$. The molecules themselves do not move through the material, but their oscillating dipole moments collectively exhibit wave motion, just as waves in pools of water are formed by the up and down motion of the water. From here, the description of the process gets complicated and in many ways beyond the scope of our present discussion. We can form a basic qualitative understanding, however, by considering the classical description of the process, which is that the dipoles, once oscillating, behave as microscopic antennas, re-radiating fields that in turn co-propagate with the applied field. Depending on the frequency, there will be some phase difference between the incident field and the radiated field at a given dipole location. This results in a net field (formed through the superposition of the two) that now interacts with the next dipole. Radiation from this dipole adds to the previous field as before, and the process repeats from dipole to dipole. The accumulated phase shifts at each location are manifested as a net slowing down of the phase velocity of the resultant wave. Attenuation of the field may also occur which, in this classical model, can be accounted for by partial phase cancellation between incident and radiated fields.

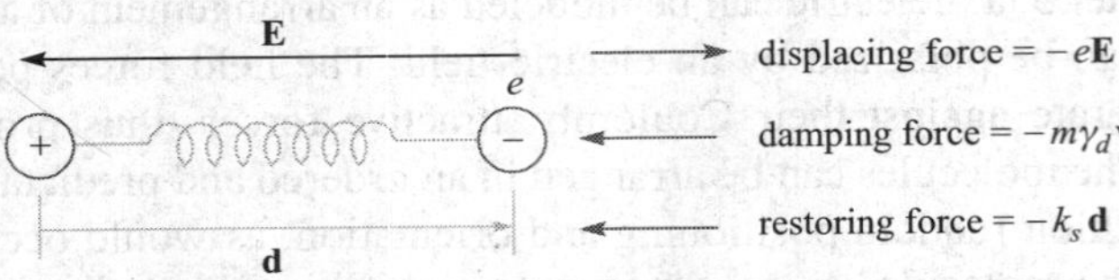

Fig. D.1 Atomic dipole model, with Coulomb force between positive and negative charge modeled by that of a spring having spring constant k_s . An applied electric field displaces the electron through distance d, resulting in dipole moment $\mathbf{p} = -e\mathbf{d}$.

In our classical model, the medium is an ensemble of identical fixed electron oscillators, in which the Coulomb binding forces on the electrons are modeled by springs that attach the electrons to the positive nuclei. We consider electrons for simplicity, but similar models can be used for any bound charged particle. Figure D.1 shows a single oscillator, located at position z in the material, and oriented along x. A uniform plane wave, assumed to be linearly polarized along x, propagates through the material in the z direction. The electric field in the wave displaces the electron of the oscillator in the x direction through a distance represented by the vector $\mathbf{d}$; a dipole moment is thus established,

$$\mathbf{p}(z, t) = -e\mathbf{d}(z, t) \tag{D.5}$$

where the electron charge, e, is treated as a positive quantity. The applied force is

$$\mathbf{F}_a(z, t) = -e\mathbf{E}(z, t) \tag{D.6}$$

We need to remember that $\mathbf{E}(z, t)$ at a given oscillator location is the *net* field, composed of the original applied field plus the radiated fields from all other oscillators. The relative phasing between oscillators is precisely determined by the spatial and temporal behavior of $\mathbf{E}(z, t)$.

The restoring force on the electron, $\mathbf{F}_r$, is that produced by the spring which is assumed to obey Hooke's law:

$$\mathbf{F}_r(z, t) = -k_s\mathbf{d}(z, t) \tag{D.7}$$

where k_s is the spring constant (not to be confused with the propagation constant). If the field is turned off, the electron is released and will oscillate about the nucleus at the *resonant frequency,* given by

$$\omega_0 = \sqrt{k_s/m} \tag{D.8}$$

where m is the mass of the electron. The oscillation, however, will be damped since the electron will experience forces and collisions from neighboring oscillators. We model these as a velocity-dependent damping force:

$$\mathbf{F}_d(z,t) = -m\gamma_d \mathbf{v}(z,t) \tag{D.9}$$

where $\mathbf{v}(z,t)$ is the electron velocity. Associated with this damping is the *dephasing* process among the electron oscillators in the system. Their relative phasing, once fixed by the applied sinusoidal field, is destroyed through collisions and dies away exponentially until a state of totally random phase exists between oscillators. The $1/e$ point in this process occurs at the *dephasing time* of the system, which is inversely proportional to the damping coefficient, γ_d (in fact it is $2/\gamma_d$). We are, of course, driving this damped resonant system with an electric field at frequency ω. We can therefore expect the response of the oscillators, measured through the magnitude of $\mathbf{d}$, to be frequency-dependent in much the same way as an RLC circuit is when driven by a sinusoidal voltage.

We can now use Newton's second law and write down the forces acting on the single oscillator of Figure D.1. To simplify the process a little we can use the complex form of the electric field:

$$\mathbf{E}_c = \mathbf{E}_0 e^{-jkz} e^{j\omega t} \tag{D.10}$$

Defining $\mathbf{a}$ as the acceleration vector of the electron, we have

$$m\mathbf{a} = \mathbf{F}_a + \mathbf{F}_r + \mathbf{F}_d$$

or

$$m\frac{\partial^2 \mathbf{d}_c}{\partial t^2} + m\gamma_d \frac{\partial \mathbf{d}_c}{\partial t} + k_s \mathbf{d}_c = -e\mathbf{E}_c \tag{D.11}$$

Note that since we are driving the system with the complex field, $\mathbf{E}_c$, we anticipate a displacement wave, $\mathbf{d}_c$, of the form

$$\mathbf{d}_c = \mathbf{d}_0 e^{-jkz} e^{-j\omega t} \tag{D.12}$$

With the waves in this form, time differentiation produces a factor of $j\omega$. Consequently (D.11) can be simplified and rewritten in phasor form:

$$-\omega^2 \mathbf{d}_s + j\omega\gamma_d \mathbf{d}_s + \omega_0^2 \mathbf{d}_s = -\frac{e}{m}\mathbf{E}_s \tag{D.13}$$

where (D.4) has been used. We now solve (D.13) for $\mathbf{d}_s$, obtaining

$$\mathbf{d}_s = \frac{-(e/m)\mathbf{E}_s}{(\omega_0^2 - \omega^2) + j\omega\gamma_d} \tag{D.14}$$

The dipole moment associated with displacement $\mathbf{d}_s$ is

$$\mathbf{p}_s = -e\mathbf{d}_s \tag{D.15}$$

The polarization of the medium is then found, assuming that all dipoles are identical. Eq. (D.1) thus becomes

$$\mathbf{P}_s = N\mathbf{p}_s$$

which, when using (D.14) and (D.15), becomes

$$\mathbf{P}_s = \frac{Ne^2/m}{(\omega_0^2 - \omega^2) + j\omega\gamma_d}\mathbf{E}_s \tag{D.16}$$

Now, using (D.3) we identify the susceptibility associated with the resonance as

$$\chi_{\text{res}} = \frac{Ne^2}{\epsilon_0 m}\frac{1}{(\omega_0^2 - \omega^2) + j\omega\gamma_d} = \chi'_{\text{res}} - j\chi''_{\text{res}} \tag{D.17}$$

The real and imaginary parts of the permittivity are now found through the real and imaginary parts of χ_{res}: Knowing that

$$\epsilon = \epsilon_0(1 + \chi_{\text{res}}) = \epsilon' - j\epsilon''$$

we find

$$\epsilon' = \epsilon_0(1 + \chi'_{\text{res}}) \tag{D.18}$$

and

$$\epsilon'' = \epsilon_0\chi''_{\text{res}} \tag{D.19}$$

The preceding expressions can now be used in Eqs. (35) and (36) in Chapter 12 to evaluate the attenuation coefficient, α, and phase constant, β, for the plane wave as it propagates through our resonant medium.

The real and imaginary parts of χ_{res} as functions of frequency are shown in Figure D.2 for the special case in which $\omega \doteq \omega_0$. Eq. (D.17) in this instance becomes

$$\chi_{\text{res}} \doteq -\frac{Ne^2}{\epsilon_0 m\omega_0\gamma_d}\left(\frac{j + \delta_n}{1 + \delta_n^2}\right) \tag{D.20}$$

where the *normalized detuning* parameter, δ_n, is

$$\frac{2}{\gamma_d}(\omega - \omega_0) \tag{D.21}$$

Key features to note in Figure D.2 include the symmetric χ''_e function, whose full-width at its half-maximum amplitude is γ_d. Near the resonant frequency, where χ''_{res} maximizes, wave attenuation maximizes as seen from Eq. (35), Chapter 11. Additionally, we see that away from resonance, attenuation is relatively weak, and the material becomes transparent. As Figure D.2 shows, there is still significant variation of χ'_{res} with frequency away from resonance, which leads to a frequency-dependent refractive index; this is expressed approximately as

$$n \doteq \sqrt{1 + \chi'_{\text{res}}} \quad \text{(away from resonance)} \tag{D.22}$$

This frequency-dependent n, arising from the material resonance, leads to phase and group velocities that also depend on frequency. Thus, group dispersion, leading to pulse-broadening effects as discussed in Chapter 13, can be directly attributable to material resonances.

Somewhat surprisingly, the classical "spring model" described here can provide very accurate predictions on dielectric constant behavior with frequency (particularly off-resonance) and can be used to a certain extent to model absorption properties. The model is insufficient, however,

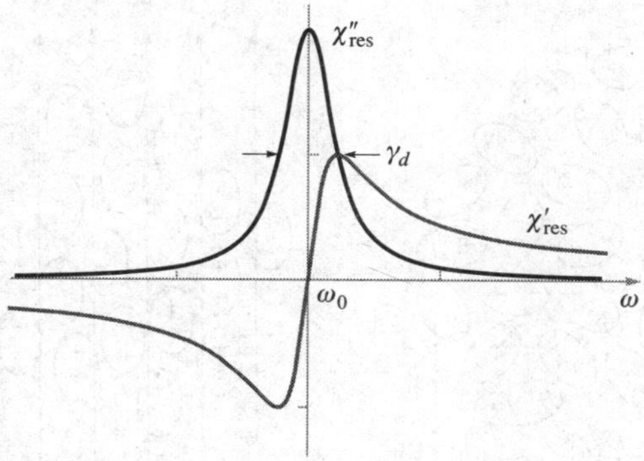

Fig. D.2 Plots of the real and imaginary parts of the resonant susceptibility, χ_{res}, as given by Eq. (D.20). The full-width at half-maximum of the imaginary part, χ''_{res}, is equal to the damping coefficient, γ_d.

when attempting to describe the more salient features of materials; specifically, it assumes that the oscillating electron can assume any one of a continuum of energy states, when in fact energy states in any atomic system are quantized. As a result, the important effects arising from transitions between discrete energy levels, such as spontaneous and stimulated absorption and emission, are not included in our classical spring system. Quantum mechanical models must be used to fully describe the medium polarization properties, but the results of such studies often reduce to those of the spring model when field amplitudes are very low.

Another way that a dielectric can respond to an electric field is through the orientation of molecules that possess permanent dipole moments. In such cases, the molecules must be free to move or rotate, and so the material is typically a liquid or a gas. Figure D.3 shows an arrangement of polar molecules in a liquid (such as water) in which there is no applied field (Figure D.3a) and where an electric field is present (Figure D.3b). Applying the field causes the dipole moments, previously having random orientations, to line up, and so a net material polarization, **P**, results. Associated with this, of course, is a susceptibility function, χ_e, through which **P** relates to **E**.

Some interesting developments occur when the applied field is time-harmonic. With field periodically reversing direction, the dipoles are forced to follow, but they do so against their natural propensity to randomize, owing to thermal motion. Thermal motion thus acts as a "restoring" force, effectively opposing the applied field. We can also think of the thermal effects as viscous forces that introduce some difficulty in "pushing" the dipoles back and forth. One might expect (correctly) that polarizations of greater amplitude in each direction can be attained at lower frequencies, since enough time is given during each cycle for the dipoles to achieve complete alignment. The polarization amplitude will weaken as the frequency increases because there is no longer enough time for complete alignment during each cycle. This is the basic description of the *dipole relaxation* mechanism for the complex permittivity. There is no resonant frequency associated with the process.

The complex susceptibility associated with dipole relaxation is essentially that of an "overdamped" oscillator, and is given by

$$\chi_{rel} = \frac{Np^2/\epsilon_0}{3k_B T(1 + j\omega\tau)} \tag{D.23}$$

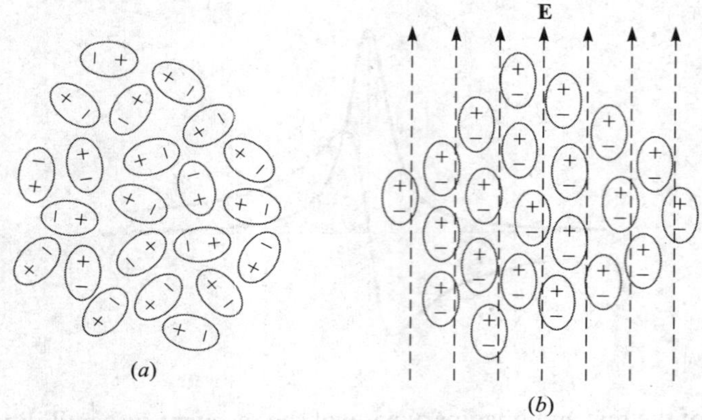

Fig. D.3 Idealized sketches of ensembles of polar molecules under conditions of (*a*) random orientation of the dipole moments, and (*b*) dipole moments aligned under the influence of an applied electric field. Conditions in (*b*) are greatly exaggerated, since typically only a very small percentage of the dipoles align themselves with the field. But still enough alignment occurs to produce measurable changes in the material properties.

where p is the permanent dipole moment magnitude of each molecule, k_B is Boltzmann's constant, and T is the temperature in degees Kelvin. τ is the thermal randomization time, defined as the time for the polarization, **P**, to relax to $1/e$ of its original value when the field is turned off. χ_{rel} is complex, and so it will possess absorptive and dispersive components (imaginary and real parts) as we found in the resonant case. The form of Eq. (D.23) is identical to that of the response of a series RC circuit driven by a sinusoidal voltage (where τ becomes RC).

Microwave absorption in water occurs through the relaxation mechanism in polar water molecules, and is the primary means by which microwave cooking is done, as discussed in Chapter 12. Frequencies near 2.5 GHz are typically used, since these provide the optimum penetration depth. The peak water absorption arising from dipole relaxation occurs at much higher frequencies, however.

A given material may possess more than one resonance and may have a dipole relaxation response as well. In such cases, the net susceptibility is found in frequency domain by the direct sum of all component susceptibilities. In general, we may write:

$$\chi_e = \chi_{\mathrm{rel}} + \sum_{i=1}^{n} \chi_{\mathrm{res}}^i \tag{D.24}$$

where χ_{res}^i is the susceptibility associated with the ith resonant frequency, and n is the number of resonances in the material. The reader is referred to the references for Chapter 11 for further reading on resonance and relaxation effects in dielectrics.

E

Answers to Odd-Numbered Problems

Chapter 1

1.1 (*a*) $0.92\mathbf{a}_x + 0.36\mathbf{a}_y + 0.4\mathbf{a}_z$ (*b*) 48.6 (*c*) $-580.5\mathbf{a}_x + 3193\mathbf{a}_y - 2902\mathbf{a}_z$

1.3 $(7.8, -7.8, 3.9)$

1.5 (*a*) $48\mathbf{a}_x + 36\mathbf{a}_y + 18\mathbf{a}_z$ (*b*) $-0.26\mathbf{a}_x + 0.39\mathbf{a}_y + 0.88\mathbf{a}_z$ (*c*) $0.59\mathbf{a}_x + 0.20\mathbf{a}_y - 0.78\mathbf{a}_z$ (*d*) $100 = 16x^2y^2 + 4x^4 + 16x^2 + 16 + 9z^4$

1.7 (*a*) (1) the plane $z = 0$, with $|x| < 2, |y| < 2$; (2) the plane $y = 0$, with $|x| < 2, |z| < 2$; (3) the plane $x = 0$, with $|y| < 2, |z| < 2$; (4) the plane $x = \pi/2$, with $|y| < 2, |z| < 2$ (*b*) the plane $2z = y$, with $|x| < 2, |y| < 2, |z| < 1$ (*c*) the plane $y = 0$, with $|x| < 2, |z| < 2$

1.9 (*a*) $0.6\mathbf{a}_x + 0.8\mathbf{a}_y$ (*b*) 53° (*c*) 26

1.11 (*a*) $(-0.3, 0.3, 0.4)$ (*b*) 0.05 (*c*) 0.12 (*d*) 78°

1.13 (*a*) $(0.93, 1.86, 2.79)$ (*b*) $(9.07, -7.86, 2.21)$ (*c*) $(0.02, 0.25, 0.26)$

1.15 (*a*) $(0.08, 0.41, 0.91)$ (*b*) $(0.30, 0.81, 0.50)$ (*c*) 30.3 (*d*) 32.0

1.17 (*a*) $(0.664, -0.379, 0.645)$ (*b*) $(-0.550, 0.832, 0.077)$ (*c*) $(0.168, 0.915, 0.367)$

1.19 (*a*) $(1/\rho)\mathbf{a}_\rho$ (*b*) $0.5\mathbf{a}_\rho$, or $0.41\mathbf{a}_x + 0.29\mathbf{a}_y$

1.21 (*a*) $-6.66\mathbf{a}_\rho - 2.77\mathbf{a}_\phi + 9\mathbf{a}_z$ (*b*) $-0.59\mathbf{a}_\rho + 0.21\mathbf{a}_\phi - 0.78\mathbf{a}_z$ (*c*) $-0.90\mathbf{a}_\rho - 0.44\mathbf{a}_z$

1.23 (*a*) 6.28 (*b*) 20.7 (*c*) 22.4 (*d*) 3.21

1.25 (*a*) $1.10\mathbf{a}_\rho + 2.21\mathbf{a}_\phi$ (*b*) 2.47 (*c*) $0.45\mathbf{a}_r + 0.89\mathbf{a}_\phi$

1.27 (*a*) 2.91 (*b*) 12.61 (*c*) 17.49 (*d*) 2.53

1.29 (*a*) $0.59\mathbf{a}_r + 0.38\mathbf{a}_\theta - 0.72\mathbf{a}_\phi$ (*b*) $0.80\mathbf{a}_r - 0.22\mathbf{a}_\theta - 0.55\mathbf{a}_\phi$ (*c*) $0.66\mathbf{a}_r + 0.39\mathbf{a}_\theta - 0.64\mathbf{a}_\phi$

Chapter 2

2.1 4.0×10^{-4} N

2.3 $21.5\mathbf{a}_x\ \mu$N

2.5 (*a*) $4.58\mathbf{a}_x - 0.15\mathbf{a}_y + 5.51\mathbf{a}_z$ (*b*) -6.89 or -22.11

2.7 $159.7\mathbf{a}_\rho + 27.4\mathbf{a}_\phi - 49.4\mathbf{a}_z$

2.9 (*a*) $(x + 1) = 0.56\,[(x + 1)^2 + (y - 1)^2 + (z - 3)^2]^{1.5}$ (*b*) 1.69 or 0.31

2.11 (*a*) $-1.63\ \mu$C (*b*) $-30.11\mathbf{a}_x - 180.63\mathbf{a}_y - 150.53\mathbf{a}_z$ (*c*) $-183.12\mathbf{a}_\rho - 150.53\mathbf{a}_z$ (*d*) -237.1

2.13 (*a*) 82.1 pC (*b*) 4.24 cm

2.15 (*a*) 3.35×10^{-2} C (*b*) 1.24×10^6 C/m^3

2.17 (*a*) $57.5\mathbf{a}_y - 28.8\mathbf{a}_z$ V/m (*b*) $23\mathbf{a}_y - 46\mathbf{a}_z$

2.19 (*a*) $7.2\mathbf{a}_x + 14.4\mathbf{a}_y$ kV/m (*b*) $4.9\mathbf{a}_x + 9.8\mathbf{a}_y + 4.9\mathbf{a}_z$ kV/m

2.21 $126\mathbf{a}_y$ μN/m

2.23 (*a*) 8.1 kV/m (*b*) −8.1 kV/m

2.25 $-3.9\mathbf{a}_x - 12.4\mathbf{a}_y - 2.5\mathbf{a}_z$ V/m

2.27 (*a*) $y^2 - x^2 = 4xy - 19$ (*b*) $0.99\mathbf{a}_x + 0.12\mathbf{a}_y$

2.29 (*a*) 12.2 (*b*) $-0.87\mathbf{a}_x - 0.50\mathbf{a}_y$ (*c*) $y = (1/5)\ln\cos 5x + 0.13$

Chapter 3

3.1 (*a*) Penny: +5 nC; nickel: 0; dime: 0; can: −5 nC (*b*) Coins: same as at start; can: −2 nC

3.3 (*a*) 0.25 nC (*b*) 9.45 pC

3.5 360 C

3.7 (*a*) 4.0×10^{-9} nC (*b*) 3.2×10^{-4} nC/m^2

3.9 (*a*) 164 pC (*b*) 130 nC/m^2 (*c*) 32.5 nC/m^2

3.11 $\mathbf{D} = 0 \quad (\rho < 1 \text{ mm})$; $D_\rho = \frac{10^{-15}}{2\pi^2\rho}[\sin(2000\pi\rho) + 2\pi[1 - 10^3\rho\cos(2000\pi\rho)]]\text{C/ m}^2$ $(1 \text{ mm} < \rho < 1.5 \text{ mm})$; $D_\rho = \frac{2.5\times10^{-15}}{\pi\rho}\text{ C/m}^2$ $(\rho > 1.5 \text{ m}m)$

3.13 (*a*) $D_r(r < 2) = 0; \quad D_r(r = 3) = 8.9 \times 10^{-9}\text{ C/m}^2; D_r(r = 5) = 6.4 \times 10^{-10}\text{ C/m}^2$
(*b*) $\rho_{s0} = -(4/9) \times 10^{-9}\text{ C/m}^2$

3.15 (*a*) $[(8\pi L)/3][\rho_1^3 - 10^{-9}]\,\mu$C where ρ_1 is in meters (*b*) $4(\rho_1^3 - 10^{-9})/(3\rho_1)\mu\text{C/m}^2$ where ρ_1 is in meters
(*c*) $D_\rho(0.8\text{ mm}) = 0; D_\rho(1.6\text{ mm}) = 3.6 \times 10^{-6}\mu\text{C/m}^2; D_\rho(2.4\text{ mm}) = 3.9 \times 10^{-6}\,\mu\text{C/m}^2$

3.17 (*a*) 0.1028 C (*b*) 12.83 (*c*) 0.1026 C

3.19 113 nC

3.21 (*a*) 8.96 (*b*) 71.67 (*c*) −2

3.23 (*b*) $\rho_{v0} = 3Q/(4\pi a^3) \quad (0 < r < a)$; $D_r = Qr/4\pi a^3$ and $\nabla\cdot\text{D} = 3Q/(4\pi a^3)$ $(0 < r < a)$; $D_r = Q/(4\pi r^2)$ and $\nabla\cdot\mathbf{D} = 0\ (r > a)$

3.25 (*a*) 17.50 C/m^3 (*b*) $5\mathbf{a}_r$ C/m^2 (*c*) 320π C (*d*) 320π C

3.27 (*a*) 1.20 mC/m^3 (*b*) 0 (*c*) -32μC/m^2

3.29 (*a*) 3.47 C (*b*) 3.47 C

3.31 −3.91 C

Chapter 4

4.1 (*a*) −12 nJ (*b*) 24 nJ (*c*) −36 nJ (*d*) −44.9 nJ
(*e*) −41.8 nJ

4.3 (*a*) 3.1 μJ (*b*) 3.1 μJ

4.5 (*a*) 2 (*b*) −2

4.7 (*a*) 90 (*b*) 82

4.9 (*a*) 8.14 V (*b*) 1.36 V

4.11 1.98 kV

4.13 576 pJ

4.15 −68.4 V

4.17 (*a*) −3.026 V (*b*) −9.678 V

4.19 .081 V

4.21 (*a*) −15.0 V (*b*) 15.0 V
(*c*) $7.1\mathbf{a}_x + 22.8\mathbf{a}_y - 71.1\mathbf{a}_z$ V/m
(*d*) 75.0 V/m
(*e*) $-0.095\mathbf{a}_x - 0.304\mathbf{a}_y + 0.948\mathbf{a}_z$
(*f*) $62.8\mathbf{a}_x + 202\mathbf{a}_y - 629\mathbf{a}_z$ pC/m^2

4.23 (*a*) $-48\rho^{-4}$ V/m (*b*) −673 pC/m^3 (*c*) −1.96 nC

4.25 (*a*) $V_p = 279.9$ V, $\mathbf{E}_p = -179.9\mathbf{a}_\rho - 75.0\mathbf{a}_\phi$ V/m, $\mathbf{D}_p = -1.59\mathbf{a}_\rho - .664\mathbf{a}_\phi$ nC/m^2, $\rho_{vp} = -443$ pC/m^3 (*b*) −5.56 nC

4.27 (*a*) 5.78 V (*b*) 25.2 V/m (*c*) 5.76 V

4.29 1.31 V

4.31 (*a*) 387 pJ (*b*) 207 pJ

4.33 (*a*) $(5 \times 10^{-6})/(4\pi r^2)\mathbf{a}_r$ C/m^2
(*b*) 2.81 J (*c*) 4.45 pF

4.35 (a) 0.779 μJ (b) 1.59 μJ

Chapter 5

5.1 (a) -1.23 MA (b) 0 (c) 0, as expected

5.3 (a) 77.4 A (b) $53.0\mathbf{a}_r$ A/m^2

5.5 (a) -178.0 A (b) 0 (c) 0

5.7 (a) mass flux density in (kg/m^2 – s) and mass density in (kg/m^3) (b) -550 g/m^3 – s

5.9 (a) 0.28 mm (b) 6.0×10^7 A/m^2

5.11 (a) $\mathbf{E} = [(9.55)/\rho l)]\mathbf{a}_\rho$ V/m, $V = (4.88)/l$ V and $R = (1.63)/l\,\Omega$, where l is the cylinder length (not given) (b) $14.64/l$ W

5.13 (a) 0.147 V (b) 0.144 V

5.15 (a) $(\rho + 1)z^2 \cos\phi = 2$ (b) $\rho = 0.10, \mathbf{E}(.10, .2\pi, 1.5) = -18.2\mathbf{a}_\rho + 145\mathbf{a}_\phi - 26.7\mathbf{a}_z$ V/m (c) 1.32 nC/m^2

5.17 (a) $\mathbf{D}(z = 0) = -(100\epsilon_0 x)/(x^2 + 4)\mathbf{a}_z$ C/m^2 (c) -0.92 nC

5.19 (a) At 0 V: $2x^2y - z = 0$. At 60 V: $2x^2y - z = 6/z$ (b) 1.04 nC/m^2 (c) $-[0.60\mathbf{a}_x + 0.68\mathbf{a}_y + 0.43\mathbf{a}_z]$

5.21 (a) 1.20 kV (b) $\mathbf{E}_p = 723\mathbf{a}_x - 18.9\mathbf{a}_y$ V/m

5.23 (a) 289.5 V (b) $z/[(x-1)^2 + y^2 + z^2]^{1.5} - z/[(x+1)^2 + y^2 + z^2]^{1.5} = 0.222$

5.25 (a) 4.7×10^{-5} S/m (b) 1.1×10^{-3} S/m (c) 1.2×10^{-2} S/m

Chapter 6

6.1 (a) 6.26 pC/m^2 (b) 1.000176

6.3 (a) $\mathbf{E} = [(144.9)/\rho]\mathbf{a}_\rho$ V/m, $\mathbf{D} = (3.28\mathbf{a}_\rho)/\rho$ nC/m^2 (b) $V_{ab} = 192$ V, $\chi_e = 1.56$ (c) $[(5.0 \times 10^{-29})/\rho]\mathbf{a}_\rho$ C $\cdot$ m

6.5 (a) 80 V/m (b) $-60\mathbf{a}_y - 30\mathbf{a}_z$ V/m (c) 67.1 V/m (d) 104.4 V/m (e) 40.0° (f) 2.12 nC/m^2 (g) 2.97 nC/m^2 (h) $2.12\mathbf{a}_x - 2.66\mathbf{a}_y - 1.33\mathbf{a}_z$ nC/m^2 (i) $1.70\mathbf{a}_x - 2.13\mathbf{a}_y - 1.06\mathbf{a}_z$ nC/m^2 (j) 54.5°

6.7 $125\mathbf{a}_x + 175\mathbf{a}_y$ V/m

6.9 (a) $\mathbf{E}_2 = \mathbf{E}_1$ (b) $W_{E1} = 45.1$ μJ, $W_{E2} = 338$ μJ

6.11 barium titanate

6.13 451 pF

6.15 (a) 3.05 nF (b) 5.21 nF (c) 6.32 nF (d) 9.83 nF

6.17 (a) 143 pF (b) 101 pF

6.19 (a) 53.3 pF (b) 41.7 pF

6.21 $K_1 = 23.0, \rho_L = 8.87$ nC/m, $a = 13.8$ m, $C = 35.5$ pF

6.23 (a) 473 nC/m^2 (b) -15.8 nC/m^2

6.25 Exact value: 57 pF/m

6.27 Exact value: $11\epsilon_0$ F/m

6.29 (b) $C \approx 110$ pF/m (c) Result would not change.

6.31 (a) 3.64 nC/m (b) 206 mA

Chapter 7

7.1 (a) -8 V (b) $8\mathbf{a}_x - 8\mathbf{a}_y - 24\mathbf{a}_z$ V/m (c) $-4xz(z^2 + 3y^2)$ C/m^3 (d) $xy^2z^3 = -4$ (e) $y^2 - 2x^2 = 2$ and $3x^2 - z^2 = 2$ (f) No

7.3 $f(x, y) = -4e^{2x} + 3x^2, V(x, y) = 3(x^2 - y^2)$

7.5 (b) $A = 112.5, B = -12.5$ or $A = -12.5, B = 112.5$

7.7 (a) -106 pC/m^3 (b) ± 0.399 pC/m^2 (depending on which side of the surface is considered)

7.9 (a) yes, yes, yes, no (b) At the 100 V surface, no for all. At the 0 V surfaces, yes, except for $V_1 + 3$. (c) Only V_2 is

7.11 (a) 33.33 V (b) $[(100)/3]\mathbf{a}_z + 50\mathbf{a}_y$ V/m

7.13 (a) 1.01 cm (b) 22.8 kV/m (c) 3.15

7.15 (*a*) $(-2.00 \times 10^4)\phi + 3.78 \times 10^3$ V
(*b*) $[(2.00 \times 10^4)/\rho]\mathbf{a}_\phi$ V/m
(*c*) $(2.00 \times 10^4\epsilon_0/\rho)\mathbf{a}_\phi$ C/m^2
(*d*) $[(2.00 \times 10^4)/\rho]$ C/m^2
(*e*) 84.7 nC
(*f*) $V(\phi) = 28.7\phi + 194.9$ V,
$\mathbf{E} = -(28.7)/\rho\mathbf{a}_\phi$ V/m,
$\mathbf{D} = -(28.7\epsilon_0)/\rho\mathbf{a}_\phi$ C/m^2,
$\rho_s = (28.7\epsilon_0)/$
ρ C/m^2, $Q_b = 122$ pC (*g*) 471 pF

7.17 (*a*) 12.5 mm (*b*) 26.7 kV/m
(*c*) 4.23 (with given $\rho_s = 1.0\,\mu$C/m^2)

7.19 (*a*) $\alpha_A = 26.57°, \alpha_B = 56.31°$ (*b*) 23.3 V

7.21 (*a*) $833.3r^{-.4}$ V (*b*) $833.3r^{-.4}$ V

7.23 71.9173 V

7.25 12.5 V

7.27 (*a*) No (*b*) Yes (*c*) Yes (*d*) No (*e*) No

7.29 38 V

7.31 40 V

7.33 (*b*) 90 V

7.35 CCW from upper left: (*a*) 48, 42, 19, 34, 19, 42, 48 (*b*) 48, 40, 19, 30, 19, 40, 48
(*c*) 47.3, 37.1, 17.4, 27.4, 17.4, 37.1, 47.3

Chapter 8

8.1 (*a*) $-294\mathbf{a}_x + 196\mathbf{a}_y\ \mu$A/m
(*b*) $-127\mathbf{a}_x + 382\mathbf{a}_y\ \mu$A/m
(*c*) $-421\mathbf{a}_x + 578\mathbf{a}_y\ \mu$A/m

8.3 (*a*) $H = \frac{I}{2\pi\rho}\left[1 - \frac{a}{\sqrt{\rho^2+a^2}}\right]\mathbf{a}_\phi$ A/m
(*b*) $1/\sqrt{3}$

8.5 $|\mathbf{H}| = \frac{I}{2\pi}\left[\left(\frac{2}{y^2+2y+5} - \frac{2}{y^2-2y+5}\right)^2 + \left(\frac{(y-1)}{y^2-2y+5} - \frac{(y+1)}{y^2+2y+5}\right)^2\right]^{1/2}$

8.7 (*a*) $0.53\mathbf{a}_x + 0.80\mathbf{a}_y + 0.27\mathbf{a}_z$
(*b*) 5.73 μA/m
(*c*) $\pm(-\mathbf{a}_x + \mathbf{a}_y - \mathbf{a}_z)/\sqrt{3}$

8.9 $-1.50\mathbf{a}_y$ A/m

8.11 2.0 A/m, 933 mA/m, 360 mA/m, 0

8.13 (*e*) $H_z(a < \rho < b) = k_b; H_z(\rho > b) = 0$

8.15 (*a*) $45e^{-150\rho}\mathbf{a}_z$ kA/m^2
(*b*) $12.6[1 - (1 + 150\rho_0)e^{-150\rho_0}]$ A
(*c*) $\frac{2.00}{\rho}[1 - (1 + 150\rho)e^{-150\rho}]$ A/m

8.17 (*a*) $2.2 \times 10^{-1}\mathbf{a}_\phi$ A/m (just inside), $2.3 \times 10^{-2}\mathbf{a}_\phi$ A/m (just outside)
(*b*) $3.4 \times 10^{-1}\mathbf{a}_\phi$ A/m
(*c*) $1.3 \times 10^{-1}\mathbf{a}_\phi$ A/m (*d*) $-1.3 \times 10^{-1}\mathbf{a}_z$ A/m

8.19 0

8.21 $530\mathbf{a}_x + 460\mathbf{a}_y - 148\mathbf{a}_z$

8.23 (*a*) $60\rho\mathbf{a}_z$ A/m^2 (*b*) 40π A (*c*) 40π A

8.25 (*a*) -259 A (*b*) -259 A

8.27 (*a*) $2(x + 2y)/z^3\mathbf{a}_x + 1/z^2\mathbf{a}_z$ A/m
(*b*) same as part (*a*) (*c*) 1/8 A

8.29 (*a*) $1.59 \times 10^7\mathbf{a}_z$ A/m^2
(*b*) $7.96 \times 10^6\rho\mathbf{a}_\phi$ A/m, $10\rho\mathbf{a}_\phi$ Wb/m^2
(*c*) as expected (*d*) $1/(\pi\rho)\mathbf{a}_\phi$ A/m, $\mu_0/(\pi\rho)\mathbf{a}_\phi$ Wb/m^2 (*e*) as expected

8.31 (*a*) 0.392 μWb (*b*) 1.49 μWb (*c*) 27 μWb

8.35 (*a*) -40ϕ A $\quad (2 < \rho < 4), 0\ (\rho > 4)$
(*b*) $40\mu_0 \ln(3/\rho)\mathbf{a}_z$ Wb/m

8.37 $[120 - (400/\pi)\phi]$ A $(0 < \phi < 2\pi)$

8.39 (*a*) $-30\mathbf{a}_y$ A/m (*b*) $30y - 6$ A
(*c*) $-30\mu_0\mathbf{a}_y$ Wb/m^2 (*d*) $\mu_0(30x - 3)\mathbf{a}_z$ Wb/m

8.41 (*a*) $-100\rho/\mu_0\mathbf{a}_\phi$ A/m, $-100\rho\mathbf{a}_\phi$ Wb/m^2
(*b*) $-\frac{200}{\mu_0}\mathbf{a}_z$ A/m^2 (*c*) -500 MA (*d*) -500 MA

8.43 $A_z = \frac{\mu_0 I}{96\pi}\left[\left(\frac{\rho^2}{a^2} - 25\right) + 98\ln\left(\frac{5a}{\rho}\right)\right]$ Wb/m

Chapter 9

9.1 (*a*) $(.90, 0, -.135)$ (*b*) $3 \times 10^5\mathbf{a}_x - 9 \times 10^4\mathbf{a}_z$ m/s (*c*) 1.5×10^{-5} J

9.3 (*a*) $.70\mathbf{a}_x + .70\mathbf{a}_y - .12\mathbf{a}_z$ (*b*) 7.25 fJ

9.5 (*a*) $-18\mathbf{a}_x$ nN (*b*) $19.8\mathbf{a}_z$ nN (*c*) $36\mathbf{a}_x$ nN

9.7 (*a*) $-35.2\mathbf{a}_y$ nN/m (*b*) 0 (*c*) 0

9.9 $4\pi \times 10^{-5}$ N/m

9.13 (*a*) $-1.8 \times 10^{-4}\mathbf{a}_y$ N · m
(*b*) $-1.8 \times 10^{-4}\mathbf{a}_y$ N · m
(*c*) $-1.5 \times 10^{-5}\mathbf{a}_y$ N · m

9.15 $(6 \times 10^{-6})[b - 2\tan^{-1}(b/2)]\mathbf{a}_y$ N · m

9.17 $\Delta w/w = \Delta m/m = 1.3 \times 10^{-6}$

9.19 (*a*) $77.6y\mathbf{a}_z$ kA/m (*b*) 5.15×10^{-6} H/m
(*c*) 4.1 (*d*) $241y\mathbf{a}_z$ kA/m
(*e*) $77.6\mathbf{a}_x$ kA/m^2
(*f*) $241\mathbf{a}_x$ kA/m^2 (*g*) $318\mathbf{a}_x$ kA/m^2

9.21 (Use $\chi_m = .003$) (*a*) 47.7 A/m
(*b*) 6.0 A/m (*c*) 0.288 A/m

9.23 (*a*) 637 A/m, 1.91×10^{-3} Wb/m^2, 884 A/m (*b*) 478 A/m, 2.39×10^{-3} Wb/m^2, 1.42×10^3 A/m
(*c*) 382 A/m, 3.82×10^{-3} Wb/m^2, 2.66×10^3 A/m

9.25 (*a*) $1.91/\rho$ A/m $(0 < \rho < \infty)$
(*b*) $(2.4 \times 10^{-6}/\rho)\mathbf{a}_\phi$ T $(\rho < .01)$,
$(1.4 \times 10^{-5}/\rho)\mathbf{a}_\phi$ T $(.01 < \rho < .02)$,
$(2.4 \times 10^{-6}/\rho)\mathbf{a}_\phi$ T $(\rho > .02)$
(ρ in meters)

9.27 (*a*) $-4.83\mathbf{a}_x - 7.24\mathbf{a}_y + 9.66\mathbf{a}_z$ A/m
(*b*) $54.83\mathbf{a}_x - 22.76\mathbf{a}_y + 10.34\mathbf{a}_z$ A/m
(*c*) $54.83\mathbf{a}_x - 22.76\mathbf{a}_y + 10.34\mathbf{a}_z$ A/m
(*d*) $-1.93\mathbf{a}_x - 2.90\mathbf{a}_y + 3.86\mathbf{a}_z$ A/m
(*e*) 102° (*f*) 95°

9.29 10.5 mA

9.31 (*a*) 2.8×10^{-4} Wb (*b*) 2.1×10^{-4} Wb
(*c*) $\approx 2.5 \times 10^{-4}$ Wb

9.33 (*a*) $23.9/\rho$ A/m (*b*) $3.0 \times 10^{-4}/\rho$ Wb/m^2
(*c*) 5.0×10^{-7} Wb
(*d*) $23.9/\rho$ A/m, $6.0 \times 10^{-4}/\rho$ Wb/m^2, 1.0×10^{-6} Wb (*e*) 1.5×10^{-6} Wb

9.35 (*a*) $20/(\pi r \sin\theta)\mathbf{a}_\phi$ A/m
(*b*) 1.35×10^{-4} J

9.37 0.17 μH

9.39 (*a*) $(1/2)wd\mu_0 K_0^2$ J/m (*b*) $\mu_0 d/w$ H/m
(*c*) $\Phi = \mu_0 d K_0$ Wb

9.41 (*a*) 33 μH (*b*) 24 μH

9.43 (*b*) $L_{\text{int}} = \frac{2W_H}{I^2}$
$$= \frac{\mu_0}{8\pi}\left[\frac{d^4 - 4a^2c^2 + 3c^4 + 4c^4\ln(a/c)}{(a^2 - c^2)^2}\right] \text{H/m}$$

Chapter 10

10.1 (*a*) $-5.33\sin 120\pi t$ V
(*b*) $21.3\sin(120\pi t)$ mA

10.3 (*a*) $-1.13 \times 10^5[\cos(3 \times 10^8 t - 1) - \cos(3 \times 10^8 t)]$ V (*b*) 0

10.5 (*a*) -4.32 V (*b*) -0.293 V

10.7 (*a*) $(-1.44)/(9.1 + 39.6t)$ A
(*b*) $-1.44[\frac{1}{61.9 - 39.6t} + \frac{1}{9.1 + 39.6t}]$ A

10.9 $2.9 \times 10^3[\cos(1.5 \times 10^8 t - 0.13x) - \cos(1.5 \times 10^8 t)]$ W

10.11 (*a*) $(\frac{10}{\rho})\cos(10^5 t)\mathbf{a}_\rho$ A/m^2
(*b*) $8\pi\cos(10^5 t)$ A (*c*) $-0.8\pi\sin(10^5 t)$ A (*d*) 0.1

10.13 (*a*) $\mathbf{D} = 1.33 \times 10^{-13}\sin(1.5 \times 10^8 t - bx)\mathbf{a}_y$ C/m^2, $\mathbf{E} = 3.0 \times 10^{-3}\sin(1.5 \times 10^8 t - bx)\mathbf{a}_y$ V/m
(*b*) $\mathbf{B} = (2.0)b \times 10^{-11}\sin(1.5 \times 10^8 t - bx)\mathbf{a}_z$T, $\mathbf{H} = (4.0 \times 10^{-6})b \sin(1.5 \times 10^8 t - bx)\mathbf{a}_z$ A/m (*c*) $4.0 \times 10^{-6}b^2\cos(1.5 \times 10^8 t - bx)\mathbf{a}_y$ A/m^2
(*d*) $\sqrt{5.0}$ m^{-1}

10.15 $\mathbf{B} = 6 \times 10^{-5}\cos(10^{10}t - \beta x)\mathbf{a}_z$T, $\mathbf{D} = -(2\beta \times 10^{-10})\cos(10^{10}t - \beta x)\mathbf{a}_y$ C/m^2, $\mathbf{E} = -1.67\beta\cos(10^{10}t - \beta x)\mathbf{a}_y$V/m, $\beta = \pm 600$ rad/m

10.17 $a = 66$ m^{-1}

10.21 (*a*) $\pi \times 10^9$sec^{-1}
(*b*) $\frac{500}{\rho}\sin(10\pi z)\sin(\omega t)\mathbf{a}_\rho$ V/m

10.23 (*a*) $\mathbf{E}_{N1} = 10\cos(10^9 t)\mathbf{a}_z$V/m
$\mathbf{E}_{t1} = (30\mathbf{a}_x + 20\mathbf{a}_y)\cos(10^9 t)$ V/m
$\mathbf{D}_{N1} = 200\cos(10^9 t)\mathbf{a}_z$ pC/m^2
$\mathbf{D}_{t1} = (600\mathbf{a}_x + 400\mathbf{a}_y)\cos(10^9 t)$ pC/m^2
(*b*) $\mathbf{J}_{N1} = 40\cos(10^9 t)\mathbf{a}_z$ mA/m^2
$\mathbf{J}_{t1} = (120\mathbf{a}_x + 80\mathbf{a}_y)\cos(10^9 t)$ mA/m^2

(*c*) $\mathbf{E}_{t2} = (30\mathbf{a}_x + 20\mathbf{a}_y)\cos(10^9 t)$ V/m
$\mathbf{D}_{t2} = (300\mathbf{a}_x + 200\mathbf{a}_y)\cos(10^9 t)$ pC/m^2
$\mathbf{J}_{t2} = (30\mathbf{a}_x + 20\mathbf{a}_y)\cos(10^9 t)$ mA/m^2
(*d*) $\mathbf{E}_{N2} = 20.3\cos(10^9 t + 5.6°)\mathbf{a}_z$ V/m
$\mathbf{D}_{N2} = 203\cos(10^9 t + 5.6°)\mathbf{a}_z$ pC/m^2
$\mathbf{J}_{N2} = 20.3\cos(10^9 t + 5.6°)\mathbf{a}_z$ mA/m^2

10.25 (*b*) $\mathbf{B} = \left(t - \frac{z}{c}\right)\mathbf{a}_y$ T
$\mathbf{H} = \frac{1}{\mu_0}\left(t - \frac{z}{c}\right)\mathbf{a}_y$ A/m
$\mathbf{E} = (ct - z)\mathbf{a}_x$ V/m
$\mathbf{D} = \epsilon_0(ct - z)\mathbf{a}_x$ C/m^2

Chapter 11

11.1 (*a*) $\gamma = 0.104 + j2.40$ m^{-1}, $\alpha = 0.104$ Np/m, $\beta = 2.40$ rad/m, $\lambda = 2.62$ m, $Z_0 = 100 - j4.0\ \Omega$
(*b*) 12.5% 2.75×10^3 degrees

11.3 (*a*) 96 pF/m (*b*) 1.44×10^8 m/s (*c*) 3.5 rad/m (*d*) $\Gamma = -0.09$, $s = 1.2$

11.5 (*a*) 83.3 nH/m, 33.3 pF/m (*b*) 65 cm

11.7 7.9 mW

11.9 (*a*) $\lambda/8$ (*b*) $\lambda/8 + m\lambda/2$

11.11 (*a*) V_0^2/R_L (*b*) $R_L V_0^2/(R\ell + R_L)^2$ (*c*) V_0^2/R_L (*d*) $(V_0^2/R_L)\exp(-2\ell\sqrt{RG})$

11.13 (*a*) 6.28×10^8 rad/s (*b*) $4\cos(\omega t - \pi z)$A
(*c*) $0.287\angle 1.28$ rad
(*d*) $57.5\exp[j(\pi z + 1.28)]$ V
(*e*) $257.5\angle 36°$ V

11.15 (*a*) 104 V (*b*) $52.6 - j123$ V

11.17 $P_{25} = 2.28$ W, $P_{100} = 1.16$ W

11.19 16.5 W

11.21 (*a*) $s = 2.62$ (*b*) $Z_L = 1.04 \times 10^3 + j69.8\ \Omega$ (*c*) $z_{\max} = -7.2$ mm

11.23 (*a*) 0.037λ or 0.74 m (*b*) 2.61 (*c*) 2.61 (*d*) 0.463λ or 9.26 m

11.25 (*a*) $495 + j290\ \Omega$ (*b*) $j98\ \Omega$

11.27 (*a*) 2.6 (*b*) $11 - j7.0$ mS (*c*) 0.213λ

11.29 $47.8 + j49.3\ \Omega$

11.31 (*a*) 3.8 cm (*b*)14.2 cm

11.33 (*a*) $d_1 = 7.6$ cm, $d = 17.3$ cm (*b*) $d_1 = 1.8$ cm, $d = 6.9$ cm

11.35 (*a*) 39.6 cm (*b*) 24 pF

11.37 $V_L = (2/3)V_0$ $\quad (l/v < t < \infty)$ and is zero for $t < l/\nu$. $I_B = (V_0/75)$ A for $0 < t < \infty$.

11.39 $\frac{l}{\nu} < t < \frac{5l}{4\nu}$: $\quad V_1 = 0.44\,V_0$
$\frac{3l}{\nu} < t < \frac{13l}{4\nu}$: $\quad V_2 = -0.15\,V_0$
$\frac{5l}{\nu} < t < \frac{21l}{4\nu}$: $\quad V_3 = 0.049\,V_0$
$\frac{7l}{\nu} < t < \frac{29l}{4\nu}$: $\quad V_4 = -0.017\,V_0$

Voltages in between these times are zero.

11.41 $0 < t < \frac{l}{2_y}$: $\quad V_L = 0$
$\frac{l}{2\nu} < t < \frac{3l}{2\nu}$: $\quad V_L = \frac{V_0}{2}t > \frac{3l}{2\nu}$:
$V_L = V_0$

Chapter 12

12.3 (*a*) 0.33 rad/m (*b*) 18.9 m (*c*) $-3.76 \times 10^3\mathbf{a}_z$ V/m

12.5 (*a*) $\omega = 3\pi \times 10^8$ sec^{-1}, $\lambda = 2$ m, and $\beta = \pi$ rad/m (*b*) $-8.5\mathbf{a}_x - 9.9\mathbf{a}_y$ A/m
(*c*) 9.08 kV/m

12.7 $\beta = 25$m^{-1}, $\eta = 278.5\ \Omega$, $\lambda = 25$ cm, $v_p = 1.01 \times 10^8$ m/s, $\epsilon_R = 4.01$, $\mu_R = 2.19$, and $\mathbf{H}(x, y, z, t) = 2\cos(8\pi \times 10^8 t - 25x)\mathbf{a}_y + 5\sin(8\pi \times 10^8 t - 25x)\mathbf{a}_z$ A/m

12.9 (*a*) $\beta = 0.4\pi$ rad/m, $\lambda = 5$ m, $v_p = 5 \times 10^7$ m/s, and $\eta = 251\ \Omega$
(*b*) $-403\cos(2\pi \times 10^7 t)$ V/m
(*c*) $1.61\cos(2\pi \times 10^{-7} t)$ A/m

12.11 (*a*) 0.74 kV/m (*b*) –3.0 A/m

12.13 $\mu = 2.28 \times 10^{-6}$ H/m, $\epsilon' = 1.07 \times 10^{-11}$ F/m, and $\epsilon'' = 2.90 \times 10^{-12}$ F/m

12.15 (*a*) $\lambda = 3$ cm, $\alpha = 0$ (*b*) $\lambda = 2.95$ cm, $\alpha = 9.24 \times 10^{-2}$ Np/m (*c*) $\lambda = 1.33$ cm, $\alpha = 335$ Np/m

12.17 $\langle \mathcal{S}_z \rangle(z = 0) = 315\mathbf{a}_z$ W/m^2, $\langle \mathcal{S}_z \rangle(z = 0.6) = 248\mathbf{a}_z$ W/m^2

12.19 (*a*) $\omega = 4 \times 10^8$ rad/s (*b*) $\mathbf{H}(\rho, z, t) = (4.0/\rho)\cos(4 \times 10^8 t - 4z)\mathbf{a}_\phi$ A/m
(*c*)$\langle \mathcal{S} \rangle =$

$(2.0 \times 10^{-3}/\rho^2)\cos^2(4 \times 10^8 t - 4z)\mathbf{a}_z$ W/m² (*d*) P = 5.7 kW

12.21 (*a*) $H_{\phi 1}(\rho) = (54.5/\rho)(10^4\rho^2 - 1)$ A/m $(.01 < \rho < .012)$, $H_{\phi 2}(\rho) = (24/\rho)$ A/m $(\rho > .012)$, $H_\phi = 0 \quad (\rho < .01\text{m})$
(*b*) $\mathbf{E} = 1.09\mathbf{a}_z$ V/m
(*c*) $\langle \mathcal{S} \rangle =$
$-(59.4/\rho)(10^4\rho^2 - 1)\mathbf{a}_\rho$ W/m² $(.01 < \rho < .012$ m$)$, $-(26/\rho)\mathbf{a}_\rho$W/m² $(\rho > 0.12$ m$)$

12.23 (*a*) $1.4 \times 10^{-3}\Omega$/m (*b*) 4.1×10^{-2} Ω/m (*c*) $4.1 \times 10^{-1}\Omega$/m

12.25 $f = 1$ GHz, $\sigma = 1.1 \times 10^5$ S/m

12.27 (*a*) 4.7×10^{-8} (*b*) 3.2×10^3 (*c*) 3.2×10^3

12.29 (*a*) $\mathbf{H}_s = (E_0/\eta_0)(\mathbf{a}_y - j\mathbf{a}_x)e^{-j\beta z}$
(*b*) $\langle \mathcal{S} \rangle = (E_0^2/\eta_0)\mathbf{a}_z$ W/m² (assuming E_0 is real)

12.31 (*a*) $L = 14.6\ \lambda$ (*b*) Left

12.33 (*a*) $\mathbf{H}_s = (1/\eta)[-18e^{j\phi}\mathbf{a}_x + 15\mathbf{a}_y]e^{-j\beta z}$ A/m (*b*) $\langle \mathcal{S} \rangle = 275$ Re $\{(1/\eta^*)\}$ W/m²

Chapter 13

13.1 0.01%

13.3 0.056 and 17.9

13.5 (*a*) 4.7×10^8 Hz (*b*) $691 + j177\ \Omega$ (*c*) −1.7 cm

13.7 (*a*) $s_1 = 1.96, s_2 = 2, s_3 = 1$ (*b*) −0.81 m

13.9 (*a*) 6.25×10^{-2} (*b*) 0.938 (*c*) 1.67

13.11 $641 + j501\ \Omega$

13.13 Reflected wave: left circular polarization; power fraction = 0.09. Transmitted wave: right circular polarization; power fraction = 0.91

13.15 (*a*) 1.2 GHz (*b*) 1.27

13.17 (*a*) $s = 2.13$ and $1 - |\Gamma|^2 = 0.87$
(*b*) $s = 1$ and $1 - |\Gamma|^2 = 1$

13.19 (*a*) $d_1 = d_2 = d_3 = 0$ or $d_1 = d_3 = 0,\ d_2 = \lambda/2$
(*b*) $d_1 = d_2 = d_3 = \lambda/4$

13.21 (*a*) Reflected power: 15%. Transmitted power: 85% (*b*) Reflected wave: s-polarized. Transmitted wave: Right elliptically polarized.

13.23 $n_0 = (n_1/n_2)\sqrt{n_1^2 - n_2^2}$

13.25 0.76

13.27 2

13.29 4.3 km

Chapter 14

14.1 $W_m =$
$\frac{\mu_0 I^2}{4\pi}\left[\frac{c^4}{(c^2-b^2)^2}\ln\left(\frac{c}{b}\right) + \frac{b^2-(3/4)c^2}{(c^2-b^2)}\right]$ J

14.3 14.2 pF/m, 0.786 μH/m, 0, 0.023 Ω/m

14.5 (*a*) 2.8 (*b*) 5.85×10^{-2}

14.7 (*a*) 5.0 (*b*) 1.6

14.9 9

14.11 (*a*) 12.8 GHz (*b*) Yes

14.14 3.32

14.16 (*a*) $\theta_{\min} = \sin^{-1}(n_3/n_1)$
(*b*) $v_{p,\max} = c/n_3$

14.18 greater than

14.20 (*a*) $-0.284\mathbf{a}_x - 0.959\mathbf{a}_z$ (*b*) 0.258

14.22 (*a*) $-j(1.5 \times 10^{-2})e^{-j1000}\mathbf{a}_z$ V/m
(*b*) $-j(1.5 \times 10^{-2})e^{-j1000}\mathbf{a}_y$ V/m
(*c*) $-j(1.5 \times 10^{-2})(\mathbf{a}_y + \mathbf{a}_z)$ V/m
(*d*) $-(1.24 \times 10^{-2})(\mathbf{a}_y + \mathbf{a}_z)$ V/m
(*e*) 1.75×10^{-2} V/m

14.24 (*a*) 85.4 A (*b*) 5.03 A

14.26 (*a*) $0.2e^{-j1000\pi}$ V/m
(*b*) $0.2e^{-j1000\pi}e^{j0.5\pi}$ V/m (*c*) 0

Chapter 15

15.1 9

15.3 1.5 ns

15.5 (*a*) 2.5 GHz $< f <$ 3.75 GHz (air-filled)
(*b*) 3.75 GHs $< f <$ 4.5 GHz (air-filled)

15.7 $a = 1.1$ cm, $b = 0.90$ cm

15.11 72 cm

Index

E

F

G

H

I

K

L

M

N

O

P

Q

R

S

T

DIVERGENCE

RECTANGULAR $\nabla \cdot \mathbf{D} = \frac{\partial D_x}{\partial x} + \frac{\partial D_y}{\partial y} + \frac{\partial D_z}{\partial z}$

CYLINDRICAL $\nabla \cdot \mathbf{D} = \frac{1}{\rho}\frac{\partial}{\partial \rho}(\rho D_\rho) + \frac{1}{\rho}\frac{\partial D_\phi}{\partial \phi} + \frac{\partial D_z}{\partial z}$

SPHERICAL $\nabla \cdot \mathbf{D} = \frac{1}{r^2}\frac{\partial}{\partial r}(r^2 D_r) + \frac{1}{r\sin\theta}\frac{\partial}{\partial \theta}(D_\theta \sin\theta) + \frac{1}{r\sin\theta}\frac{\partial D_\phi}{\partial \phi}$

GRADIENT

RECTANGULAR $\nabla V = \frac{\partial V}{\partial x}\mathbf{a}_x + \frac{\partial V}{\partial y}\mathbf{a}_y + \frac{\partial V}{\partial z}\mathbf{a}_z$

CYLINDRICAL $\nabla V = \frac{\partial V}{\partial \rho}\mathbf{a}_\rho + \frac{1}{\rho}\frac{\partial V}{\partial \phi}\mathbf{a}_\phi + \frac{\partial V}{\partial z}\mathbf{a}_z$

SPHERICAL $\nabla V = \frac{\partial V}{\partial r}\mathbf{a}_r + \frac{1}{r}\frac{\partial V}{\partial \theta}\mathbf{a}_\theta + \frac{1}{r\sin\theta}\frac{\partial V}{\partial \phi}\mathbf{a}_\phi$

CURL

RECTANGULAR $$\nabla \times \mathbf{H} = \left(\frac{\partial H_z}{\partial y} - \frac{\partial H_y}{\partial z}\right)\mathbf{a}_x + \left(\frac{\partial H_x}{\partial z} - \frac{\partial H_z}{\partial x}\right)\mathbf{a}_y + \left(\frac{\partial H_y}{\partial x} - \frac{\partial H_x}{\partial y}\right)\mathbf{a}_z$$

CYLINDRICAL $$\nabla \times \mathbf{H} = \left(\frac{1}{\rho}\frac{\partial H_z}{\partial \phi} - \frac{\partial H_\phi}{\partial z}\right)\mathbf{a}_\rho + \left(\frac{\partial H_\rho}{\partial z} - \frac{\partial H_z}{\partial \rho}\right)\mathbf{a}_\phi + \frac{1}{\rho}\left[\frac{\partial(\rho H_\phi)}{\partial \rho} - \frac{\partial H_\rho}{\partial \phi}\right]\mathbf{a}_z$$

SPHERICAL $$\nabla \times \mathbf{H} = \frac{1}{r\sin\theta}\left[\frac{\partial(H_\phi \sin\theta)}{\partial \theta} - \frac{\partial H_\theta}{\partial \phi}\right]\mathbf{a}_r + \frac{1}{r}\left[\frac{1}{\sin\theta}\frac{\partial H_r}{\partial \phi} - \frac{\partial(rH_\phi)}{\partial r}\right]\mathbf{a}_\theta + \frac{1}{r}\left[\frac{\partial(rH_\theta)}{\partial r} - \frac{\partial H_r}{\partial \theta}\right]\mathbf{a}_\phi$$

LAPLACIAN

RECTANGULAR $$\nabla^2 V = \frac{\partial^2 V}{\partial x^2} + \frac{\partial^2 V}{\partial y^2} + \frac{\partial^2 V}{\partial z^2}$$

CYLINDRICAL $$\nabla^2 V = \frac{1}{\rho}\frac{\partial}{\partial \rho}\left(\rho\frac{\partial V}{\partial \rho}\right) + \frac{1}{\rho^2}\frac{\partial^2 V}{\partial \phi^2} + \frac{\partial^2 V}{\partial z^2}$$

SPHERICAL $$\nabla^2 V = \frac{1}{r^2}\frac{\partial}{\partial r}\left(r^2\frac{\partial V}{\partial r}\right) + \frac{1}{r^2\sin\theta}\frac{\partial}{\partial \theta}\left(\sin\theta\frac{\partial V}{\partial \theta}\right) + \frac{1}{r^2\sin^2\theta}\frac{\partial^2 V}{\partial \phi^2}$$

SAMPLE ILLUSTRATIONS FROM THE CD-ROM

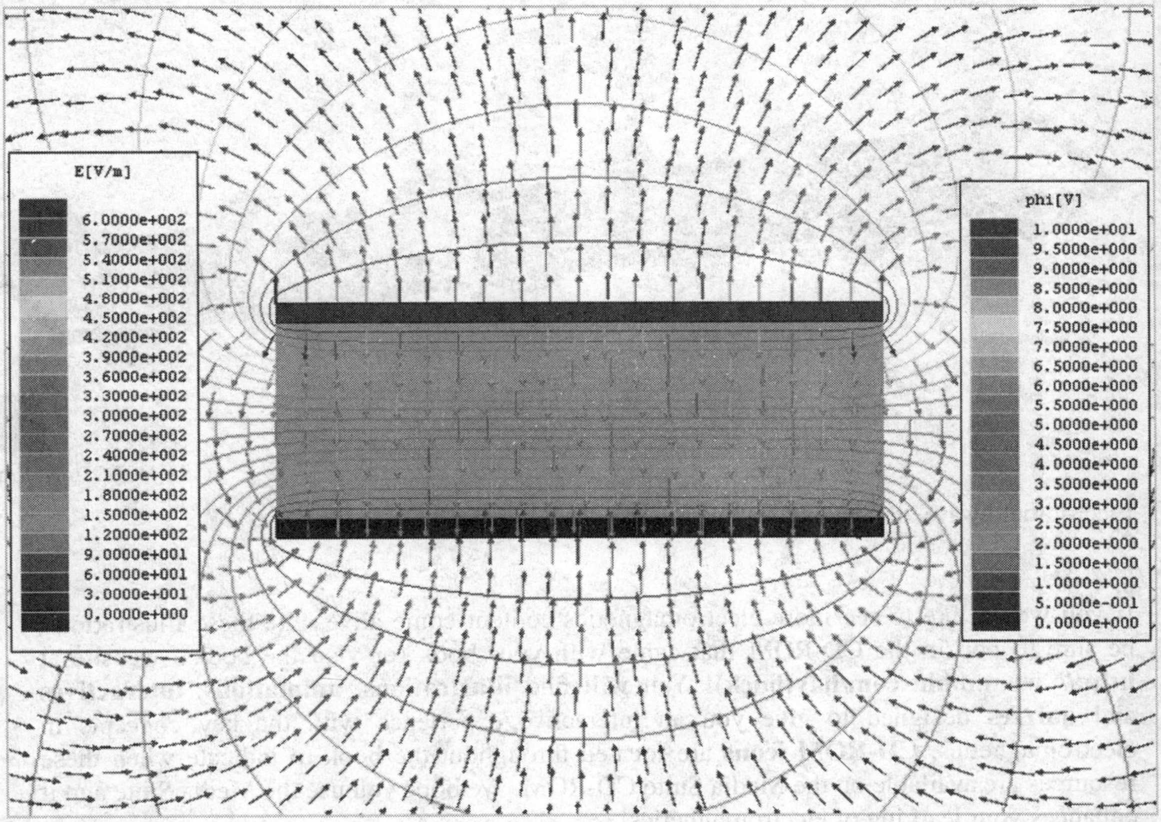

Field map in the cross-section of a parallel-plate line: equi-potential lines and E-field streamlines.

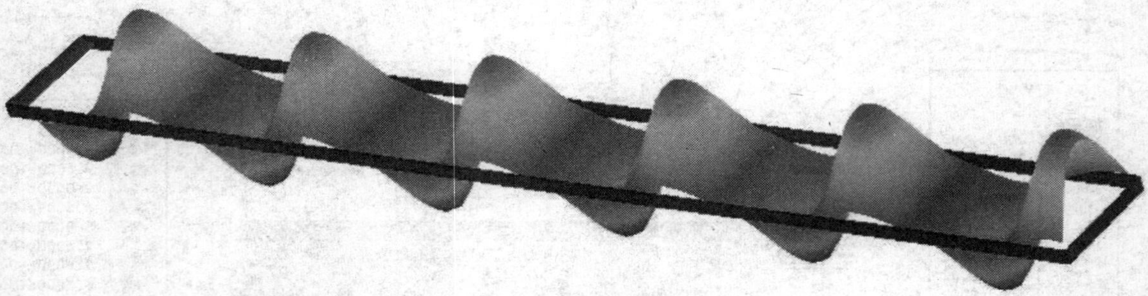

MEFiSTo-3D Pro™

Electric-field wave along a metallic waveguide: the dominant TE_{10} mode.

If you would like to see more electromagnetics content come alive, like these illustrations, be sure to pop in the CD-ROM that came with your book (or visit the book's website at **http:// www.mhhe.com/haytbuck**)! You will find **illustrations, animations, interactives** and **quizzes** designed to give you an interactive experience with the key concepts in electromagnetics. **CD-ROM icons** are located throughout the book to indicate when these resources are available on the Media Suite CD-ROM. We hope you use the Media Suite and it enhances your learning of electromagnetics!